SMALL WASTEWATER SYSTEM OPERATION AND MAINTENANCE

Volume II

First Edition

A Field Study Training Program

prepared by

Office of Water Programs
California State University, Sacramento

★★★★★★★★★★★★★★★★★★★★★★★★★★★★★★★★

Kenneth D. Kerri, Project Director

John Brady, Co-Director

★★★★★★★★★★★★★★★★★★★★★★★★★★★★★★★★

2002

NOTICE

This manual is revised and updated before each printing based on comments from persons using the manual.

FIRST EDITION

First Printing, 2002 4,000

ISBN 1-884701-22-1 (Volume II)
ISBN 1-884701-23-X (Volumes I and II)

OPERATOR TRAINING MANUALS

OPERATOR TRAINING MANUALS AND VIDEOS IN THIS SERIES are available from the Office of Water Programs, California State University, Sacramento, 6000 J Street, Sacramento, CA 95819-6025, phone: (916) 278-6142, e-mail: waterof-fice@csus.edu or FAX: (916) 278-5959, website: www.owp.csus.edu.

1. *SMALL WASTEWATER SYSTEM OPERATION AND MAINTENANCE*, 2 Volumes,
2. *OPERATION OF WASTEWATER TREATMENT PLANTS*, 2 Volumes,
3. *OPERATION AND MAINTENANCE OF WASTEWATER COLLECTION SYSTEMS*, 2 Volumes,
4. *INDUSTRIAL WASTE TREATMENT*, 2 Volumes,
5. *ADVANCED WASTE TREATMENT*,
6. *TREATMENT OF METAL WASTESTREAMS*,
7. *PRETREATMENT FACILITY INSPECTION*,
8. *WATER TREATMENT PLANT OPERATION*, 2 Volumes,
9. *SMALL WATER SYSTEM OPERATION AND MAINTENANCE*,
10. *WATER DISTRIBUTION SYSTEM OPERATION AND MAINTENANCE*, and
11. *UTILITY MANAGEMENT*.

SMALL WATER SYSTEMS OPERATION AND MAINTENANCE INFORMATION VIDEOS also are available to help the operators, managers, owners, and elected board members of small water systems. Ten videos, ranging from 15 minutes to 59 minutes, demonstrate the procedures to safely and effectively operate and maintain surface water treatment systems, groundwater treatment systems, and distribution and storage systems. Other topics covered include monitoring, managerial, financial, and emergency response procedures for small water systems. These videos are used with a *LEARNING BOOKLET* that provides additional essential information and references.

PREFACE TO THE FIRST EDITION

The purposes of this small wastewater system field study training program are to:

1. Develop new qualified small wastewater system operators,
2. Expand the abilities of existing operators, permitting better service to both their employers and the public, and
3. Prepare operators for civil service and *CERTIFICATION EXAMINATIONS.*[1]

To provide you with the knowledge and skills needed to operate and maintain small wastewater systems as efficiently and effectively as possible, experienced wastewater system operators prepared the material in each chapter of this manual.

Small wastewater systems vary from town to town and from region to region. The material contained in this program is presented to provide you with an understanding of the basic operation and maintenance aspects of your system and with information to help you analyze and solve operation and maintenance problems. This information will help you operate and maintain your system in a safe and efficient manner. Each chapter covers different types of small wastewater systems and all of these systems may not be used in your location.

Small wastewater system operation and maintenance is a rapidly advancing field. To keep pace with scientific and technological advances, the material in this manual must be revised and updated periodically. *THIS MEANS THAT YOU, THE OPERATOR, MUST RECOGNIZE THE NEED TO BE AWARE OF NEW ADVANCES AND THE NEED FOR CONTINUOUS TRAINING BEYOND THIS PROGRAM.*

The project directors are indebted to the many operators and other persons who contributed to this operator training manual. Many portions of this manual were taken from our first operator training manuals, *OPERATION OF WASTEWATER TREATMENT PLANTS*, Volumes I and II. The contributions of these pioneers in our field are greatly appreciated. Peg Hannah served as Research Editor for this project. The illustrations were prepared by Martin Garrity for other operator training manuals. Gay Kornweibel, Assistant Director of the Office of Water Programs, coordinated all the office activities involved in developing and implementing a new training course. And special thanks are well deserved by Janet Weeks, who typed manuscripts, check and rechecked facts, coordinated the printing process, and did everything else that had to be done to complete this project successfully. Thanks also to the California Rural Water Association for the support of Pat Conway and Don Snelling during the review of this material.

2002

KENNETH D. KERRI
JOHN BRADY

---

[1] *Certification Examination. An examination administered by a state agency or professional association that small wastewater system operators take to indicate a level of professional competence. In many states and provinces, certification of small wastewater system operators may be voluntary. Current trends indicate that more states, provinces, and employers will require small wastewater system operators to be "certified" in the future.*

OBJECTIVES OF THIS MANUAL

Proper installation, inspection, operation, maintenance, repair, and management of small wastewater systems have a significant impact on the operation and maintenance costs and effectiveness of the systems. The objectives of this manual are to provide small wastewater system operators with the knowledge and skills required to operate and maintain these systems safely and effectively, thus eliminating or reducing the following problems:

1. Health hazards created by the discharge of improperly treated wastewater;
2. System failures that result from the lack of proper installation, inspection, preventive maintenance, surveillance and repair programs designed to protect the public investment in these facilities;
3. Fish kills and environmental damage caused by wastewater system problems;
4. Corrosion damage to pipes, equipment, tanks, and structures in the wastewater system;
5. Complaints from the public or local officials due to the unreliability or failure of the wastewater system to perform as designed; and
6. Odors caused by insufficient and/or inadequate wastewater and sludge handling and treatment procedures.

SCOPE OF THIS MANUAL

Operators with the responsibility for operating and maintaining package wastewater treatment plants and wastewater treatment processes, which are used by smaller communities and treatment agencies, will find this manual (Volume II) very helpful. Topics covered include wastewater stabilization ponds, activated sludge, rotating biological contactors, and some alternative treatment systems. Other topics in Volume II include disinfection, chlorination and dechlorination, wastewater treatment, disposal, and reuse methods, laboratory procedures, and management of a small wastewater collection and treatment agency.

Volume I of *SMALL WASTEWATER SYSTEM OPERATION AND MAINTENANCE* focuses on septic tanks and on-site treatment and disposal systems, wastewater collection systems, and pumping or lift stations. Volume I contains information on:

1. What small wastewater system operators do,
2. Sources of wastewater and damage caused by improperly treated wastewater,
3. How to operate and maintain septic tanks and on-site wastewater treatment and disposal systems,
4. How to operate and maintain collection systems and pump stations,
5. Techniques for recognizing hazards and developing safe procedures and safety programs,
6. Procedures for setting up and carrying out an effective maintenance program, and
7. How to calculate the cost of operating a small wastewater system and develop an appropriate rate structure.

Material in both volumes furnishes you with information concerning situations encountered by most small wastewater system operators in most areas. These materials provide you with an understanding of the basic operation and maintenance concepts for small wastewater systems and with an ability to analyze and solve problems when they occur. Operation and maintenance programs for small wastewater systems will vary with the age of the system, the extent and effectiveness of previous programs, and local conditions. You will have to adapt the information and procedures in this manual to your particular situation.

Technology is advancing very rapidly in the field of operation and maintenance of small wastewater systems. This means that you, the small wastewater system operator, must be aware of new advances and recognize the need for continuous personal training reaching beyond this program. ***TRAINING OPPORTUNITIES EXIST IN YOUR DAILY WORK EXPERIENCE, FROM YOUR ASSOCIATES, AND FROM ATTENDING MEETINGS, WORKSHOPS, CONFERENCES, AND CLASSES.***

USES OF THIS MANUAL

This manual was developed to serve the needs of operators in several different situations. The format was developed to serve as a home-study or self-paced instruction course for operators in remote areas or persons unable to attend formal classes either due to shift work, personal reasons, or the unavailability of suitable classes. This home-study training program uses the concepts of self-paced instruction where you are your own instructor and work at your own speed. In order to certify that a person has successfully completed the program, an objective test is included at the end of each chapter.

Also, this manual can serve effectively as a textbook in the classroom. Many colleges and universities have used similar manuals as texts in formal classes (often taught by operators). In areas where colleges are not available or are unable to offer classes in the operation of small wastewater systems, operators and utility agencies can join together to offer their own courses using the manual.

Cities or utility agencies can use the manual in several types of on-the-job training programs. In one type of program, a manual is purchased for each operator. A senior operator or a group of operators are designated as instructors. These operators help answer questions when the persons in the training program have questions or need assistance. The instructors grade the objective tests at the end of each chapter, record scores, and notify California State University, Sacramento, of the scores when a person successfully completes this program. This approach eliminates any waiting while papers are being graded and returned by CSUS.

This manual was prepared to help operators operate and maintain their small wastewater systems. Please feel free to use the manual in the manner that best fits your training needs and the needs of other operators. We will be happy to work with you to assist you in developing your training program. Please feel free to contact:

Office of Water Programs
California State University, Sacramento
6000 J Street
Sacramento, California 95819-6025

Phone: (916) 278-6142
FAX: (916) 278-5959

e-mail: wateroffice@csus.edu

INSTRUCTIONS TO PARTICIPANTS IN HOME-STUDY COURSE

Procedures for reading the lessons and answering the questions are contained in this section.

To progress steadily through this program, you should establish a regular study schedule. For example, many operators in the past have set aside two hours during two evenings a week for study.

The study material is contained in two volumes having a total of fifteen chapters. Some chapters are longer and more difficult than others. For this reason, many of the chapters are divided into two or more lessons. The time required to complete a lesson will depend on your background, experience, and study habits. Some people might require an hour to complete a lesson and some might require three hours, but that is perfectly all right. *THE IMPORTANT THING IS THAT YOU UNDERSTAND THE MATERIAL IN THE LESSON!*

Each lesson is arranged for you to read a short section, write the answers to questions at the end of the section, check your answers against suggested answers; and then *YOU* decide if you understand the material sufficiently to continue or whether you should read the section again. You will find that this procedure is slower than reading a normal textbook, but you will remember much more when you have finished the lesson.

Some discussion and review questions are provided following each lesson in some of the chapters. These questions review the important points you have covered in the lesson. Write the answers to these questions in your notebook.

At the end of each chapter you will find an "objective test," which contains true or false, multiple-choice, or best answer types of questions. The purposes of this exam are to review the chapter and to give you experience in taking different types of exams. Mark your answers on the special answer sheet provided for each chapter. *MAIL TO THE PROGRAM DIRECTOR ONLY YOUR ANSWERS TO OBJECTIVE TESTS ON THE PROVIDED ANSWER SHEETS.* Your answer sheets will not be returned to you so you may wish to record your answers in your notebook before mailing the answer sheet to the Program Director.

The answer sheets you mail to the Program Director will be computer-graded in the following manner: true-false questions are considered as having only one possible answer so they have a value of 1 point. Best Answer questions have only one correct answer and are worth 4 points. Multiple-choice questions are worth 5 points since there may be more than one correct answer and all correct answers must be filled in for full credit. For the multiple-choice questions, the computer deducts a point for each correct answer *NOT* filled in and each *INCORRECT* response filled in. For example, if 1, 3, and 5 are the correct responses and you fill in 1 and 5, a point is deducted for failing to fill in 3. If you fill in 1, 3, 4, and 5, a point is deducted because 4 is filled in and should not be.

After you have completed the last objective test in each volume, you will find a final examination. This exam is provided for your review. The results will reveal how well you remember the material. You may wish to review the entire manual before you take the final exam. Some of the questions are essay-type questions, which are used by some states and professional associations for higher-level certification examinations. After you have completed the final examination, grade your own paper and determine the areas in which you might need additional review before your next certification or civil service examination.

You are your own teacher in this program. You could merely look up the suggested answers at the end of each chapter or copy the answers from someone else, but you would not have given yourself a chance to understand the material. Consequently, you would not be able to apply the material to the operation and maintenance of your facilities nor recall it during an examination for certification or a civil service position.

YOU WILL GET OUT OF THIS PROGRAM WHAT YOU PUT INTO IT.

SUMMARY OF PROCEDURE

A. OPERATOR (YOU)

1. Read what you are expected to learn in each chapter (the chapter objectives).
2. Read sections in the lesson.
3. Write your answers to questions at the end of each section in your notebook. You should write the answers to the questions just as you would if these were questions on a test.
4. Compare your answers with the suggested answers.
5. Decide whether to reread the section or to continue with the next section.
6. Write your answers to the discussion and review questions at the end of each lesson in your notebook.
7. Mark your answers to the objective test on the answer sheet provided by the Project Director or by your instructor.
8. Mail material to the Project Director. (Send *ONLY* your completed answer sheet.)

   Office of Water Programs
   California State University, Sacramento
   6000 J Street
   Sacramento, CA 95819-6025

B. PROJECT DIRECTOR

1. Mails answer sheet for each chapter to operator.
2. Corrects tests, answers any questions, and returns results to operators including explanations and solutions for missed questions.

C. ORDER OF WORKING LESSONS

To complete this program, you will have to work all of the chapters. You may proceed in numerical order, or you may wish to work some lessons sooner. *SAFETY IS A VERY IMPORTANT TOPIC.* Everyone working in a small wastewater system must always be safety conscious. Operators daily encounter situations and equipment that can cause a serious disabling injury or illness if the operator is not aware of the potential danger and does not exercise adequate precautions. In each chapter, *SAFE PROCEDURES ARE ALWAYS STRESSED.*

D. COMPLETION

Following successful completion of each volume in this program, a Certificate of Completion will be sent to you. If you wish, the Certificate can be sent to your supervisor, the mayor of your town, or any other official you think appropriate. Some operators have been presented their Certificate at a City Council meeting, got their picture in the newspaper, and received a pay raise. When you successfully complete the course, the Project Director will ask you how you would like your name to appear on your Certificate and if you would like the Certificate sent directly to you, or to your supervisor or other municipal official.

TECHNICAL REVIEWERS

Russ Armstrong	Dwight Lancaster	Kurt Ohlinger
Pat Conway	Martin Lopez	Don Snelling
David Dauwalder	Thurlow Morrow	Tim Sullivan
Steve Desmond	Stew Oakley	

SMALL WASTEWATER SYSTEM OPERATION AND MAINTENANCE COURSE OUTLINE

VOLUME II

SMALL WASTEWATER SYSTEM OPERATION AND MAINTENANCE COURSE OUTLINE

VOLUME I

Another similar operator training program that may be of interest to you is our manual and study program on small drinking water systems.

SMALL WATER SYSTEM OPERATION AND MAINTENANCE (Wells, Small Treatment Plants, and Rates)

COURSE OUTLINE

CHAPTER 9

WASTEWATER STABILIZATION PONDS

by

A. Hiatt

Revised by

John Brady

TABLE OF CONTENTS

Chapter 9. WASTEWATER STABILIZATION PONDS

OBJECTIVES

Chapter 9. WASTEWATER STABILIZATION PONDS

Following completion of Chapter 9, you should be able to:

1. Explain how wastewater stabilization ponds work and what factors influence and control pond treatment processes,
2. Identify the different types of ponds,
3. Place a new pond into operation,
4. Schedule and conduct normal and abnormal operation and maintenance duties,
5. Collect samples, interpret lab results, and make appropriate adjustments in pond operation,
6. Recognize factors that indicate a pond is not performing properly, identify the source of the problem, and take corrective action,
7. Develop a pond operating strategy,
8. Conduct your duties in a safe fashion,
9. Determine pond loadings,
10. Keep records for a waste treatment pond facility, and
11. Review plans and specifications for new ponds.

WORDS

Chapter 9. WASTEWATER STABILIZATION PONDS

AEROBIC (AIR-O-bick) AEROBIC

A condition in which atmospheric or dissolved molecular oxygen is present in the aquatic (water) environment.

ALGAE (AL-gee) ALGAE

Microscopic plants which contain chlorophyll and live floating or suspended in water. They also may be attached to structures, rocks, or other submerged surfaces. Algae produce oxygen during sunlight hours and use oxygen during the night hours. Their biological activities appreciably affect the pH, alkalinity, and dissolved oxygen of the water.

ALGAL (AL-gull) BLOOM ALGAL BLOOM

Sudden, massive growths of microscopic and macroscopic plant life, such as green or blue-green algae, which can, under the proper conditions, develop in lakes, reservoirs, and also in ponds.

ANAEROBIC (AN-air-O-bick) ANAEROBIC

A condition in which atmospheric or dissolved molecular oxygen is *NOT* present in the aquatic (water) environment.

APPURTENANCE (uh-PURR-ten-nans) APPURTENANCE

Machinery, appliances, structures and other parts of the main structure necessary to allow it to operate as intended, but not considered part of the main structure.

BOD (pronounce as separate letters) BOD

Biochemical **O**xygen **D**emand. The rate at which organisms use the oxygen in water or wastewater while stabilizing decomposable organic matter under aerobic conditions. In decomposition, organic matter serves as food for the bacteria and energy results from its oxidation. BOD measurements are used as a measure of the organic strength of wastes in water.

BACTERIA (back-TEAR-e-ah) BACTERIA

Bacteria are living organisms, microscopic in size, which usually consist of a single cell. Most bacteria use organic matter for their food and produce waste products as a result of their life processes.

BIOCHEMICAL OXYGEN DEMAND (BOD) BIOCHEMICAL OXYGEN DEMAND (BOD)

The rate at which organisms use the oxygen in water or wastewater while stabilizing decomposable organic matter under aerobic conditions. In decomposition, organic matter serves as food for the bacteria and energy results from its oxidation. BOD measurements are used as a measure of the organic strength of wastes in water.

BIOFLOCCULATION (BUY-o-flock-u-LAY-shun) BIOFLOCCULATION

The clumping together of fine, dispersed organic particles by the action of certain bacteria and algae. This results in faster and more complete settling of the organic solids in wastewater.

CHEMICAL OXYGEN DEMAND (COD) CHEMICAL OXYGEN DEMAND (COD)

A measure of the oxygen-consuming capacity of organic matter present in wastewater. COD is expressed as the amount of oxygen consumed from a chemical oxidant in mg/*L* during a specific test. Results are not necessarily related to the biochemical oxygen demand (BOD) because the chemical oxidant may react with substances that bacteria do not stabilize.

COLIFORM (COAL-i-form) COLIFORM

One type of bacteria. The presence of coliform-group bacteria is an indication of possible pathogenic bacterial contamination. The human intestinal tract is one of the main habitats of coliform bacteria. They may also be found in the intestinal tracts of warm-blooded animals, and in plants, soil, air, and the aquatic environment. Fecal coliforms are those coliforms found in the feces of various warm-blooded animals; whereas the term "coliform" also includes other environmental sources.

COMPOSITE (come-PAH-zit) (PROPORTIONAL) SAMPLE — COMPOSITE (PROPORTIONAL) SAMPLE

A composite sample is a collection of individual samples obtained at regular intervals, usually every one or two hours during a 24-hour time span.9.11 Each individual sample is combined with the others in proportion to the rate of flow when the sample was collected. The resulting mixture (composite sample) forms a representative sample and is analyzed to determine the average conditions during the sampling period.

DETENTION TIME — DETENTION TIME

The time required to fill a tank at a given flow or the theoretical time required for a given flow of wastewater to pass through a tank. In septic tanks this detention time will decrease as the volumes of sludge and scum increase.

DISINFECTION (dis-in-FECT-shun) — DISINFECTION

The process designed to kill or inactivate most microorganisms in wastewater, including essentially all pathogenic (disease-causing) bacteria. There are several ways to disinfect, with chlorination being the most frequently used in water and wastewater treatment plants. Compare with STERILIZATION.

DISSOLVED OXYGEN — DISSOLVED OXYGEN

Molecular (atmospheric) oxygen dissolved in water or wastewater, usually abbreviated DO.

DUCKWEED — DUCKWEED

A small, green, cloverleaf-shaped floating plant, about one-quarter inch (6 mm) across which appears as a grainy "scum" on the surface of a pond.

EFFLUENT (EF-lu-ent) — EFFLUENT

Wastewater or other liquid—raw (untreated), partially or completely treated—flowing *FROM* an interceptor, reservoir, basin, treatment process, or treatment plant.

FACULTATIVE (FACK-ul-TAY-tive) POND — FACULTATIVE POND

The most common type of pond in current use. The upper portion (supernatant) is aerobic, while the bottom layer is anaerobic. Algae supply most of the oxygen to the supernatant.

FREE OXYGEN — FREE OXYGEN

Molecular oxygen available for respiration by organisms. Molecular oxygen is the oxygen molecule, O_2, that is not combined with another element to form a compound.

GRAB SAMPLE — GRAB SAMPLE

A single sample of water collected at a particular time and place which represents the composition of the water only at that time and place.

INFLUENT (IN-flu-ent) — INFLUENT

Wastewater or other liquid—raw (untreated) or partially treated—flowing *INTO* an interceptor, reservoir, basin, treatment process, or treatment plant.

MEDIAN — MEDIAN

The middle measurement or value. When several measurements are ranked by magnitude (largest to smallest), half of the measurements will be larger and half will be smaller.

MOLECULAR OXYGEN — MOLECULAR OXYGEN

The oxygen molecule, O_2, that is not combined with another element to form a compound.

NPDES PERMIT — NPDES PERMIT

National **P**ollutant **D**ischarge **E**limination **S**ystem permit is the regulatory agency document issued by either a federal or state agency which is designed to control all discharges of potential pollutants from point sources and storm water runoff into U.S. waterways. NPDES permits regulate discharges into navigable waters from all point sources of pollution, including industries, municipal wastewater treatment plants, sanitary landfills, large agricultural feedlots and return irrigation flows.

OVERTURN — OVERTURN

The almost spontaneous mixing of all layers of water in a reservoir or lake when the water temperature becomes similar from top to bottom. This may occur in the fall/winter when the surface waters cool to the same temperature as the bottom waters and also in the spring when the surface waters warm after the ice melts. This is also called "turnover."

PARALLEL OPERATION — PARALLEL OPERATION

Wastewater being treated is split and a portion flows to one treatment unit while the remainder flows to another similar treatment unit. Also see SERIES OPERATION.

PATHOGENIC (PATH-o-JEN-ick) ORGANISMS — PATHOGENIC ORGANISMS

Bacteria, viruses, cysts, or protozoa which can cause disease (giardiasis, cryptosporidiosis, typhoid, cholera, dysentery) in a host (such as a person). There are many types of organisms which do *NOT* cause disease and which are *NOT* called pathogenic. Many beneficial bacteria are found in wastewater treatment processes actively cleaning up organic wastes.

PERCOLATION (PURR-co-LAY-shun) — PERCOLATION

The movement or flow of water through soil or rocks.

pH (pronounce as separate letters) — pH

pH is an expression of the intensity of the basic or acidic condition of a liquid. Mathematically, pH is the logarithm (base 10) of the reciprocal of the hydrogen ion activity.

$$pH = \text{Log} \frac{1}{[H^+]}$$

The pH may range from 0 to 14, where 0 is most acidic, 14 most basic, and 7 neutral. Natural waters usually have a pH between 6.5 and 8.5.

PHOTOSYNTHESIS (foe-toe-SIN-thuh-sis) — PHOTOSYNTHESIS

A process in which organisms, with the aid of chlorophyll (green plant enzyme), convert carbon dioxide and inorganic substances into oxygen and additional plant material, using sunlight for energy. All green plants grow by this process.

POPULATION EQUIVALENT — POPULATION EQUIVALENT

A means of expressing the strength of organic material in wastewater. In a domestic wastewater system, microorganisms use up about 0.2 pound of oxygen per day for each person using the system (as measured by the standard BOD test). May also be expressed as flow (100 gallons per day per person) or suspended solids (0.2 lb SS/day/person).

$$\text{Population Equivalent, persons} = \frac{\text{Flow, MGD x BOD, mg}/L \text{ x 8.34 lbs/gal}}{\text{0.2 lb BOD/day/person}}$$

RIPRAP — RIPRAP

Broken stones, boulders, or other materials placed compactly or irregularly on levees or dikes for the protection of earth surfaces against the erosive action of waves.

ROTATING BIOLOGICAL CONTACTOR (RBC) — ROTATING BIOLOGICAL CONTACTOR (RBC)

A secondary biological treatment process for domestic and biodegradable industrial wastes. Biological contactors have a rotating "shaft" surrounded by plastic discs called the "media." The shaft and media are called the "drum." A biological slime grows on the media when conditions are suitable and the microorganisms that make up the slime (biomass) stabilize the waste products by using the organic material for growth and reproduction.

SEPTIC (SEP-tick) — SEPTIC

A condition produced by anaerobic bacteria. If severe, the sludge produces hydrogen sulfide, turns black, gives off foul odors, contains little or no dissolved oxygen, and the wastewater has a high oxygen demand.

SERIES OPERATION — SERIES OPERATION

Wastewater being treated flows through one treatment unit and then flows through another similar treatment unit. Also see PARALLEL OPERATION.

SHOCK LOAD (activated sludge) — SHOCK LOAD

The arrival at a plant of a waste which is toxic to organisms in sufficient quantity or strength to cause operating problems. Possible problems include odors and bulking sludge which will result in a high loss of solids from the secondary clarifiers into the plant effluent and a biological process upset that may require several days to a week to recover. Organic or hydraulic overloads also can cause a shock load.

SHORT-CIRCUITING — SHORT-CIRCUITING

A condition that occurs in tanks or basins when some of the flowing water entering a tank or basin flows along a nearly direct pathway from the inlet to the outlet. This is usually undesirable since it may result in shorter contact, reaction, or settling times in comparison with the theoretical (calculated) or presumed detention times.

SPLASH PAD — SPLASH PAD

A structure made of concrete or other durable material to protect bare soil from erosion by splashing or falling water.

STABILIZED WASTE — STABILIZED WASTE

A waste that has been treated or decomposed to the extent that, if discharged or released, its rate and state of decomposition would be such that the waste would not cause a nuisance or odors in the receiving water.

STERILIZATION (STARE-uh-luh-ZAY-shun) — STERILIZATION

The removal or destruction of all microorganisms, including pathogenic and other bacteria, vegetative forms and spores. Compare with DISINFECTION.

STOP LOG — STOP LOG

A log or board in an outlet box or device used to control the water level in ponds and also the flow from one pond to another pond or system.

SUPERNATANT (sue-per-NAY-tent) (activated sludge) — SUPERNATANT

Liquid removed from settled sludge. Supernatant commonly refers to the liquid between the sludge on the bottom and the scum on the surface of a tank.

TOXIC (TOX-ick) — TOXIC

A substance which is poisonous to a living organism.

TOXICITY (tox-IS-it-tee) — TOXICITY

The relative degree of being poisonous or toxic. A condition which may exist in wastes and will inhibit or destroy the growth or function of certain organisms.

TRICKLING FILTER — TRICKLING FILTER

A treatment process in which the wastewater trickles over media that provide the opportunity for the formation of slimes or biomass which contain organisms that feed upon and remove wastes from the water being treated.

CHAPTER 9. WASTEWATER STABILIZATION PONDS

USED FOR TREATMENT OF WASTEWATER AND OTHER WASTES

(Lesson 1 of 3 Lessons)

9.0 USE OF PONDS

Ponds (3 to 6 feet or 1 to 2 meters deep) are often used to treat wastewater and other wastes instead of, or in addition to, conventional waste treatment processes. (See Figure 9.1 for a typical pond plant layout and Table 9.1 for the purposes of pond parts.) When discharged into ponds, wastes are treated or *STABILIZED*[1] by several natural processes acting at the same time. Heavy solids settle to the bottom where they are decomposed by bacteria. Lighter suspended material is broken down by bacteria in suspension. Some wastewater is disposed of by evaporation from the pond surface or by *PERCOLATING*[2] into the ground.

Dissolved nutrient materials, such as nitrogen and phosphorus, are used by green *ALGAE*,[3] which are actually microscopic plants floating and living in the water. The algae use carbon dioxide(CO_2) and bicarbonate to build body protoplasm. In so growing, they need nitrogen and phosphorus in their metabolism much as land plants do. Like land plants, they release oxygen and some carbon dioxide as waste products.

Ponds can serve as very effective treatment facilities. Extensive studies of their performance have led to a better understanding of the natural processes by which ponds treat wastes. Information is provided here on the natural processes and ways operators can regulate pond processes for efficient waste treatment.

TABLE 9.1 PURPOSE OF POND PARTS

Part	Purpose
Flowmeter	Measures and records flows into pond and possibly effluent.
Bar Screen	Removes coarse material from pond influent.
Pond Inlets	Distribute influent in pond.
Pond Depth and Outlet Control	Regulates outflow from pond and depth of water in pond. Allows pond to be drained for cleaning and inspection.
Outlet Baffle	Prevents scum and other surface debris from flowing to next pond or receiving waters.
Dike or Levee	Separates ponds and holds wastewater being treated in ponds.
Transfer Line	Conveys wastewater from one pond to another.
Recirculation Line	Returns pond effluent rich in algae and oxygen from second pond to first pond for seeding, dilution, and process control.
Chlorination	Applies chlorine to treated wastewater for disinfection purposes.
Chlorine Contact Basin	Provides contact time for chlorine to disinfect pond effluent.
Effluent Line	Conveys treated wastewater to receiving waters, to point of reuse (irrigation), or to land disposal site.

9.1 WASTE TREATMENT BY PONDS

Since 1958, engineers have designed and constructed a great number of ponds using research by qualified biological consultants, current scientific knowledge of ponding, and the

[1] *Stabilized Waste. A waste that has been treated or decomposed to the extent that, if discharged or released, its rate and state of decomposition would be such that the waste would not cause a nuisance or odors in the receiving water.*

[2] *Percolation (PURR-co-LAY-shun). The movement or flow of water through soil or rocks.*

[3] *Algae (AL-gee). Microscopic plants which contain chlorophyll and live floating or suspended in water. They also may be attached to structures, rocks, or other submerged surfaces. Algae produce oxygen during sunlight hours and use oxygen during the night hours. Their biological activities appreciably affect the pH, alkalinity, and dissolved oxygen of the water.*

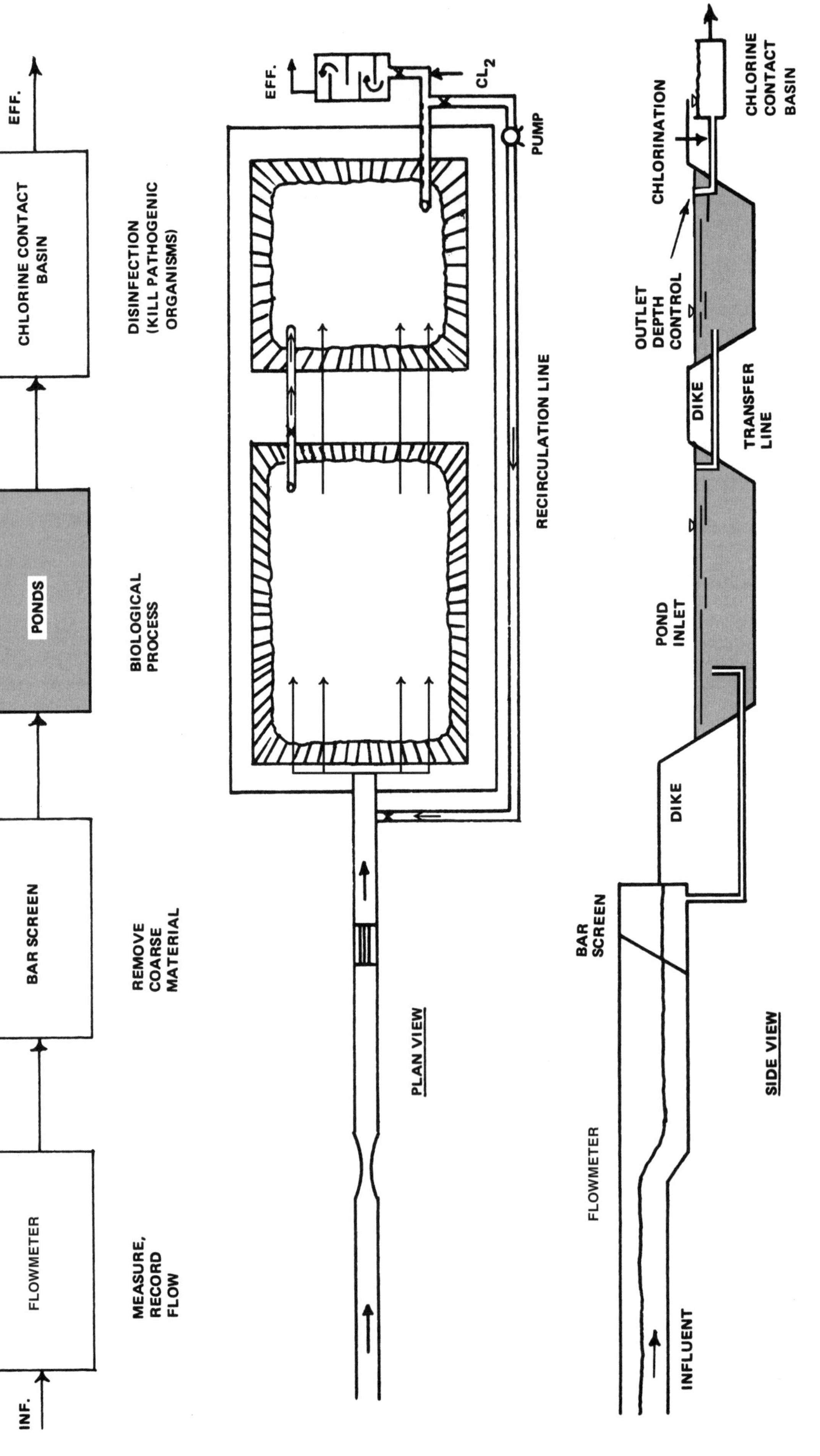

Fig. 9.1 Typical plant; ponds only

experience of past successes and failures. When operated in a knowledgeable and purposeful manner, these ponds have successfully performed a variety of functions.

As a complete process, wastewater stabilization ponds offer many advantages for smaller installations. This is true provided that land is not costly and the location is isolated from residential, commercial, and recreational areas. The advantages are that a pond:

1. Does not require expensive equipment,
2. Does not require highly trained operating personnel,
3. Is economical to construct,
4. Provides treatment that is equal or superior to some conventional processes,
5. Is a satisfactory method of treating wastewater on a temporary basis,
6. Is adaptable to changing loads,
7. Is adaptable to land application,
8. Consumes little energy,
9. Serves as a wildlife habitat,
10. Has an increased potential design life,
11. Has few short-term sludge handling and disposal problems, and
12. Is probably the most trouble-free of any treatment process when used correctly, *PROVIDED A CONSISTENTLY HIGH-QUALITY EFFLUENT IS NOT REQUIRED.*

The limitations are that a pond:

1. May produce odors,
2. Requires a large area of land,
3. Treats wastes inconsistently depending on climatic conditions,
4. May contaminate groundwaters (if the pond is not properly lined),
5. May have high suspended solids levels in the effluent due to algae,
6. May be slow to recover from *SHOCK LOADS*,[4]
7. Has limited process control capabilities, and
8. May require supplemental aeration equipment to achieve odor-free treatment.

QUESTIONS

Write your answers in a notebook and then compare your answers with those on page 45.

9.1A What conditions are critical for the use of wastewater stabilization ponds for smaller installations?

9.1B Discuss the limitations of ponds.

9.2 POND CLASSIFICATIONS AND APPLICATIONS

Treatment of wastewater using ponds, as a complete treatment process, is a method used to treat the wastewater from single-family homes as well as villages and small towns. Ponds designed to receive wastes with no prior treatment are often referred to as "raw wastewater (sewage) lagoons" or "stabilization ponds" (Figure 9.2). They may require sizable areas of land.

Ponds are quite commonly used in series (one pond following another) after a primary wastewater treatment plant to provide additional clarification, *BOD*[5] removal, and *DISINFECTION.*[6] These ponds are sometimes called *FACULTATIVE PONDS.*[7] Ponds are sometimes used in series after a *TRICKLING FILTER*[8] plant or *ROTATING BIOLOGICAL CONTACTOR (RBC).*[9] These are sometimes called "polishing ponds." Ponds placed in series with each other can provide a high-quality effluent that is acceptable for discharge into many watercourses. If the *DETENTION TIME*[10] is long enough, many ponds can meet fecal *COLIFORM*[11] standards.

A great many variations in ponds are possible due to differences in depth, operating conditions, and loadings. A bold line of distinction among different types of ponds is often impossible. Current literature generally uses three broad pond classifications: *AEROBIC,*[12] *ANAEROBIC,*[13] and facultative.

---

[4] *Shock Load. The arrival at a plant of a waste which is toxic to organisms in sufficient quantity or strength to cause operating problems. Possible problems include odors and bulking sludge which will result in a high loss of solids from the secondary clarifiers into the plant effluent and a biological process upset that may require several days to a week to recover. Organic or hydraulic overloads also can cause a shock load.*

[5] *Biochemical Oxygen Demand (BOD). The rate at which organisms use the oxygen in water or wastewater while stabilizing decomposable organic matter under aerobic conditions. In decomposition, organic matter serves as food for the bacteria and energy results from its oxidation. BOD measurements are used as a measure of the organic strength of wastes in water.*

[6] *Disinfection (dis-in-FECT-shun). The process designed to kill or inactivate most microorganisms in wastewater, including essentially all pathogenic (disease-causing) bacteria. There are several ways to disinfect, with chlorination being the most frequently used in water and wastewater treatment plants.*

[7] *Facultative (FACK-ul-TAY-tive) Pond. The most common type of pond in current use. The upper portion (supernatant) is aerobic, while the bottom layer is anaerobic. Algae supply most of the oxygen to the supernatant.*

[8] *Trickling Filter. A treatment process in which the wastewater trickles over media that provide the opportunity for the formation of slimes or biomass which contain organisms that feed upon and remove wastes from the water being treated.*

[9] *Rotating Biological Contactor (RBC). A secondary biological treatment process for domestic and biodegradable industrial wastes. Biological contactors have a rotating "shaft" surrounded by plastic discs called the "media." The shaft and media are called the "drum." A biological slime grows on the media when conditions are suitable and the microorganisms that make up the slime (biomass) stabilize the waste products by using the organic material for growth and reproduction.*

[10] *Detention Time. The time required to fill a tank at a given flow or the theoretical time required for a given flow of wastewater to pass through a tank. In septic tanks this detention time will decrease as the volumes of sludge and scum increase.*

[11] *Coliform (COAL-i-form). One type of bacteria. The presence of coliform-group bacteria is an indication of possible pathogenic bacterial contamination. The human intestinal tract is one of the main habitats of coliform bacteria. They may also be found in the intestinal tracts of warm-blooded animals, and in plants, soil, air, and the aquatic environment. Fecal coliforms are those coliforms found in the feces of various warm-blooded animals; whereas the term "coliform" also includes other environmental sources.*

[12] *Aerobic (AIR-O-bick). A condition in which atmospheric or dissolved molecular oxygen is present in the aquatic (water) environment.*

[13] *Anaerobic (AN-air-O-bick). A condition in which atmospheric or dissolved molecular oxygen is NOT present in the aquatic (water) environment.*

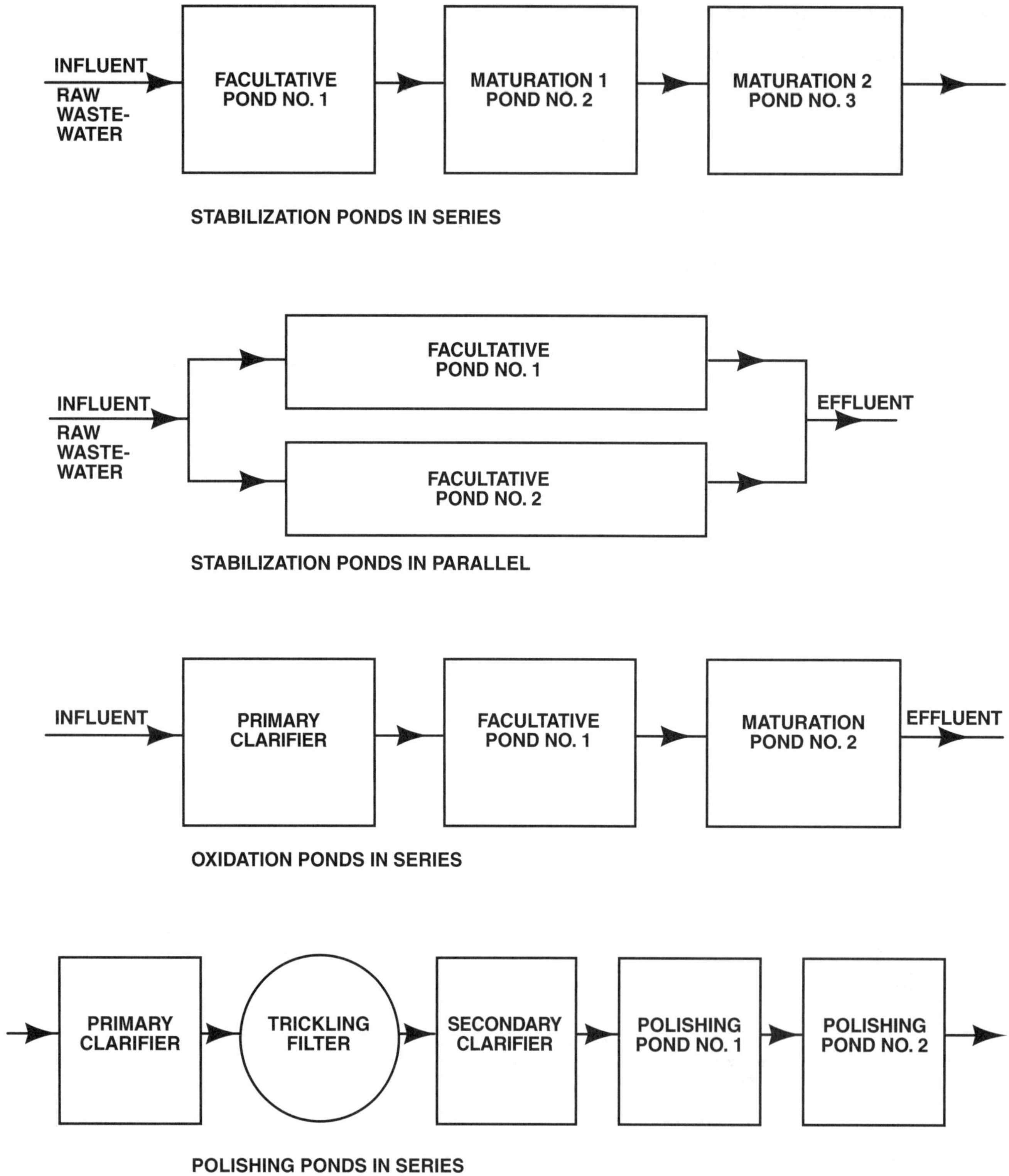

Fig. 9.2 Pond classifications

AEROBIC ponds are characterized by having dissolved oxygen distributed throughout their contents practically all of the time. They usually require an additional source of oxygen to supplement the rather minimal amount that can be diffused from the atmosphere at the water surface. The additional source of oxygen may be supplied by algae during daylight hours, by mechanical agitation of the surface, or by compressors bubbling air through the pond.

ANAEROBIC ponds, as the name implies, usually are without any dissolved oxygen throughout their entire depth. Treatment depends on fermentation of the sludge at the pond bottom. This process can be quite odorous under certain conditions, but it is highly efficient in destroying organic wastes. Anaerobic ponds are basically a pretreatment unit and they are seldom used for domestic wastewater treatment; however, some aerobic and facultative domestic-waste ponds become anaerobic when they are badly overloaded.

FACULTATIVE ponds are the most common type in current use. The upper portion (supernatant) of these ponds is aerobic, while the bottom layer is anaerobic. Algae supply most of the oxygen to the supernatant. Facultative ponds are most common because it is almost impossible to maintain completely aerobic or anaerobic conditions all the time at all depths of the pond.

Pond uses may be classified according to detention time. A pond with a detention time of less than three days will perform in ways similar to a sedimentation or settling tank. Some growth of algae will occur in the pond, but it will not have a major effect on the treatment of the wastewater.

Abundant growth of algae will be observed in ponds with detention periods from three to around twenty days, and large amounts of algae will be found in the pond effluent. In some effluents, the stored organic material may be greater than the amount in the influent. Detention times in this range merely allow the organic material to change form and delay problems until the algae settle out in the receiving waters.

Longer detention periods in ponds may provide time for the reduction of algae due to a reduction in nutrients in the water that sustain algae. Usually, this will occur in facultative ponds having anaerobic conditions on the bottom and aerobic conditions on the surface. Combined aerobic-anaerobic treatment provided by long detention periods produces good stabilization of the pond influent during treatment in the pond.

Controlled-discharge ponds are facultative ponds with long detention times of up to 180 days or longer. These ponds may discharge effluent only once (fall) or twice (fall and spring) a year.

QUESTIONS

Write your answers in a notebook and then compare your answers with those on page 45.

9.2A What is the difference between the terms "aerobic," "anaerobic," and "facultative" when applied to ponds?

9.2B How does the use of a pond vary depending on detention time?

9.3 EXPLANATION OF TREATMENT PROCESS

As mentioned in the previous section, waste disposal ponds also are classified according to their dissolved oxygen content. Oxygen in an aerobic pond is distributed throughout the entire depth practically all the time. An anaerobic pond is predominantly without oxygen most of the time because oxygen requirements are much greater than the oxygen supply. In a facultative pond, the upper portion is aerobic most of the time, whereas the bottom layer is predominantly anaerobic.

In aerobic ponds or in the aerobic layer of facultative ponds, organic matter contained in the wastewater is first converted by aerobic bacteria to carbon dioxide and ammonia, and finally, to algae in the presence of sunlight. Algae are simple one- or many-celled microscopic plants that are essential to the successful operation of both aerobic and facultative ponds.

By using sunlight through *PHOTOSYNTHESIS*,[14] the algae use the carbon dioxide in the water to produce *FREE OXYGEN*,[15] making it available to the aerobic bacteria that inhabit the pond. Each pound of algae in a healthy pond is capable of producing 1.6 pounds of oxygen on a normal summer day. Algae live on carbon dioxide and other nutrients in the wastewater. At night, when light is no longer available for photosynthesis, algae use up the oxygen by respiration and produce carbon dioxide. The alternating use and production of oxygen and carbon dioxide can result in diurnal (daily) variations of both *pH*[16] and dissolved oxygen. During the day, algae use carbon dioxide, which raises the pH, while at night they produce carbon dioxide and the pH is lowered. Algae are found in the soil, water, and air; they occur naturally in a pond without seeding and multiply greatly under favorable conditions. Figure 9.3 illustrates the role of algae in treating wastes in a pond.

---

[14] *Photosynthesis (foe-toe-SIN-thuh-sis). A process in which organisms, with the aid of chlorophyll (green plant enzyme), convert carbon dioxide and inorganic substances into oxygen and additional plant material, using sunlight for energy. All green plants grow by this process.*

[15] *Free Oxygen. Molecular oxygen available for respiration by organisms. Molecular oxygen is the oxygen molecule, O_2, that is not combined with another element to form a compound.*

[16] *pH (pronounce as separate letters). pH is an expression of the intensity of the basic or acidic condition of a liquid. Mathematically, pH is the logarithm (base 10) of the reciprocal of the hydrogen ion activity.*

$$pH = Log \frac{1}{[H^+]}$$

The pH may range from 0 to 14, where 0 is most acidic, 14 most basic, and 7 neutral. Natural waters usually have a pH between 6.5 and 8.5.

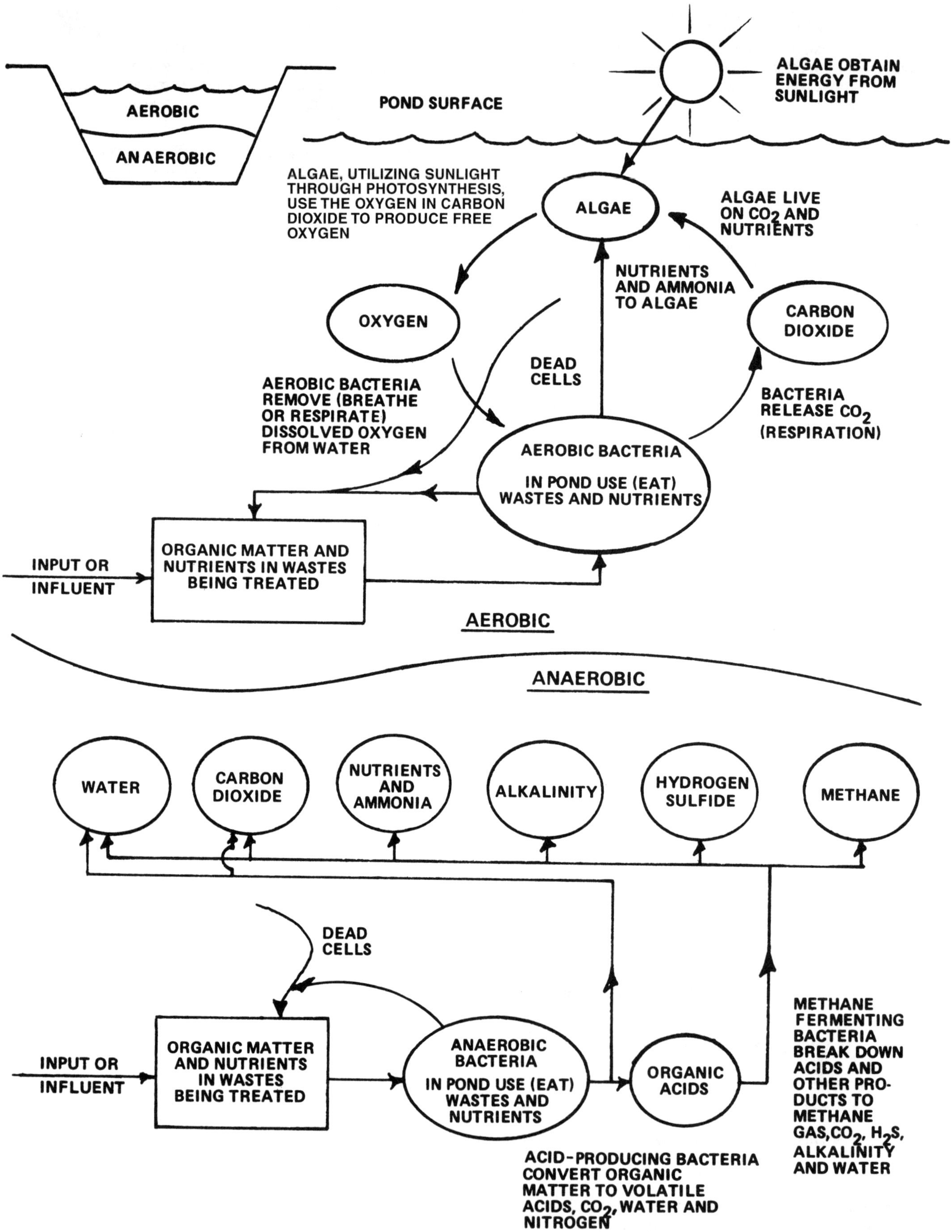

Fig. 9.3 Process of decomposition in aerobic and anaerobic layers of a pond

In anaerobic ponds or in the anaerobic layer of facultative ponds, the organic matter is first converted by a group of organisms called the "acid producers" to organic acids. In an established pond, at the same time, a group called the "methane fermenters" breaks down the organic acids produced by the first group to form methane gas, carbon dioxide, ammonia, water, and alkalinity. This process is illustrated in Figure 9.3.

In a successful facultative pond, the processes characteristic of aerobic ponds occur in the surface layers, while those similar to anaerobic ponds occur in its bottom layers.

During certain periods, sludge decomposition in the anaerobic zone is interrupted and it begins to accumulate. If sludge accumulation occurs and decomposition does not set in, it is probably due to a lack of the right bacteria, low pH, presence of substances that slow or stop the process, or a low temperature. Under these circumstances, the acid production will continue at a slower rate, but the rate of gas (methane) production slows down considerably.

Sludge storage in ponds is continuous, with small amounts stored during warm weather and larger amounts when it is cold. During low temperatures, the bacteria cannot multiply fast enough to handle the waste. When warm weather comes, the "acid producers" begin to decompose the accumulated sludge deposits built up during the winter. If the organic acid production is too great, a lowered pH will occur with the possibilities of an upset pond and resulting hydrogen sulfide odors.

Hydrogen sulfide ordinarily is not a problem in properly designed and operated ponds because it dissociates (divides) into hydrogen and hydrosulfide ions at high pH and may form insoluble metallic sulfides or sulfates. This high degree of dissociation and the formation of insoluble metallic sulfides are the reasons that ponds having a pH above 8.5 do not emit odors, even when hydrogen sulfide is present in relatively large amounts. An exception occurs in northern climates during the spring when the pH is low and the pond is just getting started; then hydrogen sulfide odors can be a problem.

All of the organic matter that finds its way to the bottom of a stabilization pond through the various processes of sludge decomposition is subject to *METHANE FERMENTATION*, provided that proper conditions exist or become established. In order for methane fermentation to occur, an abundance of organic matter must be deposited and continually converted to organic acids. An abundant population of methane bacteria must be present. They require a pH level of from 6.5 to 7.5 within the sludge, alkalinity of several hundred mg/*L* to buffer (neutralize) the organic acids (volatile acid/alkalinity relationship), and suitable temperatures. Once methane fermentation is established, it accounts for a considerable amount of the organic solids removal.

QUESTIONS

Write your answers in a notebook and then compare your answers with those on page 45.

9.3A How is oxygen produced by algae?

9.3B Where do the algae found in a pond come from?

9.3C What happens to organic matter in a pond?

9.4 POND PERFORMANCE

The treatment efficiencies that can be expected from ponds vary more than most other treatment devices. Some of the many variables are:

1. Physical Factors
   a. type of soil (permeability)
   b. surface area
   c. depth
   d. wind action
   e. sunlight
   f. temperature
   g. *SHORT-CIRCUITING*[17]
   h. inflow variations
2. Chemical Factors
   a. organic material
   b. pH
   c. solids
   d. concentration and nature of waste
3. Biological Factors
   a. type of bacteria
   b. type and quantity of algae
   c. activity of organisms
   d. nutrient deficiencies
   e. *TOXIC*[18] concentrations

The performance expected from a pond depends largely on its design. The design, of course, is determined by the waste discharge requirements or the water quality standards to be met in the receiving waters. Overall treatment efficiency may be about the same as primary treatment (only settling of solids), or it may be equivalent to the best secondary biological treatment plants. Some ponds, usually those located in hot, arid areas, have been designed to take advantage of percolation and high evaporation rates so that there is no discharge.

Depending on design, ponds can be expected to provide BOD removals from 50 to 90 percent. Facultative ponds, under normal design loads with 50 to 60 days' detention time, will usually remove approximately 90 to 95 percent of the coliform bacteria and 70 to 80 percent of the BOD load approximately 80 percent of the time. Controlled-discharge ponds

[17] *Short-Circuiting. A condition that occurs in tanks or basins when some of the flowing water entering a tank or basin flows along a nearly direct pathway from the inlet to the outlet. This is usually undesirable since it may result in shorter contact, reaction, or settling times in comparison with the theoretical (calculated) or presumed detention times.*

[18] *Toxic (TOX-ick). A substance which is poisonous to a living organism.*

with 180-day detention times can produce BOD removals from 85 to 95 percent, total suspended solids removals from 85 to 95 percent, and fecal coliform reductions up to 99 percent.

Physical sedimentation by itself has been found to remove approximately 90 percent of the suspended solids in three days. About 80 percent of the dissolved organic solids can be removed by biological action in ten days. However, in a pond with a healthy algae and bacteria population, a phenomenon known as *BIOFLOCCULATION*[19] can occur. This will remove approximately 85 percent of both suspended and dissolved solids within hours. Bioflocculation is accelerated by increased temperature, wave action, and high dissolved oxygen content.

Pond detention times are sometimes specified by regulatory agencies to ensure adequate treatment and removal of *PATHOGENIC ORGANISMS*.[20] Many agencies specify effluent or receiving water quality standards in terms of *MEDIAN*[21] and maximum MPN (Most Probable Number) values that should not be exceeded. In critical water use areas, chlorination or other means of disinfection can be used to further reduce the coliform level (see Chapter 12, "Disinfection and Chlorination"). In many areas the pond effluent must not contain any chlorine residual; therefore, the effluent must be dechlorinated.

A pond is generally regarded as not fulfilling its function when it creates a visual or odor nuisance, or leaves a high BOD, solids, grease, or coliform-group bacteria concentration in the discharge (unless it was designed to be anaerobic in the first stages and aerobic in later ponds for final treatment).

QUESTIONS

Write your answers in a notebook and then compare your answers with those on page 45.

9.4A What biological factors influence the treatment efficiency of a pond?

9.4B What is bioflocculation?

9.4C What factors indicate that a pond is not fulfilling its function (operating properly)?

END OF LESSON 1 OF 3 LESSONS
on
WASTEWATER STABILIZATION PONDS

Please answer the discussion and review questions next.

DISCUSSION AND REVIEW QUESTIONS

Chapter 9. WASTEWATER STABILIZATION PONDS

(Lesson 1 of 3 Lessons)

At the end of each lesson in this chapter, you will find some discussion and review questions. The purpose of these questions is to indicate to you how well you understand the material in the lesson. Write the answers to these questions in your notebook before continuing.

1. Where is the chlorine contact basin located in a typical plant with ponds only?
2. When wastewater flows through different treatment processes in a plant, where might ponds be located?
3. Why are most ponds "facultative ponds"?
4. Where does the oxygen come from that is produced by algae in a pond?
5. What is photosynthesis?
6. What are the three types of factors that may influence pond performance?

[19] *Bioflocculation (BUY-o-flock-u-LAY-shun). The clumping together of fine, dispersed organic particles by the action of certain bacteria and algae. This results in faster and more complete settling of the organic solids in wastewater.*

[20] *Pathogenic (PATH-o-JEN-ick) Organisms. Bacteria, viruses, cysts, or protozoa which can cause disease (giardiasis, cryptosporidiosis, typhoid, cholera, dysentery) in a host (such as a person). There are many types of organisms which do NOT cause disease and which are NOT called pathogenic. Many beneficial bacteria are found in wastewater treatment processes actively cleaning up organic wastes.*

[21] *Median. The middle measurement or value. When several measurements are ranked by magnitude (largest to smallest), half of the measurements will be larger and half will be smaller.*

CHAPTER 9. WASTEWATER STABILIZATION PONDS

(Lesson 2 of 3 Lessons)

9.5 STARTING THE POND

One of the most critical periods of a pond's life is the time that it is first placed in operation. If at all possible, at least one foot (0.3 m) of water should be in the pond before wastes are introduced. The water should be added to the pond in advance to prevent odors developing from waste solids exposed to the atmosphere. Thus, a source of water should be available when starting a pond.

A good practice is to start ponds during the warmer part of the year because a shallow starting depth allows the contents of the pond to cool too rapidly if nights are cold. Generally speaking, the warmer the pond contents, the more efficient the treatment processes.

ALGAL BLOOMS[22] normally will appear from seven to twelve days after wastes are introduced into a pond, but it generally takes at least sixty days to establish a thriving biological community. A definite green color is evidence that a flourishing algae population has been established. After this length of time has elapsed, bacterial decomposition of bottom solids will usually become established. This is generally indicated by bubbles coming to the surface near the pond inlet where most of the sludge deposits occur. Although the bottom is anaerobic, travel of the gas through the aerobic surface layers generally prevents odor release.

Wastes should be discharged to the pond intermittently during the first few weeks with constant monitoring of the pH. The pH in the pond should be kept above 7.5 if possible. Initially, the pH of the bottom sludge will be below 7 due to the digestion of the sludge by acid-producing bacteria. If the pH starts to drop, discharge to the pond should be diverted to another pond or diluted with makeup water (water from another source) if another pond is not available until the pH recovers. A high pH is essential to encourage a balanced anaerobic fermentation (bacterial decomposition) of bottom sludge. This high pH also is a good indicator of high algal activity since removal of the carbon dioxide from the water in algal metabolism tends to keep the pH high. A continuing low pH indicates acid production, which will cause odors. Soda ash (sodium carbonate) may be added to the influent to a pond to increase the pH.

QUESTIONS

Write your answers in a notebook and then compare your answers with those on pages 45 and 46.

9.5A Why should at least one foot of fresh water cover the pond bottom before wastes are introduced?

9.5B Why should ponds be started during the warmer part of the year if at all possible?

9.5C What does a definite green color in a pond indicate?

9.5D When bubbles are observed coming to the pond surface near the inlet, what is happening in the pond?

9.6 DAILY OPERATION AND MAINTENANCE

BECAUSE PONDS ARE RELATIVELY SIMPLE TO OPERATE, THEY ARE PROBABLY NEGLECTED MORE THAN ANY OTHER TYPE OF WASTEWATER TREATMENT PROCESS. Many of the complaints that arise regarding ponds are the result of neglect or poor housekeeping. Following are listed the day-to-day operational and maintenance duties that will help to ensure peak treatment efficiency and to present your plant to its neighbors as a well-run waste treatment facility. If problems develop in a pond, refer to this section and Section 9.10, "Review of Plans and Specifications," as troubleshooting guides.

9.60 Scum Control

Scum accumulation is a common characteristic of ponds and is usually greatest in the spring when the water warms and vigorous biological activity resumes. Ordinarily, wind action will break up scum accumulations and cause them to settle; however, in the absence of wind or in sheltered areas, other means must be used. If scum is not broken up, it will dry on top and become crusted. Not only is the scum more difficult to break up then, but a species of blue-green algae is apt to become established on the scum. This can give rise to disagreeable odors. If scum is allowed to accumulate, it can reach proportions where it cuts off a significant amount of sunlight from the pond. When this happens, the production of

[22] *Algal (AL-gull) Bloom. Sudden, massive growths of microscopic and macroscopic plant life, such as green or blue-green algae, which can, under the proper conditions, develop in lakes, reservoirs, and also in ponds.*

oxygen by algae is reduced and odor problems can result. Rafts of scum cause a very unsightly appearance in ponds and can quite likely become a source of botulism that will have a devastating effect on any waterfowl and shore birds that may be attracted to the facility.

Many methods of breaking up scum have been used, including agitation with garden rakes from the shore, jets of water from pumps or tank trucks, and the use of outboard motors on boats in large ponds. Outboard motors should be of the air-cooled type to avoid plugging the cooling system with algae and scum. Scum is broken up most easily if it is attended to promptly.

9.61 Odor Control

Eventually, odors probably will come from a wastewater treatment pond no matter what kind of process is used. Most odors are caused by overloading (see Section 9.109 to determine pond loading) or poor housekeeping practices and can be remedied by taking corrective measures. If a pond is overloaded, stop loading and divert influent to other ponds, if available, until the odor problem stops. Then gradually start loading the pond again. Once a pond develops odor problems, it is more apt to cause trouble than other ponds.

There are times, such as when unexpected shutdowns of mechanical equipment occur, that pond processes may be upset and cause odors. For these unexpected occurrences, it is strongly advised that a careful plan for emergency odor control be available. Odors usually occur during the spring warm-up in colder climates because biological activity has been reduced during cold weather. When the water warms, microorganisms become active and use up all of the available dissolved oxygen, creating anaerobic conditions in which odors may be produced.

There are several suggested ways to reduce odors in ponds. These ways include recirculation from aerobic units, the use of floating aerators, and heavy chlorination. Recirculation from an aerobic pond to the inlet of an anaerobic pond (one part recycle flow to six parts influent flow) will reduce or eliminate odors. If a recirculation program is used: (1) the recirculation rate must not raise the flow rate through the pond to a level that will cause a pond overflow, and (2) the recirculation water should be drawn from the surface of the source pond to ensure the highest possible dissolved oxygen levels in the recirculated water. Usually, floating aeration and chlorination equipment are too expensive to have standing idle waiting for an odor problem to develop. Small gasoline-powered pumps have been used to recirculate pond contents and help ponds recover. These pumps also have been used to spray pond water on the surface of the pond to add additional oxygen to the top layer of a pond to assist recovery. Odor-masking chemicals also have been promoted for this purpose and have some uses for concentrated sources of specific odors. However, in almost all cases, process procedures of the type mentioned previously are preferable. In any event, it is poor procedure to wait until the emergency arises to plan for odor control. Often, several days are needed to receive delivery of materials or chemicals, if they are required. Try to have possible alternate methods of control ready to go in case they are needed.

In some areas, sodium nitrate has been added to ponds as a source of oxygen for microorganisms rather than sulfate compounds, thus preventing odors. To be effective, sodium nitrate must be dispersed throughout the water in the pond. Once mixed in the pond, it acts very quickly because many common organisms (facultative groups) can use the oxygen in nitrate compounds in the absence of dissolved oxygen. The amount of sodium nitrate depends on the oxygen demand. Some operators dose their ponds successfully between 0.3 and 0.4 mg/*L* of sodium nitrate. Liquid sodium hypochlorite or chlorine solution is a faster acting solution, but not necessarily the best chemical because it will interfere with biological stabilization of the wastes.

A different odor-control strategy will be needed if the ponds have already frozen over in the winter but are expected to develop odor problems when the ice thaws in the spring and the water in the pond turns over. In this situation, some operators spread sodium nitrate on the ice surface before thaw so that the chemical will be available to supply oxygen when needed.

9.62 Weed and Insect Control

Weed control is an essential part of good housekeeping and is not a difficult task with modern herbicides and soil sterilants. Weeds around the edge are most objectionable because they allow a sheltered area for mosquito breeding and scum accumulation. In most average ponds, there has been little need for mosquito control when edges are kept free of weed growth. Weeds also can hinder pond circulation. Aquatic weeds, such as tules, will grow in depths shallower than three feet (1 m), so an operating pond level of at least this depth is necessary. Tules may emerge singly or be well scattered, but they should be removed promptly by hand because they will quickly multiply from the root system. One of the best methods for control of undesirable vegetation is achieved by a daily practice of close inspection and immediate removal of the young plants (including roots). Effective weed control strategies include using *RIPRAP*[23] or concrete on the pond banks and maintaining two to three inches (5 to 8 cm) of gravel on the roads.

Suspended vegetation, such as *DUCKWEED*,[24] usually will not flourish if the pond is exposed to a clean sweep of the wind. Dike vegetation control is aided by regular mowing and use of a cover grass that will crowd out undesirable growth. Because emergent weed growth will occur only when sunlight is able to reach the pond bottom, the single best preventive measure against emergent growth is to maintain a water depth of at least three feet (1 m). Due to greater water clarity, the amount of sunlight reaching the bottom will be greater in secondary or final ponds than in primary ponds. Because shallow water promotes growth, there will likely always be a battle to keep emergent weeds from becoming well established around lagoon banks.

Whenever emergent or suspended weeds are being pulled from the lagoon, protective gear such as waterproof gloves, boots, and goggles should be worn to reduce the chance of infection from pathogens that may be present in the water. Pulled weeds should be buried to prevent odor and insect problems. Although most stabilization ponds are no deeper than five feet (1.7 m), there is still sufficient depth to drown a

---

[23] *Riprap. Broken stones, boulders, or other materials placed compactly or irregularly on levees or dikes for the protection of earth surfaces against the erosive action of waves.*

[24] *Duckweed. A small, green, cloverleaf-shaped floating plant, about one-quarter inch (6 mm) across which appears as a grainy "scum" on the surface of a pond.*

person, especially if that person gets caught in a sticky clay liner. Using the buddy system and approved flotation devices will greatly increase the safety level when performing any pond maintenance, especially when using a boat or mowing the dike. Control measures include:

Emergent Weeds

1. Keep the water level above three feet (1 m).
2. Pull out new (first-year) growth by hand.
3. Drown the weeds by raising the water level.
4. Lower the water level, cut the weeds or burn them with a gas burner, and raise the water level (cut and drown).
5. Use riprap (large broken stones or concrete) along the bank. If riprap is used and growth continues, herbicides will be the only alternative.
6. Install a pond liner.
7. Use herbicides as a last resort.

Suspended Vegetation

1. Keep the pond exposed to a clean sweep of the wind.
2. A few ducks may be used to eat light growth of duckweed.
3. Small ponds may be skimmed with rakes or boards. This may have to be repeated.
4. Excessive growth can be mechanically harvested.
5. Use herbicides as a last resort.

Dike Vegetation

1. Mow regularly during the growing season. Dike slopes may be cut using sickle bars or weed-eater equipment. When using heavy equipment, mowers designed especially for cutting slopes are preferable. Any tractor used on the dike should have a low center of gravity. The tractor must be provided with an approved roll-over protective structure and seat belts if within certain use/weight classifications. Check with your local safety regulatory agency for applicable rules.
2. Seed or reseed slopes with desirable grasses that will form a thick and somewhat impenetrable mat.
3. Use herbicides as a last resort.

Herbicides are a last resort in vegetation control for ponds, not only because of the obvious hazards facing the operator, but also because of dangers presented to the biological growth in the pond and receiving stream (if discharging). Care must be taken to follow mixing, application, and storage directions exactly. Safety for the operator and for any other persons who might come in contact with the herbicide is of utmost importance. Proper protective gear and warnings (verbal and written) are vital. Also, desirable vegetation could be killed by herbicide drift on a windy day or by other improper applications.

Availability, formulation, trade names, and federal clearance of state registration for certain herbicides may change. Do not use stored herbicides without checking on their current status of approval.

Some herbicides are effective only on certain plants, and herbicides are most effective at specific stages in the life of the plant. Cattails, for instance, are usually sprayed when the shoots are two to three feet (0.7 to 1 m) tall and cattails are developing.

Perhaps the most important factor in the selection of the proper herbicide is site compatibility. In addition to the other important instructions on the product label, a comprehensive listing of approved application sites is usually provided. This list should be checked to verify that the product is the best choice for each specific application. Herbicides that are effective on many types of undesirable lagoon vegetation and that are approved for both pond and river (discharging lagoon) applications are often the logical choice. Some areas require an herbicide application license and complete reporting of the type and amount of material used. Verify with your regulatory agency that you can be in compliance with applicable regulations before attempting an herbicide application program.

Duckweed[25] consists of tiny green plants that float on the surface of a pond. Duckweed stops sunlight penetration and hinders surface aeration, thus reducing dissolved oxygen in the pond. Rakes, water sprays, or pushing a board with a boat can be used to move the duckweed into a corner of a pond where it can be physically removed. Another approach is to install a surface discharge in the downwind corner of a pond. The prevailing winds will blow the duckweed into the corner and the duckweed will be removed by the surface discharge. The duckweed can be captured by a wire mesh bucket placed in the effluent line after the surface outlet. The bucket will need to be removed and cleaned as the duckweed accumulates.

Mosquitoes will breed in sheltered areas of standing water where there is vegetation or scum to which the egg rafts of the female mosquito can become attached. These egg rafts are fragile and will not withstand the action of disturbed water surfaces such as caused by wind action or normal currents. Keeping the water edge clear of vegetation and keeping any scum broken up will normally give adequate control. Shallow, isolated pools left by the receding pond level should be drained or sprayed with a larvacide. Mosquito fish (*Gambusia*) are an effective means of controlling mosquitoes. A stocking density

[25] *Material in this section was adapted from an article entitled "Control of Undesirable Lagoon Vegetation," by Doug Matheny; it appeared in the Oklahoma Water Pollution Control Association's DRIP, Spring 1988.*

of 2,500 fish per acre has been reported. However, stocking levels will depend on the local situation and environmental conditions, as well as the desire to sustain the fish population.

Any of several tiny shrimp-like animals may infest the pond from time to time during the warmer months of the year (March-November). There are two general types of microcrustaceans: (1) those that have the appearance of crayfish without claws, and (2) those that have the appearance of fleas. During periods of large "blooms," these organisms will appear as large orange or red cloud-like patterns in the pond. They can prove to be a valuable part of the treatment process, especially in polishing ponds. Microcrustaceans live on algae and at times will appear in such numbers as to almost clear the pond of algae. During the more severe infestations, there will be a sharp drop in the dissolved oxygen of the pond, accompanied by a lowered pH. This is a temporary condition because the microcrustaceans will outrun the algae supply and there will be a mass die-off of the animals, followed by a rapid greening up of the pond. If the pond is operated on a batch basis, the period when the algae concentration is low may be a good time for release of water due to the low suspended solids values.

Ordinarily there should be no great concern about these infestations because they soon balance themselves; however, in the case of a heavily loaded pond, a sustained low dissolved oxygen content may give rise to obnoxious odors. In that event, any of several commercial sprays can be used to control the shrimp-like animals.

Chironomid midges are often produced in wastewater ponds in sufficient numbers to be serious nuisances to nearby residential areas, farm workers, recreation sites, and industrial plants. When emerging in large numbers, they may also create traffic hazards. At present the only satisfactory control is through the use of insecticides. Control measures are time consuming and may be difficult, particularly if there is a discharge to a receiving stream. If possible, lower the level in the ponds (one pond at a time) enough to contain a day's inflow before applying an insecticide. Holding the insecticide for at least one day will kill more insects and reduce the effect of the insecticide on receiving waters. For better results, insecticides should be applied on a calm day and any recirculation pumps should be stopped.

CAUTION: Before attempting to apply any insecticide or pesticide, contact your local official in charge of approving pesticide applications. This person can tell you which chemicals may be applied, the conditions of application, and safe procedures.

9.63 Levee Maintenance

Levee slope erosion caused by wave action or surface runoff from precipitation is probably the most serious maintenance problem. If allowed to continue, it will result in a narrowing of the levee crown, which will make accessibility with maintenance equipment most difficult.

If the levee slope is composed of easily erodible material, one long-range solution is the use of bank protection such as stone riprap or broken concrete rubble. Good bank-protection materials include small pieces of broken street materials, curbs, gutters, and also bricks and other suitable materials from building demolition. Also a semi-porous plastic sheet can be used with riprap. The sheet allows the two-way movement of air and water but prevents the movement of soils. The sheet also discourages weed growth and digging by crayfish (crawdads). Geosynthetics are an effective means of levee erosion control.

Portions of the pond levee or dike not exposed to wave action should be planted with a low-growing, spreading grass to prevent erosion by surface runoff. Native grasses may naturally seed the levees or local highway departments may be consulted for suitable grasses to control erosion. If necessary, grass may have to be mowed to prevent it from becoming too high. Do not allow large grazing animals (cows, horses) to control vegetation because they may damage the levees near the waterline and possibly complicate erosion problems. Goats have proven effective in controlling vegetation and grasses.

Plants or grasses with long roots, such as willows and alfalfa, should not be allowed to grow on levees because they may damage the levees and possibly cause levee failure and costly repair. Burrowing animals such as muskrats, badgers, squirrels, and gophers also may cause levees to fail. Remove these animals from levees as soon as possible and repair their burrowed holes immediately. Local agricultural departments may be able to assist with the development of a rodent control program.

Levee tops should be crowned so that rainwater will drain over the side in a sheet flow. Otherwise the water may flow a considerable distance along the levee crown and gather enough flow to cause erosion when it finally spills over the side and down the slope. If the levees are to be used as roadways during wet weather, they should be paved or well graveled.

If seepage or leakage from the ponds appears on the outside of levees, ask your engineer or a qualified person to investigate and solve this problem before further levee damage occurs.

9.64 Headworks and Screening

Be sure to clean the bar screen as frequently as necessary. The screen should be inspected at least once or twice a day with more frequent visits during storm periods. Screenings should be disposed of daily in a sanitary manner, such as by burial, to avoid odors and fly breeding. One method of disposal is to place screenings in garbage cans and request that your local garbage service dispose of the screenings at a sanitary landfill disposal site.

Many pond installations have grit chambers at the headworks to protect raw wastewater lift pumps or prevent plugging of the influent lines. There are many types of grit removal equipment. Grit removed by the various types of mechanical equipment or by manual means will usually contain small amounts of organic matter and should therefore be disposed of in a sanitary manner. Disposal by burial is the most common method. Disposal methods for screenings and grit must be approved by the permitting regulatory agency.

QUESTIONS

Write your answers in a notebook and then compare your answers with those on page 46.

9.6A Why should scum not be allowed to accumulate on the surface of a pond?

9.6B How can scum accumulations be broken up?

9.6C What are the causes of odors from a pond?

9.6D What precaution would you take to be prepared for an odor problem that might develop?

9.6E Why are weeds objectionable in and around ponds?

9.6F How can weeds be controlled and removed in and around ponds?

9.6G Why should insects be controlled?

9.6H Why should a pond be lowered before an insecticide is applied?

9.65 Some Operating Hints

1. Anaerobic ponds should be covered and isolated for odor control and followed by aerobic ponds. Floating polystyrene planks can be used to cover anaerobic ponds and can be painted white for protection from the sun. These will help to confine odors and heat and tend to make the anaerobic ponds more efficient.

2. Placing *PONDS IN SERIES*[26] tends to cause the first pond to become overloaded and may never allow it to recover; the overload may be carried to the next pond in series. Feeding *PONDS IN PARALLEL*[26] allows the distribution of the incoming load evenly between units. Whether ponds are operated in series or in parallel should depend on the loading situation and *NPDES PERMIT*[27] requirements.

   When operating ponds in series, the accumulation of solids in the first pond may become a serious problem after a long period of use. Periodically the flow should be routed around the first pond. This pond should then be drained and the solids removed and disposed of in an approved manner. A special permit from the regulatory agency may be required before burying solids removed from a pond. Also, the solids may be removed by dredging or pumping.

3. A large amount of recirculation, say 25 to 100 percent, can be very helpful. This allows the algae and other aerobic organisms to become thoroughly mixed with incoming raw wastewater. At the same time, good oxygen transfer can be attained by passing the incoming water over a deck and down steps or other type of aerator. However, this procedure can cause heat loss.

4. Heavy chlorination at the recirculation point can assist in odor control, but it will probably interfere with treatment.

5. As with any treatment process, it is necessary to measure the important water quality indicators (DO and solids) at frequent, regular intervals and plot them so that you have some idea of the direction the process is taking in time to take corrective action when necessary. During normal pond operation, the algae produced, dissolved oxygen (DO), and solids will be highest in the summer and during daylight hours. Dissolved oxygen and solids will be lowest during the winter and at night.

6. When solids start floating to the surface of a pond during the spring or fall *OVERTURN*,[28] the pond may have to be taken out of service and cleaned. Measurement of the sludge depth on the bottom of a pond also will indicate when a pond should be cleaned. Usually, ponds are cleaned when the wet sludge is over one foot (0.3 m) deep.

7. Before applying insecticides or herbicides, be sure to check with appropriate authorities regarding the regulation of their use and the long-term effects of the pesticide you plan to use. Do not apply pesticides that may be toxic to organisms in the receiving waters.

8. Algae in effluent. Unfortunately, there is little an operator can do by changing pond operating procedures to effectively remove algae from a pond effluent. The best approach is to operate ponds in series and to draw off the effluent just below the surface by use of a good baffling arrangement.

   If the algae must be removed from the effluent of a pond, additional facilities may be designed and constructed. Techniques for eliminating algae from pond effluents are described in Section 9.11, "Eliminating Algae From Pond Effluents." Suitable effluents have been obtained under favorable field conditions. Algae removal processes include microscreening, slow sand filtration, dissolved air flotation, and algae harvesting. In final ponds that are operated in series with periodic discharges, alum may be added in doses of less than 20 mg/*L* to improve effluent quality

---

[26]

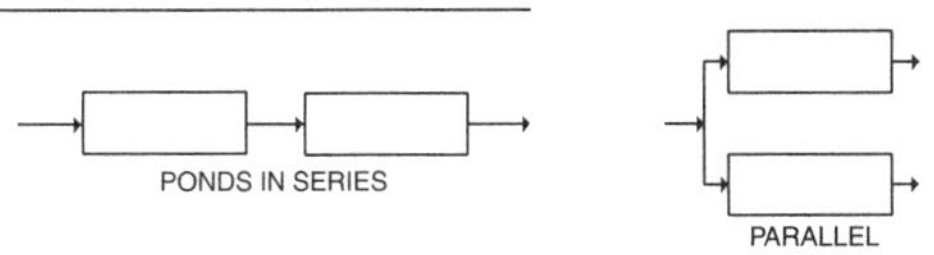

[27] *NPDES Permit.* ***N**ational **P**ollutant **D**ischarge **E**limination **S**ystem permit is the regulatory agency document issued by either a federal or state agency which is designed to control all discharges of potential pollutants from point sources and storm water runoff into U.S. waterways. NPDES permits regulate discharges into navigable waters from all point sources of pollution, including industries, municipal wastewater treatment plants, sanitary landfills, large agricultural feedlots and return irrigation flows.*

[28] *Overturn. The almost spontaneous mixing of all layers of water in a reservoir or lake when the water temperature becomes similar from top to bottom. This may occur in the fall/winter when the surface waters cool to the same temperature as the bottom waters and also in the spring when the surface waters warm after the ice melts. This is also called "turnover."*

before discharge. Also, algae have been removed from the effluent of polishing ponds by the use of agricultural "pod" type pressure filters and the use of straw.

9.66 Abnormal Operation

Abnormal operation occurs when ponds are overloaded because the BOD loads and/or flows are too high. Excessive BOD levels can occur when influent loads exceed design capacity due to population increases, industrial growth, or illegal industrial discharges. Under these conditions, new facilities must be constructed or the BOD loading must be reduced at the source. Repeated wide fluctuations in BOD loads over short time periods will also interfere with pond performance and can create a nearly constant state of abnormal operation.

Another type of overloading can occur when too much flow is diverted to one pond. This can happen when an operator accidentally feeds one pond more than the other or when a pipe opening is blocked by rags, solids, or grit due to low pipe velocities, and thus too much flow is diverted to another pond. When this happens and the overloaded pond starts producing odors, take the pond out of service and divert flows to the other ponds until the overloaded pond recovers. Hopefully, the ponds in service will not become overloaded. Also, be sure to remove the rags, solids, or grit that caused the overloading and inspect the other pipes to prevent this problem from happening again in the other ponds. Another source of plugging or blocking of pipes carrying water from one pond to another can be naturally occurring algae. Pipes can be cleaned on a scheduled basis using a high-velocity cleaner or hydro-jet used to clean sewers.

Usually, ponds do not become overloaded during storms and periods of high runoff because there is not a significant increase in the BOD loading on the ponds.

Large amounts of brown or black scum on the surface of a pond is an indication that the pond is overloaded. Scum on the surface of a pond often leads to odor problems. The best way to control scum is to take corrective action as soon as possible (see Section 9.60, "Scum Control").

During winter conditions, the pond can become covered with ice and snow. Sunlight is no longer available to the algae and oxygen cannot enter the water from the atmosphere. Without dissolved oxygen available for aerobic decomposition, anaerobic decomposition of the solids occurs. Anaerobic decomposition takes place slowly because of the low temperatures. By keeping the pond surface at a high level, a longer detention time will be obtained and heat losses will be minimized. During the period of ice cover, odorous gases formed by anaerobic decomposition accumulate under the ice and are dissolved into the wastewater being treated.

Some odors may be observed in the spring just after the ice cover breaks up because the pond is still in an anaerobic state and some of these dissolved gases are being released. Melting of ice in the spring provides dilution water with a high oxygen content, thus the ponds usually become facultative in a few days after breakup of the ice if they are not organically (BOD) overloaded.

9.67 Batch Operation

Some ponds do not discharge continuously. These ponds may discharge only once (fall) or twice (fall and spring) a year. Discharges should be made only when necessary and, if possible, during the nonrecreational season when flows are high in the receiving waters.

If your pond is allowed to discharge intermittently (controlled discharge), you must work closely with your pollution control agency and be sure that you are in compliance with the National Pollutant Discharge Elimination System (NPDES) permit. Before and during the discharge, samples should be collected from the pond being emptied and from the receiving waters both upstream and downstream from the point of discharge. Samples are usually analyzed for DO, BOD, pH, total suspended solids, and coliform-group bacteria.

Ponds should not be emptied too quickly. If a pond is emptied too quickly, the wet side slope along the shoreline may slide into the pond. Usually, ponds are emptied in two weeks or less, depending on how much water is to be discharged. Normally, 1.0 to 1.5 feet (0.3 to 0.45 m) of water is left in the bottom of the pond.

9.68 Controlled Discharge

A common operational modification to aerated and facultative lagoons is the controlled-discharge pond mode in which pond discharge is prohibited during the winter months in cold climates and/or during the peak algal growth periods in all climates. In this approach, each cell in the system is isolated and then discharged sequentially. Sufficient storage capacity is provided in the lagoon system to allow wastewater storage during winter months, peak algal growth periods, or receiving stream low-flow periods. In the Great Lakes states of Michigan, Minnesota, and Wisconsin, as well as in the province of Ontario, Canada, many pond systems are designed to discharge in the spring and/or fall when water quality effects are minimized. As a secondary benefit, operational costs are lower than for a continuous-discharge lagoon because of reduction in laboratory monitoring requirements and the need for less operator control.

A similar modification is called a hydrograph controlled-release lagoon (HCRL). Water is retained in the pond until flow volume and conditions in the receiving stream are adequate for discharge, thus eliminating the need for costly additional treatment. An HCRL system has three principal components: a stream-flow monitoring system, a storage cell, and an effluent discharge system. The stream-flow monitoring system measures the flow rate in the stream and transmits the data to the effluent discharge system. The effluent discharge system consists of a controller and a discharge structure. The controller operates a discharge device, such as a motor-driven sluice gate, through which wastewater is discharged from the storage lagoon; however, these tasks can be manually performed.

Key considerations for HCRLs include:

- An effluent release model keyed to flow (as measured by depth) in the receiving stream is needed,
- Outlets should be capable of drawing from different depths to ensure best quality, and
- Storage cells should be sized by use of water-balance equations.

9.69 Shutting Down a Pond

Ponds may be shut down for short periods of time without any problems developing. For example, if flow to a pond must be stopped to repair a pipe or a valve, no precautionary procedures are necessary. If a pond is full and received no flows for a long period of time, start up the pond with caution and gradually increase the load. If the full load is applied immediately, the pond may become overloaded because the microorganism population in the pond is low and insufficient to treat the load.

Stop all flow to a pond when emptying it to remove bottom deposits, repair inlet or outlet structures, or repair levees. Drain by use of discharge valves or pump water from one pond to other ponds. Try to feed the other ponds remaining in service equally to prevent them from becoming overloaded. Frequently, there is a time lag between the overloading of a pond and the development of problems. Therefore, watch the other ponds and lab results closely for any signs (odors, low pH, low DO, drop in alkalinity, loss of green color) of potential problems developing.

9.610 Operating Strategy

To prevent ponds from developing odors or discharging an effluent in violation of the NPDES permit requirements, develop a plan to keep the ponds operating as intended.

A. Maintain constant water elevations in the ponds.

If the NPDES permit allows the discharge of a pond into receiving waters, keep a constant water level to help maintain constant loadings. When the water surface elevation starts to drop, look for the following possible causes:

1. Discharge valve open too far or a *STOP LOG*[29] is missing;
2. Levees leaking due to animal burrows, cracks, soil settlement, or erosion; and
3. Inlet lines plugged or restricted and causing wastewater to back up into the collection system.

When the water surface starts to rise, look for:

1. Discharge valve closed or lines plugged, and
2. Sources of infiltration.

NOTE: Under some conditions you may not want to maintain constant water levels in your ponds. For example, you may allow the water surface to fluctuate to:

1. Control shoreline aquatic vegetation,
2. Control mosquito breeding and burrowing rodents,
3. Handle fluctuating inflows, and
4. Regulate discharge (continuous, intermittent, or seasonal).

B. Distribute inflow equally to ponds.

All ponds designed to receive the inflow should receive the same hydraulic and organic (BOD) loadings.

C. Keep pond levees or dikes in good condition.

Proper maintenance of pond levees can be a time-saving activity. Regularly inspect levees for leaks and erosion and correct any problems before they become serious. If erosion is a problem at the waterline, install riprap. Do not allow weeds to grow along the waterline and keep weeds on the levee mowed. If insect larvae are observed on the pond surface, spray with an appropriate insecticide before problems develop.

D. Observe and test pond condition.

Daily visual observations can reveal if a pond is treating the wastewater properly. The pond should be a deep green color indicating a healthy algae population. Scum and floating weeds should not be present, but if they are present, remove them to allow sunlight to reach the algae in the pond.

Once or twice a week, tests should be conducted to determine pond dissolved oxygen level, pH, and temperature. Sampling should be conducted at the same time of day to ensure comparable results. Effluent dissolved oxygen should be measured at this time also. Other effluent tests should be conducted at least weekly and include BOD, suspended solids, dissolved solids, coliform-group bacteria, and chlorine residual. If ponds are operated on a batch or controlled-discharge basis, these effluent tests will have to be determined only during periods of discharge.

During warm summer months, algae populations tend to be high and may cause high suspended solids concentrations in the effluent. An advantage of ponds in arid regions is that this is also a period of high evaporation rates. Under these conditions, effluent flows may drop to almost zero or may be stopped. In the fall and winter when the weather is cool and sunlight is reduced, the algae population in ponds is also reduced, thus resulting in reduced suspended solids. This situation could allow ponds to meet effluent requirements during this period.

If test results reveal that certain water quality indicators (such as DO, BOD, pH, or suspended solids) are tending to move in the wrong direction, try to identify the cause and take corrective action. Remember that ponds are a biological process and that changes resulting from corrective action may not occur until a week or so after changes were made.

E. Baffles.

Baffle walls placed properly in aerated ponds can ensure maximum efficiency by directing the flow along a path that will cause all of the water to pass through the area of influence of each aerator. This pattern of baffles eliminates short-circuiting through the pond.

F. Troubleshooting.

The Troubleshooting Guide on pages 24 through 25 provides you with step-by-step procedures to follow when problems develop in a pond.

[29] *Stop Log. A log or board in an outlet box or device used to control the water level in ponds and also the flow from one pond to another pond or system.*

TROUBLESHOOTING GUIDE FOR PONDS

(Adapted from *PERFORMANCE EVALUATION AND TROUBLESHOOTING AT MUNICIPAL WASTEWATER TREATMENT FACILITIES,* Office of Water Program Operations, US EPA, Washington DC.)

Indicator/Observation	Probable Cause	Check or Monitor	Solution
1. Poor quality effluent.	1a. Mixing/agitation equipment failure.	1a. Monitor surface aerators, rotors, or aeration equipment, and DO of pond water.	1a. 1. Restart out of service mixing or aeration units. 2. Increase operating time. 3. Increase recycle flow from effluent to influent. 4. Provide additional mixing (small boat and outboard motor for one hour for every two hours of daylight, or at least three times a day).
	1b. Organic overload.	1b. Monitor influent laboratory data: BOD, SS, DO, pH, temperature.	1b. 1. Increase or start recirculation of effluent to influent. 2. Mix pond contents hourly by surface aerators or outboard motor on small boat at least three times per day. 3. Increase run cycle on surface aeration equipment to maintain at least 1.0 mg/*L* DO. 4. Add chemicals to help reduce pond load by prechlorination for BOD reduction, add sodium nitrate or hydrogen peroxide for oxygen input.
	1c. Excessive turbidity from scum mats.	1c. Floating mats of scum or sludge on surface and corners of pond.	1c. 1. Break up scum mats. a. water sprays b. poles or rakes c. mixing equipment d. outboard motor 2. Check pond influent for excess grease or scum. 3. If seasonal temperature change, rising sludge from pond may have to be broken up daily by mixing methods in item 1c.1. above.
	1d. Blockage of light by excessive plant growth (tules, reeds, and/or grasses) near dikes.	1d. Visual inspection for weed growth in and near ponds.	1d. 1. Remove plant growth. 2. Schedule regular herbicide applications to levees and dikes as a last resort.
	1e. Low temperature.	1e. Monitor air and pond water temperature.	1e. 1. Freezing weather, raise pond levels to increase water depth. 2. Reduce recirculation rates. 3. If possible, operate ponds in series.
	1f. Toxic material in influent.	1f. Color change, low DO, low pH for no apparent reason.	1f. 1. Sample and try to identify toxic material. If the evidence of the toxic material can be verified in time, isolate the toxic material in one pond until it can be identified and then detoxified or neutralized. 2. Increase recirculation from effluent to influent. 3. Increase surface aeration or pond mixing times. 4. Implement new or enforce existing sewer-use ordinances. 5. See 1b. above.
	1g. Loss of pond volume caused by sludge accumulation.	1g. Sludge depth.	1g. Remove sludge.

Indicator/Observation	Probable Cause	Check or Monitor	Solution
1. Poor quality effluent. (continued)	1h. Process chemicals from illegal drug laboratories dumped into rural area manholes.	1h. Surveillance of rural area manholes.	1h. Install locking devices on manhole to prevent access by unauthorized persons. Isolate "shocked" ponds and gradually recycle polished effluents.
2. Low dissolved oxygen in ponds.	2a. Low algal growth.	2a. Visual inspection, pond not green, low DO during daylight hours, none from sunup until noon. Odor.	2a. 1. Recirculate last pond effluent to inlet of first pond. 2. Same as 1b.
	2b. Excessive scum accumulation.	2a. Pond surface conditions.	2b. 1. Break up and sink re-suspended scum. 2. Skim off grease balls and scum. Dispose of in landfill.
3. Odors.	3a. Anaerobic conditions or spring and fall turnovers.	3a. Monitor pond loading, BOD, SS, pH, DO, and temperature.	3a. 1. Limit organic load by diverting influent flows to several ponds. 2. Isolate pond and remove excess sludge solids. 3. Same as 1b.
	3b. Hydrogen sulfide in pond influent.	3b. Monitor total dissolved H_2S.	3b. 1. Check collection system. 2. Pretreat with chlorine or pre-aeration. 3. Aerate plant influent in small pond at least 30 minutes before applying to other ponds.
4. Inability to maintain sufficient liquid.	4a. Leakage.	4a. 1. Seepage around dikes. 2. Sampling wells located on the outside perimeter of the pond(s).	4a. 1. Apply bentonite clay to the pond water to seal the leak. 2. If sampling well analysis indicates contamination, the lagoon/pond site may require lining.
	4b. Excessive evaporation or percolation.	4b. Detention time in pond is probably too long.	4b. Divert land drainage or stream flow into pond.
5. Insect generation.	5a. Layers of scum and excessive plant growth in sheltered portions of the pond.	5a. Visual inspection. Mosquitoes.	5a. 1. Weed and scum removal. 2. Application of approved insecticides or larvicides.
	5b. Shallow pools of standing water outside pond.	5b. Visual inspection.	5b. Cut vegetation outside pond and fill in potholes that collect standing water nearby.
6. Levee erosion.	6a. Windy conditions.	6a. Visual inspection.	6a. 1. Riprap levee. 2. Construct a wind barrier around pond.
	6b. Excessive surface aerator operating time.	6b. Aerator operating time.	6b. Reduce aerator operating time if DO levels allow.
7. Excessive weeds and tule growth.	7a. Pond too shallow.	7a. Visual inspection for weeds in the area.	7a. Deepen all pond areas to at least 3 feet.
	7b. Inadequate maintenance program to control vegetation.	7b. Maintenance program.	7b. 1. Correct program deficiency. 2. Install pond lining. 3. Herbicide program as a last resort.
	7c. Poor circulation.	7c. Visual inspection of flow characteristics.	7c. Fluctuate pond level.
8. Animals burrowing into the dikes.	8. Burrowing animals (gophers, squirrels, crayfish).	8. Visual inspection.	8. 1. Alter pond level several times in rapid succession. 2. Remove animals as soon as possible. 3. Provide riprap with semi-porous sheet on levee slopes.
9. Groundwater contamination.	9. Leakage through bottom and/or sides of pond.	9. Seepage around pond dikes.	9. Apply bentonite clay to pond water to seal leak.

QUESTIONS

Write your answers in a notebook and then compare your answers with those on page 46.

9.6I Why are the contents of ponds recirculated?

9.6J When do ponds that are operated on a batch basis discharge?

9.6K What factors would you consider when developing an operating strategy for ponds?

9.7 AERATION SYSTEMS AND EQUIPMENT

9.70 Surface Aerators (Table 9.2 and Figures 9.4 and 9.5)

Surface aerators have been used in two types of applications:

1. To provide additional air for ponds during the night, during cold weather, or for overloaded ponds; and
2. To provide a mechanical aeration device for ponds operated as aerated lagoons. Aerated lagoons operate like activated sludge aeration tanks without returning any settled activated sludge.

TABLE 9.2 PURPOSE OF AERATOR PARTS

Part	Purpose
PLATFORM AERATOR COMPONENTS	
Aerator	Introduces oxygen into pond.
Electric Motor	Provides energy to drive aerator.
Drive Reduction	Converts torque from motor to
Gear Box	drive impeller.
Draft Tube	Conveys bottom contents of pond to surface for aeration.
Discharge Guide	Regulates spray patterns for oxygen transfer to water.
Jacking Screw	Adjusts aerator impeller level in water to regulate oxygen transfer and motor loading amps.
FLOATING AERATOR COMPONENTS (additional)	
Pontoons or Floats	Provide platform for motor.
Guy Wires	Maintain aerator positions in ponds and are anchored to pond levee.
Power Cable	Conveys power to motor.
Impeller (aerator)	Pumps water into air to be aerated.

In both cases, the aerators are operated by time clocks with established ON/OFF cycles. Laboratory tests on the dissolved oxygen in a pond indicate the time period for ON and OFF cycles to maintain aerobic conditions in the surface layers of the pond. Adjustments in the ON/OFF cycles are necessary when changes occur in the quantity and quality of the influent and seasonal weather conditions. Some experienced operators have correlated their lab test results to pond appearance and regulate the ON/OFF cycles using the following rule: "If the pond has foam on the surface, reduce the operating time of the aerator; and if there is no evidence of foam on the pond surface, increase the operating time of the aerator." If there is a trace of foam on the surface, the operating time is satisfactory.

Surface aerators may be either stationary or floating. Maintenance of surface aerators should be conducted in accordance with manufacturer's recommendations. Always turn off, tag, and lock out electric current when repairing surface aerators. Special precautions may be necessary to handle problems with icing and winter maintenance. Overhead guy wires have been used to prevent aerators from turning over when iced up. The operator should be aware that if a power failure occurs, even for a short time, the pond surface will freeze rapidly and possibly damage mechanical aerators.

9.71 Aeration Tubing

Another method of aerating ponds is by the use of air compressors connected to plastic tubes placed across the pond bottom. Holes are drilled in the plastic tubes, which serve as diffusers to disperse air in the pond. With this method of aeration, diffusers can become plugged and create maintenance problems. If an aeration system becomes plugged due to deposits of carbonate, try reducing the aeration.

Aeration tubing requires an effective maintenance program. The type of maintenance depends on the type of aeration system and problems encountered.

1. Acid cleaning. Some aeration tubing systems require cleaning on a weekly basis. If excessive hardness in the wastewater leads to bicarbonate deposits, the deposits can be removed by the use of an acid wash. This can be done using an industrial solution of hydrogen chloride, commonly known as muriatic acid. This liquid acid solution is a strong acid and requires careful handling.
2. High-pressure air purging. Some aerated lagoon facilities are capable of applying 90 psi (620 kPa) of compressed air to the system just after the blowers and the air manifold. The air manifold valves and blowers should be off. This is not the pressure level that will reach the lagoon tubing because of the friction losses in the tubing system. High-pressure cleaning will greatly deform the slits in the tubing and break off deposits farther back in the tubing. High-pressure air may be applied for three to five minutes once a year.
3. Slit deformation. The aeration tubing slits may be allowed to deform by shutting off the air once a day for about 20 minutes. The weight of the water above the tubing will cause the slits to be deformed inward, thus breaking up the deposits. However, this procedure may allow some suspended material and/or sediment to enter the air line. This material can plug the air line when the air is turned on again.
4. Removal of tubing and cleaning. A very thorough but time-consuming job is the drawdown of the lagoon and removal of the tubing for cleaning. The tubing can be flexed by hand or with special equipment to break up deposits and the inner surfaces of the tubing may be blown out with high-pressure air. You may be able to avoid draining the lagoon to clean the tubing by using a boat, retrieving one end of a line of tubing, and manually flexing each section of tubing.
5. Blowouts. Blowouts of aeration tubing and boils caused by slit enlargements must be repaired. The damaged area must be cut out and a special coupling inserted and clamped in place. The lagoon water level may have to be lowered so the line can be hooked and drawn to the surface for repair.

Fig. 9.4 Surface aerator
(Permission of EIMCO)

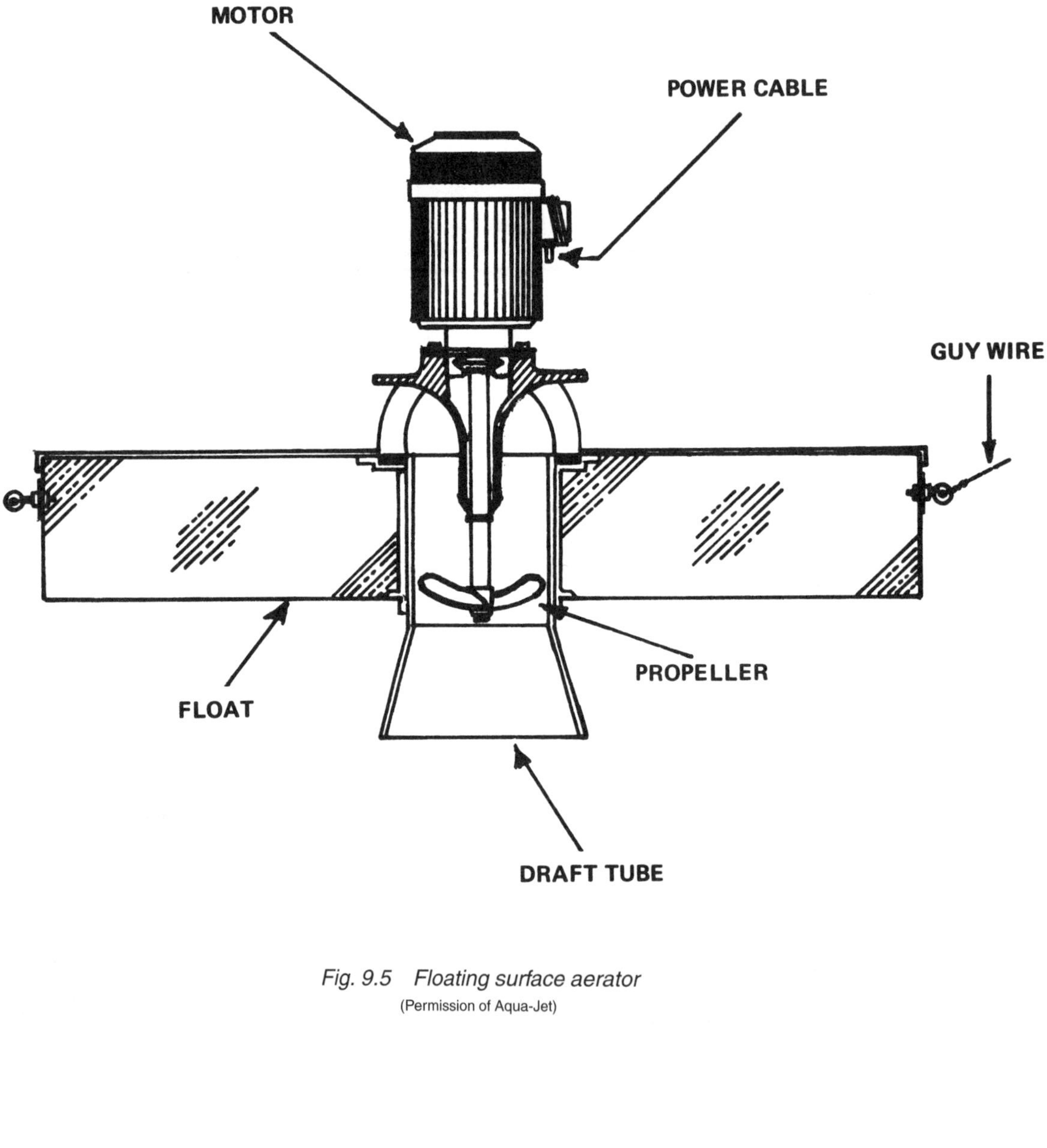

Fig. 9.5 Floating surface aerator
(Permission of Aqua-Jet)

Acknowledgment

Material in this section was provided by Jon Jewett.

9.8 SAMPLING AND ANALYSIS

9.80 Importance

Probably the most important sampling that can be accomplished easily by any operator is routine pH and dissolved oxygen analysis. A good practice is to take pH, temperature, and dissolved oxygen tests several times a week, and occasionally during the night. These values should be recorded because they will serve as a valuable record of performance and for monitoring reports. Sampling should be conducted at the same time of day to ensure comparable results. However, the time of day should be varied occasionally for the tests using *GRAB SAMPLES*[30] so that the operator becomes familiar with the pond's characteristics at various times of the day. Usually, the pH and dissolved oxygen will be lowest just at sunrise. Both will get progressively higher as the day goes on, reaching their highest levels in late afternoon.

Be very careful to avoid getting any atmospheric oxygen into the sample taken to measure dissolved oxygen. This is most necessary when samples are taken in the early morning or if the dissolved oxygen in the pond is low from overloading. If possible, measure the dissolved oxygen with an electric meter and probe, being careful not to allow the membrane on the end of the probe to be exposed to the atmosphere during actual DO measurement of the water sample.

Ponds often have clearly developed individuality, each being a biological community that is unique. Apparently identical adjacent ponds receiving the same influent in the same amount often have a different pH and a different dissolved oxygen content at any given time. One pond may generate considerable scum while its neighbor doesn't have any scum. For this reason, *EACH* pond should be tested routinely for pH and dissolved oxygen. Such testing may indicate an unequal loading because of the internal clogging of influent or distribution lines that might not be apparent from visual inspection. Tests also may indicate differences or problems that are being created by a buildup of solids or solids recycling.

When an operator becomes familiar with operating a pond, the results of some of the chemical tests can be related to visual observations. A deep green sparkling color generally indicates a high pH and a satisfactory dissolved oxygen content. A dull green color or lack of color generally indicates a declining pH and a lowered dissolved oxygen content. A gray color indicates the pond is being overloaded or that it is not working properly.

9.81 Frequency and Location of Lab Samples

The frequency of testing and expected ranges of test results vary considerably from pond to pond, but you should establish those ranges within which your pond functions properly. Test results will also vary during the hours of the day. Table 9.3 summarizes the typical tests, locations, and frequency of sampling. (See Chapter 14, "Laboratory Procedures," for detailed testing procedures.)

TABLE 9.3 FREQUENCY AND LOCATION OF LAB SAMPLES[a]

Test	Frequency[b]	Location	Common Range
pH[c]	Weekly	Pond	7.5+
Dissolved Oxygen (DO)[c]	Weekly	Pond	4 – 12 mg/*L*
		Effluent	4 – 12 mg/*L*
Temperature	Weekly	Pond	
BOD[d]	Weekly	Influent	100 – 300 mg/*L*
		Effluent	20 – 50 mg/*L*
Coliform-Group Bacteria	Weekly	Effluent	MPN >24,000/100 m*L* (unchlorinated)
Chlorine Residual	Daily	Effluent	0.5 – 2.0 mg/*L*
Suspended Solids[e]	Weekly	Influent	100 – 350 mg/*L*
		Effluent	40 – 80 mg/*L*
Dissolved Solids	Weekly	Influent	250 – 800 mg/*L*

[a] Consistent sampling times are required for pond operational control and regulatory monitoring reports to establish logical interpretations of lab results and pond performance. Suggested sampling times are 1100 or 1300.

[b] Tests may be less frequent for ponds with long detention times (greater than 100 days).

[c] pH values above 9.0 and DO levels over 15 mg/*L* are not uncommon.

[d] Contact your regulatory agency to determine whether effluent samples should be filtered to remove algae before testing. If the samples must be filtered, the agency will recommend the proper procedures.

[e] Effluent suspended solids consist of algae, microorganisms, and other suspended matter.

Samples should always be collected from the same point or location. Raw wastewater samples for pond influent tests may be collected either at the wet well of the influent pump station or at the inlet control structure. Samples of pond effluent should be collected from the outlet control structure or from a well-mixed point in the outfall channel. Pond samples may be taken from the four corners of the pond. The samples should be collected from a point eight feet (2.5 m) out from the water's edge and one foot (0.3 m) below the water surface. Be careful; try not to stir up material from the pond bottom. Do not collect pond samples during or immediately after high winds or storms because solids will be stirred up after such activity.

BOD should be measured on a weekly basis. Samples should be taken during the day at low flow, medium flow, and high flow. The average of these three tests will give a reasonable indication of the organic load of the wastewater being treated. If it is suspected that the BOD varies sharply during the day or from day to day, or if unusual circumstances exist, the sampling frequency should be increased to obtain a clear definition of the variations. If the pond DO level is supersaturated (higher than maximum theoretical DO levels), the sample must be aerated to remove the excess oxygen before the BOD test is performed. A typical data sheet for a plant consisting mainly of ponds is provided in the Appendix at the end of this chapter. Figure 9.6 shows the influent and effluent BOD and suspended solids values from an actual pond.

[30] *Grab Sample. A single sample of water collected at a particular time and place which represents the composition of the water only at that time and place.*

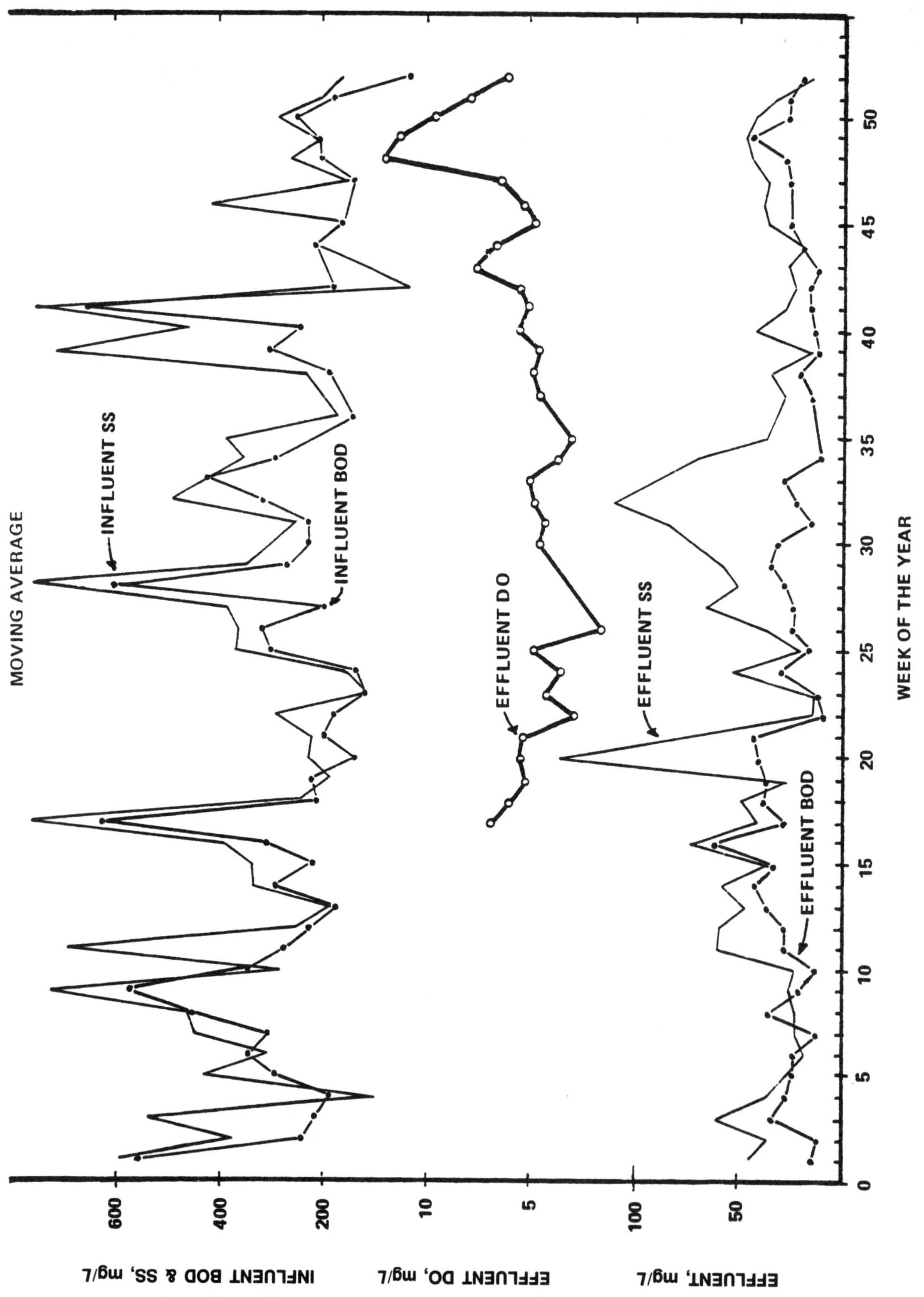

Fig. 9.6 Pond influent and effluent BOD and suspended solids values and effluent DO values

A grab sample is a single sample. Grab samples are used to measure temperature, pH, dissolved oxygen, and chlorine residual. These tests must be performed immediately after the sample is collected in order to obtain accurate results. *COMPOSITE SAMPLES*[31] of pond influent or effluent are collected by gathering individual samples at regular intervals over a selected period of time. The individual samples are then mixed together in proportion to the flow at the time of sampling. Pond samples may be composited by mixing equal portions from the four corners of the pond. Composite samples should be placed in a refrigerator or ice chest as soon as possible after they are collected. BOD and suspended solids are measured using composite samples.

Tests of pH, DO, and temperature are important indicators of the condition of the pond, whereas BOD, coliform, and solids tests measure the efficiency of the pond in treating wastes. BOD is also used to calculate the loading on the pond.

In order to estimate the organic loading on the pond, the operator must have some knowledge of the biochemical oxygen demand (BOD) of the waste and the approximate average daily flow. Influent BOD and solids will vary with time of day, day of week, and season, but a pond is a good equalizer if not overloaded.

Tests for pond alkalinity can provide helpful information to an operator. After you have determined "normal" alkalinity levels for your pond, a sudden change in alkalinity of 10 to 20 mg/*L* or more may indicate that a problem is developing. A change in alkalinity may be a warning that the pH of the pond could change in a day or two if corrective action is not taken. If a trend in changes in alkalinity continues in one direction for two or three days, the cause of this change should be identified. For useful results, the alkalinity test should be performed every day.

9.82 Expected Treatment Efficiencies[32]

Table 9.4 is provided as a guide to indicate probable removal efficiencies of typical ponds.

TABLE 9.4 EXPECTED RANGES OF REMOVAL BY PONDS

Test	Detention Time	Expected Removal
BOD		50 to 90%
BOD (facultative pond)	50 to 60 days	70 to 80%[a]
Coliform Bacteria (facultative pond)	50 to 60 days	90 to 95%
Suspended Solids	After 3 days	90%
Dissolved Organic Solids	After 10 days	80%

[a] Expected removal approximately 80 percent of the time with poorer removals during the remainder of the time.

The calculation of pond BOD removal efficiency is figured in terms of the percentage of BOD removed.

EXAMPLE

The influent BOD to a series of ponds is 300 mg/*L*, and the effluent BOD is 60 mg/*L*. What is the efficiency of BOD removal?

$$\text{BOD Removal, \%} = \frac{(\text{In} - \text{Out})}{\text{In}} \times 100\%$$

$$= \frac{(300\ mg/L - 60\ mg/L)}{300\ mg/L} \times 100\%$$

$$= \frac{240\ mg/L}{300\ mg/L} \times 100\%$$

$$= 0.80 \times 100\%$$

$$= 80\%$$

Throughout this manual, Calculator Math procedures developed by T. M. Morrow also will be illustrated and used by permission. Many operators have found this procedure helpful because it outlines the procedures followed to solve arithmetic problems when using a calculator.

$$\text{BOD Removal} = (300\ mg/L - 60\ mg/L) \times 100\% \div 300\ mg/L = 80\%$$

9.83 Response to Poor Pond Performance

See Section 9.6, "Daily Operation and Maintenance," especially Section 9.65, "Some Operating Hints."

QUESTIONS

Write your answers in a notebook and then compare your answers with those on page 46.

9.7A Surface aerators have been used in what two types of applications in ponds?

9.8A Why should the pH, temperature, and dissolved oxygen be measured in a pond?

9.8B If the color of a pond is dull green, gray, or colorless, what is happening in the pond?

9.8C Influent BOD to a series of ponds is 200 mg/*L*. If the BOD in the effluent of the last pond is 40 mg/*L*, what is the BOD removal efficiency?

9.9 SAFETY

Even though a pond has little mechanical equipment, there still are hazards. Catwalks should have guardrails and nonskid walking surfaces. Headworks and any enclosed *APPURTENANCES*[33] should be well ventilated to prevent dangerous gas accumulations.

[31] *Composite (come-PAH-zit) (Proportional) Sample. A composite sample is a collection of individual samples obtained at regular intervals, usually every one or two hours during a 24-hour time span. Each individual sample is combined with the others in proportion to the rate of flow when the sample was collected. The resulting mixture (composite sample) forms a representative sample and is analyzed to determine the average conditions during the sampling period.*

[32] $\textit{Waste Removal, \%} = \frac{(In - Out)}{In} \times 100\%$

[33] *Appurtenance (uh-PURR-ten-nans). Machinery, appliances, structures and other parts of the main structure necessary to allow it to operate as intended, but not considered part of the main structure.*

WARNING

AN OPERATOR SHOULD <u>ALWAYS</u> BE ACCOMPANIED BY A HELPER WHEN PERFORMING ANY TASK THAT IS <u>DANGEROUS</u> SINCE POND LOCATIONS ARE USUALLY QUITE ISOLATED. IMMEDIATE AID MIGHT BE NEEDED TO PREVENT SERIOUS INJURY OR LOSS OF LIFE.

Be very careful when removing debris from channels and ponds. Do not attempt to lift too much. Make certain you have secure footing so you won't slip and fall. Never stand in a boat or lean over too far to one side, for you could fall into the pond and also possibly tip over the boat. Always wear a life jacket when in a boat.

Electrical wires and electrical equipment are always a source of potential danger. Exercise caution when cutting weeds or removing vegetation such as trees next to electrical wires. Electrical wires in damp areas can be especially dangerous. Be careful when spraying weeds around electrical wires and equipment because the spray could act as a conductor. Always turn off, tag, and lock out electric current when repairing surface aerators and other equipment operated by electricity.

When applying pesticides or herbicides, be sure they are approved by the appropriate officials for your specific use and *FOLLOW THE DIRECTIONS EXACTLY.* This includes following the directions for using the proper procedures for mixing or preparing the solution, applying the solution, disposing of any excess solution and the containers, and cleaning up before you go home. Protective clothing, gloves, and respiratory protection may be required depending on the product, product strength, application method, and wind conditions. Not only can you kill the target pest or weed, but carelessness can harm nearby grasses, plants, trees, fish, birds, and even people, including yourself. Also, failure to follow directions may result in harm to the algae and microorganisms in the pond. Many areas require specific applicator training/certification. Check with the local regulatory agency.

Extreme care must be exercised when working with wastewater to avoid infections and diseases. *Clostridium* is a group of obligate bacterium that are normal inhabitants of the intestinal tract of almost all animals. In the presence of free oxygen, these bacteria produce highly resistant intercellular spores. Under anaerobic conditions where nutrients and moisture are present, as in puncture wounds or severely smashed tissue, the spores can "germinate" allowing the organisms to resume their growth and reproduction. During growth and reproduction, the organisms produce toxins that cause severe diseases.

There are two members of the *clostridia* group that present the greatest peril to people working in the wastewater treatment industry. They are *clostridia tetani* and *clostridia gangrene.* The tetanus toxin attacks the central nervous system and eventually interferes with breathing and may cause death, if not treated. Gangrene toxin causes the tissue to rot in the area of the wound, and can cause death, if untreated. Fortunately, there is an inoculation that can protect against tetanus, and gangrene infections can be treated with antibiotics.

Adequate precautions should be observed by applying first-aid procedures to all cuts and scrapes and always washing before eating or smoking. In all cases of injury involving puncture or severe tissue damage, rapid medical attention must be given to prevent severe damage or death.

Fences should surround ponds to keep unauthorized persons and animals out of the pond area. They should be located in such a manner that they will not interfere with mechanical or hand maintenance of levee slopes.

QUESTIONS

Write your answers in a notebook and then compare your answers with those on page 46.

9.9A What safety devices should be provided on catwalks over ponds?

9.9B Why should an operator be accompanied by a helper when performing any dangerous task?

END OF LESSON 2 OF 3 LESSONS
on
WASTEWATER STABILIZATION PONDS

Please answer the discussion and review questions next.

DISCUSSION AND REVIEW QUESTIONS

Chapter 9. WASTEWATER STABILIZATION PONDS

(Lesson 2 of 3 Lessons)

Write the answers to these questions in your notebook before continuing. The question numbering continues from Lesson 1.

7. Why should water be introduced into a new pond before any wastewater?
8. Why is good housekeeping an important factor in operating a properly functioning pond?
9. Why may chlorine compounds or chlorine solution not be the best method of odor control in a pond?
10. What precautions should be taken when applying an insecticide?
11. What lab tests measure the condition of a pond?
12. Estimate the BOD removal efficiency of a series of ponds if the influent BOD is 250 mg/*L* and the effluent BOD is 50 mg/*L*.
13. Why should fences be installed around ponds?

CHAPTER 9. WASTEWATER STABILIZATION PONDS

(Lesson 3 of 3 Lessons)

9.10 REVIEW OF PLANS AND SPECIFICATIONS

A careful review of the plans and specifications of a proposed pond can provide the operator an opportunity to suggest design improvements and changes before construction that will allow the operator and the ponds to do a better job. Guidelines for reviewing plans and specifications are provided in this section. If you are having trouble with an existing pond or ponds, check the items discussed in this section for help in locating possible sources of problems.

9.100 Location

The general considerations for the location of other types of wastewater treatment plants also apply to the location of ponds. Isolation should be as great as can be economically provided. Ponds should be isolated to prevent nuisances such as odors and insects associated with ponds from disturbing residential, commercial, and recreational neighbors, as well as to prevent possible traffic hazards from insects. Attention to the direction of prevailing winds with due regard for present and projected downwind residential, commercial, and recreational development is of utmost importance.

Winds can have both favorable and unfavorable impacts on ponds. Winds are desirable in terms of blowing surface scum and weeds not rooted in the pond bottom to one side of the pond where they can be removed. Also, winds can be helpful by mixing the contents of a pond, such as DO, algae, and incoming wastes. An undesirable aspect of winds is the creation of waves that can erode the pond levee. Both of these factors should be considered when selecting the location of the ponds and the arrangement or length of the ponds. If high winds are to be expected in the area where ponds will be constructed, try to arrange the ponds so the winds will blow across the short width of the pond rather than the length in order to reduce levee erosion caused by waves.

9.101 Chemistry of Waste

Before the design of any pond is undertaken, it should be determined whether the waste to be treated contains any toxic constituents that may interfere with growth of algae or bacteria. Certain wastes, such as dairy products and wine products, are difficult to treat because of their low pH. Any processing waste must be carefully investigated before one can be certain that it can be successfully treated by ponds. Some process wastes contain powerful fungicides and disinfectants that may have a great inhibitive effect on the biological activity in a pond. Other wastes may have nutrient deficiencies that could inhibit the growth of desirable types of algae. Some industrial wastes are difficult to treat due to high levels of starch, sugars, fats, oils, grease, or cellulose.

In addition, some natural water supplies have a high sulfur content or other chemicals that limit the possibility of desired sludge decomposition.

9.102 Headworks and Screening

A headworks with a bar screen is desirable to remove rags, bones, and other large objects that might lodge in pipes or control structures.

A trash shredder is a luxury that may not be warranted. Any material that gets past an adequate bar screen will, in all probability, not harm the influent pump. Any fecal matter will be pulverized when going through the pump.

Rotating screens can be very effective in removing grit, rags, plastic, and rubber goods from influent flow. Any material that can be kept out of the ponds will help enhance the ability of the pond to treat the wastewater and increase the useful life of the facility.

9.103 Flow Measuring Devices

An influent measuring device should be installed to give a direct reading on the daily volume of wastes that are introduced into the ponds. This information is used to determine pond detention time and also, along with a BOD measurement of the influent, is required to estimate the organic loading on the pond. Comparison of influent and effluent flow rates is necessary for estimating percolation and evaporation losses. A 24-hour flow rate recording instrument will show flow patterns that can be useful in showing a possible need for process adjustments.

A measuring device provides basic data for prediction of future plant expansion needs or for detecting unauthorized or abnormal flows. Reliable, well-kept records on flow volume help justify budgets and greatly assist an engineer's design of a plant expansion or new installation.

9.104 Inlet and Outlet Structures

Inlet structures should be simple, foolproof, and constructed of standard manufactured materials and devices so that replacement parts are readily available. Telescoping friction-fit tubes (see Figure 9.7) for regulating spill or discharge height

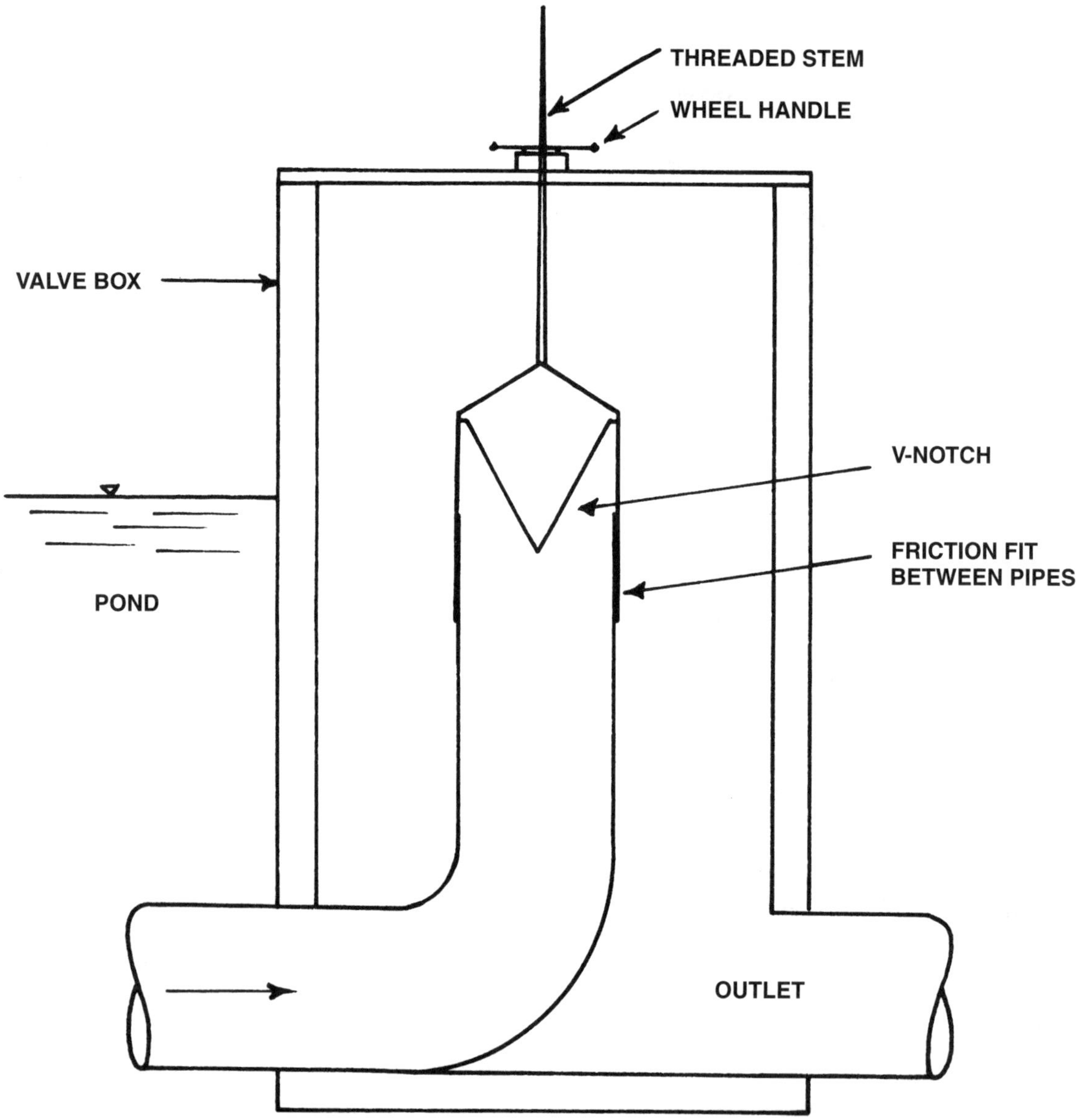

Fig. 9.7 Telescoping friction-fit tubes for regulating discharge. These tubes must be exercised regularly to prevent becoming stuck.

should be avoided because a biological growth may become attached and prevent the tubes from telescoping if they are not cleaned regularly. Occasional dosages of hypochlorite solution can effectively discourage growths. Also, the formation of ice can prevent adjustment of the tubes. If freezing is a problem, place a polyethylene floating ring sprayed with urea around the friction tube of the telescopic valve to prevent freeze-ups. This device will act as a floating baffle to keep scum and floating debris from clogging up the tube and entering the effluent.

The inlet structure must provide the ability to direct the flow to any or all of the available ponds. This flexibility will make it possible to isolate ponds for maintenance or process control.

Figure 9.1 (page 10) shows four inlets to one pond. With this type of installation, usually only one inlet valve is open at a time. If all the valves were open, velocities in the pipes might become too low and solids would settle out in the pipes. To overcome low velocities, close all but one valve or recycle pond effluent back to the inlet. When all inlet valves are open, the load is more evenly distributed throughout the pond, but requires sufficient recycle flow to maintain the desired velocities in the inlet lines.

A submerged inlet will minimize the occurrence of floating material and will help conserve the heat of the pond by introducing the warmer wastewater into the depths of the pond. Warm wastewater introduced at the bottom of a cold water mass will channel to the surface and spread unless it is promptly and vigorously mixed with cold water. Warm wastewater spilled onto the surface of the pond will spread out in a thin layer on the surface and not contribute to the warmth of the lower regions of the pond where heat is needed for bacterial decomposition. Inlet and outlet structures should be so located in relation to each other to minimize possible short-circuiting.

Outlet structures (Figures 9.8 and 9.9) should consist of a baffled and submerged pipe inlet to prevent scum and other floating surface material from leaving the pond. The actual level of the pond and rate of outflow can be controlled by the use of flash boards in the outlet structure. A rowboat may be used when access to the outlet baffle (shown in Figure 9.8) is required.

Valves that have stems extending into the stream flow should be avoided. Stringy material and rags will collect, form an obstruction, and may render the valve inoperative.

Free overfalls (Figure 9.10) at the outlet should be avoided to minimize release of odors, foaming, and gas entrapment, which may hamper pipe flows. Free overfalls should be converted to submerged outfalls if they are causing nuisances and other problems.

If a pond has a surface outlet, floating material can be kept out of the effluent by building a simple baffle around the outlet. The baffle can be constructed of wood or other suitable material. This baffle should be securely supported or anchored.

9.105 Levees

The selection of the steepness of the levee slope must depend on several variables. A steep slope erodes more quickly from wave wash unless the levee material is of a rocky nature or else protected by riprap. However, a steep slope minimizes waterline weed growth. Equipment operation and the performance of routine maintenance are more difficult on steep slopes. A gentle slope will erode the least from wave wash. Also, it is easier to operate equipment and easier to perform routine maintenance on a gentle slope. However, waterline weed growth will have a much greater opportunity to flourish.

Ensure that provisions are made to adequately compact and/or seal the levee banks to prevent leaking. Pipes passing through levees should be as close to horizontal as possible to reduce the possibility of leaks. Also cut-off walls should be installed when pipes pass through levees to prevent leakage around the outside of the pipes. Proper compaction and sealing is necessary around pipes, *SPLASH PADS*,[34] cleanouts, valves, inlet and outlet control structures, and also recirculation, transfer, and drain lines. Once a leak develops, stopping it can be very difficult.

The top of the levee should be at least ten feet (3.1 m) wide to allow for maintenance vehicles. Provisions should be made for a rounded or sloping top to allow for drainage. Pave or gravel the levee surface if it will be used as a roadway during wet weather.

9.106 Pond Depths

An observed phenomenon of lightly loaded, shallow secondary ponds and tertiary ponds is that they are apt to become infested with filamentous algae and mosses that not only limit the penetration of sunlight into the pond, but hamper circulation of the pond's contents and clog inlet and outlet structures. When the loading is increased, this condition improves because these algae and mosses require relatively clean water (low nutrients) for their environment.

Pond depths of four feet (1.2 m) or more allow a greater conservation of heat from the incoming wastes. This encourages biological activity as the ratio between pond volume and pond area is more favorable. In facultative ponds, depths over four feet (1.2 m) provide physical storage for dissolved oxygen accumulated during the day. The stored dissolved oxygen carries over through the night when no oxygen is released by the algae, unless floating algae and poor circulation keep all the oxygen near the surface. This physical storage of DO is very important during the colder months when nights are long.

A pond operating depth of at least three feet (1 m) is recommended to prevent tule and cattail growth. Weeds that emerge along the shoreline can be effectively controlled by spraying with any of several products available.

9.107 Fencing and Signs

The pond area must be surrounded by a fence capable of keeping livestock out and discouraging trespassing. A gate wide enough to allow mowing equipment and other maintenance vehicles to enter the pond facility should be provided. All access gates should have locks.

Signs should be posted along the fence around the ponds to indicate the nature of the facility and to forbid trespassing. The signs should not be more than 300 feet (90 m) apart, but the spacing should also comply with local penal codes governing trespass.

[34] *Splash Pad. A structure made of concrete or other durable material to protect bare soil from erosion by splashing or falling water.*

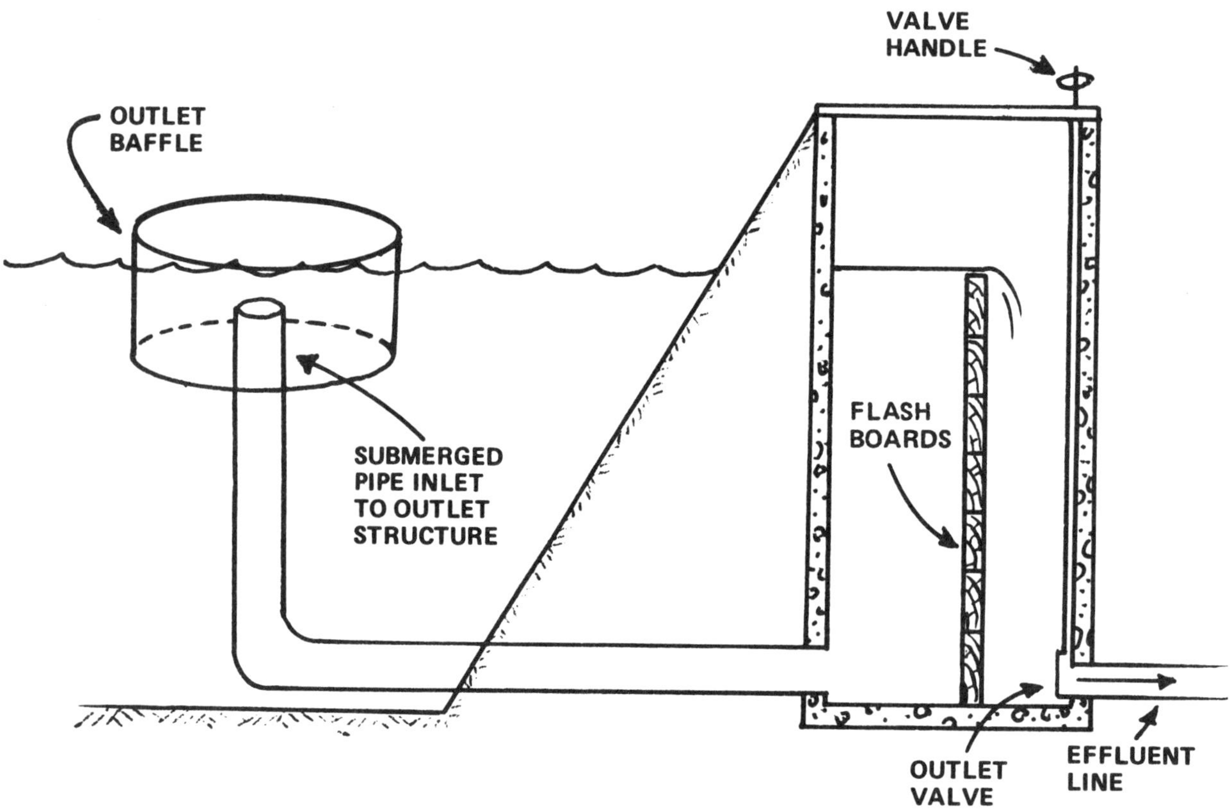

Fig. 9.8 Pond outlet structure

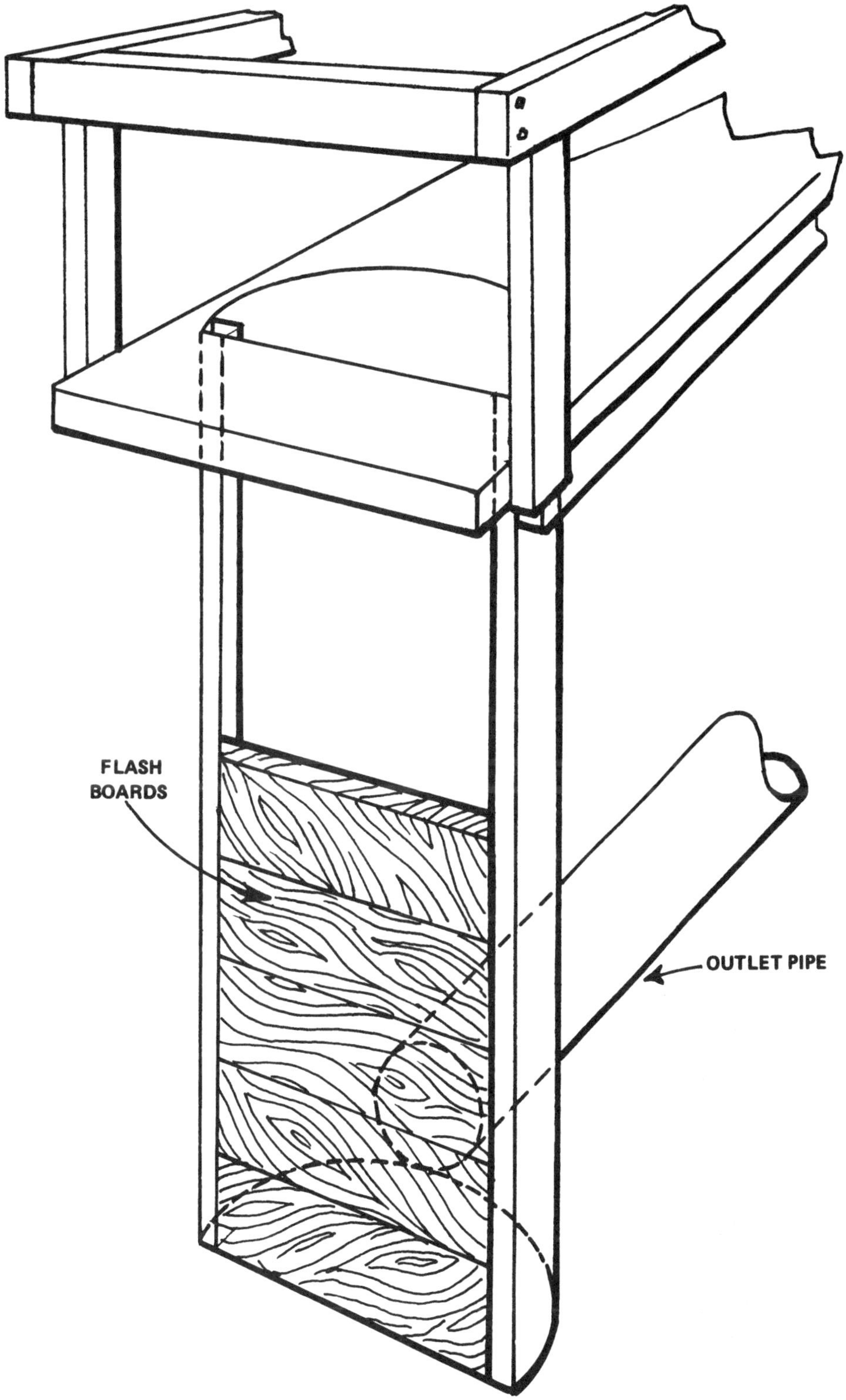

Fig. 9.9 Pond outlet control structure

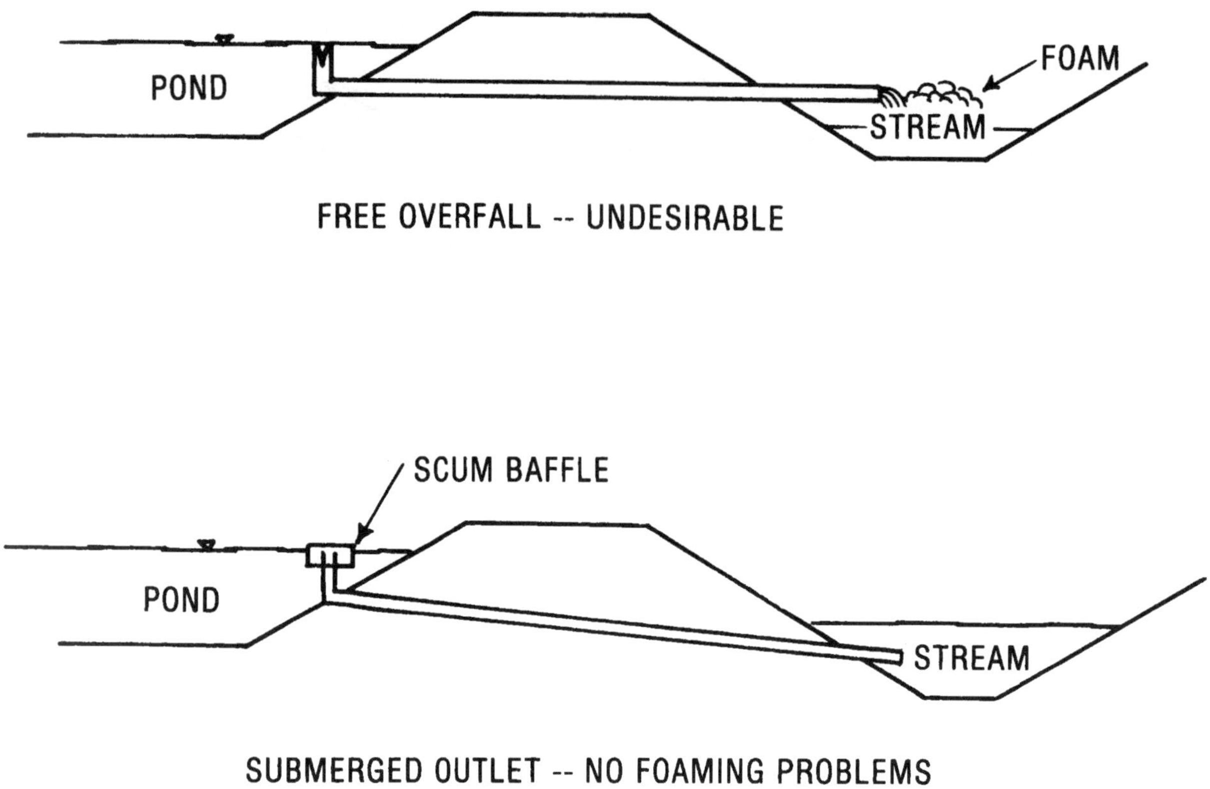

Fig. 9.10 Free overfall and submerged outlet

9.108 Surface Aerators

Provisions must be made for easy access and sufficient space for maintenance and repair of fixed aerators. Alternate anchor points should be installed in order to move floating aerators. Also, be sure the electrical cables are long enough to permit easy movement of the aerator and large enough to handle anticipated loads.

QUESTIONS

Write your answers in a notebook and then compare your answers with those on pages 46 and 47.

9.10A Why are some wastes not easily treated by ponds?

9.10B Why should the influent to a pond be metered?

9.10C Why should the inlet to a pond be submerged?

9.10D Why should the outlet be submerged?

9.10E Why should free overfalls be avoided?

9.10F How could problems created by a surface outlet be reduced or corrected?

9.10G What is the minimum recommended pond operating depth?

9.109 Pond Loading (English and Metric Calculations)

The waste loading on a pond is generally spoken of in relation to its area, and may be stated in several different ways:

- lbs of BOD per day per acre = lbs BOD/day/acre (this is called "organic loading");
- inches (or feet) of depth added per day (this is called "hydraulic loading" or "overflow rate"); or
- persons (or population served) per acre (this is called "population loading").

Detention time is related directly to pond hydraulic loading, which is actually the rate of inflow of wastewater. It may be expressed as million gallons per day (MGD), or as the number of acre-inches per day or acre-feet per day (one acre-foot is equal to 43,560 cubic feet). We must know the pond volume in order to determine detention time; this is most easily computed on an acre-foot basis.

Detention Time

$$\text{Detention (in days)} = \frac{\text{Pond Volume, ac-ft}}{\text{Influent Rate, ac-ft/day}}$$

This equation does not take into consideration water that may be lost through evaporation or percolation. Detention time may vary from 30 to 120 days, depending on the treatment requirements to be met. Controlled-discharge ponds may have minimum detention times of 180 days. Ponds whose discharges are disposed of by land application may have 210-day minimum detention times.

Population Loading

Loading calculated on a population-served basis is expressed simply as:

$$\text{No. of Persons per Acre} = \frac{\text{Population Served, persons}}{\text{Surface Area of Pond, ac}}$$

The population loading may vary from 50 to 500 persons per acre, depending on many local factors.

Hydraulic Loading

The hydraulic loading or overflow rate is expressed as:

$$\text{Inches per Day} = \frac{\text{Inflow, ac-in per day}}{\text{Average Pond Area, ac}}$$

The hydraulic loading may vary from half an inch to several inches per day, depending on the organic load of the influent.

NOTE: If the wastewater inflow rate is known in million gallons per day (MGD), it can be converted to an equivalent number of acre-inches per day as follows:

Inflow, acre-inches per day = Inflow, MGD × 36.8[35]

If the pond detention time is known, the hydraulic loading can also be calculated, as follows:

$$\text{Inches per Day} = \frac{\text{Depth of Pond, in}}{\text{Detention Time, days}}$$

Organic Loading

The organic loading is expressed as:

$$\text{Organic Load, lbs BOD per Day per Acre} = \frac{(\text{BOD, mg}/L)(\text{Flow, MGD})(8.34\text{ lbs/gal})^{36}}{\text{Pond Area, ac}}$$

Typical organic loadings may range from 10 to 50 lbs BOD per day per acre. Recirculation will help an overloaded pond.

EXAMPLE CALCULATIONS

NOTE TO OPERATORS: If you have difficulty following the work shown in Example 1 below, you should refer to Section 9.16, "Ponds—Math Supplement," page 44, for further details. If you have no trouble, continue with the lesson.

EXAMPLE 1

Use of the pond loading formulas can be illustrated by examining a typical situation. The following data should be obtained so that all the calculations can be performed:

Essential Data

1. Depth of Pond = 4 feet
2. Width of Pond
   - Bottom = 412 feet
   - Water Surface = 428 feet
   - Average Width = 420 feet
3. Length of Pond
   - Bottom = 667 feet
   - Water Surface = 683 feet
   - Average Length = 675 feet
4. Side Slopes
   (2 ft horizontal to 1 ft vertical) = 2:1

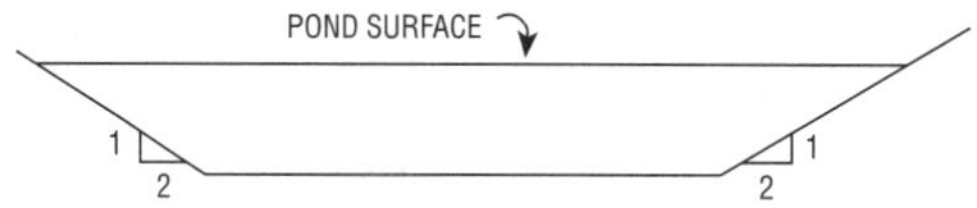

5. Influent Flow = 0.2 million gallons per day
   = 0.2 MGD
   = 200,000 gallons per day
6. Influent BOD = 200 mg/*L*
   = 200 lbs BOD per million lbs of wastewater
7. Population = 2,000 persons

To calculate the loading criteria, first determine the pond area and volume.

I. POND AREA, ACRES

$$\text{Avg Pond Area, ac} = \frac{(\text{Avg Width, ft})(\text{Avg Length, ft})}{43{,}560\text{ sq ft/ac}}$$

$$= \frac{(420\text{ ft})(675\text{ ft})}{43{,}560\text{ sq ft/ac}}$$

$$= 6.51\text{ ac}$$

II. POND VOLUME, ACRE-FEET

$$\text{Volume, ac-ft} = (\text{Avg Area, ac})(\text{Depth, ft})$$

$$= (6.51\text{ ac})(4\text{ ft})$$

$$= 26.04\text{ ac-ft (say 26 ac-ft)}$$

Convert flow rate from gallons per day to ac-ft/day.

$$\text{Flow Rate, ac-ft/day} = \frac{200{,}000\text{ gal}}{\text{day}} \times \frac{\text{cu ft}}{7.48\text{ gal}} \times \frac{\text{ac}}{43{,}560\text{ sq ft}}$$

Or

$$\text{Flow Rate, ac-ft/day} = 200{,}000\text{ GPD} \div 7.48\text{ gal/cu ft} \div 43{,}560\text{ sq ft/ac} = 0.61\text{ ac-ft/day}$$

III. LOADING CRITERIA

NOTE TO OPERATORS: Details for calculations in the remainder of Example 1 have not been given. If you have trouble, go to the end of this lesson and study Section 9.16, "Ponds—Math Supplement," for details, and try to apply them to this section.

1. *DETENTION TIME*

$$\text{Detention Time days} = \frac{\text{Pond Volume, ac-ft}}{\text{Flow Rate, ac-ft/day}}$$

$$= \frac{26\text{ ac-ft}}{0.61\text{ ac-ft/day}}$$

$$= 42.6\text{ days}$$

Or

$$\text{Detention Time, days} = 26\text{ ac-ft} \div 0.61\text{ ac-ft/day} = 42.6\text{ days}$$

---

[35] $1\text{ MGD} = \frac{1{,}000{,}000\text{ gal}}{\text{day}} \times \frac{1\text{ cu ft}}{7.48\text{ gal}} \times \frac{1\text{ ac}}{43{,}560\text{ sq ft}} \times \frac{12\text{ in}}{1\text{ ft}} = 36.8\,\frac{\text{ac-in}}{\text{day}}$

[36] *Recall lbs/day = (Conc, mg/M mg)(M gal/day)(8.34 lbs/gal)*

2. *POPULATION LOADING*

$$\text{Number of Persons per Acre} = \frac{\text{Population Served by Sewer System, persons}}{\text{Surface Pond Area, ac}}$$

$$= \frac{(2{,}000 \text{ Persons})(43{,}560 \text{ sq ft/ac})}{(428 \text{ ft})(683 \text{ ft})}$$

$$= 298 \text{ Persons/ac}$$

Or

$$\text{Number of Persons per Acre} = 2{,}000 \text{ Persons} \times 43{,}560 \text{ sq ft/ac} \div 428 \text{ ft} \div 683 \text{ ft}$$

$$= 298 \text{ Persons/acre}$$

NOTE: If there is a significant waste flow from industry mixed in with the domestic waste, an adjustment must be made to take the industrial waste into consideration. This is usually done by analyzing the industrial waste and converting it to a *POPULATION EQUIVALENT.*[37]

3. *HYDRAULIC LOADING (OVERFLOW RATE)*

$$\text{Inches per Day} = \frac{\text{Depth of Pond, in}}{\text{Detention Time, days}}$$

$$= \frac{(\text{Depth, 4 ft})(12 \text{ in/ft})}{42.6 \text{ days}}$$

$$= 1.13 \text{ in/day}$$

Or

$$\text{Inches per Day} = 4 \text{ ft} \times 12 \text{ in/ft} \div 42.6 \text{ days} = 1.13 \text{ in/day}$$

4. *ORGANIC LOADING*

$$\text{Organic Load, lbs BOD/day/ac} = \frac{(\text{BOD, mg}/L)(\text{Flow, MGD})(8.34 \text{ lbs/gal})}{\text{Area, ac}}$$

$$= \frac{200 \text{ lbs}}{\text{M lbs}} \times \frac{0.2 \text{ M gal}}{\text{day}} \times \frac{8.34 \text{ lbs}}{\text{gal}} \times \frac{1}{6.5 \text{ ac}}$$

$$= 51 \text{ lbs BOD/day/ac}$$

Or

$$\text{Organic Load, lbs BOD/day/ac} = 200 \text{ mg}/L \times 0.2 \text{ MGD} \times 8.34 \text{ lbs/gal} \div 6.5 \text{ ac}$$

$$= 51 \text{ lbs BOD/day/ac}$$

EXAMPLE 2

NOTE TO OPERATORS: Details for calculations in Example 2 have not been given. If you have trouble, go back and study the procedures for Example 1 and try to apply them to Example 2.

Suppose that a small wastewater treatment plant must be completely shut down for major repairs that will require several months of work. Enough vacant land is near the plant to enable 16 acres of temporary ponds to be constructed as raw wastewater stabilization ponds. Determine if this is feasible, given the following data:

Influent Rate = 1 MGD

Influent BOD = 150 mg/*L*
= 150 lbs BOD per million lbs of wastewater

Pond Area = 16 acres

$$\text{Average Operating Depth} = 42 \text{ inches} = \frac{42 \text{ in}}{12 \text{ in/ft}} = 3.5 \text{ ft}$$

Assume that at least a 60-day detention period (average time the wastewater must take to flow through the pond for disinfection) is desired for bacterial die-off.

Assume that the organic loading (BOD) should not exceed 50 lbs per day per acre.

Calculate what the *WASTE DETENTION TIME* would be in the pond:

One acre-foot = 325,829 gallons

$$\text{Pond Volume, ac-ft} = \text{Pond Area, ac} \times \text{Pond Depth, ft}$$

$$= 16 \text{ ac} \times 3.5 \text{ ft}$$

$$= 56 \text{ ac-ft}$$

$$\text{Influent Flow Rate, ac-ft/day} = 1{,}000{,}000 \text{ gal per day}$$

$$= \frac{1{,}000{,}000 \text{ GPD}}{325{,}829 \text{ gal/ac-ft}}$$

$$= 3.07 \text{ ac-ft per day}$$

$$\text{Detention Time, days} = \frac{56 \text{ ac-ft}}{3.07 \text{ ac-ft per day}}$$

$$= 18.2 \text{ days}$$

Or

$$\text{Detention Time, days} = 56 \text{ ac-ft} \div 3.07 \text{ ac-ft/day} = 18.2 \text{ days}$$

Thus the detention time would not be sufficient to satisfy requirements. Increasing the depth to 5 feet would help.

Calculate the organic loading:

$$\text{Organic Loading, lbs BOD/day} = (\text{BOD, mg}/L)(\text{Flow, MGD})(8.34 \text{ lbs/gal})$$

$$= (150 \text{ mg}/L)(1 \text{ MGD})(8.34 \text{ lbs/gal})$$

$$= 1{,}251 \text{ lbs BOD per day}$$

Or

$$\text{Organic Loading, lbs BOD/day} = 150 \text{ mg}/L \times 1 \text{ MGD} \times 8.34 \text{ lbs/gal}$$

$$= 1{,}251 \text{ lbs BOD/day}$$

[37] *Population Equivalent. A means of expressing the strength of organic material in wastewater. In a domestic wastewater system, microorganisms use up about 0.2 pound of oxygen per day for each person using the system (as measured by the standard BOD test). May also be expressed as flow (100 gallons per day per person) or suspended solids (0.2 lb SS/day/person).*

$$\text{Population Equivalent, persons} = \frac{\text{Flow, MGD} \times \text{BOD, mg}/L \times 8.34 \text{ lbs/gal}}{0.2 \text{ lb BOD/day/person}}$$

The organic loading per acre of pond would be:

$$\text{Organic Loading, lbs BOD/day/ac} = \frac{\text{Loading, lbs BOD/day}}{\text{Area, ac}}$$

$$= \frac{\text{1,251 lbs BOD per day}}{\text{16 ac}}$$

$$= \text{78.2 lbs BOD/day/ac}$$

Or

$$\text{Organic Loading, lbs BOD/day/ac} = \text{1,251 lbs BOD/day} \div \text{16 ac}$$

$$= \text{78.2 lbs BOD/day/ac}$$

Therefore, the organic loading would exceed the desired maximum of 50 lbs BOD/day/acre.

QUESTION

Write your answers in a notebook and then compare your answers with those on page 47.

9.10H Given a pond receiving a flow of 2.0 MGD from 20,000 people. Influent BOD is 180 mg/*L*. Pond area is 24 acres, and the average operating depth is four feet. Determine the detention time, organic loading, population loading, and hydraulic loading.

EXAMPLE METRIC CALCULATIONS

EXAMPLE 1

Use of the pond loading formulas can be illustrated by examining a typical situation. The following data should be obtained so that all calculations can be performed:

Essential Data

1. Depth of Pond = 2.0 meters
2. Width of Pond
   - Bottom = 125 meters
   - Water Surface = 133 meters
   - Average Width = 129 meters
3. Length of Pond
   - Bottom = 200 meters
   - Water Surface = 208 meters
   - Average Length = 204 meters
4. Side Slopes
   (2 m horizontal to 1 m vertical) = 2:1

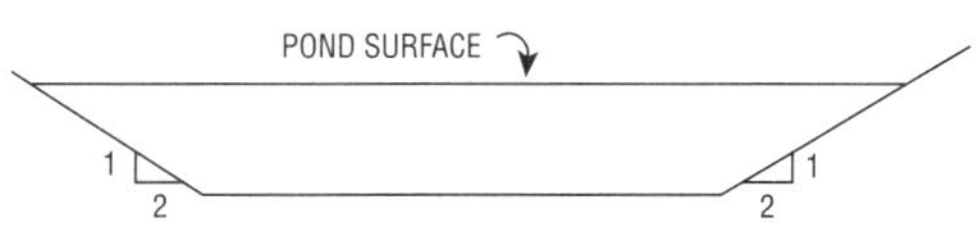

5. Influent Flow = 800 cubic meters per day
6. Influent BOD = 200 mg/*L*
7. Population = 2,000 persons

To calculate the loading criteria, first determine the pond area and volume.

I. POND AREA, SQUARE METERS

$$\text{Pond Area, sq m} = \text{(Avg Width, m)(Avg Length, m)}$$

$$= \text{(129 m)(204 m)}$$

$$= \text{26,316 sq m}$$

II. POND VOLUME, CUBIC METERS

$$\text{Volume, cu m} = \text{(Avg Area, sq m)(Depth, m)}$$

$$= \text{(26,316 sq m)(2 m)}$$

$$= \text{52,632 cu m}$$

III. LOADING CRITERIA

1. *DETENTION TIME*

$$\text{Detention Time, days} = \frac{\text{Pond Volume, cu m}}{\text{Flow Rate, cu m/day}}$$

$$= \frac{\text{52,632 cu m}}{\text{800 cu m/day}}$$

$$= \text{66 days}$$

Or

$$\text{Detention Time, days} = \text{52,632 cu m} \div \text{800 cu m} = \text{66 days}$$

2. *POPULATION LOADING*

$$\text{Number of Persons per square meter} = \frac{\text{Population Served by Sewer System, persons}}{\text{Surface Pond Area, sq m}}$$

$$= \frac{\text{(2,000 Persons)}}{\text{(133 m)(208 m)}}$$

$$= \text{0.072 Persons/sq m}$$

Or

$$\text{Number of Persons per square meter} = \text{2,000 Persons} \div \text{133 m} \div \text{208 m}$$

$$= \text{0.072 Persons/sq m}$$

NOTE: If there is a significant waste flow from industry mixed in with the domestic waste, an adjustment must be made to take the industrial waste into consideration. This is usually done by analyzing the industrial waste and converting it to a population equivalent.

3. *HYDRAULIC LOADING (OVERFLOW RATE)*

$$\text{Centimeters per Day} = \frac{\text{Depth of Pond, cm}}{\text{Detention Time, days}}$$

$$= \frac{\text{(Depth, 2 m)(100 cm/m)}}{\text{66 days}}$$

$$= \text{3.03 cm/day}$$

Or

$$\text{Centimeters per Day} = \text{2 m} \times \text{100 cm/m} \div \text{66 days}$$

$$= \text{3.03 cm/day}$$

4. *ORGANIC LOADING*

$$\text{Organic Load, gm BOD/day/sq m} = \frac{(\text{BOD, mg}/L)(\text{Flow, cu m/day})(1\text{ gm}/1{,}000\text{ mg})(1{,}000\ L/\text{cu m})}{\text{Area, sq m}}$$

$$= \frac{200\text{ mg}/L \times 800\text{ cu m/day} \times 1\text{ gm}/1{,}000\text{ mg} \times 1{,}000\ L/\text{cu m}}{26{,}316\text{ sq m}}$$

$$= 6.1\text{ gm BOD/day/sq m}$$

Or

$$\text{Organic Load, gm BOD/day/sq m} = 200\text{ mg}/L \times 800\text{ cu m/day} \times 1{,}000\ L/\text{cu m} \div 1{,}000\text{ mg/gm} \div 26{,}316\text{ sq m}$$

$$= 6.1\text{ gm BOD/day/sq m}$$

EXAMPLE 2

Suppose that a small wastewater treatment plant must be completely shut down for major repairs that will require several months of work. Enough vacant land is near the plant to enable 65,000 square meters of temporary ponds to be constructed as raw wastewater stabilization ponds. Determine if this is feasible, given the following data:

Influent Rate	= 4,000 cu m/day
Influent BOD	= 150 mg/*L*
Pond Area	= 65,000 square meters
Average Operating Depth	= 1.2 meters

Assume that at least a 60-day detention period (average time the wastewater must take to flow through the pond for disinfection) is desired for bacterial die-off.

Assume that the organic loading (BOD) should not exceed 5 grams per day per square meter.

Calculate what the *WASTE DETENTION TIME* would be in the pond:

$$\text{Pond Volume, cu m} = \text{Pond Area, sq m} \times \text{Pond Depth, m}$$

$$= 65{,}000\text{ sq m} \times 1.2\text{ m}$$

$$= 78{,}000\text{ cu m}$$

$$\text{Detention Time, days} = \frac{\text{Pond Volume, cu m}}{\text{Influent Flow, cu m/day}}$$

$$= \frac{78{,}000\text{ cu m}}{4{,}000\text{ cu m/day}}$$

$$= 19.5\text{ days}$$

Or

$$\text{Detention Time, days} = 78{,}000\text{ cu m} \div 4{,}000\text{ cu m} = 19.5\text{ days}$$

Thus the detention time would not be sufficient to satisfy requirements. Increasing the depth to 3.5 meters would help.

Calculate the organic loading:

$$\text{Organic Loading, gm BOD/day} = (\text{BOD, mg}/L)(\text{Flow, cu m/day})(1\text{ gm}/1{,}000\text{ mg})(1{,}000\ L/\text{cu m})$$

$$= (150\text{ mg}/L)(4{,}000\text{ cu m/day})(1\text{ gm}/1{,}000\text{ mg})(1{,}000\ L/\text{cu m})$$

$$= 600{,}000\text{ gm BOD per day}$$

Or

$$\text{Organic Loading, gm BOD/day} = 150\text{ mg}/L \times 4{,}000\text{ cu m/day} \times 1{,}000\ L/\text{cu m} \div 1{,}000\text{ mg/gm}$$

$$= 600{,}000\text{ gm BOD/day}$$

The organic loading per square meter of pond area would be:

$$\text{Organic Loading, gm BOD/day/sq m} = \frac{\text{Loading, gm BOD/day}}{\text{Area, sq m}}$$

$$= \frac{600{,}000\text{ gm BOD/day}}{65{,}000\text{ sq m}}$$

$$= 9.2\text{ gm BOD/day/sq m}$$

Or

$$\text{Organic Loading, gm BOD/day/sq m} = 600{,}000\text{ gm BOD/day} \div 65{,}000\text{ sq m}$$

$$= 9.2\text{ gm BOD/day/sq m}$$

Therefore, the organic loading would exceed the desired maximum of 5 gm BOD/day/sq m.

9.11 ELIMINATING ALGAE FROM POND EFFLUENTS

9.110 Treatment Methods

Algae are usually present in the effluent from ponds with continuous discharges. Algae can create undesirable impacts on the receiving waters in terms of a loss of aesthetic values, increased turbidity, suspended solids, and biochemical oxygen demand, and also the development of nuisance conditions. In most areas, the algae in the effluent increase the suspended solids concentration to the point that the NPDES effluent limitation on total suspended solids is exceeded, thus resulting in a permit violation.

Algal growths in polishing ponds can be greatly reduced by the use of microcrustaceans. When a population is established, a balance develops between the crustaceans and the algal growth, resulting in a clear, algae-free water.

Researchers have attempted to develop cost-effective treatment processes for removing algae from pond effluents. These efforts have included the use of centrifuges, chemical coagulation, filtration, microstraining, magnetic separation, and ultrafiltration. Although some of these processes have been very effective, none of these techniques have achieved a high level of acceptance due to the increased treatment costs associated with the processes.

9.111 Duckweed Systems

The Lemna Duckweed System is a patented process that uses aquatic duckweed plants for wastewater treatment. This system is used effectively as a polishing pond after a conventional wastewater treatment pond. The duckweed covering the polishing pond's surface prevents the penetration of sunlight and causes the algae to die and settle out of the wastewater being treated. Plastic grids (approximately 10 ft by 10 ft square) are placed on the surface of the pond to prevent the wind from blowing all the duckweed to one side of the pond. The population of duckweed within each grid reproduces and must be harvested on a regular basis for the system to be effective.

Duckweed plants are capable of removing phosphorus and nitrogen from the water. With sufficient detention time (greater than 30 days) and intensive harvesting, significant nutrient removal may be possible. Duckweed needs a water temperature of 50°F (10°C) or greater to be effective. If the water temperature drops below 50°F, duckweed will recover when the water temperature increases to 50°F.

The key considerations for duckweed systems include:

- A 30-day retention time, shallow pond depth, and frequent harvesting (every 1 to 3 days at peak season) if nutrient removal is desired,
- A plan must be in place for disposing of the harvested plants (that is, processing on site, transportation, and ultimate disposal),
- Post-aeration of the effluent may be necessary to meet dissolved oxygen (DO) requirements, and
- Capability to draw effluent from several levels and thereby avoid high algal concentrations near the water surface during discharge.

9.112 Intermittent Sand Filters

The intermittent sand filter is an outdoor, gravity-actuated, slow-rate filtration system that capitalizes on the availability of land area. It is a biological and physical wastewater treatment mechanism consisting of an underdrained bed of granular material, usually sand. The filter surface is flooded intermittently with lagoon effluent at intervals that permit the surface to drain between applications. It is recommended that the flow be directed to one filter for 24 hours. That filter is then allowed to drain and dry for one to two days and the flow is directed to an adjacent filter. It is preferable to have three filter beds where good operation and treatment can be accomplished over a three-day cycle. The system can remove suspended solids (algae) and BOD as well as convert ammonia to nitrate-nitrogen.

The preferred filter design for lagoon upgrading is as follows:

- Three cells in series, all 30 to 36 inches (76 to 91 cm) deep,
- Progressively finer sands (for example, effective sizes of 0.72 mm, 0.40 mm, and 0.17 mm),
- Effluent quality of 10 mg/*L* of BOD_5 and SS, and
- Complete nitrification expected, except under extremely cold conditions.

This configuration is considered to provide the most effective approach (in terms of achieving the most favorable effluent quality attainable) for lagoon upgrading.

For additional information about the operation and maintenance of sand filters, see *SMALL WASTEWATER SYSTEM OPERATION AND MAINTENANCE*, Volume I, Chapter 5, "Wastewater Treatment and Effluent Disposal Methods."

A 28-minute instructional video on intermittent sand filters is available from Orenco Systems, Inc. The video provides installation and operational guidelines as well as routine and preventive maintenance procedures. The video is available from Orenco Systems, Inc., 814 Airway Avenue, Sutherlin, OR 97479. Price, $16.00 (quantity discounts available).

9.113 Land Treatment

Another approach to improving effluent quality from waste stabilization ponds is to apply the pond effluent to an overland-flow treatment system followed by a constructed wetland wastewater treatment system.

Overland-flow systems consist of grass planted on slowly permeable soil and slopes typically from 2 to 8 percent. The slope lengths range from 100 to 200 feet (30 to 60 m). The application rate is expressed in inches per day and depends on treatment requirements and local conditions. Application periods usually range from 6 to 12 hours per day. BOD removals are in the 75 to 90 percent range and suspended solids in the 50 to 90 percent range. For additional information on overland-flow treatment systems, see *ADVANCED WASTE TREATMENT*, Chapter 8, "Wastewater Reclamation and Recycling," in this series of operator training manuals.

If higher levels of effluent quality are desired, effluent from an overland-flow system can be treated by constructed wetlands. Wetlands systems often consist of two cells and they can be of either the free surface flow (FSF) type or the submerged flow (SF) type. The vegetation in the wetland system may consist mainly of cattails. If the cattails die off in cold winter months, the organic debris is partially burnt and the remaining unburned material is removed to prevent plugging of the drains. For additional information on constructed wetlands, see Chapter 13, "Alternative Wastewater Treatment, Disposal, and Reuse Methods."

9.114 Straw

Research in England has proven that blue-green algae can be controlled by the active agent produced during the decomposition of barley straw. Wheat straw works, though not as well as barley straw. Straw should be applied to the final wastewater stabilization pond twice a year, once in the autumn and once again in the spring before algal growth starts. Recommended applications are about one-third ounce of straw per 1.33 cubic yards of water in the pond (a 3-foot-deep pond that is 20 feet by 50 feet would need about 2 pounds of straw, more if the water in the pond changes quickly). A dose of 5 grams of straw per cubic meter is also suggested. Water quality indicators should be monitored to ensure desired performance of the straw.

QUESTIONS

Write your answers in a notebook and then compare your answers with those on page 47.

9.11A What are the undesirable effects of algae on the receiving waters?

9.11B How does the Lemna Duckweed System eliminate algae from pond effluents?

9.11C What types of land treatment systems are used to improve effluent quality from waste stabilization ponds?

9.12 ARITHMETIC ASSIGNMENT

Turn to the Arithmetic Appendix at the back of this manual. Work the example problems in Section A.20, "*WASTEWATER STABILIZATION PONDS*," and check the arithmetic using your calculator. You should be able to get the same answers.

9.13 ACKNOWLEDGMENT

Liberal use has been made of the many papers presented by Professor W. J. Oswald of the University of California at Berkeley on the subject of the treatment of wastes by ponding.

9.14 ADDITIONAL READING

1. *"MOP 11,"* Chapter 23, "Natural Biological Processes."* *OPERATION OF MUNICIPAL WASTEWATER TREATMENT PLANTS* (MOP 11). Obtain from Water Environment Federation (WEF), Publications Order Department, 601 Wythe Street, Alexandria, VA 22314-1994. Order No. M05110. Price to members, $99.00; nonmembers, $148.00; plus shipping and handling.
2. *"NEW YORK MANUAL,"* page 5-43, "Stabilization Ponds."* *MANUAL OF INSTRUCTION FOR WASTEWATER TREATMENT PLANT OPERATORS* (two-volume set) distributed in New York by the New York State Department of Health, Office of Public Health Education, Water Pollution Control Board. Distributed outside of New York State by Health Education Services, PO Box 7126, Albany, NY 12224. Price for the two-volume set, $20.00, plus $4.50 shipping and handling. Make checks payable to Health Education Services.
3. *"TEXAS MANUAL,"* Chapter 16, "Stabilization Ponds."* *MANUAL OF WASTEWATER TREATMENT.* Obtain from Texas Water Utilities Association, 1106 Clayton Lane, Suite 101 East, Austin, TX 78723-1093. Price to members, $25.00; nonmembers, $35.00; plus $3.50 shipping and handling.
4. *STABILIZATION POND OPERATION AND MAINTENANCE.* Obtain from Minnesota Pollution Control Agency, 520 Lafayette Road, St. Paul, MN 55155-4194. Price, $15.00.
5. *OPERATIONS MANUAL—STABILIZATION PONDS,* U.S. Environmental Protection Agency. Obtain from ERIC/CSMEE, 1929 Kenny Road, Columbus, OH 43210-1080. Order No. U015. EPA No. 430-9-77-012. Price, $12.10, plus $5.50 shipping and handling.

* Depends on edition.

9.15 METRIC CALCULATIONS

Refer to Section 9.109, "Pond Loading (English and Metric Calculations)."

END OF LESSON 3 OF 3 LESSONS
on
WASTEWATER STABILIZATION PONDS

Please answer the discussion and review questions on page 45 before continuing with the Objective Test. (You may skip Section 9.16 unless you would like to see the arithmetic calculations for Example 1 in this chapter.)

9.16 PONDS—MATH SUPPLEMENT (Details of Example Calculations)

References: See the Appendix, "How to Solve Wastewater Treatment Plant Arithmetic Problems," at the end of this manual.

Solution: Example 1

I. Pond Area, sq ft = (Average Width, ft)(Average Length, ft)

A. Ca culate average width.

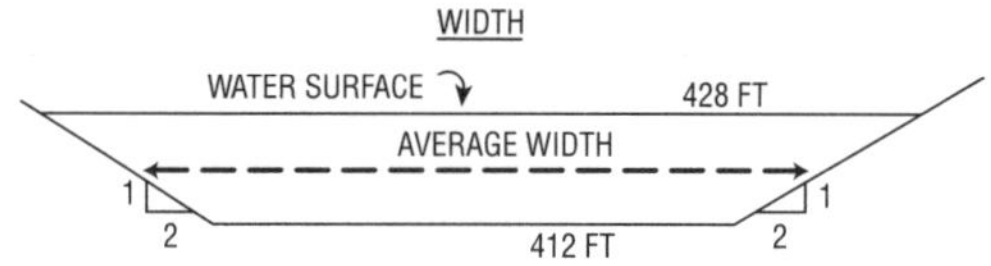

$$\text{Average Width, ft} = \frac{\text{Water Surface Width, ft} + \text{Bottom Width, ft}}{2}$$

$$= \frac{428\text{ ft} + 412\text{ ft}}{2}$$

$$= \frac{840\text{ ft}}{2}$$

$$= 420\text{ ft}$$

B. Calculate average length.

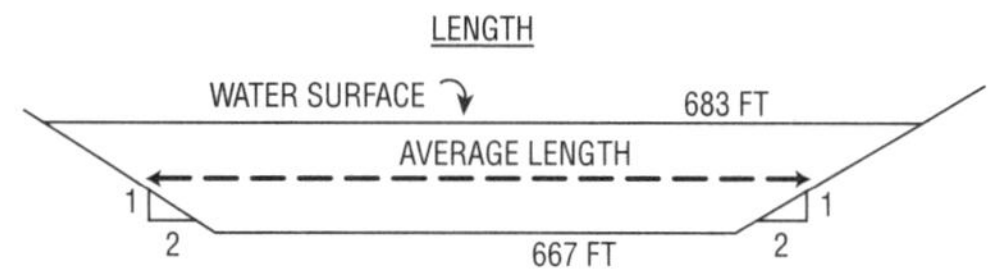

$$\text{Average Length, ft} = \frac{\text{Water Surface Length, ft} + \text{Bottom Length, ft}}{2}$$

$$= \frac{683\text{ ft} + 667\text{ ft}}{2}$$

$$= \frac{1{,}350\text{ ft}}{2}$$

$$= 675\text{ ft}$$

C. Calculate pond area.

$$\text{Area, sq ft} = (\text{Average Width, ft})(\text{Average Length, ft})$$

$$= (420\text{ ft})(675\text{ ft})$$

$$= 283{,}500\text{ sq ft}$$

Units: When we multiply ft by ft, we obtain square feet or ft^2.

$$\text{Area, acres} = \frac{\text{Area, sq ft}}{43{,}560\text{ sq ft/ac}}$$

$$= \frac{283{,}500\text{ sq ft}}{43{,}560\text{ sq ft/ac}}$$

$$= 6.5\text{ ac}$$

Units: The sq ft on top (numerator) and bottom (denominator) cancel out, and the /acre on the bottom shifts to the top (numerator).

Our result is 6.508 acres, but we will round off our answer to the nearest tenth (0.1), or 6.5. This is sufficient accuracy.

II. Calculate pond volume.

$$\text{Pond Volume, ac-ft} = (\text{Area, ac})(\text{Depth, ft})$$

$$= (6.51\text{ ac})(4\text{ ft})$$

$$= 26.04\text{ ac-ft}$$

DISCUSSION AND REVIEW QUESTIONS

Chapter 9. WASTEWATER STABILIZATION PONDS

(Lesson 3 of 3 Lessons)

Write the answers to these questions in your notebook before continuing with the Objective Test on page 48. The question numbering continues from Lesson 2.

14. Why is it desirable for a pond to be isolated from neighbors?
15. How can scum be prevented from leaving a pond?
16. How can erosion of levee slopes be controlled?
17. What methods are commonly used to eliminate algae from pond effluents?

A pond receives an inflow of 0.01 MGD from 100 people. The pond is 150 feet long, 150 feet wide, and 4 feet deep. Influent BOD is 200 mg/*L*. Determine the following loading criteria:

18. Detention time in days;
19. Population loading in persons per acre;
20. Hydraulic loading in inches per day; and
21. Organic loading in pounds of BOD applied per day per acre.

SUGGESTED ANSWERS

Chapter 9. WASTEWATER STABILIZATION PONDS

ANSWERS TO QUESTIONS IN LESSON 1

Answers to questions on page 11.

9.1A Wastewater stabilization ponds are advantageous for small installations where land is not costly and the location is isolated from residential, commercial, and recreational areas.

9.1B The limitations of a pond are that it may:

1. Produce odors,
2. Require a large amount of land,
3. Treat wastes inconsistently depending on climatic conditions,
4. Contaminate groundwaters, if not properly lined,
5. Leave high levels of suspended solids in the effluent,
6. Be slow to recover from shock loads,
7. Have limited process control capabilities, or
8. Require supplemental aeration equipment to achieve odor-free treatment.

Answers to questions on page 13.

9.2A *AEROBIC* ponds have DO distributed throughout the pond; *ANAEROBIC* ponds do not contain any DO. Most ponds are facultative and have aerobic (have DO) conditions on the surface and are anaerobic (no DO) on the bottom.

9.2B The use of a pond will vary depending on the detention period. Ponds with detention times less than three days will act like sedimentation tanks. In ponds with a detention period from three to twenty days, the organic material in the influent will be converted to algae, and high concentrations of algae will be found in the effluent. Ponds with longer detention periods provide time for the reduction of algae and a better effluent due to a reduction in nutrients in the water that sustain algae.

Answers to questions on page 15.

9.3A Algae produce oxygen from the oxygen in the carbon dioxide molecule (CO_2) through photosynthesis.

9.3B Algae simply appear in a pond on their own without seeding. They are found in soil, water, and air and multiply under favorable conditions.

9.3C Organic matter in a pond is converted to carbon dioxide and ammonia and, finally, in the presence of sunlight, to algae. The organic matter in anaerobic bottom sections is first converted by a group of organisms called "acid producers" to carbon dioxide, nitrogen, and organic acids. Next, a group called the "methane fermenters" breaks down the acids and other products of the first group to form methane gas. Another end product of organic reduction is water.

Answers to questions on page 16.

9.4A Biological factors influencing the treatment efficiency of a pond include the type of bacteria, type and quantity of algae, activity of organisms, nutrient deficiencies, and toxic concentrations.

9.4B Bioflocculation is the clumping together of fine, dispersed organic particles into settleable solids by the action of certain bacteria and algae.

9.4C A pond is not functioning properly when it creates a visual or odor nuisance, or leaves a high BOD, solids, grease, or coliform-group bacteria concentration in the effluent unless it was designed to be anaerobic in the first stages and aerobic in later ponds for final treatment.

ANSWERS TO QUESTIONS IN LESSON 2

Answers to questions on page 17.

9.5A At least one foot of water should cover the pond bottom before wastes are introduced to prevent decomposing solids from being exposed and causing odor problems.

9.5B Ponds should be started during the warmer months because higher temperatures are associated with efficient treatment processes.

9.5C A definite green color in a pond indicates a flourishing algae population and is a good sign.

9.5D When bubbles are observed coming to the pond surface near the inlet, this indicates that the solids that settled to the bottom are being decomposed anaerobically by bacterial action.

Answers to questions on page 21.

9.6A Scum should not be allowed to accumulate on the surface of a pond because it is unsightly, may prevent sunlight from reaching the algae, and an odor-producing species of algae may develop on the scum. Also, scum can become a source of botulism.

9.6B Scum accumulations may be broken up with rakes, jets of water, or by use of outboard motors.

9.6C Most odors are caused in ponds by overloading or poor housekeeping.

9.6D To prepare for an odor problem, a careful plan for emergency odor control must be developed. For example, an odor control chemical should be available before an odor problem develops. Sodium nitrate or a floating aerator will help control odors and improve treatment of the wastewater.

9.6E Weeds are objectionable in and around ponds because they provide a shelter for the breeding of mosquitoes, promote scum accumulation, and also hinder pond circulation.

9.6F Weeds may be controlled by removing them manually or by using herbicides and soil sterilants.

9.6G Insects should be controlled because they may, in sufficient numbers, be a serious nuisance to nearby residential areas, farm workers, recreation sites, industrial plants, and drivers on highways.

9.6H A pond should be lowered before the application of an insecticide to improve the destruction of insects and reduce the effect of the insecticide on the receiving waters by holding the wastewater at least one day. Lowering of the pond also will dry up weeds and insects.

Answers to questions on page 26.

9.6I The contents of ponds are recirculated to allow algae and other aerobic organisms to become thoroughly mixed with incoming raw wastewater.

9.6J Ponds that are operated on a batch basis may discharge only once (fall) or twice (fall and spring) a year.

9.6K To develop an operating strategy for ponds, consider the following factors:

1. Maintain constant water elevations in the ponds;
2. Distribute inflow equally to ponds;
3. Keep pond levees or dikes in good condition; and
4. Observe and test pond condition on a routine schedule and time for dissolved oxygen, pH, and temperature.

Answers to questions on page 31.

9.7A Surface aerators have been used (1) to provide additional air for ponds during the night, during cold weather, and for overloaded ponds, and (2) to provide air for ponds operated as aerated lagoons.

9.8A pH, temperature, and dissolved oxygen should be measured to provide a record of pond performance and to indicate the status (health) of the pond and whether corrective action is or may be necessary. DO may be expected to be low in the morning and increase with sunlight hours.

9.8B When a pond turns dull green, gray, or colorless, generally the pH and dissolved oxygen have dropped too low. This condition may be caused by overloading or lack of circulation.

9.8C Influent BOD to a series of ponds is 200 mg/*L*. If the BOD in the effluent of the last pond is 40 mg/*L*, what is the BOD removal efficiency?

$$\text{BOD Removal, \%} = \frac{(\text{In} - \text{Out})}{\text{In}} \times 100\%$$

$$= \frac{(200 \text{ mg}/L - 40 \text{ mg}/L)}{200 \text{ mg}/L} \times 100\%$$

$$= \frac{160 \text{ mg}/L}{200 \text{ mg}/L} \times 100\%$$

$$= 0.80 \times 100\%$$

$$= 80\%$$

Answers to questions on page 32.

9.9A Catwalks over ponds should have guardrails and nonskid walking surfaces.

9.9B An operator should be accompanied by a helper when performing any dangerous task because immediate aid might be needed to prevent serious injury or loss of life.

ANSWERS TO QUESTIONS IN LESSON 3

Answers to questions on page 38.

9.10A Some wastes are not easily treated by ponds because they contain substances with interfering concentrations that hinder algal or bacterial growth.

9.10B The influent to a pond should be metered to justify budgets, indicate unexpected fluctuations in flows that may cause upsets, and provide data for future expansion when necessary. Influent flows also are used to determine pond detention time and to determine pond loadings.

9.10C The inlet to a pond should be submerged to distribute the heat of the influent as much as possible and to minimize the occurrence of floating material.

9.10D The outlet of a pond should be submerged to prevent the discharge of floating material.

9.10E Free overfalls should be avoided to minimize odors, foaming, and gas entrapment, which may hamper the flow of water in pipes. They are generally controlled with pipes at the outfall.

9.10F The discharge of floating material over a surface outlet may be corrected by constructing a baffle around the outlet.

9.10G The minimum recommended pond operating depth is three feet (one meter). At shallower depths, aquatic weeds become a nuisance and pond performance is apt to be irregular.

Answers to question on page 41.

9.10H Given: Flow = 2.0 MGD
Population = 20,000 people
Influent BOD = 180 mg/*L*
Pond Area = 24 acres
Average Depth = 4 feet

Find: Detention Time
Organic Loading
Population Loading
Hydraulic Loading

Determine pond volume in acre-feet.

$$\text{Volume, ac-ft} = (\text{Avg Area, ac})(\text{Depth, ft})$$
$$= (24\text{ ac})(4\text{ ft})$$
$$= 96\text{ ac-ft}$$

Convert flow rate from MGD to ac-ft per day.

$$\text{Flow, ac-ft/day} = \frac{2{,}000{,}000\text{ gal}}{\text{day}} \times \frac{\text{cu ft}}{7.48\text{ gal}} \times \frac{\text{acre}}{43{,}560\text{ sq ft}}$$
$$= 6.1\text{ ac-ft/day}$$

Determine detention time in days.

$$\text{Detention Time, days} = \frac{\text{Pond Volume, ac-ft}}{\text{Flow Rate, ac-ft/day}}$$
$$= \frac{96\text{ ac-ft}}{6.1\text{ ac-ft/day}}$$
$$= 15.7\text{ days}$$

Calculate organic loading in pounds of BOD per day per acre.

$$\text{Organic Loading, lbs BOD/day/ac} = \frac{(\text{BOD, mg}/L)(\text{Flow, MGD})(8.34\text{ lbs/gal})}{\text{Area, ac}}$$
$$= \frac{180\text{ lbs}}{\text{M lbs}} \times \frac{2.0\text{ MG}}{\text{day}} \times \frac{8.34\text{ lbs}}{\text{gal}} \times \frac{1}{24\text{ ac}}$$
$$= 125\text{ lbs BOD/day/ac}$$

Estimate the population loading in persons per acre.

$$\text{Population Loading, persons/ac} = \frac{\text{Population, persons}}{\text{Surface Area, ac}}$$
$$= \frac{20{,}000\text{ persons}}{24\text{ ac}}$$
$$= 833\text{ persons/ac}$$

Calculate the hydraulic loading in inches per day.

$$\text{Hydraulic Loading, in/day} = \frac{\text{Pond Depth, in}}{\text{Detention Time, days}}$$
$$= \frac{(4\text{ ft})(12\text{ in/ft})}{15.7\text{ days}}$$
$$= \frac{48}{15.7}$$
$$= 3.06\text{ in/day}$$

Answers to questions on page 43.

9.11A Algae can create undesirable effects on the receiving waters in terms of a loss of aesthetic values, increased turbidity, suspended solids, and biochemical oxygen demand; algae can also cause the development of nuisance conditions.

9.11B The Lemna Duckweed System eliminates algae from pond effluents by covering the pond's surface with duckweed, thus preventing the penetration of sunlight and causing algae to die and settle out of the wastewater being treated.

9.11C Land treatment systems used to improve effluent quality from stabilization ponds consist of applying the effluent to an overland-flow treatment system followed by a constructed wetland wastewater treatment system.

OBJECTIVE TEST

Chapter 9. WASTEWATER STABILIZATION PONDS

Please write your name and mark the correct answers on the answer sheet, as directed at the end of Chapter 1, Volume I. There may be more than one correct answer to each multiple-choice question.

True-False

1. Algae are microscopic animals floating and living in water.
   1. True
   2. False
2. When ponds are operated in a knowledgeable and purposeful manner, they can perform a variety of functions.
   1. True
   2. False
3. Usually, floating aeration and chlorination equipment are too expensive to have standing idle waiting for an odor problem to develop.
   1. True
   2. False
4. Plants or grasses with long roots should be allowed to grow on levees to prevent erosion.
   1. True
   2. False
5. Heavy chlorination at the recirculation point of a pond can assist with treatment, but will probably interfere with odor control.
   1. True
   2. False
6. Even though a pond has little mechanical equipment, there are still safety hazards.
   1. True
   2. False

Best Answer (Select only the closest or best answer.)

7. What is the purpose of a pond flowmeter?
   1. Distributes influent in pond
   2. Measures and records flows into pond
   3. Regulates outflow from pond
   4. Removes coarse material from pond influent
8. What is the purpose of a pond recirculation line?
   1. Conveys treated wastewater to receiving waters
   2. Conveys wastewater from one pond to another
   3. Prevents scum and other surface debris from flowing to next pond
   4. Returns pond effluent rich in algae and oxygen from second pond to first pond
9. How do algae produce free oxygen in ponds?
   1. By allowing microorganisms to produce oxygen
   2. By converting oxygen from nitrate
   3. By transferring oxygen from the atmosphere
   4. By using sunlight through photosynthesis
10. What happens to oxygen in a pond at night?
    1. Algae produce additional oxygen at night
    2. Algae use up the oxygen by respiration
    3. Oxygen levels increase due to lack of algal activity
    4. Oxygen levels remain constant
11. Why does the pH increase in a pond during the day?
    1. Algae produce carbon dioxide
    2. Algae use carbon dioxide
    3. Carbon dioxide is added from the air
    4. Influent wastewaters contain higher levels of acid
12. Why is duckweed undesirable in wastewater stabilization ponds?
    1. Attracts ducks
    2. Causes supersaturation
    3. Presents control problems
    4. Stops sunlight penetration
13. Why do ponds *NOT* become organically overloaded during storms and periods of high runoff?
    1. Because the higher flows facilitate recirculation
    2. Because the lower temperatures increase microbial activity
    3. Because there are lower waste discharges during these times
    4. Because there is *NOT* a significant increase in BOD loading
14. When will the pH and dissolved oxygen be lowest in a pond?
    1. Around midnight
    2. Around noon
    3. Just at sunrise
    4. Just at sunset
15. What is the desired color of a properly operating pond?
    1. Dark green color
    2. Deep blue color
    3. Dull green color
    4. Gray color
16. Why should telescoping friction-fit tubes for regulating discharge be exercised regularly?
    1. To prevent becoming stuck
    2. To prevent formation of scum mats
    3. To prevent organic overload
    4. To prevent weed growth along shoreline

Multiple Choice (Select all correct answers.)

17. What are the advantages of a pond?
    1. Adaptable to changing loads
    2. Consumes little energy
    3. Has few sludge handling and disposal problems
    4. Protects groundwater
    5. Requires a large area of land
18. How can oxygen be supplied to ponds?
    1. By adding sodium hypochlorite to the pond influent
    2. By algae during daylight hours
    3. By compressors bubbling air through the pond
    4. By mechanical agitation of the surface
    5. By promoting pond overturn
19. Which variables influence the treatment efficiency of ponds?
    1. Concentration of waste
    2. Nutrients
    3. pH
    4. Temperature
    5. Type and quantity of algae
20. Which day-to-day operational and maintenance duties should be performed to ensure peak pond treatment efficiency?
    1. Insect control
    2. Levee maintenance
    3. Odor control
    4. Scum control
    5. Weed control
21. Which methods have been used to break up scum on the surface of ponds?
    1. Agitation with garden rakes
    2. Chemical sprays
    3. Jets of water from pumps
    4. Lawn mowers
    5. Vacuum cleaners
22. Why are weeds around the edge of a pond objectionable?
    1. Allow scum accumulation
    2. Allow sheltered area for mosquito breeding
    3. Contribute to evapotranspiration
    4. Provide habitat for water fowl
    5. Provide source of food for mosquito fish
23. How can mosquitoes be controlled in ponds?
    1. Import frog populations
    2. Keep any scum broken up
    3. Keep water edge clear of vegetation
    4. Plant mosquito fish
    5. Swat mosquitoes as they hatch
24. What are the most serious causes of pond levee slope erosion?
    1. Burrowing animals
    2. Excessive vegetation on slopes
    3. Surface runoff from precipitation
    4. Tule growth along shoreline
    5. Wave action
25. Which items could be probable causes of poor quality pond effluent?
    1. Blockage of light by excessive plant growth near dikes
    2. Excessive turbidity from scum mats
    3. Low temperature
    4. Organic overload
    5. Toxic material in influent
26. What are possible solutions to a pond odor problem?
    1. Aerate plant influent
    2. Increase or start recirculation from effluent to influent
    3. Inspect collection system for sources of excessive organic loads
    4. Limit organic loads
    5. Pretreat with chlorine
27. Which tests are used to measure the efficiency of a pond in treating wastes?
    1. BOD (biochemical oxygen demand)
    2. DO (dissolved oxygen)
    3. pH
    4. Solids
    5. Temperature
28. How can winds have a favorable impact on a pond?
    1. Blow surface scum to one side of pond for removal
    2. Blow weeds that are *NOT* rooted to the pond bottom to one side of pond for removal
    3. Create waves that can erode pond levee
    4. Increase evaporation rates by wave action
    5. Mix contents of pond
29. How are the loadings on a pond calculated?
    1. Coliform loading
    2. Detention loading
    3. Hydraulic loading
    4. Organic loading
    5. Population loading
30. How can algae in a pond effluent create undesirable impacts on receiving waters?
    1. Development of nuisance conditions
    2. Increased BOD
    3. Increased suspended solids
    4. Increased turbidity
    5. Loss of aesthetic values
31. How can receiving waters be protected from algae in pond effluents? By use of
    1. Activated sludge treatment
    2. Duckweed
    3. Land treatment
    4. Sand filters
    5. Trickling filters

End of Objective Test

APPENDIX

Monthly Data Sheet

CLEANWATER, U.S.A.

WATER POLLUTION CONTROL PLANT

MONTHLY RECORD __________ 20__ OPERATOR: __________

DATE	DAY	WEATHER	TIME OF VISIT	FLOW (MGD) EFFLUENT	TEMP. °F NO. 1 POND	TEMP. °F NO. 2 POND	pH NO. 1 POND	pH NO. 2 POND	D.O. NO. 1 POND	D.O. NO. 2 POND	B.O.D. INFLUENT	B.O.D. EFFLUENT	CL_2 FEED RATE LB./DAY	CL_2 RESIDUAL Mg./L.	E-COLI. M.P.N./100	MIXING HRS. NO. 1 POND	MIXING HRS. NO. 2 POND	REMARKS
1	T												35					
2	W	OVERCAST	9:30 A.M.	0.025	68	70	7.6	7.9	0.8	1.4	183	32	35	3.2	62	-0-	-0-	EFFL. SS=45 mg/l
3	T	CLEAR	2:00 P.M.	0.038	76	79	7.7	8.4	1.3	8.7			35	1.4				
4	F												"					
5	S												"					
6	S												"					
7	M	CLEAR	10:00 A.M.	0.026	70	74	7.6	8.1	1.2	7.6			"					
8	T												"					
9	W	CLEAR	10:30 A.M.	0.027	69	73	7.6	8.1	1.1	7.8	158	29	35	0.8	2400	-0-	-0-	EFFL. SS=53 mg/l
10	T												"					
11	F	CLEAR HOT	3:15 P.M.	0.039	77	80	7.6	8.3	1.5	11.2			"					
12	S												"					
13	S												"					
14	M	CLEAR	9:30 A.M.	0.023	70	75	7.8	8.5	1.9	7.6			"					
15	T												40					INCREASED CL_2 FEED RATE TO 40#/DAY
16	W	CLEAR	1:30 P.M.	0.031	75	79	7.8	8.4	1.7	10.6	162	33	40	4.2	700	-0-	-0-	EFFL. SS=37 mg/l
17	T												40					
18	F	CLEAR	3:30 P.M.	0.037	79	85	7.8	8.6	2.1	14.5			40	2.5	23			
19	S												"					
20	S												40	2.8				
21	M												"					DECLARED HOLIDAY - NO PLANT CHECK
22	T	CLEAR	9:00 A.M.	0.020	71	76	7.6	8.4	1.0	7.8			40	5.6				
23	W	CLEAR	10:30 A.M.	0.028	72	78	7.7	8.4	1.0	8.1	178	30	40	5.0	6	-0-	-0-	EFFL. SS=48 mg/l
24	T												"					
25	F	CLEAR	2:15 P.M.	0.035	80	87	7.8	8.5	1.3	11.4			40	3.8				
26	S												"					
27	S												"					
28	M	CLEAR	10:00 A.M.	0.026	73	79	7.7	8.5	1.1	8.6			40	4.7				CL_2 FEED RATE BACK TO 35#/DAY
29	T												35					
30	W	CLEAR	4:15 P.M.	0.038	81	88	7.8	8.5	1.8	15.3	168	35	35	1.1	6	-0-	-0-	EFFL. SS=57 mg/l
31																		
MAX.				0.039	81	88	7.8	8.6	2.1	15.3	183	35	40	5.6	2400	-0-	-0-	
MIN.				0.020	68	70	7.6	7.9	0.8	7.6	162	29	35	0.8	6	-0-	-0-	
AVG.		CLEAR		0.030	73	78	7.7	8.3	1.3	9.2	169	31	37.2	2.7	62	-0-	-0-	

SUMMARY DATA	
% REMOVAL B.O.D.	81.6 %
LBS. B.O.D. / ACRE / DAY	52.8
DETENTION TIME — DAYS	86.7

PONDS OPERATED IN SERIES ALL MONTH.

MIXERS NOT OPERATED DURING MONTH.

COST DATA	
MAN DAYS 4 x 2.5 HRS PER VISIT PAYROLL	245.52
POWER PURCHASED	69.60
OTHER UTILITIES (GAS, H_2O)	NONE
GASOLINE, OIL, GREASE	NONE
CHEMICALS & SUPPLIES	138.60
MAINTENANCE	NONE
VEHICLE COSTS	57.40
OTHER	NONE
TOTAL	$ 511.12
OPERATING COST/CAPITA/MO.	$ 1.68
COST/1000 GAL. TREATED	$ 0.55

FLOW METER:
LAST 232602
1st 231672
TOTAL 0.930 MG.

ELECTRIC METER:
LAST 5978
1st 5817
MULT. 40 x 161 = 6440 KWH

CHAPTER 10

ACTIVATED SLUDGE

Package Plants and Oxidation Ditches

by

John Brady

TABLE OF CONTENTS

Chapter 10. ACTIVATED SLUDGE

PACKAGE PLANTS AND OXIDATION DITCHES

LESSON 3

OBJECTIVES

Chapter 10. ACTIVATED SLUDGE

PACKAGE PLANTS AND OXIDATION DITCHES

Following completion of Chapter 10, you should be able to:

1. Explain the principles of the activated sludge process and the factors that influence and control the process,
2. Inspect a new activated sludge facility for proper installation,
3. Place a new activated sludge process into service,
4. Schedule and conduct operation and maintenance duties,
5. Collect samples, interpret lab results, and make appropriate adjustments in treatment processes,
6. Recognize factors that indicate an activated sludge process is not performing properly, identify the source of the problem, and take corrective action,
7. Conduct your duties in a safe fashion,
8. Determine aerator loadings and understand the application of different loading guidelines,
9. Keep records for an activated sludge plant, and
10. Review plans and specifications for an activated sludge plant.

WORDS

Chapter 10. ACTIVATED SLUDGE

PACKAGE PLANTS AND OXIDATION DITCHES

ABSORPTION (ab-SORP-shun) ABSORPTION

The taking in or soaking up of one substance into the body of another by molecular or chemical action (as tree roots absorb dissolved nutrients in the soil).

ACTIVATED SLUDGE (ACK-ta-VATE-ed sluj) ACTIVATED SLUDGE

Sludge particles produced in raw or settled wastewater (primary effluent) by the growth of organisms (including zoogleal bacteria) in aeration tanks in the presence of dissolved oxygen. The term "activated" comes from the fact that the particles are teeming with bacteria, fungi, and protozoa. Activated sludge is different from primary sludge in that the sludge particles contain many living organisms which can feed on the incoming wastewater.

ACTIVATED SLUDGE (ACK-ta-VATE-ed sluj) PROCESS ACTIVATED SLUDGE PROCESS

A biological wastewater treatment process which speeds up the decomposition of wastes in the wastewater being treated. Activated sludge is added to wastewater and the mixture (mixed liquor) is aerated and agitated. After some time in the aeration tank, the activated sludge is allowed to settle out by sedimentation and is disposed of (wasted) or reused (returned to the aeration tank) as needed. The remaining wastewater then undergoes more treatment.

ADSORPTION (add-SORP-shun) ADSORPTION

The gathering of a gas, liquid, or dissolved substance on the surface or interface zone of another material.

AERATION (air-A-shun) LIQUOR AERATION LIQUOR

Mixed liquor. The contents of the aeration tank including living organisms and material carried into the tank by either untreated wastewater or primary effluent.

AERATION (air-A-shun) TANK AERATION TANK

The tank where raw or settled wastewater is mixed with return sludge and aerated. The same as aeration bay, aerator, or reactor.

AEROBES AEROBES

Bacteria that must have molecular (dissolved) oxygen (DO) to survive. Aerobes are aerobic bacteria.

AEROBIC BACTERIA (AIR-O-bick back-TEAR-e-ah) AEROBIC BACTERIA

Bacteria which will live and reproduce only in an environment containing oxygen which is available for their respiration (breathing), namely atmospheric oxygen or oxygen dissolved in water. Oxygen combined chemically, such as in water molecules (H_2O), cannot be used for respiration by aerobic bacteria.

AEROBIC (AIR-O-bick) DIGESTION AEROBIC DIGESTION

The breakdown of wastes by microorganisms in the presence of dissolved oxygen. This digestion process may be used to treat only waste activated sludge, or trickling filter sludge and primary (raw) sludge, or waste sludge from activated sludge treatment plants designed without primary settling. The sludge to be treated is placed in a large aerated tank where aerobic microorganisms decompose the organic matter in the sludge. This is an extension of the activated sludge process.

AGGLOMERATION (a-GLOM-er-A-shun) AGGLOMERATION

The growing or coming together of small scattered particles into larger flocs or particles which settle rapidly. Also see FLOC.

AIR LIFT PUMP AIR LIFT PUMP

A special type of pump. This device consists of a vertical riser pipe submerged in the wastewater or sludge to be pumped. Compressed air is injected into a tail piece at the bottom of the pipe. Fine air bubbles mix with the wastewater or sludge to form a mixture lighter than the surrounding water which causes the mixture to rise in the discharge pipe to the outlet. An air-lift pump works like the center stand in a percolator coffee pot.

ANAEROBES | ANAEROBES

Bacteria that do not need molecular (dissolved) oxygen (DO) to survive.

ANAEROBIC (AN-air-O-bick) | ANAEROBIC

A condition in which atmospheric or dissolved molecular oxygen is *NOT* present in the aquatic (water) environment.

ANODE (an-O-d) | ANODE

The positive pole or electrode of an electrolytic system, such as a battery. The anode attracts negatively charged particles or ions (anions).

ANOXIC (an-OX-ick) | ANOXIC

A condition in which atmospheric or dissolved molecular oxygen is *NOT* present in the aquatic (water) environment and nitrate is present. Oxygen deficient or lacking sufficient oxygen. The term is similar to ANAEROBIC.

ASHING | ASHING

Formation of an activated sludge floc in a clarifier effluent that is well oxidized and floats on the water surface (has the appearance of gray ash).

BOD (pronounce as separate letters) | BOD

Biochemical **O**xygen **D**emand. The rate at which organisms use the oxygen in water or wastewater while stabilizing decomposable organic matter under aerobic conditions. In decomposition, organic matter serves as food for the bacteria and energy results from its oxidation. BOD measurements are used as a measure of the organic strength of wastes in water.

BOD_5 | BOD_5

BOD_5 refers to the five-day biochemical oxygen demand. The total amount of oxygen used by microorganisms decomposing organic matter increases each day until the ultimate BOD is reached, usually in 50 to 70 days. BOD usually refers to the five-day BOD or BOD_5.

BACTERIAL (back-TEAR-e-al) CULTURE | BACTERIAL CULTURE

In the case of activated sludge, the bacterial culture refers to the group of bacteria classified as AEROBES, and FACULTATIVE organisms, which covers a wide range of organisms. Most treatment processes in the United States grow facultative organisms which use the carbonaceous (carbon compounds) BOD. Facultative organisms can live when oxygen resources are low. When "nitrification" is required, the nitrifying organisms are OBLIGATE AEROBES (require oxygen) and must have at least 0.5 mg/*L* of dissolved oxygen throughout the whole system to function properly.

BATCH PROCESS | BATCH PROCESS

A treatment process in which a tank or reactor is filled, the wastewater (or other solution) is treated or a chemical solution is prepared, and the tank is emptied. The tank may then be filled and the process repeated. Batch processes are also used to cleanse, stabilize or condition chemical solutions for use in industrial manufacturing and treatment processes.

BIOASSAY (BUY-o-ass-SAY) | BIOASSAY

(1) A way of showing or measuring the effect of biological treatment on a particular substance or waste, or

(2) A method of determining the relative toxicity of a test sample of industrial wastes or other wastes by using live test organisms such as fish.

BIOMASS (BUY-o-MASS) | BIOMASS

A mass or clump of organic material consisting of living organisms feeding on the wastes in wastewater, dead organisms and other debris. Also see ZOOGLEAL MASS.

BULKING (BULK-ing) | BULKING

Clouds of billowing sludge that occur throughout secondary clarifiers and sludge thickeners when the sludge does not settle properly. In the activated sludge process, bulking is usually caused by filamentous bacteria or bound water.

BURPING | BURPING

"Billowing" solids that result when the settling tank sludge blanket becomes too deep (occupies too much volume in the bottom of the tank) and solids are swept by the water up and out over the effluent weirs in the discharged effluent.

COD (pronounce as separate letters) | COD

Chemical **O**xygen **D**emand. A measure of the oxygen-consuming capacity of organic matter present in wastewater. COD is expressed as the amount of oxygen consumed from a chemical oxidant in mg/*L* during a specific test. Results are not necessarily related to the biochemical oxygen demand (BOD) because the chemical oxidant may react with substances that bacteria do not stabilize.

CATHODIC (ca-THOD-ick) PROTECTION — CATHODIC PROTECTION

An electrical system for prevention of rust, corrosion, and pitting of steel and iron surfaces in contact with water, wastewater or soil. A low-voltage current is made to flow through a liquid (water) or a soil in contact with the metal in such a manner that the external electromotive force renders the metal structure cathodic. This concentrates corrosion on auxiliary anodic parts which are deliberately allowed to corrode instead of letting the structure corrode.

CENTRIFUGE — CENTRIFUGE

A mechanical device that uses centrifugal or rotational forces to separate solids from liquids.

COAGULATION (co-AGG-you-LAY-shun) — COAGULATION

The clumping together of very fine particles into larger particles (floc) caused by the use of chemicals (coagulants). The chemicals neutralize the electrical charges of the fine particles, allowing them to come closer and form larger clumps. This clumping together makes it easier to separate the solids from the water by settling, skimming, draining or filtering.

COMMINUTOR (com-mih-NEW-ter) — COMMINUTOR

A device used to reduce the size of the solid chunks in wastewater by shredding (comminuting). The shredding action is like many scissors cutting or chopping to shreds all the large influent solids material in the wastewater.

CONTACT STABILIZATION — CONTACT STABILIZATION

Contact stabilization is a modification of the conventional activated sludge process. In contact stabilization, two aeration tanks are used. One tank is for separate reaeration of the return sludge for at least four hours before it is permitted to flow into the other aeration tank to be mixed with the primary effluent requiring treatment. The process may also occur in one long tank.

DENITRIFICATION (dee-NYE-truh-fuh-KAY-shun) — DENITRIFICATION

(1) The anoxic biological reduction of nitrate nitrogen to nitrogen gas.

(2) The removal of some nitrogen from a system.

(3) An anoxic process that occurs when nitrite or nitrate ions are reduced to nitrogen gas and nitrogen bubbles are formed as a result of this process. The bubbles attach to the biological floc in the activated sludge process and float the floc to the surface of the secondary clarifiers. This condition is often the cause of rising sludge observed in secondary clarifiers or gravity thickeners. Also see NITRIFICATION.

DIFFUSED-AIR AERATION — DIFFUSED-AIR AERATION

A diffused-air activated sludge plant takes air, compresses it, and then discharges the air below the water surface of the aerator through some type of air diffusion device.

DIFFUSER — DIFFUSER

A device (porous plate, tube, bag) used to break the air stream from the blower system into fine bubbles in an aeration tank or reactor.

DISSOLVED OXYGEN — DISSOLVED OXYGEN

Molecular (atmospheric) oxygen dissolved in water or wastewater, usually abbreviated DO.

ENDOGENOUS (en-DODGE-en-us) RESPIRATION — ENDOGENOUS RESPIRATION

A situation where living organisms oxidize some of their own cellular mass instead of new organic matter they adsorb or absorb from their environment.

F/M RATIO — F/M RATIO

See FOOD/MICROORGANISM RATIO.

FACULTATIVE (FACK-ul-TAY-tive) — FACULTATIVE

Facultative bacteria can use either dissolved molecular oxygen or oxygen obtained from food materials such as sulfate or nitrate ions. In other words, facultative bacteria can live under aerobic, anoxic, or anaerobic conditions.

FILAMENTOUS (FILL-a-MEN-tuss) ORGANISMS — FILAMENTOUS ORGANISMS

Organisms that grow in a thread or filamentous form. Common types are *Thiothrix* and *Actinomycetes*. A common cause of sludge bulking in the activated sludge process.

FLOC — FLOC

Clumps of bacteria and particles or coagulants and impurities that have come together and formed a cluster. Found in aeration tanks, secondary clarifiers and chemical precipitation processes.

FOOD/MICROORGANISM RATIO FOOD/MICROORGANISM RATIO

Food to microorganism ratio. A measure of food provided to bacteria in an aeration tank.

$$\frac{\text{Food}}{\text{Microorganisms}} = \frac{\text{BOD, lbs/day}}{\text{MLVSS, lbs}}$$

$$= \frac{\text{Flow, MGD} \times \text{BOD, mg/}L \times 8.34\text{ lbs/gal}}{\text{Volume, MG} \times \text{MLVSS, mg/}L \times 8.34\text{ lbs/gal}}$$

or by calculator math system

$$= \text{Flow, MGD} \times \text{BOD, mg/}L \div \text{Volume, MG} \div \text{MLVSS, mg/}L$$

or metric

$$= \frac{\text{BOD, kg/day}}{\text{MLVSS, kg}}$$

$$= \frac{\text{Flow, M}L\text{/day} \times \text{BOD, mg/}L \times 1\text{ kg/M mg}}{\text{Volume, M}L \times \text{MLVSS, mg/}L \times 1\text{ kg/M mg}}$$

Commonly abbreviated F/M Ratio.

HEADER HEADER

A large pipe to which the ends of a series of smaller pipes are connected. Also called a MANIFOLD.

LINEAL (LIN-e-al) LINEAL

The length in one direction of a line. For example, a board 12 feet long has 12 lineal feet in its length.

MANIFOLD MANIFOLD

A large pipe to which the ends of a series of smaller pipes are connected. Also called a HEADER.

MECHANICAL AERATION MECHANICAL AERATION

The use of machinery to mix air and water so that oxygen can be absorbed into the water. Some examples are: paddle wheels, mixers, or rotating brushes to agitate the surface of an aeration tank; pumps to create fountains; and pumps to discharge water down a series of steps forming falls or cascades.

MICROORGANISMS (MY-crow-OR-gan-IS-zums) MICROORGANISMS

Very small organisms that can be seen only through a microscope. Some microorganisms use the wastes in wastewater for food and thus remove or alter much of the undesirable matter.

MIXED LIQUOR MIXED LIQUOR

When the activated sludge in an aeration tank is mixed with primary effluent or the raw wastewater and return sludge, this mixture is then referred to as mixed liquor as long as it is in the aeration tank. Mixed liquor also may refer to the contents of mixed aerobic or anaerobic digesters.

MIXED LIQUOR SUSPENDED SOLIDS (MLSS) MIXED LIQUOR SUSPENDED SOLIDS (MLSS)

Suspended solids in the mixed liquor of an aeration tank.

MIXED LIQUOR VOLATILE SUSPENDED SOLIDS (MLVSS) MIXED LIQUOR VOLATILE SUSPENDED SOLIDS (MLVSS)

The organic or volatile suspended solids in the mixed liquor of an aeration tank. This volatile portion is used as a measure or indication of the microorganisms present.

NPDES PERMIT NPDES PERMIT

National **P**ollutant **D**ischarge **E**limination **S**ystem permit is the regulatory agency document issued by either a federal or state agency which is designed to control all discharges of potential pollutants from point sources and storm water runoff into U.S. waterways. NPDES permits regulate discharges into navigable waters from all point sources of pollution, including industries, municipal wastewater treatment plants, sanitary landfills, large agricultural feedlots and return irrigation flows.

NITRIFICATION (NYE-truh-fuh-KAY-shun) NITRIFICATION

An aerobic process in which bacteria change the ammonia and organic nitrogen in wastewater into oxidized nitrogen (usually nitrate). The second-stage BOD is sometimes referred to as the "nitrogenous BOD" (first-stage BOD is called the "carbonaceous BOD"). Also see DENITRIFICATION.

OBLIGATE AEROBES | OBLIGATE AEROBES

Bacteria that must have atmospheric or dissolved molecular oxygen to live and reproduce.

OXIDATION (ox-uh-DAY-shun) | OXIDATION

Oxidation is the addition of oxygen, removal of hydrogen, or the removal of electrons from an element or compound. In wastewater treatment, organic matter is oxidized to more stable substances. The opposite of REDUCTION.

POLYMER (POLY-mer) | POLYMER

A long chain molecule formed by the union of many monomers (molecules of lower molecular weight). Polymers are used with other chemical coagulants to aid in binding small suspended particles to larger chemical flocs for their removal from water.

PROTOZOA (pro-toe-ZOE-ah) | PROTOZOA

A group of motile microscopic organisms (usually single-celled and aerobic) that sometimes cluster into colonies and generally consume bacteria as an energy source.

REAGENT (re-A-gent) | REAGENT

A pure chemical substance that is used to make new products or is used in chemical tests to measure, detect, or examine other substances.

REDUCTION (re-DUCK-shun) | REDUCTION

Reduction is the addition of hydrogen, removal of oxygen, or the addition of electrons to an element or compound. Under anaerobic conditions (no dissolved oxygen present), sulfur compounds are reduced to odor-producing hydrogen sulfide (H_2S) and other compounds. The opposite of OXIDATION.

RISING SLUDGE | RISING SLUDGE

Rising sludge occurs in the secondary clarifiers of activated sludge plants when the sludge settles to the bottom of the clarifier, is compacted, and then starts to rise to the surface, usually as a result of denitrification, or anaerobic biological activity that produces carbon dioxide and/or methane.

SECCHI (SECK-key) DISC | SECCHI DISC

A flat, white disc lowered into the water by a rope until it is just barely visible. At this point, the depth of the disc from the water surface is the recorded Secchi disc transparency.

SEPTIC (SEP-tick) | SEPTIC

A condition produced by anaerobic bacteria. If severe, the sludge produces hydrogen sulfide, turns black, gives off foul odors, contains little or no dissolved oxygen, and the wastewater has a high oxygen demand.

SEQUENCING BATCH REACTOR (SBR) | SEQUENCING BATCH REACTOR (SBR)

A form of activated sludge treatment process.

SHOCK LOAD (activated sludge) | SHOCK LOAD

The arrival at a plant of a waste which is toxic to organisms in sufficient quantity or strength to cause operating problems. Possible problems include odors and bulking sludge which will result in a high loss of solids from the secondary clarifiers into the plant effluent and a biological process upset that may require several days to a week to recover. Organic or hydraulic overloads also can cause a shock load.

SHORT-CIRCUITING | SHORT-CIRCUITING

A condition that occurs in tanks or basins when some of the flowing water entering a tank or basin flows along a nearly direct pathway from the inlet to the outlet. This is usually undesirable since it may result in shorter contact, reaction, or settling times in comparison with the theoretical (calculated) or presumed detention times.

SLUDGE AGE | SLUDGE AGE

A measure of the length of time a particle of suspended solids has been retained in the activated sludge process.

$$\text{Sludge Age, days} = \frac{\text{Suspended Solids Under Aeration, lbs or kg}}{\text{Suspended Solids Added, lbs/day or kg/day}}$$

SOLIDS CONCENTRATION | SOLIDS CONCENTRATION

The solids in the aeration tank which carry microorganisms that feed on wastewater. Expressed as milligrams per liter of Mixed Liquor Volatile Suspended Solids (MLVSS, mg/*L*).

STABILIZED WASTE — STABILIZED WASTE

A waste that has been treated or decomposed to the extent that, if discharged or released, its rate and state of decomposition would be such that the waste would not cause a nuisance or odors in the receiving water.

STAFF GAGE — STAFF GAGE

A ruler or graduated scale used to measure the depth or elevation of water in a channel, tank, or stream.

STANDARDIZE — STANDARDIZE

To compare with a standard.

(1) In wet chemistry, to find out the exact strength of a solution by comparing it with a standard of known strength. This information is used to adjust the strength by adding more water or more of the substance dissolved.

(2) To set up an instrument or device to read a standard. This allows you to adjust the instrument so that it reads accurately, or enables you to apply a correction factor to the readings.

STEP-FEED AERATION — STEP-FEED AERATION

Step-feed aeration is a modification of the conventional activated sludge process. In step aeration, primary effluent enters the aeration tank at several points along the length of the tank, rather than all of the primary effluent entering at the beginning or head of the tank and flowing through the entire tank in a plug flow mode.

SUPERNATANT (sue-per-NAY-tent) (activated sludge) — SUPERNATANT

Liquid removed from settled sludge. Supernatant commonly refers to the liquid between the sludge on the bottom and the scum on the surface of a tank.

TOC (pronounce as separate letters) — TOC

Total **O**rganic **C**arbon. TOC measures the amount of organic carbon in water.

TURBIDITY (ter-BID-it-tee) — TURBIDITY

The cloudy appearance of water caused by the presence of suspended and colloidal matter. In the waterworks field, a turbidity measurement is used to indicate the clarity of water. Technically, turbidity is an optical property of the water based on the amount of light reflected by suspended particles. Turbidity cannot be directly equated to suspended solids because white particles reflect more light than dark-colored particles and many small particles will reflect more light than an equivalent large particle.

TURBIDITY METER — TURBIDITY METER

An instrument for measuring and comparing the turbidity of liquids by passing light through them and determining how much light is reflected by the particles in the liquid. The normal measuring range is 0 to 100 and is expressed as Nephelometric Turbidity Units (NTUs).

WET CHEMISTRY — WET CHEMISTRY

Laboratory procedures used to analyze a sample of water using liquid chemical solutions (wet) instead of, or in addition to, laboratory instruments.

ZOOGLEAL (ZOE-glee-al) MASS — ZOOGLEAL MASS

Jelly-like masses of bacteria found in both the trickling filter and activated sludge processes. These masses may be formed for or function as the protection against predators and for storage of food supplies. Also see BIOMASS.

CHAPTER 10. ACTIVATED SLUDGE

PACKAGE PLANTS AND OXIDATION DITCHES

(Lesson 1 of 3 Lessons)

10.0 THE ACTIVATED SLUDGE PROCESS

10.00 Definitions

ACTIVATED SLUDGE (Figure 10.1). Activated sludge consists of sludge particles produced in raw or settled wastewater (primary effluent) by the growth of organisms (including zoogleal bacteria) in aeration tanks in the presence of dissolved oxygen. The term "activated" comes from the fact that the particles are teeming with bacteria, fungi, and protozoa.

ACTIVATED SLUDGE PROCESS (Figure 10.2). The activated sludge process is a biological wastewater treatment process that uses suspended growth *MICROORGANISMS*[1] to speed up decomposition of wastes. When activated sludge is added to wastewater, the microorganisms feed and grow on waste in the wastewater. As the organisms grow and reproduce, more and more waste is removed, leaving the wastewater partially cleaned. To function efficiently, the mass of organisms (*SOLIDS CONCENTRATION*[2]) needs a steady balance of food (*FOOD/MICROORGANISM RATIO*[3]) and oxygen.

10.01 Wastewater Treatment by Activated Sludge

The activated sludge wastewater treatment process can be used as a standard wastewater treatment system removing 93 to 98 percent of the organic carbon (carbonaceous portion) as measured by the biochemical oxygen demand test. Or, the process can be used to pretreat industrial wastes containing very high *BOD*[4] concentrations to remove a portion of the waste load before being discharged to a collection system or other treatment system. The activated sludge process can also be operated to remove specific nutrients not removed in standard wastewater treatment; specifically, the removal of phosphorus and nitrogen which can create undesirable effects on receiving water when discharged from a wastewater treatment plant.

The components required for an activated sludge wastewater treatment plant depend on the quantity of wastewater to be treated, the makeup of the wastewater, and whether the wastewater has received any pretreatment before entering the activated sludge system. The common feature of all activated sludge systems is that there must be a tank or basin where the following functions can occur:

- Aeration/mixing: the purpose is to mix the suspended growth organisms (activated sludge) with the incoming wastewater so they can obtain food, and to supply dissolved oxygen to the *AEROBIC*[5] and *FACULTATIVE*[6] organisms. This function takes place in what is called an aeration tank or basin, reactor, oxidation cell, or oxidation ditch.

---

[1] *Microorganisms (MY-crow-OR-gan-IS-zums). Very small organisms that can be seen only through a microscope. Some microorganisms use the wastes in wastewater for food and thus remove or alter much of the undesirable matter.*

[2] *Solids Concentration. The solids in the aeration tank which carry microorganisms that feed on wastewater. Expressed as milligrams per liter of Mixed Liquor Volatile Suspended Solids (MLVSS, mg/L).*

[3] *Food/Microorganism Ratio. Food to microorganism ratio. A measure of food provided to bacteria in an aeration tank.*

$$\frac{\text{Food}}{\text{Microorganisms}} = \frac{\text{BOD, lbs/day or kg/day}}{\text{MLVSS, lbs or kg}}$$

[4] *BOD (pronounce as separate letters). **B**iochemical **O**xygen **D**emand. The rate at which organisms use the oxygen in water or wastewater while stabilizing decomposable organic matter under aerobic conditions. In decomposition, organic matter serves as food for the bacteria and energy results from its oxidation. BOD measurements are used as a measure of the organic strength of wastes in water.*

[5] *Aerobic Bacteria (AIR-O-bick back-TEAR-e-ah). Bacteria which will live and reproduce only in an environment containing oxygen which is available for their respiration (breathing), namely atmospheric oxygen or oxygen dissolved in water. Oxygen combined chemically, such as in water molecules (H_2O), cannot be used for respiration by aerobic bacteria.*

[6] *Facultative (FACK-ul-TAY-tive). Facultative bacteria can use either dissolved molecular oxygen or oxygen obtained from food materials such as sulfate or nitrate ions. In other words, facultative bacteria can live under aerobic, anoxic, or anaerobic conditions.*

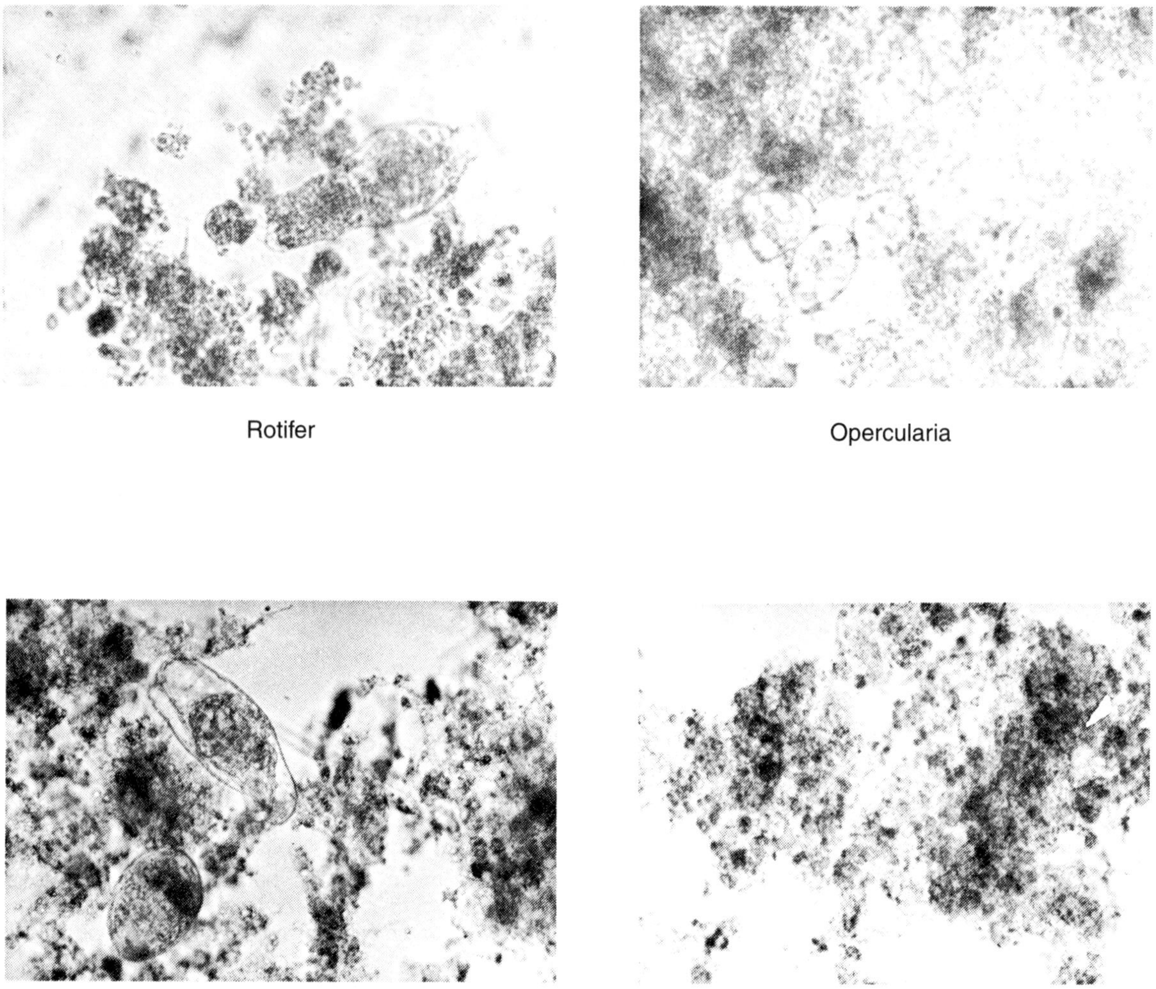

Rotifer

Opercularia

Rotifer and opercularia

Zoogleal mass

Fig. 10.1 Microorganisms in activated sludge

For information on the use of microorganisms to operate the activated sludge process, see Chapter 2, "Activated Sludge," in *ADVANCED WASTE TREATMENT.* Also see Section 11.10, "Microbiology for Activated Sludge," by Paul V. Bohlier in Chapter 11 of *OPERATION OF WASTEWATER TREATMENT PLANTS*, Volume II.

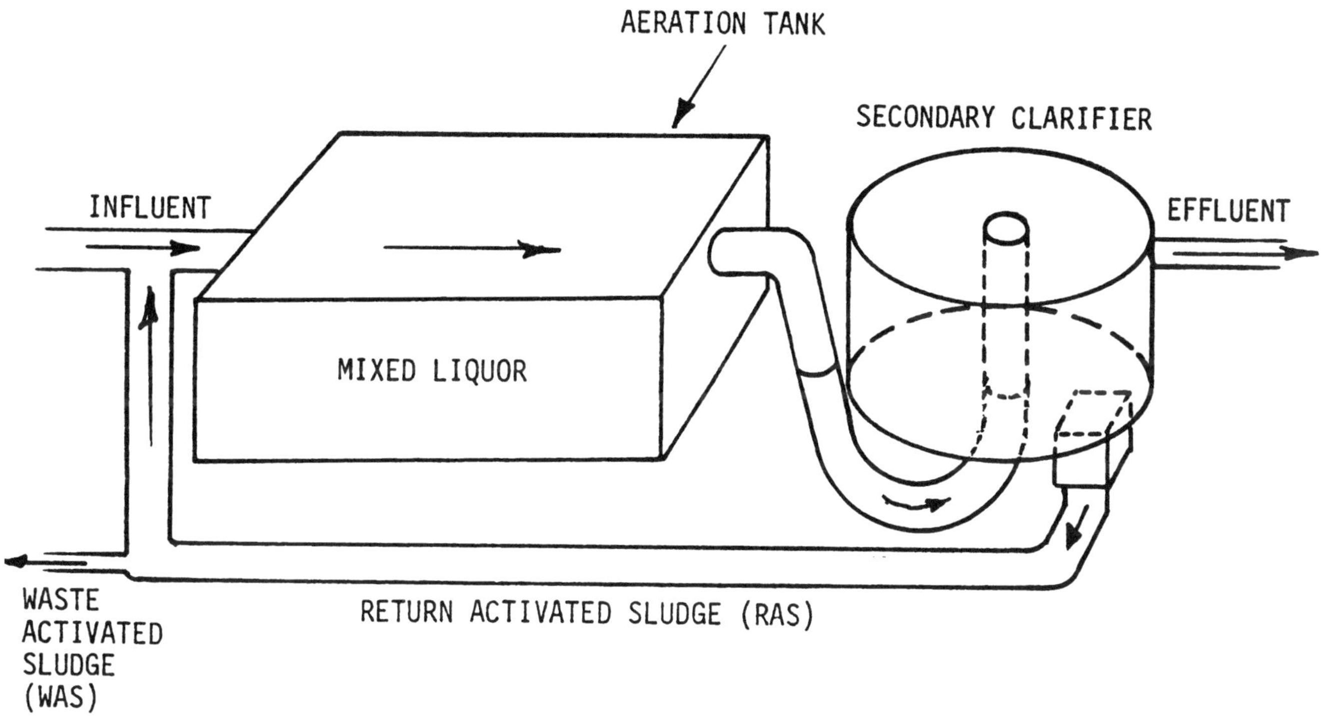

Fig. 10.2 Activated sludge process

- Settling: after spending a prescribed amount of time in the aeration/mixing mode, the activated sludge is provided a quiescent (quiet) period. During this period, the organisms tend to clump together and form large particles (called *FLOC*[7]) that will normally settle out of the wastewater and form a blanket layer of solids on the bottom of the tank or basin. The remaining clear liquid above the activated sludge blanket is the supernatant (effluent). This settling function takes place in a secondary clarifier, settling tank, or settler.
- Decant: the settled activated sludge culture is separated from the clarified wastewater. The decant stage may vary slightly depending on the layout of the treatment system. If the entire process of aeration/mixing, settling, and decant occurs in a single tank, the decant stage occurs in the following manner.

  After the quiescent period in which solids flocculation/separation has occurred, the supernatant layer of treated wastewater (effluent) is removed from the tank, to a level just above the settled activated sludge blanket. Once the treated effluent has been removed, the tank is again filled with untreated wastewater and the process of mixing/aeration and settling is repeated. This type of activated sludge system is referred to as a batch reactor or, if the system is specifically designed and automated to perform the above functions, it is called a sequencing batch reactor (SBR).

  In conventional activated sludge plants, aeration/mixing and settling occur in separate tanks, with the wastewater continuously flowing through them. In continuous-flow systems, the treated wastewater (effluent) leaves the settling tank from the surface or supernatant zone above the activated sludge blanket in the tank. At the same time, a portion of the activated sludge blanket is removed from the bottom of the settling tank and returned to the aeration or reactor tank where it will be mixed with incoming wastewater to be treated and supplied with dissolved oxygen by the aeration system.

In a typical activated sludge plant for a community with a conventional gravity wastewater collection system, the plant process sequence and flow diagram could be similar to Figures 10.3 and 10.4.

Figures 10.5 and 10.6 show the typical process sequence and flow diagram for an activated sludge plant designed for a small community that is using septic tanks. In this example, the septic tank effluent is collected and routed to a wastewater treatment facility by means of a STEP (Septic Tank Effluent Pump) or SDGS (Small-Diameter Gravity Sewer) system.

In a small wastewater treatment facility using the activated sludge process as the treatment mode, the physical components would most likely be one of the three systems that this chapter will discuss:

1. Standard-rate activated sludge,
2. Package plants or field-fabricated plants, and
3. Oxidation ditches.

Sequencing batch reactors (SBRs) are a special modification of the activated sludge process that is used by some small systems. For details on how to operate and maintain sequencing batch reactors, see the training manual on *OPERATION OF WASTEWATER TREATMENT PLANTS*, Volume II, Chapter 11, "Activated Sludge," Section 11.9, "Sequencing Batch Reactors (SBRs)."

10.02 Process Description

Secondary treatment in the form of an activated sludge process (Figures 10.3 and 10.4) is aimed at *OXIDATION*[8] and removal of soluble or finely divided suspended materials that were not removed by previous treatment. Aerobic organisms do this in a few hours as wastewater flows through an aeration tank. The organisms *STABILIZE*[9] soluble or finely divided suspended solids by converting the solids to bacteria and protozoa bodies that are removed by sedimentation in a clarifier. Some ongoing organism growth and reproduction releases a portion of the waste in the form of carbon dioxide, water, and sulfate and nitrate compounds. The remaining solids are changed to a form that can be settled and removed as sludge during sedimentation.

After the aeration period, the wastewater is routed to a secondary settling tank for a liquid-organism (water-solids) separation. Settled organisms in the final clarifier are in a deteriorating condition due to lack of oxygen and food and should be returned to the aeration tank as quickly as possible. The remaining clarifier effluent is usually chlorinated, dechlorinated, and discharged from the plant.

Conversion of dissolved and suspended material to settleable solids is the main objective of high-rate activated sludge processes, while low-rate processes stress oxidation. In the activated sludge process, the biochemical oxidation carried out by living organisms is stressed. The same organisms also are effective in converting substances to settleable solids if the plant is operated properly.

When wastewater enters the aeration tanks, it is mixed with the activated sludge to form a mixture of sludge, carrier water, and influent solids. These solids come mainly from the discharges from homes, factories, and businesses. The activated sludge that is added contains many different types of helpful living organisms that were grown during previous contact with wastewater. These organisms are the workers in the treatment process. They use the incoming wastes for food and as a source of energy for their life processes and for the reproduction of more organisms. The activated sludge also

[7] *Floc. Clumps of bacteria and particles or coagulants and impurities that have come together and formed a cluster. Found in aeration tanks, secondary clarifiers and chemical precipitation processes.*

[8] *Oxidation (ox-uh-DAY-shun). Oxidation is the addition of oxygen, removal of hydrogen, or the removal of electrons from an element or compound. In wastewater treatment, organic matter is oxidized to more stable substances. The opposite of reduction.*

[9] *Stabilized Waste. A waste that has been treated or decomposed to the extent that, if discharged or released, its rate and state of decomposition would be such that the waste would not cause a nuisance or odors in the receiving water.*

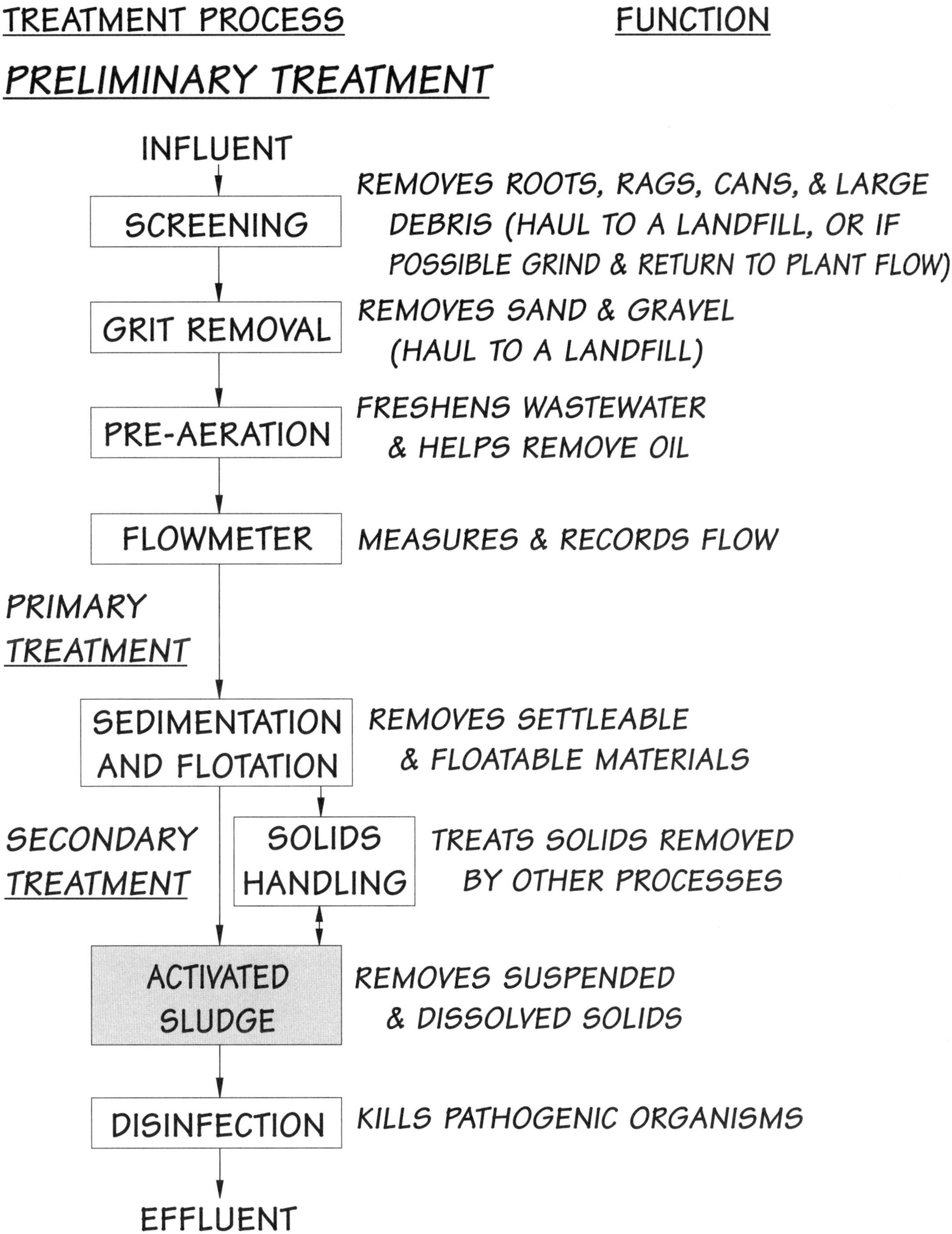

Fig. 10.3 Flow diagram of a typical plant

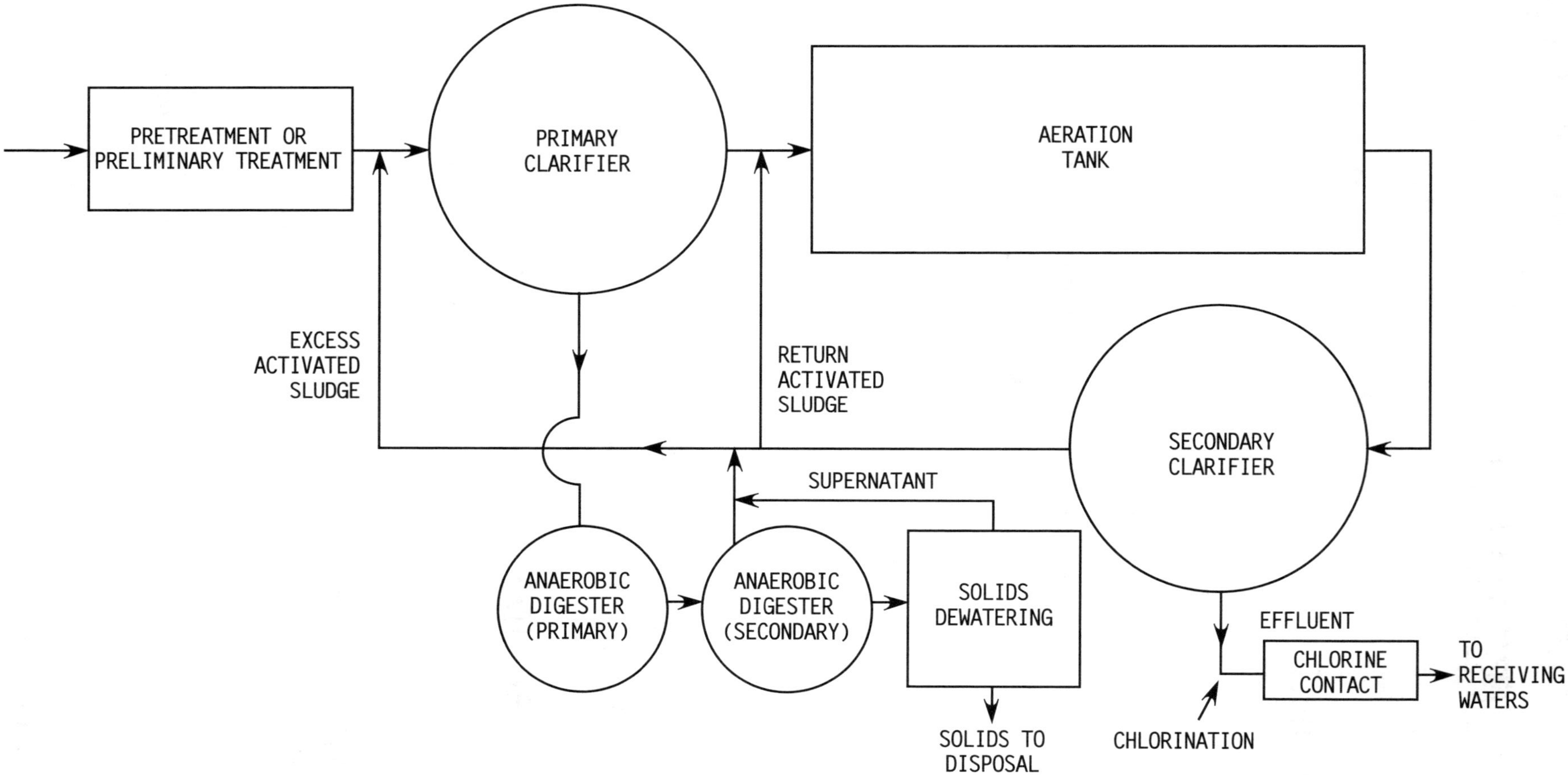

Fig. 10.4 Plan layout of a typical activated sludge plant

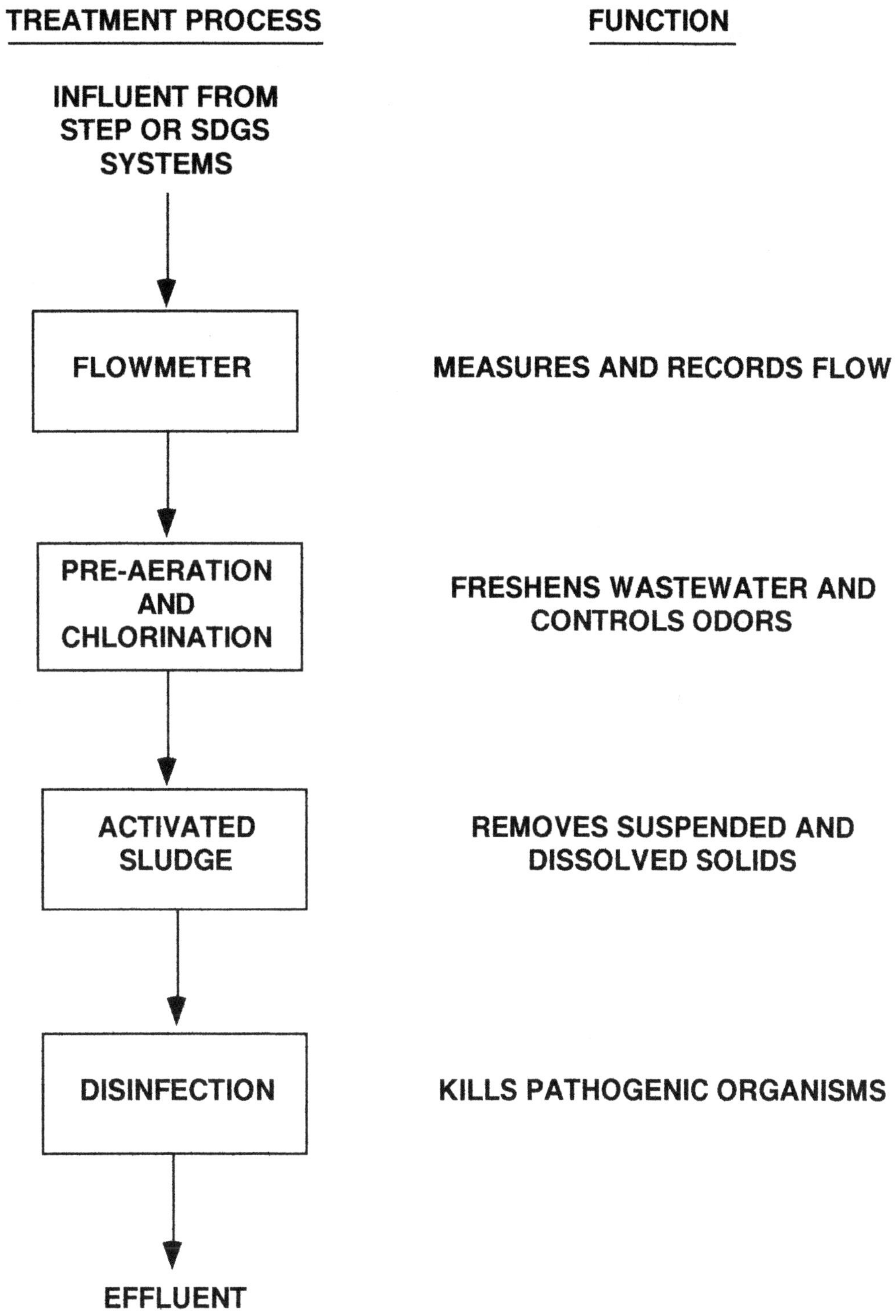

Fig. 10.5 Flow diagram of a typical activated sludge plant treating wastewater from STEP or SDGS system

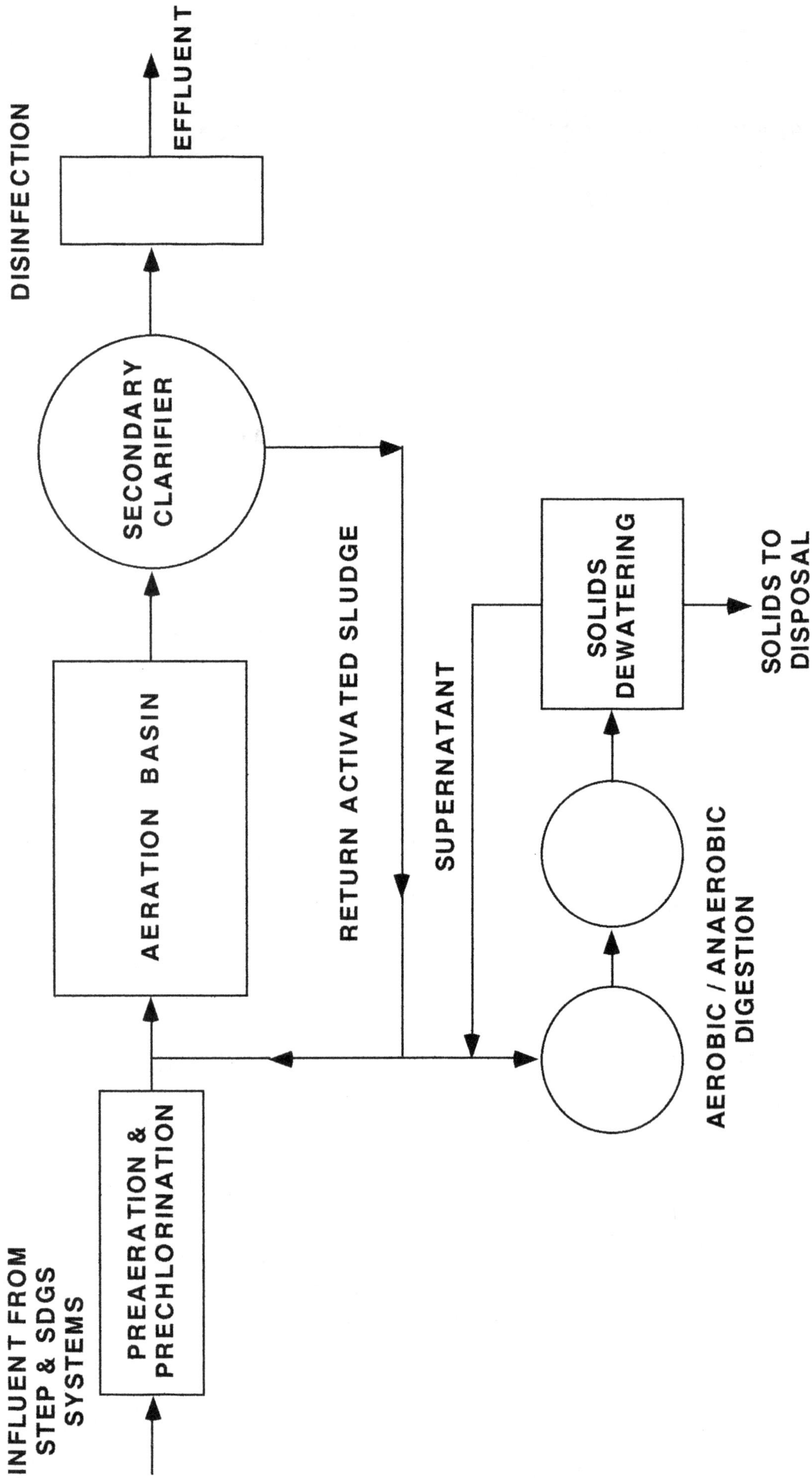

Fig. 10.6 Plant layout of a typical activated sludge plant for STEP and SDGS systems

forms a lacy network, or floc mass, that entraps materials not used as food.

Some types of organisms will require a long time to use the available food in the wastewater at a given waste concentration; however, the organisms will compete with each other in the use of available food (waste) and thereby shorten the time factor and increase the portion of waste stabilized. The ratio of food to organisms is a primary control in the activated sludge process.

Organism populations tend to increase with waste (food) load and time spent in the aeration tank. Under favorable conditions, the operator will remove the excess organisms to maintain the required number of active worker organisms for effective waste treatment. Therefore, removal of organisms from the treatment process (sludge wasting) is a very important control technique.

Oxygen, usually supplied from air, is needed by the living organisms as they oxidize wastes to obtain energy for growth. Insufficient oxygen will slow down aerobic organisms, make facultative organisms work less efficiently, and favor production of foul-smelling intermediate products of decomposition and incomplete reactions. Adding air also encourages mixing in the aerator.

An increase in organisms in an aeration tank will require greater amounts of oxygen. More food in the influent encourages more organism activity and more oxidation; consequently, more oxygen is required in the aeration tank. An excess of oxygen is required for complete waste stabilization. Therefore, the dissolved oxygen (DO) content in the aeration tank is an essential control test. Some minimum level of oxygen must be maintained to favor the desired type of organism activity to achieve the necessary treatment efficiency. If the DO in the aeration tank is too low, *FILAMENTOUS ORGANISMS*[10] will thrive and the sludge floc will not settle in the secondary clarifier. Therefore, the proper DO level must be maintained so that solids will settle properly and the plant effluent will be clear.

Flows must be distributed evenly among two or more similar treatment units. If the plant is equipped with a splitter box or a series of boxes, it will be necessary to check periodically to verify that the flow is being split as intended.

Activated sludge solids concentrations in the aerator and the secondary clarifier should be determined by the operator for process control purposes. Solids are in a deteriorating condition as long as they remain in the secondary clarifier. Controlling the depth of sludge blanket in the secondary clarifier and the concentration of solids in the aerator is very important for successful wastewater treatment. *CENTRIFUGE*[11] tests will give a quick estimate of solids concentrations and locations in the units. Precise solids tests should be made periodically for comparison with centrifuge solids tests. Before any changes are made in the mode of operation, precise solids measurements should be obtained. Settleability tests show the degree and volume of solids settling that may be obtained in a secondary clarifier; however, visual plant checks show what is actually happening.

Primary clarifiers are designed to remove material that settles to the bottom of the tank or floats to the water surface. Activated sludge helps this process along by collecting and *AGGLOMERATING*[12] the tiny particles in the primary effluent or raw wastewater so that they will settle better. If for some reason the organisms fail to make this change in the soluble solids, then the secondary clarifier effluent quality will not be satisfactory. For the activated sludge process to work properly, the operator must control the number of organisms, the dissolved oxygen in the aeration tanks, and the treatment time. When these factors are under proper control, the organisms will convert soluble solids and agglomerate the fine particles into a floc mass.

A floc mass is made up of millions of organisms (10^{12} to 10^{18}/100 m*L* in a good activated sludge), including bacteria, fungi, yeast, protozoa, and worms. When a floc mass is returned to the aerator from the final clarifier, the organisms grow as a result of taking food from the inflowing wastewater. The surface of the floc mass is irregular and promotes the transfer of wastewater pollutants into the solids by means of mechanical entrapment, absorption, adsorption, or adhesion. Many substances not used as food also are transferred to the floc mass, thus improving the quality of the plant effluent.

Material taken into the floc mass is partially oxidized by the organisms to form cell mass and oxidation products. Ash or inorganic material (silt and sand) taken in by the floc mass increase the density of the mass. Mixing the contents of the aerator causes the floc masses to bump into each other and form larger clumps. Eventually, these masses become heavy enough to settle to the bottom of the secondary clarifier where they can be removed easily. This sludge now contains most of the organisms and waste material that had been mixed in the wastewater.

---

[10] *Filamentous (FILL-a-MEN-tuss) Organisms. Organisms that grow in a thread or filamentous form. Common types are Thiothrix and Actinomycetes. A common cause of sludge bulking in the activated sludge process.*

[11] *Centrifuge. A mechanical device that uses centrifugal or rotational forces to separate solids from liquids.*

[12] *Agglomeration (a-GLOM-er-A-shun). The growing or coming together of small scattered particles into larger flocs or particles which settle rapidly.*

The next step in the activated sludge process is removal of sludge from the secondary clarifier. Some of the material is converted and released to the atmosphere in the form of stripped gases (carbon dioxide or volatile gases not converted and released in the aeration tank). That leaves water and sludge solids. A certain amount of the solids (return activated sludge) will be returned to the aerator to treat incoming wastewater. The operator must pump these solids to the aerator. The rest of the activated sludge must be wasted or removed and disposed of so that it does not continue in the plant flow. After the sludge solids have been removed from the final clarifier, the treated wastewater (clarifier effluent) moves to advanced waste treatment processes and/or the disinfection process.

The successful operation of an activated sludge plant requires the operator to be aware of the many factors influencing the process and to check them repeatedly. To keep the organisms working in the activated sludge, the operator *MUST* provide a suitable environment. High concentrations of acids, bases, and other toxic substances are undesirable and may kill the working organisms. Uneven flows of wastewater may cause overfeeding, starvation, and other problems that upset the activated sludge process. Failure to supply enough oxygen can cause an unfavorable environment, which results in decreased organism activity.

While successful operation of an activated sludge plant involves an understanding of many factors, actual control of the process as outlined in this section is relatively simple. *CONTROL CONSISTS OF MAINTAINING THE PROPER SOLIDS (FLOC MASS OR ORGANISMS) CONCENTRATION IN THE AERATOR FOR THE WASTE (FOOD) INFLOW BY ADJUSTING THE WASTE SLUDGE PUMPING RATE AND REGULATING THE OXYGEN (AIR) SUPPLY TO MAINTAIN A SATISFACTORY LEVEL OF DISSOLVED OXYGEN IN THE PROCESS.*

QUESTIONS

Write your answers in a notebook and then compare your answers with those on page 111.

10.0A What is the purpose of the activated sludge process in treating wastewater?

10.0B What is a stabilized waste?

10.0C Why is air added to the aeration tank in the activated sludge process?

10.0D What happens to the air requirement in the aeration tank when the strength (BOD) of the incoming wastewater increases?

10.0E What factors could cause an unsuitable environment for the activated sludge process in an aeration tank?

10.1 REQUIREMENTS FOR CONTROL

Effective control of the activated sludge process depends on the operator's ability to interpret and adjust several interrelated factors. Some of these factors are:

1. Effluent quality requirements.
2. Wastewater flow, concentration, and characteristics of the wastewater received.
3. Amount of activated sludge (containing the working organisms) to be maintained in the process relative to inflow.
4. Amount of oxygen required to stabilize wastewater oxygen demands and to maintain a satisfactory level of dissolved oxygen to meet organism requirements.
5. Equal division of plant flow and waste load between duplicate treatment units (two or more clarifiers or aeration tanks).
6. Transfer of the pollutional material (food) from the wastewater to the floc mass (solids or active worker organisms) and separation of the solids from the treated wastewater.
7. Effective control and disposal of in-plant residues (solids, scums, and supernatants) to accomplish ultimate disposal in a nonpolluting manner.
8. Provisions for maintaining a suitable environment for the work force of living organisms treating the wastes to keep them healthy and happy.

Effluent quality requirements may be stated by your regulatory agency in terms of percentage removal of wastes. Current regulations frequently specify the allowable quantities of wastes that may be discharged. These quantities are based on flow and concentrations of significant items such as solids, oxygen demand (*BOD_5*[13]), coliform bacteria, nitrogen, and oil, as specified by your regulatory agencies in your *NPDES PERMIT*.[14]

The effluent quality requirements in your NPDES permit usually determine what kind of activated sludge operation you

[13] *BOD_5. BOD_5 refers to the five-day biochemical oxygen demand. The total amount of oxygen used by microorganisms decomposing organic matter increases each day until the ultimate BOD is reached, usually in 50 to 70 days. BOD usually refers to the five-day BOD or BOD_5.*

[14] *NPDES Permit. **N**ational **P**ollutant **D**ischarge **E**limination **S**ystem permit is the regulatory agency document issued by either a federal or state agency which is designed to control all discharges of potential pollutants from point sources and storm water runoff into U.S. waterways. NPDES permits regulate discharges into navigable waters from all point sources of pollution, including industries, municipal wastewater treatment plants, sanitary landfills, large agricultural feedlots and return irrigation flows.*

can use and how tightly you must control the process. For example, if an effluent containing 50 mg/*L* of suspended solids and BOD_5 is satisfactory, a high-rate activated sludge process will probably meet your needs. If the limit is 10 mg/*L*, the high-rate process alone would not be suitable. If a high degree of treatment is required, very close process control and additional treatment after the activated sludge process may be needed. Today, secondary treatment plants are expected to remove 85 percent of the BOD and provide an effluent with a 30-day average BOD of less than 30 mg/*L*.

The treatment plant operator has little control over the makeup or amount of influent coming into the treatment plant. However, municipal ordinances may limit or forbid industries from dumping substances or wastes into the collection system if they could seriously damage treatment facilities or create safety hazards. Even with these laws to protect the collection system, some type of inspection program may be necessary (see *PRETREATMENT FACILITY INSPECTION* in this series of training manuals). It may be necessary to work out some additional plans for disposal, pretreatment, or controlled discharge to be sure that substances harmful to your treatment plant are diluted before they enter the plant.

QUESTIONS

Write your answers in a notebook and then compare your answers with those on page 111.

10.1A What two different ways may effluent quality requirements be stated by regulatory agencies?

10.1B How might harmful industrial waste discharges be regulated to protect an activated sludge process?

END OF LESSON 1 OF 3 LESSONS
on
ACTIVATED SLUDGE

Please answer the discussion and review questions next.

DISCUSSION AND REVIEW QUESTIONS

Chapter 10. ACTIVATED SLUDGE

(Lesson 1 of 3 Lessons)

At the end of each lesson in this chapter you will find some discussion and review questions. The purpose of these questions is to indicate to you how well you understand the material in the lesson. Write the answers to these questions in your notebook before continuing.

1. Some activated sludge plants do not have a primary clarifier. True or False?
2. Define activated sludge.
3. Sketch the activated sludge process.
4. Why should activated sludge in the final clarifier be returned to the aeration tank as soon as possible?
5. Define facultative bacteria.
6. How can the operator control the activated sludge process?

CHAPTER 10. ACTIVATED SLUDGE

PACKAGE PLANTS AND OXIDATION DITCHES

(Lesson 2 of 3 Lessons)

10.2 ACTIVATED SLUDGE PACKAGE PLANTS

10.20 Purpose of Activated Sludge Package Plants

The activated sludge treatment process is widely used in small systems or communities because it provides high-quality biological treatment when operated properly in the extended aeration mode of the activated sludge process.

This type of plant comes in many sizes and arrangements, but basically there are just two to four compartments made from one large tank. Figure 10.7 shows a two-compartment plant; Figure 10.8 shows a three-compartment plant. These plants are fabricated from steel with all of the necessary components at an environmental equipment supplier's factory. Table 10.1 lists the typical parts of an activated sludge plant and briefly describes their purposes.

The complete unit is delivered to the community's treatment site by truck. The plant may be set above or on the ground on a prepared pad. Most installations are placed in excavations with the top of the tank six inches to a foot (15 to 30 cm) above normal ground level to prevent surface water inflow and to provide better access for operational personnel to perform routine cleaning and maintenance. Before backfilling the excavation around the tank, several items must be checked because they are important to the plant's operation and life span.

1. The tank must be absolutely level in both directions; one side or one end lower than the other will hinder proper operation of the aeration equipment or cause *SHORT-CIRCUITING*[15] through the tank's compartments, thus creating process problems.
2. The tank's exterior corrosion protection material should be checked for any damage that may have occurred during transport or placement. Scratches from cables or equipment used in placement of the tank should be re-coated with the proper protective material.

TABLE 10.1 PURPOSES OF PARTS IN ACTIVATED SLUDGE PLANT

Part	Purpose
1. Bar screen	Catches rags and large debris to prevent entry into the aeration tank.
2. Aeration tank	Provides detention time during which activated sludge microorganisms treat wastewater.
3. Blower	Supplies air to the plant. Air is used for mixing, oxygenation, and pumping by air lift action.
4. Diffusers	Transfers oxygen from air to water for respiration by microorganisms. Also keeps the contents of the aeration tank mixed and moving.
5. Settling tank	Allows activated sludge to be separated from the water being treated. Clear effluent leaves the plant and settled activated sludge is either returned to the aeration tank or wasted to the aerobic digester to maintain the desired F/M ratio.
6. Air lifts	Pumps that move settled activated sludge from one compartment to another by the use of air.
7. Weir	Controls flow from one compartment to another.
8. Air filter	Filters inlet air to blowers. Protects blowers from internal damage or excessive wear from dirt and prevents diffusers from clogging internally.
9. Time clock	Controls operation of blower. In normal operation, the blower is operated several times per hour; it is rarely run continuously unless an operational problem requires continuous operation.
10. Froth control	Some plants are equipped with effluent pumps to supply water pressure to spray heads along the perimeter (outer edges) of the aeration tank to control foam or froth.

[15] *Short-Circuiting. A condition that occurs in tanks or basins when some of the flowing water entering a tank or basin flows along a nearly direct pathway from the inlet to the outlet. This is usually undesirable since it may result in shorter contact, reaction, or settling times in comparison with the theoretical (calculated) or presumed detention times.*

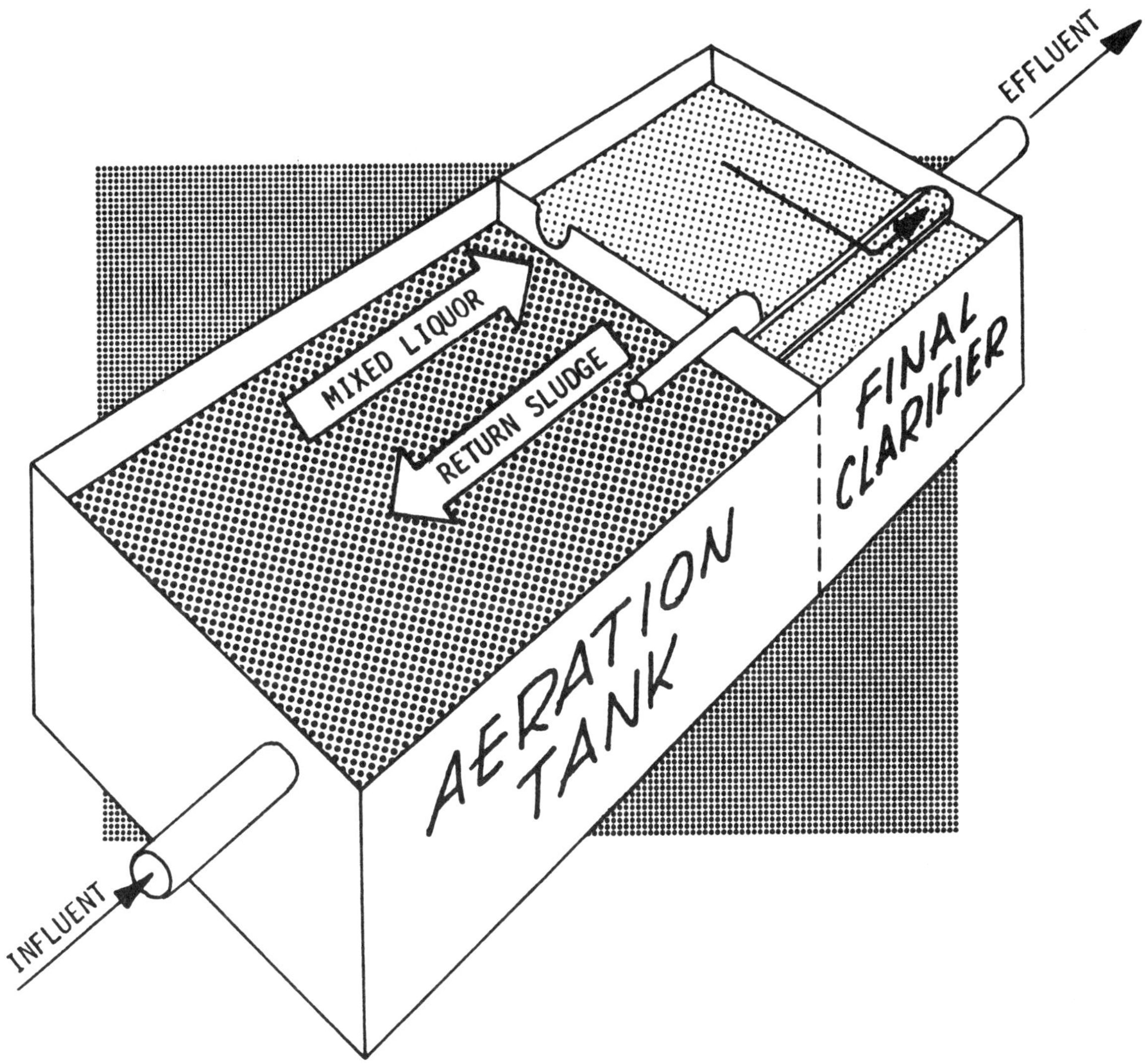

Fig. 10.7 Package plant (two compartments)

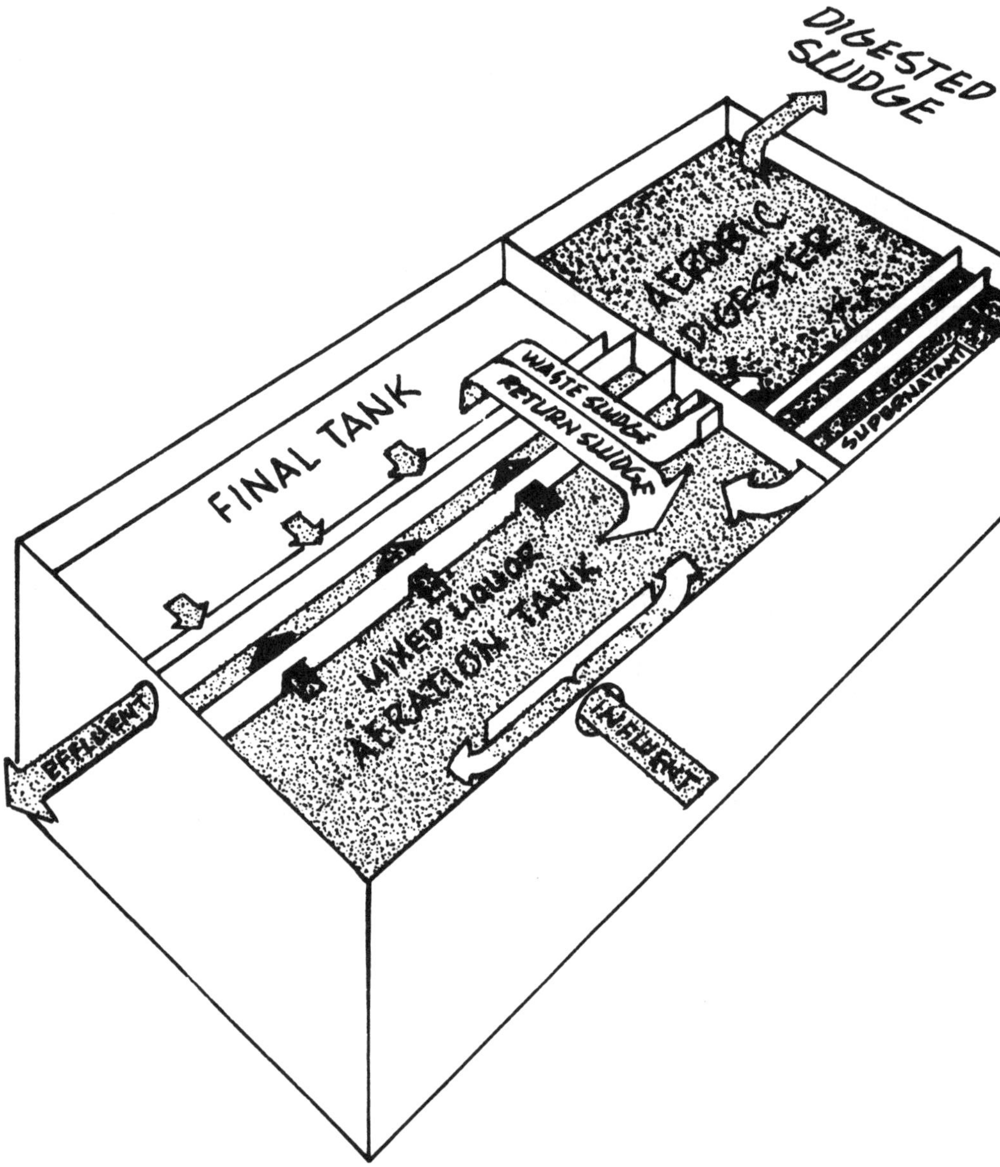

Fig. 10.8 Package plant (three compartments)

3. Most installations are protected by *CATHODIC PROTECTION*.[16] Check to be sure the *ANODES*[17] are properly placed and that the electrical leads are well bonded to the tank and anodes.

4. Buried tanks must be placed on a foundation of sand bedding material.

5. The space between the sides of the tank and the walls of the excavation should be filled using Class A material.

After backfilling, the soil around the plant should be sloped away from the tank to provide good drainage and to allow for easy access during wet weather. After the plant is placed in its proper location, remove all debris, piping supports, and equipment tie downs that were installed in the compartments by the manufacturer for shipping purposes. In most areas it is good practice to fill the package plant with water as soon as possible after the space between the excavation walls and the plant's walls have been backfilled with soil and compacted to prevent flotation of the plant from high groundwater. If the plant ever is floated, it must be re-excavated and leveled again for the plant to function properly.

The main parts of an activated sludge package plant are the two to four compartments that provide the following process functions:

- Aeration tank or reactor—this is usually the largest compartment of the plant. Its purpose is to provide oxygen and mixing energy to the activated sludge and incoming wastewater.

- Secondary clarifier or settler—this compartment is next to the aeration tank and receives flow (mixed liquor) from the aeration tank. The settling compartment is normally one-third the size of the aeration tank and is provided with a hopper on the bottom. The tank provides the quiescent period to the mixed liquor to separate the activated sludge from wastewater by settling. The wastewater effluent flows over a weir and normally leaves the plant to either be further treated or disinfected and discharged. The settled activated sludge is returned to the aeration tank.

All activated sludge package plants have at least the two compartments described above; some plants are provided with a third compartment, the aerobic digester.

- Aerobic digester—this compartment is about the same size as the clarifier compartment. The aerobic digester serves three functions, solids concentration, solids stabilization, and microorganism storage, all of which are very beneficial to the process and the operator. During the time excess activated sludge solids are being held in the clarifier compartment, the concentration of solids increases because the clarified supernatant is frequently drawn off (decanted). In addition, microorganisms in the tank continue to feed on and digest solid material in the wastewater, thereby reducing some of the solids through *ENDOGENOUS RESPIRATION*.[18]

  The clarifier compartment also provides a reserve supply of activated sludge solids that may be returned to the aeration tank as seed culture in cases of plant upsets caused by hydraulic overload, toxic discharge, or an unexpected high-organic load (shock load). The availability of this reserve sludge reduces the time required to establish a new culture from days to a few hours.

- Some plants also have a fourth compartment. Depending on the specific needs of the particular treatment plant, the compartment may be located at the beginning (inlet) or effluent end of the plant. When the tank is located at the plant inlet, it could serve as a rock or debris trap, a primary clarifier, or a pre-aeration or other pretreatment unit. If the tank is located at the effluent end of the plant, it may be used as a holding chamber to chemically dose post-treatment systems, as a post-aeration chamber before discharge, or as a disinfection chamber for the treated effluent.

10.200 Use of Activated Sludge Package Plants

The type of plant described in the previous section was a pre-fabricated activated sludge treatment unit. Other systems are sometimes constructed using separate pre-built tanks for the various functions; the individual tanks in these systems are connected with pipelines or channels to convey wastewater from one unit process to another. In other instances, small plants are constructed on site to handle slightly larger wastewater flows or provide a system for a specific activated sludge process. Some examples of the plant configurations available are shown in Figure 10.9.

Most activated sludge package plants are designed to operate in the extended aeration mode. Conventional activated sludge plants are normally operated with a *SLUDGE AGE*[19] ranging from three to seven days and a *MIXED LIQUOR SUSPENDED SOLIDS (MLSS)*[20] concentration, a measure of the microorganisms in the biological culture, of 2,000 to 3,000 mg/*L*. Extended aeration package plants operate with a sludge age ranging from ten to twenty days. At this sludge age, Mixed Liquor Suspended Solids (MLSS) are maintained at a concentration of 2,000 to 6,000 mg/*L*. The longer sludge age and higher MLSS concentration makes this process mode very reliable and less likely to be upset by shock loads or operational errors. The operation of these plants is similar to the operation of any other activated sludge plant. A high-quality effluent requires attention, understanding, and good plant operation.

---

[16] *Cathodic (ca-THOD-ick) Protection. An electrical system for prevention of rust, corrosion, and pitting of steel and iron surfaces in contact with water, wastewater or soil. A low-voltage current is made to flow through a liquid (water) or a soil in contact with the metal in such a manner that the external electromotive force renders the metal structure cathodic. This concentrates corrosion on auxiliary anodic parts which are deliberately allowed to corrode instead of letting the structure corrode.*

[17] *Anode (an-O-d). The positive pole or electrode of an electrolytic system, such as a battery. The anode attracts negatively charged particles or ions (anions).*

[18] *Endogenous (en-DODGE-en-us) Respiration. A situation where living organisms oxidize some of their own cellular mass instead of new organic matter they adsorb or absorb from their environment.*

[19] *Sludge Age. A measure of the length of time a particle of suspended solids has been retained in the activated sludge process.*

$$\text{Sludge Age, days} = \frac{\text{Suspended Solids Under Aeration, lbs or kg}}{\text{Suspended Solids Added, lbs/day or kg/day}}$$

[20] *Mixed Liquor Suspended Solids (MLSS). Suspended solids in the mixed liquor of an aeration tank.*

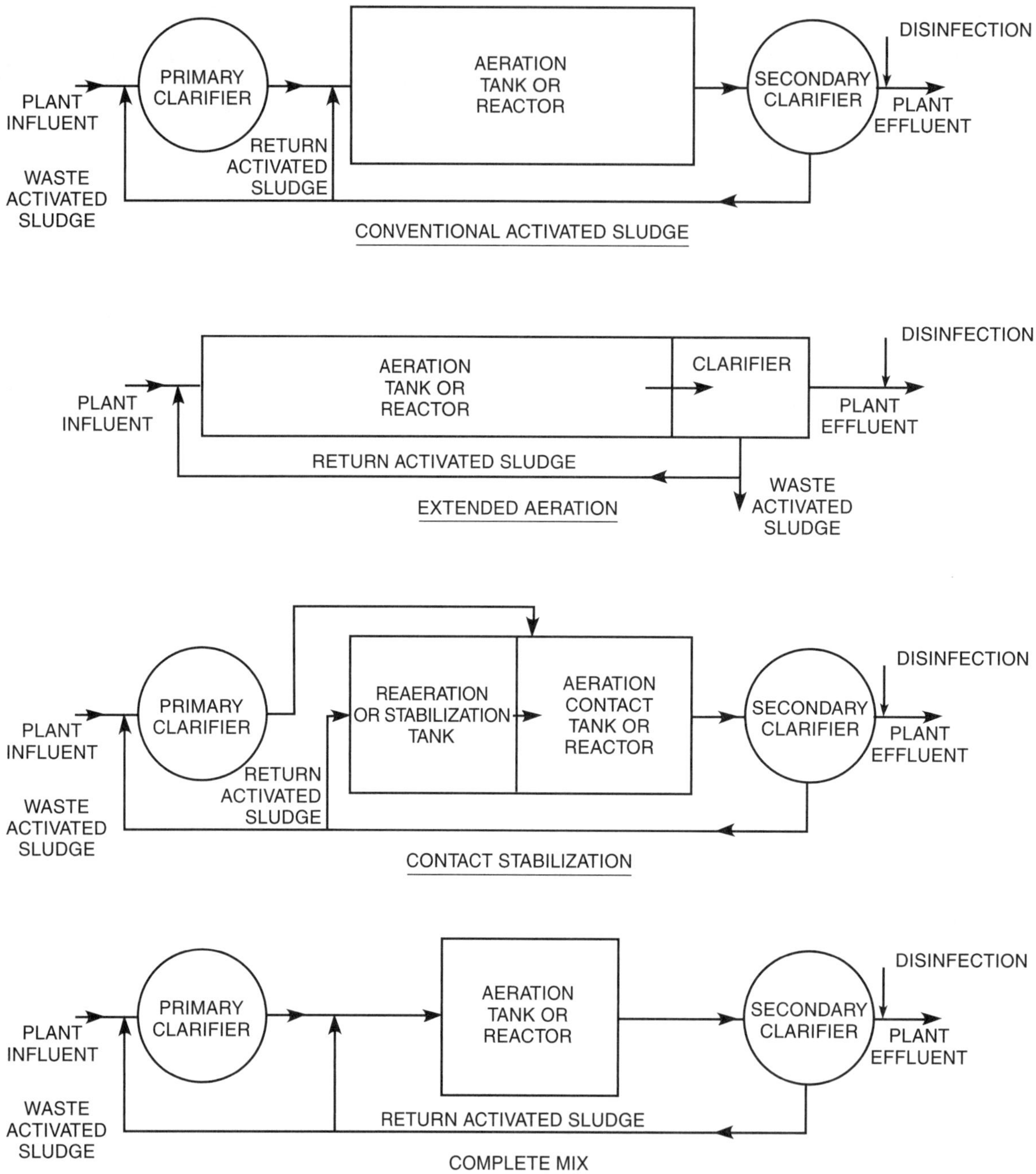

NOTE: ALL FOUR PLANTS HAVE AERATION HEADERS RUNNING THE LENGTH OF THE AERATION TANK. COMPLETE MIX ALSO HAS AERATION HEADERS RUNNING ACROSS THE AERATION TANK AT REGULAR INTERVALS.

Fig. 10.9 Types of package aeration plants

You may be assigned to operate a small activated sludge plant. You may not have laboratory facilities and may not be supplied with equipment for simple tests such as dissolved oxygen (DO), pH, or settleable solids. These tests should be run occasionally, at least three to four times a year. If necessary, the tests can be run by an outside laboratory or a local high school or college instructor may be able to assist you in conducting periodic analyses. If you do not have access to a laboratory, this manual will provide you with basic procedures for performing some simple tests that will help you effectively operate your plant. The tests described in this manual will not provide precise information; however, the results will be a useful guide for operating your plant if the tests are performed the same way each time, they are run at least once a week, and the results are recorded for comparison with previous test results. For details on how to perform the settleability and suspended solids tests, see Chapter 14, "Laboratory Procedures."

10.201 Package Plant Treatment Process Modes (Figure 10.9)

The most common modes of operation for activated sludge treatment processes are extended aeration, contact stabilization, and complete mix. All three of these process modes are essentially modifications of the conventional activated sludge process. The name of each mode basically describes the structural arrangement of the aeration tank(s) as well as the various arrangements of process streams that are used to provide process flexibility. Realistically, almost all package activated sludge plants are of the extended aeration type. They have long solids retention times, high mixed liquor suspended solids, and low food/microorganism ratios.

Extended Aeration

Extended aeration is similar to conventional activated sludge except that the organisms are retained in the aeration tank longer and do not get as much food. The organisms get less food because there are more of them to feed. Mixed liquor suspended solids concentrations range from 2,000 to 6,000 mg/*L*. In addition to consuming the incoming food, the organisms also consume any stored food in the dead organisms. The new products are carbon dioxide, water, and a biologically inert residue. Extended aeration does not produce as much waste sludge as other processes because much of the material is oxidized to soluble compounds and/or gases. However, wasting still is necessary to maintain proper control of the process.

Contact Stabilization

Contact stabilization is similar to conventional activated sludge except that the capture of the waste material and the digestion of that material by the organisms are accomplished in two different aeration tanks. In both the conventional and the contact stabilization modes, the activated sludge flows to the clarifier to be separated from the wastewater. The organisms can *ADSORB*[21] the waste material on their cell walls in only fifteen to thirty minutes, but it takes several hours to *ABSORB*[22] the material through the cell walls. In conventional activated sludge, adsorption and absorption both take place in one tank; therefore, the wastewater has to remain in the tank for a longer time. In the contact stabilization mode, however, the settled organisms are moved to another separate aeration tank (called a stabilization or re-aeration tank) where they digest their food and then are returned hungry to the original aeration tank (contact tank) ready to eat more food. The mixed liquor suspended solids (MLSS) concentration of the contact tank should be maintained around 1,500 to 2,000 mg/*L*. If the MLSS concentration gets too high, the sludge that the microorganisms form is disposed of or "wasted" as in conventional activated sludge. In most package plants, contact stabilization is not used and the adsorption/oxidation is achieved in one tank.

Complete Mix

In an ideal complete-mix activated sludge plant, the contents of the aeration tank are completely mixed; that is, the MLSS are uniformly distributed throughout the entire aeration tank. To ensure that this is achieved, special arrangements are often used to *UNIFORMLY DISTRIBUTE THE INFLUENT AND WITHDRAW THE EFFLUENT FROM THE AERATION TANK.* Attention to the tank shape and to intensive mixing is important. There are some means that the operator may use to evaluate the degree to which a particular process operates in the complete-mix mode. First and foremost, the contents of the tank should be as uniform as possible throughout the tank. This can be confirmed by measurements of DO and suspended solids. If the tank is thoroughly mixed, these measurements should be nearly uniform. If the DO and/or suspended solids measurements are far from uniform, incomplete mixing is indicated and the operator should work with an engineer or the package plant manufacturer to determine the cause. The settleability of the complete-mix sludges is generally well within the accepted range of normal operation. The MLSS in the aeration tank ranges from 2,000 to 5,000 mg/*L*.

Most package plants have one influent line and one influent port. The aeration compartment has one aeration *HEADER*[23] (Figure 10.10) and the tank contents are completely mixed in a very short time. A limitation of the complete-mix mode of operation is that the process may be more susceptible to short-circuiting.

10.202 Aeration Methods

Two methods are commonly used to supply oxygen from the air to the bacteria: *MECHANICAL AERATION* and *DIFFUSED AERATION.*

Mechanical aeration devices agitate the water surface in the aerator to cause spray and waves by paddle wheels (Figure 10.11), mixers, rotating brushes (Figure 10.12), or some other method of splashing water into the air or air into the water where oxygen can be absorbed. Mechanical aerators in the tank tend to be lower in installation and maintenance costs. Usually they are more versatile in terms of mixing, production of surface area of bubbles, and oxygen transfer per unit of applied power.

---

[21] *Adsorption (add-SORP-shun). The gathering of a gas, liquid, or dissolved substance on the surface or interface zone of another material.*

[22] *Absorption (ab-SORP-shun). The taking in or soaking up of one substance into the body of another by molecular or chemical action (as tree roots absorb dissolved nutrients in the soil).*

[23] *Header. A large pipe to which the ends of a series of smaller pipes are connected. Also called a manifold.*

Air headers (package plant)

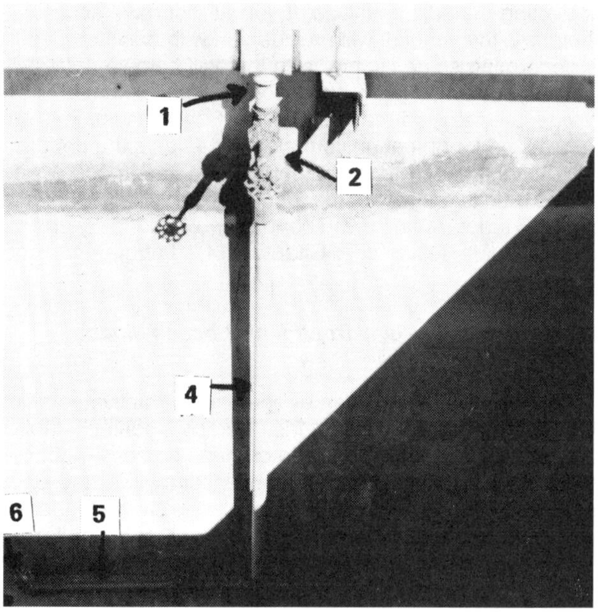

Air header (grit chamber)

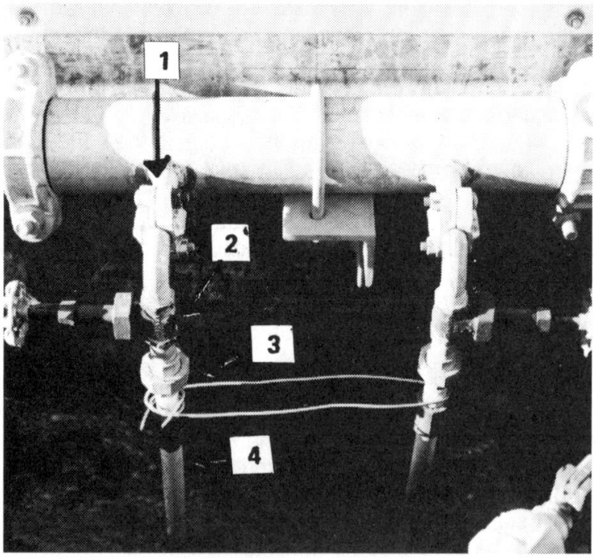

Air header (channel aeration)

1. distribution system connector fitting
2 header valve—regulating and isolation
3. union
4. riser pipe
5. horizontal air header
6. diffuser

Fig. 10.10 Fixed headers

Fig. 10.11 Mechanical aeration device

(Permission of INFILCO INC.)

Brush rotor

Brush rotor

Outboard bearing assembly

Brush rotor drive motor and gear reducer assembly

Fig. 10.12 Brush rotor used in oxidation ditches

Diffused-air systems are more common to manufactured package activated sludge plants. A device called a diffuser (Figure 10.13) is used to break up the air stream from the blower system into fine bubbles in the mixed liquor. The smaller the bubble, the greater the oxygen transfer due to the greater surface area of rising air bubbles surrounded by water. Unfortunately, fine bubbles will tend to regroup into larger bubbles while rising unless they are broken up by suitable mixing energy and turbulence, which is caused by the volume and pressure of the air being supplied from the compressor to the diffusers.

There are a number of devices (Figure 10.14) that can be used to diffuse (break up) an air stream into small bubbles to obtain a high efficiency of oxygen transfer from the air bubbles into the water. To obtain this highest possible oxygen transfer efficiency, the diffuser material must be very fine, such as in a Carborundum plate diffuser or a tightly wound diffuser. Unfortunately, these types of diffusers require an air supply that is very, very clean. Any dust particles will clog the pores of the diffuser internally, which will create back pressure on the air system and reduce oxygen transfer to the water. These diffusers also plug very quickly if for any reason the air supply to the diffuser is disrupted. When this occurs, activated sludge liquor may flow back into the diffusers. For these reasons, diffusers that generate larger air bubbles and are less efficient in oxygen transfer (3 to 5 percent of oxygen in air) are often used because they do not require as frequent cleaning or other maintenance.

Most diffused-air package plants are equipped with a diffuser known as a sparger, disk, or comb (Figure 10.15). These diffusers are placed on air pipes that are connected to the main air manifold from the air blowers. These pipes may be equipped with knuckle joints so that an array of diffusers may be mechanically lifted from the aeration tank for maintenance work such as cleaning or replacement.

Another option is to install a pipe union on the air main near the top of the aeration tank. Another pipe union is used to attach a one-inch diameter pipe on which two to four diffusers have been installed. This arrangement (Figure 10.10) makes it possible to disconnect the pipe to raise the bottom portion of the pipe and diffusers out of the aeration tank for maintenance.

All diffusers for mixing and oxygen transfer in an aeration tank are located approximately 18 inches (46 cm) or less above the bottom of the tank to facilitate rolling the contents of the tank to provide the mixing action needed. All the diffusers in the aeration tank must be positioned at the same elevation or, if they are used for other purposes in the tank, they must have different air pressures supplying them.

QUESTIONS

Write your answers in a notebook and then compare your answers with those on page 111.

10.2A How many compartments are there in a typical package extended aeration plant? What is the purpose of each compartment?

10.2B What are some common characteristics of all package aeration plants?

10.2C Why should package plants have cathodic protection?

10.21 Pre-Start Check-Out

While the tank is empty, check the following:

1. Tank must be level from one end to the other.
2. If the tank is constructed of metal, is cathodic protection required for corrosion control?
3. Condition of paint on exterior and interior.
4. Removal of all rocks and debris from both compartments.
5. If equipped with *COMMINUTOR*[24] or grinder, check lubrication.
6. Check aeration device:
   a. Lubrication of motors
   b. Direction of rotation
   c. Mechanical aeration—proper agitator depth
   d. Compressor, if diffused air
      (1) Air filter and oil bath
      (2) Air header and valves
      (3) Air lift tubes and valves
      (4) Diffusers installed
      (5) Swing header lifts easily and free
      (6) Fixed headers or air drop legs secure, unions tight, and isolation valves operate freely
7. Record and file the following data:
   a. Plant model and serial number
   b. Two copies of plant operation and maintenance manual
   c. Nameplate data from equipment
      (1) Comminutor
      (2) Comminutor motor
      (3) Aeration motor
      (4) Compressor or agitator
      (5) Amperage at design load
      (6) Oils and greases specified for each unit
      (7) Sizes and types of blower inlet air filters
      (8) If blower belt driven, size of belts; if multiple sheaves (two or more belts driving blower from motor), belts must be ordered in matched sets. If one belt breaks, all belts must be replaced at the same time.
8. Check influent gate or valve for proper operation.

[24] *Comminutor (com-mih-NEW-ter). A device used to reduce the size of the solid chunks in wastewater by shredding (comminuting). The shredding action is like many scissors cutting or chopping to shreds all the large influent solids material in the wastewater.*

Fig. 10.13 Air diffuser

(Courtesy Paul Hallbach, National Training Center, Water Quality Office/EPA)

NOTE: Diffuser does not appear to be level. Most of the air is leaving from the higher left side which has a lower head of water over it. Diffuser should be level to ensure even distribution of air leaving the diffuser.

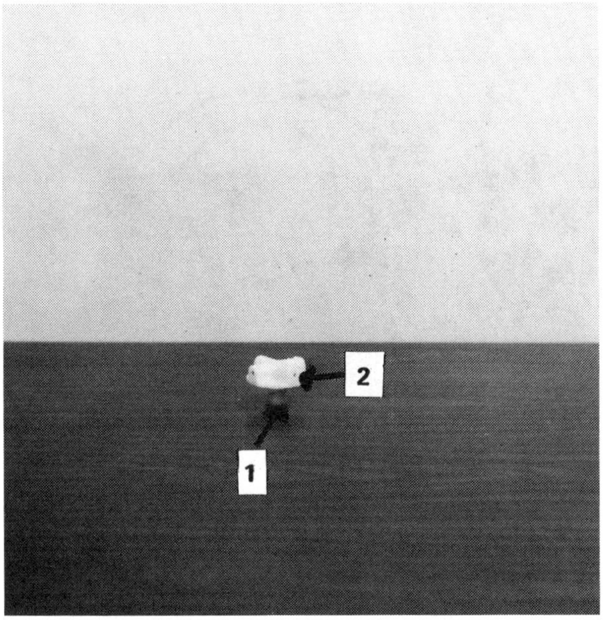

1. threaded stub for header mounting
2. air outlet orifices (four)

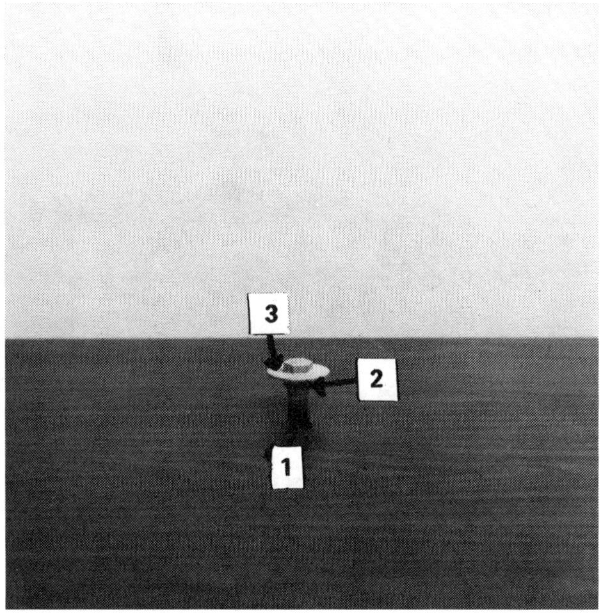

1. threaded stub for header mounting
2. air outlet orifices (four)
3. air deflector crown

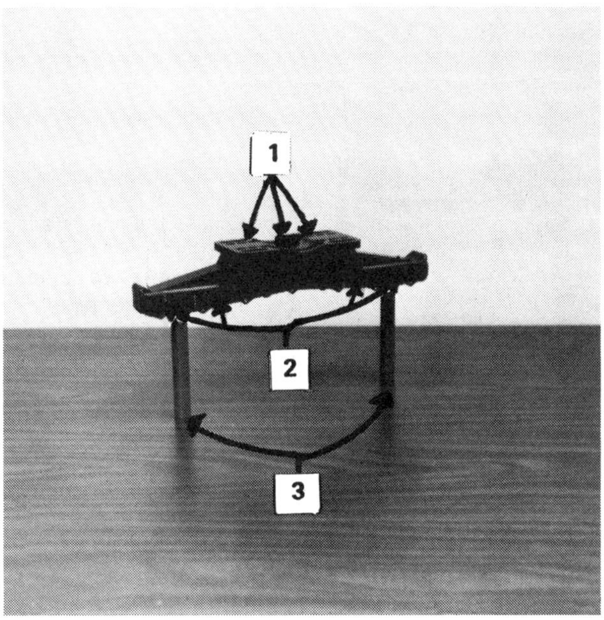

1. air inlet and header mounting surfaces
2. air outlet orifices (eight)
3. plunge tubes

Fig. 10.14 Coarse bubble diffusers

Air headers

Air distribution piping and air headers

Swing header

1. distribution system connector fittings
2. header valve—regulating and isolation
3. double pivot upper swing joint
4. upper riser pipe
5. lower riser pipe
6. pivot elbow
7. leveling tee
8. horizontal air header
9. air blowoff leg
10. hoist

Fig. 10.15 Swing headers

10.22 Starting the Plant

If the plant has not been previously filled with fresh water, the aeration blower should be started and air introduced to the diffusers prior to the introduction of wastewater. Some types of diffusers could become plugged if wastewater enters the tank before air is coming out of the diffusers. In a package plant the aeration blower is intended to operate intermittently. The diffusers installed are ones not likely to be affected by wastewater flowing back into the diffusers when the blower is idle. If the plant is the diffused-air type with air lifts for return sludge, the air line valve to the air lifts will have to be closed until the settling compartment is filled. Otherwise all the air will attempt to go to the empty compartment and none will go to the diffusers. Once the settling compartment is filled from the overflow from the aeration tank, the air lift valves may be opened. They will have to be adjusted to return a constant stream of water and solids to the aeration tank. This adjustment is usually two to three turns open on the air valve to each air lift. If the plant is equipped with an aerobic digestion compartment, the valves on the air headers in that compartment should also be closed. There will be no excess sludge to digest for some time so there is no need to aerate that compartment. If the plant is started in cold weather or if mosquito breeding is a problem, the air valves to the diffusers should be opened slightly to obtain a slight roll of the water to prevent ice formation or to reduce insect breeding.

You may have several options in starting a new plant. It is going to require a week or more to establish an activated sludge culture. The quicker a culture is established, the sooner the plant will produce an acceptable effluent. The fastest start would be to acquire seed sludge from a nearby activated sludge plant. A septic tank pumper tank truck can be used to transport a thousand or fifteen hundred gallons of concentrated return sludge from another wastewater treatment plant and dump it in your aeration compartment (with the aeration system running). Aerate the new sludge for an hour, then admit raw wastewater slowly into the aeration tank. Once the clarifier compartment is filled, open the air valves on the air lift return lines or start the return sludge pumps. During the start-up mode it is good practice to return as much settled activated sludge as possible back into the aeration tank, even when it has a very low concentration of MLSS. The need is to get the organisms back into an aerobic environment with a food source as soon as possible. However, care must be taken to be sure that the combined flow of RAS (Return Activated Sludge) and raw influent does not exceed the design capacity (shorten the detention time) of the clarifier. If the total flow is too high, the sludge will not have time to settle out and will flow out of the plant.

During the first few days of start-up, the aeration blower should be operated on a continuous basis to adequately supply oxygen and mixing during the start-up period. After several days of operation and when you can observe at least 20 percent by volume of MLSS in a jar test, then you can program the time clock to gradually reduce blower operation. The blower should never be off for longer than one hour. In the aeration compartment during normal operation the blower should operate for 10 to 15 minutes, then turn off for 5 to 10 minutes, and then turn back on for 10 to 15 minutes to repeat the cycle. Blower operating time is such that the aeration tank dissolved oxygen (DO) level during blower operation is 2 to 3 mg/*L*, and the mixed liquor DO does not drop below 1.0 mg/*L* of DO during the OFF cycle of the blower. If the DO drops below 1.0 mg/*L*, the blower must be operated longer or more frequently.

If you don't have access to seed activated sludge for start-up then you must raise your own activated sludge culture. If there were existing treatment facilities before the new plant was installed, that is, if the plant was replacing a leach field or some other treatment method, then continue to use that facility. Start the new plant by admitting a small amount of raw wastewater (filling the aeration and clarifier tanks only). Run the aeration equipment and return sludge system on a continuous recycle basis. The next day add additional raw wastewater (20 percent of the aeration tank volume) at two-hour intervals. After the third day, the plant should be capable of receiving the full flow. The effluent will not be of the highest quality until sufficient MLSS have been established in the aeration tank.

In cases where there are no prior wastewater treatment facilities, then the flow is admitted to the new plant and the highest return sludge rates possible are maintained. Once sufficient MLSS are established, return flows are reduced to start concentrating solids in the clarifier to obtain a higher concentration of solids in the return sludge.

The last two methods described above usually will result in development of a white foam on the aeration tank during start-up. An accelerated growth of filamentous organisms on the tank walls and weirs also will appear. The effluent will have a milky appearance. All of these problems will occur while the F/M ratio is too high during the start-up period; that is, while the MLSS concentration is low, meaning there is more food than the organisms can digest.

Foam can be controlled by the froth sprays (if the tank is so equipped) or by use of a water hose with a sprinkler. The slime growths can be brushed or hosed off the walls and weirs. Once an activated sludge culture is established, the effluent will become clear and lose the milky or cloudy appearance and the foam and filamentous growths should disappear.

See Chapter 14, "Laboratory Procedures," for procedures to determine solids, DO, and pH.

10.23 Operation of Aeration Equipment

In conventional activated sludge plants equipped with diffused air, aeration equipment is operated continuously to reduce clogging of the diffusers. In mechanically aerated plants, the aeration equipment may or may not be operated on a continuous basis. This depends on the process mode and the waste being treated. For economic efficiency, small activated sludge package plants with blowers and air diffusers quite frequently are operated intermittently; again, blower operating time will depend on the process mode and the waste being treated.

Intermittent blower operation extends equipment life and is more economical than continuous operation in areas having high electrical rates. The prime requirement is that DO levels be maintained in the aeration tank as described previously. All valves to the aeration tank diffusers should be kept wide open. Only air valves on the air lifts or diffusers in the aerobic digestion tank are throttled for control purposes.

The blower discharge air manifold is equipped with a pressure relief valve that generally is set at two psig (pounds per square inch gage pressure) above the blower manifold discharge pressure. The purpose of the relief valve is to prevent excessive loading on the blower, but if air valves are throttled, restricting air flow, and air is vented through the pressure relief valve, it is not an efficient operation. The pressure relief valve may open on blower start-up if there is wastewater in the diffusers or drop pipes. It may take a minute or so for the water to be expelled from the pipes and diffusers, but the relief valve should not continue to vent after start-up. If the relief valve does continue to vent, check all diffuser control valves for open position and check for blocked or plugged diffusers. If both of these items are OK, check the air relief valve setting with a pressure gage and adjust the springs or weights to let the valve open at two psig above normal manifold discharge pressure.

Blowers are controlled by a time clock. In units where two blowers are used to supply air, the clock alternates the RUN and OFF operation of the units. Some plants are equipped with two time clocks. This is especially needed if the plant is subject to intermittent loading, such as wastewater flow from a school. One clock is programmed for the school week when students are in attendance and the other clock is programmed for the weekends when little or no flow is discharged to the plant. The second clock provides a backup for the first clock in case of failure. Installing two clocks also eliminates the need for an operator to visit the plant on weekends to manually program a single clock for weekend or weekday operation. Clocks can also control the operation of other devices such as pumps, chlorinators, lighting, and other process units.

You can judge how well the aeration equipment is working by the appearance of the water in the settling compartment and the effluent that goes over the weir. If the water is murky or cloudy and the aeration compartment has a rotten egg odor (H_2S), not enough air is being supplied. The air supplied (aeration rate) should be increased. If the water is clear in the settling compartment, the aeration rate is probably sufficient. If you have a DO probe or lab equipment to measure the DO, try to maintain a DO level of around 2 mg/*L* throughout the aeration tank. Try to measure the DO at different locations in the aeration tank as well as from top to bottom.

QUESTIONS

Write your answers in a notebook and then compare your answers with those on page 111.

10.2D Why should the valve to the air lift pumps be closed until the settling compartment is filled?

10.2E What would you do if the water in the aeration compartment was murky or cloudy and the aeration compartment had a rotten egg odor (H_2S)?

10.24 Wasting Sludge

Many older package plants did not come equipped with facilities for wasting sludge. The reason was that the extended aeration system was considered capable of stabilizing the aeration tank suspended solids to a level that was acceptable for release to the receiving waters. Today, a higher degree of treatment and effluent quality is required. Therefore, some sludge solids must be removed (wasted) from the system routinely to achieve optimum plant performance. For best results (in terms of effluent quality), waste about five percent of the solids each week during summer operation to prevent excessive solids *BURPING*.[25]

To waste the excess activated sludge, the approximate amount of MLSS in the aeration tank should be determined. If laboratory equipment is not available, a simple test using a quart jar can be performed. Fill the jar with mixed liquor as it flows from the aeration tank into the settling tank. Let the jar sit without mixing for 30 minutes. If at least 50 percent of the jar is sludge after 30 minutes, then a portion of the sludge should be wasted to the aerobic digester. Mixed liquor solids in the aeration tank should never be less than 20 percent in the jar test to have a good, healthy activated sludge culture.

If the plant is not equipped with an aerobic digester, sludge wasting should be postponed until the jar test indicates a well-compacted sludge volume of at least 70 percent. Once that amount of MLSS is accumulated, sludge wasting should occur. Arrange for a septic tank pump truck to haul off the wasted activated sludge.

Continuously aerate the aeration tank for four to six hours before the arrival of the septic tank pumper. During this aeration period, turn off the return sludge air lifts or pumps so that sludge will become more concentrated in the settling tank. The septic tank pumper should pump the waste activated sludge out of the settling hoppers and dispose of it at a larger treatment plant, in a sanitary landfill, or some other approved site.

In plants equipped with an aerobic digester, waste activated sludge can be returned to the digester. However, the mixed liquor solids in the aeration tank should not be reduced by more than 10 percent per day to prevent causing a system overload (shock loading) in the digester.

Another method of sludge wasting is to turn off the return pumps or air lifts for one hour and continue to let the rest of the plant function. After one hour of not returning sludge, use a portable pump to pump about five percent of the waste solids from the settling compartment to a sand or soil drying bed. The amount of solids pumped is determined by measuring the depth of the sludge blanket and then lowering it five percent. Record the pumping time and weekly waste solids for this time period if results are satisfactory.

If your package activated sludge plant does not have an aerobic digester, applying waste activated sludge to drying beds may cause odor problems. If odors from waste sludge drying beds are a problem, consider the following solutions:

1. Waste the excess activated sludge into an aerated holding tank. This tank can be pumped out and the sludge disposed of in a manner approved by the regulatory agency. If aerated long enough, the sludge could be applied to drying beds.
2. The excess or waste activated sludge can be removed by a septic tank pumper and disposed of in a manner approved by the regulatory agency.
3. Arrange for disposal of the excess activated sludge at a larger treatment plant nearby.

[25] *Burping. "Billowing" solids that result when the settling tank sludge blanket becomes too deep (occupies too much volume in the bottom of the tank) and solids are swept by the water up and out over the effluent weirs in the discharged effluent.*

Annually, check the bottom of the hoppers for rocks, sticks, and grit deposits by probing or dewatering the tank. Also check the tail pieces of the air lifts to be sure that they are clear of rags and rubber goods and in proper working condition.

The frequency and amount of wasting may be revised after several months of operation by examining:

1. The amount of carryover of solids in the effluent,
2. The depth to which the solids settle in the aeration compartment when the aeration device is off (should be greater than one-third of the distance from top to bottom),
3. The appearance of floc and foam in the aeration compartment (color, settleability, foam makeup) and excess solids on the water surface in the tank, and
4. The results of laboratory testing (refer to Section 10.26, "Laboratory Testing for Small Activated Sludge Plants").

A white, fluffy foam indicates low solids content in the aerator while a brown, leathery foam suggests high solids concentrations. If you notice high effluent solids levels at the same time each day, the solids loading may be too great for the final clarifier. Excessive solids indicate that the mixed liquor suspended solids concentration is too high for the flows and more solids should be wasted. Sometimes a loss of solids over the weirs is caused by a combination of high solids and high flows. Turning off foam sprays and hoses can increase the settling time and may allow the sludge to settle better.

10.25 Operation of Activated Sludge Package Plants

These small plants are intended to function in the following ranges:

1. Plant Flow	5,000 to 100,000 gallons/day (19,000 to 380,000 liters/day) 0.005 to 0.1 MGD (0.019 to 0.38 M*L*D)
2. BOD Loading	10 to 30 lbs BOD/day per 1,000 cubic feet (200 to 500 kg BOD/day per 1,000 cubic meters) of aeration tank
3. F/M Ratio	0.05 to 0.3 lb BOD/day/lb MLVSS
4. MLSS Concentration	2,000 to 6,000 mg/*L*
5. Sludge Age	20 to 35 days
6. Dissolved Oxygen in Aeration Tank	1.0 to 3.0 mg/*L*
7. BOD Removal, %	85 to 90%
8. Suspended Solids Removal, %	90 to 95%

10.250 Normal Operation

Package activated sludge plants should be checked every day. Each visit should include the following:

1. Check the appearance of aeration tank and final clarification compartment.
2. Check aeration unit for proper operation and lubrication.
3. Check return sludge line for proper operation. If the air lift is not flowing properly, briefly close the outlet valve to force the air to go down and out the tail piece. Usually this will blow it out and clear any obstructions. Re-open the discharge valve and adjust to desired return sludge flow. *CAUTION:* Back flushing the air lift will disturb the settled sludge blanket in the clarifier hopper when the tail piece is cleared and this may cause solids to be discharged from the clarifier into the plant effluent. Back flushing should be very brief. If repeated clogging of the air lift occurs, the settling tank should be dewatered and the debris removed from the hopper.
4. Check comminuting device for lubrication and operation. Clean bar screen by removing captured rags and debris, hose down screen and inlet to aeration tank.
5. Hose down aeration tank and final clarification compartment. Remove any accumulations of sludge from the walls and hoppers of the clarifier tank with a squeegee. Sludge deposits left for long periods of time without oxygen will become *SEPTIC*[26] and produce odors caused by anaerobic decomposition.These septic deposits in either the aeration tank or the clarifier may be harmful to the activated sludge process.
6. Brush weirs when necessary to remove algae and captured materials.
7. Skim off grease and other floating material such as plastic and rubber goods. Dispose of this material, along with captured bar screen debris, at an approved site.
8. Perform necessary maintenance and operational adjustments to other process units if used; that is, chlorination/disinfection equipment. Check plant discharge for proper appearance, grease, or material of wastewater origin that is not desirable.

10.251 Abnormal Operation

Remember that changing conditions or abnormal conditions can upset the microorganisms in the aeration tank. As the temperature changes from season to season, the activity of the organisms speeds up or slows down. Also, the flows and waste (food as measured by BOD and suspended solids) in the plant influent change seasonally. All of these factors require the operator to gradually adjust aeration rates, return sludge rates, and wasting rates. Abnormal conditions may consist of high flows or high solids concentrations as a result of storms or weekend loads. These problems require the operator and the plant to be prepared and to do the best possible job with available facilities.

Toxic wastes such as pesticides, detergents, solvents, or high or low pH levels can upset or kill the microorganisms in the aeration tank. Plant effluent usually does not deteriorate until after the toxic substance has passed through the plant. To correct problems from toxic substances, try to locate the source and prevent future discharges.

---

[26] *Septic (SEP-tick). A condition produced by anaerobic bacteria. If severe, the sludge produces hydrogen sulfide, turns black, gives off foul odors, contains little or no dissolved oxygen, and the wastewater has a high oxygen demand.*

If the microorganisms in the aeration tank have been killed by toxic wastes, try to build up the microorganisms as if you were working with a new plant. If the plant is equipped with an aerobic digester, use a portion of the sludge solids in the digester compartment. Turn off the air to the aerobic digester diffusers, let the sludge solids settle out, decant the liquid above the sludge blanket, and pump the settled sludge back into the aeration tank. This sludge will introduce a large number of organisms required to treat the wastewater and will permit the plant to return rapidly to normal operation (this is one of the largest benefits of an activated sludge plant having aerobic digestion). Operate the blower continuously for several hours to supply dissolved oxygen to the aeration tank. Once the activated sludge culture is well established (may require several days), intermittent blower operation can be resumed. Any time solids are lost out of the plant for any reason, this method should permit rapid recovery of the process.

An outstanding example of a toxic substance added by operators is the use of chlorine for odor control (prechlorination). Chlorine is a toxicant and should not be allowed to enter the activated sludge process in uncontrolled amounts because it is not selective with respect to the types of organisms it will kill or inactivate. Although chlorine is used in larger activated sludge plants for the control of *BULKING*[27] sludge, its use in package plants is not recommended. It may kill the organisms that should be retained as workers. Chlorine is effective in disinfecting the plant effluent *AFTER* treatment by the activated sludge process.

10.252 Troubleshooting

When problems develop in the activated sludge process, try to identify the cause of the problem, and select the best possible solution. Table 10.2 lists a number of problems that may be encountered when operating an activated sludge package plant and suggests the probable cause of each problem, what should be checked or monitored, and possible solutions to the problem. Remember that the activated sludge process is a biological process and may require from three days to a week or longer to show any response to the proper corrective action. Allow seven or more days for the process to stabilize after making a change in the treatment process.

1. Solids in the effluent

   a. If the effluent appears turbid (muddy or cloudy), the return activated sludge pumping rate is out of balance. Try increasing the return sludge rate. Also, consider the possible presence of something toxic to the microorganisms or the possibility of a hydraulic overload washing out some of the solids.

   b. If the activated sludge is not settling in the clarifier (sludge bulking), several possible factors could be causing this problem. Look for too low a solids level in the system; low dissolved oxygen concentrations in the aeration tank; strong, stale septic influent; high grease levels in the influent; or alkaline wastes from a laundry.

   c. If the solids level is too high in the sludge compartment of the secondary clarifier, solids will appear in the effluent. Try increasing the return sludge pumping rate. Remember, excessive return flows can cause a hydraulic overload on the clarifier.

   d. If odors are present and the aeration tank mixed liquor appears black as compared with the usual brown color, try increasing the aeration rates and look for septic dead spots.

   e. If light-colored floating sludge solids are observed on the clarifier surface, try reducing the aeration rates. Try to maintain the dissolved oxygen at around 2 mg/*L* throughout the entire aeration tank.

   f. Long sludge retention times in the clarifier will result in anoxic/anaerobic conditions that will release gases that will be trapped in the sludge and cause it to float.

2. Odors

   a. If the effluent is turbid and the aeration tank mixed liquor appears black as compared with the usual brown color, try increasing aeration rates and look for septic dead spots.

   b. If clumps of black solids appear on the clarifier surface, try increasing the return sludge rate. Also, be sure the sludge return line is not plugged and that there are no septic dead spots around the edges or elsewhere in the clarifier.

   c. Examine the method of wasting and disposing of waste activated sludge to be sure this is not the source of the odors.

   d. Poor housekeeping could result in odors. Do not allow solids to accumulate or debris removed from the wastewater to sit around the plant in open containers.

3. Foaming/Frothing

   Foaming is usually caused by too low a solids level, while frothing is caused by too long a solids retention time.

   a. If too much activated sludge was wasted, reduce wasting rate.

   b. If over-aeration caused excessive foaming, reduce aeration rates.

   c. If the plant is recovering from overload or septic conditions, allow time for recovery.

   d. If frothing is occurring and building up on the clarifier tank, skim off the floating froth and dispose of it with grease and other debris at a landfill. Do not return the floating froth from the clarifier back to the aeration tank. This seeds the culture and prolongs the problem of frothing.

   Foaming can be controlled by water sprays or commercially available defoaming agents until the cause is corrected by reducing or stopping wasting and building up solids levels in the aeration tank.

To learn more about the operation of an activated sludge process under both normal and abnormal conditions, you may refer to another training manual in this series, *OPERATION OF WASTEWATER TREATMENT PLANTS*, Volume II, Chapter 11, "Activated Sludge." There you will also find a troubleshooting guide for activated sludge plants.

---

[27] *Bulking (BULK-ing). Clouds of billowing sludge that occur throughout secondary clarifiers and sludge thickeners when the sludge does not settle properly. In the activated sludge process, bulking is usually caused by filamentous bacteria or bound water.*

TABLE 10.2 ACTIVATED SLUDGE PACKAGE PLANT TROUBLESHOOTING GUIDE

Indicator/ Observation	Probable Cause	Check or Monitor	Solution
Rocks, mud, debris	Broken collection main	Check collection system	Repair broken main.
	Open manhole	Check collection system	Replace manhole cover—add riser ring to cone to raise cover above existing grade to prevent future drainage.
		Check aeration tank for rocks, mud, debris	Remove debris from aeration tank.
Rags (mops), large pieces of plastic in aeration tank	Bar screen	Check bar screen for rags and debris	Clean bar screen more frequently.
		Check screen spacing	Reduce bar openings to capture more material.
	Comminutor	Comminutor operation	Comminutor operating in right direction—replace broken combs on comminutor screen.
		Comminutor cutters	Cutters dull, replace.
Aeration tank No mixing Low DO	Blower off, does not run	Blower operation	Check power supply to time clock and blower motors. Re-set main power breaker.
		Time clock	Program time clock to correct run times.
		Blower motor	Check motor for operation. Replace if burned out.
	Blower motor runs	Blower drive Belts—Broken	a. Broken belt. Replace. If more than one belt on drive, replace all belts.
		Blower drive Belts—Loose	b. Belts loose, slipping. Adjust belt tension between motor and blowers.
		Burned	c. Check sheave alignment. Realign motor and blower.
		Direct drive or gear reducer	Check coupling.
			Check gear reducer output. If low, replace.
	Blower operation normal	Blower inlet air filter dirty	Clean filter or replace filter cartridge.
	Air manifold pressure lower than normal	Blower discharge check valve	Check valve not opening. Remove obstruction, or replace valve.
		Blower is hot	Check blower lubricating oil level and/or adjust oil to proper operating level.
			Check blower enclosure ventilation screens for obstructions, restricting cooling air.
		Blower worn out	Replace blower.
Large air boil	Broken pipe/diffuser	Check drop pipe and diffusers	Replace pipe/replace diffuser valves, missing diffuser.
Air valve malfunction	Air manifold pressure higher than normal	Air valves	Repair air valves.
	Air relief valve venting	Air valves	Open all air diffuser valves in aeration tank to full open.

Indicator/ Observation	Probable Cause	Check or Monitor	Solution
Air valve malfunction (continued)		Diffusers plugged	Remove drop pipes and diffusers. Clean diffusers or replace. Flush drop pipes with clean water.
			Rag mops or plastic wrapped around diffusers restricts air flow. Remove rags. Check comminutor or bar screen.
High DO	Blower operating time	Check time clock	Program time clock for shorter operating time.
No return sludge flow	Air valve closed to air lift	Operate valve to air lift	Open air control valve so return pipe runs at least 1/4 full.
	Tail pipe of air lift plugged	No return with air control valve full open	Tap on air line to tail pipe. Back flush by closing several air valves for 30 seconds on aeration tank diffuser drop pipes.
	Debris in settling tank hopper	Sound hopper (measure sludge depth in hopper)	Dewater settling tank. Remove debris from tank and hopper. Clean air lift pipes.
Return has low solids, almost clear liquid	Low solids content	Check MLSS concentration	Build up solids. If aerobic digester, transfer solids to aeration tank.
	Solids amount normal—sludge blanket bridged	Sound settling hoppers if sludge blanket bridged or coned	Squeegee or probe settling hopper, breaking bridge or channel so solids can be removed.
Corrosion	Electrolysis	Sacrificial anodes	Check anodes. If absorbed, replace with new anodes.
Rusting of metal pipes, tuberculations	Exposed pipes	Protective coating	Repaint metal surfaces, particularly those under water in aeration tanks.
Low DO in aeration tank	Blower not operating long enough	DO level in aeration tank	Check DO hourly.
		Blower tank clock setting	Increase blower run times.
Septic sludge	Insufficient aeration	Blower operation	Increase blower run time.
	Return sludge rate too low	Sound sludge blanket for depth and solids condition	Increase return sludge rate.
	Shock load/toxic waste	Sample influent to plant—solids, pH, temperature	Stop sludge wasting.
		Heavy organic load—septic tank pumped into collection system	Increase aeration time.
			Increase return rate.
White foam on aeration tank	Low MLSS	Run jar test	If necessary, add solids from aerobic digester.
Dark, leathery froth	High MLSS	Run jar test, if more than 70% settled sludge	Waste sludge to aerobic digester.
Odors	Low DO, shock load	Run DO test, check pH	Increase aeration.
Scum on clarifier	Skimmer operation	Check surface skimmers	Check air flow to surface skimmer, clean inlets. Hand skim tank. Dispose with screenings.
Clarifier effluent ashing Sludge clumps, dark brown 1/4-1" in size, rise to surface	Too much solids in system	Run jar test—over 70% settled sludge	Waste sludge to normal level concentration.

Indicator/ Observation	Probable Cause	Check or Monitor	Solution
Pin floc Pin head size of floc, turbid effluent	Sludge too old	Jar test 50% settles in 5 minutes, rapid settling of granular particles	Waste some of the return sludge.
Rising sludge or clumping Large patches of sludge rise to clarifier surface	Sludge is denitrifying	Jar test—settled sludge rises to surface in two hours after settling	Increase return sludge rate. Decrease blower operating time slightly.
Bulking sludge does not settle	Low DO, pH, septicity, grease, industrial waste, F/M ratio out of proportion	Run test on influent, mixed liquor	Increase aeration. Adjust pH to near 7.0. Reduce return sludge rate slightly.
Straggler floc—fluffy cream sludge floc lost over weir in effluent	Sludge age too low or too high	Jar test—clear effluent	Increase MLSS.
		Jar test MLSS dark brown, scummy foam	Decrease MLSS by wasting sludge.

10.253 Shutdown

Shutdown procedures depend on whether the plant is being shut down because of operational problems, for maintenance and repairs, or for the off-season such as would be the case in a resort area. Activated sludge microorganisms die quickly from lack of oxygen if the aeration system is out of service for even a short time. Whenever the tank must be drained, try to determine the groundwater level. A high groundwater level can float a tank and cause considerable damage to structures and pipes. Diffusers should be cleaned before the tank is returned to service. If the package plant is shut down during the off-season, "moth ball" the equipment to prevent damage from weather and moisture. Exact procedures will depend on location and climate.

10.254 Operational Strategy

This section provides a brief summary of the basic concepts in the operation and control of the activated sludge process as it relates to both package plants and oxidation ditches. The following list outlines items that you, the operator, must consider in the day-to-day operation of your treatment plant.

A log book should be kept by the operator with notations entered as to conditions, activities, and changes at the plant. Any equipment changes or adjustments should be recorded, such as adjustments of operating times (time clock settings at various durations during the year) and return sludge rates (air valve position, pipe volume). All such information will be a valuable resource you can refer to in the future, especially if problems develop and corrective actions have been taken. Flow conditions, temperature, rain, snow, and test results should all be recorded. The more information you keep, the easier it is to solve a similar problem. Maintenance records are also required and should include lubrication, oil changes, filter changes, belt replacement, equipment repair, and painting. Keep accurate and complete records. By reviewing recorded data, the answers to the following questions can help you develop an operating strategy to improve plant performance.

1. Do the influent flow characteristics vary significantly during the year? Is your activated sludge process operation adequate to provide suitable treatment for these variations?
2. Is adequate pretreatment and collection system monitoring being practiced to avoid downstream mechanical or process failure?
3. Are routine solids tests performed (centrifuge, settleability, depth of blanket, visual observations) with results plotted on graphs to assist you in determining if a change in the process mode or operation is necessary?
4. Is suitable aeration time and mixing being provided to allow adequate oxidation, conversion, and floc formation of the solids?
5. Is adequate sludge wasting being practiced to properly maintain a favorable food to microorganism balance throughout the system?
6. If an increase or decrease in organisms occurs, is the oxygen level adjusted accordingly to maintain proper solids settling and production of a clear final effluent?
7. Does the return sludge flow rate produce a high concentration of solids with the minimum amount of water being returned to the aerator?

8. Do you visit your plant on a regular basis to observe process conditions, check equipment for proper operation, lubricate and maintain equipment, and clean process tanks and related equipment?
9. When a problem develops in your activated sludge process, do you refer to Section 10.252, "Troubleshooting"?
10. Do you avoid injuries by avoiding hazards and following safe procedures?
11. Before leaving your plant for the day, do you make a final and detailed check of the equipment for proper operation? Do you ensure that flow rates are set properly, that flow gates are set for possible storm and/or high-flow conditions, that timer-controlled equipment and equipment alarms are set correctly, that equipment is stored properly, and that buildings and gates are securely locked?

10.26 Laboratory Testing for Small Activated Sludge Plants

Testing of the plant wastestreams is very important. The dissolved oxygen level may be determined with a DO analyzer. This is an instrument that gives an immediate reading of the dissolved oxygen level when a probe is inserted into the flow stream or into a grab sample. DO analyzers require periodic standardization. The probes may become coated with grease or oil, membranes may be damaged, or a substance toxic to the probe will give an erroneous reading. Most DO analyzers are delicate instruments that must be handled with care.

Dissolved oxygen may also be determined by *WET CHEMISTRY.*[28] Certain *REAGENTS*[29] must be mixed and *STANDARDIZED*[30] to perform accurate testing. The wet chemistry procedures are not difficult but care must be exercised in making the reagents and performing the test. When wet chemistry (Winkler method) is used in mixed liquor, an inhibitor must be added in the procedure to stop the organisms from using oxygen during the test; otherwise, inaccurate readings will be obtained. Dissolved oxygen kits that have prepackaged reagents are available for running the test.

pH is a useful test but not really a necessity unless an alkaline or acid waste is being treated. The pH can be determined by simple color comparator units or electronic instruments.

Chlorine concentrations may be determined by color comparator kits or wet chemistry methods.

Suspended solids determinations provide information as to plant loadings coming in and solids being discharged in the effluent. The suspended solids test is also used to determine mixed liquor solids concentrations as well as return sludge concentrations. Accurate suspended solids information is determined by filtering a precise volume of well-mixed sample, drying the filtered residue, and very accurately weighing the residue.

The centrifuge test is a method for approximating the suspended solids in the mixed liquor or return sludge (not on the influent or effluent). Some centrifuges are operated by an electric motor that spins the sample tubes; other units are turned by a hand-operated crank. A well-mixed sample is taken and two centrifuge tubes are filled to the upper fill mark on each tube. The tubes are placed in a centrifuge and spun at a given speed for a set time. A hand-operated centrifuge is usually cranked at 1,800 RPM for one minute. After the centrifuge stops spinning, the tubes are removed and the solids portion is read on the tube. The amount of the tube that the solids occupy in m*L* or in percent can be converted to a comparison parts per million (ppm) or milligrams per liter (mg/*L*) that would approximate a Gooch crucible test by using the graph like the one shown in Figure 10.16.

Another way to measure suspended solids is the jar test method described in Chapter 14, "Laboratory Procedures." Note, however, that the jar test method is only an approximation. This method can be used for plant control of mixed liquor or return sludge, but it cannot supply accurate enough information for NPDES reports or other reports requiring accurate information.

BOD (biochemical oxygen demand) is a wet chemistry *BIOASSAY*[31] test. It requires the accurate measurement of dissolved oxygen at the beginning and end of the test period. A DO probe or wet chemistry is used for this determination. The BOD test must also be accurately performed and incubated at a precise temperature (20°C) for a prescribed time period, usually five days. Even under ideal conditions, the BOD test is an approximation.

Depending on the treatment plant and the degree of treatment provided, other tests may be required, such as settleable solids, grease, nitrogen, total Kjeldahl nitrogen (TKN), and phosphorus. After disinfection, the effluent is tested for coli-

[28] *Wet Chemistry. Laboratory procedures used to analyze a sample of water using liquid chemical solutions (wet) instead of, or in addition to, laboratory instruments.*

[29] *Reagent (re-A-gent). A pure chemical substance that is used to make new products or is used in chemical tests to measure, detect, or examine other substances.*

[30] *Standardize. To compare with a standard. (1) In wet chemistry, to find out the exact strength of a solution by comparing it with a standard of known strength. This information is used to adjust the strength by adding more water or more of the substance dissolved. (2) To set up an instrument or device to read a standard. This allows you to adjust the instrument so that it reads accurately, or enables you to apply a correction factor to the readings.*

[31] *Bioassay (BUY-o-ass-SAY). (1) A way of showing or measuring the effect of biological treatment on a particular substance or waste, or (2) A method of determining the relative toxicity of a test sample of industrial wastes or other wastes by using live test organisms such as fish.*

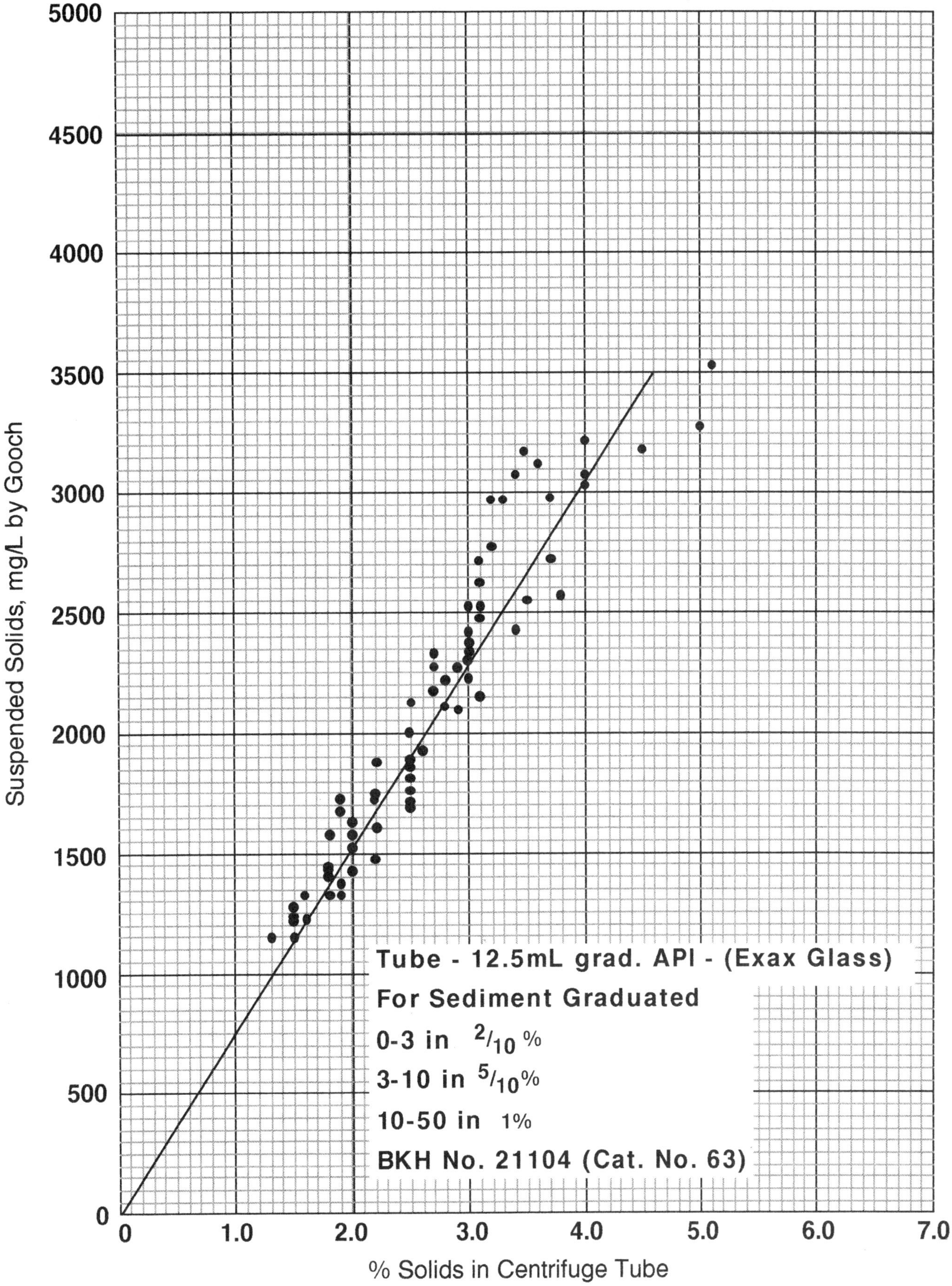

Fig. 10.16 Estimation of suspended solids using centrifuge

form organisms. The *SECCHI DISC*[32] test is used to measure the clarity of the plant effluent.

Testing for solids condition may be accomplished by the settling test. Using a quart jar, take a sample from the aeration compartment after the aeration device has been operating for ten to fifteen minutes and fill the jar to the top. Let the jar stand and watch the floc form and settle to the bottom of the jar. At the end of thirty minutes, the jar should be approximately half full of the settled solids, or slightly less, and have a chocolate brown color with clear water above it. The solids should have a curled or granular appearance. If the solids do not settle half way and the water above them is cloudy or murky in appearance, a longer aeration period, more air, or solids wasting is needed. If the solids settle to less than a quarter of the jar's depth and the water above the solids is murky or cloudy, no wasting of solids should be done and the solids level (MLSS or Mixed Liquor Suspended Solids) in the aerator should be allowed to increase.

If the solids settle to the bottom of the jar with a clear liquor on top and stay down one hour and come up in two hours, this is an indication of good operation. Solids should never be allowed to remain in the settling compartment longer than two hours. If the solids should rise in one hour, this is an indication usually of too much air, or too many solids. Make a slight adjustment to reduce the air to the aeration compartment or increase the return sludge rate.

Another possible cause of solids rising in one hour could be that there are not enough solids under aeration. When this happens the sludge will rise because of the high respiration (breathing) rate of the overtaxed organisms in the settling compartment. Under these circumstances you would want to try to increase or build up the mixed liquor suspended solids. To identify the cause of problems and select the proper solution, you must keep good records and observe what is happening in your plant.

The final clarifier should be equipped with a scum baffle. A properly operated plant will produce some light, oxidized floc that will float to the surface of the settling compartment. A scum baffle will prevent this flow from leaving the compartment in the plant effluent. The better the treatment, the more likely scum froth will develop, unless the unit is septic.

The frequency of performing each test will depend on how the plant is operated, how it performs, and the regulatory requirements for monitoring the discharge.

QUESTIONS

Write your answers in a notebook and then compare your answers with those on page 111.

10.2F If it becomes necessary to waste sludge, how much and when should it be wasted?

10.2G How frequently should a package plant be visually checked by an operator?

10.2H What should you do if you take a sample in a jar from the aeration compartment and after 30 minutes

a. solids do not settle to the bottom half of the jar?

b. solids settle to the bottom and then float to the top?

10.27 Safety

Operators of wastewater treatment plants have one of the worst safety records in the United States. If you are the operator of a small package plant, you must be extremely careful. Frequently, you may be the only person at the plant and there will be no one nearby to come to your rescue or help you if you are hurt or seriously injured. *THEREFORE, PRACTICE SAFETY, AVOID HAZARDOUS CONDITIONS, AND USE SAFE PROCEDURES.* Prepare and rehearse a safety plan for all hazards that exist at your plant.

You can be killed by toxic gases (hydrogen sulfide), electric shock, or drowning, or suffocate in an oxygen-deficient atmosphere. Slippery surfaces, falls, and attempting to lift heavy objects can cause serious injury. Cuts and bruises can lead to infection, and pathogens (disease-causing bacteria) in wastewater can make you sick. Chlorine and dangerous lab chemicals can blind you or seriously burn your skin. Whenever you must enter a confined space, additional personnel will be required. All work in confined spaces must be done in full compliance with your safety regulatory agency's requirements. Do electrical troubleshooting *ONLY* if you are qualified and authorized to do so. If you must work alone doing routine operation and maintenance, phone your office at regular intervals and have someone check on you if you fail to report. You can avoid injuries if you avoid hazards and follow safe procedures.

10.28 Maintenance

Maintenance of equipment in package plants should follow the manufacturer's instructions. Items requiring attention include:

1. Plant cleanliness. Wash down tank walls, weirs, and channels to reduce the collection of odor-causing materials,
2. Aeration equipment:
   a. Air blowers and air diffusion units, and
   b. Mechanical aerators,
3. Air-lift pumps,
4. Scum skimmer,
5. Sludge scrapers,
6. Froth spray system,
7. Weirs, gates, and valves, and
8. Raw wastewater pumps.

[32] *Secchi (SECK-key) Disc. A flat, white disc lowered into the water by a rope until it is just barely visible. At this point, the depth of the disc from the water surface is the recorded Secchi disc transparency.*

10.29 Additional Reading

1. *PACKAGE TREATMENT PLANTS—OPERATIONS MANUAL*, U.S. Environmental Protection Agency. Obtain from ERIC/CSMEE, 1929 Kenny Road, Columbus, OH 43210-1080. Order No. U012. EPA No. 430-9-77-005. Price, $14.16, plus $5.50 shipping and handling.

   NOTE: This is an outstanding publication and if you are the operator of a package activated sludge plant, you should obtain this Operations Manual.

2. *OPERATOR'S POCKET GUIDE TO ACTIVATED SLUDGE—PART II: PROCESS CONTROL AND TROUBLESHOOTING.* Obtain from WSBC Civil Engineers, Inc., Attn: Pocket Guide Department, 11490 Westheimer, Suite 600, Houston, TX 70777. Price, $3.50, plus $2.00 shipping and handling. (Quantity discounts available.)

QUESTIONS

Write your answers in a notebook and then compare your answers with those on page 111.

10.2I How can operators avoid being injured?

10.2J What safety precautions should be taken if you must work alone?

END OF LESSON 2 OF 3 LESSONS
on
ACTIVATED SLUDGE

Please answer the discussion and review questions next.

DISCUSSION AND REVIEW QUESTIONS

Chapter 10. ACTIVATED SLUDGE

(Lesson 2 of 3 Lessons)

Write the answers to these questions in your notebook before continuing. The question numbering continues from Lesson 1.

7. What is a temporary way to control foaming in the aeration compartment of a package plant?
8. How would you operate the aeration device in a package plant?
9. How would you waste sludge in a package plant?
10. What items can cause problems for the operator of a package activated sludge plant?

CHAPTER 10. ACTIVATED SLUDGE

PACKAGE PLANTS AND OXIDATION DITCHES

(Lesson 3 of 3 Lessons)

10.3 OXIDATION DITCHES

10.30 Use of Oxidation Ditches

10.300 Flow Path for Oxidation Ditches

The oxidation ditch (Figure 10.17 and Table 10.3) is a modified form of the activated sludge process and is usually operated in the extended aeration mode.

TABLE 10.3 PURPOSES OF PARTS IN AN OXIDATION DITCH

Part	Purpose
Oxidation Ditch	Provides detention time where activated sludge microorganisms treat wastewater.
Rotor	Causes surface aeration which transfers oxygen from air to water for respiration by microorganisms. Also keeps the contents of the ditch mixed and moving.
Level Control Weir	Regulates how deep rotors sit in the flow of wastewater. This affects the amount of oxygen dissolved in or transferred to the water being treated. Overflow goes to final settling tank.
Final Settling Tank	Allows activated sludge to be separated from water being treated. Clear effluent leaves plant and settled activated sludge is either returned to oxidation ditch or wasted.
Return Sludge Pump	Returns settled activated sludge from final settling tank to oxidation ditch or to excess sludge handling facilities.
Excess Sludge Handling Facilities (not shown)	Treat waste activated sludge for ultimate disposal.

The main parts of the oxidation ditch are the aeration basin, which generally consists of two channels placed side by side and connected at the ends to produce one continuous loop of the wastewater flow, a brush rotor assembly (see Figure 10.12, page 80), settling tank, return sludge pump, and excess sludge handling facilities.

There is usually no primary settling tank or grit removal system used in this process. Inorganic solids such as sand and grit are captured in the oxidation ditch and removed during sludge wasting or cleaning operations. The raw wastewater passes directly through a bar screen to the ditch. The bar screen is necessary for the protection of the mechanical equipment such as rotor and pumps. Comminutors or barminutors may be installed after the bar screen or instead of a bar screen. The oxidation ditch forms the aeration basin and here the raw wastewater is mixed with previously formed active organisms. The rotor is the aeration device that transfers the necessary oxygen into the liquid for microbial life and keeps the contents of the ditch mixed and moving. The velocity of the liquid in the ditch must be maintained to prevent settling of solids, normally 1.0 to 1.5 feet per second (0.3 to 0.45 m/sec). The ends of the ditch are well rounded to prevent eddying and dead areas, and the outside edges of the curves are given erosion protection.

The mixed liquor flows from the ditch to a clarifier for separation. The clarified water passes over the effluent weir and is chlorinated. Plant effluent is discharged to either a receiving stream, percolation ditches, or a subsurface disposal or leaching system. In approved systems, the treated water may be used for irrigation of hay crops, golf courses, or roadside landscaping. If irrigation is practiced, the water is usually delivered to a holding pond and then pumped to the irrigation site as needed. In some areas, the regulatory agency may require that all runoff irrigation water be captured and recycled. The settled sludge is removed from the bottom of the clarifier by a pump and is returned to the ditch or wasted. Scum that floats to the surface of the clarifier is removed and either returned to the oxidation ditch for further treatment or disposed of by burial.

Since the oxidation ditch is operated as a closed system, the amount of volatile suspended solids will gradually increase. It will periodically become necessary to remove some sludge from the process. Wasting of sludge lowers the MLSS (Mixed Liquor Suspended Solids) concentration in the ditch and keeps the microorganisms more active. Control of sludge concentration by wasting of excess sludge is one of the reasons for the high reductions possible by this process. Excess sludge may be dried directly on sludge drying beds or stored in a holding tank or in sludge lagoons for later disposal to larger treatment plants or approved sanitary landfills.

The basic process design results in simple, easy operation. A high mixed liquor suspended solids (MLSS) concentration, usually between 2,000 to 6,000 mg/*L* (some plants carry 6,000 to 8,000 mg/*L* MLSS), is carried in the ditch and the plant may be capable of handling shock and peak loads without upsetting plant operation. There is no foam problem after solids buildup, as experienced with other types of activated sludge plants. Cold weather operation has less effect on plant efficiency than other processes because of the large number of

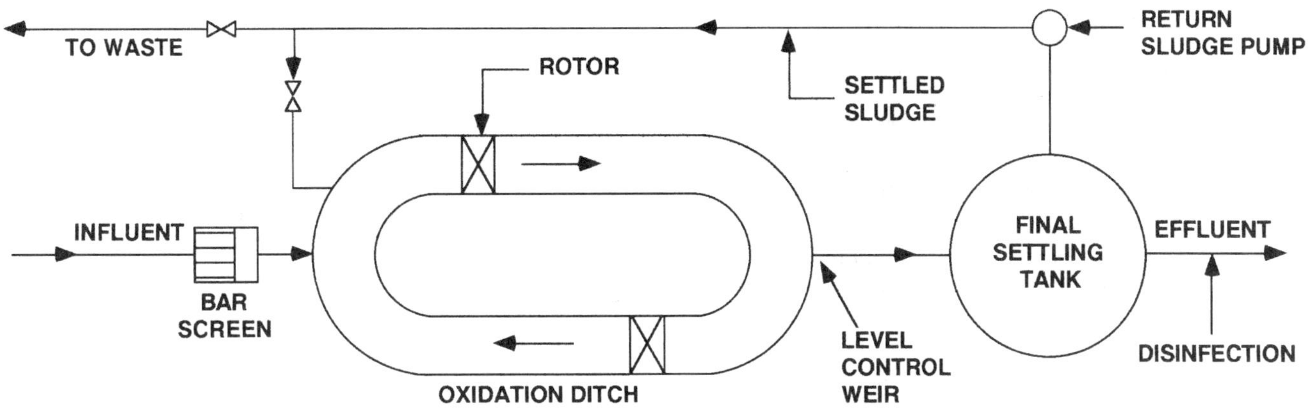

Fig. 10.17 Oxidation ditch plant

(Source: "Oxidation Ditch" prepared by William L. Berk for the New England Regional Wastewater Institute, South Portland, Maine 04106. August 1970)

microorganisms in the ditch. However, in some hot climates, excessive foam can occur when the MLSS is high. Also, the solids will have poor settling characteristics with solids being lost over the weirs. Lowering the MLSS usually corrects these problems.

Operating plants in the United States are achieving BOD removals of about 90 percent and as high as 98 percent.

10.301 Description of Oxidation Ditches

1. Plant flow	0.2 to 20.0 MGD (750 to 75,000 cu m/day)
2. BOD loading	10 to 50 lbs BOD/day per 1,000 cubic feet of ditch (200 to 800 kg/day per 1,000 cu m)
3. F/M ratio[33]	0.03 to 0.1 lb BOD/day/lb MLVSS
4. MLSS concentration	2,000 to 6,000 mg/*L*
5. Sludge age	20 to 35 days
6. Ditch detention time	3 to 24 hours
7. Minimum velocity	1.0 fps (0.3 mps)
8. DO levels	0.5 to 3.0 mg/*L*
9. Liquid level	3.0 to 7.0 feet (1 to 2 m)

See Section 10.35, "Operational Guidelines," for procedures to calculate these items.

10.31 Safety

Lost time, injury, and even death are the results of not being concerned with applying the rules of safety to all activities involved in operating and maintaining a plant. Practicing "safety" is not just knowing what to do; it is a lifestyle. Personnel must not only acquire this "lifestyle," but must also know what to do if an accident occurs. Refer to Volume I, Chapter 3, "Safety," for more information.

Some safety precautions that should be observed at all times when working in a treatment plant include:

1. Wear safety shoes with steel toes, shanks, and soles that retard slipping. Cork-inserted composition soles provide the best traction for all-around use.
2. Slippery algal growths should be scrubbed off and washed away whenever they appear.
3. Keep all areas clear of spilled oil or grease. Use soap and water, not gasoline or solvents, for cleaning.
4. Wear gloves when working with equipment or while in direct contact with wastewater.
5. Do not leave tools, equipment, and materials where they could create a safety hazard.
6. Adequate lighting should be provided for night work and in areas with limited existing lighting.
7. Ice conditions in winter require spiked shoes and the sanding of icy areas if the ice cannot be thawed away with wash water.

---

[33] *Food/Microorganism Ratio. A measure of food provided to bacteria in an aeration tank.*

$$\frac{\text{Food}}{\text{Microorganisms}} = \frac{\text{BOD, lbs/day or kg/day}}{\text{MLVSS, lbs or kgs}}$$

$$= \frac{\text{Flow, MGD} \times \text{BOD, mg/L} \times 8.34\text{ lbs/gal}}{\text{Ditch Volume, MG} \times \text{SS, mg/L} \times \text{Volatile Portion, decimal} \times 8.34\text{ lbs/gal}}$$

or by calculator math system

$$= \text{Flow, MGD} \times \text{BOD, mg/L} \div \text{Ditch Volume, MG} \div \text{SS, mg/L} \div \text{Volatile Portion, decimal}$$

8. Remove only sections of handrails, deck plates, or grating necessary for the immediate job. Removed sections should be properly stored out of the way and properly secured against falling into tanks. The area should be barricaded to prevent unauthorized personnel entry and possible injury.
9. Do not walk on top of the oxidation ditch sidewalls; you may slip and fall into the ditch.
10. Prepare an emergency response plan for each hazard in the plant. Outline each response action to be taken. Rehearse the response plan.

QUESTIONS

Write your answers in a notebook and then compare your answers with those on pages 111 and 112.

10.3A Why are the ends of oxidation ditches well rounded?

10.3B List the major components of an oxidation ditch treatment process.

10.3C Why should operators avoid walking on top of oxidation ditch sidewalls?

10.32 Start-Up

Prior to start-up, prepare a punch list form to be used to record any deficiencies that may be revealed. The punch list is a good way to identify problems when everyone is observing the deficiencies. Make no mechanical changes or repairs until you have signed off on the project. Any work may invalidate warranties on the equipment.

There are two primary objectives of start-up. One is to make certain that all mechanical equipment is operating properly. The second is to develop a proper microbial floc (activated sludge) in the oxidation ditch. Floc development is essential for the plant to succeed in reducing the quantities of polluting materials in the raw wastewater.

The start-up procedures presented here should be used along with the manufacturer's start-up procedures for all components of the plant. The plant operator, the contractor, engineer, equipment manufacturer's representative, and regulatory agency's representative should all be present at the start-up to respond to problems that may be identified.

During start-up of the plant, some construction may still be in progress. Special care should be given to ensure that all safety procedures are followed at all times.

10.320 Pre-Start Inspection

Check all equipment to verify that it is properly installed and secured in place.

Check the inlet structure for debris and clean all debris from the structure prior to start-up.

Clean all debris from the ditch structure prior to start-up. Check the walkways to ensure there is no debris that can later fall into the channel. Also inspect the influent and effluent lines to be sure that they are free of debris.

If you are preparing to start a rotor that is a new installation, one that has not been in operation for some time, or one that has just been overhauled, a thorough inspection check should be made prior to starting the rotor to prevent damage to the rotor and injury to personnel.

Rotor checklist.

1. The ON/OFF switch is OFF and the main breaker is locked out and tagged.
2. Motor secured to gear reducer.
3. Gear reducer assembly secured to the mounting platform.
4. Rotor cylinder shaft secured to reducer coupling bolts.
5. Rotor blades and teeth secured to the cylinder.
6. Driver, rotor, bearings, and stand(s) properly aligned.
7. Rotor turns with a reasonable pull on the cylinder.
8. Rotor anti-rotation screws properly adjusted.
9. Proper oil type and quantity in the reducer.
10. All bearings greased with the proper lubricant.
11. All lube oil line fittings tight.
12. Gear reducer housing air vent "open."
13. All bolts tight.
14. All tools and foreign material are clear of the rotor assembly and inlet piping.
15. Safety guards over moving parts properly installed and secure.

Following completion of the pre-start checks, the rotor assembly is ready to start. Be sure all personnel are clear of the rotor assembly. Turn the main power breaker ON. The ON/OFF switch should now be positioned to ON and the rotor will start. Activate the various switches in sequence, starting at the main breaker and working to the local ON/OFF switch. The main breaker should never be used for an ON/OFF switch.

If the rotor assembly is part of a new oxidation ditch installation, the following additional steps should be taken:

1. Check the rotation of the rotor and run the rotor for at least one hour (see "bump start" below).
2. Start the rotor(s) with the rotor(s) out of the water.
3. Check and record the motor amperage and voltage on each phase.
4. Recheck the rotor support bearing(s) and drive alignment and realign if necessary according to manufacturer's installation drawings.
5. Tighten all nuts, bolts, and set screws to prepare the rotor for regular operation. Check and record the items noted under Section 10.330, "Normal Operation, Records, Mechanical."

NOTE: When starting a new or recently overhauled rotor assembly, a "bump start" is recommended *BEFORE* a routine full start and run is attempted.

A "bump start" allows operation of the rotor assembly for 2 to 3 seconds and is accomplished by briefly positioning the ON/OFF switch to ON and immediately returning it to the OFF position.

This short run-time will allow you to determine if the rotor assembly operates freely and properly, operates in the proper rotation, and will avoid extensive damage to the rotor assembly if it is not properly installed.

Make sure the adjustable weir operates freely and does not bind. Set the weir at the proper elevation in accordance with the O & M manual or the manufacturer's instructions.

The clarifier structure and piping should be inspected and cleared of all debris. All control gates and valves should be checked for smooth operation and proper seating.

The return sludge and waste sludge systems should be examined for leakage and all valves should be operated one complete cycle and set for normal operation. Pumps should be manually operated with liquid in them to check proper operation. One pump should be operated and checked for vibration, excessive noise, or overheating, and the amperage reading should be recorded. The same procedure should then be repeated for the second pump with the first pump shut off. If your plant has waste sludge treatment facilities (holding tanks, lagoons, or drying beds), operate the necessary valves to allow pumping to this part of the plant. Operate each pump manually and note the discharge. Shut off both pumps and return the valves to the normal flow position for returning the settled sludge back to the oxidation ditch.

10.321 Plant Start-Up

Start to fill the ditch with water or wastewater. If possible, add a water tank or two full of healthy seed activated sludge from a nearby plant. Divert all wastewater to be treated into the ditch. If the ditch was initially filled with wastewater and one or two days were required to fill the ditch during hot weather, odor problems could develop. Start the rotors when the water level reaches the bottom of the rotor blades.

During plant start-up, a dark gray color of the developing MLSS may be seen. A dark gray color usually indicates a lack of bacterial buildup in the mixed liquor. If this condition continues for more than several days, check your return sludge system to see that it is operating properly.

Do not start discharging to the clarifier when the water reaches the rotors. Allow the wastewater to continue to fill the ditch and to be treated. When the water level approaches the maximum submergence of the rotors and the peak motor load, then start allowing some of the water to be discharged to the clarifier. During start-up, an unstable clarifier effluent may result due to the inadequate biological treatment. As this effluent is generally the discharged product (as final effluent), chlorination is to be used to reduce health hazards on the receiving waters during this time. State regulatory agencies should be contacted to ensure that the receiving waters will not be harmed as a result of heavily chlorinating the plant effluent. Effluent dechlorination using sulfur dioxide may be required.

During the period of start-up, wastewater testing procedures should be initiated as soon as possible. The actual flow rates should be recorded and also the incoming BOD and *COD*[34] levels.

Building up of the MLSS concentration is the most important activity during the start-up process. At least three and possibly up to fifteen days are required to build up the MLSS concentration. In the event the actual MLSS concentration cannot be determined on a daily basis, you should at least record daily the results of the 30-minute sludge settleability tests. (See Section 10.26, "Laboratory Testing for Small Activated Sludge Plants," and Chapter 14, "Laboratory Procedures," for details on how to conduct sludge settleability tests.) During this period, maintain the highest possible return activated sludge rate.

Dissolved oxygen (DO) should be measured at a sampling location approximately 15 feet (4.5 m) upstream from the rotor. Until the desired MLSS concentration is reached and the 30-minute sludge settleability volume reaches 20 percent, a minimum DO concentration of 2.0 mg/*L* should be kept in the ditch. Following this period, a lesser DO concentration may be desired, but never less than 0.5 mg/*L* at a point 15 feet (4.5 m) upstream of the rotors.

Following start-up, when the plant has stabilized, the solids should settle rapidly in the clarifier leaving a clear, odorless, and stable effluent. The solids should look like particles, golden to rich dark brown in color, with sharply defined edges.

You should not expect immediate results from the start-up procedures. Plant start-up takes time, sometimes over a month if a "seed" activated sludge from another treatment plant is not available. Also, some conditions may occur during start-up that would, under normal conditions, indicate a poorly operating process such as light foaming in the ditch or a cloudy supernatant in the settleable solids tests. These conditions should only be temporary if the information in this section is applied properly.

QUESTIONS

Write your answers in a notebook and then compare your answers with those on page 112.

10.3D What are the two primary objectives of start-up?

10.3E What items or structures should be inspected during the pre-start inspection?

10.3F During plant start-up, when should the rotors be started?

10.3G When an oxidation ditch is operating properly, how should the activated sludge solids appear?

---

[34] *COD (pronounce as separate letters). **C**hemical **O**xygen **D**emand. A measure of the oxygen-consuming capacity of organic matter present in wastewater. COD is expressed as the amount of oxygen consumed from a chemical oxidant in mg/L during a specific test. Results are not necessarily related to the biochemical oxygen demand (BOD) because the chemical oxidant may react with substances that bacteria do not stabilize.*

10.33 Operation

10.330 Normal Operation

Process controls and operation of an oxidation ditch are similar to the activated sludge process. To obtain maximum performance efficiency, the following control methods must be maintained:

1. Proper food supply (measured as BOD or COD and no toxicants) for the microorganisms;
2. Proper DO levels in the oxidation ditch;
3. Proper ditch environment (no toxicants, and sufficient microorganisms to treat the wastes);
4. Proper ditch detention time to treat the wastes by control of the adjustable weir; and
5. Proper water/solids separation in the clarifier.

PROPER FOOD SUPPLY FOR THE MICROORGANISMS

Influent flows and waste characteristics are subject to limited control by the operator. Municipal ordinances may prohibit discharge to the collection system of materials damaging to treatment structures or human safety. Control over wastes dumped into the collection system requires a pretreatment facility inspection program to ensure compliance. Alternate means of disposal, pretreatment, or controlled discharge of significantly damaging wastes may be required in order to permit dilution to an acceptable level by the time the waste arrives at the treatment plant. For waste discharge control methods, refer to Chapter 4, "Preventing and Minimizing Wastes at the Source," in *INDUSTRIAL WASTE TREATMENT*, Volume I, or the *PRETREATMENT FACILITY INSPECTION* manual in this series of operator training manuals.

PROPER DO LEVELS

Proper operation of the process depends on the rotor assembly supplying the right amount of oxygen to the waste flow in the ditch. For the best operation, a DO concentration of 0.5 to 2.0 mg/*L* should be maintained just upstream (15 feet or 4.5 m) of the rotors. Over-oxygenation wastes power and excessive DO levels can cause a pinpoint floc to form that does not settle and is lost over the weir in the settling tank. Control of rotor oxygenation is achieved by adjusting the ditch outlet level control weir.

The level or elevation of the rotors is fixed but the deeper the rotors sit in the water, the greater the transfer of oxygen from the air to the water (greater DO). The ditch outlet level control weir regulates the level of water in the oxidation ditch.

PROPER ENVIRONMENT

The oxidation ditch process with its long-term aeration basin is designed to carry MLSS concentrations of 2,000 to 6,000 mg/*L*. This provides a large organism mass in the system.

Performance of the ditch and ditch environment can be evaluated by conducting a few simple tests and general observations. The color and characteristics of the floc in the ditch as well as the clarity of the effluent should be observed and recorded daily. Typical tests are settleable solids, DO upstream of the rotor, pH, and residual chlorine in the plant effluent. These test procedures are outlined in Chapter 14, "Laboratory Procedures." Laboratory tests such as BOD, COD, suspended solids, volatile solids, total solids, and microscopic examinations should be performed periodically by the plant operator or an outside laboratory. These test procedures are outlined in Chapter 16, "Laboratory Procedures and Chemistry," in *OPERATION AND MAINTENANCE OF WASTEWATER TREATMENT PLANTS*, Volume II. The results will aid you in determining the actual operating efficiency and performance of the process.

Oxidation ditch solids are controlled by regulating the return sludge rate and waste sludge rate. Remember that solids continue to deteriorate as long as they remain in the clarifier. Adjust the return sludge rate to return the microorganisms in a healthy condition from the final settling tank to the oxidation ditch. If dark solids appear in the settling tank, either the return sludge rate should be increased (solids remaining too long in clarifier) or the DO levels are too low in the oxidation ditch.

Adjusting the waste sludge rate regulates the solids concentration (number of microorganisms) in the oxidation ditch. If the surface of the ditch has a white, crisp foam, reduce the sludge wasting rate. Some plants have operated successfully with an MLSS of 6,000 to 8,000 mg/*L* and very little wasting of solids. If the surface has a thick, dark foam, increase the wasting rate. Waste activated sludge may be removed from the ditch by pumping to a sludge holding tank, to sludge drying beds, to sludge lagoons, or to a tank truck. Ultimate disposal may be to larger treatment plants.

Remember that this is a biological treatment process and several days may be required before the process responds to operation changes. Make your changes slowly, be patient, and observe and record the results. Allow seven or more days for the process to stabilize after making a change. For additional information on the regulation of the process, see Section 10.26, "Laboratory Testing for Small Activated Sludge Plants," which discusses the use of the settling test for operational control. For more details on controlling the activated sludge process, see *OPERATION AND MAINTENANCE OF WASTEWATER TREATMENT PLANTS*, Volume II, Chapter 11, "Activated Sludge."

PROPER TREATMENT TIME AND FLOW VELOCITIES

Treatment time is directly related to the flow of wastewater and is controlled by an adjustable weir. Velocities in the ditch should be maintained at 1.0 to 1.5 feet per second (0.3 to 0.45 m/sec) to prevent the deposition of floc. With this in mind, the ditch contents should travel the complete circuit of the ditch, or from rotor to rotor, every 3 to 6 minutes. If the rotors are operated by time clocks (30 minutes off and 30 minutes on, for example), the velocities in the ditch must be sufficient to resuspend any settled material. Settled sludge not resuspended forms sludge deposits that can become anaerobic and be toxic to the activated sludge organisms.

PROPER WATER/SOLIDS SEPARATION

MLSS that have entered and settled in the secondary clarifier are continuously removed from the clarifier and returned to the oxidation ditch or to the waste sludge handling facility. Usually, all sludge formed by the process and settled in the clarifier is returned to the ditch, except when wasting sludge. Scum that is captured on the surface of the clarifier also is removed from the clarifier and either returned to the oxidation ditch for further treatment or disposed of by burial. All methods of disposal of waste sludge, screenings, foam, or other debris must be approved by the regulatory agency.

OBSERVATIONS

Some aspects of the operation of your oxidation ditch plant can be controlled and adjusted with the help of some general observations. General observations of the plant are important to help you determine whether or not your oxidation ditch is operating as intended. These observations include color of the mixed liquor in the ditch, odor at the plant site, and clarity of the ditch and sedimentation tank surfaces.

- Color. You should note the color of the mixed liquor in the ditch daily. A properly operating oxidation ditch plant mixed liquor should have a medium to rich dark brown color. If the MLSS, following proper start-up, changes color from a dark brown to a light brown and the MLSS appear to be thinner than before, the sludge waste rate may be too high, which may cause the plant to lose efficiency in removing waste materials. By decreasing sludge waste rates before the color lightens too much, you can ensure that the plant effluent quality will not deteriorate due to low MLSS concentrations.

  If the MLSS becomes black, the ditch is not receiving enough oxygen and has gone "anaerobic." The oxygen output of the rotors must be increased to eliminate the black color and return the process to normal aerobic operation. This is done by increasing the submergence level of the rotor.

- Odor. When the oxidation ditch plant is operating properly, there will be little or no odor. Odor, if detected, should have an earthy smell. If an odor other than this is present, you should check and determine the cause. Odor similar to rotten eggs indicates that the ditch may be going anaerobic, requiring more oxygen or a higher ditch velocity to prevent deposition of solids. The color of the MLSS could be black if this were the case.

  Odor may also be a sign of poor housekeeping. Grease and solids buildup on the edge of the ditch or sedimentation tank will go anaerobic and cause odors. With an oxidation ditch, odors are much more often caused by poor housekeeping than poor operation. Be sure to verify that any odors are not being generated in the collection system.

- Clarity. In a properly operating oxidation ditch a "layer" of "clear" water or supernatant is usually visible a few feet upstream from the rotor. The depth of this relatively clear water may vary from almost nothing to as much as two or more inches (5 cm) above the mixed liquor. The clarity will depend on the ditch velocity and the settling characteristics of the activated sludge solids. Flocculated particles may be seen in the clarified water.

  Two other good indications of a properly operating oxidation ditch are the clarity of the settling tank water surface and the oxidation ditch surface free of foam buildup. Foam buildup in the ditch (normally not enough to be a nuisance) is usually caused by an insufficient MLSS concentration. Most frequently, foam buildup is only seen during plant start-up and will gradually disappear.

  Clarity of the effluent from the secondary clarifier discharged over the weirs is the best indication of plant performance. A very clear effluent shows that the plant is achieving excellent pollutant removals. A cloudy effluent often indicates a problem with the plant operation.

- Foam. Heavy foam can build up if the MLSS is too high and the biological activity is operating at a high level of respiration.

RECORDS

Daily, or as scheduled, the items listed below should be checked and a record made of these checks. Accurate records are essential and invaluable in evaluating rotor efficiency and in establishing normal operating conditions.

Check:

ITEM	CONDITION
Mechanical	
1. Motor	High or uneven amperage
2. Motor	High temperature
3. Motor	Unusual noise
4. Motor	Operating hours
5. Gear reducer	Unusual bearing or gear noise
6. Gear reducer	Proper oil level (use sight glass or dipstick to check)
7. Gear reducer housing	Air vent pipe "open"
8. Outboard bearing	Unusual bearing noise
9. Rotor	Unusual noise or vibration
10. Rotor	Remove any debris caught in rotor blades such as rags, weeds, or plastic goods.
Operation	
11. DO concentration	Use portable DO probe or perform Winkler lab test. DO should be taken approximately 15 feet (4.5 m) *UPSTREAM* of the rotor(s). Maintain 0.5 to 2.0 mg/*L* at this point.
12. Ditch velocity	1.0 to 1.5 fps (0.3 to 0.45 mps)
13. Ditch water level	Read *STAFF GAGE*[35] in ditch. Convert this reading to brush aerator submergence.
14. Ditch surface condition	Neither white foam nor a heavy brown scum
15. Mixed Liquor Suspended Solids concentration (MLSS)	2,000 to 6,000 mg/*L*

[35] *Staff Gage. A ruler or graduated scale used to measure the depth or elevation of water in a channel, tank, or stream.*

FINAL PLANT SURVEY

Before leaving for the day, one final inspection should be made around the plant. The following questions may help in seeing that you have left the plant in a condition so that it will be operating properly the next time it is attended.

1. Are any pieces of equipment operating poorly? Will they need to be checked before the next scheduled day of operator attendance (hot bearings, loose belts)?
2. Are return sludge rates set at the correct level?
3. Are flowmeters clean and operating?
4. Are inlet gates set properly in case of high flows?
5. Has the rotor level of submergence been set properly?
6. If some equipment is time-clock controlled, are the time clocks set?
7. If remote alarms are used to warn someone about equipment failures, are these set properly?
8. Is equipment stored and locked so as to prevent vandalism?
9. Are outside lights on or set to come on?

PERFORMANCE

Figure 10.18 shows the performance of an actual oxidation ditch.

QUESTIONS

Write your answers in a notebook and then compare your answers with those on page 112.

10.3H How is the DO in the oxidation ditch regulated?

10.3I What should be the velocity in the ditch?

10.3J What observations should be made daily to help indicate the performance of the oxidation ditch?

10.331 Abnormal Operation

In making the routine checks that are noted under Section 10.330, "Normal Operation, Records, Mechanical," you may occasionally find some conditions to be abnormal. Serious damage to the rotor assembly may result if the abnormal conditions are not corrected as soon as possible.

Rotor assembly operation is *ESSENTIAL* for the efficient operation of the oxidation ditch. Loss of rotor-generated mixing and air entrainment to the mixed liquor in the ditch for an extended period of time will turn your oxidation ditch activated sludge process into an upset, bulking activated sludge process.

Listed in Table 10.4 are some abnormal conditions, possible causes, and your response to the conditions that will aid you in the safe and efficient operation of the rotor.

TABLE 10.4 ABNORMAL BRUSH ROTOR OPERATION

Item	Abnormal Condition	Possible Cause	Operator Response
Motor	Motor "off"	Ambient temperature in switchboard panel or room too high. Tripped breakers.	Check for overloading. Reduce temperature with fans. Reset breakers.
		Degree of rotor submergence results in excessive amperage draw. Tripped breakers.	Adjust rotor submergence. Reset breakers.
		Motor shorted or burned out.	Repair motor.
	Motor "hot" to the touch	Dry bearings.	Lubricate bearings.
		Excessive grease.	Remove grease and grease properly.
		Worn bearings.	Replace bearings.
		Excessive amperage draw (motor load).	Adjust rotor submergence.
Gear Reducer	Grinding, chipping, whirring, or whining noise	Low oil level.	Drain and/or refill. Check for leakage and make repairs.
		Gear drive part(s) partially or totally worn.	Replace worn parts.
Bearings	Grinding or bumping noise	Insufficient bearing grease.	Grease bearing(s) routinely.
		Worn bearing(s).	Replace bearing(s).

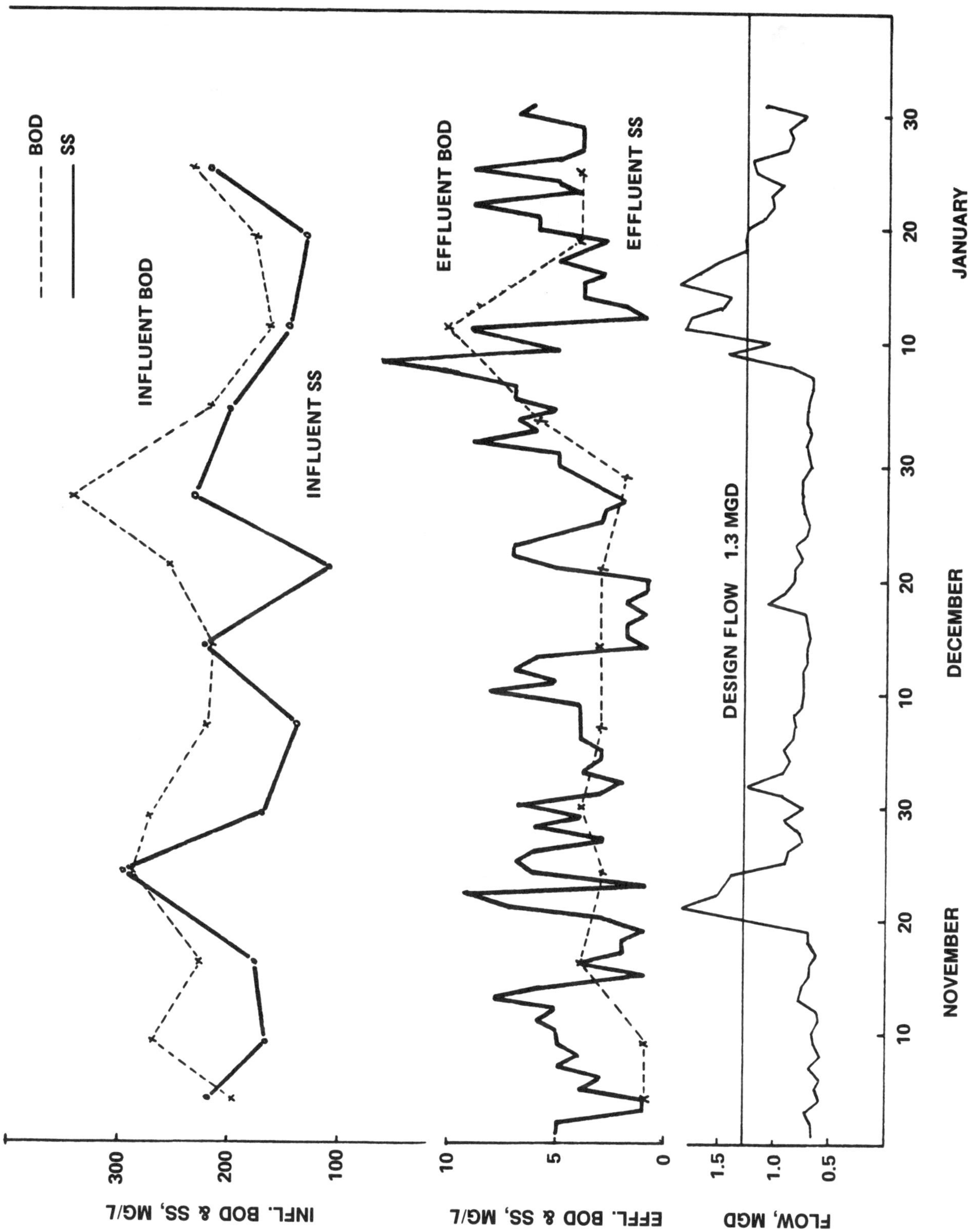

Fig. 10.18 Actual performance of an oxidation ditch

Environmental factors that affect the wastewater treatment process include temperature and precipitation. The wastewater temperature affects the activity of the microorganisms. During cold weather, this reduced activity might lower the efficiency of the treatment system. Besides biological effects of temperature, the flocculation and sedimentation of the mixed liquor solids are not as effective at lower temperatures.

Ice buildup will hinder or stop altogether the proper operation of mechanical parts such as rotors and sludge scraper mechanisms. Chunks of ice may develop that float in the ditch and eventually enter the area of the rotor assembly. Unless adequate safeguards are provided, serious damage to the rotor *WILL* result. Some of the safeguards that *MUST* be considered are:

1. The oxidation ditch in cold weather areas *SHOULD BE OPERATED FOR THE MAXIMUM DETENTION TIME PRACTICAL* in order to conserve as much heat in the wastewater as possible. This action will inhibit ice chunk formation. Set the effluent control weir at the highest possible level.
2. The splashing and/or spraying action produced by the rotor will allow ice to form on the rotor assembly. Serious consideration must be given to covering the whole rotor assembly with a structure made of wood, fiberglass, or other suitable material that will bridge over the ditch. Installations of this type generally are not heated.

NOTE: If it is necessary to shut the rotor off for maintenance, *ICE WILL FORM ON THE ROTOR*. Prior to restarting the rotor, hose it off with water to thaw the ice. Otherwise, the ice could cause vibrations that can damage the rotor.

CAUTION: Ice conditions in winter require spiked shoes and sanding icy areas if the ice cannot be thawed away with wash water.

In some treatment plants, very heavy rains or snow melts cause flows to the treatment plant to exceed three to four times design flow. This is generally accompanied by a weaker wastewater in terms of BOD and COD due to the dilution effect of the storm water. These hydraulic overloads may exceed the capacity of the clarifier to settle sludge solids properly. When this occurs, extremely high BOD, COD, and suspended solids concentrations will be discharged to the receiving stream in the final effluent and possible process upsets can occur if corrective action is not taken. Coagulant chemicals such as alum, ferric chloride, and *POLYMERS*[36] can be added to the final settling tank to assist in settling solids during abnormal conditions. When chemicals such as alum are used, the volume of return sludge is increased and the pH of the sludge may be reduced.

Another method of preventing hydraulic overloads from causing this discharge of high BOD, COD, and suspended solids concentrations is to shut down one or more rotor assemblies in the ditch. This will allow the ditch to act as a large settling tank, keeping the MLSS from flowing into the clarifier where they could be washed out. When the plant flow decreases to more normal levels, the rotor(s) can then be restarted to resume normal operation.

Other than the above corrective actions, without plant modifications there is not much the operator can do to offset the changes in treatment efficiency caused by temperature changes and high flows during abnormal conditions. However, ice buildup can be controlled by frequent observation and removal during cold periods. For persistent cold weather problems, construction of a *LIGHTWEIGHT* building over the ditch and clarifier may be more economical in the long run than fighting ice. The normal wastewater temperature will generally supply sufficient heat inside a building to prevent ice from forming. However, the moist atmosphere over the water surface may lead to "fogging," excess condensation, and even ice buildup on uninsulated surfaces inside the building.

Usually, it is necessary to vary the amount of MLSS in the ditch as seasons change. Because the microorganisms are not as active in winter at low temperatures, the MLSS will need to be higher in the winter than in the summer if complete *NITRIFICATION*[37] is desired.

10.332 Shutdown

Shutdown of the oxidation ditch may be necessary for emergency repairs. Try to schedule any shutdowns so the activated sludge microorganisms will be without air for the shortest possible time period. Problems such as odors and a loss of the microorganism culture can start within two hours after the rotors have been shut off. The microorganism culture may start to deteriorate within 10 minutes, but can recover quickly if the downtime does not exceed four hours. If possible, try to keep one rotor operating at all times.

To prevent injury to personnel, shut down the rotor assembly whenever any maintenance function is performed.

1. Turn the local ON/OFF switch to OFF.
2. Turn any control panel or motor control center ON/OFF switches to the OFF position.
3. Turn the main power breaker OFF.
4. Lock out and tag the main power breaker in the OFF position.

Maintenance work may now be performed.

If the oxidation ditch treats seasonal loads and is shut down during the off-season, protect equipment from the weather and moisture.

[36] *Polymer (POLY-mer). A long chain molecule formed by the union of many monomers (molecules of lower molecular weight). Polymers are used with other chemical coagulants to aid in binding small suspended particles to larger chemical flocs for their removal from water.*

[37] *Nitrification (NYE-truh-fuh-KAY-shun). An aerobic process in which bacteria change the ammonia and organic nitrogen in wastewater into oxidized nitrogen (usually nitrate). The second-stage BOD is sometimes referred to as the "nitrogenous BOD" (first-stage BOD is called the "carbonaceous BOD").*

10.333 Troubleshooting

Refer to Section 10.252, "Troubleshooting." The problems of package aeration plants and the solutions are very similar to those of oxidation ditches.

If floatables appear in the final settling tank, examine the baffle around the level control weir. Since oxidation ditches do not have primary clarifiers, plastic goods and other floatables can be a problem if the baffle is not properly adjusted.

10.334 Operational Strategy

Refer to Section 10.254, "Operational Strategy." The items considered in the operational strategy for package aeration plants are the same as those considered for oxidation ditches.

10.34 Maintenance

10.340 Housekeeping

A general cleanup each day at the plant is important. This not only gives you a more pleasant place to work, but also helps your plant perform better.

Daily cleanup usually includes removing and disposing of debris that may have accumulated on the bar screen; removing grease and scum from the surface of the clarifier; and washing or brushing down the ditch and clarifier weirs and walls.

10.341 Equipment Maintenance

Regularly scheduled equipment maintenance must be performed. As boring as some equipment manufacturers' instruction manuals may be, *READ THEM!* Learn what they say. You should check each piece of equipment daily to see that it is functioning properly. You may have very few mechanical devices in your oxidation ditch plant, but they are all important. The rotors and pumps should be inspected to see that they are operating properly. If pumps are clogged, the obstructions should be removed. Listen for unusual noises. Check for loose bolts. Look for evidence of excess heat, such as burnt paint. Be aware of unusual odors that may indicate overheated devices or switches. Uncovering a mechanical problem in its early stages could prevent a costly repair or replacement at a later date.

Lubrication should also be performed with a fixed operating schedule and properly recorded. Follow the lubrication and maintenance instructions furnished with each piece of equipment. If you are unable to find the instructions, write to the manufacturer for a new set. Make sure that the proper lubricants are used. Overlubrication is wasteful and reduces the effectiveness of lubricant seals and may cause overheating of bearings or gears.

Painting should be done periodically. In addition to beautifying your plant, it gives a good protective coating on all iron and/or metal surfaces and will prolong the life of the metal.

NEVER PAINT OVER EQUIPMENT IDENTIFICATION TAGS! You will regret it if you do.

You should make it your business to know the manufacturer and local supplier of every piece of equipment in your plant. Knowing the manufacturer and how to quickly contact the manufacturer may save important time and money when equipment breakdowns occur.

1. Motors (See *OPERATION OF WASTEWATER TREATMENT PLANTS*, Volume II, Chapter 15, "Maintenance," for additional details.)

   Motors should be greased after about 2,000 hours of operation or as often as conditions and/or the manufacturer recommends. The motor must be stopped when greasing begins.

   Remove the filler and drain plugs, free the grease holes of any hardened grease, add new grease through the filler hole until it starts to come out of the drain hole. Start the motor and let it run for about 15 minutes to expel any excess grease. Stop the motor and install the filler and drain plugs.

   Rotor assembly motors are generally very much exposed to a high degree of moisture. For this reason the motor should be checked at least yearly by an electrician to be sure all parts are in good working condition.

2. Gear Reducer

   Generally, all new oil-lubricated equipment has a "break-in" period of about 400 hours. After this time, the oil should be drained from the gear reducer, the unit flushed, and new oil added. This procedure removes fine metal particles that have worn off the internal components as a result of the initial close tolerances as the equipment was "breaking-in." If large quantities of fine metal particles are found after the "break-in" period, the manufacturer should be consulted. A small magnet can be used to check the sediment in the old oil. Iron particles will be attracted to the magnet.

   Oil in the gear reducers should be checked to see if water has entered the unit. If water is in the oil, it will cause the oil to have a cloudy tan or cocoa brown color.

3. Bearings

   a. Gear Reducer Bearings

      These bearings are generally greased twice per week *WHILE THE ROTOR ASSEMBLY IS IN OPERATION* to ensure proper distribution of the lubricant.

      Extension tubes or pipes are usually attached to the bearing cap. A grease fitting is then installed on the other end of the tube or pipe. This allows for safety when greasing the bearings while the rotor is operating by extending the grease fitting away from the hazardous area of the operating rotor. This grease fitting modification may be used in other equipment applications where lubrication of running equipment would present a safety hazard.

   b. Rotor Intermediate and Outboard Bearings

      These bearings are generally lubricated *DAILY WHILE THE ROTOR ASSEMBLY IS IN OPERATION.*

      In most bearing lubrication applications, overlubrication is wasteful and reduces the effectiveness of lubricant seals and may cause overheating of bearings. Rotor intermediate and outboard bearings *ARE GENERALLY THE EXCEPTION.* As a rule, these bearings cannot be overlubricated.

      Some rotor bearings are protected by neoprene seals that retain lubricant and keep out moisture. The rotor shaft may have "slingers" near the bearing and a detachable shield may cover the bearing. Be sure you do not overlubricate and destroy the seals.

4. Lubrication

   Lubrication of this equipment in locations that have extreme weather variations is critical. Oil and grease must be changed, using the proper type and grade for the expected weather conditions, as determined by the equipment manufacturer.

   NOTE: The proper type and grade of oil is a necessity. If it is too thin or too thick, it will interfere with proper functioning of bearings and gears.

   Bearing grease changes may be accomplished by flushing the grease housing with a 30 to 90 weight oil. The oil is added in the same manner as the grease.

   Lubricant suppliers can provide you with a list of oils and greases that are equal to the product recommended by the equipment manufacturer.

QUESTIONS

Write your answers in a notebook and then compare your answers with those on page 112.

10.3K What problems can be caused by ice during cold weather?

10.3L Why are more microorganisms (higher MLSS) needed in the ditch during cold weather than during warm weather?

10.3M Why is a general cleanup each day at the plant important?

10.35 Operational Guidelines

10.350 English System

Section 10.301, "Description of Oxidation Ditches," lists operational guidelines for an oxidation ditch plant. This section outlines the procedures for you to follow to calculate the guidelines for your oxidation ditch. The first step is to draw a sketch of your oxidation ditch, to obtain the important dimensions, and to record the appropriate flow and wastewater characteristics.

1. Oxidation ditch dimensions.
   a. Length, ft = 200 ft
   b. Bottom, ft = 8 ft
   c. Height, ft = 8 ft
   d. Depth, ft = 4 ft
   e. Radius, ft = 28 ft
   f. Slope = 2

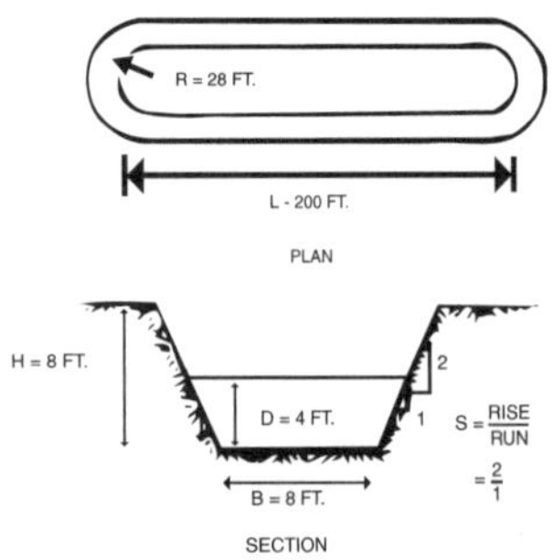

2. Flow and waste characteristics.
   a. Flow, MGD = 0.2 MGD
   b. BOD, mg/*L* = 200 mg/*L*
   c. Influent SS, mg/*L* = 200 mg/*L*
   d. MLSS, mg/*L* = 4,000 mg/*L*
   e. Volatile Matter in MLSS = 70%

3. Calculate oxidation ditch volume in cubic feet and million gallons.

   a. Determine cross-sectional area.

   Find average width of water in ditch, W in feet.

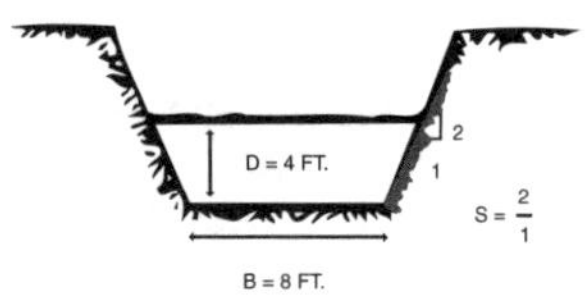

$$\text{W, ft} = B + \frac{D}{S}$$

$$= 8 \text{ ft} + \frac{4 \text{ ft}}{2}$$

$$= 8 \text{ ft} + 2 \text{ ft}$$

$$= 10 \text{ ft}$$

$$\text{Area, sq ft} = W\ D$$

$$= 10 \text{ ft} \times 4 \text{ ft}$$

$$= 40 \text{ sq ft}$$

   b. Determine the centerline length of the ditch. Find the length around the two ends.

$$\text{Ends L, ft} = 2\,\pi\,R$$

$$= 2 \times 3.14 \times 28 \text{ ft}$$

$$= 176 \text{ ft}$$

$$\text{Total L, ft} = \text{Ends L, ft} + 2\text{ L, ft}$$

$$= 176 \text{ ft} + (2 \times 200 \text{ ft})$$

$$= 576 \text{ ft}$$

   c. Calculate ditch volume.

$$\text{Vol, cu ft} = \text{Length, ft} \times \text{Area, sq ft}$$

$$= 576 \text{ ft} \times 40 \text{ sq ft}$$

$$= 23{,}040 \text{ cu ft}$$

$$= 23.04 \cdot (1{,}000 \text{ cu ft})$$ (read as 23.04 one thousand cubic feet)

$$\text{Vol, gal} = 23{,}040 \text{ cu ft} \times 7.48 \text{ gal/cu ft}$$

$$= 172{,}339 \text{ gal}$$

$$= 0.172 \text{ MG}$$

4. Calculate the BOD loading in pounds BOD per day and pounds BOD per day per 1,000 cu ft of ditch volume.

$$\text{BOD Loading, lbs BOD/day} = \text{Flow, MGD} \times \text{BOD, mg}/L \times 8.34 \text{ lbs gal}$$

$$= 0.2 \text{ MGD} \times 200 \text{ mg}/L \times 8.34 \text{ lbs/gal*}$$

$$= 334 \text{ lbs BOD/day}$$

$$\text{BOD Loading, lbs BOD/day/ 1,000 cu ft} = \frac{\text{BOD, lbs BOD/day}}{\text{Ditch Vol, 1,000 cu ft}}$$

$$= \frac{334 \text{ lbs BOD/day}}{23.04 \bullet (1{,}000 \text{ cu ft})}$$

$$= 14 \text{ lbs BOD/day/1,000 cu ft}$$

* Remember that 1 liter = 1,000,000 mg and that

$$\frac{\text{lbs}}{\text{day}} = \frac{\text{Mil Gal}}{\text{day}} \times \frac{\text{mg}}{\text{Mil mg}} \times \frac{\text{lbs}}{\text{gal}} = \frac{\text{lbs}}{\text{day}}$$

5. Determine the Food/Microorganism Ratio.

   a. Microorganisms are measured as pounds of Mixed Liquor Volatile Suspended Solids under aeration in the oxidation ditch. VM means Volatile Matter.

$$\text{MLVSS, lbs} = \text{Vol, MG} \times \text{MLSS, mg}/L \times \text{VM} \times 8.34 \text{ lbs/gal}$$

$$= 0.172 \text{ MG} \times 4{,}000 \text{ mg}/L \times 0.70 \times 8.34 \text{ lbs/gal}$$

$$= 4{,}017 \text{ lbs MLVSS}$$

   b. Calculate F/M Ratio.

$$\frac{F}{M}, \text{ lbs BOD/day/lb MLVSS} = \frac{\text{BOD, lbs BOD/day}}{\text{MLVSS, lbs}}$$

$$= \frac{334 \text{ lbs BOD/day}}{4{,}017 \text{ lbs MLVSS}}$$

$$= 0.08 \text{ lb BOD/day/lb MLVSS}$$

6. Determine the Sludge Age, days.

   a. Calculate pounds of solids under aeration.

$$\text{Aeration Solids, lbs} = \text{Vol, MG} \times \text{MLSS, mg}/L \times 8.34 \text{ lbs/gal}$$

$$= 0.172 \text{ MG} \times 4{,}000 \text{ mg}/L \times 8.34 \text{ lbs/gal}$$

$$= 5{,}738 \text{ lbs}$$

   b. Calculate solids fed to ditch, lbs/day.

$$\text{Solids Added, lbs/day} = \text{Flow, MGD} \times \text{Infl SS, mg}/L \times 8.34 \text{ lbs/gal}$$

$$= 0.2 \text{ MGD} \times 200 \text{ mg}/L \times 8.34 \text{ lbs/gal}$$

$$= 334 \text{ lbs/day}$$

   c. Calculate Sludge Age, days.

$$\text{Sludge Age, days} = \frac{\text{Solids Under Aeration, lbs}}{\text{Solids Added, lbs/day}}$$

$$= \frac{5{,}738 \text{ lbs}}{334 \text{ lbs/day}}$$

$$= 17 \text{ days}$$

7. Calculate ditch detention time, hours.

$$\text{Detention Time, hours} = \frac{\text{Ditch Volume, MG} \times 24 \text{ hr/day}}{\text{Flow, MGD}}$$

$$= \frac{0.172 \text{ MG} \times 24 \text{ hr/day}}{0.2 \text{ MGD}}$$

$$= 20.6 \text{ hours}$$

Or

$$\text{Detention Time, hours} = 0.172 \text{ MG} \times 24 \text{ hr/day} \div 0.2 \text{ MGD}$$

$$= 20.6 \text{ hours}$$

10.351 Metric System

This section will show you how to calculate operational criteria using the metric system. To start, you must collect the same basic information.

1. Oxidation ditch dimensions.

   a. Length, m = 60 meters

   b. Bottom, m = 2.5 meters

   c. Height, m = 2.5 meters

   d. Depth, m = 1.2 meters

   e. Radius, m = 9 meters

   f. Slope = 2

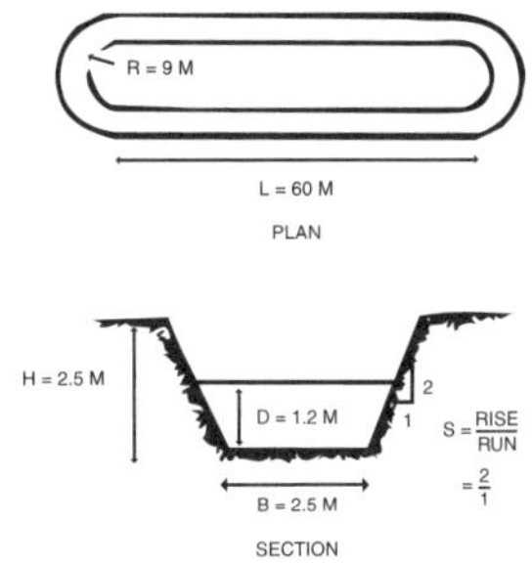

2. Flow and waste characteristics.

   a. Flow, cu m/day = 750 cu m/day

   b. BOD, mg/*L* = 200 mg/*L*

   c. Influent SS, mg/*L* = 200 mg/*L*

   d. MLSS, mg/*L* = 4,000 mg/*L*

   e. Volatile Matter in MLSS = 70%

3. Calculate oxidation ditch volume in cubic meters.

   a. Determine cross-sectional area.

      Find average width of water in ditch, W in meters.

$$\text{W, m} = B + \frac{D}{S}$$

$$= 2.5 \text{ m} + \frac{1.2 \text{ m}}{2}$$

$$= 3.1 \text{ m}$$

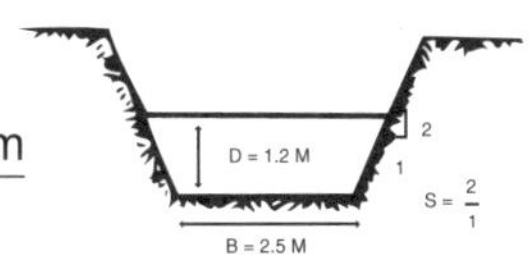

$$\text{Area, sq m} = W\,D$$

$$= 3.1 \text{ m} \times 1.2 \text{ m}$$

$$= 3.72 \text{ sq m}$$

   b. Determine the centerline length of the ditch. Find the length around the two ends.

$$\text{Ends L, m} = 2\,\pi\,R$$

$$= 2 \times 3.14 \times 9 \text{ m}$$

$$= 56.5 \text{ m}$$

$$\text{Total L, m} = \text{Ends L, m} + 2 \text{ L, m}$$

$$= 56.5 \text{ m} + (2 \times 60 \text{ m})$$

$$= 176.5 \text{ m}$$

c. Calculate ditch volume.

$$\text{Vol, cu m} = \text{Length, m} \times \text{Area, sq m}$$
$$= 176.5 \text{ m} \times 3.72 \text{ sq m}$$
$$= 657 \text{ cu m}$$
$$= 0.657 \bullet (1{,}000 \text{ cu m})$$ (read as 0.657 one thousand cubic meters)

4. Calculate the BOD loading in kg BOD per day and kg BOD per day per 1,000 cu m of ditch volume.

$$\text{BOD Loading, kg BOD/day} = \text{Flow, } \frac{\text{cu m}}{\text{day}} \times \text{BOD } \frac{\text{mg}}{L} \times \frac{1 \text{ kg}}{1{,}000{,}000 \text{ mg}} \times \frac{1{,}000\ L}{1 \text{ cu m}}$$
$$= 750 \frac{\text{cu m}}{\text{day}} \times 200 \frac{\text{mg}}{L} \times \frac{1 \text{ kg}}{1{,}000{,}000 \text{ mg}} \times \frac{1{,}000\ L}{1 \text{ cu m}}$$
$$= 150 \text{ kg BOD/day}$$

$$\text{BOD Loading, kg BOD/day/1,000 cu m} = \frac{\text{BOD, kg BOD/day}}{\text{Ditch Volume, 1,000 cu m}}$$
$$= \frac{150 \text{ kg BOD/day}}{0.657 \bullet (1{,}000 \text{ cu m})}$$
$$= 228 \text{ kg BOD/day/1,000 cu m}$$

5. Determine the Food/Microorganism Ratio.

a. Calculate kilograms of MLVSS under aeration in the oxidation ditch.

$$\text{MLVSS, kg} = \text{Vol, cu m} \times \text{MLSS, } \frac{\text{mg}}{L} \times \text{VM} \times \frac{1 \text{ kg}}{1{,}000{,}000 \text{ mg}} \times \frac{1{,}000\ L}{1 \text{ cu m}}$$
$$= 657 \text{ cu m} \times 4{,}000 \frac{\text{mg}}{L} \times 0.70 \times \frac{1 \text{ kg}}{1{,}000{,}000 \text{ mg}} \times \frac{1{,}000\ L}{1 \text{ cu m}}$$
$$= 1{,}840 \text{ kg MLVSS}$$

b. Calculate F/M Ratio.

$$\frac{F}{M}, \text{ kg BOD/day/kg MLVSS} = \frac{\text{BOD, kg/day}}{\text{MLVSS, kg}}$$
$$= \frac{150 \text{ kg BOD/day}}{1{,}840 \text{ kg MLVSS}}$$
$$= 0.08 \text{ kg BOD/day/kg MLVSS}$$

6. Determine the Sludge Age, days.

a. Calculate kilograms of solids under aeration.

$$\text{Aeration Solids, kg} = \text{Vol, cu m} \times \text{MLSS, mg/}L \times \frac{1 \text{ kg}}{1{,}000{,}000 \text{ mg}} \times \frac{1{,}000\ L}{1 \text{ cu m}}$$
$$= 657 \text{ cu m} \times 4{,}000 \text{ mg/}L \times \frac{1 \text{ kg}}{1{,}000{,}000 \text{ mg}} \times \frac{1{,}000\ L}{1 \text{ cu m}}$$
$$= 2{,}628 \text{ kg}$$

b. Calculate solids fed to ditch, kg/day.

$$\text{Solids Added, kg/day} = \text{Flow, cu m/day} \times \text{Infl SS, mg/}L \times \frac{1 \text{ kg}}{1{,}000{,}000 \text{ mg}} \times \frac{1{,}000\ L}{1 \text{ cu m}}$$
$$= 750 \text{ cu m/day} \times 200 \text{ mg/}L \times \frac{1 \text{ kg}}{1{,}000{,}000 \text{ mg}} \times \frac{1{,}000\ L}{1 \text{ cu m}}$$
$$= 150 \text{ kg/day}$$

c. Calculate Sludge Age, days.

$$\text{Sludge Age, days} = \frac{\text{Solids Under Aeration, kg}}{\text{Solids Added, kg/day}}$$
$$= \frac{2{,}628 \text{ kg}}{150 \text{ kg/day}}$$
$$= 17.5 \text{ days}$$

7. Calculate ditch detention time, hours.

$$\text{Detention Time, hours} = \frac{\text{Ditch Volume, cu m} \times 24 \text{ hr/day}}{\text{Flow, cu m/day}}$$
$$= \frac{657 \text{ cu m} \times 24 \text{ hr/day}}{750 \text{ cu m/day}}$$
$$= 21 \text{ hours}$$

QUESTION

Write your answers in a notebook and then compare your answers with those on page 112.

10.3N Determine the BOD loading on an oxidation ditch in:

1. Pounds of BOD/day, and
2. Kilograms of BOD/day.

The inflow is 0.8 MGD and the influent BOD is 250 mg/*L*.

1 MGD = 3,785 cu m/day

10.4 REVIEW OF PLANS AND SPECIFICATIONS

As an operator, you can be very helpful to design engineers in pointing out some design features that would make your job easier. This section indicates some of the items that you should look for when reviewing plans and specifications for expansion of existing facilities or construction of new package plants and oxidation ditches.

10.40 Package Plants

1. Is the plant designed for implementation of modifications to the activated sludge process (adequate flexibility to accommodate future treatment requirements)?
2. Are adequate standby units (equipment) designed into the system?
3. Are ladders, railings, and walkways provided to allow safe, easy access to equipment, pipes, and valves for normal operation, routine maintenance, or repair?
4. Are adequate remote and local controls provided for the mechanical equipment?
5. Are the equipment and related instrumentation designed to operate at low flows and loads common in the early stages of plant operation?
6. Are adequate dewatering systems provided to permit rapid servicing of submerged equipment?
7. Is the chlorination facility flexible enough to allow for prechlorination and are adequate control devices provided?
8. Is the treatment plant's total connected horsepower adequate to allow operation of all equipment in parallel?
9. Are flow equalization facilities provided to handle high flows during wet weather or industrial discharges?
10. Is adequate support equipment provided to allow easy and safe removal of aeration diffusers?
11. Is a laboratory provided with appropriate equipment for conducting at least minimum process control tests?
12. Are adequate sludge drying beds provided with consideration given to wet weather conditions?
13. Is a sludge transfer pump provided to transfer sludge to the drying beds or solids handling facilities?
14. Are the beds designed for easy removal of dried sludge and are there provisions for proper disposal of the sludge that will comply with regulatory agency requirements?
15. Is there a provision for waste or digested sludge liquid storage in place of or in addition to sludge drying beds?
16. Is standby or auxiliary power provided?

10.41 Oxidation Ditches

1. Influent and return activated sludge should enter an oxidation ditch just upstream of a rotor assembly to afford immediate mixing with mixed liquor in the channel.
2. Effluent should exit the oxidation ditch upstream of the rotor and far enough upstream from the injection of the influent and return activated sludge to prevent short-circuiting.
3. Water level in the aeration channel should be controlled by an adjustable weir. In calculating weir height or "set point," use maximum raw flow plus maximum recirculated flow to prevent excessive rotor immersions.
4. Walkways with railings must be provided across the aeration channel to provide access to the rotor for maintenance. The normal location is upstream of the rotor. Location should be such to prevent spray from the rotor on the walkway. Approved flotation devices should be provided at strategic locations if accidental entry into the channel is possible.
5. Horizontal baffles, placed within 15 feet (4.5 m) downstream of the rotor, should be used on all basins with water depths over 6 feet (1.8 m) to provide proper mixing of the entire depth of the basin.
6. In a single oxidation ditch or aeration channel, the rotor drive assembly should be on the outboard side for ease of access.
7. The ditch should be constructed with some type of lining. Consideration should be given to the most economical means of lining available in the particular plant location. As typical liners, consideration could be given to gunite or shotcrete, poured concrete, asphalt, precast concrete, or tile liners.
8. All drive and gear assemblies should be elevated out of the water and safe and easy access should be provided for maintenance.
9. Standby or auxiliary power must be provided to operate critical equipment.
10. Floating aerators and related equipment should be provided in case of rotor failure in order to maintain adequate treatment.
11. Secondary clarifier.
    a. Surface Rate: 600 gal/day/sq ft (24 cu m/day/sq m) based upon plant design flow. Where wide variations in plant hydraulic loading are expected, care should be taken to limit maximum instantaneous surface rate to 1,200 gal/day/sq ft (49 cu m/day/sq m).
    b. Solids Loading: Normal oxidation ditch operation and surface loading rates will prevent excessive solids loading to the final clarifier. However, if complete nitrification is required 12 months per year, the maximum instantaneous solids loading must not exceed 35 lbs/day/sq ft (150 kg/day/sq m).
    c. Detention Time: Three hours based on plant design flow. In no case, however, should the final clarifier have a side water depth less than eight feet (2.4 m).
12. Effluent disposal.
    a. Receiving Stream: The discharge pipe should be located where flood water will not flow back into the plant if there is a power outage or pump failure. A tide gate or check valve can be installed in the effluent line to prevent backflows. Be sure the outlet in the receiving waters is submerged to reduce foam and scum problems.

b. Percolation Ditch: Design should be based on a percolation rate of gallons per *LINEAL*[38] foot of ditch. Provisions must be made to prevent blowing sand and/or dirt from entering the ditch.

13. Winter climates.

a. Provisions should be made to maximize detention times in order to conserve as much heat as possible in the wastewater.

b. Is it practical to cover the oxidation ditch? If not, consider a large and a small ditch or a single ditch that can be modified to use only half its total capacity.

c. Be sure that all equipment requiring normal maintenance is housed and/or heated. This will extend the useful life of the equipment and facilitate service work. Changing a gear box, repairing a pipe, or installing electrical components becomes a major task at very low temperatures.

d. The rotor assembly should be provided with a lightweight cover to prevent rotor icing.

e. Equip the clarifier with a lightweight tarpaulin to keep heavy snow out of the clarifier and to reduce freezing problems.

f. A subsurface disposal pipeline and percolation field should be provided for effluent disposal in winter months. A percolation ditch will freeze over in the winter.

QUESTIONS

Write your answers in a notebook and then compare your answers with those on page 112.

10.4A What provisions should be made to permit rapid servicing of submerged equipment?

10.4B What items would you check when reviewing the location of a walkway intended to provide access to the rotor for maintenance?

10.5 ARITHMETIC ASSIGNMENT

Turn to the Arithmetic Appendix at the back of this manual and work the example problems in Section A.21, *ACTIVATED SLUDGE (OXIDATION DITCHES)*; check the arithmetic using your calculator.

10.6 METRIC CALCULATIONS

Refer to Section 10.35, "Operational Guidelines," for metric calculations.

END OF LESSON 3 OF 3 LESSONS
on
ACTIVATED SLUDGE

Please answer the discussion and review questions before continuing with the Objective Test.

DISCUSSION AND REVIEW QUESTIONS

Chapter 10. ACTIVATED SLUDGE

(Lesson 3 of 3 Lessons)

Write the answers to these questions in your notebook before continuing with the Objective Test on page 113. The question numbering continues from Lesson 2.

11. What happens to inorganic solids such as grit, sand, and silt that enter an oxidation ditch plant?

12. During plant start-up, what does a dark gray color in the mixed liquor indicate and what would you do if this condition persists for more than several days?

13. During plant start-up, when should water be discharged to the clarifier?

14. How can an operator determine if the sludge wasting rate should be increased or decreased?

15. What would you do if solids were in the effluent of the final settling tank during high flows caused by storms or during high influent solids levels caused by the cleaning of the collection system sewers?

[38] *Lineal (LIN-e-al). The length in one direction of a line. For example, a board 12 feet long has 12 lineal feet in its length.*

SUGGESTED ANSWERS

Chapter 10. ACTIVATED SLUDGE

PACKAGE PLANTS AND OXIDATION DITCHES

ANSWERS TO QUESTIONS IN LESSON 1

Answers to questions on page 70.

10.0A The purpose of the activated sludge process in treating wastewater is to oxidize and remove soluble or finely divided suspended materials that were not removed by previous treatment.

10.0B A stabilized waste is a waste that has been treated or decomposed to the extent that, if discharged or released, its rate and state of decomposition would be such that the waste would not cause a nuisance or odors.

10.0C Air is added to the aeration tank in the activated sludge process to provide oxygen to sustain the living organisms as they oxidize wastes to obtain energy for growth. The application of air also encourages mixing in the aerator.

10.0D Air requirements increase when the strength (BOD) of the incoming wastewater increases because more food (wastes) encourages biological activity (reproduction and respiration).

10.0E Factors that could cause an unsuitable environment for the activated sludge process in an aeration tank include:

1. High concentrations of acids, bases, and other toxic substances,
2. Uneven flows of wastewater that cause overfeeding or starvation,
3. Failure to supply enough oxygen, and
4. Failure to maintain F/M ratio.

Answers to questions on page 71.

10.1A Effluent quality requirements may be stated by regulatory agencies in terms of percentage removal of wastes or allowable quantities of wastes that may be discharged.

10.1B Harmful industrial waste discharges may be regulated by municipal ordinances preventing discharge to the collection system, requiring pretreatment, or controlling the discharge in order to protect an activated sludge process.

ANSWERS TO QUESTIONS IN LESSON 2

Answers to questions on page 81.

10.2A A typical package extended aeration plant may have either two or three compartments. The purpose of each compartment is:

1. Aeration. Aerate and mix waste to be treated with activated sludge.
2. Clarification and Settling. Allow activated sludge to be separated from water being treated. Clarified effluent leaves plant and settled activated sludge is returned to aeration tank to treat more wastewater.
3. Aerobic digestion (optional). Treatment of waste activated sludge.

10.2B Common characteristics of all package aeration plants include long solids retention times, high mixed liquor suspended solids, and low food/microorganism ratios.

10.2C Package plants should have cathodic protection to prevent rust, corrosion, and pitting of steel and iron surfaces in contact with water, wastewater, or soil.

Answers to questions on page 86.

10.2D The valve to the air lift pumps must be closed until the settling compartment is filled or all of the air will attempt to flow out the air lifts and no air will flow out the diffusers.

10.2E If the water in the aeration compartment is murky or cloudy and the aeration compartment has a rotten egg odor (H_2S), you should increase the aeration rate.

Answers to questions on page 94.

10.2F For best results, in terms of effluent quality, waste about five percent of the solids each week during warm weather operation.

10.2G A package plant should be visually checked by an operator every day.

10.2H a. If the solids do not settle in the jar, the aeration rate should be increased and the MLSS concentration checked for proper level.

b. If the solids settle and then float to the surface, the aeration rate should be reduced a little each day until the solids settle properly. Also, reduce the wasting rate to increase the MLSS content.

Answers to questions on page 95.

10.2I Operators can avoid being injured on the job by practicing safety, by avoiding hazardous conditions, and by using safe procedures.

10.2J If you must work alone, phone your office at regular intervals and have someone check on you if you fail to report.

ANSWERS TO QUESTIONS IN LESSON 3

Answers to questions on page 98.

10.3A The ends of oxidation ditches are well rounded to prevent eddying and dead areas.

10.3B The major components of an oxidation ditch treatment process include:

1. Oxidation ditch,
2. Rotor,
3. Level control weir,
4. Final settling tank,
5. Return sludge pump, and
6. Excess sludge handling facilities.

10.3C Operators should avoid walking on top of oxidation ditch sidewalls to avoid slipping and falling into the ditch.

Answers to questions on page 99.

10.3D The two primary objectives of start-up are to:

1. Make certain that all mechanical equipment is operating properly, and
2. Develop a proper microbial floc (activated sludge) in the oxidation ditch.

10.3E The following items or structures should be inspected during the pre-start inspection.

1. Inlet structure,
2. Oxidation ditch structure,
3. Rotor,
4. Adjustable weir,
5. Clarifier structure and piping, and
6. Return sludge and waste sludge systems.

10.3F During plant start-up, start the rotors when the water level in the ditch reaches the bottom of the rotor blades.

10.3G When an oxidation ditch is operating properly, the activated sludge solids should look like particles, golden to rich dark brown in color, with sharply defined edges.

Answers to questions on page 102.

10.3H The DO in the oxidation ditch is regulated by raising or lowering the ditch outlet level control weir. This weir controls the level of water in the ditch, which in turn regulates the degree of submergence of the rotors. The greater the submergence, the more oxygen is transferred from the air.

10.3I The ditch velocity should be maintained between 1.0 and 1.5 fps (0.3 to 0.45 mps).

10.3J Daily observations of ditch color, odors, lack of foam on the aerator surface, and settling tank clarity help indicate the performance of the oxidation ditch.

Answers to questions on page 106.

10.3K During cold weather, ice buildup will hinder or stop altogether the proper operation of mechanical parts such as rotors and sludge scraper mechanisms.

10.3L More microorganisms are needed to treat the same amount of wastes during cold weather than during warm weather because the colder the water, the less active the microorganisms.

10.3M A general cleanup each day provides a more pleasant place to work and also improves plant performance.

Answers to question on page 108.

10.3N Determine the BOD loading on an oxidation ditch in:

1. Pounds of BOD/day, and
2. Kilograms of BOD/day.

The inflow is 0.8 MGD and the influent BOD is 250 mg/*L*.

1 MGD = 3,785 cu m/day

Known		Unknown
Flow, MGD	= 0.8 MGD	1. BOD Loading, lbs BOD/day
BOD, mg/*L*	= 250 mg/*L*	2. BOD Loading, kg BOD/day
1 MGD	= 3,785 cu m/day	

1. Determine BOD loading, lbs BOD/day.

$$\text{BOD Loading, lbs BOD/day} = \text{Flow, MGD} \times \text{BOD, mg}/L \times 8.34\ \text{lbs/gal}$$

$$= 0.8\ \text{MGD} \times 250\ \text{mg}/L \times 8.34\ \text{lbs/gal}$$

$$= 1{,}668\ \text{lbs BOD/day}$$

2. Determine BOD loading, kg BOD/day.

a. Convert flow from MGD to cubic meters per day.

$$\text{Flow, cu m/day} = \text{Flow, MGD} \times \frac{3{,}785\ \text{cu m/day}}{1\ \text{MG}}$$

$$= \frac{0.8\ \text{MGD} \times 3{,}785\ \text{cu m/day}}{1\ \text{MGD}}$$

$$= 3{,}028\ \text{cu m/day}$$

b. Calculate BOD loading, kg BOD/day.

$$\text{BOD Loading, kg BOD/day} = \text{Flow, cu m/day} \times \text{BOD, mg}/L \times \frac{1\ \text{kg}}{1{,}000{,}000\ \text{mg}} \times \frac{1{,}000\ L}{1\ \text{cu m}}$$

$$= 3{,}028\ \text{cu m/day} \times 250\ \text{mg}/L \times \frac{1\ \text{kg}}{1{,}000{,}000\ \text{mg}} \times \frac{1{,}000\ L}{1\ \text{cu m}}$$

$$= 757\ \text{kg BOD/day}$$

Or

$$\text{BOD Loading, kg BOD/day} = 3{,}028\ \text{cu m/day} \times 250\ \text{mg}/L \div 1{,}000{,}000\ \text{mg/kg} \times 1{,}000\ L/\text{cu m}$$

$$= 757\ \text{kg BOD/day}$$

Answers to questions on page 110.

10.4A Adequate dewatering systems should be provided to permit rapid servicing of submerged equipment.

10.4B When reviewing the location of a walkway intended to provide access to the rotor for maintenance, check:

1. Distance from walkway to rotor,
2. Normal location is upstream from rotor, and
3. Spray from the rotor will not fall on the walkway.

OBJECTIVE TEST

Chapter 10. ACTIVATED SLUDGE

PACKAGE PLANTS AND OXIDATION DITCHES

Please write your name and mark the correct answers on the answer sheet, as directed at the end of Volume I, Chapter 1. There may be more than one correct answer to each multiple-choice question.

True-False

1. Activated sludge solids are in a deteriorating condition as long as they remain in the secondary clarifier.
   1. True
   2. False
2. High concentrations of acids, bases, and other toxic substances are undesirable and may kill the working organisms.
   1. True
   2. False
3. The treatment plant operator has significant control over the makeup or amount of influent coming into the plant.
   1. True
   2. False
4. In small activated sludge package plants, the aeration blower is intended to operate continuously.
   1. True
   2. False
5. Some sludge must be wasted routinely to allow optimum performance of small activated sludge package plants.
   1. True
   2. False
6. Plant effluent deteriorates as soon as a toxic substance reaches an activated sludge plant.
   1. True
   2. False
7. Chlorine is effective in disinfecting the plant effluent after treatment by the activated sludge process.
   1. True
   2. False
8. The oxidation ditch is a modified form of oxidation ponds.
   1. True
   2. False
9. Operators walk on top of the oxidation ditch sidewalls when maintaining the ditch.
   1. True
   2. False
10. Wastewater flows and waste characteristics from homes are measured before being discharged to sewers.
    1. True
    2. False
11. Usually, it is necessary for the operator to vary the amount of MLSS in the ditch as seasons change.
    1. True
    2. False
12. Uncovering a mechanical problem in its early stages could prevent a costly repair or replacement at a later date.
    1. True
    2. False

Best Answer (Select only the closest or best answer.)

13. What is a primary control in the activated sludge process?
    1. Depth of activated sludge
    2. Influent flow rate
    3. Ratio of food to organisms
    4. Regulation of toxic chemicals
14. What is a common cause of sludge bulking in the activated sludge process?
    1. Filamentous bacteria
    2. High dissolved oxygen (DO) in the aeration tank
    3. Pin floc
    4. Properly settling solids
15. How are blowers controlled in a small package activated sludge plant?
    1. Mixed Liquor Volatile Suspended Solids (MLVSS)
    2. pH
    3. Temperature
    4. Time clock
16. What does a white, fluffy foam on the surface of an aeration tank indicate?
    1. High dissolved oxygen content in the aerator
    2. High solids content in the aerator
    3. Low dissolved oxygen content in the aerator
    4. Low solids content in the aerator
17. How can problems caused by toxic substances be corrected?
    1. Add detoxifying chemical
    2. Change to more tolerant microorganisms
    3. Locate source and prevent future discharges
    4. Provide pretreatment
18. What could be the probable cause of pin floc?
    1. Low DO
    2. Septicity
    3. Sludge is denitrifying
    4. Sludge is too old

19. What is a Secchi disc used to measure?
    1. Clarity of plant effluent
    2. Settleable solids of plant effluent
    3. Suspended solids of plant effluent
    4. Turbidity of plant effluent
20. How can solids condition be determined in the activated sludge treatment process?
    1. Dissolved oxygen determination
    2. Microorganism count
    3. pH measurement
    4. Settling test
21. How should areas of spilled oil or grease be cleaned?
    1. Algae
    2. Gasoline
    3. Soap and water
    4. Solvents
22. What is the most important activity during the start-up process of an oxidation ditch?
    1. Adjusting pH to optimum level
    2. Building up of the MLSS concentration
    3. Introducing acclimated microorganisms to the process
    4. Maintaining near saturation dissolved oxygen levels
23. What should an operator do if the surface of an oxidation ditch has a thick, dark foam?
    1. Decrease the brush rotor depth
    2. Decrease the wasting rate
    3. Increase the brush rotor depth
    4. Increase the wasting rate
24. How should an oxidation ditch be operated during cold weather?
    1. Maximum buildup of ice
    2. Maximum detention time practical
    3. Minimum detention time practical
    4. Minimum levels of microorganisms

Multiple Choice (Select all correct answers.)

25. What types of waste constituents can be removed by the activated sludge process?
    1. Heavy metals
    2. Nitrogen
    3. Organic carbon (BOD)
    4. Phosphorus
    5. Toxic wastes
26. What problems are created in the activated sludge process by insufficient oxygen?
    1. Creates demand for excess use of aeration equipment
    2. Favors production of foul-smelling products of decomposition
    3. Makes facultative organisms work less efficiently
    4. Requires less operator attention to process performance
    5. Slows down activity of aerobic organisms
27. What must the operator control for the activated sludge process to work properly?
    1. Dissolved oxygen in the aeration tanks
    2. Influent organic loading
    3. Number of organisms in aeration tank
    4. Temperature of influent wastewater
    5. Treatment time
28. What items should be checked *BEFORE* backfilling the excavation around a new activated sludge package plant?
    1. Tank dissolved oxygen level is adequate for microorganisms
    2. Tank exterior must be checked for corrosion protection material
    3. Tank has sufficient microorganisms to treat incoming wastes
    4. Tank must be absolutely level in both directions
    5. Tank must have proper level of water
29. What methods are commonly used to supply oxygen from the air to the bacteria in the activated sludge process?
    1. Algae aeration
    2. Atmosphere aeration
    3. Diffused aeration
    4. Mechanical aeration
    5. Surface aeration
30. How can an operator confirm that enough air is being supplied by the aeration equipment to the aeration tank?
    1. Aeration compartment has a rotten egg odor
    2. Measure DO with a probe
    3. Water is clear in settling compartment
    4. Water is cloudy
    5. Water is murky
31. Where can wasted activated sludge from a small package plant be disposed of?
    1. At a larger treatment plant
    2. In a sanitary landfill
    3. In an injection well
    4. On agricultural crops
    5. To surface waters
32. What types of toxic wastes can upset or kill the microorganisms in the aeration tank?
    1. Detergents
    2. High pH levels
    3. Low pH levels
    4. Pesticides
    5. Solvents
33. What could be the probable causes of rocks, mud, and debris in the influent of an activated sludge package plant?
    1. Broken collection main
    2. CCTV inspection of sewers
    3. Flood waters
    4. Mudslide
    5. Open manhole
34. What could be the probable cause of a large air boil in an aeration tank?
    1. Air manifold pressure higher than normal
    2. Air relief valve venting
    3. Broken pipe/diffuser
    4. Excessively high influent water temperature
    5. Mixer paddle broken
35. What could be the probable causes of no return sludge flow?
    1. Activated sludge fails to settle
    2. Air valve closed to air lift
    3. All sludge is being wasted
    4. Debris in settling tank hopper
    5. Tail pipe of air lift plugged

36. How can rusting of metal pipes be prevented in a package activated sludge plant?
    1. Apply protective coatings
    2. Keep debris from flowing in pipes
    3. Maintain sacrificial anodes
    4. Prevent excessive flow velocities
    5. Use iron-free water

37. What could be the cause of odors at a small package activated sludge plant?
    1. Aeration rates too low
    2. Improper wasting or disposal of waste activated sludge
    3. Poor housekeeping
    4. Return sludge rate too low
    5. Septic dead spots in clarifier

38. What could an operator do to improve performance if the results of an activated sludge aeration tank settling test indicate that the solids do not settle half way and the water above the solids is cloudy or murky in appearance?
    1. Add more air
    2. Adjust pH
    3. Increase aeration period
    4. Stop wasting solids
    5. Waste additional solids

39. What kinds of safety hazards may the operator of a small package plant encounter on the job?
    1. Confined spaces
    2. Drowning
    3. Electric shock
    4. Pathogenic organisms
    5. Toxic gases

40. What are the purposes of an oxidation ditch rotor?
    1. Allows activated sludge to be separated from water
    2. Causes surface aeration
    3. Keeps contents of ditch mixed and moving
    4. Provides detention time for microorganisms
    5. Regulates level of wastewater in ditch

41. How can excess sludge from an oxidation ditch be disposed of properly?
    1. Discharged on to land
    2. Discharged to surface waters
    3. Dried directly on sludge drying beds
    4. Stored for later disposal in approved sanitary landfills
    5. Stored for later disposal at larger treatment plants

42. When starting a new or recently overhauled rotor assembly, why is a "bump start" recommended *BEFORE* a routine full start and run is attempted? Allows operator to determine if rotor assembly operates
    1. At a high level of oxygen transfer
    2. Freely
    3. In a manner that will *NOT* destroy beneficial microorganisms
    4. In proper rotation
    5. Properly

43. What items should be checked when starting a new pump?
    1. Amperage reading recorded
    2. Excess noise
    3. Overheating
    4. Vibration
    5. Water temperature

44. For an oxidation ditch to obtain maximum performance efficiency, which process control methods must be maintained?
    1. Proper DO levels in the oxidation ditch
    2. Proper ditch detention time to treat the wastes by control of the adjustable weir
    3. Proper ditch environment (no toxicants and sufficient microorganisms to treat the wastes)
    4. Proper food supply (measured as BOD or COD and no toxicants) for the microorganisms
    5. Proper water/solids separation in the clarifier

45. What happens when over-oxygenation occurs in an oxidation ditch?
    1. Microorganisms become overly active
    2. Pinpoint floc forms
    3. Waste of power
    4. Wastes settle on bottom of ditch
    5. Wastewater overflows the ditch walls

46. What could be the cause of dark solids in the settling tank following an oxidation ditch?
    1. Brush rotor is too low in the water
    2. DO levels are too low
    3. Ditch levels are too low
    4. Influent flow is too low
    5. Return sludge rate is too low

47. What types of general observations should be made by the operator of an oxidation ditch?
    1. Clarity of the ditch
    2. Clarity of the sedimentation tank surfaces
    3. Color of the mixed liquor in the ditch
    4. Influent wastewater BOD
    5. Odor at the plant site

48. What could be possible causes of abnormal brush rotor operation?
    1. Excessive grease
    2. Gear drive part(s) partially or totally worn
    3. Insufficient bearing grease
    4. Low oil level
    5. Worn bearings

End of Objective Test

CHAPTER 11

ROTATING BIOLOGICAL CONTACTORS

by

Richard Wick

Revised by

John Brady

TABLE OF CONTENTS

Chapter 11. ROTATING BIOLOGICAL CONTACTORS

OBJECTIVES

Chapter 11. ROTATING BIOLOGICAL CONTACTORS

Following completion of Chapter 11, you should be able to:

1. Describe a rotating biological contactor and the purpose of each important part,
2. Start up and operate a rotating biological contactor,
3. Operate a rotating biological contactor under abnormal conditions,
4. Shut down and restart a rotating biological contactor,
5. Maintain and troubleshoot a rotating biological contactor,
6. Safely perform the duties of the operator of a rotating biological contactor,
7. Review the plans and specifications for a rotating biological contactor, and
8. Calculate the hydraulic and organic loadings on a rotating biological contactor.

WORDS

Chapter 11. ROTATING BIOLOGICAL CONTACTORS

BIODEGRADABLE (BUY-o-dee-GRADE-able) BIODEGRADABLE

Organic matter that can be broken down by bacteria to more stable forms which will not create a nuisance or give off foul odors is considered biodegradable.

COMPOSITE (come-PAH-zit) (PROPORTIONAL) SAMPLE COMPOSITE (PROPORTIONAL) SAMPLE

A composite sample is a collection of individual samples obtained at regular intervals, usually every one or two hours during a 24-hour time span. Each individual sample is combined with the others in proportion to the rate of flow when the sample was collected. The resulting mixture (composite sample) forms a representative sample and is analyzed to determine the average conditions during the sampling period.

DENITRIFICATION (dee-NYE-truh-fuh-KAY-shun) DENITRIFICATION

(1) The anoxic biological reduction of nitrate nitrogen to nitrogen gas.

(2) The removal of some nitrogen from a system.

(3) An anoxic process that occurs when nitrite or nitrate ions are reduced to nitrogen gas and nitrogen bubbles are formed as a result of this process. The bubbles attach to the biological floc and float the floc to the surface of the secondary clarifiers. This condition is often the cause of rising sludge observed in secondary clarifiers or gravity thickeners. Also see NITRIFICATION.

GRAB SAMPLE GRAB SAMPLE

A single sample of water collected at a particular time and place which represents the composition of the water only at that time and place.

MPN (pronounce as separate letters) MPN

MPN is the **M**ost **P**robable **N**umber of coliform-group organisms per unit volume of sample water. Expressed as a density or population of organisms per 100 m*L* of sample water.

NAMEPLATE NAMEPLATE

A durable metal plate found on equipment which lists critical operating conditions for the equipment.

NEUTRALIZATION (new-trall-i-ZAY-shun) NEUTRALIZATION

Addition of an acid or alkali (base) to a liquid to cause the pH of the liquid to move toward a neutral pH of 7.0.

NITRIFICATION (NYE-truh-fuh-KAY-shun) NITRIFICATION

An aerobic process in which bacteria change the ammonia and organic nitrogen in wastewater into oxidized nitrogen (usually nitrate). The second-stage BOD is sometimes referred to as the "nitrogenous BOD" (first-stage BOD is called the "carbonaceous BOD"). Also see DENITRIFICATION.

PYROMETER (pie-ROM-uh-ter) PYROMETER

An apparatus used to measure high temperatures.

SEPTIC TANK SEPTIC TANK

A system sometimes used where wastewater collection systems and treatment plants are not available. The system is a settling tank in which settled sludge and floatable scum are in intimate contact with the wastewater flowing through the tank and the organic solids are decomposed by anaerobic bacterial action. Used to treat wastewater and produce an effluent that flows into a subsurface leaching (filtering and disposal) system where additional treatment takes place. Also referred to as an INTERCEPTOR; however, the preferred term is *SEPTIC TANK*.

SEPTIC TANK EFFLUENT PUMP (STEP) SYSTEM SEPTIC TANK EFFLUENT PUMP (STEP) SYSTEM

A facility where effluent is pumped from a septic tank into a pressurized collection system which may flow into a gravity sewer, treatment plant, or subsurface leaching system.

SLOUGHED SLOUGHED

Biological or biomass growths that break off from the rotating biological contactor (RBC) media. The "sloughed" biological growth is suspended in the RBC effluent and is later removed in the secondary clarifier as sludge.

SMALL-DIAMETER GRAVITY SEWER (SDGS) SYSTEM SMALL-DIAMETER GRAVITY SEWER (SDGS) SYSTEM

A type of collection system in which a series of septic tanks discharge effluent by gravity, pump, or siphon to a small-diameter wastewater collection main. The wastewater flows by gravity to a lift station, a manhole in a conventional gravity collection system, or directly to a wastewater treatment plant.

SOLUBLE BOD SOLUBLE BOD

Soluble BOD is the BOD of water that has been filtered in the standard suspended solids test. The soluble BOD is a measure of food for microorganisms that is dissolved in the water being treated.

SUPERNATANT (sue-per-NAY-tent) SUPERNATANT

Liquid removed from settled sludge. Supernatant commonly refers to the liquid between the sludge on the bottom and the scum on the surface of an anaerobic digester or a septic tank (interceptor). This liquid from an anaerobic digester is usually returned to the influent wet well or to the primary clarifier. If the liquid (supernatant) is from a septic tank, the liquid is discharged by gravity or by a pump to a wastewater collection system.

TOXIC SUBSTANCES TOXIC SUBSTANCES

Substances which are poisonous to living organisms and restrict their ability to treat wastes.

TRICKLING FILTER TRICKLING FILTER

A treatment process in which the wastewater trickles over media that provide the opportunity for the formation of slimes or biomass which contain organisms that feed upon and remove wastes from the water being treated.

CHAPTER 11. ROTATING BIOLOGICAL CONTACTORS

11.0 DESCRIPTION OF ROTATING BIOLOGICAL CONTACTORS

Rotating biological contactors (RBCs) are a secondary biological treatment process for domestic and *BIODEGRADABLE*[1] industrial wastes. For rotating biological contactors to treat wastewater efficiently, the wastewater must first receive at least primary treatment (flow through a primary clarifier) where the major portion of the settled solids are removed for treatment and disposal. For wastewater transported in conventional gravity sewers or pressure sewers where grinder pumps discharge to the wastewater collection system, treatment is commonly accomplished by routing the wastewater flow through a wastewater treatment facility as shown in Figure 11.1. Figure 11.2 indicates a typical flow route through a wastewater treatment facility where wastewater flows from septic tanks (interceptors) and is discharged to small-diameter gravity sewers, or from septic tank effluent pump system sewers. The difference is that septic tanks in small systems perform similarly to primary clarifiers in larger systems. Figure 11.3 illustrates how rotating biological contactors are used in small systems with septic tanks, collection systems, treatment, disinfection, and disposal systems.

The major requirement for efficient RBC performance is that some form of sedimentation (primary treatment) has been provided to the wastewater for removal of settleable and floatable solids before reaching the rotating biological contactors (reactors) for further treatment.

Additional preliminary treatment or pretreatment may or may not be needed. The makeup of the wastewater, type of collection system used to convey the wastewater to the treatment facility, and other factors such as length of travel time of the flow through the system to the treatment plant, temperature of the wastewater, and condition of the collection system, will dictate pretreatment requirements.

Biological contactors have a rotating "shaft" surrounded by plastic discs called the "media." The shaft and media are called the "drum" (Figures 11.4 and 11.5). A biological slime grows on the media when conditions are suitable. This process is very similar to a trickling filter where the biological slime grows on rock or other media and settled wastewater (primary clarifier effluent) is applied over the media. With rotating biological contactors, the biological slime or film grows on the surface of the plastic disc media. The slime is rotated into the settled wastewater and then into the atmosphere to provide oxygen for the organisms (Figure 11.4). The wastewater being treated either flows parallel to the rotating shaft or perpendicular to the shaft as it flows from stage to stage or tank to tank.

The rotating biological contactor process uses several plastic media drums. The plastic disc media are made of high-density plastic circular sheets usually 12 feet (3.6 m) in diameter. These sheets are bonded and assembled onto horizontal shafts up to 25 feet (7.5 m) in length. Spacing between the sheets provides the hollow (void) space for distribution of wastewater and air (Figures 11.5 and 11.6). The media provide the required surface area for the slimes containing the active biological growth. Concrete or coated steel tanks usually hold the wastewater being treated. The media rotate at about 1.5 RPM while approximately 40 percent of the media surface is immersed in the wastewater (Figure 11.6).

As the drum rotates, the media pick up a thin layer of wastewater, which flows over the biological slimes on the discs. Organisms living in the slimes use organic matter from the wastewater for food and dissolved oxygen from the air, thus removing wastes from the water being treated. As the attached slimes pass through the wastewater, some of the slimes are *SLOUGHED*[2] from the media as the media rotates downward into the wastewater being treated. The effluent with the sloughed slimes flows to the secondary clarifier where the slimes are removed from the effluent by settling. Figure 11.7 shows the location of a rotating biological contactor process in a wastewater treatment plant. The process is located in the same position as the trickling filter or activated sludge aeration basin. Usually, the process operates on a "once-through" scheme, with no recycling of effluent or sludge, which makes it a simple process to operate. Figure 11.8 shows a treatment plant without a primary clarifier because the solids and floatables have been removed by septic tanks (interceptors).

---

[1] *Biodegradable (BUY-o-dee-GRADE-able). Organic matter that can be broken down by bacteria to more stable forms which will not create a nuisance or give off foul odors is considered biodegradable.*

[2] *Sloughed. Biological or biomass growths that break off from the rotating biological contactor (RBC) media. The "sloughed" biological growth is suspended in the RBC effluent and is later removed in the secondary clarifier as sludge.*

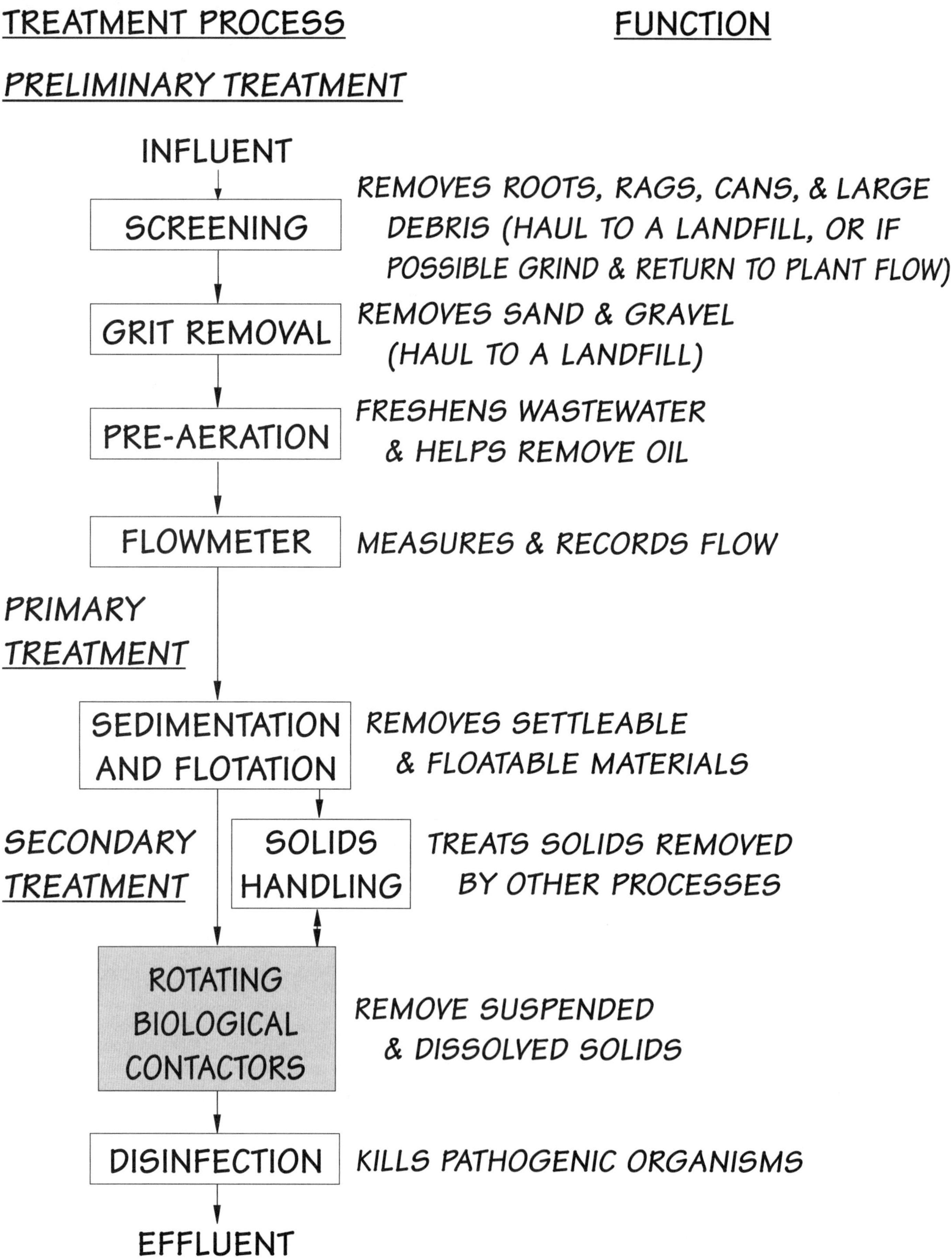

Fig. 11.1 Flow diagram of treatment plant

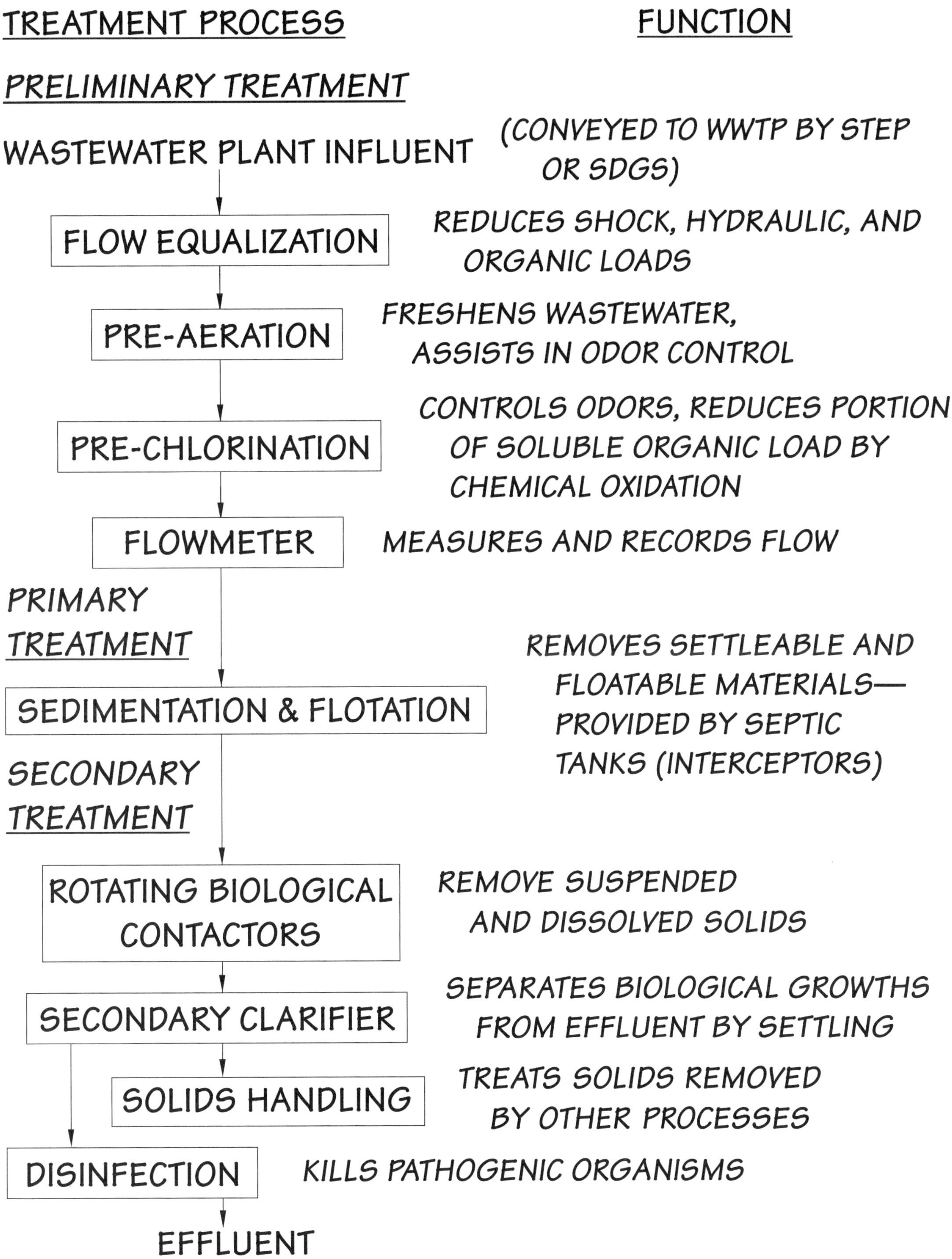

Fig. 11.2 Flow diagram of treatment plant treating wastewater from community served by septic tanks (interceptors)

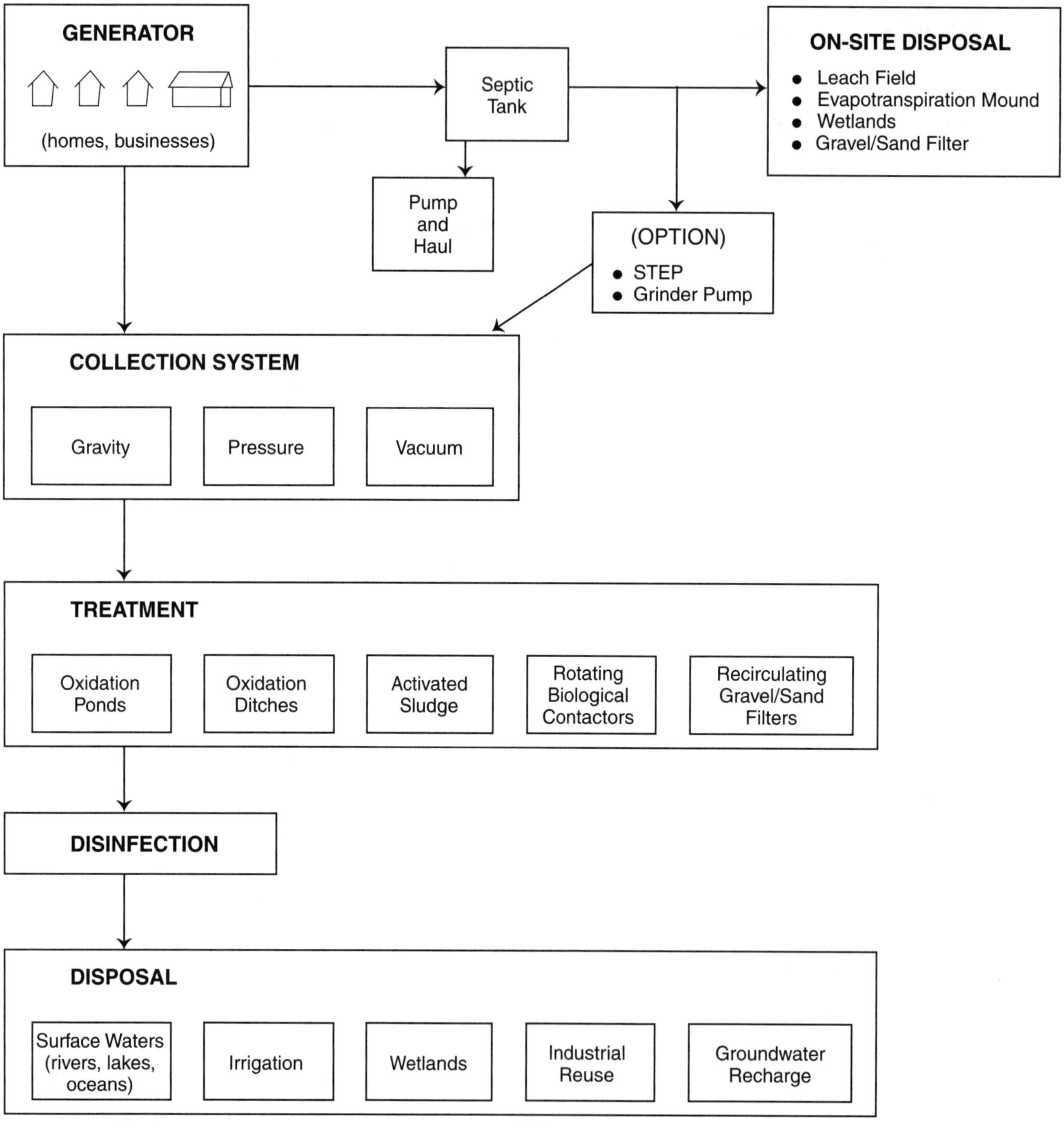

Fig. 11.3 Flow diagram of small collection, treatment, and disposal systems

Fig. 11.4 Rotating biological contactors
(Permission of Autotrol Corporation)

Fig. 11.5 Plastic disc media and biological contactor drum
(Permission of Autotrol Corporation)

Media cross section

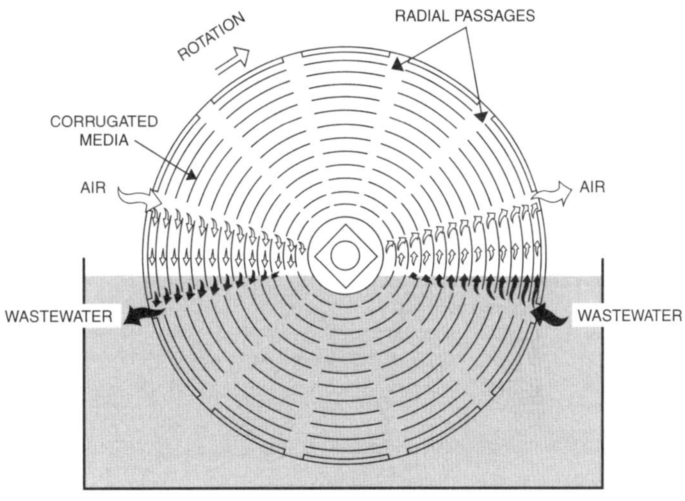

End-view sketch illustrates exchange of air and wastewater

Fig. 11.6 Sections of the plastic disc media

(Permission of Autotrol Corporation)

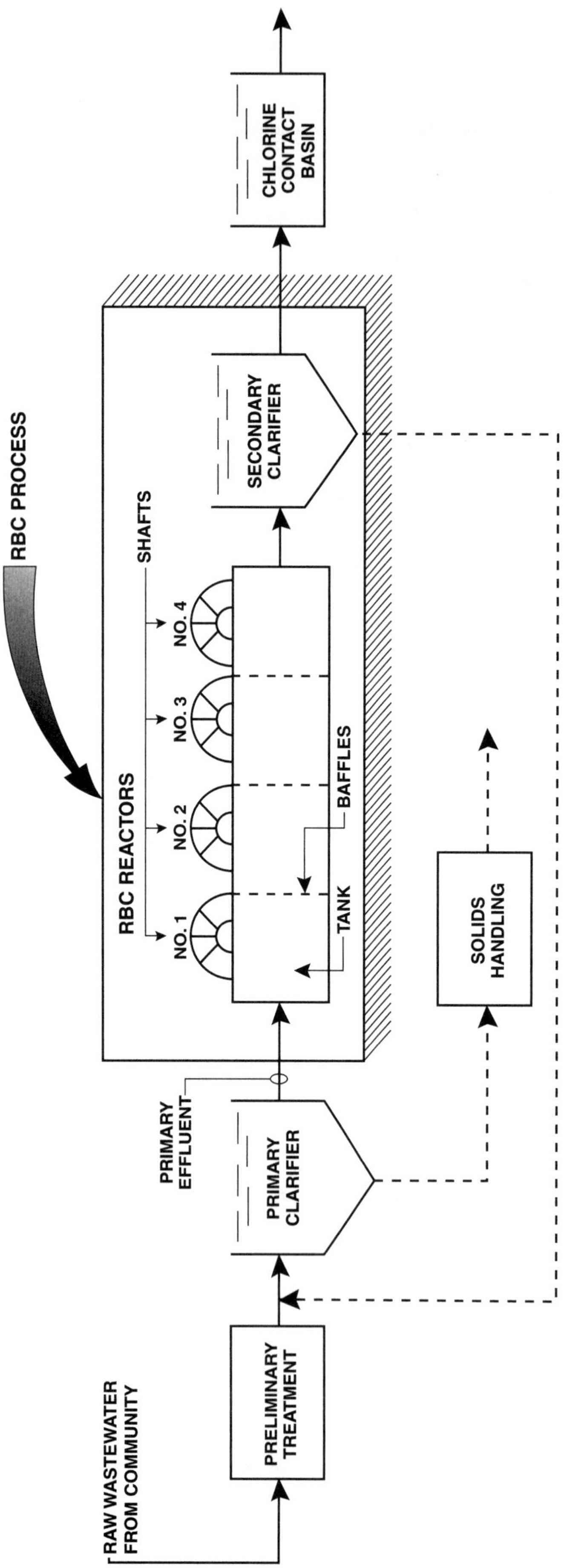

Fig. 11.7 Typical rotating biological contactor (reactor) treatment plant

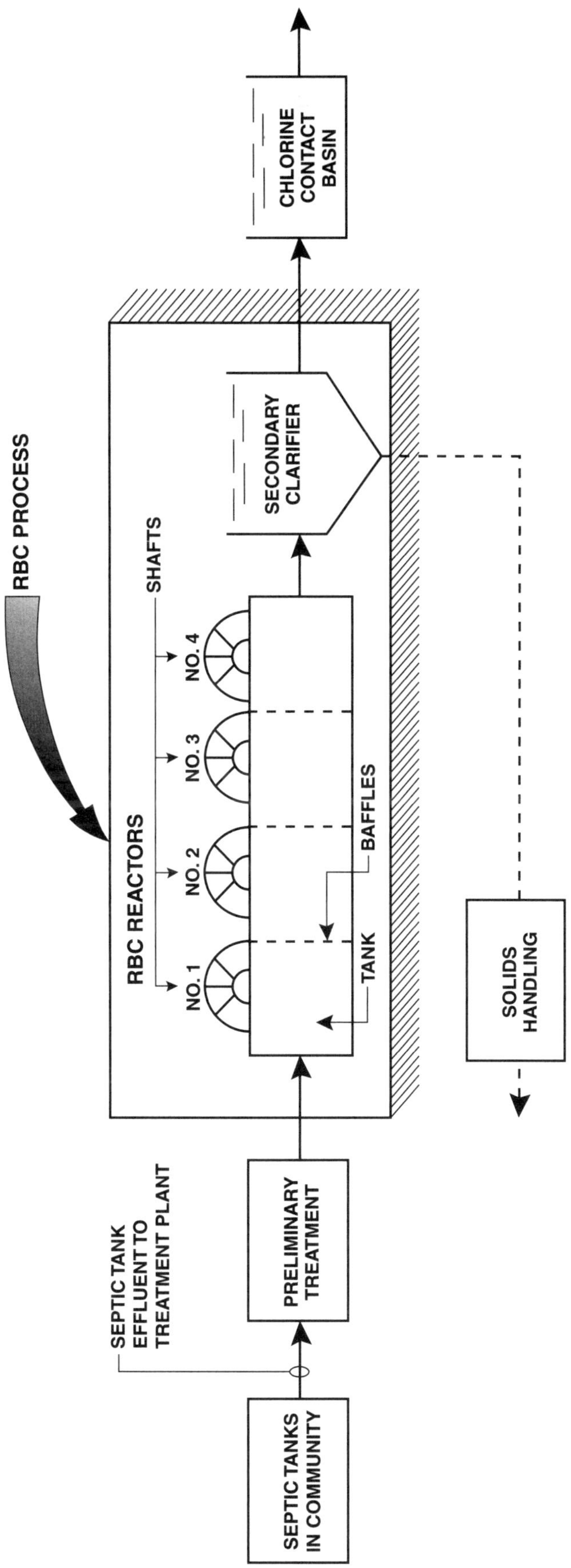

Fig. 11.8 Typical rotating biological contactor (reactor) treatment plant treating wastewater from septic tanks (interceptors)

The major parts of the RBC process are listed in Table 11.1 along with their purposes. The concrete or steel tanks are commonly shaped to conform to the general shape of the media. This shape eliminates dead spots where solids could settle out and cause odors and septic conditions.

TABLE 11.1 PURPOSE OF PARTS OF A ROTATING BIOLOGICAL CONTACTOR

Part	Purpose
Concrete or Steel Tank Divided Into Bays (Sections) by Baffles (Bulkheads)	Tank. Holds the wastewater being treated and allows the wastewater to come in contact with the organisms on the discs.
	Bays and baffles. Prevent short-circuiting of wastewater.
Orifice or Weir Located in Baffle	Controls flow from one stage to the next stage or from one bay to the next bay.
Rotating Media	Provide support for organisms. Rotation provides food (from wastewater being treated) and air for organisms.
Cover Over Contactor	Protects organisms from severe fluctuations in the weather, especially freezing. Also contains odors.
Drive Assembly	Rotates the media.
Influent Lines With Valves	Influent lines. Transport wastewater to be treated to the rotating biological contactor bays.
	Influent valves. Regulate influent to contactor and also isolate contactor for maintenance.
Effluent Lines With Valves	Effluent lines. Convey treated wastewater from the contactor to the secondary clarifier.
	Effluent valves. Regulate effluent from the contactor and also isolate contactor for maintenance.
Underdrains	Allow for removal of solids that may settle out in tank.

The rotating biological contactor process is usually divided into multiple stages (up to four) with three possible layouts (Figure 11.9). Each stage is separated by a removable baffle, concrete wall, or cross-tank bulkhead. Wastewater flow can be either parallel or perpendicular to the shaft. Each bulkhead or baffle has an underwater orifice or hole to permit flow from one stage to the next. Each section of media between bulkheads acts as a separate stage of treatment.

Staging is used to maximize the effectiveness of a given amount of media surface area by separating competing biomass populations. Organisms on the first-stage media are exposed to high levels of BOD and reduce the BOD at a high rate. As the BOD levels decrease from stage to stage, the rate at which the organisms can remove BOD decreases and *NITRIFICATION*[3] starts.

Treatment plants requiring four or more shafts of media usually are arranged so that each shaft serves as an individual stage of treatment. The shafts are arranged so that the flow is perpendicular to the shafts (Figure 11.9, Layout No. 3). The media should move against the direction of flow to provide better treatment of the wastewater. Plants with fewer than four shafts are usually arranged with the flow parallel to the shaft (Figure 11.9, Layout No. 1).

Rotating biological contactors are covered for several reasons relating to climatic conditions:

1. Protect biological slime growths from freezing,
2. Prevent intense rains from washing off some of the slime growth,
3. Stop exposure of media to direct sunlight to prevent growth of algae,
4. Avoid exposure of media to sunlight, which may cause the media to become brittle, and
5. Provide protection for operators from sun, rain, snow, or wind while maintaining equipment.

Fiberglass covers in the shape of the media are easily removed for maintenance. In some areas, the rotating biological contactors are covered by a building. In other areas only a roof is placed over the media for protection against sunlight. The type of cover depends on climatic conditions.

Three types of drive assemblies are used to rotate the shafts supporting the media:

1. Motor with chain drive (Figure 11.10),
2. Motor with direct shaft drive, and
3. Air drive (Figure 11.11).

The first type of drive assembly consists of a motor and belt or chain drive. The second type of drive system consists of a motor, belt drive, and speed reducer mounted directly to the RBC shaft. The third type of drive unit consists of plastic cups attached to the outside of the media (Figure 11.11). A small air header below the edge of the media releases air into the cups. The air in the cups creates a buoyant force, which then makes the shaft turn. With any of the three types of drive assemblies, the mainshaft is supported by two main bearings.

Individual units are usually provided with influent and effluent line valving to allow isolation for maintenance reasons. Usually, the units are not shut down during low-flow conditions because power consumption is minimal and, as the flows decrease, the percent of BOD removal increases.

[3] *Nitrification (NYE-truh-fuh-KAY-shun). An aerobic process in which bacteria change the ammonia and organic nitrogen in wastewater into oxidized nitrogen (usually nitrate). The second-stage BOD is sometimes referred to as the "nitrogenous BOD" (first-stage BOD is called the "carbonaceous BOD").*

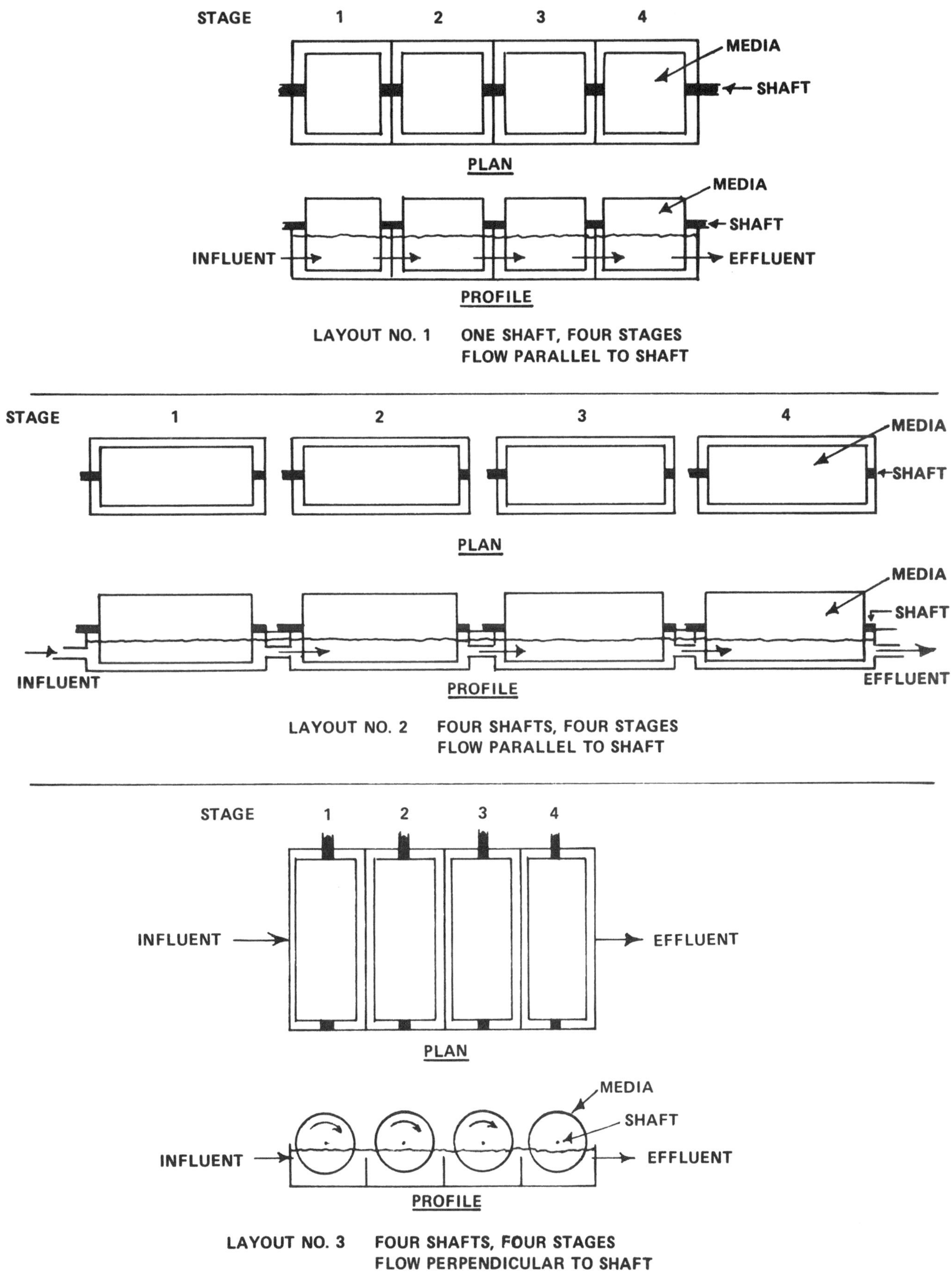

Fig. 11.9 Possible rotating biological contactor layouts

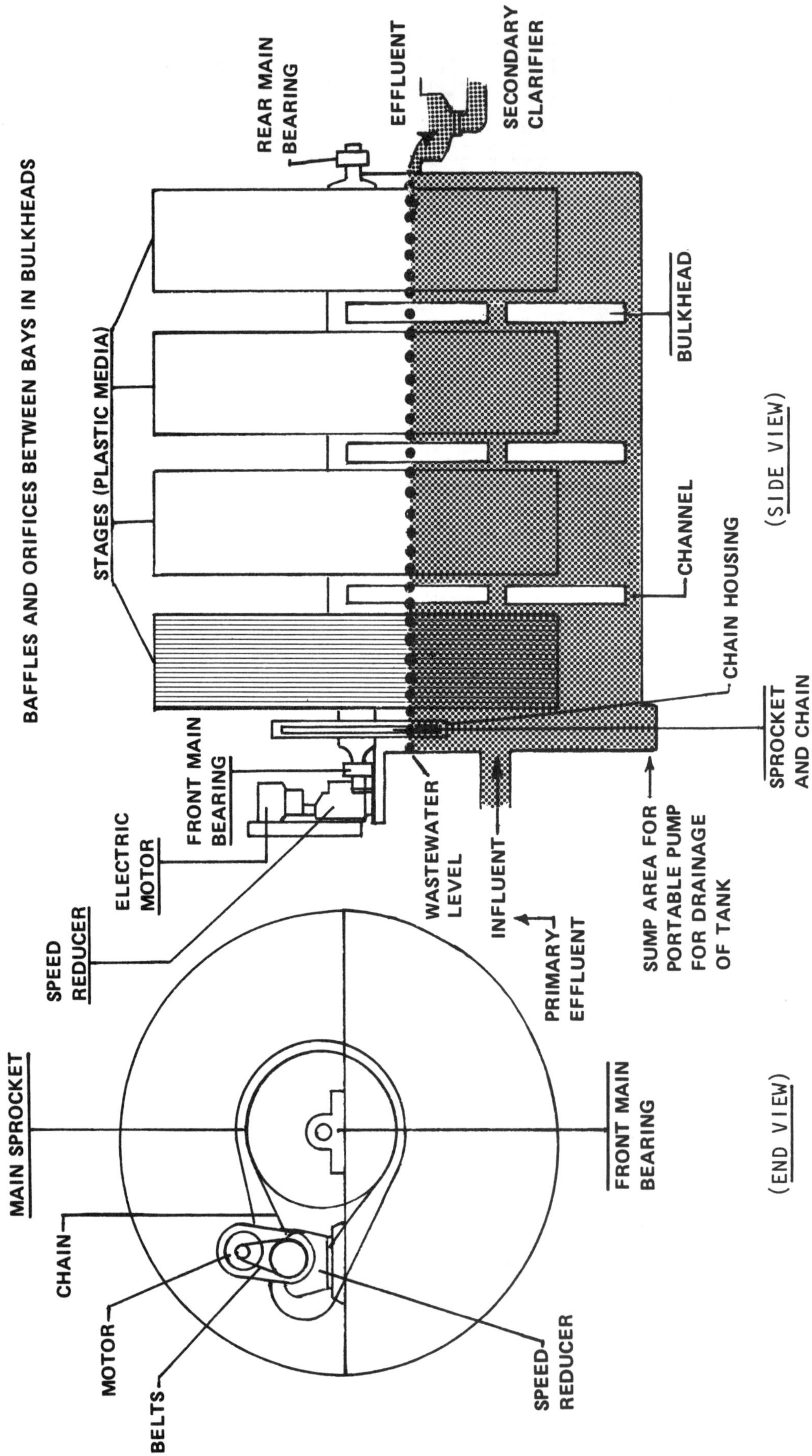

Fig. 11.10 Motor with chain drive unit

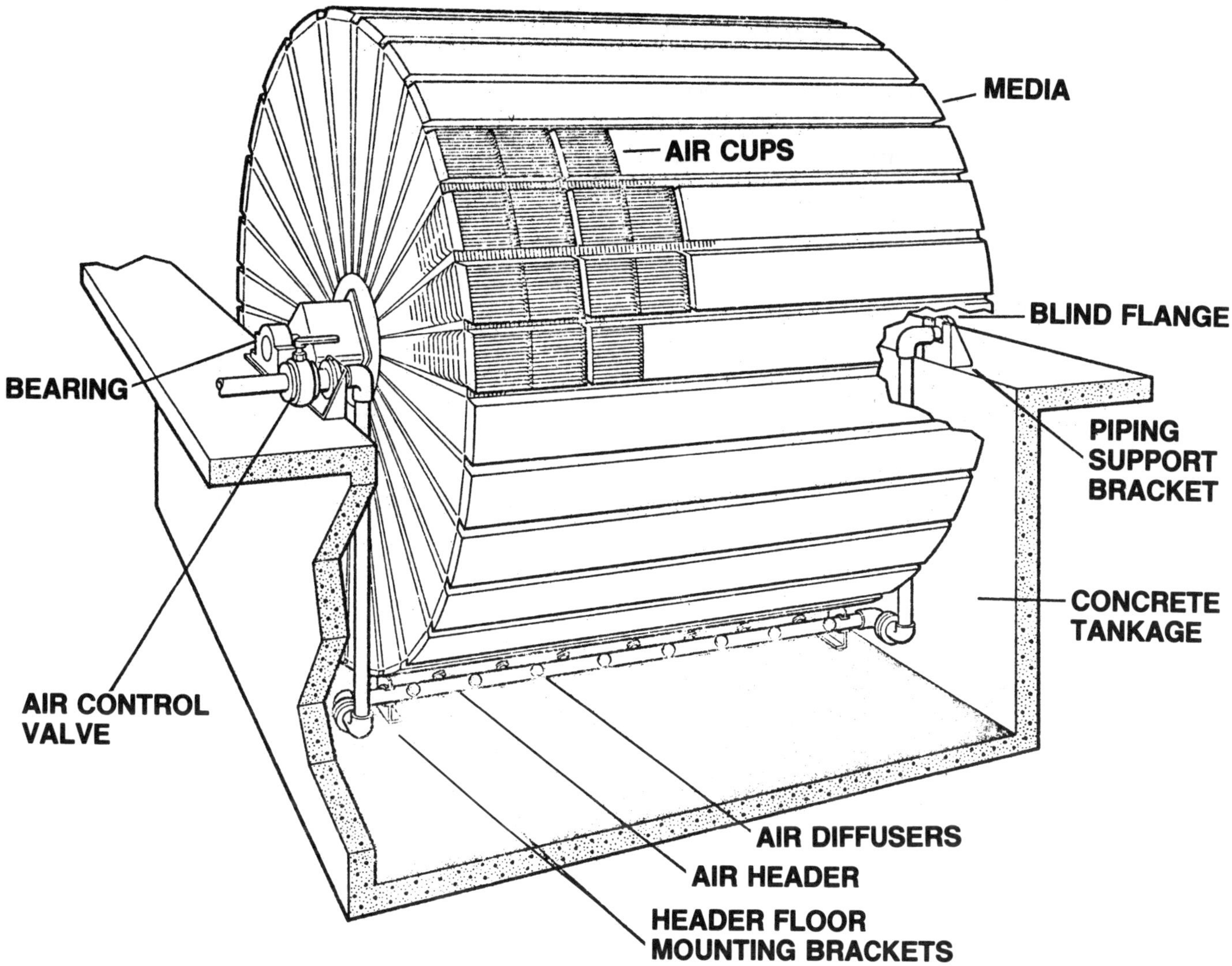

MEDIA

Media with polyethylene air cups attached to the outer perimeter capture air as it is released from the air header.

AIR CUPS

Air cups, attached to the media, capture air, which in turn rotates the media.

AIR CONTROL VALVE

Butterfly control valve controls inlet air supply to each unit.

AIR HEADER

Lightweight headers that carry the air through the system run the length of the media assembly and are easily removable for cleaning.

HEADER FLOOR MOUNTING BRACKETS

Brackets secure header to the floor of the tank.

AIR DIFFUSER

Coarse-bubble air diffusers distribute air from the header into the air cups and media.

PIPING SUPPORT BRACKET

Bracket on each end of the header holds unit in place.

Fig. 11.11 Air drive unit

(Permission of Autotrol Corporation)

QUESTIONS

Write your answers in a notebook and then compare your answers with those on page 153.

11.0A How does a rotating biological contactor (RBC) treat wastewater?

11.0B Why is the RBC process divided into multiple stages?

11.0C What is the purpose of a cover over the RBC unit?

11.0D What are the three types of drive units?

11.1 PROCESS OPERATION

Plants have been designed to treat flows ranging from 18,000 GPD to 50 MGD (70,000 liters/day to 200 Million liters/day), however, the majority of plants treat flows of less than 5 MGD (20 M*L*D). Typical operating and performance characteristics are as follows:

Characteristic	Range
Hydraulic Loading[4]	
BOD Removal	1.5 to 6 GPD/sq ft[5] (60 to 240 liters per day/sq m)[5]
Nitrogen Removal	1.5 to 1.8 GPD/sq ft[5] (60 to 72 liters per day/sq m)[5]
Organic Loading[4]	
Soluble BOD[6,7]	2.5 to 4 lbs BOD/day/1,000 sq ft[5] (12 to 20 kg BOD/day/1,000 sq m)[5]
BOD Removal	80 to 95 percent
Effluent Total BOD	10 to 30 mg/*L*
Effluent Soluble BOD	5 to 15 mg/*L*
Effluent NH_3-N	1 to 10 mg/*L*
Effluent NO_3-N	2 to 7 mg/*L*

See Section 11.5, "Loading Calculations," for procedures showing how to calculate the hydraulic and organic loadings on rotating biological contactors. Both hydraulic and organic loads to a rotating biological contactor should be evaluated. Experience indicates that the organic load (especially to the first stage) often controls the performance of the RBC. Performance by rotating biological contactors also is affected by hydraulic loadings and temperatures below 55°F (13°C).

Advantages of rotating biological contactors over *TRICKLING FILTERS*[8] include the elimination of the rotating distributor, the elimination of problems caused by ponding on the media, and reduction of problems with filter flies. More efficient use of the media is achieved due to the even or uniform rotation of the media into the wastewater being treated. Another advantage of RBCs is the lack of anaerobic conditions in properly designed facilities as compared with anaerobic conditions found in the bottom of trickling filters.

A limitation of the process, as compared with trickling filters, is the lack of flexibility due to the absence of provisions for recirculation; however, in most installations recirculation is not needed but may improve effluent quality when used. Rotating biological contactors are more sensitive to industrial waste shock loading impacts than trickling filters. Care must be taken to ensure that industrial organic loadings to the RBCs do not cause low dissolved oxygen conditions in the treatment system.

11.10 Pretreatment Requirements

Rotating biological contactors are usually preceded by preliminary treatment processes consisting of screening and grit removal, and by primary settling. Grit and large organic matter, if not removed, can settle beneath the drums and form sludge deposits that reduce the effective tank volume, produce septic conditions, physically damage the media, and possibly stall the unit.

Some rotating biological contactor plants have aerated flow equalization tanks between the primary clarifiers and the rotating biological contactors. Flow equalization tanks may be installed to equalize or balance highly fluctuating flows and to allow for the dilution of strong wastes and neutralization of highly acidic or alkaline wastes. These equalization tanks are capable of reducing shock loads and providing pre-aeration to the RBC.

11.11 Start-Up

Prior to plant start-up, become familiar with and understand the contents of the plant O & M manual. If you have any questions, be sure to ask the design engineer or the manufacturer's representative. Both of these persons should instruct the operator on the proper operation of the plant and maintenance of the equipment. If it is possible, try to visit an operating system and talk to an experienced RBC operator. This operator's experience and knowledge could be very helpful during start-up and also during operation.

---

[4] *Hydraulic and organic loadings depend on influent flow, influent soluble BOD, effluent BOD, temperature, and surface area of plastic media. Manufacturers provide charts converting flow to hydraulic and organic loadings for their media.*

[5] *All areas refer to media surface area.*

[6] *Soluble BOD. Soluble BOD is the BOD of water that has been filtered in the standard suspended solids test. The soluble BOD is a measure of food for microorganisms that is dissolved in the water being treated.*

[7] *BOD loading for first stage or first shafts only.*

[8] *Trickling Filter. A treatment process in which the wastewater trickles over media that provide the opportunity for the formation of slimes or biomass which contain organisms that feed upon and remove wastes from the water being treated.*

11.110 Pre-Start Checks for New Equipment

Before starting any equipment or allowing any wastewater to enter the treatment process, record pertinent data including motor data and the sizes of drive belts, chains, and bearings. Also, check the following items before starting any equipment:

1. *TIGHTNESS*

   Inspect the following parts for tightness in accordance with manufacturer's recommendations.

   a. Anchor bolts
   b. Mounting studs
   c. Bearing caps and set screws (check any torque limitations)
   d. Locking collars
   e. Jacking screws
   f. Roller chain (be sure chain is properly aligned)
   g. Media (unbalanced media may cause slippage)
   h. Belts (use matched sets on multiple-belt drives)

2. *LUBRICATION*

   Be sure the following parts have been correctly lubricated with proper lubricants in accordance with manufacturer's recommendations.

   a. Mainshaft bearings
   b. Roller chain
   c. Speed reducer

3. *CLEARANCES*

   a. Between media and tank wall
   b. Between media and baffles or cover support beams
   c. Between chain casing and media
   d. Between roller chain, sprockets, and chain casing

4. *ALIGNMENT*

   Check drive unit alignment to drum shaft.

5. *FLOW CONTROL EQUIPMENT*

   a. Influent and effluent gates
   b. Any weirs for proper level
   c. Drain plugs and/or ports

6. *ELECTRICAL CONTROLS*

   a. Check all breakers
   b. Check START/STOP switches for proper operation

7. *SAFETY*

   a. Be sure safety guards are properly installed over chains and other moving parts.
   b. Be sure safety rails and toe plates are properly installed.

11.111 Procedure for Starting Unit

Actual start-up procedures for a new unit should be in your plant O & M manual provided by the manufacturer. A typical starting procedure is outlined below.

1. Switch on power, allow shaft to rotate one turn, turn off the power, lock out, and tag the main breaker. Inspect and correct if necessary during this revolution:

   a. Movement of chain casing.
   b. Unusual noises.
   c. Direction of media rotation. Where wastewater flow is parallel to the rotating media shaft, the direction of rotation is not critical. If the wastewater flow is perpendicular to the rotating media shaft, the media should be moving through the wastewater against the direction of flow (see Figure 11.9).

2. Switch on power and allow shaft to rotate for 15 minutes. Inspect the following:

   a. Chain drive sprocket alignment.
   b. Noises in bearings, chain drives, and drive package.
   c. Motor amperage. Compare with *NAMEPLATE*[9] value.
   d. Temperature of mainshaft bearing (by hand) and drive package pillow block bearing. If too hot for the hand, use a *PYROMETER*[10] or thermometer. Temperature should not exceed 200°F (93°C).
   e. Tightness of shaft bearing cap bolts. Tighten to manufacturer's recommended torque.
   f. Determine number of revolutions per minute for drum and record for future reference.

3. Open inlet valve and allow wastewater to fill the tank (all four stages if in one tank). Open the outlet valve to allow water to flow through the tank. Make inspections listed in steps 1 and 2 again while drum is rotating. Shut off power, lock out, and tag the main breaker to make any corrections.

4. Check the relationship between the clarifier inlet and the rotating biological contactor outlet for hydraulic balance. This means that you want to be sure the tank containing the biological contactor will not overflow, possibly causing equipment damage.

5. Initial reading of hydraulic or electronic load cell, if so equipped. Record operating weight while unit is rotating.

6. Operate supplemental or air drive air and check for uniform distribution and proper blower operation.

7. See Section 11.20 for break-in maintenance instructions, which start after eight hours of operation.

Encourage the development of biological slimes by regulating the flow rate and strength of the wastewater applied to nearly constant levels by the use of recirculation, if available. Maintaining the wastewater temperatures at 55°F (15°C) or higher during start-up will help. The best rotating speed is one that will shear off excess growth. The media on the drum is usually operated at about 1.5 RPM while approximately 40 percent of the media surface is immersed in the wastewater being treated.

Allow one to two weeks for an even growth of biological slimes (biomass) to develop on the surface of the media with normal strength wastewater. After start-up, a slimy growth (biomass) will appear. During the first week, excessive sloughing will occur naturally. This sloughing is normal and the sloughed material is soon replaced with a fairly uniform, shaggy, brown-to-gray appearing biomass with very few or no bare spots.

---

[9] *Nameplate. A durable metal plate found on equipment which lists critical operating conditions for the equipment.*

[10] *Pyrometer (pie-ROM-uh-ter). An apparatus used to measure high temperatures.*

Follow the same start-up procedures whether a plant is starting at less than design flow or at full design flow. Start-up during cold weather takes longer because the organisms in the slime growth (biomass) are not as active and require more time to grow and reproduce.

The RBC works best when operating at or near design BOD and hydraulic loadings. If the plant is currently over-designed to handle expansion of the system or population growth, it may be possible to operate a portion of units at the beginning of start-up. If all units are not needed (based on loading calculations), the operating units should be switched between running and not running RBC trains to protect them from deterioration.

11.12 Operation

Rotating biological contactor treatment plants are not difficult to operate and produce a good effluent provided the operator properly and regularly performs the duties of inspecting the equipment, testing the influent and effluent, observing the media, maintaining the equipment, and taking corrective action, when necessary.

11.120 Inspecting Equipment

This treatment process has relatively few moving parts. There is a drive train to rotate the shaft and there are bearings upon which the shaft rotates. Check the following items when inspecting equipment.

1. Feel the outer housing of the shaft bearing to see if it is running hot. Use a pyrometer or thermometer, if temperature is too hot for your hand. If the temperature exceeds 200°F (93°C), the bearings may need to be replaced. Also, check for proper lubrication and be sure the shaft is properly aligned. The longer the shaft, the more critical the alignment. See the manufacturer's manual for procedures to check alignment and allowable tolerances.
2. Listen for unusual noises in motor bearings. Locate the cause of unusual noises and correct it.
3. Feel motors to determine if they are running hot. If hot, determine cause and correct. If hot, check motor nameplate for permissible degrees of temperature rise (typically 40°C). If temperature rise is greater than the nameplate rating, check the following items:
    a. Amperage draw on each leg,
    b. Bearing lubrication and temperature, and
    c. Motor alignment to drive unit.
4. Look around drive train and shaft bearing for oil spills. If oil is visible, check oil levels in the speed reducers and chain drive system. Also, look for damaged or worn out gaskets or seals. Clean up leaking oil for reasons of safety and appearance.
5. Inspect chain drive for alignment, tightness, and sprocket wear.

6. Inspect belts for proper tension.
7. Be sure all guards over moving parts and equipment are in place and properly installed.
8. Clean up any spills, messes, or debris.

Damage to the equipment will be minimized if required maintenance is done as soon as the need is identified. The RBC unit will not repair itself.

11.121 Testing Influent and Effluent

Wastewater analysis is required to monitor overall plant and process performance. Because there are few process control functions to be performed, only a minimal analysis is required to monitor and report daily performance. To determine if the rotating biological contactors are operating properly, you should measure (1) BOD, (2) suspended solids, (3) pH, and (4) dissolved oxygen (DO). Performance is best monitored by analysis of a 24-hour *COMPOSITE SAMPLE*[11] for BOD and suspended solids on a daily basis. DO and pH should be measured using *GRAB SAMPLES*[12] at specific times. Actual frequency of tests may depend on how often you need the results for plant control and also how often your NPDES permit requires you to sample and analyze the plant effluent.

Sampling points should include

1. Plant influent: BOD and Suspended Solids (SS),
2. Plant effluent: BOD and SS,
3. RBC effluent: BOD, SS, and DO, and
4. Secondary clarifier effluent: BOD, SS, and DO.

BOD, SS, and DO tests performed on samples taken at or after each drum provide useful information on how well the system is working.

DISSOLVED OXYGEN

The DO in the wastewater being treated beneath the rotating media will vary from stage to stage. A plant designed to treat primary effluent for BOD and suspended solids removal will usually have 0.5 to 1.0 mg/*L* DO in the first stage. The DO level should increase to 1 to 3 mg/*L* at the end of the first

---

[11] *Composite (come-PAH-zit) (Proportional) Sample. A composite sample is a collection of individual samples obtained at regular intervals, usually every one or two hours during a 24-hour time span. Each individual sample is combined with the others in proportion to the rate of flow when the sample was collected. The resulting mixture (composite sample) forms a representative sample and is analyzed to determine the average conditions during the sampling period.*

[12] *Grab Sample. A single sample of water collected at a particular time and place which represents the composition of the water only at that time and place.*

stage. A plant designed for nitrification to convert ammonia and organic nitrogen compounds to nitrate usually will have four stages and DO levels of 4 to 8 mg/*L*. The difference between an RBC unit designed for BOD removal and one designed for nitrification is the design flow applied per square foot of media surface area is lower for nitrification. DO in the first stage of a nitrification unit will be more than 1 mg/*L* DO and often as high as 2 to 3 mg/*L*.

EFFLUENT VALUES

Typical BOD, suspended solids, and ammonia and nitrate effluent values for rotating biological contactors depend on NPDES permit requirements and design effluent values. As flows increase, effluent values increase, thus lowering treatment constituent removal efficiencies because a greater flow is applied to each square foot of media while the time the wastewater is in contact with the slime growths is reduced. Also, the greater the levels of BOD, suspended solids, and nitrogen in the influent, the greater the levels in the plant effluent. Figure 11.12 shows typical influent and effluent values for a rotating biological contactor. The influent and effluent data plotted are seven-day moving averages, which smooth out daily fluctuations and reveal trends. Procedures for calculating moving averages are explained in Chapter 18, "Analysis and Presentation of Data," in *OPERATION OF WASTEWATER TREATMENT PLANTS*, Volume II.

If analysis of samples reveals a decrease in process efficiency, look for three possible causes:

1. Reduced wastewater temperatures,
2. Unusual variations in flow and/or organic loadings, and
3. High or low pH values (less than 6.5 or greater than 8.5).

Once the cause of the problem has been identified, possible solutions can be considered and the problem corrected.

TEMPERATURE

Wastewater temperatures below 55°F (13°C) will result in a reduction of biological activity and in a decrease in BOD or organic material removal. Not much can be done by the operator except to wait for the temperatures to increase again. Under severe conditions, provisions can be made to heat the building, the air inside the RBC unit cover, or the RBC unit influent.

Solar heat can be used effectively to maintain temperature in buildings and enclosures without drying out the biological slime growths. Ceilings should be kept low to use available heat effectively. If existing buildings have high ceilings, large vaned fans can be mounted on the ceilings to direct heat downward.

INFLUENT VARIATIONS

When large daily influent flow and/or organic (BOD) variations occur, a reduction in process efficiency is likely to result. Before corrective steps are taken, the exact extent of the problem and resulting change in process efficiency must be determined. In most cases, when the influent flow and/or organic peak loads are less than three times the daily average values during a 24-hour period, little decrease in process efficiency will result.

In treatment plants where the influent flow and/or organic loads exceed design values for a sustained period, the effluent BOD and suspended solids must be measured regularly during this period to determine if corrective action is required.

During periods of severe organic overload, the bulkhead or baffle between stages one and two may be removed. This procedure provides a larger amount of media surface area for the first stage of treatment. If the plant is continuously overloaded and the effluent violates the NPDES permit requirements, additional treatment units should be installed. A possible short-term solution to an organic overload problem might be the installation of facilities to recycle effluent; however, this would cause a greater increase of any hydraulic overload. Supplemental aeration is an effective method of controlling excess biomass and improving treatment efficiency.

pH

Every wastewater has an optimum pH level for best treatability. Domestic wastewater pH varies between 6.5 and 8.5 and will have little effect on organic waste removal efficiency. However, if this range is exceeded at any time (due to industrial waste discharges, for example), a decrease in efficiency is likely.

To adjust the pH toward 7.0, either pre-aerate the influent or add chemicals. If the pH is too low, add sodium bicarbonate or lime.[13] If the pH is too high, add acetic acid. The amount of chemical to be added depends on the characteristics of the water and can best be determined by adding chemicals to samples in the lab and measuring the change in pH. Before chemicals are used to adjust the pH, run tests in the lab to determine how much of each chemical being considered will be needed to raise the pH to the desired level. Don't guess how much chemical may be needed. Always wear appropriate safety gear (goggles and/or face shield, impervious gloves, protective clothing) when handling chemicals.

When dealing with nitrification, pH and alkalinity are very critical. The pH should be kept as close as possible to a value of 8.4 when nitrifying. The alkalinity level in the raw wastewater should be maintained at a level at least 7.1 times the influent ammonia concentration to allow the reaction to go to completion without adversely affecting the microorganisms. Effluent alkalinity of 50 to 80 mg/*L* is required to avoid pH depression in the effluent. Sodium bicarbonate can be used to increase both the alkalinity and pH.

Another cause of pH variations could be the addition of *SUPERNATANT*[14] from an aerobic or anaerobic digester. The supernatant should be tested for pH and suspended solids. Without testing the supernatant, you will not know what kind of load you're placing on the rest of the plant. Sometimes, it's best to return supernatant at low flows to the plant. Caution should be taken to avoid overloading the process. If the supernatant pH is too low, supernatant could be drawn off during high flows when these flows can be used for dilution and *NEUTRALIZATION*.[15]

---

[13] *Lime is relatively insoluble and yields less alkalinity control per pound added than sodium bicarbonate. Lime often settles in the bottom of the tanks and basins and eventually needs to be removed.*

[14] *Supernatant (sue-per-NAY-tent). Liquid removed from settled sludge. Supernatant commonly refers to the liquid between the sludge on the bottom and the scum on the surface of an anaerobic digester or a septic tank (interceptor). This liquid from an anaerobic digester is usually returned to the influent wet well or to the primary clarifier. If the liquid (supernatant) is from a septic tank, the liquid is discharged by gravity or by a pump to a wastewater collection system.*

[15] *Neutralization (new-trall-i-ZAY-shun). Addition of an acid or alkali (base) to a liquid to cause the pH of the liquid to move toward a neutral pH of 7.0.*

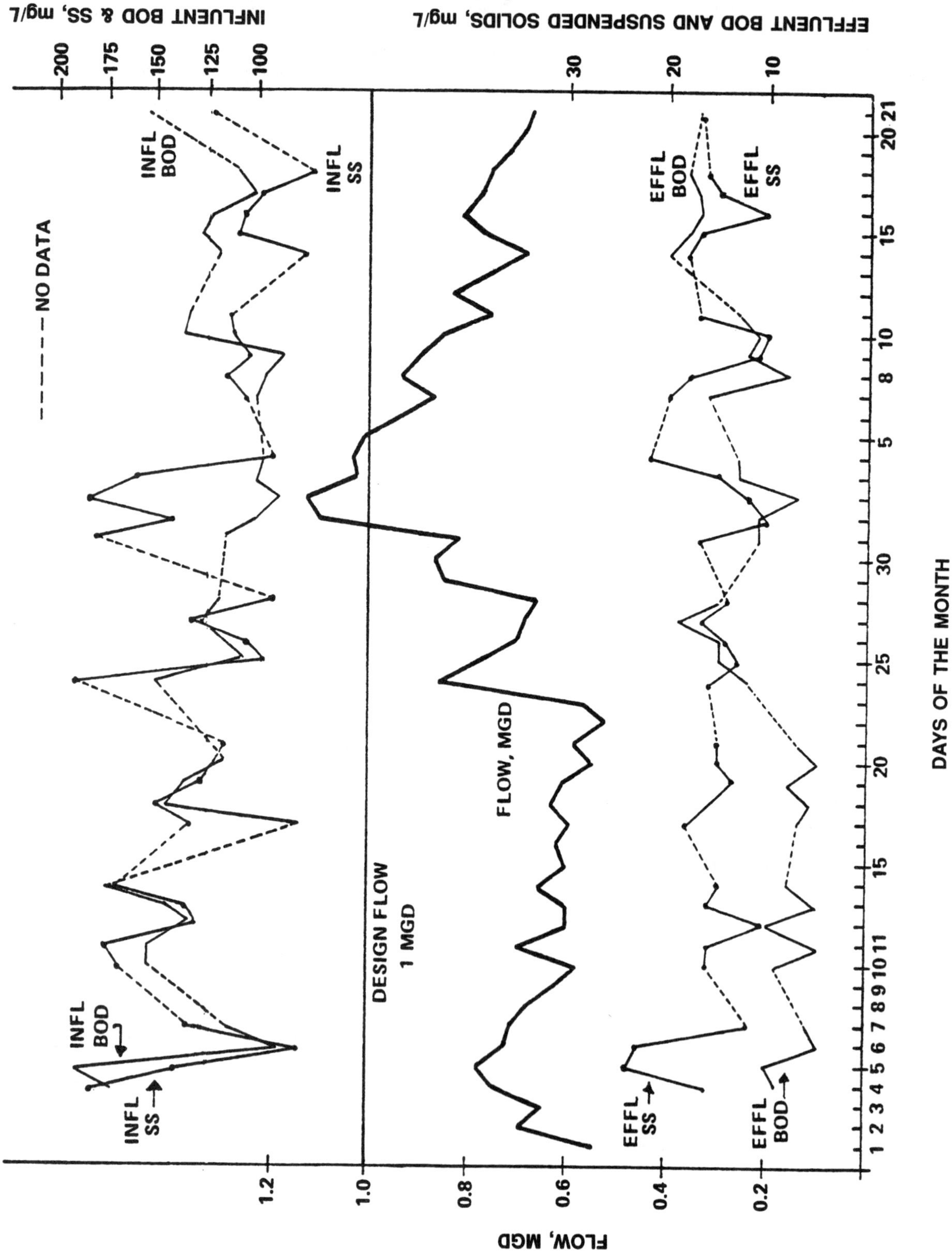

NOTE: Influent data (top lines) are for conventional wastewater before any preliminary or primary treatment. Typical wastewater flowing from septic tanks will have 40 to 100 mg/L BOD and 30 to 80 mg/L suspended solids.

Fig. 11.12 Typical BOD and suspended solids values for an RBC unit

11.122 Observing the Media

Rotating biological contactors use bacteria and other living organisms growing on the media to treat wastes. Because of this, you can use your senses of sight and smell to identify problems. The slime growth or biomass should have a brown-to-gray color, no algae present, a shaggy appearance with a fairly uniform coverage, and very few or no bare spots. The odor should not be offensive, and certainly there should be no sulfide (rotten egg) smells.

BLACK APPEARANCE

If the appearance becomes black and odors that are not normal do occur, then this could be an indication of solids or BOD overloading. These conditions would probably be accompanied by low DO in the plant effluent. Compare previous influent suspended solids and BOD values with current test results to determine if there is an increase. To solve this problem, place another rotating biological contactor unit in service, if possible, operate or install supplemental air, or try to pre-aerate the influent to the RBC unit. Also, review the operation of the primary clarifiers and sludge digesters to be sure they are not the source of the overload.

WHITE APPEARANCE

A white appearance on the disc surface also might be present during high loading conditions. This might be due to a type of bacteria that feeds on sulfur compounds. The overloading could result from industrial discharges containing sulfur compounds upon which certain sulfur-loving bacteria thrive and produce a white slime biomass. Corrective action consists of placing another RBC unit in service, operating supplemental air, or trying to pre-aerate the influent to the unit. During periods of severe organic or sulfur overloading, remove the bulkhead or baffle between stages one and two. Prechlorination of plant influent also will control sulfur-loving bacteria.

Another cause of overloading may be sludge deposits that have been allowed to accumulate in the bottom of the bays. To remove these deposits, drain the bays, wash the sludge deposits out, and return the unit to service. Be sure the orifices in the baffles between the bays are clear.

The biomass on the surface of the media and ends of the drums is made up of many different types of organisms. These organisms can be observed using a microscope that has at least 400 times magnification. Organism observations should be made whenever there is a change in the appearance of the biomass or a change in the effluent quality. These organism observations can be related to DO at the point where the samples are taken.

SLOUGHING

If severe sloughing or loss of biomass occurs after the start-up period and process difficulty arises, the causes may be due to the influent wastewater containing *TOXIC SUBSTANCES*[16] that kill the organisms in the biomass or restrict their ability to treat wastes. Microscopic observation of the biomass will indicate whether the organisms are dead or alive. If dead, then this indicates the presence of a toxic substance. To solve this problem caused by toxicity, steps must be taken to eliminate the toxic substance even though this may be very difficult and costly. Biological processes will never operate properly as long as they attempt to treat toxic wastes. Until the toxic substance can be located and eliminated, loading peaks should be dampened (reduced) and a diluted, uniform concentration of the toxic substance allowed to reach the media in order to minimize harm to the biological culture. While the corrections are made at the plant, dampening may be accomplished by regulating inflow to the plant. Be careful not to flood any homes or overflow any low manholes. Toxic wastes may be diluted in the plant using plant effluent. Try to pump any toxic wastes to emergency storage to prevent the toxic substances from being discharged to receiving waters.

Another problem that could cause loss of biomass is an unusual variation in flow and/or organic loading. In small communities, one cause may be high flow during the day and near zero flow at night. During the day, the biomass is receiving food and oxygen and starts growing; then the night flow drops to near zero—available food is reduced and nearly stops. The biomass starts sloughing off again due to lack of food. When diurnal swings (day-night variations) occur, investigate the possibility of taking a portion of the system out of service at night and then putting it back on line in the morning. If the system has been designed with this degree of flexibility, it is a good way to treat load variations.

Possible solutions to sloughing of the biomass due to excessive variations in plant flow and/or organic loading include throttling peak conditions and recycling from the secondary clarifier or RBC effluent during low flows. Be very careful when throttling plant inflows that low-elevation homes are not flooded or that manholes do not overflow. Usually, RBC units do not have provisions for any recycling from the secondary clarifier. If low flows at night are creating operational problems due to lack of organic matter, a possible solution is the installation of a pump to recirculate water from the secondary clarifier. If recirculation is provided, try to maintain a hydraulic loading rate of greater than 1.0 to 1.5 GPD/sq ft (40 to 60 liters per day/sq m). A flow equalization tank can be used to provide fairly continuous or even flows.

11.123 Weighing the Biomass

The shaft and media of an RBC are designed to carry biomass, but it is possible that the design load carrying capacity of the shaft and media could be exceeded by excessive biomass growth, which may lead to shaft failure. Refer to the Appendix at the end of this chapter for a description of the equipment and procedures that can be used to determine the total operating weight on the RBC shaft.

11.124 Control of Snails

Snails are not a problem in RBCs used to remove carbonaceous biochemical oxygen demand (CBOD) because the growth of "bugs" (microbes) removing CBOD is high and the microbial slime (treating the wastewater) consumed by snails is quickly replaced by new growth. Snails can be a problem where the RBC is expected to remove nitrogenous biochemical oxygen demand (NBOD) because snails remove slow-growing nitrifying bacteria and interfere with nitrification. Snail shells are a problem when they clog tanks, pipes, and pumps.

Chlorination is commonly used to control snails on RBCs. One approach is to take an RBC train (the RBCs treating a particular slug or flow of wastewater) off line. Add chlorine to the water to a concentration of 60 to 70 mg/*L*, rotate the RBC in the superchlorinated solution for two to three days, and then return the RBC to service. Chlorination also can be used to

[16] *Toxic Substances. Substances which are poisonous to living organisms and restrict their ability to treat wastes.*

control filamentous microorganisms. The superchlorinated water can be dechlorinated before it is discharged from the facility by applying sulfur dioxide. When the RBC is returned to service, it will take a period of time for the biomass to recover and reach full treatment capacity.

Another approach to control snails is to increase the pH to 10. A pH of 10 will kill snails without harming the microbial growth on the RBCs. The pH can be increased by adding caustic soda (sodium hydroxide) or soda ash (sodium carbonate) and maintaining the RBC exposure for eight hours. Operators may have to increase the pH to 10 every one to two months to control the snails.

Also, if secondary solids are being recycled to the head of the plant, snail eggs and snails may be recycled along with the solids. Some operators reduce this transfer of snails by sending the solids from the secondary clarifier directly to anaerobic digesters.

11.125 Operational Troubleshooting

Indicators of possible rotating biological contactor process operational problems, probable causes, and solutions are summarized in Table 11.2.

The top ten things operators can do to improve RBC plant performance are as follows:

- Split influent flows to RBCs accurately,
- Adjust staging on RBCs,
- Aerate RBCs,
- Inspect/clean RBC tank solids deposits,
- Optimize secondary sludge removals,
- Control/minimize/equalize in-plant returns, for example, digester supernatant, thickener overflows, belt filter press precoat chemicals, and filter backwash. Avoid BOD and ammonia spikes to RBCs,
- Control equalization tank more efficiently,
- Recirculate biological sludge to aerated RBCs,
- Recirculate RBC effluent to primary anaerobic treatment tank (septic tank) for denitrification and alkalinity recovery, and
- Recirculate biological sludge to aerated equalization tank.

11.13 Abnormal Operation

Abnormal operating conditions may develop under the following circumstances:

1. High or low flows,
2. High or low solids loading, and
3. Power outages.

When your plant must treat high or low flows or solids (organic) loads, abnormal conditions develop as the treatment efficiency drops. For solutions to these problems, refer to Section 11.12, "Operation," and Table 11.2. One advantage of RBC units is the fact that high flows usually do not wash the slime growths off the media; consequently the organisms are present and treating the wastewater during and after the high flows.

A power outage requires the operator to take certain precautions to protect the equipment and the slime growths while no power is available. If the power is off for less than eight hours, nothing needs to be done. If the power outage lasts longer than eight hours, the RBC shaft needs to be turned about one-half turn every eight to ten hours. The shaft can be rotated at shorter intervals and smaller increments, if desired, but must equal the same distance in relation to time as previously noted (that is, one-half turn every eight to ten hours). Turning prevents all the slime growth from accumulating on the bottom portion of the plastic disc media. Before attempting to turn the shaft, lock out and tag the power in case the outage ends abruptly. To turn the shaft, *REMOVE THE BELT GUARD USING EXTREME CARE.* Turn the shaft by using the spokes on the pulley or a strap wrench on the shaft. *DO NOT EXPOSE YOUR FINGERS OR HANDS TO THE PINCH POINT BETWEEN THE BELT(S) AND THE PULLEY(S).* Place a wedge-shaped block between the belts and belt pulley or sheave to hold the shaft and media in the desired location. Actually, the shaft is very delicately balanced and easy to rotate. Do not try to weld handles or brackets to the shaft to facilitate turning because this will throw the shaft off balance.

WARNING: If the shaft starts to roll back to its original position before you get the block properly inserted, do not try to stop the shaft. Let it roll back and stop. If you try to stop the shaft from rolling back, you could injure yourself and also damage the belts and the belt's pulley or sheave.

Gently spray water on the slime growth that is not submerged frequently enough to keep the biomass moist whenever the drum is not rotating.

If the power outage lasts longer than 12 hours, more than normal sloughing will occur from the media when the unit is placed back in service. When the sloughing becomes excessive, increase the sludge pumping rate from the secondary clarifier.

11.14 Shutdown and Restart

The rotating biological contactor may be stopped by turning off the power to the drive package. If the process is to be stopped for longer than eight hours, follow the precautions listed in Section 11.13, "Abnormal Operation," when a power outage occurs. Do not allow one portion of the media to be submerged in the wastewater being treated for more than eight hours. Occasionally, spray the media not submerged to prevent the slime growth from drying out whenever the drum is not rotating.

If the shaft cannot be rotated, or if freezing temperatures exist and the flow is stopped, the tank must be drained or

TABLE 11.2 TROUBLESHOOTING GUIDE—ROTATING BIOLOGICAL CONTACTORS

(Adapted from *PERFORMANCE EVALUATION AND TROUBLESHOOTING AT MUNICIPAL WASTEWATER TREATMENT FACILITIES*, Office of Water Program Operations, US EPA, Washington, DC.)

Indicator/Observation	Probable Cause	Check or Monitor	Solution
1. Decreased treatment efficiency.	1a. Organic overload.	1a. Check peak organic loads—BOD, SS, DO, pH, temperature.	1a. 1. Improve pretreatment of plant. 2. Place another RBC in service, if available. 3. Remove bulkhead between stages 1 and 2 for larger first stage. 4. Recycle effluent as a possible short-term solution. 5. Install or operate supplemental aeration.
	1b. Hydraulic overload.	1b. Check peak hydraulic loads—if less than twice the daily average, should not be the cause.	1b. 1. Flow equalization; eliminate source of excessive flow. 2. Balance flows between reactors. 3. Store peak flows in collection system, monitor possible overflows of collection system.
	1c. pH too high or too low.	1c. Desired range is 6.5 - 8.5 for secondary treatment; 8 - 8.5 for nitrification.	1c. 1. Eliminate source of undesirable pH or add acid or base to adjust pH. When nitrifying, maintain alkalinity at 7 times the influent NH_3 concentration. 2. Sodium bicarbonate can be used to increase both pH and alkalinity.
	1d. Low wastewater temperatures.	1d. Temperatures less than 55°F (13°C) will reduce efficiency.	1d. 1. Cover RBC to contain heat of wastewater. 2. Heat influent to unit or building.
2. Excessive sloughing of biomass from discs.	2a. Toxic materials in influent.	2a. Determine material and its source.	2a. 1. Eliminate toxic material if possible—if not, use flow equalization to reduce variations in concentration so biomass can acclimate. Monitor industrial contributors for flow variations and toxic discharges. 2. Recycle effluent for dilution.
	2b. Excessive pH variations.	2b. pH below 5 or above 10 can cause sloughing.	2b. Eliminate source of pH variations or maintain control of influent pH.
	2c. Unusual variation in flow and/or organic loading.	2c. Influent flow rate(s) and organic strength.	2c. Eliminate/reduce variations by throttling peak conditions and recycling from the secondary clarifier or RBC effluent during low flows. 2d. Monitor industrial contributors for flow variations and also toxic discharges.
3. Development of white biomass over most of disc area.	3a. Septic influent or high H_2S concentrations.	3a. Influent odor.	3a. Pre-aerate wastewater or add sodium nitrate or hydrogen peroxide or place another RBC unit in service. Prechlorination of influent will also control sulfur-loving bacteria. If STEP or SDGS systems, check septic tanks for excessive solids buildup. If excessive solids, pump solids out of septic tank.
	3b. First stage is overloaded organically.	3b. Organic loading on first stage.	3b. 1. Improve pretreatment of plant. 2. Place another RBC in service, if available. 3. Adjust baffles between first and second stages to increase total surface area in first stage. 4. Install or operate supplemental aeration.
4. Solids accumulating in reactors.	4a. Inadequate pretreatment.	4a. Determine if solids are grit or organic.	4a. Remove solids from reactors and provide improved grit removal or primary settling.

pumped out so that the water level remains below the RBC media to prevent equipment damage. A portable sump pump may be used to drain the tank. A sump is usually located at the end of the unit by the motor. Pump the water either to the primary clarifier or to the inlet end of an RBC unit in operation. A trough running the full length of the tank allows the solids to be pumped out. While the tank is empty, inspect for cracks and any other damage and make necessary repairs.

Try to keep the slime growths moist to minimize sloughing and a reduction in organism activity when the process starts again. A loss in process efficiency can result if the slimes are washed off the media. *DO NOT WASH THE SLIME GROWTH OFF THE MEDIA* because you will be washing away the organisms that treat the wastewater. If the unit is to be out of service for longer than one day, the slimes may be washed off the media to prevent the development of odor problems.

Restart rotation by applying power to the drive unit. Before applying power, inspect the shaft and drive unit for possible interference from such items as tools or bulkheads. If slippage occurs from an unbalanced media, inspect and adjust alignment and tension. Check the lubrication of the shaft bearings and the drive assembly and bearings.

11.15 Decommissioning RBCs for Extended Shutdowns

There are a variety of situations in which RBCs are taken out of service for months each year. Examples are summer camps, RV parks, ski resorts, marinas, and other seasonally operated facilities. Some larger plants may have standby capacity or seasonal effluent limits so RBCs are treatment alternatives that are put into and taken out of service at various times of the year. To shut down an RBC for an extended period, use the following procedures.

1. Biofilm thickness on the RBC media should be reduced (or minimized) before extended or seasonal shutdowns to prevent accumulation of air-dried solids in the media or basin during the shutdown period. Reduction of biofilm can be accomplished in varying degrees as follows:

   - Stop wastewater flow to the RBC while continuing rotation of the RBC in idle wastewater. The process will eventually deplete nutrients in the wastewater and the biofilm will starve and slough off.
   - Stop wastewater flow to the RBC, shut off the RBC, drain the tank, fill the basin with "clean" water (or secondary effluent), and restart the RBC. The biofilm will starve and eventually slough off.
   - Use chlorine (maintain a chlorine residual of 4 mg/*L*) to strip the biomass or add sodium hydroxide (maintain a pH of 10 to 12) water in the basin to chemically kill the biofilm and produce sloughing. Chemical cleaning is normally not indicated for routine seasonal shutdowns.

   Use of water is more effective than use of wastewater during cleaning. The effectiveness of the biomass reduction procedure can be assessed visually or determined with load cell equipment (see the Appendix at the end of this chapter). **In all cases, intermittent scour with supplemental aeration will accelerate biomass removal.**

NOTE: Under no circumstances should an RBC be operated for more than a few revolutions if there is a normal wet biofilm growth on the media and the basin is not filled to normal operating level. RBC design relies on buoyancy of the tank liquid to support a portion of the loaded shaft.

2. After biomass reduction, drain or pump out the RBC tank, flush, wash, or otherwise remove any sludge or solids from the basin. The tank should remain empty and clean during duration of shutdown; do not allow an accumulation of incidental water (which might freeze) or debris.
3. Protect the RBC from sunlight.
4. During shutdown, the following precautions are recommended:
   - If in a dusty environment, cover drive and RBC shaft bearings with plastic or similar material to prevent dust and dirt accumulation.
   - Check lubrication of motor, reducer, and bearings on shutdown and every six months.
   - At one-month intervals, after removing dust cover material, using the drive, rotate the RBC for five minutes. While doing so, on three-month intervals, lubricate the RBC shaft bearings.
   - Maintain a rust-preventive coating on RBC shaft ends and bearings.
   - Refer to the manufacturer's O & M manual for lubrication and process information on equipment restart.

RBCs have the advantage of being convenient to take out of or put back into service. During initial start-up, it takes several weeks to reach biomass equilibrium because the new media tends to be very smooth and there is no existing biomass population. However, establishing a biomass on a previously operated RBC occurs more quickly. Accordingly, when taking an RBC out of service for a season, it is neither necessary nor advantageous to get back to squeaky clean media.

QUESTIONS

Write your answers in a notebook and then compare your answers with those on page 153.

11.1A Why should debris, grit, and suspended solids be removed before the wastewater being treated reaches the RBC unit?

11.1B List the major items for a pre-start check.

11.1C What are the main operational duties for an operator of an RBC unit?

11.2 MAINTENANCE

Rotating biological contactors have few moving parts and require minor amounts of preventive maintenance. Chain drives, belt drives, sprockets, rotating shafts, and any other moving parts should be inspected and maintained in accordance with manufacturers' instructions or your plant's O & M manual. All exposed parts, bearing housing, shaft ends, and bolts should be painted or covered with a layer of grease to prevent rust damage. Motors, speed reducers, and all other metal parts should be painted for protection. Do not paint over nameplate data.

Maintenance also includes the repair or replacement of broken parts. A preventive maintenance program that keeps equipment properly lubricated and adjusted to help reduce wear and breakage requires less time and money than a program that waits for breakdowns to occur before taking any action. The frequency of inspection and lubrication is usually provided by manufacturer's instructions and also may be found in the plant O & M manual. The following sections indicate a typical maintenance program for a rotating biological contactor treatment process. More details can be found in a plant O & M manual.

11.20 Break-In Maintenance

AFTER 8 HOURS OF OPERATION

1. Recheck tightening torque of capscrews in all split-tapered bushings in the drive package.
2. Visually inspect hubs and capscrews for general condition and possibility of rubbing against an obstruction.
3. Inspect belt drive (drive package) and tighten as needed.

AFTER 24 HOURS OF OPERATION

1. Inspect all chain drives.

AFTER 40 HOURS OF OPERATION

1. Inspect all belt drives in drive packages.

AFTER 100 HOURS OF OPERATION

1. Change oil in speed reducer. Use manufacturer's recommended lubricants.
2. Clean magnetic drain plug in speed reducer. If a large amount of metal particles is found in the magnetic plug, open the gear box and look for chipped or broken gear teeth. If broken or damaged teeth are found, contact the engineer or contractor that is responsible for repair of equipment under warranty.
3. Check all capscrews in split-tapered bushings and setscrews in drive package output sprocket and bearing for tightness.
4. Inspect belt drive of drive package.

AFTER 3 WEEKS OF OPERATION

1. Change oil in chain casing. Be sure oil level is at or above the mark on the dipstick. Use manufacturer's recommended lubricants.

11.21 Preventive Maintenance Program

Interval	Procedure
Daily	1. Check for hot shaft and bearings. If hot, check bearing lubrication and alignment. Replace bearings, if temperature exceeds 200°F (93°C), if there are still problems.
Daily	2. Listen for unusual noises in shaft and bearings. Identify cause of noise and correct, if necessary.
4 wk.	3. Inspect all chain drives.
4 wk.	4. Inspect mainshaft bearings and drive bearings.
3 mo.	5. Grease the mainshaft bearings and drive bearings. Use manufacturer's recommended lubricants. Add grease slowly while shaft rotates. When grease begins to ooze from the housing, the bearings contain the correct amount of grease. Add six full strokes where bearings cannot be seen. Lubricant providers can supply a list of "or equal" products that meet specifications. If you use an alternate brand lubricant, have the provider give you a letter of assurance of an "or equal" product.
3 mo.	6. Check for unusual odors that may indicate overheated painted surfaces or hot wire insulation. Find the cause of any odors.
3 mo.	7. Change oil in chain casing. Use manufacturer's recommended lubricants. Be sure oil level is at or above the mark on the dipstick, or to a level that covers the lowest point on the shaft sprocket.
3 mo.	8. Inspect belt drive.
6 mo.	9. Apply a generous coating of general purpose grease to mainshaft stub ends, mainshaft bearings, and end collars.
6 mo.	10. Change oil in speed reducer. Use manufacturer's recommended lubricants. If speed reducer is a variable drive unit, operate unit through full range once per week to prevent locking of disc drive sheaves.
6 mo.	11. Clean magnetic drain plug in speed reducer. Check for excess metallic particles.
6 mo.	12. Purge the grease in the double-sealed shaft seals of the speed reducer by removing the plug located 180 degrees from the grease fitting on both the input and output seal cages. Pump grease into the seal cages and then replace the plug. Use manufacturer's recommended grease.
12 mo.	13. Grease motor bearings. Use manufacturer's recommended grease. To grease motor bearings, stop motor and remove drain plugs. Inject new grease with pressure gun until all old grease has been forced out of the bearing through the grease drain. Run motor until all excess grease has been expelled. This may require up to several hours running time for some motors. Replace drain plugs.

11.22 Housekeeping

Properly designed systems have sufficient turbulence to prevent solids or sloughed slime growths from settling out on the bottom of the bays. If grease balls appear on the water surface in the bays, they should be removed with a dip net or screen device.

If the media comes apart on welded media systems, squeeze the two unbonded sections together with a pair of pliers. Take another pair of pliers and force a heated nail through the media. The heat from the nail will melt the plastic and make a plastic weld between the two sections of media.

11.23 Troubleshooting Guide

11.230 Roller Chain Drive

Trouble	Probable Cause	Corrective Action
1. Noisy Drive	1a. Moving parts rub stationary parts.	1a. Tighten and align casing and chain. Remove dirt or other interfering matter.
	1b. Chain does not fit sprockets.	1b. Replace with correct parts.
	1c. Loose chain.	1c. Maintain a taut chain at all times.
	1d. Faulty lubrication.	1d. Lubricate properly.
	1e. Misalignment or improper assembly.	1e. Correct alignment and assembly of the drive.
	1f. Worn parts.	1f. Replace worn chain or bearings. Reverse worn sprockets before replacing.
2. Rapid Wear	2a. Faulty lubrication.	2a. Lubricate properly.
	2b. Loose or misaligned parts.	2b. Align and tighten entire drive.
3. Chain Climbs Sprockets	3a. Chain does not fit sprockets.	3a. Replace chain or sprockets.
	3b. Worn-out chain or worn sprockets.	3b. Replace chain. Reverse or replace sprockets.
	3c. Loose chain.	3c. Tighten.
4. Stiff Chain	4a. Faulty lubrication.	4a. Lubricate properly.
	4b. Rust or corrosion.	4b. Clean and lubricate.
	4c. Misalignment or improper assembly.	4c. Correct alignment and assembly of the drive.
	4d. Worn-out chain or worn sprockets.	4d. Replace chain. Reverse or replace sprockets.
5. Broken Chain or Sprockets	5a. Shock or overload.	5a. Repair or replace broken parts. Add more RBC units or remove baffles between stages 1 and 2.
	5b. Wrong size chain, or chain that does not fit sprockets.	5b. Replace chain. Reverse or replace sprockets.
	5c. Rust or corrosion.	5c. Replace parts. Correct corrosive conditions.
	5d. Misalignment.	5d. Correct alignment.
	5e. Interferences.	5e. Make sure no solids interfere between chain and sprocket teeth. Loosen chain if necessary for proper clearance over sprocket teeth.

11.231 Belt Drive

Trouble	Probable Cause	Corrective Action
1. Excessive edge wear	1. Misalignment or non-rigid centers.	1. Check alignment and/or reinforcement mounting.
2. Jacket wear on pressure-face side of belt tooth*	2. Bent flange.	2. Straighten flange.
3. Excessive jacket wear between belt teeth (exposed tension members)*	3. Excessive overload and/or excessive belt tightness.	3. Reduce installation tension and/or increase drive load-carrying capacity.
4. Cracks in Neoprene backing	4a. Excessive installation tension.	4a. Reduce installation tension.
	4b. Exposure to excessively low temp. (below –30°F or –35°C).	4b. Eliminate low temperature condition or consult factory for proper belt construction.
5. Softening of Neoprene backing	5. Exposure to excessive heat (+200°F or 93°C) and/or oil.	5. Eliminate high temperature and oil condition or consult factory for proper belt construction.
6. Tensile or tooth shear failure*	6a. Small or sub-minimum diameter pulley.	6a. Increase pulley diameter.
	6b. Belt too narrow.	6b. Increase belt width.
7. Excessive pulley tooth wear (on pressure face and/or OD)*	7a. Excessive overload and/or excessive belt tightness.	7a. Reduce installation tension and/or increase drive load-carrying capacity.
	7b. Insufficient hardness of pulley material.	7b. Surface-harden pulley or use harder material.
8. Unmounting of flange	8a. Incorrect flange installation.	8a. Reinstall flange correctly.
	8b. Misalignment.	8b. Correct alignment.
9. Excessive drive noise	9a. Misalignment.	9a. Correct alignment.
	9b. Excessive installation tension.	9b. Reduce tension.
	9c. Sub-minimum pulley diameter.	9c. Increase pulley diameter.
10. Tooth shear*	10a. Less than six teeth in mesh (TIM).	10a. Increase TIM or use next smaller pitch.
	10b. Excessive load.	10b. Increase drive load-carrying capacity.
11. Apparent belt strength	11. Reduction of center distance or non-rigid mounting.	11. Re-tension drive and/or reinforce mounting.
12. Cracks or premature wear at belt tooth root*	12. Improper pulley groove top radius.	12. Regroove or install new pulley.
13. Tensile break	13a. Excessive load.	13a. Increase load-carrying capacity of drive.
	13b. Sub-minimum pulley diameter.	13b. Increase pulley diameter.

* Pertains to a timing belt system only. Recent systems use a V-belt drive.

11.232 V-Belt Drive

Trouble	Probable Cause	Corrective Action
1. Short belt life	1a. Spin burns from belt slipping on driver sheave under stalled load conditions or when starting.	1a. Tension belts.
	1b. Gouges or extreme cover wear caused by belts rubbing on drive.	1b. Eliminate obstruction or realign drive to provide clearance.
	1c. High ambient temperature.	1c. Use gripnotch belts. Provide ventilation. Shield belts.
	1d. Grease or oil on belts.	1d. Check for leaky bearings. Clean belts and sheaves.
	1e. Worn sheaves.	1e. Replace sheaves.
2. Belts turn over in grooves	2a. Damaged cord section in belts. Frayed or gouged belts.	2a. Replace belts.
	2b. Excessive vibration.	2b. Tension belts. Replace belts if damaged.
	2c. Flat idler pulley misaligned.	2c. Realign idler.
	2d. Worn sheaves.	2d. Replace sheaves.
	2e. Sheave misalignment.	2e. Realign drive.

11.232 V-Belt Drive (continued)

Trouble	Probable Cause	Corrective Action
3. Belts squeal	3a. High starting load. Belts not tensioned properly. Excessive overload.	3a. Tension drive or redesign and replace drive.
	3b. Insufficient arc of contact.	3b. Increase center distance or use gripnotch belts.
4. Belt breakage	4a. Foreign material in drive.	4a. Provide drive guard.
	4b. Belts damaged during installation.	4b. Follow V-drive manufacturer's installation instructions.
	4c. Shock or extreme overload.	4c. Eliminate overload cause or redesign drive.
5. Belt stretch beyond take-up	5a. Worn sheaves.	5a. Replace sheaves.
	5b. Under-designed drive.	5b. Redesign and replace drive.
	5c. Take-up slipped.	5c. Reposition take-up.
	5d. Drive excessively tensioned.	5d. Properly tension drive.
	5e. Damaged cord section during installation.	5e. Replace belts and properly install.
6. Excessive vibration	6a. Damaged belt cord section.	6a. Replace belts.
	6b. Loose belts.	6b. Tension drive.
	6c. Belts improperly tensioned.	6c. Tension drive with slack of each belt on the same side of drive.
7. Belts too long at installation	7a. Insufficient take-up.	7a. Use shorter belts.
	7b. Drive improperly set up.	7b. Recheck driver and driven machine setup.
	7c. Wrong size belts.	7c. Use correct size belts.
8. Belts too short at installation	8a. Insufficient take-up.	8a. Use longer belts.
	8b. Drive improperly set up.	8b. Recheck driver and driven machine setup.
	8c. Wrong size belts.	8c. Use correct size belts.
9. Belts mismatched at installation	9a. Belts matched by code number only.	9a. Replace belts with machine-matched belts.
	9b. Old belts and new belts used together on same drive.	9b. Replace with new belts.
	9c. Different brand name belts used on the same drive.	9c. Replace with set of machine-matched belts.
	9d. Driver and driven shafts not parallel.	9d. Realign drive.
	9e. Worn sheaves.	9e. Replace sheaves.
10. Belts mismatched after service	10a. Belts improperly tensioned, causing more stretch of some belts than others.	10a. Replace belts and tension drive with slack of each belt on the same side of the drive.
	10b. Old belts and new belts used together on the same drive.	10b. Replace with new belts.
	10c. Different brand name belts used on the same drive.	10c. Replace with set of machine-matched belts.
	10d. Driver and driven shafts shifted from parallel.	10d. Realign drive.
	10e. Belt cord section damaged during installation.	10e. Replace belts and install properly.

11.233 Bearings and Motors

Trouble	Probable Cause	Corrective Action
1. Shaft bearings running hot or failing. If temperatures exceed 200°F (93°C), the bearings may need to be replaced.	1a. Inadequate lubrication.	1a. 1. Lubricate bearings per manufacturer's instructions. 2. Check tightness (torque) and alignment of bearings.
	1b. Shaft misalignment.	1b. Align shaft properly. The longer the shaft, the more critical the alignment.
2. Motors running hot.	2a. Inadequate maintenance.	2a. 1. Lubricate per manufacturer's instructions. 2. Maintain correct oil level and oil viscosity in speed reducer.
	2b. Improper chain drive alignment.	2b. Align properly.
	2c. Excessive ambient temperature.	2c. Control/lower ambient temperature.

11.234 RBC Shaft

Trouble	Probable Cause	Corrective Action
1. Noisy bearing	1a. Rollers or bearing race damaged.	1a. Replace bearing cartridge.
	1b. No grease between bearing cartridge housing and pillow block housing.	1b. Lubricate outside diameter of cartridge housing.
2. Hot bearing	2a. No grease in bearing.	2a. See item 4 below.
	2b. Bearing rollers and/or cups and cones damaged.	2b. Replace bearing cartridge assembly.
3. Bearing leaks grease	3. Grease escapes between the seal and locking collar when lubricating the bearing or when grease expands due to an increase in ambient temperature.	3. NONE—This is normal.
4. Excessive grease usage (more than 4 oz. of grease is required to lubricate the bearing)	4. Worn or damaged seals and/or seal lip contact area on bearing.	4. Replace seals. Inspect bearing rollers, races, and seal lip contact area. If worn or damaged, replace bearing cartridge assembly.
5. Evidence of water in bearing (purged grease is milky or rusty)	5a. Water trapped between locking collar and seal.	5a. Wipe clean and relubricate.
	5b. Worn or damaged seals and/or seal lip contact area on bearings.	5b. Replace seals. Inspect bearing rollers, races, and seal lip contact area. If worn or damaged, replace bearing cartridge assembly.
6. Excessive shaft weight and biomass thickness	6. Organic loading too high.	6. Decrease organic loading.

11.3 SAFETY

Any equipment with moving parts or electrical components should be considered a potential safety hazard. *ALWAYS SHUT OFF THE POWER TO UNIT, TAG THE SWITCH, AND LOCK THE MAIN BREAKER IN THE "OFF" POSITION BEFORE WORKING ON A UNIT.*

11.30 Slow-Moving Equipment

Slow-moving equipment does not appear dangerous. Unfortunately, moving parts such as the sprockets, chains, belts, and sheaves can cause serious injury by tearing and/or crushing your hands or legs.

11.31 Wiring and Connections

Wiring and connections should be inspected regularly for potential hazards such as loose connections and bare wires. Again, always shut off, tag, and lock out the main breaker before working on a unit.

11.32 Slippery Surfaces

Caution must be taken on slippery surfaces. Falls can result in serious injuries. Any spilled oil or grease must be cleaned up immediately. If covers over the media allow sufficient space for walkways, condensed moisture on surfaces will create slippery places. If the temperature of the air within the enclosure can be kept several degrees above the temperature of the wastewater, condensation is significantly reduced. This condensation cannot be avoided completely so walk carefully at all times. The application of a nonskid material on walkways can greatly reduce the potential for slips and falls.

11.33 Infections and Diseases

Precautions must be taken to prevent infections in cuts or open wounds and illnesses from waterborne diseases. After working on a unit, always wash your hands before smoking or eating. *GOOD PERSONAL HYGIENE MUST BE PRACTICED BY ALL OPERATORS AT ALL TIMES.*

11.4 REVIEW OF PLANS AND SPECIFICATIONS

When reviewing plans and specifications, be sure the following items are included in the design of rotating biological contactors.

1. Enclosure to protect biomass from freezing temperature. Enclosure constructed of suitable corrosion-resistant materials and has windows or louvered structures in sides for ventilation. Forced ventilation is not necessary.
2. Heating. A source of heat is helpful during winter operation to minimize the corrosion caused by condensation and to improve operator comfort. If the temperature of the air within the enclosure is kept several degrees above the temperature of the wastewater, condensation is significantly reduced. Ceilings should be kept low to effectively use available heat.
3. Recirculation. Provisions for recirculating secondary effluent and also secondary sludge will allow for flexibility and can improve performance.
4. Supplemental aeration. Better RBC media systems are designed to accommodate the benefits of air scour and biomass control afforded by supplemental aeration. Consider installation of headers and diffusers with future blower hookup if the cost of the blower is not warranted during initial installation.
5. Provisions should be made to catch and hold oil that is drained from the gear boxes and other units.

QUESTIONS

Write your answers in a notebook and then compare your answers with those on page 153.

11.2A How often should maintenance be performed on an RBC unit during start-up?

11.3A List some possible safety hazards operators encounter when working around RBC units.

11.5 LOADING CALCULATIONS

11.50 Typical Loading Rates

Hydraulic and organic loadings on rotating biological contactors depend on influent soluble BOD, effluent total and soluble BOD requirements, and wastewater temperature. The loading rates given here are for "typical" values and values for your plant could be different.

Characteristic	**Range**
Hydraulic Loading	
BOD Removal	1.5 to 6 GPD/sq ft (60 to 240 liters per day/sq m)
Nitrogen Removal	1.5 to 1.8 GPD/sq ft (60 to 72 liters per day/sq m)
Organic Loading	
Soluble BOD	2.5 to 4 lbs BOD/day/1,000 sq ft (12 to 20 kg BOD/day/1,000 sq m)
Total BOD	6 to 8 lbs BOD/day/1,000 sq ft (30 to 40 kg BOD/day/1,000 sq m)

11.51 Computing Hydraulic Loading

Hydraulic loading on a rotating biological contactor is the amount (gallons) of wastewater per day that flows past the rotating media. To calculate the hydraulic loading, we must know:

1. Gallons per day treated by the rotating biological contactor, and
2. The surface area of the media in square feet.

EXAMPLE 1

A rotating biological contactor treats a flow of 3.5 MGD. The surface area of the media is 1,000,000 square feet (provided by manufacturer, see plant O & M manual). What is the hydraulic loading in GPD/sq ft?

Known		**Unknown**
Flow, MGD	= 3.5 MGD	Hydraulic Loading, GPD/sq ft
Surface Area, sq ft	= 1,000,000 sq ft	

Calculate the hydraulic loading in gallons per day of wastewater per square foot of surface media.

$$\text{Hydraulic Loading, GPD/sq ft} = \frac{\text{Flow, gal/day}}{\text{Surface Area, sq ft}}$$

$$= \frac{3{,}500{,}000 \text{ GPD}}{1{,}000{,}000 \text{ sq ft}}$$

$$= 3.5 \text{ GPD/sq ft}$$

Or

$$\text{Hydraulic Loading, GPD/sq ft} = 3{,}500{,}000 \text{ GPD} \div 1000{,}000 \text{ sq ft}$$

$$= 3.5 \text{ GPD/sq ft}$$

11.52 Computing Organic (BOD) Loading

Organic loadings on rotating biological contactors are based on soluble BOD. Soluble BOD is the BOD of water that has been filtered in the standard suspended solids test. If soluble BOD information is not available, soluble BOD may be estimated on the basis of the total BOD and the suspended solids as follows:

$$\text{Soluble BOD, mg/}L = \text{Total BOD, mg/}L - \text{Suspended BOD, mg/}L$$

where

$$\text{Suspended BOD, mg/}L = \text{K} \times \text{Suspended Solids, mg/}L$$

Therefore,

$$\text{Soluble BOD, mg/}L = \text{Total BOD, mg/}L - (\text{K} \times \text{Suspended Solids, mg/}L)$$

where

K = 0.5 to 0.7 for most domestic wastewaters.

EXAMPLE 2

The rotating biological contactor in Example 1 treats an influent with a total BOD of 200 mg/*L* and suspended solids of 250 mg/*L*. Assume a K value of 0.5 to calculate the soluble

BOD. What is the organic loading in pounds of soluble BOD per day per 1,000 square feet of media surface?

Known		Unknown
Flow, MGD	= 3.5 MGD	Organic Loading, lbs BOD/day/1,000 sq ft
Surface Area, sq ft	= 1,000,000 sq ft	
Total BOD, mg/*L*	= 200 mg/*L*	
SS, mg/*L*	= 250 mg/*L*	
K	= 0.5	

1. Estimate the soluble BOD treated by the rotating biological contactor.

$$\text{Soluble BOD, mg}/L = \text{Total BOD, mg}/L - (\text{K} \times \text{Suspended Solids, mg}/L)$$

$$= 200\text{ mg}/L - (0.5 \times 250\text{ mg}/L)$$

$$= 200\text{ mg}/L - 125\text{ mg}/L$$

$$= 75\text{ mg}/L$$

2. Determine the soluble BOD applied to the rotating biological contactor in pounds of soluble BOD per day.

$$\text{BOD Applied, lbs BOD/day} = (\text{Soluble BOD, mg}/L)(\text{Flow, MGD})(8.34\text{ lbs/gal})$$

$$= (75\text{ mg}/L)(3.5\text{ MGD})(8.34\text{ lbs/gal})$$

$$= 2{,}189\text{ lbs Soluble BOD/day}$$

Or

$$\text{BOD Applied, lbs/day} = 75\text{ mg}/L \times 3.5\text{ MGD} \times 8.34\text{ lbs/gal}$$

$$= 2{,}189\text{ lbs Soluble BOD/day}$$

3. Calculate the organic loading in pounds of soluble BOD per day per 1,000 square feet of media surface. Since the media surface area is 1,000,000 square feet, there are 1,000 • 1,000 square feet of surface area.

$$\text{Organic BOD Loading, lbs BOD/day/1,000 sq ft} = \frac{\text{Soluble BOD Applied, lbs BOD/day}}{\text{Surface Area of Media (in 1,000 sq ft)}}$$

$$= \frac{2{,}189\text{ lbs Soluble BOD/day}}{1{,}000 \bullet 1{,}000\text{ sq ft}}$$

$$= 2.2\text{ lbs BOD/day/1,000 sq ft}$$

Or

$$\text{Organic BOD Loading, lbs BOD/day/1,000 sq ft} = 2{,}189\text{ lbs Soluble BOD/day} \div 1{,}000\text{ sq ft}$$

$$= 2.2\text{ lbs Soluble BOD/day/1,000 sq ft}$$

11.53 Typical Loading Rates (Metric)

The next three sections show typical loading rates for rotating biological contactors and how to calculate the hydraulic and organic loading using the metric system.

Characteristic	Range
Hydraulic Loading	
BOD Removal	0.06 to 0.24 cu m/day/sq m
Nitrogen Removal	0.06 to 0.07 cu m/day/sq m
Organic Loading	
Soluble BOD	15 to 25 gm BOD/day/sq m
Total BOD	36 to 50 gm BOD/day/sq m

11.54 Computing Hydraulic Loading (Metric)

Hydraulic loading on a rotating biological contactor is the amount (cubic meters) of wastewater per day that flows past the rotating media.

EXAMPLE 3

A rotating biological contactor treats a flow of 15,000 cubic meters per day. The surface area of the media is 100,000 square meters (provided by the manufacturer). What is the hydraulic loading in cu m/day/sq m?

Known		Unknown
Flow, cu m/day	= 15,000 cu m/day	Hydraulic Loading, cu m/day/sq m
Surface Area, sq m	= 100,000 sq m	

Calculate the hydraulic loading in cubic meters per day of wastewater per square meter of surface media.

$$\text{Hydraulic Loading, cu m/day/sq m} = \frac{\text{Flow, cu m/day}}{\text{Surface Area, sq m}}$$

$$= \frac{15{,}000\text{ cu m/day}}{100{,}000\text{ sq m}}$$

$$= 0.15\text{ cu m/day/sq m}$$

11.55 Computing Organic (BOD) Loading (Metric)

EXAMPLE 4

The rotating biological contactor in Example 3 treats an influent with a soluble BOD of 75 mg/*L*. What is the organic loading in grams of soluble BOD per day per square meter of media surface?

Known		Unknown
Flow, cu m/day	= 15,000 cu m/day	Organic Loading, gm BOD/day/sq m
Surface Area, sq m	= 100,000 sq m	
Soluble BOD, mg/*L*	= 75 mg/*L*	

1. Determine the soluble BOD applied to the rotating biological contactor in grams of soluble BOD per day.

$$\text{BOD Applied, gm/day} = (\text{BOD, mg}/L)(\text{Flow, cu m/day})(1{,}000\text{ kg/cu m})(1{,}000\text{ gm/kg})$$

$$= \frac{75\text{ mg}}{1{,}000{,}000\text{ mg}} \times \frac{15{,}000\text{ cu m}}{\text{day}} \times \frac{1{,}000\text{ kg}}{\text{cu m}} \times \frac{1{,}000\text{ gm}}{\text{kg}}$$

$$= 1{,}125{,}000\text{ gm BOD/day}$$

Or

$$\text{BOD Applied, gm/day} = 75\text{ mg} \times 15{,}000\text{ cu m/day} \times 1{,}000\text{ kg/cu m} \times 1{,}000\text{ gm/kg} \div 1{,}000{,}000\text{ mg}$$

$$= 1{,}125{,}000\text{ gm BOD/day}$$

2. Calculate the organic loading in grams of soluble BOD per day per square meter of media surface.

$$\text{Organic BOD Loading, gm BOD/day/sq m} = \frac{\text{Soluble BOD Applied, gm BOD/day}}{\text{Surface Area of Media, sq m}}$$

$$= \frac{1{,}125{,}000\text{ gm Soluble BOD/day}}{100{,}000\text{ sq m}}$$

$$= 11.2\text{ gm BOD/day/sq m}$$

QUESTIONS

Write your answers in a notebook and then compare your answers with those on pages 153 and 154.

A rotating biological contactor treats a flow of 2.0 MGD with an influent soluble BOD of 100 mg/*L*. The total surface area of the media is 500,000 square feet.

11.5A What is the hydraulic loading in gallons per day per square foot of media surface?

11.5B What is the organic loading in pounds of soluble BOD per day per 1,000 square feet of surface area?

11.5C If the first stage surface area is 250,000 square feet, what is the soluble BOD loading on the first stage only?

11.6 ARITHMETIC ASSIGNMENT

Turn to the Arithmetic Appendix at the back of this manual and work all the problems in Section A.22, *ROTATING BIOLOGICAL CONTACTORS*. Check the arithmetic with your calculator.

11.7 ACKNOWLEDGMENTS

Jerry Greene, Autotrol Corporation, reviewed this section and provided helpful suggestions, photographs, drawings, and procedures. His contributions are greatly appreciated.

Mark Gehring with US Filter, John R. Harrison, and Mark Lambert also reviewed this revised chapter and their comments and suggestions are sincerely appreciated.

11.8 ADDITIONAL READING

1. *MOP 11*, Chapter 21, "Rotating Biological Contactors."*
2. *NEW YORK MANUAL*, Chapter 5, "Secondary Treatment."*
3. *AEROBIC FIXED-GROWTH REACTORS*. Obtain from Water Environment Federation (WEF), Publications Order Department, 601 Wythe Street, Alexandria, VA 22314-1994. Order No. P09107. Price to members, $62.00; nonmembers, $92.00; plus shipping and handling.

* Depends on edition.

11.9 METRIC CALCULATIONS

Refer to Section 11.5, "Loading Calculations," for metric calculations.

END OF CHAPTER 11
on
ROTATING BIOLOGICAL CONTACTORS

Please answer the discussion and review questions before continuing with the Objective Test.

DISCUSSION AND REVIEW QUESTIONS

Chapter 11. ROTATING BIOLOGICAL CONTACTORS

Write the answers to these questions in your notebook before continuing with the Objective Test on page 154. The purpose of these questions is to indicate to you how well you understand the material in the chapter.

1. Describe the rotating biological contactor (RBC) process and discuss how it works.
2. What RBC equipment would you inspect on a regular basis and how would you do it for an RBC unit?
3. What water quality indicators would you test for in the effluent from an RBC treatment plant?
4. How do the slime growths (biomass) on the plastic media look under (a) normal conditions, and (b) abnormal conditions?
5. What factors can cause a decrease in the process efficiency of an RBC unit and how can these problems be corrected?

SUGGESTED ANSWERS

Chapter 11. ROTATING BIOLOGICAL CONTACTORS

Answers to questions on page 136.

11.0A A rotating biological contactor treats wastewater by allowing the slime growths to develop on the surface of plastic media. The slime growths contain organisms that remove the organic materials from the wastewater.

11.0B The RBC process is divided into multiple stages to increase the effectiveness of a given amount of media surface area. Organisms on the first-stage media are exposed to high levels of BOD and reduce the BOD at a high rate. As the BOD levels decrease from stage to stage, the rate at which the organisms can remove BOD decreases and nitrification starts.

11.0C The purpose of a cover over the RBC unit is to protect organisms from severe changes in weather, especially freezing. Other reasons include preventing heavy rains from washing off slime and protecting operators from weather while maintaining equipment. Covers also contain odors and prevent the growth of algae on the media.

11.0D The three types of drive units are:

1. Motor with chain drive,
2. Motor with direct shaft drive, and
3. Air drive.

Answers to questions on page 144.

11.1A Debris, grit, and suspended solids should be removed to prevent sludge deposits forming beneath the drum and the tank bottom. These sludge deposits can reduce the effective tank volume, produce septic conditions, scrape the slimes from the media, and possibly stall the unit.

11.1B The major items on a pre-start check include:

1. Tightness of bolts and parts,
2. Lubrication of equipment,
3. Clearances for moving parts,
4. Drive unit alignment to drum shaft,
5. Flow control equipment,
6. Electrical controls, and
7. Proper installation of safety guards.

11.1C The main operational duties for an operator of an RBC unit include:

1. Inspecting equipment,
2. Testing influent and effluent,
3. Observing the media,
4. Maintaining the equipment, and
5. Correcting any problems.

Answers to questions on page 150.

11.2A During start-up, maintenance should be performed on an RBC unit:

1. After 8 hours of operation,
2. After 24 hours of operation,
3. After 40 hours of operation,
4. After 100 hours of operation, and
5. After 3 weeks of operation.

11.3A Possible safety hazards to operators working around RBC units include:

1. Moving parts and equipment,
2. Electrical power,
3. Bare electrical wires and loose connections,
4. Slippery surfaces,
5. Infections in cuts and open wounds, and
6. Illnesses from waterborne diseases.

Answers to questions on page 152.

11.5A, 11.5B, and 11.5C

A rotating biological contactor treats a flow of 2.0 MGD with an influent soluble BOD of 100 mg/*L*. The total surface area of the media is 500,000 square feet.

Known		Unknown	
Flow, MGD	= 2.0 MGD	7.5A	Hydraulic Loading, GPD/sq ft
Soluble BOD, mg/*L*	= 100 mg/*L*	7.5B	Organic Loading, lbs BOD/day/1,000 sq ft
Surface Area, sq ft	= 500,000 sq ft	7.5C	First Stage BOD Loading, lbs BOD/day/1,000 sq ft
Surface Area, sq ft (First Stage)	= 250,000 sq ft		

11.5A Calculate the hydraulic loading in gallons per day of wastewater per square foot of media surface.

$$\text{Hydraulic Loading, GPD/sq ft} = \frac{\text{Flow, gal/day}}{\text{Surface Area, sq ft}}$$

$$= \frac{2{,}000{,}000 \text{ gal/day}}{500{,}000 \text{ sq ft}}$$

$$= 4 \text{ GPD/sq ft}$$

11.5B 1. Determine the soluble BOD applied to the rotating biological contactor in pounds of soluble BOD per day.

$$\text{BOD Applied, lbs BOD/day} = (\text{Soluble BOD, mg}/L)(\text{Flow, MGD})(8.34 \text{ lbs/gal})$$

$$= (100 \text{ mg}/L)(2.0 \text{ MGD})(8.34 \text{ lbs/gal})$$

$$= 1{,}668 \text{ lbs Soluble BOD/day}$$

Or

$$\text{BOD Applied, lbs BOD/day} = 100 \text{ mg}/L \times 2.0 \text{ MGD} \times 8.34 \text{ lbs/gal}$$

$$= 1{,}668 \text{ lbs Soluble BOD/day}$$

2. Calculate the organic loading in pounds of soluble BOD per day per 1,000 square feet of media surface.

$$\text{Organic BOD Loading, lbs BOD/day/1,000 sq ft} = \frac{\text{Soluble BOD Applied, lbs BOD/day}}{\text{Surface Area of Media (in 1,000 sq ft)}}$$

$$= \frac{\text{1,668 lbs Soluble BOD/day}}{500 \bullet \text{1,000 sq ft}}$$

$$= \text{3.3 lbs BOD/day/1,000 sq ft}$$

Or

$$\text{Organic BOD Loading, lbs BOD/day/1,000 sq ft} = \text{1,668 lbs Soluble BOD/day} \; \square \; \text{500 sq ft}$$

$$= \text{3.3 lbs BOD/day/1,000 sq ft}$$

11.5C Calculate the organic loading on the first stage in pounds of soluble BOD per day per 1,000 square feet of first stage surface area.

$$\text{Organic BOD Loading, lbs BOD/day/1,000 sq ft} = \frac{\text{Soluble BOD Applied, lbs BOD/day}}{\text{First Stage Surface Area (in 1,000 sq ft)}}$$

$$= \frac{\text{1,668 lbs Soluble BOD/day}}{250 \bullet \text{1,000 sq ft}}$$

$$= \text{6.6 lbs BOD/day/1,000 sq ft}$$

NOTE: Soluble BOD loading on first stage should be less than 4 lbs BOD/day/1,000 sq ft.

OBJECTIVE TEST

Chapter 11. ROTATING BIOLOGICAL CONTACTORS

Please write your name and mark the correct answers on the answer sheet, as directed at the end of Volume I, Chapter 1. There may be more than one correct answer to each multiple-choice question.

True-False

1. Organisms living in the slimes on RBCs use organic matter from the wastewater for food and dissolved oxygen from the air, thus removing wastes from the water being treated.
   1. True
   2. False
2. As the flows to a rotating biological contactor decrease, the percent of BOD removal decreases.
   1. True
   2. False
3. If the wastewater flow is perpendicular to the rotating media shaft, the media should be moving through the wastewater in the direction of flow.
   1. True
   2. False
4. Organisms living in slime growths on RBC media are present and treating the wastewater during and after high flows.
   1. True
   2. False
5. RBC process efficiency can be increased by washing the slimes off the media.
   1. True
   2. False
6. Always shut off power to a rotating biological contactor unit, tag the switch, and lock the main breaker in the OFF position before working on a unit.
   1. True
   2. False

Best Answer (Select only the closest or best answer.)

7. What makes the rotating biological contactors a simple process to operate?
   1. No algae to require DO and pH adjustments
   2. No recycling of effluent or sludge
   3. No sludge to return
   4. No sludge to waste
8. Why is staging used in the rotating biological contactor process?
   1. To keep the wastewater flowing uniformly
   2. To maximize the effectiveness of a given amount of media
   3. To minimize the settling of solids in each tank
   4. To prevent short-circuiting
9. What is the purpose of rotating biological contactor media?
   1. Control flow from one stage to next stage
   2. Hold wastewater being treated
   3. Protect organisms from weather
   4. Provide support for organisms
10. What operational problem could be created by a possible short-term solution to an organic overload problem by the installation of facilities to recycle effluent?
    1. Could require additional monitoring
    2. Effluent might be toxic
    3. Greater increase of any hydraulic overload
    4. May need to add more operational staff
11. How can snails be controlled on RBC media?
    1. Chlorination
    2. Decrease the pH to 4
    3. Scrape snails off media
    4. Wash snails off media

12. How do properly designed rotating biological contactor systems prevent solids or sloughed slime growths from settling out on the bottom of the bays?
    1. Coagulation and flocculation
    2. Precipitation
    3. Scrapers
    4. Turbulence

13. How can grease balls be removed from the water surface in RBC bays?
    1. Bioaugmentation
    2. DAFT unit
    3. Dip net
    4. Sprays

Multiple Choice (Select all correct answers.)

14. Why are rotating biological contactors covered?
    1. Avoid exposure of media to sunlight, which may cause the media to become brittle
    2. Prevent intense rains from washing off some of the slime growths
    3. Protect biological slime growths from freezing
    4. Provide protection for operators from sun, rain, snow, or wind while maintaining equipment
    5. Stop exposure of media to direct sunlight to prevent growth of algae

15. What problems can be caused by grit and large organic matter not removed by pretreatment? Formation of sludge deposits that
    1. Cause increase in pH
    2. Generate nitrogen gas
    3. Produce septic conditions
    4. Reduce effective tank volume
    5. Scrape slimes from media

16. Before starting any rotating biological contactor equipment or allowing any wastewater to enter the treatment process, which items should be checked?
    1. Clearances
    2. Effluent quality
    3. Lubrication
    4. Safety
    5. Tightness

17. What does an operator need to do to produce a good effluent from a rotating biological contactor treatment plant? Operator must properly and regularly perform the duties of
    1. Inspecting the equipment
    2. Maintaining the equipment
    3. Observing the media
    4. Taking corrective action when necessary
    5. Testing the influent and effluent

18. To determine if biological contactors are operating properly, which items should be measured or tested in the influent and effluent?
    1. BOD
    2. Coliforms
    3. pH
    4. Suspended solids
    5. Temperature

19. Which items are measured using composite samples?
    1. BOD
    2. DO
    3. pH
    4. Suspended solids
    5. Temperature

20. When dealing with nitrification, which items are very critical?
    1. Alkalinity
    2. BOD
    3. Coliforms
    4. pH
    5. Suspended solids

21. What could cause excess sloughing or increased loss of biomass from a rotating biological contactor media?
    1. Changes in slime growth reproduction dynamics
    2. Influent dissolved oxygen fluctuations
    3. Seasonal weather patterns
    4. Toxic or inhibitory substances
    5. Unusual variation in flow and/or organic loading

22. An RBC has decreased treatment efficiency due to an organic overload. What are possible solutions to this organic overload?
    1. Addition and/or operation of supplemental aeration
    2. Chlorinate influent
    3. Improve pretreatment of plant
    4. Place another RBC in service, if available
    5. Remove bulkhead between stages 1 and 2 for larger first stage

23. How can rust damage be prevented?
    1. Brush away any rust
    2. Cover with layer of grease
    3. Mop
    4. Paint
    5. Sweep

24. What could be the probable cause of rapid wear on a roller chain drive?
    1. Faulty lubrication
    2. Loose parts
    3. Misaligned parts
    4. Rust or corrosion
    5. Worn parts

25. What could be the probable cause of excessive drive noise on a belt drive?
    1. Excessive installation tension
    2. Inadequate lubrication
    3. Misalignment
    4. Sub-minimum pulley diameter
    5. Tooth shear

26. Which items should be checked when reviewing plans and specifications for rotating biological contactors?
    1. Enclosure constructed of suitable corrosion-resistant material
    2. Enclosure has windows or louvered structures in sides for ventilation
    3. Enclosure to protect biomass
    4. Provisions for recirculation to improve performance
    5. Source of heat to minimize corrosion caused by condensation in winter

END OF OBJECTIVE TEST

APPENDIX

QUANTITATIVE BIOMASS ASSESSMENT AND CORRECTIVE ACTIONS

(Adapted from operations manual, courtesy of US Filter, Envirex Products)

The shaft and media of a rotating biological contactor (RBC) are designed to carry biomass. It is possible that the design load carrying capacity of the shaft and media could be exceeded by excessive biomass growth, which may lead to shaft failure. The bearing load cell and hydraulic weighing equipment will provide a determination of the total weight on the shaft.

The operation of the RBC weighing apparatus is used in conjunction with the load cell bearing, which is located on the expansion bearing end (idle end) of the RBC unit. A load cell kit is used on an expansion-type bearing. **The kit is to be installed in the field prior to setting the bearings on the base plates**.

The RBC weighing apparatus consists of a hand-operated hydraulic pump, pressure gage, flexible hose, and fittings required to connect the hose to the quick-disconnect coupler attached to the bearing housing (Figure 11.13).

The bearing load cell is the most accurate method of determining biomass weight. By weighing the RBC in operation, shaft weight, media weight, biomass, and any entrapped water are measured. Data have been developed to convert a psi (pounds per square inch) reading obtained from the hydraulic weighing device to the total operating weight of the shaft.

SET-UP PROCEDURE

The RBC unit is weighed while it is rotating.

Loosen the nuts securing the bearing housing to the base plate until they can be turned by hand. Snug up each nut finger-tight, then back off one (1) turn. This limits the bearing movement.

Connect the hose to the coupler located on the bearing housing. With the relief valve open, stroke the pump a few times to purge air from the pump cylinder. Close the valve and proceed.

Apply pressure with slow, steady strokes of the hand pump. The pressure will rise steadily until the bearing is lifted off the base. With continued stroking of the pump, the pressure should remain constant until the bearing reaches the limiting nuts, after which it increases sharply. Stop pumping before the bearing reaches the nuts to prevent O-ring blow-by.

If the bearing is against the limiting nuts (nuts cannot be rotated by hand), release the pressure in small increments until the nuts can be rotated by hand.

PRESSURE READING

Observe the pressure required to keep the bearing floating between the bearing base and the limiting nuts.

Most RBCs are not perfectly balanced and a pressure swing will be observed during rotation. This pressure range should be recorded.

If the load cell bearing alternately bottoms out and hits the limiting nuts while the RBC is rotating, there is a large volume of air trapped in the pressure system. In this case, the extreme pressure readings are inaccurate and only the relatively stable mean pressure (average) should be recorded.

If one side of the bearing housing touches the limiting nuts before the other side, release the pressure and re-pressurize the system. (If, after several attempts, one side of the bearing continues to touch the limiting nuts before the other side, record the pressure reading and note this condition.)

If the load cell bearing settles as it rotates, this is due to a loss in system pressure. The pump must be continually stroked to keep the bearing floating to record an accurate pressure reading.

Pressure readings can be converted to RBC weight from the curve(s) supplied by the manufacturer. Note the load limitation(s).

RE-SETTING BEARING

Open relief valve slowly to lower bearing. Check bearing to ensure that it is level and properly aligned. Bearing housing should rest evenly on base plate. Bearing cartridge should be centered in housing.

Disconnect hose from coupler. Re-torque nuts securing bearing housing to base plate to 160 lbs/ft. (Threads must be clean, dry, and free from oil and grease.)

FREQUENCY FOR LOAD CELL READINGS

Readings should be taken twice a month and should become a part of permanent plant records. Overloading condition will require corrective action. Contact the RBC manufacturer if the following suggested corrective measures are not effective.

CONTINUAL OVERLOAD CONDITION MAY SHORTEN EQUIPMENT LIFE AND MAY DECREASE PROCESS EFFICIENCY. IN ADDITION, SUSTAINED OVERLOADED CONDITION WILL VOID THE EQUIPMENT WARRANTY. SHAFT, MEDIA, AND/OR SUPPORT FATIGUE FAILURES MAY OCCUR. MEDIA CLOGGING MAY ALSO RESULT.

CAUSES OF EXCESSIVE BIOMASS

- High stage of system SBOD loadings
- Extended stage detention times
- Cold wastewater temperature

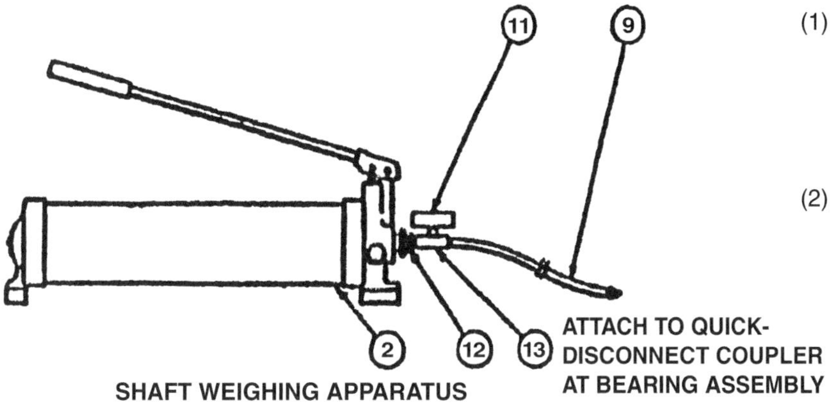

NOTES

(1) Field Note: Install “O” ring, quick-disconnect coupler, and piston before setting shaft on bearing base. Do not twist “O” ring when installing it on piston. Lubricate “O” ring with clean oil or light grease before assembling piston into bearing base.

(2) 1.9 liters (2 quarts) hydraulic oil required.

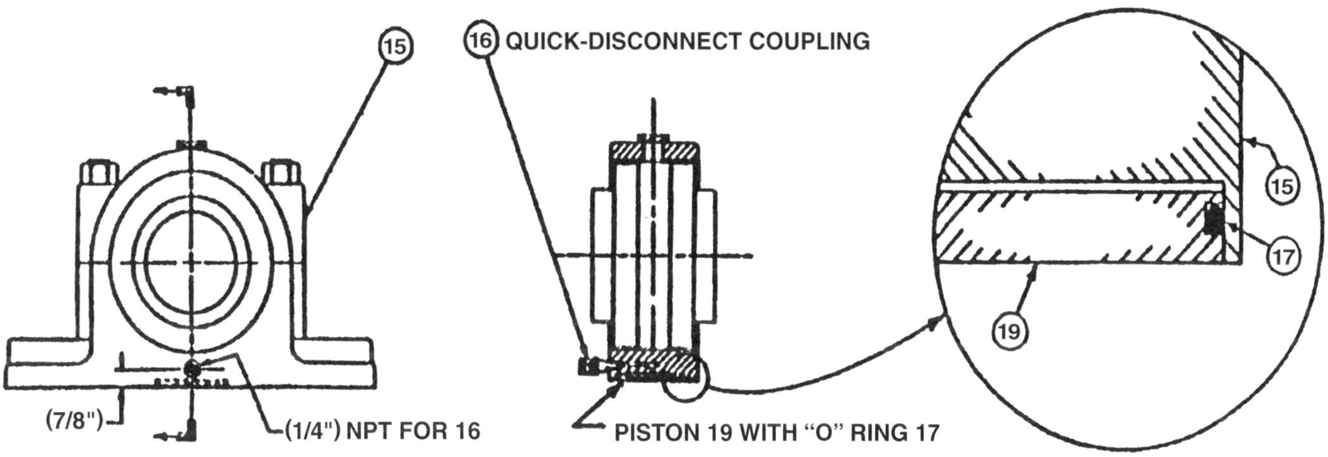

LEGEND

Item	Description
2	Pump, hydraulic
9	Hose with male coupler
11	Gage
12	Hex, nipple
13	Pipe, tee
15	Pillow block bearing
16	Quick-disconnect coupler with dust cap
17	“O” ring
19	Piston

Fig. 11.13 RBC shaft weighing apparatus installation diagram

(Courtesy of US Filter Envirex Products)

- Specific organic make-up of SBOD
- Precipitation of inorganic salts, for example, lime precipitates as $CaCO_3$
- Polymer coagulation (from upstream or side stream wastes, for example, belt press or centrifuge returns)
- De-emulsification of solvents carrying FOGs (fats, oil, and grease)

OVERLOAD CORRECTION

The most common causes for excessive shaft weight are:

1. Organic loading rate in excess of design, and
2. Improper staging of RBCs.

Total shaft weight reduction can be achieved by either changing the organic loading or increasing the shearing forces on the biomass. This can be accomplished by one of the following methods:

1. Changing the staging configuration,
2. Step feeding the influent,
3. Recirculation (organic load dilution),
4. Diffused aeration,
5. Starvation, or
6. Chemical stripping.

1. Changing the Staging Configuration

Increasing the media surface area in a given stage by changing the baffle location will reduce the overall organic load and subsequent biomass weight for that stage. This method is limited to installations with removable baffles and it should be remembered that increasing the area of a stage will either decrease the size or number of the following stages, which will change downstream loading, removal, and biomass.

2. Step Feeding

Feeding a portion of the RBC influent to other than the influent stages will have essentially the same effect as changing baffle configuration.

3. Recirculation

Organic load dilution can be affected by recirculation through the RBC(s) of RBC effluent mixed liquor or secondary clarifier effluent. Although organic loading is not decreased, the strength of the influent waste is decreased, which can have a positive influence on biomass thickness reduction in the initial stage(s). This method also has the effect of decreasing the size and number of following stages, which will change the downstream loading, removal, and biomass.

4. Diffused Aeration

This addition enhances the biological process while stripping biomass. Most effective when applied intermittently at high rates, for example, 10 to 12 cfm/ft of RBC media length. This tool represents a long-term solution for controlling periods of excessive biomass growth.

5. Starvation

Requires RBC be taken out of service so may be impractical on single- or two-shaft plant. Rate of decay is only about 10 percent that of biomass growth.

6. Chemical Stripping

Caustic most effective but dangerous to use and difficult disposal. RBC must be taken out of service and treatment, particularly nitrification, will not recover for weeks.

Combinations of the above approaches may be tried until a proven method can be selected. In any event, this procedure should allow the operator to control biofilm to maintain a relatively thin, uniform biofilm layer. Failure to use this method or some other method of positive biofilm control may have undesirable consequences, in some cases.

CHAPTER 12

DISINFECTION AND CHLORINATION

by

John Brady

Much of this text is taken from Chapter 10, "Disinfection and Chlorination," *OPERATION OF WASTEWATER TREATMENT PLANTS*, Volume I, by Leonard W. Hom, revised by Tom Ikesaki, with a special section by J. L. Beals.

TABLE OF CONTENTS

Chapter 12. DISINFECTION AND CHLORINATION

OBJECTIVES

Chapter 12. DISINFECTION AND CHLORINATION

Following completion of Chapter 12, you should be able to:

1. Explain the principles of wastewater disinfection and problems associated with wastewater disinfection,
2. Control the chlorination process to obtain the desired effluent disinfection,
3. Handle chlorine products safely,
4. Inspect new chlorination facilities for proper installation,
5. Schedule and conduct chlorination operation and maintenance duties,
6. Recognize factors that indicate the chlorination process is not performing properly, identify the source of the problem, and take corrective action,
7. Conduct your duties in a safe fashion,
8. Determine chlorine dosages,
9. Explain applications and limitations of uses of chlorine other than for disinfection, and
10. Keep records of chlorination operation.

NOTE: This chapter does not contain information on the safe operation and maintenance of gaseous chlorinators or sulfur dioxide dechlorinators. For information on these units, see *OPERATION OF WASTEWATER TREATMENT PLANTS,* Volume I, Chapter 10, "Disinfection and Chlorination."

WORDS

Chapter 12. DISINFECTION AND CHLORINATION

AMPEROMETRIC (am-PURR-o-MET-rick) AMPEROMETRIC

A method of measurement that records electric current flowing or generated, rather than recording voltage. Amperometric titration is a means of measuring concentrations of certain substances in water.

BACTERIA (back-TEAR-e-ah) BACTERIA

Bacteria are living organisms, microscopic in size, which usually consist of a single cell. Most bacteria use organic matter for their food and produce waste products as a result of their life processes.

BREAKOUT OF CHLORINE BREAKOUT OF CHLORINE

A point at which chlorine leaves solution as a gas because the chlorine feed rate is too high. The solution is saturated and cannot dissolve any more chlorine. The maximum strength a chlorine solution can attain is approximately 3,500 mg/*L*. Beyond this concentration molecular chlorine is present which will break out of solution and cause "off-gassing" at the point of application.

BREAKPOINT CHLORINATION BREAKPOINT CHLORINATION

Addition of chlorine to water or wastewater until the chlorine demand has been satisfied. At this point, further additions of chlorine result in a residual that is directly proportional to the amount of chlorine added beyond the breakpoint.

CHLORAMINES (KLOR-uh-means) CHLORAMINES

Compounds formed by the reaction of hypochlorous acid (or aqueous chlorine) with ammonia.

CHLORINATION (KLOR-uh-NAY-shun) CHLORINATION

The application of chlorine to water or wastewater, generally for the purpose of disinfection, but frequently for accomplishing other biological or chemical results.

CHLORINE CONTACT CHAMBER CHLORINE CONTACT CHAMBER

A baffled basin that provides sufficient detention time of chlorine contact with wastewater for disinfection to occur. The minimum contact time is usually 30 minutes. (Also commonly referred to as basin or tank.)

CHLORINE DEMAND CHLORINE DEMAND

Chlorine demand is the difference between the amount of chlorine added to wastewater and the amount of residual chlorine remaining after a given contact time. Chlorine demand may change with dosage, time, temperature, pH, and nature and amount of the impurities in the water.

Chlorine Demand, mg/*L* = Chlorine Applied, mg/*L* – Chlorine Residual, mg/*L*

CHLORINE REQUIREMENT CHLORINE REQUIREMENT

The amount of chlorine which is needed for a particular purpose. Some reasons for adding chlorine are reducing the number of coliform bacteria (Most Probable Number), obtaining a particular chlorine residual, or oxidizing some substance in the water. In each case a definite dosage of chlorine will be necessary. This dosage is the chlorine requirement.

CHLORINE RESIDUAL CHLORINE RESIDUAL

The concentration of chlorine present in water after the chlorine demand has been satisfied. The concentration is expressed in terms of the total chlorine residual, which includes both the free and combined or chemically bound chlorine residuals.

CHLORINE RESIDUAL ANALYZER CHLORINE RESIDUAL ANALYZER

An instrument used to measure chlorine residual in water or wastewater. This instrument also can be used to control the chlorine dose rate.

CHLORORGANIC (klor-or-GAN-ick) CHLORORGANIC

Organic compounds combined with chlorine. These compounds generally originate from, or are associated with, living or dead organic materials.

COLIFORM (COAL-i-form) COLIFORM

One type of bacteria. The presence of coliform-group bacteria is an indication of possible pathogenic bacterial contamination. The human intestinal tract is one of the main habitats of coliform bacteria. They may also be found in the intestinal tracts of warm-blooded animals, and in plants, soil, air, and the aquatic environment. Fecal coliforms are those coliforms found in the feces of various warm-blooded animals; whereas the term "coliform" also includes other environmental sources.

COMBINED AVAILABLE CHLORINE COMBINED AVAILABLE CHLORINE

The total chlorine, present as chloramine or other derivatives, that is present in a water and is still available for disinfection and for oxidation of organic matter. The combined chlorine compounds are more stable than free chlorine forms, but they are somewhat slower in disinfection action.

COMBINED AVAILABLE CHLORINE RESIDUAL COMBINED AVAILABLE CHLORINE RESIDUAL

The concentration of residual chlorine that is combined with ammonia, organic nitrogen, or both in water as a chloramine (or other chloro derivative) and yet is still available to oxidize organic matter and help kill bacteria.

COMBINED CHLORINE COMBINED CHLORINE

The sum of the chlorine species composed of free chlorine and ammonia, including monochloramine, dichloramine, and trichloramine (nitrogen trichloride). Dichloramine is the strongest disinfectant of these chlorine species, but it has less oxidative capacity than free chlorine.

DNA (pronounce as separate letters) DNA

Deoxyribo**n**ucleic **a**cid. A chemical that encodes genetic information that is transmitted between generations of cells.

DECHLORINATION (dee-KLOR-uh-NAY-shun) DECHLORINATION

The removal of chlorine from the effluent of a treatment plant. Chlorine needs to be removed because chlorine is toxic to fish and other aquatic life.

DISINFECTION (dis-in-FECT-shun) DISINFECTION

The process designed to kill or inactivate most microorganisms in wastewater, including essentially all pathogenic (disease-causing) bacteria. There are several ways to disinfect, with chlorination being the most frequently used in water and wastewater treatment plants. Compare with STERILIZATION.

EDUCTOR (e-DUCK-ter) EDUCTOR

A hydraulic device used to create a negative pressure (suction) by forcing a liquid through a restriction, such as a Venturi. An eductor or aspirator (the hydraulic device) may be used in the laboratory in place of a vacuum pump. As an injector, it is used to produce vacuum for chlorinators. Sometimes used instead of a suction pump.

ELECTROLYSIS (ee-leck-TRAWL-uh-sis) ELECTROLYSIS

The decomposition of material by an outside electric current.

END POINT END POINT

Samples of water or wastewater are titrated to the end point. This means that a chemical is added, drop by drop, to a sample until a certain color change (blue to clear, for example) occurs. This is called the *END POINT* of the titration. In addition to a color change, an end point may be reached by the formation of a precipitate or the reaching of a specified pH. An end point may be detected by the use of an electronic device such as a pH meter. The completion of a desired chemical reaction.

ENZYMES (EN-zimes) ENZYMES

Organic substances (produced by living organisms) which cause or speed up chemical reactions. Organic catalysts and/or biochemical catalysts.

FREE AVAILABLE CHLORINE FREE AVAILABLE CHLORINE

The amount of chlorine available in water. This chlorine may be in the form of dissolved gas (Cl_2), hypochlorous acid (HOCl), or hypochlorite ion (OCl^-), but does not include chlorine combined with an amine (ammonia or nitrogen) or other organic compound.

FOOT VALVE FOOT VALVE

A special type of check valve located at the bottom end of the suction pipe on a pump. This valve opens when the pump operates to allow water to enter the suction pipe but closes when the pump shuts off to prevent water from flowing out of the suction pipe.

GAS/LIQUID — GAS/LIQUID

Gaseous/Liquid. Gaseous/liquid chlorination refers to the fact that free chlorine is delivered to small treatment plants in containers (150-pound chlorine cylinders) that hold liquid chlorine with a free chlorine gas above the liquid in the container. The release of chlorine gas from the liquid chlorine surface depends on the temperature of the liquid and the pressure of the chlorine gas on the liquid surface. As chlorine gas is removed from the container, the gas pressure drops and more liquid chlorine becomes chlorine gas.

HEPATITIS (HEP-uh-TIE-tis) — HEPATITIS

Hepatitis is an inflammation of the liver caused by an acute viral infection. Yellow jaundice is one symptom of hepatitis.

HYPOCHLORINATION (HI-poe-KLOR-uh-NAY-shun) — HYPOCHLORINATION

The application of hypochlorite compounds to water or wastewater for the purpose of disinfection.

HYPOCHLORINATORS (HI-poe-KLOR-uh-NAY-tors) — HYPOCHLORINATORS

Chlorine pumps, chemical feed pumps or devices used to dispense chlorine solutions made from hypochlorites such as bleach (sodium hypochlorite) or calcium hypochlorite into the water being treated.

HYPOCHLORITE (HI-poe-KLOR-ite) — HYPOCHLORITE

Chemical compounds containing available chlorine; used for disinfection. They are available as liquids (bleach) or solids (powder, granules, and pellets) in barrels, drums, and cans. Salts of hypochlorous acid.

MPN (pronounce as separate letters) — MPN

MPN is the **M**ost **P**robable **N**umber of coliform-group organisms per unit volume of sample water. Expressed as a density or population of organisms per 100 m*L* of sample water.

MSDS — MSDS

See MATERIAL SAFETY DATA SHEET (MSDS).

MATERIAL SAFETY DATA SHEET (MSDS) — MATERIAL SAFETY DATA SHEET (MSDS)

A document which provides pertinent information and a profile of a particular hazardous substance or mixture. An MSDS is normally developed by the manufacturer or formulator of the hazardous substance or mixture. The MSDS is required to be made available to employees and operators whenever there is the likelihood of the hazardous substance or mixture being introduced into the workplace. Some manufacturers are preparing MSDSs for products that are not considered to be hazardous to show that the product or substance is *NOT* hazardous.

MICRON (MY-kron) — MICRON

µm, Micrometer or Micron. A unit of length. One millionth of a meter or one thousandth of a millimeter. One micron equals 0.00004 of an inch.

NITROGENOUS (nye-TRAH-jen-us) — NITROGENOUS

A term used to describe chemical compounds (usually organic) containing nitrogen in combined forms. Proteins and nitrate are nitrogenous compounds.

OXIDIZING AGENT — OXIDIZING AGENT

Any substance, such as oxygen (O_2) or chlorine (Cl_2), that will readily add (take on) electrons. When oxygen or chlorine is added to wastewater, organic substances are oxidized. These oxidized organic substances are more stable and less likely to give off odors or to contain disease-causing bacteria. The opposite is a REDUCING AGENT.

PARASITIC (PAIR-a-SIT-tick) BACTERIA — PARASITIC BACTERIA

Parasitic bacteria are those bacteria which normally live off another living organism, known as the "host."

PATHOGENIC (PATH-o-JEN-ick) ORGANISMS — PATHOGENIC ORGANISMS

Bacteria, viruses, cysts, or protozoa which can cause disease (giardiasis, cryptosporidiosis, typhoid, cholera, dysentery) in a host (such as a person). There are many types of organisms which do *NOT* cause disease and which are *NOT* called pathogenic. Many beneficial bacteria are found in wastewater treatment processes actively cleaning up organic wastes.

POSTCHLORINATION — POSTCHLORINATION

The addition of chlorine to the plant discharge or effluent, *FOLLOWING* plant treatment, for disinfection purposes.

POTABLE (POE-tuh-bull) WATER — POTABLE WATER

Water that does not contain objectionable pollution, contamination, minerals, or infective agents and is considered satisfactory for drinking.

PRECHLORINATION — PRECHLORINATION

The addition of chlorine in the collection system serving the plant or at the headworks of the plant *PRIOR TO* other treatment processes mainly for odor and corrosion control. Also applied to aid disinfection, to reduce plant BOD load, to aid in settling, to control foaming in Imhoff units and to help remove oil.

RNA (pronounce as separate letters) — RNA

Ribo**n**ucleic **a**cid. A chemical that provides the structure for protein synthesis (building up).

REAGENT (re-A-gent) — REAGENT

A pure chemical substance that is used to make new products or is used in chemical tests to measure, detect, or examine other substances.

REDUCING AGENT — REDUCING AGENT

Any substance, such as base metal (iron) or the sulfide ion (S^{2-}), that will readily donate (give up) electrons. The opposite is an OXIDIZING AGENT.

RESIDUAL ANALYZER, CHLORINE — RESIDUAL ANALYZER, CHLORINE

See CHLORINE RESIDUAL ANALYZER.

RESIDUAL CHLORINE — RESIDUAL CHLORINE

The concentration of chlorine present in water after the chlorine demand has been satisfied. The concentration is expressed in terms of the total chlorine residual, which includes both the free and combined or chemically bound chlorine residuals.

ROTAMETER (RODE-uh-ME-ter) — ROTAMETER

A device used to measure the flow rate of gases and liquids. The gas or liquid being measured flows vertically up a tapered, calibrated tube. Inside the tube is a small ball or bullet-shaped float (it may rotate) that rises or falls depending on the flow rate. The flow rate may be read on a scale behind or on the tube by looking at the middle of the ball or at the widest part or top of the float.

SAPROPHYTIC (SAP-row-FIT-ick) ORGANISMS — SAPROPHYTIC ORGANISMS

Organisms living on dead or decaying organic matter. They help natural decomposition of the organic solids in wastewater.

SEPTICITY (sep-TIS-it-tee) — SEPTICITY

Septicity is the condition in which organic matter decomposes to form foul-smelling products associated with the absence of free oxygen. If severe, the wastewater produces hydrogen sulfide, turns black, gives off foul odors, contains little or no dissolved oxygen, and the wastewater has a high oxygen demand.

SET POINT — SET POINT

The position at which the control or controller is set. This is the same as the desired value of the process variable. For example, a thermostat is set to maintain a desired temperature.

SPORE — SPORE

The reproductive body of an organism which is capable of giving rise to a new organism either directly or indirectly. A viable (able to live and grow) body regarded as the resting stage of an organism. A spore is usually more resistant to disinfectants and heat than most organisms. Gangrene and tetanus bacteria are common spore-forming organisms.

STERILIZATION (STARE-uh-luh-ZAY-shun) — STERILIZATION

The removal or destruction of all microorganisms, including pathogenic and other bacteria, vegetative forms and spores. Compare with DISINFECTION.

TITRATE (TIE-trate) — TITRATE

To *TITRATE* a sample, a chemical solution of known strength is added drop by drop until a certain color change, precipitate, or pH change in the sample is observed (end point). Titration is the process of adding the chemical reagent in small increments (0.1 – 1.0 milliliter) until completion of the reaction, as signaled by the end point.

TOTAL CHLORINE | TOTAL CHLORINE

The total concentration of chlorine in water, including the combined chlorine (such as inorganic and organic chloramines) and the free available chlorine.

TOTAL CHLORINE RESIDUAL | TOTAL CHLORINE RESIDUAL

The total amount of chlorine residual (value for residual chlorine, including both free chlorine and chemically bound chlorine) present in a water sample after a given contact time.

V-NOTCH WEIR | V-NOTCH WEIR

A triangular weir with a "V" notch calibrated in gallons per minute readings. The weir can be placed in a pipe or open channel. As the flow passes through the "V", the depth of water flowing over the weir can be measured and converted to a flow in gallons per minute.

CHAPTER 12. DISINFECTION AND CHLORINATION

(Lesson 1 of 3 Lessons)

12.0 NEED FOR DISINFECTION

12.00 Removal or Inactivation of Disease-Causing Organisms

Homes, hospitals, and industrial facilities all discharge liquid and solid waste materials into the wastewater system. Diseases from human discharges may be transmitted by wastewater. Typical waterborne disease-causing microorganisms include bacteria, viruses, and parasites. These microorganisms are commonly referred to as *PATHOGENIC* (disease-causing) *ORGANISMS*.[1] These microorganisms can cause the following types of illnesses and infections:

Bacteria

Salmonellosis
Shigellosis
Typhoid Fever
Cholera
Paratyphoid
Bacillary Dysentery
Anthrax
Gangrene
Tetanus

Internal Parasites

Amoebic Dysentery
Giardiasis
Cryptosporidium
Various Roundworms (nematodes)
Various Tapeworms (cestodes)

Viruses

Polio
Infectious Hepatitis

NOTE: To date there is no record of an operator becoming infected by the HIV virus or developing AIDS due to on-the-job duties.

Disease-producing microorganisms are potentially present in all wastewaters. These microorganisms must be removed or killed before treated wastewater can be discharged to the receiving waters and, in some instances, subsurface ground injection. The purpose of disinfection is to destroy or inactivate pathogenic microorganisms and thus prevent the spread of waterborne diseases.

Wastewater treatment processes usually remove some of the pathogenic microorganisms through the following processes:

1. Physical removal through sedimentation and filtration,
2. Natural die-away or die-off of microorganisms in an unfavorable environment, and
3. Destruction by chemicals introduced for treatment purposes.

The effectiveness of treatment processes in killing or removing microorganisms depends on the type of treatment process and the hydraulic and organic loadings. Typical levels of microorganism destruction or removal by various processes are summarized below:

Treatment Process	Microorganism Removal, %
Screening	10 – 20
Grit Channel	10 – 25
Primary Sedimentation and/or Septic Tank	25 – 75
Rotating Biological Contactors	90 – 95
Activated Sludge (SBRs)	90 – 98
Oxidation Lagoons	80 – 95
Gravel/Sand Filters, Single Pass	90 – 95
Recirculating Gravel/Sand Filters	95 – 99
Disinfection	98 – 99

Although pathogenic microorganisms are reduced in number by the various treatment processes listed and by natural die-away or die-off in unfavorable environments, many microorganisms still remain. To ensure that essentially all pathogenic microorganisms are destroyed in the effluents that are surface discharged from community wastewater treatment plants, disinfection is practiced.

In the case of a rural home using a septic tank with on-site subsurface effluent disposal, a specified distance must be maintained between the wastewater effluent disposal area and a water source such as a domestic water well, water

[1] *Pathogenic (PATH-o-JEN-ick) Organisms. Bacteria, viruses, cysts, or protozoa which can cause disease (giardiasis, cryptosporidiosis, typhoid, cholera, dysentery) in a host (such as a person). There are many types of organisms which do NOT cause disease and which are NOT called pathogenic. Many beneficial bacteria are found in wastewater treatment processes actively cleaning up organic wastes.*

supply line, stream, or lake. These distances are established by the local health or environmental control agency and are based on the type of soil and distance to the water table at the site to prevent the possibility of the spread of waterborne diseases. The purpose of established distances is to provide a zone (area) for filtering action as the wastewater passes through the soil.

12.01 Disinfection

Two terms you should understand are "disinfection" and "sterilization." *DISINFECTION* is the destruction or inactivation of pathogenic microorganisms, while *STERILIZATION* is the destruction of all microorganisms.

The main objective of disinfection is to prevent the spread of disease by protecting:

1. Public water supplies,
2. Receiving waters used for recreational purposes, and
3. Shellfish growing areas.

Disinfection is effective because pathogenic microorganisms are more sensitive to destruction by chemicals than nonpathogens. Many nonpathogenic microorganisms are classified as *SAPROPHYTES*[2] and are essential to wastewater treatment processes. Since chlorine gas or hypochlorite solutions are the most widely used chemical for disinfection, this chapter will deal primarily with the principles and practices of chlorine disinfection (Figures 12.1 and 12.2). Chlorination for disinfection purposes results in killing essentially all of the pathogens in the plant effluent. No attempt is made to *STERILIZE* wastewater because it is unnecessary and impractical. Many other sensitive organisms in contact with chlorine are destroyed too. Chlorine is a nonselective killer. Organisms are affected by chlorine on the basis of their sensitivity and growth rate and also on the concentration of chlorine and exposure time. This is why dechlorination is practiced to protect fish and other aquatic organisms from being harmed by chlorinated effluents that are discharged to receiving waters. For information on effluent dechlorination, see *OPERATION OF WASTEWATER TREATMENT PLANTS*, Volume I, Chapter 10, "Disinfection and Chlorination."

One of the main uses of chlorine in wastewater treatment is for disinfection. Chlorine is relatively easy to obtain and cheap to manufacture. Even at relatively low dosages, chlorine is extremely effective.

Today, people are living in closer contact with wastewater than ever before. Wastewater effluent may be used for irrigating lawns, parks, cemeteries, freeway planting, golf courses, college campuses, athletic fields, and other public areas. Recreational lakes used for boating, swimming, water skiing, fishing, and other water sports are frequently made up partially and, in a few cases, solely of treated effluents. As public contact has increased and diluting waters have decreased or declined in quality, it has become obvious that more consideration must be given to disinfection practices.

QUESTIONS

Write your answers in a notebook and then compare your answers with those on page 207.

12.0A What is the purpose of disinfection? Why is this important?

12.0B How are pathogenic bacteria destroyed or removed from water?

12.0C Why are wastes not sterilized?

12.0D Why is chlorine used for disinfection?

12.02 How Is Disinfection Achieved?

12.020 Filtration

There are a number of permeable membranes that can be used as filters to remove a large portion of pathogens. These membranes must be very tight and are required to remove particles sized in the *MICRON*[3] range. When the openings in the membrane are kept small they are subject to fouling by oils, greases, and other particulate matter found in wastewater effluents. For wastewater treatment systems that have any significant flows or treat flows that are of poor quality, this method is not practical and is seldom used.

The most widely practiced filtration method is the percolation of wastewater effluents through soil. Specific soils make an excellent filtering media that retains small micro-size particles, yet permits the treated wastewater to pass through. For small-flow systems with the proper soil types and sufficient land area, percolation through soil is an economical and efficient disinfection and disposal system.

In small wastewater treatment systems consisting of individual septic tanks using leach field disposal of effluent, the removal of pathogens can be achieved by filtration. In treated wastewater effluent from these systems, the pathogenic organisms have been greatly reduced by the septic tank due to settling of the solids and decomposition of the organic matter through biological oxidation. The septic tank environment is not an ideal environment for the pathogens due to adverse temperatures and the presence of other organisms that will prey upon the pathogens during the detention time of the wastewater in the septic tank. Any pathogens that survive and remain in the septic tank effluent discharged to a subsurface disposal system, such as a leach field, are further subjected to a harsh environment. As the wastewater moves through the soil, the pathogens are filtered out, trapped in the soil, and die or become food for some other plant or organism. If the soil barrier that the wastewater flows through retains the water long enough before it enters into another water source, the pathogen numbers are so greatly reduced that the remaining few are unlikely to create any public health problem. Not only are the pathogens reduced in number but many other constituents, such as minerals or small quantities of contaminants such as some heavy metals, are removed from the wastewater as it filters through the soil.

---

[2] *Saprophytic (SAP-row-FIT-ick) Organisms. Organisms living on dead or decaying organic matter. They help natural decomposition of the organic solids in wastewater.*

[3] *Micron (MY-kron). µm, Micrometer or Micron. A unit of length. One millionth of a meter or one thousandth of a millimeter. One micron equals 0.00004 of an inch.*

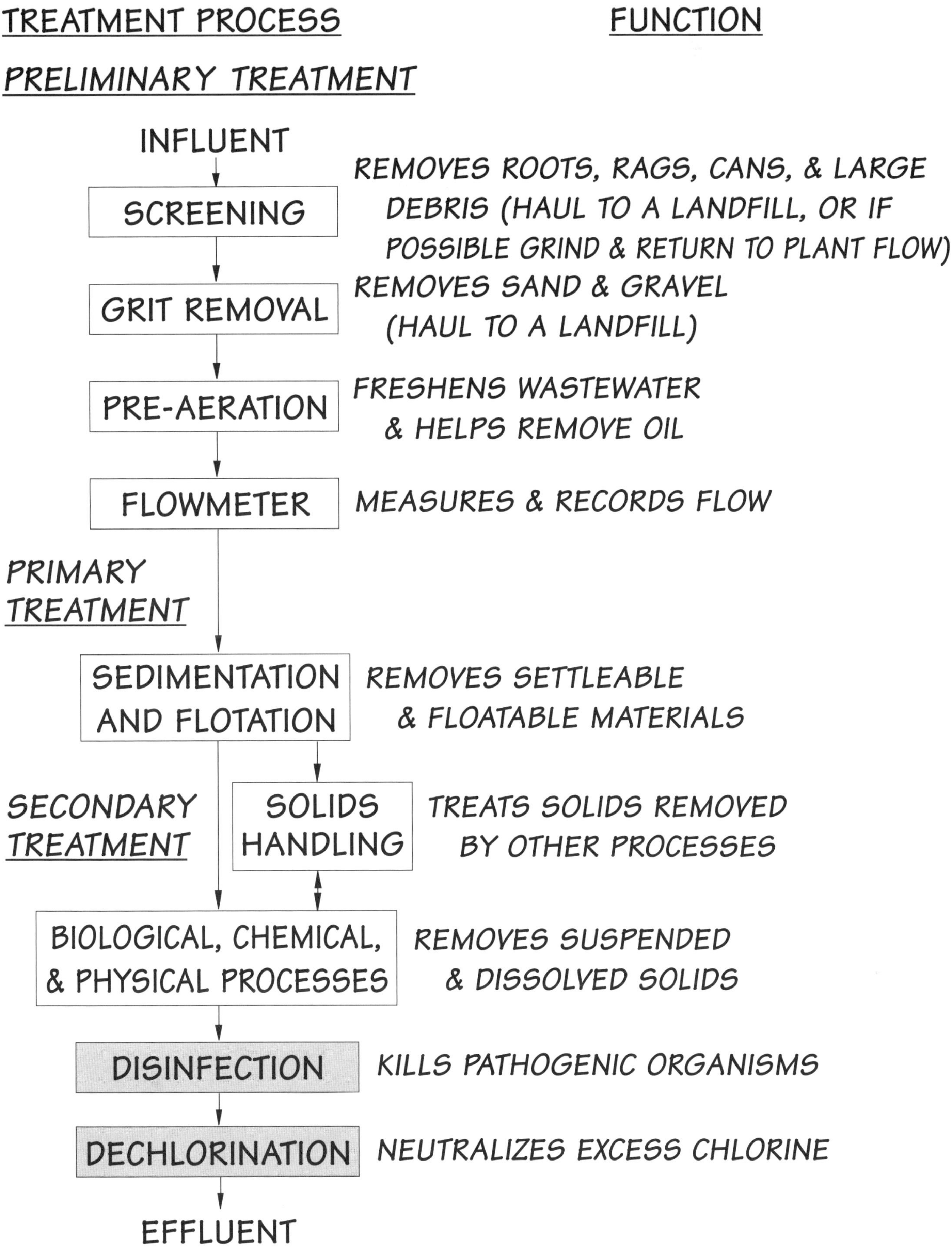

Fig. 12.1 Typical flow diagram of wastewater treatment plant

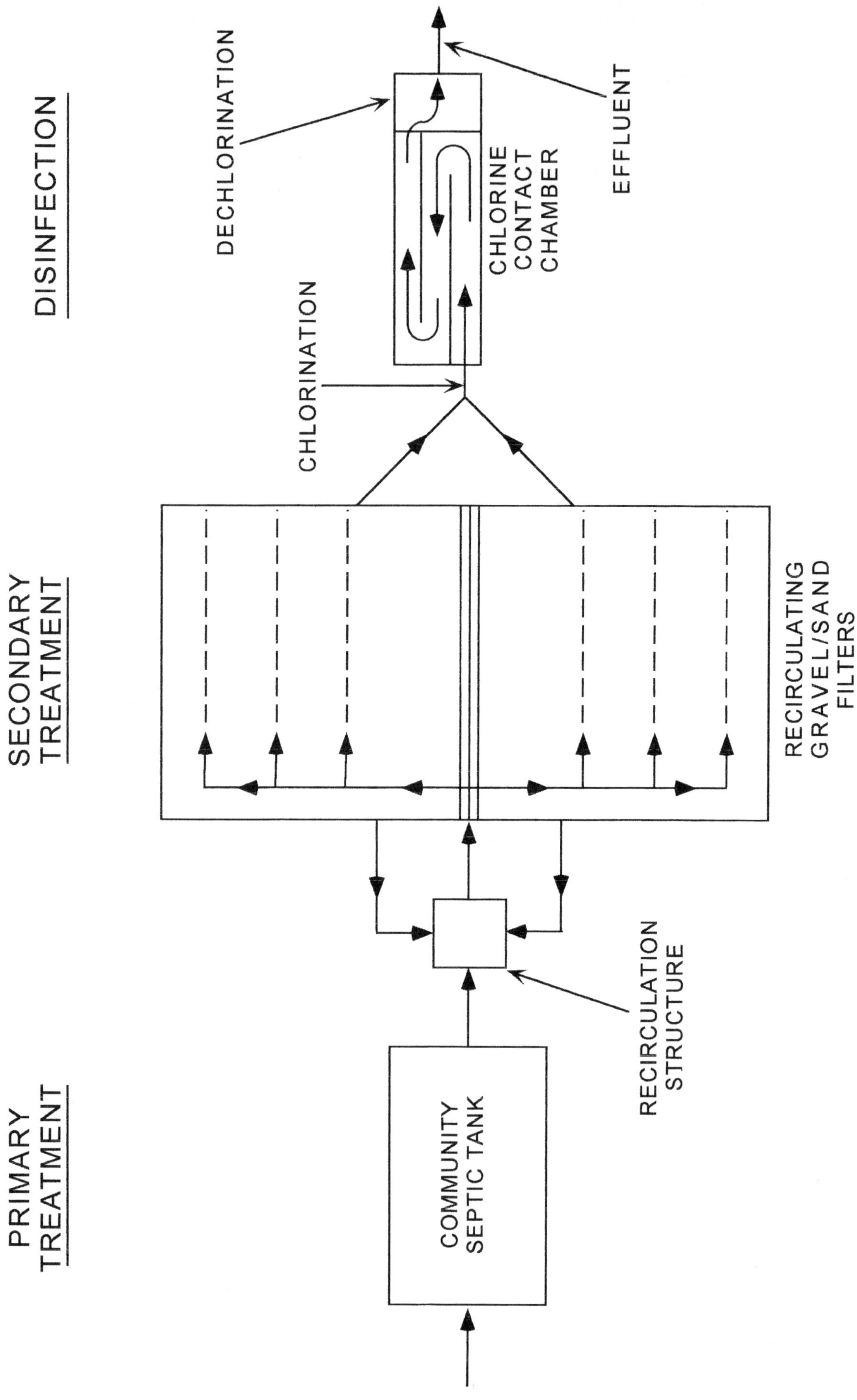

Fig. 12.2 Typical flow diagram of a small wastewater treatment facility

12.021 Chemicals

A number of chemicals can be used to disinfect wastewater effluents; the most commonly used is chlorine. Chlorine is available as free chlorine (*GAS/LIQUID*[4] in the same container) or as a *HYPOCHLORITE*,[5] which may be obtained as a liquid or a dry product in granules, pellets, and tablets. Iodine, bromine, potassium permanganate, and hydrogen peroxide have been tried but in most instances given up in favor of chlorine. Chlorination has been the most economical process for disinfection of *POTABLE WATER*[6] and wastewater.

Large wastewater treatment plants commonly use free chlorine (gas/liquid) for disinfection and application in other in-plant processes. Smaller plants typically use hypochlorite systems (liquid solution containing chlorine) for safety reasons.

Wastewater treatment facilities that are described in this manual typically use a hypochlorite disinfection system of tablet chlorinators or a liquid solution pumping system. If the plant is slightly larger, discharging effluent in excess of 100,000 gallons per day, and has a well-trained operator, cylinder-mounted gas chlorinators may be used. Only the hypochlorite disinfection systems will be discussed in detail in this manual.

12.022 Other Methods

- **Ultraviolet Light (UV)**

  Ultraviolet light (UV) has been used to disinfect potable water supplies and, in more instances, to disinfect wastewater effluents. UV is generally not used for disinfection of effluents other than very high quality tertiary effluents because the equipment used requires a high maintenance routine when disinfecting lower quality effluents. The effluent must flow through the apparatus in a narrow stream, usually through a glass tube with the UV source of quartz lamps radiating through the glass into the effluent. If the effluent has a high suspended solids content (in excess of 8 mg/*L*) or contains any grease or oil, it will foul the tube and restrict the UV transmittance to the effluent, thus reducing pathogen kill and discharging an unacceptable effluent. In some systems, it is necessary to clean the glass in the UV unit every hour, which makes it inappropriate for use except in rare cases where it is more acceptable than other methods. For additional information on disinfection using ultraviolet light, see Section 12.51, "Ultraviolet Disinfection."

- **Ozone**

  Disinfection using ozone has been attempted in limited applications and again chlorine has been found more acceptable in these applications. Limitations of using ozone include the fact that ozone must be generated on site and it can't be stored due to ozone instability. Also, energy requirements are high and costly.

- **High Temperature**

  The boiling of water at 220°F (127°C) for twenty minutes at 20 pounds pressure, as in an autoclave, will provide sterilization of the autoclave contents. Raising the effluent temperature to 212°F (100°C) for a few minutes does have pathogen reduction capability, but again the energy requirement and equipment maintenance favor the use of chlorination systems.

12.03 Reactions of Chlorine Solutions With Impurities in Wastewater

Wastewater contains a great number of complex substances and chemicals and many of these substances have a significant effect on wastewater chlorination. A few of the major impurities will be discussed in the following sections.

12.030 Inorganic Reducing Materials

Chlorine reacts rapidly with many reducing agents and more slowly with others. These reactions complicate the use of chlorine for disinfecting purposes. One of the most widely known inorganic reducing materials is hydrogen sulfide. Hydrogen sulfide reacts with chlorine to form sulfuric acid or elemental sulfur, depending on the concentration of hydrogen sulfide, pH, and temperature.

Equation 1

$$\text{Hydrogen Sulfide} + \text{Free Chlorine} + \text{Water} \rightarrow \text{Sulfuric Acid} + \text{Hydrochloric Acid}$$

$$H_2S + 4Cl_2 + 4H_2O \rightarrow H_2SO_4 + 8HCl$$

Equation 2

$$\text{Water} + \text{Free Chlorine} \rightarrow \text{Hypochlorous Acid} + \text{Hydrogen Sulfide} \rightarrow \text{Elemental Sulfur} + \text{Hydrochloric Acid} + \text{Water}$$

$$H_2O + Cl_2 \rightarrow HOCl + H_2S \rightarrow S^0_{\downarrow} + HCl + H_2O$$

One part of hydrogen sulfide (H_2S) takes about 8.5 parts of chlorine in equation 1 and 2.2 parts of chlorine in equation 2. Since these reactions occur before a *CHLORINE RESIDUAL*[7] occurs, the demand caused by the inorganic salts must be satisfied first before any disinfection can take place. Ferrous iron, manganese, and nitrite are examples of other inorganic reducing agents that react with chlorine.

12.031 Reaction With Ammonia (NH_3)

Since ammonia is present in all domestic wastewaters, the reaction of ammonia with chlorine is of great significance. When chlorine is added to waters containing ammonia, the ammonia reacts with hypochlorous acid (HOCl) to form monochloramine, dichloramine, and trichloramine. The formation of

---

[4] *Gas/Liquid. Gaseous/Liquid. Gaseous/liquid chlorination refers to the fact that free chlorine is delivered to small treatment plants in containers (150-pound chlorine cylinders) that hold liquid chlorine with a free chlorine gas above the liquid in the container. The release of chlorine gas from the liquid chlorine surface depends on the temperature of the liquid and the pressure of the chlorine gas on the liquid surface. As chlorine gas is removed from the container, the gas pressure drops and more liquid chlorine becomes chlorine gas.*

[5] *Hypochlorite (HI-poe-KLOR-ite). Chemical compounds containing available chlorine; used for disinfection. They are available as liquids (bleach) or solids (powder, granules, and pellets) in barrels, drums, and cans. Salts of hypochlorous acid.*

[6] *Potable (POE-tuh-bull) Water. Water that does not contain objectionable pollution, contamination, minerals, or infective agents and is considered satisfactory for drinking.*

[7] *Chlorine Residual. The concentration of chlorine present in water after the chlorine demand has been satisfied. The concentration is expressed in terms of the total chlorine residual, which includes both the free and combined or chemically bound chlorine residuals.*

these *CHLORAMINES*[8] depends on the pH of the solution and the initial chlorine-ammonia ratio.

Ammonia	+ Hypochlorous Acid	→	Chloramine	+ Water
NH_3	+ HOCl	→	NH_2Cl (monochloramine)	+ H_2O
NH_2Cl	+ HOCl	→	$NHCl_2$ (dichloramine)	+ H_2O
$NHCl_2$	+ HOCl	→	NCl_3 (trichloramine)	+ H_2O

In general, at the pH levels that are usually found in wastewater (pH 6.5 to 7.5), monochloramine and dichloramine exist together. At pH levels below 5.5, dichloramine exists by itself. Below pH 4.0, trichloramine is the only compound found.

The mono- and dichloramine forms have definite disinfection powers and are of interest in the measurement of chlorine residuals. Dichloramine has a more effective disinfecting power than monochloramine.

If enough chlorine is added to react with the inorganic compounds and *NITROGENOUS*[9] compounds, then this chlorine will react with organic matter to produce *CHLORORGANIC*[10] compounds or other combined forms of chlorine, which have slight disinfecting action. Then, if enough chlorine is added to react with all the above compounds, any additional chlorine will exist as *FREE AVAILABLE CHLORINE*,[11] which has the highest disinfecting action (Figure 12.3). This situation rarely exists in wastewater that contains nitrogenous compounds. The term *BREAKPOINT CHLORINATION*[12] refers to the breakpoint shown on Figure 12.3.

The exact mechanism of this disinfection action is not fully known. In some theories, chlorine is considered to exert a direct action against the bacterial cell, thus destroying it. Another theory is that the toxic character of chlorine inactivates the *ENZYMES*[13] upon which the living microorganisms are dependent for utilizing their food supply. As a result, the organisms die of starvation. From the point of view of wastewater treatment, the mechanism of the action of chlorine is much less important than its effects as a disinfecting agent.

The demand by inorganic and organic materials is referred to as "chlorine demand." The chlorine that remains in combined forms having disinfecting properties plus any free chlorine is referred to as "residual chlorine." The sum of the chlorine demand and the chlorine residual is the chlorine dose.

Chlorine Dose = Chlorine Demand + Chlorine Residual

where

Chlorine Residual = Combined Chlorine Forms + Free Chlorine

12.04 Hypochlorination[14] of Wastewater

Hypochlorination is a very common method of wastewater disinfection for small systems. Although chlorination with gaseous chlorine can cost less and have a greater disinfection efficiency than hypochlorination, the safety aspects usually make hypochlorination the preferred approach. Hypochlorination is safer because of the ease of handling hypochlorite as compared with the safety hazards of handling chlorine liquid or gas. When the pH of the wastewater is increased, the hypochlorite ion (OCl^-) formation is increased. Hypochlorite ion is a less efficient disinfectant than the hypochlorous acid (HOCl) that is formed at lower pH levels.

Hypochlorination will raise the pH of the wastewater being treated. The rise in pH will decrease the effectiveness of the hypochlorite, thereby requiring a higher dosage. This, in turn, will increase the pH even more.

12.05 Factors Influencing Disinfection

Both *CHLORINE ADDITION* and *CONTACT TIME* are essential for effective killing or inactivation of pathogenic microorganisms. Experimental determination of the best combination of combined residual chlorine and contact time is necessary to ensure both proper chlorine dose and minimum use of chlorine. Changes in pH affect the disinfection ability of chlorine and the operator must re-examine the best combination of chlorine addition and contact time when the pH fluctuates. Critical factors influencing disinfection are summarized as follows:

1. Injection point and method of mixing to get disinfectant in contact with wastewater being disinfected.

[8] *Chloramines (KLOR-uh-means). Compounds formed by the reaction of hypochlorous acid (or aqueous chlorine) with ammonia.*

[9] *Nitrogenous (nye-TRAH-jen-us). A term used to describe chemical compounds (usually organic) containing nitrogen in combined forms. Proteins and nitrate are nitrogenous compounds.*

[10] *Chlororganic (klor-or-GAN-ick). Organic compounds combined with chlorine. These compounds generally originate from, or are associated with, living or dead organic materials.*

[11] *Free Available Chlorine. The amount of chlorine available in water. This chlorine may be in the form of dissolved gas (Cl_2), hypochlorous acid (HOCl), or hypochlorite ion (OCl^-), but does not include chlorine combined with an amine (ammonia or nitrogen) or other organic compound.*

[12] *Breakpoint Chlorination. Addition of chlorine to water or wastewater until the chlorine demand has been satisfied. At this point, further additions of chlorine result in a residual that is directly proportional to the amount of chlorine added beyond the breakpoint.*

[13] *Enzymes (EN-zimes). Organic substances (produced by living organisms) which cause or speed up chemical reactions. Organic catalysts and/or biochemical catalysts.*

[14] *Hypochlorination (HI-poe-KLOR-uh-NAY-shun). The application of hypochlorite compounds to water or wastewater for the purpose of disinfection.*

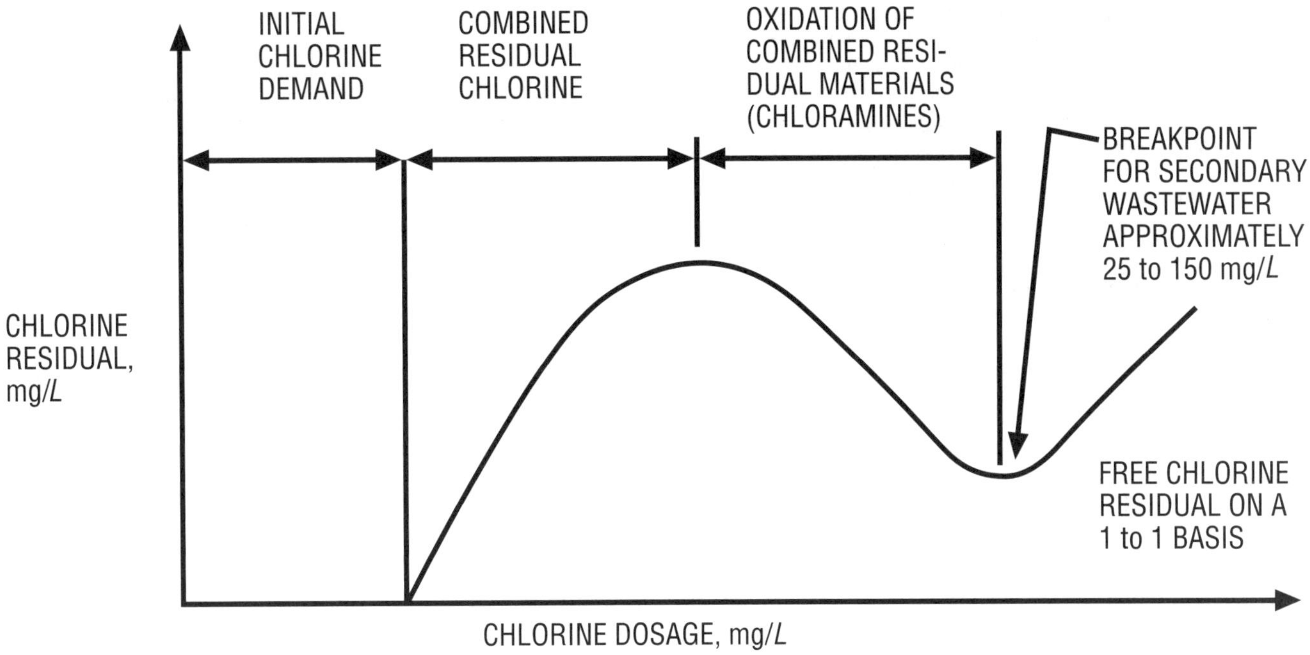

Fig. 12.3 Breakpoint chlorination curve

2. Design (shape) of *CHLORINE CONTACT CHAMBERS.*[15] Contact chambers or basins are designed in various sizes and shapes. Rectangular or square contact chambers often allow short-circuiting and consequently reduced contact times. Baffles often are installed in these basins to increase mixing action, to obtain better distribution of disinfectant, and to reduce short-circuiting, which in turn increases contact time. Long, narrow channels or pipelines have proven to be good contact chambers.
3. Contact time. With good initial mixing, the longer the contact time, the better the disinfection. Most chlorine contact basins are designed to provide a contact time of 30 minutes. In general, extending the chlorine contact time is more effective than increasing the chlorine dose to improve disinfection.
4. Effectiveness of upstream treatment processes. The lower the suspended solids and dissolved organic content of the wastewater, the better the disinfection.
5. Temperature. The higher the temperature, the more rapid the rate of disinfection.
6. Dose rate and type of chemical. Normally, the higher the dose rate, the more rapid the disinfection rate. The form or type of chemical also influences the disinfection rate.
7. pH. The lower the pH, the better the disinfection.
8. Numbers and types of organisms. The greater the organisms' concentrations, the longer the time required for disinfection. Bacterial cells are killed quickly and easily, but bacterial *SPORES*[16] are extremely resistant.

To produce effective disinfecting action, sufficient chlorine must be added after the chlorine demand is satisfied to produce a chlorine residual that will persist through the contact period.

QUESTIONS

Write your answers in a notebook and then compare your answers with those on page 207.

12.0E What does chlorine produce when it reacts with organic matter?

12.0F How is the chlorine dosage determined?

12.0G How much chlorine must be added to wastewater to produce disinfecting action?

12.06 Estimating Chlorine Requirements

The quantity of chlorine-demanding substances differs from plant to plant; therefore, the amount of chlorine that has to be

[15] *Chlorine Contact Chamber. A baffled basin that provides sufficient detention time of chlorine contact with wastewater for disinfection to occur. The minimum contact time is usually 30 minutes. (Also commonly referred to as basin or tank.)*

[16] *Spore. The reproductive body of an organism which is capable of giving rise to a new organism either directly or indirectly. A viable (able to live and grow) body regarded as the resting stage of an organism. A spore is usually more resistant to disinfectants and heat than most organisms. Gangrene and tetanus bacteria are common spore-forming organisms.*

added to ensure proper disinfection also differs. You must remember that the chlorine requirement or chlorine dose will vary with wastewater flow, time of contact between the wastewater effluent and chlorine, temperature of the effluent, pH, and major waste constituents such as hydrogen sulfide, ammonia, and amount of dead and living[17] organic matter. The amount of chlorine required to satisfy the chlorine-demanding substances is called the *CHLORINE DEMAND.* This demand is equal to the chlorine dose minus the chlorine residual.

Chlorine Demand = Chlorine Dose − Chlorine Residual

To determine the chlorine dosage rate for a particular wastewater, start with the suggested chlorine dosage listed in Table 12.1. Increase or decrease this dosage to achieve the desired chlorine residual and level of disinfection (meet effluent coliform discharge requirements).

TABLE 12.1 TYPICAL CHLORINE DOSE RANGES FOR VARIOUS APPLICATIONS

Application	Dosage Range, mg/*L*
Collection Systems	
Slime Control	1 – 15
Corrosion Control	2 – 9[a]
Odor Control	2 – 9[a]
Treatment	
Grease Removal	1 – 10
BOD Reduction	0.5 – 2[b]
Inorganic Compounds	2 – 12
Filter Ponding	1 – 10
Filter Flies	0.1 – 0.5
Activated Sludge	
Bulking	1 – 10
Digester	
Supernatant	20 – 150
Foaming	2 – 20
Disinfection	
Septic Tank Effluent	5 – 20
Rotating Biological Contactor Effluent[c]	3 – 20
Recirculating Gravel/Sand Filter Effluent	2 – 5
Oxidation Pond Effluent	5 – 20
Wetlands Effluent	2 – 10

[a] per mg/*L* H_2S (Hydrogen sulfide gas forms with water vapor forming sulfuric acid, which causes corrosion of concrete in manholes, concrete pipe lines, and concrete structures.)
[b] per mg/*L* BOD destroyed
[c] after secondary clarification

The chlorine residual is determined by one of several laboratory tests. The method of choice must be one of those approved by state and federal water pollution control agencies, otherwise the results will not be recognized. These tests are discussed in the laboratory section (Chapter 14, "Laboratory Procedures") of this manual. The amount of chlorine residual that one should maintain is determined by the desired microorganism population. The microorganism population is usually specified by state or federal NPDES permit requirements. The microorganism population usually is estimated by determining the *MPN*[18] of *COLIFORM*[19] group organisms present. This determination does not test for individual pathogenic microorganisms, but uses the coliform group of organisms as the indicator organism. Coliform organisms are found in domestic wastewaters. See Chapter 16 in *OPERATION OF WASTEWATER TREATMENT PLANTS*, Volume II, for the laboratory test to determine the MPN of coliforms present.

Calculations to determine the chlorine dosage and chlorine demand are illustrated in the following example problem.

EXAMPLE 1

A hypochlorinator is set to feed a dose of 4 pounds of chlorine per 24 hours. The wastewater flow rate is 0.050 MGD, and the chlorine, as measured by the chlorine residual test after thirty minutes of contact time, is 0.5 mg/*L*. Find the chlorine dosage and chlorine demand in mg/*L*.

$$\text{Chlorine Feed or Dose, mg/}L = \frac{\text{Pounds of Chlorine}}{\text{Million Pounds of Water}}$$

$$= \frac{\text{4 lbs Chlorine/day}}{\text{(0.050 MG/day)(8.34 lbs/gal)}}$$

$$= \frac{\text{80 lbs Chlorine/MG}}{\text{8.34 lbs/gal}}$$

$$= \text{9.5 lbs Chlorine/million pounds water}$$

$$= \text{9.5 ppm (Parts Per Million parts)}$$

$$= 9.5 \text{ mg/}L$$

Or

$$\text{Chlorine Feed or Dose, mg/}L = \text{4 lbs Chlorine/day} \div \text{0.050 MGD} \div \text{8.34 lbs/gal}$$

$$= 9.5 \text{ mg/}L$$

$$\text{Chlorine Demand, mg/}L = \text{Chlorine Dose, mg/}L - \text{Chlorine Residual, mg/}L$$

$$= 9.5 \text{ mg/}L - 0.5 \text{ mg/}L$$

$$= 9.0 \text{ mg/}L$$

The objective of disinfection is the destruction or inactivation of pathogenic organisms, and the ultimate measure of the effectiveness is the bacteriological test result. The measure-

[17] *An example of living organic matter would be ALGAE that might be present in the effluent from a wastewater treatment system using oxidation ponds.*

[18] *MPN (pronounce as separate letters). MPN is the **M**ost **P**robable **N**umber of coliform-group organisms per unit volume of sample water. Expressed as a density or population of organisms per 100 mL of sample water.*

[19] *Coliform (COAL-i-form). One type of bacteria. The presence of coliform-group bacteria is an indication of possible pathogenic bacterial contamination. The human intestinal tract is one of the main habitats of coliform bacteria. They may also be found in the intestinal tracts of warm-blooded animals, and in plants, soil, air, and the aquatic environment. Fecal coliforms are those coliforms found in the feces of various warm-blooded animals; whereas the term "coliform" also includes other environmental sources.*

ment of residual chlorine does supply a tool for practical control. If the residual chlorine value commonly effective in most wastewater treatment plants does not yield satisfactory bacteriological kills in a particular plant, the residual chlorine that does must be determined and used as a control in that plant. In other words, the 0.5 mg/*L* residual chlorine, while generally effective, is not a rigid standard but a guide that may be changed to meet local requirements.

One special case would be the use of chlorine in the effluent from a plant serving a tuberculosis hospital. Studies have shown that a residual of at least 2.0 mg/*L* should be maintained in the effluent from this type of institution, and that detention time should be at least two hours at the average rate of flow instead of the thirty minutes normally used as a basis of design. Two-stage chlorination may be particularly effective in this case.

Table 12.1 provides a list of chlorine dosages that are a reasonable guideline to produce chlorine residual adequate for applications indicated for domestic wastewaters. Individual plants may require higher or lower dosages, depending on the type and amount of suspended and dissolved organic compounds in the chlorinated sample.

QUESTIONS

Write your answers in a notebook and then compare your answers with those on page 207.

12.0H What factors influence chlorine demand?

12.0I Calculate the chlorine demand of treated domestic wastewater if:

Flow Rate = 0.06 MGD (60,000 GPD)

Hypochlorinator = 5 lbs of Chlorine per 24 hours

Residual = 0.4 mg/*L* after thirty minutes

12.0J What is the objective of disinfection?

12.0K How is the effectiveness of the chlorination process for a particular plant determined?

END OF LESSON 1 OF 3 LESSONS
on
DISINFECTION AND CHLORINATION

Please answer the discussion and review questions next.

DISCUSSION AND REVIEW QUESTIONS

Chapter 12. DISINFECTION AND CHLORINATION

(Lesson 1 of 3 Lessons)

At the end of each lesson in this chapter you will find some discussion and review questions. The purpose of these questions is to indicate to you how well you understand the material in the lesson. Write the answers to these questions in your notebook before continuing.

1. Why must wastewater be disinfected?
2. Why is chlorination used to disinfect wastewater?
3. To improve disinfection, which is more effective—increasing the chlorine dose or extending the chlorine contact time?
4. What constituents in wastewater are mainly responsible for the chlorine demand?
5. How do suspended and dissolved organic compounds in an effluent affect disinfection?

CHAPTER 12. DISINFECTION AND CHLORINATION

(Lesson 2 of 3 Lessons)

12.1 POINTS OF CHLORINE APPLICATION (Figure 12.4)

12.10 Collection System Chlorination

One of the primary benefits of collection system chlorination is preventing the deterioration of structures by inhibiting the production of hydrogen sulfide (H_2S) through biological decomposition. Other benefits include odor and *SEPTICITY*[20] control, and possibly BOD reduction to decrease the load imposed on the wastewater treatment processes. In some instances, the maximum benefit may result from a single application of chlorine at a point on the main intercepting sewer before the junction of all feeder sewer lines. In others, several applications at more than one point on the main intercepting sewer or at the upper ends of the feeder lines may prove most effective. Due to high costs, *CHLORINATION SHOULD BE CONSIDERED AS A TEMPORARY OR EMERGENCY MEASURE IN MOST CASES*, with emphasis being placed on proper design of the system. Aeration also is effective in controlling septic conditions in collection systems. Although many problems result from improper design or design for future capacity requirements, the need for hydrogen sulfide protection exists under the best of conditions in some locations.

12.11 Prechlorination

"Prechlorination" is defined as the addition of chlorine to wastewater at the entrance to the treatment plant, ahead of settling units and prior to the addition of other chemicals. In addition to its application for aiding disinfection and odor control at this point, prechlorination is applied to reduce plant BOD load, as an aid to settling, to control foaming in Imhoff units, and to help remove oil. Current trends are away from prechlorination to collection system aeration or other chemical treatment for control of odors.

12.12 Plant Chlorination

Chlorine is added to wastewater during treatment by other processes and the specific point of application is related to the results desired. The purpose of plant chlorination may be for control and prevention of odors, corrosion, sludge bulking, digester foaming, filter ponding, filter flies, and as an aid in sludge thickening. Here again, chlorination should be an emergency measure. Extreme care must be exercised when applying chlorine because it can interfere with or inhibit biological treatment processes.

12.13 Chlorination Before Filtration

More stringent discharge requirements are causing many agencies to provide filtration for effluent solids removal before discharge (for additional information, refer to *ADVANCED WASTE TREATMENT*, Chapter 4, "Solids Removal From Secondary Effluents," in this series of operator training manuals). The better designs provide a means of chlorinating the water before application to the filters to kill algae and large biological organisms. Prechlorination tends to prevent the development of biological growths that might cause short-circuiting or excessive backwashing in the filter media. Postchlorination of the effluent from the filter would be in addition to this application.

12.14 Postchlorination

"Postchlorination" is defined as the addition of chlorine to municipal or industrial wastewater following other treatment processes. This point of application should be before a chlorine contact chamber and after the final settling unit in the treatment plant. The most effective place for chlorine application for disinfection is after treatment and on a well-clarified effluent. Postchlorination is used primarily for disinfection. As a result of chlorination for disinfection, some reduction in BOD may be observed; however, chlorination is rarely practiced solely for the purpose of BOD reduction.

QUESTIONS

Write your answers in a notebook and then compare your answers with those on page 207.

12.1A What is the purpose of up-sewer chlorination?

12.1B Where should chlorine be applied in sewers?

12.1C What are the main reasons for prechlorination?

12.1D Why might chlorine be added to wastewater during treatment by other processes?

12.1E What is the objective of postchlorination?

12.2 CHLORINATION PROCESS CONTROL

12.20 Chlorinator Control

The control of chlorine to points of application is accomplished by two basic methods in small wastewater systems.

[20] *Septicity (sep-TIS-it-tee). Septicity is the condition in which organic matter decomposes to form foul-smelling products associated with the absence of free oxygen. If severe, the wastewater produces hydrogen sulfide, turns black, gives off foul odors, contains little or no dissolved oxygen, and the wastewater has a high oxygen demand.*

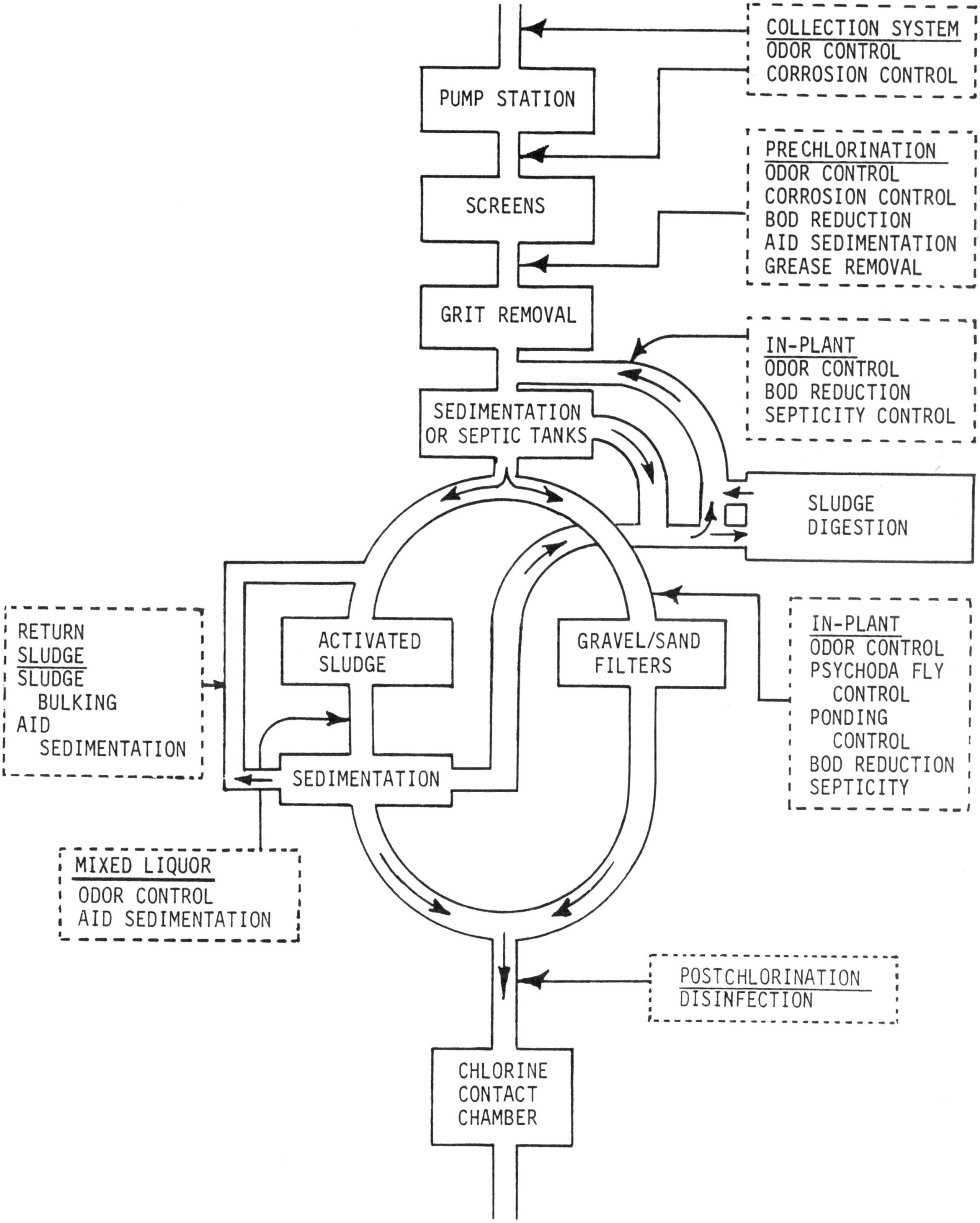

Fig. 12.4 Points of chlorine application

12.200 Manual Control

Feed-rate adjustment and starting and stopping of equipment are done by hand.

12.201 Start-Stop Control

- **Tablet *HYPOCHLORINATORS***[21] (Figure 12.5)

  The feed rate of the chlorine into the effluent is controlled in two ways: (1) by the number of tubes that are in service holding the tablets. The number of tubes in use is established by the desired chlorine residual to accomplish the desired disinfection. Once this is determined, only those tubes must be filled with tablets. (2) The actual chlorine dose to the wastewater is controlled automatically by a *V-NOTCH WEIR*[22] to control flow and depth of wastewater through the tablet feeder housing. The flow through the feeder housing may be intermittent or continuous; any time flow passes through the feeder housing chlorine will be applied.

- **Solution Feed Hypochlorinators**

  Chlorine feed rate is adjusted by hand through adjustment on the chemical feed pump (hypochlorinator). The feed pump may be started manually or automatically by a flow or level switch which activates the feed pump motor when effluent flows to the effluent chlorine contact chamber.

The selection of control methods should be based on treatment costs and treatment results (required or desired). Normally, a waste discharger must meet a disinfection standard. A manual system, which would meet the maximum requirements and overchlorinate at minimum requirement periods, might be used. It is not unusual for a plant to have maximum chlorine residual effluent requirements because of irrigation and/or marine life tolerances. In these cases, the uncontrolled or haphazard application of chlorine cannot be considered, no matter how large the added cost for controlling chlorine application or dose.

A specified chlorine residual level may be required at some point downstream from the best residual control sample point. In this case, a *RESIDUAL ANALYZER*[23] should be used to monitor and record residuals at this point. It may also be used to change the control *SET POINT*[24] of the controlling residual analyzer.

12.21 *CHLORINE REQUIREMENTS*[25] and Residuals

Ultimate control of dosage for disinfection depends upon the results desired, that is, the bacterial level or concentration acceptable or permissible at the point of discharge. Determination of chlorine requirements and residuals according to the current edition of HACH's *WATER ANALYSIS HANDBOOK* (see Section 12.9, reference 1, for ordering information) is a good method of control. Operators must remember that the chlorine requirement or chlorine dose will vary with wastewater flow, time of contact, temperature, pH, and major waste constituents such as hydrogen sulfide, ammonia, and amount of dead and living organic matter.

The chlorine requirement for various flow rates and contact times can be determined on either a plant or a laboratory scale. If the determination is made on a laboratory scale, you should expect the plant requirements to be somewhat higher. This is to be expected since the mixing and actual contact times can be more carefully controlled in the laboratory. It is preferable that the determinations be made by both methods and the results compared. If the chlorine requirement, as determined by full plant tests, is significantly greater than that determined by laboratory testing, a wastage of chlorine is indicated. The two major causes of a large discrepancy between laboratory and plant results are poor mixing at the point of chlorine injection and short-circuiting in the contact chamber. Either problem can usually be solved at a relatively small expense as compared to the savings that can be achieved by the reduced use of chlorine.

QUESTIONS

Write your answers in a notebook and then compare your answers with those on page 207.

12.2A Selection of a chlorination feed- (dose-) rate control method is based on what two main factors?

12.2B Control of chlorine dosage depends on the bacterial ________________ desired.

12.22 Hypochlorinator Feed Rate and Control

Chlorine for disinfection and other purposes is provided in most very small plants by the use of hypochlorite compounds. In the dry form (as granules, pellets, or tablets), the amount of chlorine delivered depends on the type of hypochlorite. For example, HTH (high test hypochlorite) contains approximately 65 percent chlorine, by weight, and chlorinated lime contains 34 percent.

Hypochlorite in liquid form is also available and many small plant operators prefer this form of chlorine because it eliminates the need for mixing the dry compounds and storing in a separate solution container for application. For very small systems, hypochlorite is available in the form of laundry bleach and contains 5.25 percent chlorine. For small systems, the liquid hypochlorite is shipped in plastic 30- or 50-gallon drums at a 12.5 percent chlorine concentration. The drums are so designed that the hypochlorinator feed pump sits directly on top of the drum with a suction tube equipped with a flow indicator and level switch attached to the pump to withdraw the premixed solution for direct application.

Manufacturers of hypochlorite compounds define available chlorine as the amount of gaseous chlorine required to make the equivalent hypochlorite chlorine. If you prepare a hypo-

---

[21] *Hypochlorinators (HI-poe-KLOR-uh-NAY-tors). Chlorine pumps, chemical feed pumps or devices used to dispense chlorine solutions made from hypochlorites such as bleach (sodium hypochlorite) or calcium hypochlorite into the water being treated.*

[22] *V-Notch Weir. A triangular weir with a "V" notch calibrated in gallons per minute readings. The weir can be placed in a pipe or open channel. As the flow passes through the "V", the depth of water flowing over the weir can be measured and converted to a flow in gallons per minute.*

[23] *Residual Analyzer, Chlorine. An instrument used to measure chlorine residual in water or wastewater. This instrument also can be used to control the chlorine dose rate.*

[24] *Set Point. The position at which the control or controller is set. This is the same as the desired value of the process variable. For example, a thermostat is set to maintain a desired temperature.*

[25] *Chlorine Requirement. The amount of chlorine which is needed for a particular purpose. Some reasons for adding chlorine are reducing the number of coliform bacteria (Most Probable Number), obtaining a particular chlorine residual, or oxidizing some substance in the water. In each case a definite dosage of chlorine will be necessary. This dosage is the chlorine requirement.*

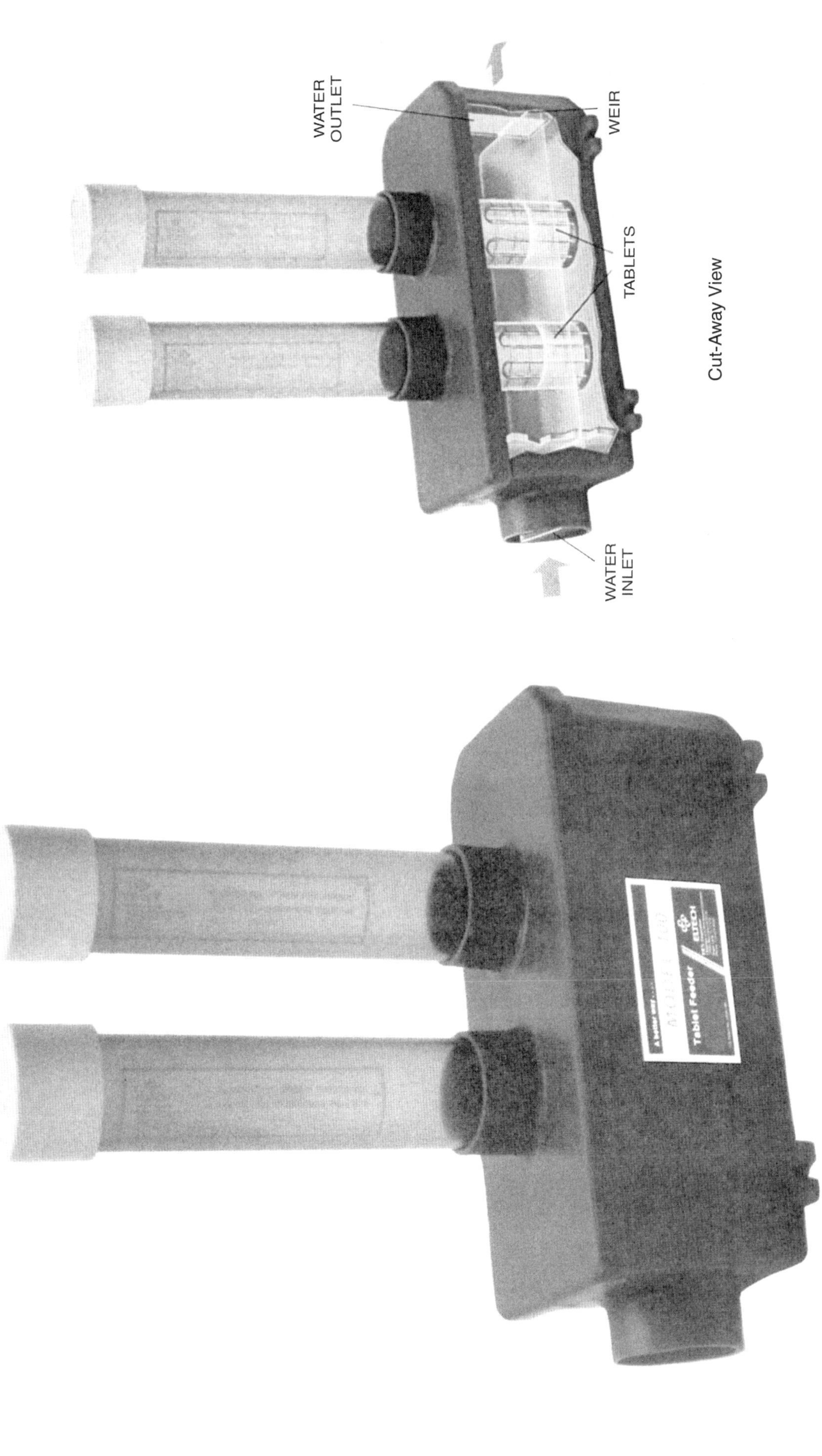

Fig. 12.5 Chlorine tablet feeder wastewater disinfecting system
(Permission of ELTECH International Corporation)

chlorite solution for disinfection and immediately measure the chlorine residual, you will find the chlorine residual about half of the expected value based on the manufacturer's amount of available chlorine. When hypochlorite is mixed with water, approximately half of the chlorine forms hydrochloric acid (HCl) and the other half forms hypochlorite (OCl^-) in the chlorine residual that you measured.

EXAMPLE 2

A wastewater requires a chlorine feed rate of 17 lbs per day. How many pounds of chlorinated lime will be required to provide the needed chlorine? Assume each pound of chlorinated lime contains 0.34 lb of available chlorine.

$$\text{Chlorinated Lime Feed Rate, lbs/day} = \frac{\text{Chlorine Required, lbs/day}}{\text{Portion of Chlorine in lbs of Hypochlorite}}$$

$$= \frac{17 \text{ lbs/day}}{0.34}$$

$$= 50 \text{ lbs/day}$$

Or

$$= \frac{17 \text{ lbs Chlorine/day}}{0.34 \text{ lb Chlorine/lb Hypochlorite}}$$

$$= 50 \text{ lbs/day}$$

Hypochlorinators can be installed on both small and large systems. Usually, the larger systems use gas/liquid chlorine because of lower costs. However, some very large cities use hypochlorite because it is safer. Treatment plants serving these cities are located in highly populated areas where escaping chlorine gas could threaten the lives of many people.

The items in feed systems of hypochlorinators and their construction are discussed below.

1. Solution storage. The storage container is made of corrosion-resistant materials such as plastic or other similar materials.
2. Solution piping, which is high-density, chemical-resistant PVC (polyvinyl chloride).
3. Diffusers that are made of the same material as the piping systems.
4. Valves and *EDUCTORS*[26] are made of PVC.
5. Pumps are made of corrosion-resistant materials. Epoxy-lined systems have been successful.
6. Flowmeters. These meters can be constructed using a Hastelloy C straight-through metal tube *ROTAMETER*[27] with the float position determined magnetically and the flow rate transmitted either electrically or pneumatically.
7. Chlorine residual analyzers of the *AMPEROMETRIC*[28] type are commonly installed.
8. Automatic controls that consist of a hypochlorite flow controller, recorder and totalizer, a ratio control station, and the necessary electronic signal converters.

The feed system can usually be operated automatically or by manual control. The operation of the hypochlorite system will usually cost twice as much as a gas/liquid chlorine system. Maintenance of the hypochlorinator system requires more operator hours than do the gas/liquid chlorine systems.

QUESTIONS

Write your answers in a notebook and then compare your answers with those on pages 207 and 208.

12.2C If you prepare a hypochlorite solution for disinfection and immediately measure the chlorine residual, why will you find the residual about half of the expected value based on the manufacturer's amount of available chlorine?

12.2D How many pounds of HTH (high test hypochlorite) should be used per day by a hypochlorinator dosing a flow of 0.55 MGD at 12 mg/*L* of chlorine? Assume 0.65 portion of chlorine in a pound of HTH.

12.23 Appurtenances Necessary To Apply Chlorine Solutions to Discharge Lines, Diffusers, and Mixing

12.230 Solution Discharge Lines (Convey Chlorine Solution to Application Point)

Solution discharge lines are made from a variety of materials depending on the requirements of service. Two primary requirements are that it must be resistant to the corrosive effects of chlorine solution and it must be of adequate size to carry the required flows. Additional considerations are pressure conditions, flexibility (if required), resistance to external corrosion and stresses when underground or passing through structures, ease and tightness of connection, and the adaptability to field fabrication or alteration.

Development of plastics has contributed greatly to chemical solution transmission. Polyvinyl chloride (PVC) pipe and black polyethylene flexible tubing have all but eliminated the use of rubber hose. Both are generally less expensive than rubber and both outlast rubber in normal service. The use of rubber hose is almost exclusively limited to applications where flexibility is required or where extremely high back pressures exist.

PVC and polyethylene can be field fabricated and altered. PVC should be Schedule 80 to limit its tendency to partially collapse under vacuum conditions. Schedule 80 PVC may be threaded and assembled with ordinary pipe tools or may be installed using solvent-welded fittings. PVC pipe greater than one inch (2.5 cm) may be manufactured with bell and spigot joints.

Rubber-lined steel pipe has been used for many years where resistance to external stresses is required. It cannot be field fabricated or altered and is thus somewhat restricted in application. PVC lining of steel pipe has not yet become economically competitive, but other plastics that can readily compete with rubber lining and are adaptable to field fabrication and alteration have been developed.

---

[26] *Eductor (e-DUCK-ter). A hydraulic device used to create a negative pressure (suction) by forcing a liquid through a restriction, such as a Venturi. An eductor or aspirator (the hydraulic device) may be used in the laboratory in place of a vacuum pump. As an injector, it is used to produce vacuum for chlorinators. Sometimes used instead of a suction pump.*

[27] *Rotameter (RODE-uh-ME-ter). A device used to measure the flow rate of gases and liquids. The gas or liquid being measured flows vertically up a tapered, calibrated tube. Inside the tube is a small ball or bullet-shaped float (it may rotate) that rises or falls depending on the flow rate. The flow rate may be read on a scale behind or on the tube by looking at the middle of the ball or at the widest part or top of the float.*

[28] *Amperometric (am-PURR-o-MET-rick). A method of measurement that records electric current flowing or generated, rather than recording voltage. Amperometric titration is a means of measuring concentrations of certain substances in water.*

Never use neoprene hose to carry chlorine solutions because it will become hard and brittle in a short time. If you have questions regarding the material or equipment needed to handle the chlorine solutions being used, be sure to consult with your chemical supplier.

12.231 Chlorine Solution Diffusers (Introduce Chlorine Solution Into Flow To Be Chlorinated)

Chlorine diffusers are normally constructed of the same materials used for solution lines. Their design is an extremely important part of a chlorination program. This importance is almost completely related to the mixing of the chlorine solution with the wastewater being treated; however, strength and flexibility also must be given consideration. In most *CIRCULAR, FILLED* conduits flowing at 0.25 ft/sec (0.08 m/sec) or greater, a solution injected at the center of the pipeline will mix with the entire flow after flowing ten pipe diameters downstream. Mixing in open channels can be accomplished by the use of a hydraulic jump (Figure 12.6) or by sizing diffuser orifices so that a high velocity (about 16 ft/sec or 4.8 m/sec) is produced at the diffuser discharge. This accomplishes two things: (1) introducing a pressure drop to get equal discharge from each orifice, and (2) imparting sufficient energy to the surrounding wastewater to complete the mixing. Generally speaking, a diffuser should be supplied for each two to three feet (0.6 to 1 meter) of channel depth. Diffusers should be placed across the width of the channel rather than in the direction of flow. Mixing also can be achieved by the use of mechanical mixers specifically designed for this purpose.

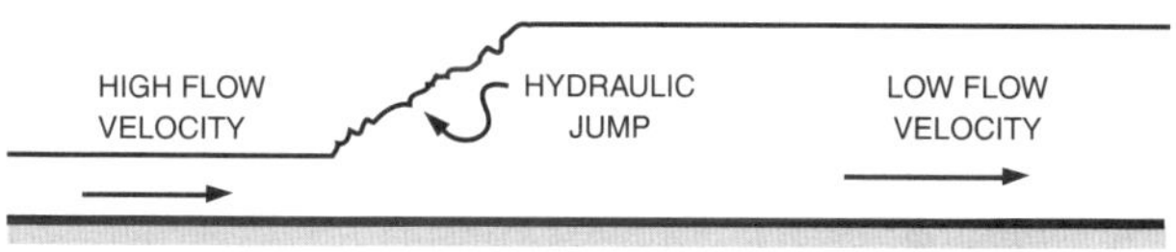

Fig. 12.6 Hydraulic jump

12.232 Mixing

Mixing as well as the speed of mixing are extremely important ahead of a chlorine contact chamber or basin or a residual sampling point. Since a contact chamber is usually designed for low velocity, little mixing occurs after wastewater enters it. Mixing must be achieved before the wastewater enters the contact chamber. The same is true for a chlorine residual sampling point. If good mixing does not occur somewhere upstream from the sampling point, erratic results will be obtained by the residual chlorine analyzing system.

QUESTIONS

Write your answers in a notebook and then compare your answers with those on page 208.

12.2E Chlorine solution discharge lines may be made of __________ or __________.

12.2F Why does little mixing of the chlorine solution with wastewater occur in chlorine contact basins?

12.24 Measurement of Chlorine Residual

12.240 Sampling

Samples to determine chlorine residuals must be collected at the discharge (effluent) end of the chlorine contact structure, whether it is a specifically built tank or an effluent pipeline that has provided the required chlorine/effluent contact time (usually at least 30 minutes) as established by the regulatory agency. The chlorine residual determination must be made as quickly as possible to prevent chlorine concentration changes in the sample. Samples that wait and are exposed to high temperatures (above 68°F or 20°C), sunlight, or periods of time longer than 15 minutes can degrade and lower the chlorine concentration of the sample.

Where appropriate, for example, in larger plants or for critical discharge receiving waters, continuous-flow automatic chlorine residual analyzers are used, as shown in Figure 12.7. Chlorinated effluent is continuously pumped to the analyzer for analysis and automatically recorded for the record to satisfy NPDES permit requirements.

Manual (grab) sampling is more common in smaller wastewater treatment plants, where the operator collects the sample, performs the analysis, records the results and, if required, corrects the chlorine dosage feed rate to maintain a chlorine residual necessary to meet disinfection requirements.

12.241 Which Method To Measure Chlorine Residual?

If you are chlorinating your wastewater plant's effluent, it is because it was mandated in the plant's NPDES permit by your controlling environmental agency. Normally, you must report at some prescribed frequency the chlorine residual concentrations in the effluent being discharged, probably along with several other water quality indicators. The method of determining chlorine residual concentrations will most likely be prescribed by the NPDES permit. There are several methods that can be used to make chlorine measurements analytically and the specific procedures for making those determinations are spelled out precisely in HACH's *WATER ANALYSIS HANDBOOK*. The methods listed in the *WATER ANALYSIS HANDBOOK* have been approved, reviewed, and updated, if necessary. This book is a standard reference for examination of potable water and wastewater (see Section 12.9, reference 1, for ordering information). Chapter 14, "Laboratory Procedures," also lists procedures for measuring chlorine residuals. Your local environmental agency has the historical background data for your area. You may recall from the previous Sections 12.02 through 12.031 that certain substances interfere with the chlorine. This interference can also distort chlorine residual determinations if certain methods and reagents are used. Consequently, one method may provide a more accurate determination than another. In some instances, regulatory agencies permit the use of a less accurate method than another due to local conditions, if slightly higher or lower chlorine concentrations will not adversely affect the environment of that area.

Table 12.2, "How To Select the Right Chlorine Procedure," provides an excellent outline to determine which procedure is best for you when measuring chlorine residuals in your plant's effluent. HACH's *WATER ANALYSIS HANDBOOK* contains the information needed to measure chlorine residuals, including apparatus, reagents, and procedures. In an emergency, a swimming pool or hot tub chlorine residual analyzer can be used, but this method is not considered accurate for treated effluents from a wastewater treatment plant due to waste matter that interferes with the test.

12.3 CHLORINE SAFETY PROGRAM

12.30 Purpose of Safety Program

Every good safety program begins with cooperation between the employee and the employer. The employee must take an active part in the overall program. The employee must

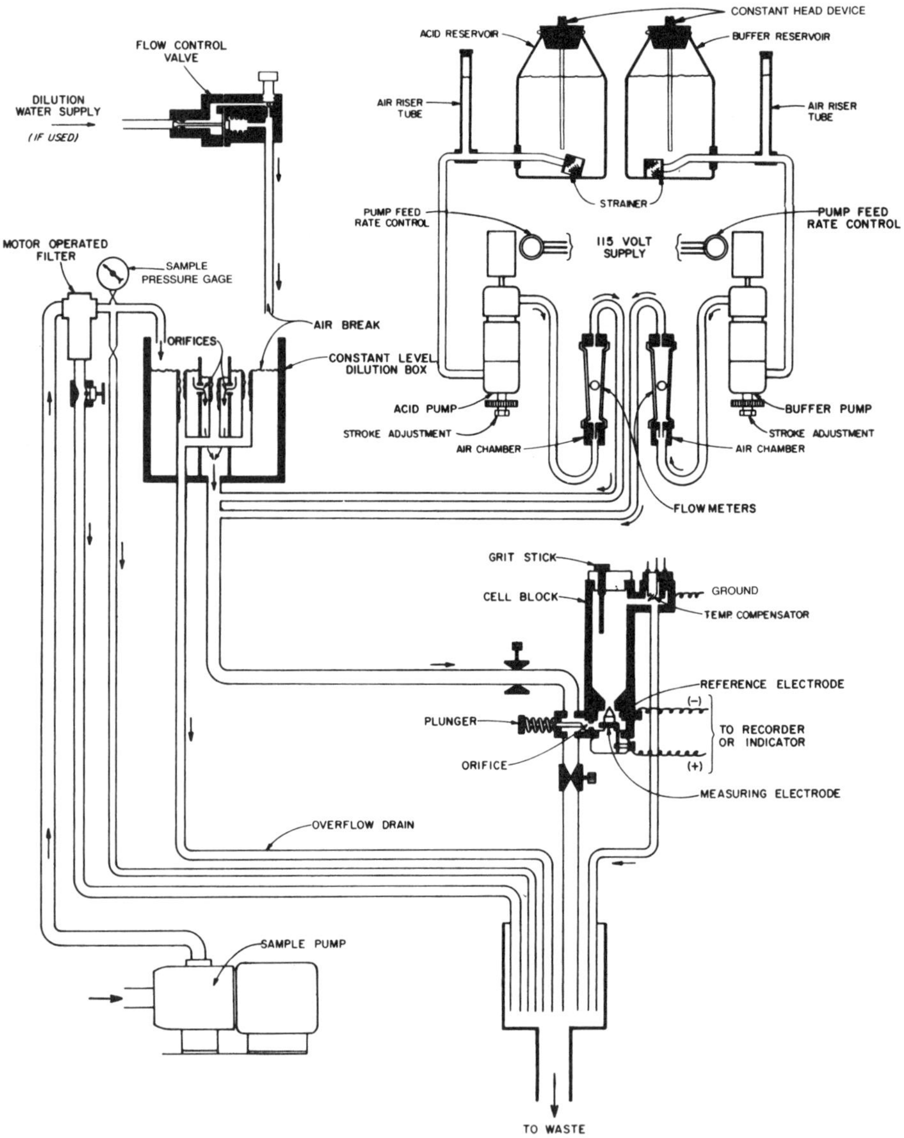

Fig. 12.7 Automatic chlorine residual analyzer

(Permission of Wallace & Tiernan Division, Penwalt Corporation)

(A continuous-flow chlorine residual analyzer also is usually equipped with 24-hour recording chart, not shown.)

TABLE 12.2 HOW TO SELECT THE RIGHT CHLORINE PROCEDURE[a]

Because there are several chlorine procedures, the following decision tree will help you select the appropriate procedure for your application.

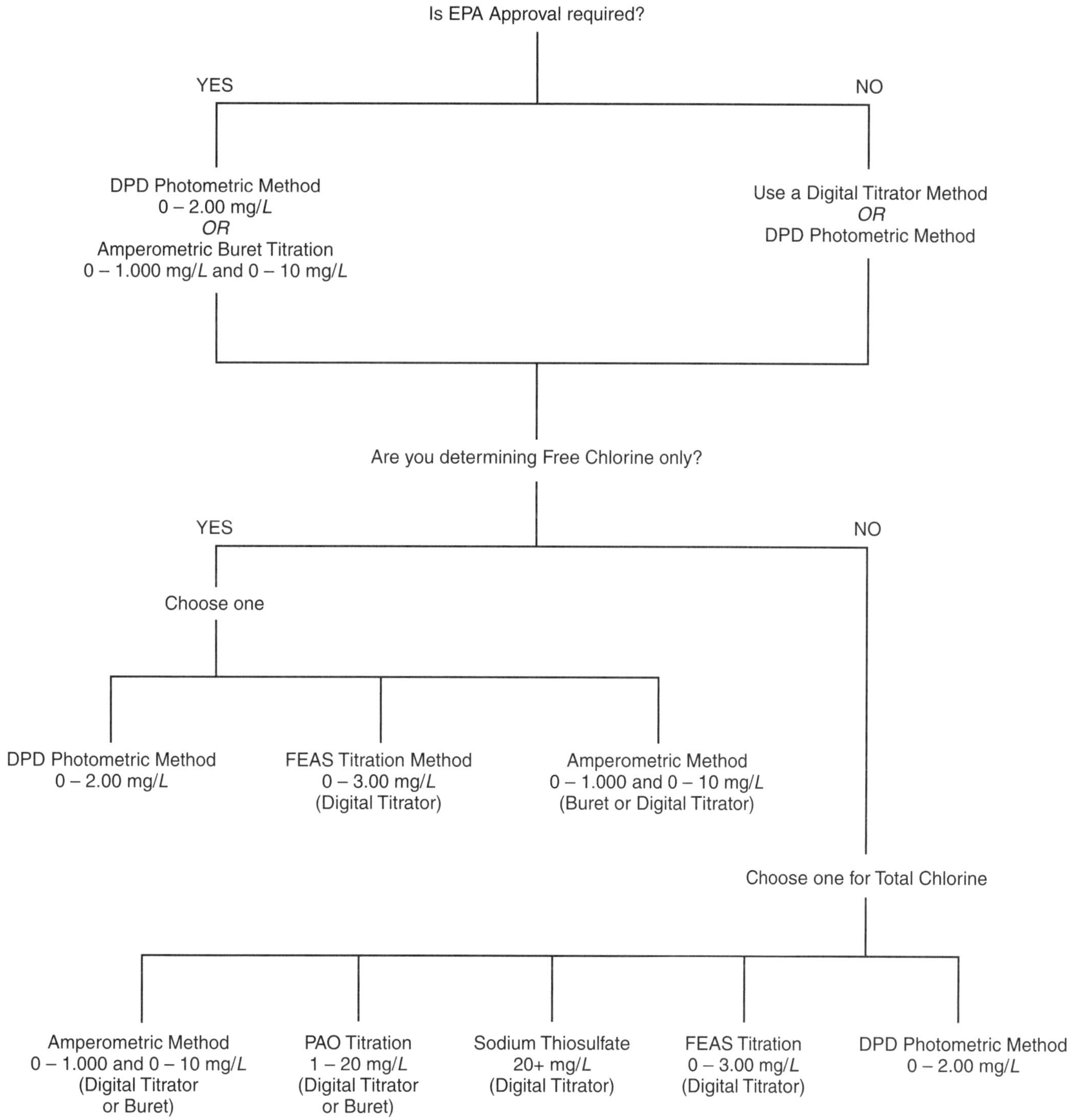

[a] HACH *WATER ANALYSIS HANDBOOK*

be responsible and should take all necessary steps to prevent accidents. This begins with the attitude that as good an effort as possible must be made by everyone. Safety is everyone's responsibility. The employer also must take an active part by supporting safety programs. There must be funding to purchase equipment and to enforce safety regulations required by OSHA and state industrial safety programs. The following elements should be included in all safety programs.

1. Establishment of a formal safety program,
2. Written rules,
3. Periodic hands-on training using safety equipment,
4. Establishment of a maintenance and calibration program for safety devices and equipment, and
5. Tours of the treatment facility for police and fire department personnel to locate hazardous areas and share chlorine safety information.

All persons handling chlorine should be thoroughly aware of its hazardous properties. Personnel should know the location and use of the various pieces of protective equipment and be instructed in safety procedures. In addition, an emergency procedure should be established and each individual should be instructed how to follow the procedures. An emergency checklist also should be developed and available. For additional information on this topic, see the Chlorine Institute's *CHLORINE MANUAL* (see Section 12.9, "Additional Reading and Contacts," reference 2, for ordering information).

12.31 Chlorine Safety and First-Aid Measures

MILD CASES

Whenever you have a mild case of chlorine exposure (which does happen from time to time around chlorination equipment), you should first leave the contaminated area. Move slowly, breathe lightly without exertion, remain calm, keep warm, and resist coughing. Notify other operators and have them repair the leak immediately.

If clothing has been contaminated, remove as soon as possible. Otherwise, the clothing will continue to give off chlorine gas, which will irritate the body even after leaving the contaminated area. Immediately wash off area affected by chlorine. Shower and put on clean clothes.

If victim has slight throat irritation, immediate relief can be accomplished by drinking milk. A mild stimulant such as hot coffee or hot tea is often used for coughing. Drinking spirits of peppermint also will help reduce throat irritation. See a physician.

EXTREME CASES

1. Follow established emergency procedures.
2. Always use proper safety equipment. Do not enter area without a self-contained breathing apparatus.
3. Remove patient from affected area immediately. Call a physician and begin appropriate treatment immediately.
4. First aid:
   a. Remove contaminated clothes to prevent clothing giving off chlorine gas, which will irritate the body.
   b. Keep patient warm and cover with blankets, if necessary.
   c. Place patient in a comfortable position on back.
   d. If breathing is difficult, administer oxygen if equipment and trained personnel are available.
   e. If breathing seems to have stopped, begin artificial respiration immediately. Mouth-to-mouth resuscitation or any of the approved methods may be used.

   f. EYES!

      If even a small amount of chlorine gets into the eyes, they should be flushed immediately with large amounts of lukewarm water so that all traces of chlorine are flushed from the eyes (at least 15 minutes). Hold the eyelids apart forcibly to ensure complete washing of all eye and lid tissues.
5. See a physician.

12.32 Hypochlorite Safety and First-Aid Measures

Hypochlorite does not present the hazards that gaseous chlorine does and therefore is safer to handle, but certain precautions are required because hypochlorite is an oxidizing agent. As an oxidizer, it will corrode metals and should not be stored in storage containers not specified to hold the solution.

Dry hypochlorite compounds are normally shipped in pails or drums with size depending upon the supplier.

Containers of hypochlorite should be stored in a cool, dry area away from other combustibles such as gasoline, solvents, oils, greases, rags, or papers.

When handling the hypochlorite, wear eye protection, such as goggles or face shields, chemical apron, and chemical gloves.

Do not inhale container vapors. After use, securely close the hypochlorite container to protect the remaining material from loss of chlorine and deterioration.

In case of fire apply liberal quantities of water.

Operators using liquid hypochlorite solutions must wear protective safety equipment including safety glasses or a face shield for eye protection as well as an apron and gloves. Without proper protection, operators can suffer severe skin burns from splashing of liquid solutions during handling or spraying due to a leaking pipe, tubing, or pressure fitting. Eyes should be flushed with eyewash and abundant amounts of water and a physician should be consulted. Skin contamination should be rinsed with water. If there is any indication of a chemical burn or rash, a physician should be consulted.

If hypochlorite is spilled, splashed, or sprayed on your skin or clothing, the water immediately starts to evaporate. This evaporation causes an increase in the chlorine and caustic concentrations until at the time of complete loss of water, almost pure sodium hydroxide (caustic) is left on your skin or clothing. Splashed clothing should be removed and washed.

When spills occur, wash with large volumes of water. The solution is messy to handle. Hypochlorite causes damage to your eyes and skin upon contact. Immediately wash affected areas thoroughly with water. Consult a physician if the area appears burned. Hypochlorite solutions are very corrosive. Hypochlorite compounds are nonflammable; however, they can cause fires when they come in contact with organics or other easily oxidizable substances.

12.33 Operator Safety Training

Training is a concern to everyone, especially when your safety and perhaps your life are concerned. Every utility agency should have an operator chlorine safety training program that introduces new operators to the program and updates previously trained operators. As soon as a training session ends, obsolescence begins. People will forget what they have learned if they don't use and practice their knowledge and skills. Operator turnover can dilute a well-trained staff. New equipment and also new techniques and procedures can dilute the readiness of trained operators. An ongoing training program can consist of a monthly luncheon seminar, a monthly safety bulletin that is to be read by every operator, and a presentation by outside speakers who can be brought in to reinforce and refresh specific elements of a safety training program. Safety training should include a scheduled review of all Material Safety Data Sheet (MSDS) information.

The American Water Works Association (AWWA) is an excellent source of training materials for safety programs, including a series of 52 lectures on common water utility safety practices (see Section 12.9, "Additional Reading and Contacts," reference 3, for ordering information).

12.34 CHEMTREC (800) 424-9300

Safely handling chemicals used in daily wastewater treatment is an operator's responsibility. However, if a situation ever gets out of hand, there are emergency teams that will respond with help anywhere there is an emergency. If an emergency does develop in your plant and you need assistance, call CHEMTREC (Chemical Transportation Emergency Center) for assistance. CHEMTREC will provide immediate advice for those at the scene of an emergency and then quickly alert experts whose products are involved for more detailed assistance and appropriate follow-up.

CHEMTREC'S EMERGENCY TOLL-FREE PHONE NUMBER IS (800) 424-9300.

QUESTIONS

Write your answers in a notebook and then compare your answers with those on page 208.

12.2G Why must the chlorine residual be measured as quickly as possible?

12.3A What would you do if hypochlorite came in contact with your hand?

END OF LESSON 2 OF 3 LESSONS
on
DISINFECTION AND CHLORINATION

Please answer the discussion and review questions next.

DISCUSSION AND REVIEW QUESTIONS

Chapter 12. DISINFECTION AND CHLORINATION

(Lesson 2 of 3 Lessons)

Write the answers to these questions in your notebook before continuing. The question numbering continues from Lesson 1.

6. Where might chlorine be applied in the treatment of wastewater?
7. Where is chlorine usually applied for disinfection purposes?
8. Calculate the hypochlorinator dose (lbs per 24 hours) to treat a waste with a chlorine demand of 12 mg/*L*, when a chlorine residual of 2 mg/*L* is desired, if the flow is 1 MGD.
9. What is the ultimate control of chlorine dosage for disinfection?
10. Why must the chlorine solution be well mixed with the wastewater?

CHAPTER 12. DISINFECTION AND CHLORINATION

(Lesson 3 of 3 Lessons)

12.4 OPERATION AND MAINTENANCE OF CHLORINATION EQUIPMENT

12.40 Disinfection Systems

This manual is intended for the small wastewater treatment plant. The following presentation of chlorination equipment is based on disinfection systems that would be the most reliable, safest, and easiest to maintain for the small system operator. Consistency of adequate disinfection of an effluent is the major concern.

This manual is not endorsing any particular manufacturer or supplier, only the method used to achieve NPDES permit requirements that would be established for the small system. For this reason, the order of presentation will be tablet chlorinators and liquid hypochlorinators.

For those plants with flows greater than what can be effectively disinfected with these systems, information is available in *OPERATION OF WASTEWATER TREATMENT PLANTS*, Volume I, Chapter 10, "Disinfection and Chlorination," that describes the operation and maintenance of systems using 150-pound containers to railroad tank cars containing chlorine.

12.41 Tablet Chlorinators

12.410 Description

The tablet-type chlorinator (refer to Figure 12.5) is a relatively maintenance-free method of chlorinating the discharge from a small wastewater treatment plant. Chlorine tablets are added in tubes that extend into a box through which the effluent passes. As the wastewater depth increases, more of the tablet is submerged thus allowing more chlorine to dissolve to match the increased flow rate. Use of this type of chlorinator is advantageous since it does not have any moving parts or electrical requirements and it does not require mixing of solutions.

The tablet chlorinator may be blocked by sticks, leaves, grease balls, and other matter, which would cause some additional chlorine tablets to dissolve as the liquid depth increases in the box. A blockage of this type may cause the wastewater to back up in preceding processes. Sometimes the tablets will also catch on the sides of the tubes and not drop to dissolve, usually due to solids buildup on the walls or improper placement of tablets in the tubes. Therefore, maintenance such as cleaning weirs, adding tablets, and cleaning and checking the tubes for "caught" tablets is necessary. Gloves, aprons, and eye protection should be worn when handling chlorine tablets.

There are several models of tablet chlorinators. Selection of a model should be based on the flow of water to be disinfected. For higher flows, more tubes are provided so that more tablets may dissolve. An adjustable weir is used in four-tube feeders to control depth of flow in the contact box. The bottom of the tablet chlorinator should be four inches above the liquid level in the contact chamber. Shims are sold with some new tablet chlorinators to prevent overchlorination at low flows. Older tablet chlorinators can be shimmed with approximately ¼-inch (0.6 cm) thick circular discs to prevent overchlorination.

QUESTIONS

Write your answers in a notebook and then compare your answers with those on page 208.

12.4A List the different types of chlorine feeders described in this chapter.

12.4B What type of debris can block flow to tablet chlorinators?

12.411 Tablet Chlorine Feeder

12.4110 OPERATION

Tablet chlorine feeders operate automatically without operator attention for as long as sixty days, depending on plant flow and the required chlorine residual. After the initial adjustment, operation is simple, automatic, and continuous.

The entire plant flow of treated wastewater passes into the unit through the inlet adaptor or pipe. As the stream of water flows past the feed tubes containing the chlorine tablets, active chlorine is released into the wastewater by the dissolving action of the wastestream in contact with the tablets.

At the outlet end, a weir (selected to match plant capacity) controls the height of the water level in the feeder, which actually controls the chlorine dose and residual in the wastewater, regardless of surges in wastewater flow entering the feeder. As the incoming wastewater flow rate increases, the wastewater level in the unit rises, immersing a greater number of chlorine tablets. When the incoming flow rate decreases, the wastewater level in the unit drops, exposing fewer tablets to the water. Since the amount of chlorine dissolved depends on the number of tablets immersed in the water, the initial chlorine concentration remains constant, regardless of the water level in the feeder.

From the feeder, the chlorinated wastewater flows into the chlorine contact tank where it is held for a sufficient time to permit effective pathogen killing action. The actual number of tubes to be filled with chlorine tablets and the weir to be used are determined by the average daily flow rate through the plant and the required chlorine residual.

After the system has been in operation for approximately one hour, samples should be taken from the effluent end of the chlorine contact tank for chlorine residual analysis. Although the chlorine residual may be determined by a variety of test methods, the preferred and recommended method is the starch-iodide procedure described in HACH's *WATER ANALYSIS HANDBOOK*.

Take several samples for chlorine analysis between the first and second hours of operation. This will allow time for the chlorine residual to reach a state of equilibrium in the detention tank. When two or three chlorine residuals determined 15 minutes apart are consistent with each other, the system may be considered stable. Occasional samples taken in the same way may be used to check the chlorine delivery at any time. If the results of the chlorine residual testing, or any other test, indicate that the unit is not providing sufficient chlorine, the necessary adjustments may be made by changing the weir size or feed tubes. (See Section 12.4112, "Troubleshooting.")

If desired, bacteriological analysis for *Escherichia coli* (*E. coli*) may be performed on the effluent from the detention tank. Optional analyses that may be run on the effluent are BOD, COD, pH, suspended solids, nitrite, ammonia, and dissolved oxygen. Tests for all these are described in HACH's *WATER ANALYSIS HANDBOOK*.

12.4111 MAINTENANCE

Tablet chlorine feeders require very little maintenance. Refilling of the feed tubes and an occasional on-stream cleaning of the feeder are the only maintenance procedures required.

Refilling the tubes is done on a schedule based on plant flow and weir size. The refill intervals are at the option of the operator. The tubes should always be filled to the top. This ensures the longest possible periods of unattended operation.

Occasional cleaning of the feeder to remove accumulated residues may be required. This will depend on the quality of the influent passing through the unit. Usually, once every six to twelve weeks is sufficient. Solids accumulated around the feed tubes are removed by pulling the tubes a few inches off the bottom of the feeder and raising the weir plate to permit the water to rush out the exit end of the dissolver. This action, if repeated a few times, will flush out most solids. Fibrous materials may occasionally stick to the bottom of the feed tubes. Use of a rod or gloved hand, combined with the shearing force of channeled water in the feeder, usually is sufficient to remove these materials.

Once a year the feed tubes may require removal of internal scale buildup by simply scraping the inside surfaces of the tubes.

When working with chlorine solutions and chlorine handling equipment, *ALWAYS* wear gloves, eye protection, and an apron.

12.4112 TROUBLESHOOTING

Incorrect amounts of chlorine delivered may be due to several causes. Some of the potential problems and their remedies are outlined below.

- **Insufficient Chlorine**

  If the chlorine residual is consistently too low in the chlorine contact tank, use an additional feed tube or a smaller size weir. Samples taken at the effluent end of the unit can be used to determine whether sufficient chlorine is being delivered into the chlorine contact tank.

  If the feed tubes do not reach the bottom of the unit, insufficient chlorine will be released into the wastewater. Check the feed tubes for contact with the bottom of the unit.

  If improperly loaded, the tablets can jam, causing the stack to remain suspended above the water level. This causes low dosage levels. Check to be sure that all the tablets are flat in the stack. If one or more tablets are jammed, a hard rap on the tube will loosen the bound tablets.

  A gross hydraulic overload will cause a lower than desired chlorine residual. Tablet chlorine feeders are designed to accommodate 100 to 200 percent over average flow. This level may be exceeded temporarily by abnormal conditions, such as heavy rains, and nothing should be changed if this is the case. However, if the overload into the plant becomes permanent due to the influent from additional sanitary installations from homes, businesses, or other sources, the additional flow must be compensated for by using a larger weir and/or more feed tubes to maintain the desired chlorine level.

- **Overchlorination**

  Overchlorination is due to excessive consumption of the chlorine tablet. The chlorine residual may be reduced by removing one or more feed tubes, adding spacers, and/or using a larger weir.

- **Plugging**

  If the flow through the weir is restricted, the water level will rise and cause backup in the unit exposing more tablets to the wastewater flow. The most common obstacles are leaves and sticks or wastewater solids. Remove these immediately whenever found.

  Under most conditions of flow, there is sufficient shearing action by the water to keep the slots in the feed tubes from becoming blocked. If the slots become obstructed, they may be cleared as described in Section 12.4111, "Maintenance," or by removing the feed tubes from the unit feeder and clearing the obstructing materials by hand.

12.4113 TABLET HANDLING AND STORAGE

Tablet chlorine is an oxidizing agent. It should be stored in a cool, dry area away from oxidizable material such as rags, paper, kerosene, or other combustible materials. Do not inhale container vapors. After use, securely close the container to protect the remaining material. In case of fire, apply liberal quantities of water.

12.4114 DECHLORINATION

Some effluents must be dechlorinated to protect aquatic life (fish) from toxic chlorine residuals. Sodium sulfite tablets can be used to dechlorinate water using equipment and procedures similar to using chlorine tablets to chlorinate wastewater. For additional information on dechlorination, see *OPERATION OF WASTEWATER TREATMENT PLANTS*, Volume I, Chapter 10, Section 10.8, "Dechlorination."

NEVER MIX or allow *CONTACT* of chlorine with sulfite tablets. This is a potentially *EXPLOSIVE* mixture.

12.4115 ACKNOWLEDGMENT

Material in this section was obtained from the ELTECH International Corporation's *INSTRUCTION*. Permission to use this material is sincerely appreciated.

QUESTIONS

Write your answers in a notebook and then compare your answers with those on page 208.

12.4C How are chlorine doses adjusted to flow changes with a chlorine tablet disinfection system?

12.4D How is chlorine released to wastewater by tablet chlorinators?

12.4E Why should the tubes always be filled to the top with chlorine tablets?

12.4F What happens if the feed tubes do not reach the bottom of the unit?

12.42 Disinfection of Wastewater Effluent by Hypochlorination

12.420 Sources of Hypochlorite

Recent technological developments have provided communities with three options for selecting a hypochlorination system:

1. On-site generation of hypochlorite,
2. Purchase of prepared hypochlorite liquid solutions ready for use, or
3. Purchase of dry chlorine compounds to be mixed into a solution for dispensing.

For the small wastewater treatment facility all systems are acceptable; however, there are several advantages with on-site chlorine/hypochlorite generation.

1. The hypochlorite solution is generated as needed. This provides a fresh solution that does not lose its strength when the disinfection system is not in use. This reduces the operational problem of readjusting dosage rates to meet a required chlorine residual.
2. The on-site generation system is safer for operators because they do not have to handle and transfer prepared liquid hypochlorite solutions or mix dry chlorine compounds to prepare a hypochlorite solution to be dispensed into the system. The only raw chemical that the operator is required to handle is salt. The on-site generated hypochlorite is 0.8 percent sodium hypochlorite, which may be only half the strength of pre-mixed or purchased hypochlorite solutions; however, pre-mixed or purchased hypochlorite solutions tend to lose strength during storage.

12.421 On-Site Chlorine/Hypochlorite Generation

On-site chlorine/hypochlorite generation systems are considered safer than other chlorination systems because they eliminate the need to store and handle dangerous gas/liquid chlorine or commercial hypochlorite solutions.

On-site systems use salt, water, and electricity to produce sodium hypochlorite. No liquid or gaseous chlorine is ever produced in the system, thus eliminating potentially dangerous storage and handling problems.

The production of hypochlorite is very simple. Salt is dissolved in water to form a concentrated brine solution. This solution is diluted and passed through a reaction tank containing cells where *ELECTROLYSIS*[29] takes place (see Figures 12.8 and 12.9). The final product solution contains approximately 0.8 percent sodium hypochlorite and 2 percent unreacted salt.

Consult the manufacturer's O & M manual for installation, start-up, operation, maintenance, and troubleshooting procedures.

NOTE: A similar system using salt has been developed to generate chlorine gas on site as needed.

12.422 Chlorine Chemicals for Hypochlorination Systems

Hypochlorite systems are safer to operate than gaseous/liquid chlorine systems. However, hypochlorite systems can be very time consuming for the operator.

Operation and maintenance of the hypochlorination system requires handling or the preparation of hypochlorite solutions. Dry chlorine products must be mixed with water to produce hypochlorite solutions. Solutions of known strength are prepared by handling dry material that must be accurately weighed and transferred to another container for mixing with a given volume of water. The material requires mixing to dissolve the dry chlorine into the water. Some compounds require continuous mixing to keep the product in solution. For this reason, many operators of small systems prefer to purchase prepared solutions of strengths up to 15 percent chlorine by volume. Hypochlorite liquid may be obtained in one-gallon, 30-gallon, or 50-gallon plastic containers. Liquid solutions, if not stored properly, or if kept for long periods, particularly if opened, will lose strength, reducing their effectiveness (chlorine strength).

The loss of hypochlorite solution strength presents another problem, particularly if opened containers are stored for a long time, or in intermittent flow applications such as disinfecting the effluent from a school where the system is shut down for a weekend or a holiday period and the chlorine solution is not used. An even more serious problem with some hypochlorination systems is unreliable flow of hypochlorite solution to provide continuous, adequate disinfection. This problem has caused a number of wastewater treatment plants to switch to other more reliable disinfection methods.

Determining an adequate pump setting to meet disinfection using the available hypochlorite solution requires running the pump, checking the chlorine residual, and readjusting the pump to achieve the adequate dose rate. This may require several adjustments and testing the chlorine residual in the effluent. There is no established guide to inform the operator what pump stroke will meet the required dose rate and desired chlorine residual. Hypochlorinator pumping units are prone to losing prime, plugging of solution tubing lines from deposits of chemicals being pumped, plugging of filters, leakage in pump heads preventing solution flow through the valves of the me-

[29] *Electrolysis (ee-leck-TRAWL-uh-sis). The decomposition of material by an outside current.*

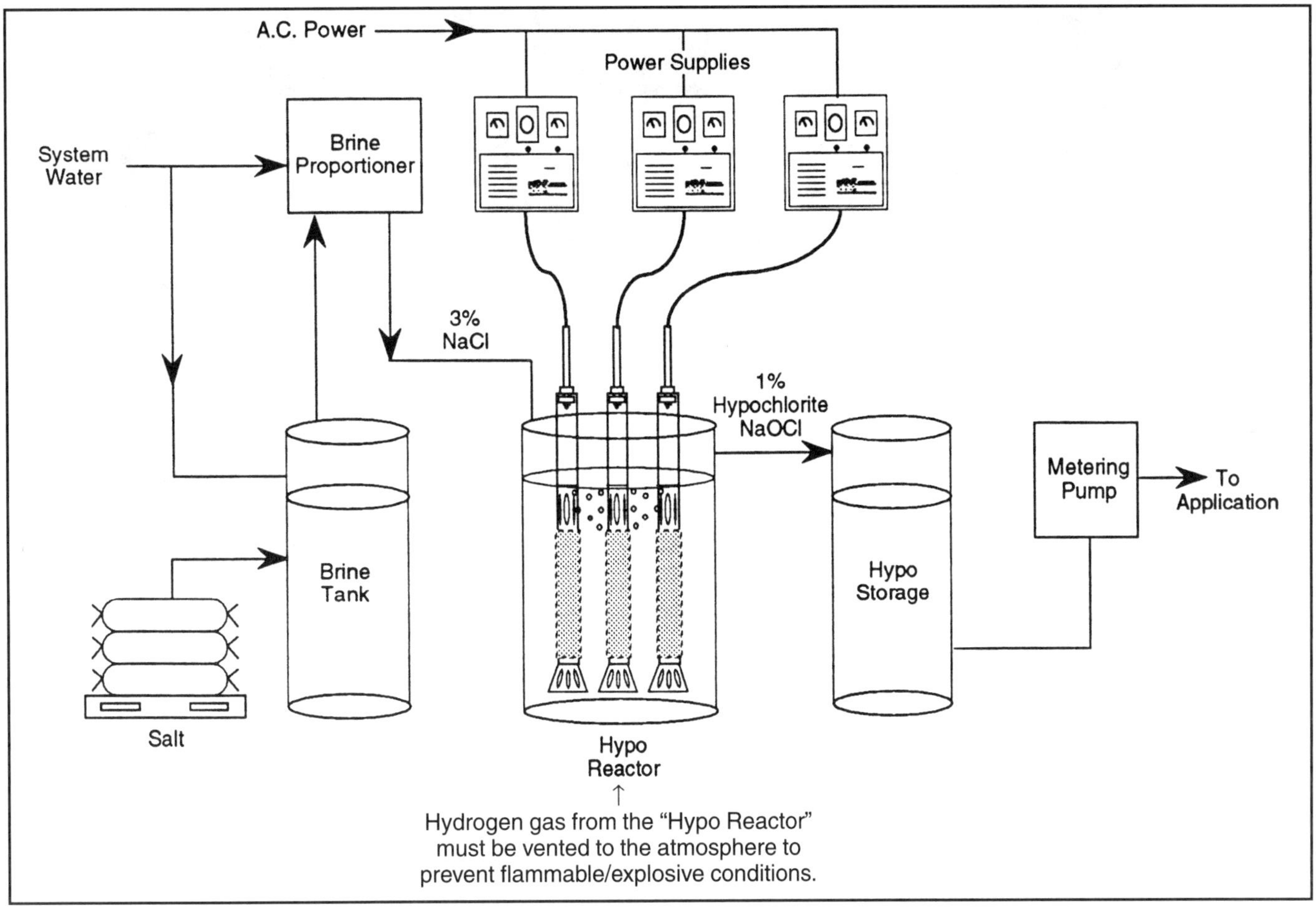

Fig. 12.8 On-site generation of hypochlorite
(Permission of Chemical Services Company)

tering pump, air lock, plugged valves, stripped gear trains, and ruptured diaphragms.

12.423 Components of a Hypochlorination System

Figure 12.10 is a diagram of a basic hypochlorination system. Hypochlorinators used on small wastewater systems are very simple and relatively easy to install. A typical wall-mounted installation is shown in Figure 12.11. The metering pump can be table mounted, wall mounted, or it can be mounted on a special plastic lid directly over a container holding solution. On a table- or wall-mounted unit, the rigid suction line is often replaced with a flexible suction line placed into a 30- or 50-gallon prepared hypochlorite solution. This eliminates the need to move the pumping system to a new container, and many small system operators prefer this arrangement.

When the solution container is nearly empty, a new container is placed beside the used one and the metering pump suction line is transferred over to the full container. When sufficient solution has been pumped out of the new container, the remaining contents of the old container can be transferred carefully into the container in service.

NOTE: Calcium hypochlorite (HTH) contains a significant amount of insoluble material that can plug pumps, pipes, and fittings. To reduce this problem, many operators mix the HTH powder or granules in a separate mixing tank and allow the insoluble solids to settle out. The remaining chlorine solution (supernatant) is then used to make up the hypochlorite feed solution in a separate tank.

12.424 Functions of Components

1. Storage container for solution. Premixed solutions are delivered in suitable drums or containers and the solution may be pumped directly from them. When mixing dry compounds to make solutions, at least two containers must be available, one to hold solutions previously mixed that are being used, and one available to mix a new batch of solution when required. Containers should be large enough to hold a two or three days' supply of hypochlorite solution. Use only containers made of materials compatible with hypochlorite solution. Never use empty containers that have held other chemicals.

2. Mixers are required when mixing and dissolving dry compounds to prepare a solution, and in some instances other chemical solutions may require continuous mixing to maintain the chemical in suspension. Purchased, prepared solutions usually do not require mixing.

3. Metering pump. There are several suppliers of chemical application pumping units. For the unit that is installed for your facility the proper operation and maintenance procedures should be in your plant O & M manual. If not, contact the supplier and request the operation and maintenance manuals for your particular unit.

Brine Tank Hypo Reactor Hypo Storage

Fig. 12.9 On-site generation of hypochlorite facility

(Permission of Chemical Services Company)

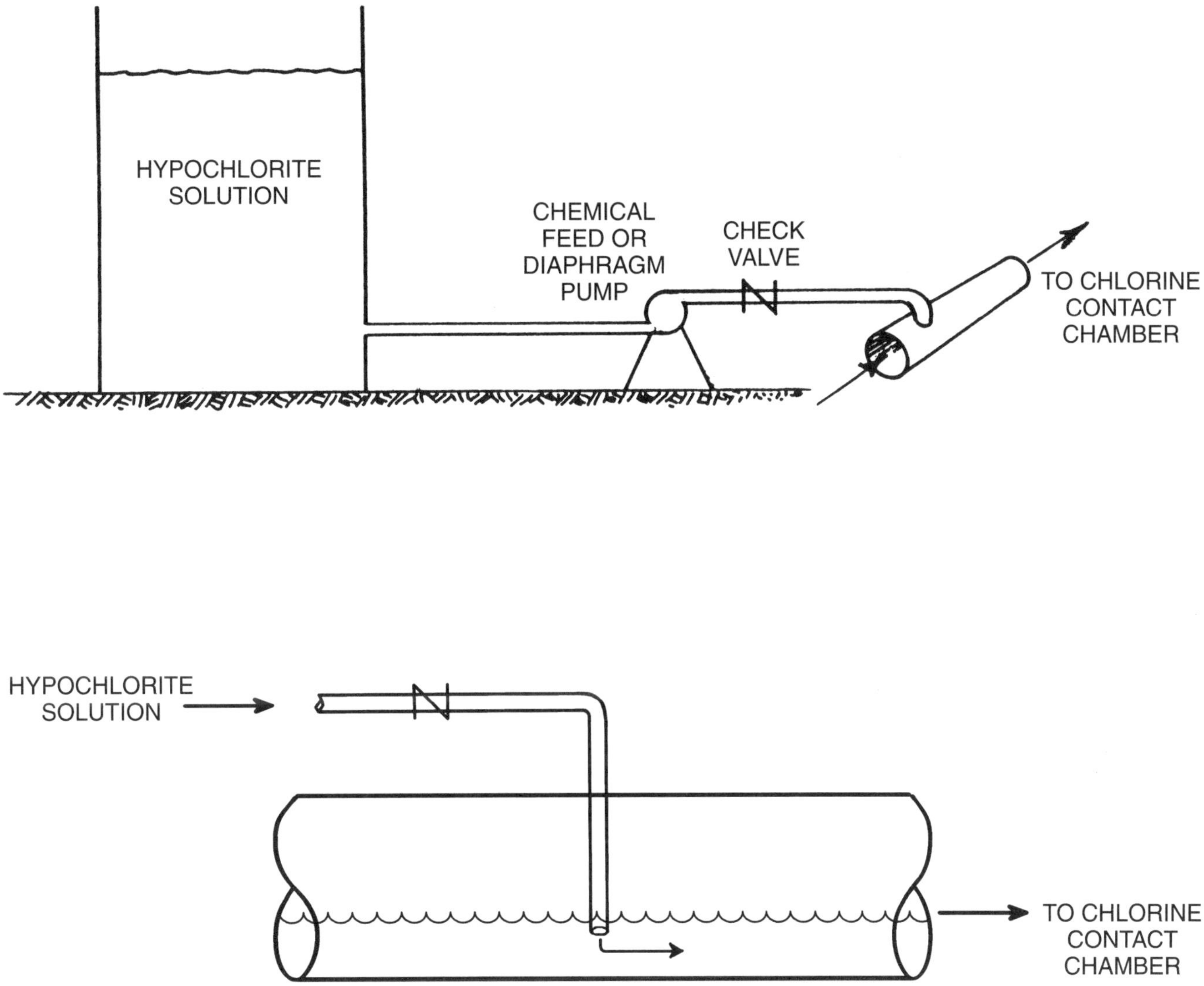

The hypochlorite solution application could be a weighted hose or a rigid pipe discharging into an open channel.

Fig. 12.10 Hypochlorinator direct pumping system

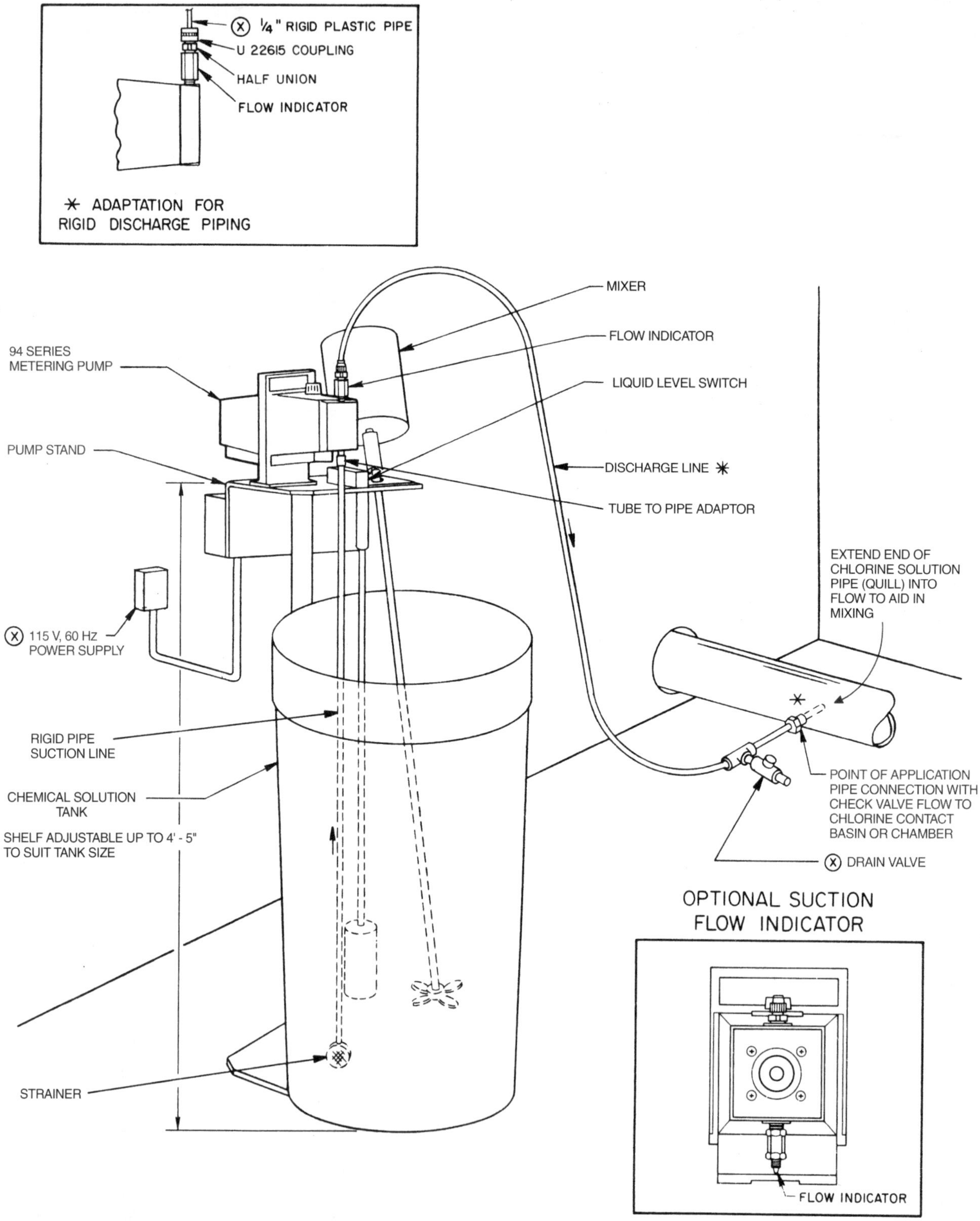

Fig. 12.11 Typical hypochlorinator installation

(Permission of Wallace & Tiernan Division, Penwalt Corporation)

The following are components with the metering pump shown in Figure 12.11.

1. Suction line equipped with a strainer/foot valve.
2. Metering pump.
3. Flow indicator.
4. Metering pump discharge line.
5. Drain valve.
6. Connection (Figure 12.12) with check valve.

   The connection is designed for installation in a ½-inch pipe tap. This may be in the bottom or side of a pipe or channel or in a tee in a line too small to tap. The point of application must be where the solution will be thoroughly mixed with all of the wastewater flowing through the pipe or system. For proper dispersion of solution, the end of the solution tube must be extended into the effluent pipe or channel approximately one-third of the diameter of the pipe. In performing routine annual maintenance, the exact replacement of the tube is critical. The chlorine solution mixes with the wastewater being treated as the flow moves to the chlorine contact chamber or basin.

7. Float switch (liquid level switch).

   The liquid level switch will activate an alarm. The alarm warns the operator to replenish the hypochlorite supply before the pump loses prime.

8. Spray guard.

   For operator protection when pumping hazardous fluids, use a spray guard covering the pump head (Figure 12.13).

NOTE: Sometimes a chemical metering pump will lose its prime and not pump any chemical. This problem is often caused by brittle cracking of the tubing from the chemical container at the point where the tubing connects to the pump. Cutting off about half an inch (1.3 cm) of the cracked tubing and reattaching the tubing to the pump can solve this problem.

Another common cause of loss of chemical pumping capacity is plugged *FOOT VALVES*[30] and strainers.

QUESTIONS

Write your answers in a notebook and then compare your answers with those on page 208.

12.4G What are the major advantages of on-site chlorine/hypochlorite generation?

12.4H What do on-site hypochlorite systems use to produce sodium hypochlorite?

12.4I What is a major problem with sodium hypochlorite systems?

12.43 Hypochlorinator Operation

12.430 Description

NOTE: The following material was adapted from pages 11 and 12, "Manual of Instruction for Package Plant Operators," prepared by West Virginia Department of Health, Office of Environmental Health Services, January 1992.

Liquid hypochlorinators are generally positive displacement pumps used to meter a chlorine solution into the influent end of the contact basin. A stock solution is made by mixing either sodium hypochlorite (5.25 to 7 percent by weight free available chlorine (bleach) or 12.5 percent industrial strength) or calcium hypochlorite (67 to 70 percent free available chlorine) with water. Calcium hypochlorite is not recommended for use, but if used, the stock solution should be mixed in a container other than the feed container so the insoluble material can settle out. The solution should be prepared every two or three days. If a greater strength solution is mixed, the solution may lose its strength and thus affect the chlorine feed rate. Normally, a week's supply of hypochlorite should be in storage and available for preparing hypochlorite solutions. Store the hypochlorite in a cool, dark place. Sodium hypochlorite can lose from two to four percent of its available chlorine content per month at room temperatures. Therefore, manufacturers recommend a maximum shelf life of 60 to 90 days.

The amount of chlorine feed may be varied by:

1. Diluting the stock solution,
2. Increasing or decreasing the feed rate, or
3. Time clock control.

The time clock control should only be used in certain situations such as schools where load patterns are predictable.

The hypochlorinator should be installed in a locked, insulated housing. A strip heater or heat lamp is required within the housing to prevent freezing of the stock solution. A check should also be made of the tubing and containers to ensure that they are suitable for use with high chlorine concentrations. Some materials (fiberglass) will deteriorate if exposed to concentrations in excess of one percent or 10,000 mg/*L*. An inlet strainer on a float should be installed on the suction side of the hypochlorinator. The float allows the strainer to withdraw chlorine from the upper portion of the stock solution container where fewer solids are found.

Calcium hypochlorite must be stored in closed containers away from any oils, greases, or fuels. Hypochlorites are strong oxidizing agents and will support and accelerate the burning of these substances. Hypochlorite has caused the rapid combustion (explosion) of a plastic bucket when ignited by a spark from a nearby refrigerator.

12.431 Start-Up

1. Solution. Chemical solutions have to be made up, or prepared solutions opened and placed in a container to be pumped.
2. Electrical. Lock out the circuit while making an inspection of an electrical circuit. Look for frayed wires. Normally, no repairs or adjustments are required. Turn the power back on.
3. Priming the pump. If the pump has been off for an extended period, the pump may have to be primed. If the pump discharges to the atmosphere, the pump will be self-priming. Having the valves wet may speed the priming process. Attempting to prime against pressure will prevent priming. Priming is fastest with the pump feed rate selector set at its highest output. *CAUTION:* Make any necessary adjustments of the pump output control selector only while the pump is running. Never adjust the pump while it is off because damage to the pump will occur.

---

[30] *Foot Valve. A special type of check valve located at the bottom end of the suction pipe on a pump. This valve opens when the pump operates to allow water to enter the suction pipe but closes when the pump shuts off to prevent water from flowing out of the suction pipe.*

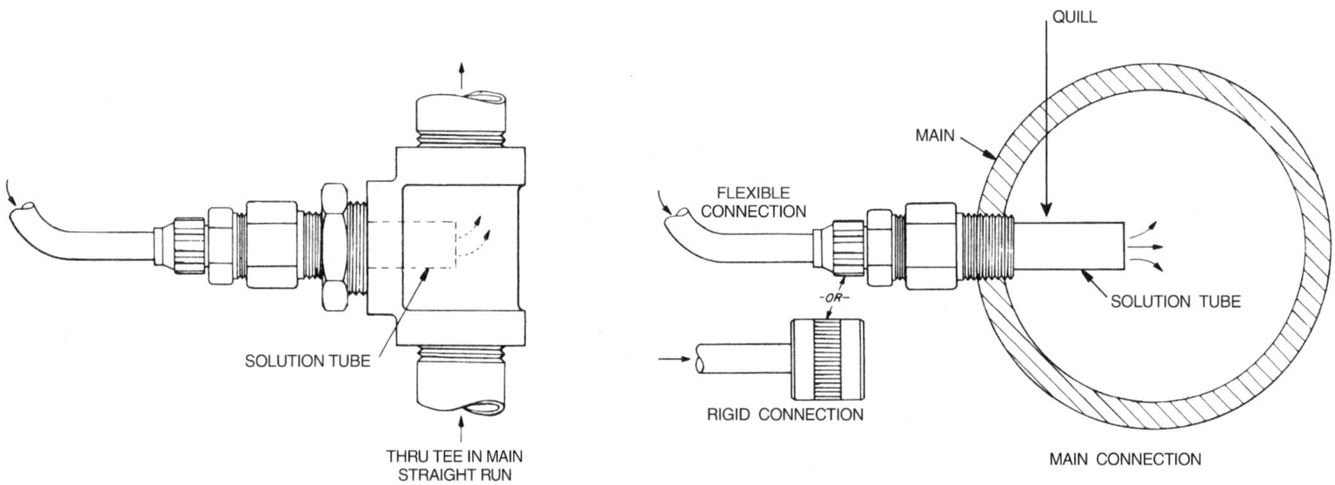

Fig. 12.12 Pipe connection

(Permission of Wallace & Tiernan Division, Penwalt Corporation)

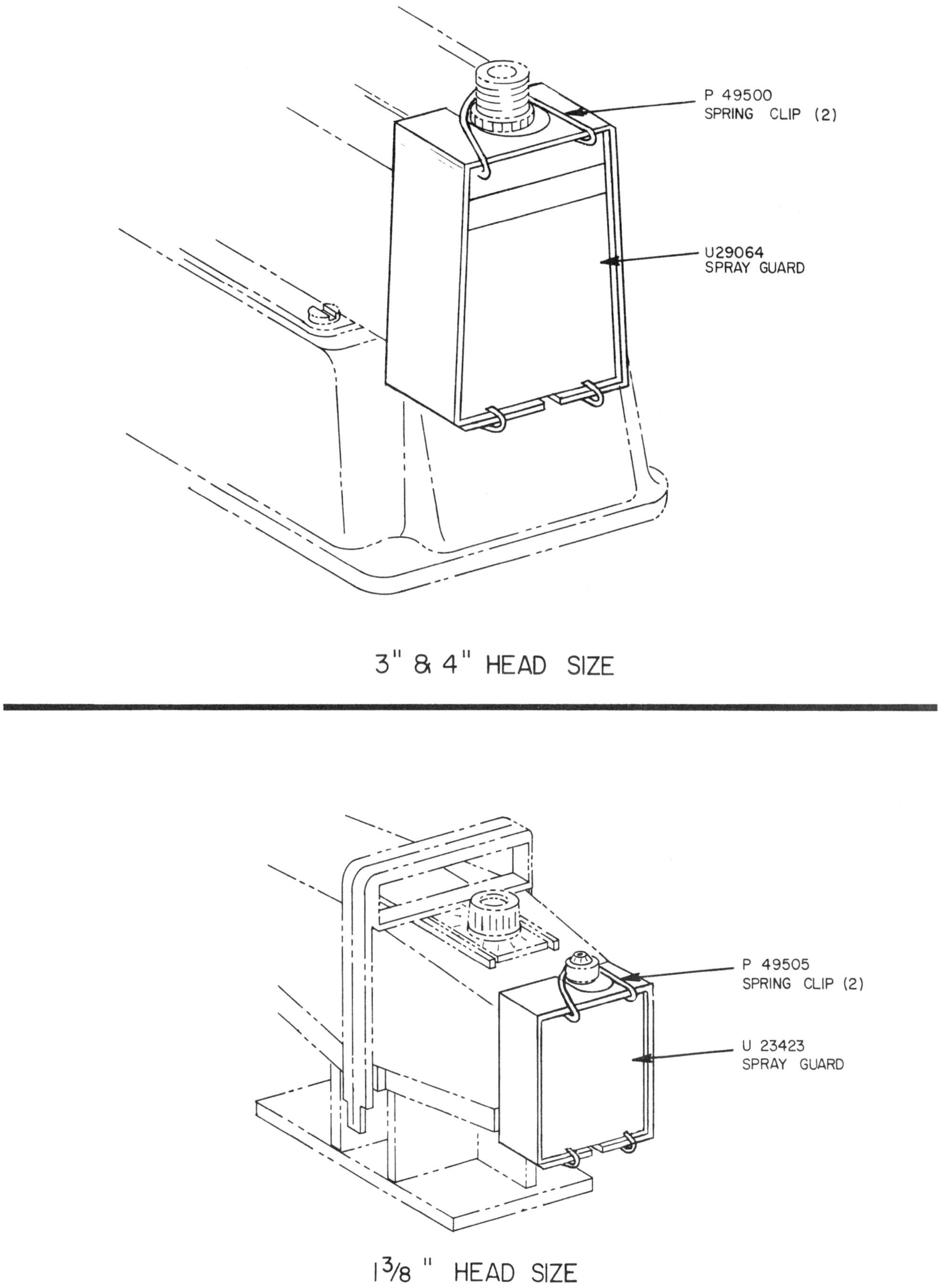

Fig. 12.13 Metering pump spray guard
(Permission of Wallace & Tiernan Division, Penwalt Corporation)

Prime the pump as follows:

a. Place a suitable container under the discharge drain valve to catch any liquid when performing step b.

b. If a pressure connection with a check valve is used, open the discharge drain valve. Or, if a discharge shut-off valve is used, shut that valve off and open the discharge drain valve. *WARNING:* To avoid possible severe personal injury, ensure that the spray guard is in place before starting the pump.

c. Start the pump. (Make pump adjustment to maximum output.)

d. When liquid is seen discharging from the drain valve, the pump is primed.

e. Return pump output selector back to the position at which it previously operated. Then stop the pump.

f. Close the drain valve and open the discharge shut-off valve (if used).

4. Restart the pump for normal operation.

5. Check the chlorine residual in the effluent. Recommended chlorine residual should be 0.5 mg/*L* at the discharge of the chlorine contact structure. Adjust the chemical feed as required.

NOTE: The feed rate is adjusted by the selector handle or knob on the pump housing. It is suggested that solution strength be adjusted so that the pump may be operated at a dial setting of 6 or 7. This permits adjustment in either direction, if required, by a temporary change in chlorine demand. Remember, the selector may be readjusted only when the pump is running. Since capacity is affected somewhat by operating pressure, the dial reading should be considered a reference point and not an exact indication of pump delivery. Once the dial setting has been established for the desired residual, an adjustment should not be necessary unless other conditions such as hypochlorite solution strength, chlorine demand of the effluent, or pump discharge pressure should change.

12.432 Shutdown

1. Short duration.

a. Turn the pump off.

b. When making any repairs, lock out and tag the electric circuit supplying the pump, or pull the pump motor plug from its electrical outlet socket.

2. Long duration, particularly in cold weather.

a. Disconnect and drain the suction and discharge lines.

b. If the anti-siphon spring is not used, turn the pump upside down to drain.

c. If the anti-siphon spring is used, remove the discharge half union and attached parts before inverting the pump to drain.

3. Failure of hypochlorinator feed pump.

a. Replace damaged pump with standby unit. Repair damaged pump or obtain replacement.

12.433 Normal Operation and Preventive Maintenance

Normal operation of the hypochlorination process requires routine observation and preventive maintenance.

DAILY

1. Inspect the building to make sure only authorized personnel have been there.
2. Read and record the level of the solution tank at the same time every day.
3. Check the chlorine residual of the effluent. Adjust the feed rate as required.

WEEKLY

1. Clean the building.
2. Replace the chemicals and wash the chemical storage tank. Try to have a 15- to 30-day supply of chlorine in storage for future needs. When preparing hypochlorite solutions, prepare only enough for a two- or three-day supply.

MONTHLY

1. Check the operation of the check valve.
2. Perform any required preventive maintenance.
3. Cleaning.

Commercial sodium hypochlorite solutions (such as Clorox) contain an excess of caustic (sodium hydroxide or NaOH). When this solution is diluted with water containing calcium and also carbonate alkalinity, the resulting solution becomes supersaturated with calcium carbonate. This calcium carbonate tends to form a coating on the poppet valves in the solution feeder or in the tubing of finger and diaphragm pumps. The coated valves will not seal properly and the feeder will fail to feed properly.

Use the following procedure to remove the carbonate scale:

a. Take a one-quart (one-liter) Mason jar and fill it half full of tap water.

b. Place one fluid ounce (20 m*L*) of 30 to 37 percent hydrochloric acid (swimming pool acid) in the jar, or use white vinegar.

ALWAYS ADD ACID TO WATER, NEVER THE REVERSE.[31]

c. Complete filling the jar with tap water.

d. Place the suction hose of the hypochlorinator in the jar and pump the entire contents of the jar through the system.

e. Return the suction hose to the hypochlorite solution tank and resume normal operation.

You can prevent the formation of the calcium carbonate coatings by obtaining the dilution water from an ordinary home water softener.

---

[31] *Water poured into a beaker or container of acid produces a violent chemical reaction generating heat and splattering of the chemical, which can cause serious injury to operators.*

NORMAL OPERATION CHECKLIST

1. Check chemical usage. Record the solution level and the number of hours of pump operation. Calculate the amount of chemical solution used and compare it with the desired feed rate.
2. Determine if every piece of equipment is operating.
3. Inspect the lubrication of the equipment.
4. Check the building for any possible problems.
5. Clean up the area.

FORMULAS

To determine the actual chlorine dose of wastewater effluent being treated by either a chlorinator or a hypochlorinator, we need to know the gallons of wastewater treated and the pounds of chlorine used to disinfect the wastewater.

1. Calculate the pounds of wastewater disinfected.

   Wastewater, lbs = (Plant Effluent, gallons)(8.34 lbs/gal)

2. To calculate the volume of hypochlorite used from a container, we need to know the dimensions of the container.

   $$\text{Volume, gallons} = \frac{(0.785)(\text{Diameter, in})^2(\text{Depth, ft})(7.48\text{ gal/cu ft})}{144\text{ sq in/sq ft}}$$

3. To calculate the pounds of chlorine used to disinfect wastewater, we need to know the gallons of hypochlorite used and the percent available chlorine in the hypochlorite solution.

   $$\text{Chlorine, lbs} = (\text{Hypochlorite, gal})(8.34\text{ lbs/gal})\left(\frac{\text{Hypochlorite,\%}}{100\%}\right)$$

4. To determine the actual chlorine dose in milligrams per liter, we divide the pounds of chlorine used by the millions of pounds of wastewater treated. Pounds of chlorine per million pounds of water is the same as parts per million or milligrams per liter.

   $$\text{Chlorine Dose, mg}/L = \frac{\text{Chlorine Used, lbs}}{\text{Wastewater Treated, Million lbs}}$$

12.434 Abnormal Operation and Troubleshooting

1. Inform your supervisor of the problem.
2. If the hypochlorinator malfunctions, it should be repaired immediately. See the shutdown operation (Step 3 in Section 12.432).
3. Solution tank level.
   a. If too low: Check the adjustment of the pump. Check the hour meter of the pump.
   b. If too high: Check the chemical pump. Check the hour meter of the pump.
4. Determine if the chemical pump is not operating.

   TROUBLESHOOTING GUIDELINES

   a. Check the electrical connection.
   b. Check the circuit breaker.
   c. Check for stoppages in the flow lines.

   CORRECTIVE MEASURES

   a. Stop the plant effluent discharge, if possible.
   b. Check for a blockage in the solution tank.
   c. Check the operation of the check valve.
   d. Check the connection of the suction hose to the pump.
   e. Check the electric circuits.
   f. Replace the chemical feed pump with another pump while repairing the defective unit.
5. The solution is not being pumped to the chlorine contact diffuser.

   TROUBLESHOOTING GUIDELINES

   a. Check the solution level.
   b. Check for blockages in the solution line.

12.435 Routine Maintenance

Hypochlorinators on small systems are normally small. Review the manufacturer's suggestions for maintenance requirements.

Suggested maintenance tasks are outlined below.

MONTHLY

1. Check the strainer for clogging.
2. Check flex lines for chemical buildup and for cuts, cracks, and also loose connections at fittings.
3. Check vat or container condition. Look for cracks and worn bottoms from moving container.
4. Inventory chemical supplies on hand.
5. Clean and check operation of the check valve.
6. Clean and check the pump head.
7. Check volume of pump output.

SEMIANNUALLY

1. Check and clean the chlorine diffuser.
2. Review the pump spare parts inventory and reorder as necessary to maintain adequate repair stock.

QUESTIONS

Write your answers in a notebook and then compare your answers with those on page 208.

12.4J How are hypochlorite solutions prevented from freezing?

12.4K Why should the hypochlorinator pump spare parts inventory be reviewed?

12.5 DISINFECTION SYSTEMS

12.50 Restricted Applications of Chlorine for Disinfection

All plants are required to disinfect their effluent before discharge to surface waters, and to be effective most effluents must maintain a chlorine residual of at least 0.5 mg/*L*. Excessive chlorine residual is not required because it is costly and can be toxic to aquatic life in the receiving streams. Chlorination as a form of disinfection may be restricted on certain chlorine-limited streams or trout streams. Dechlorination, usually using sulfur dioxide, may be used in these cases. Sulfur dioxide usually depletes the dissolved oxygen (DO) necessitating post-aeration and may lower the effluent pH. Use of ultraviolet disinfection on chlorine-limited streams may be required. For information on dechlorination, see Chapter 10, "Disinfection and Chlorination" in *OPERATION OF WASTEWATER TREATMENT PLANTS*, Volume I, in this series of operator training manuals.

12.51 Ultraviolet Disinfection (UV) (See Figure 12.14)

Common ultraviolet (UV) units either encase UV bulbs in a quartz sleeve or convey wastewater effluent through Teflon tubes surrounded by UV bulbs. UV light is able to pass through both clear Teflon and quartz with a minimal loss of energy. UV light disinfects by damaging the *DNA*[32] and *RNA*[33] of organisms. The effectiveness of UV disinfection depends on the amount of UV energy absorbed by the organism. The amount of UV energy available to be absorbed by the organism is diminished by distance from the UV source, UV absorbing constituents in the wastewater, the cleanliness of the transmission surface of the UV unit (usually Teflon or quartz), and the suspended solids concentration in the effluent. Organisms covered or shielded by suspended solids may be very resistant to UV energy due to shielding.

UV disinfection is not suitable for effluents with high turbidity, greases, oils, suspended solids, dyes, organics, or other UV-absorbing constituents. UV disinfection is applicable to most high-quality secondary treatment effluents with total suspended solids (TSS) and biochemical oxygen demand (BOD) less than 20 mg/*L*. Disinfection efficiency is related to degree of treatment of TSS and BOD.

Quartz or Teflon surfaces exposed to wastewater must be kept clean. There are several methods to clean the fouled surfaces, for example, mechanical wipers, ultrasound, or chemical cleaning with high-pressure spray wash. To protect quartz lamps from breaking, removable screens are usually installed ahead of the unit to prevent debris from entering the system.

Properly operated and maintained UV systems are as reliable as the quality of effluent to be disinfected. UV systems will not perform well if upstream process units fail, causing the effluent to be high in suspended solids or other wastewater constituents that can absorb UV light. Low-intensity UV bulbs must be changed every nine to twelve months. UV systems constantly require energy. Improperly matched lamps and ballasts either will not work or will have much shorter life cycles. The life cycle of the ballasts may be greatly reduced by excessive heat. It is important to have adequate ventilation for the power panel housing the ballasts.

UV disinfection systems must be provided with adequate access and proper handling equipment to allow quick and safe removal, cleaning, repair, and replacement of parts and equipment.

UV disinfection systems do not produce halogenated organic compounds (such as trihalomethanes) as does chlorine. Since UV disinfection does not use chemicals, there is not the possibility of chemical spills or releases. UV light must be shielded to prevent eye damage to O & M staff and plant visitors.

12.52 Chlorination Troubleshooting (See Table 12.3 and Table 12.4)

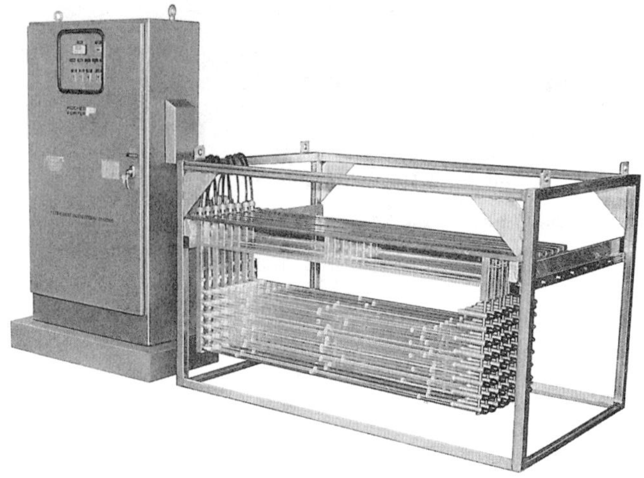

UV lamp assembly

(Reproduced with permission of Fischer & Porter (Canada) Limited)

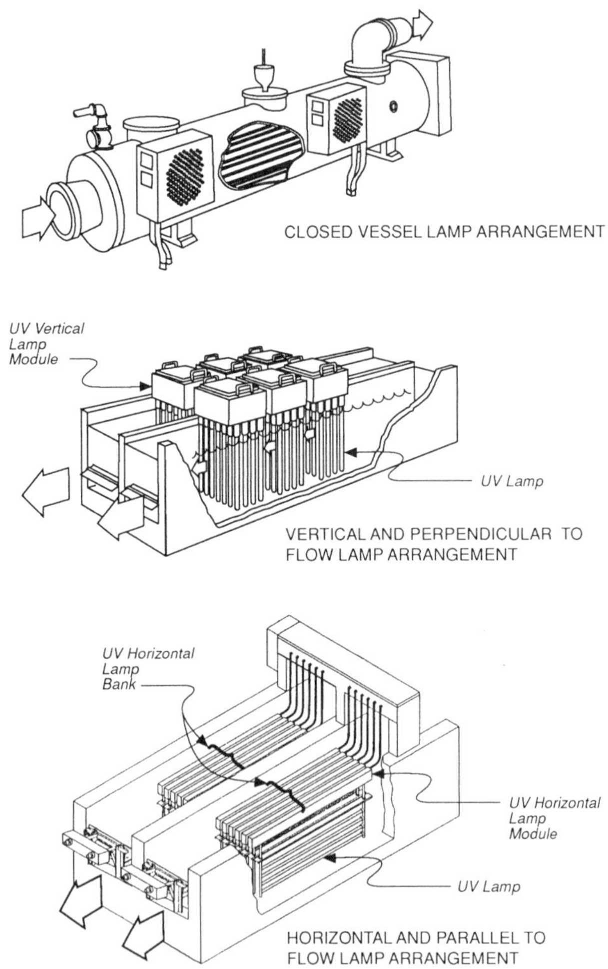

Typical UV lamp configurations

(Source: "Ultraviolet Disinfection," by CH2M Hill, reproduced with permission of CH2M Hill)

Fig. 12.14 Ultraviolet disinfection systems

---

[32] *DNA (pronounce as separate letters).* ***D****eoxyribo****n****ucleic* ***a****cid. A chemical that encodes genetic information that is transmitted between generations of cells.*

[33] *RNA (pronounce as separate letters).* ***R****ibo****n****ucleic* ***a****cid. A chemical that provides the structure for protein synthesis (building up).*

TABLE 12.3 TROUBLESHOOTING GUIDE—CHLORINATION

Indicator/ Observation	Probable Cause	Check or Monitor	Solution	
			Type of Chlorination Equipment	
			Tablet Feeder	Pump System
1. Coliform count fails to meet required NPDES permit	1a. Chlorine residual too low	1a. 1. Chlorine residual	1a. 1. Increase chlorine feed rate a. Check tubes for tablets b. Tablets reach bottom of tubes c. Weir is holding proper water level in feeder box d. Effluent flow higher than usual. Add tablets to additional tube or adjust box water depth by installing higher weir	1a. 1. Increase chlorine feed rate a. Check hypochlorite solution tank for material to be pumped b. If pump functions normally, adjust pump feed selector to a higher rate of feed
		2. 30 minutes after increasing feed rate check chlorine residual in effluent	2. a. If chlorine residual is OK, leave as is b. If chlorine residual is still low, increase box water depth or add more tubes and/or tablets	2. If chlorine residual is still low and feed rate is already in high range: a. Check pump discharge for correct output (see Section 12.431, item 3 (a-f) b. If pump discharges solution but low volume, check strainer for solids clogging in hypochlorite solution tank and pump suction line c. Check pump delivery system (see **Table 12.4** for solutions for different types of pumps)
	1b. Chlorine residual too low. Chlorine demand keeps increasing	1b. Check for solids buildup in chlorine contact chamber	1b. Clean chlorine contact chamber	1b. Clean chlorine contact chamber
	1c. Short-circuiting in chlorine contact chamber (basin)	1c. 1. Detention time through chlorine contact chamber. Run dye test. Dump dye into influent to chlorine contact chamber. Monitor time dye enters until first sign of dye in effluent leaving contact chamber	1c. 1. If less than thirty (30) minutes contact time, install baffling to create a serpentine flow through basin or add mixing to reach outer edge of chamber to achieve even distribution of flow through tank	1c. 1. If less than thirty (30) minutes contact time, install baffling to create a serpentine flow through basin or add mixing to reach outer edge of chamber to achieve even distribution of flow through tank
		2. Hypochlorite solution available		2. Provide hypochlorite solution to be pumped
		3. Suction strainer fouled		3. Clean strainer

TABLE 12.3 TROUBLESHOOTING GUIDE—CHLORINATION (continued)

Indicator/ Observation	Probable Cause	Check or Monitor	Solution: Type of Chlorination Equipment — Tablet Feeder	Solution: Type of Chlorination Equipment — Pump System
1. Coliform count fails to meet required NPDES permit (continued)	1c. Short-circuiting in chlorine contact chamber (basin) (continued)	1c. 4. Suction line plugged		1c. 4. Clean line
		5. Old hypochlorite solution		5. Increase pump feed rate or change to fresher hypochlorite solution
		6. Cold hypochlorite solution thick or freezing		6. Heat solution tank, if possible agitate solution tank contents
		7. Discharge of chemical to effluent		7. Check valve stuck closed, discharge line plugged or broken: a. Provide high discharge pressure at pump b. Broken. No excessive pressure at pump
		8. High chlorine demand of effluent due to upstream influences of spills or dumps of toxic waste or other upstream treatment processes		8. Correct upstream condition, feed maximum chlorine rate, divert effluent flow to temporary storage, for example, empty tank, pond, or control influent flow by storing in collection system without flooding
	1d. No chlorine residual in effluent	1d. Check chlorination feed system 1. Pump not running	1d. Check 1a. 1., items a. – d.	1d. Electrical system 1. Circuit breaker tripped 2. Time clock not operating 3. Pump motor burned out or pump not working
2. Chlorine residual too high in plant effluent to meet requirements	2a. Chlorine feed rate too high	2a. Chlorine residual	2a. Reduce chlorine feed rate 1. Check tablet feeder box for debris or grease on weir raising water level on tubes. Clean weir 2. Weir clean, flow normal, more than one tube in service: remove tablets from one tube 3. One tube in service: install smaller weir to increase flow through box, reducing water contact time with tablets	2a. Reduce chlorine feed rate 1. Lower pump feed rate control knob to a lower setting 2. Reset time clock for shorter pumping cycle 3. Dilute hypochlorite solution to make a weaker chemical dosage

TABLE 12.3 TROUBLESHOOTING GUIDE—CHLORINATION (continued)

Indicator/ Observation	Probable Cause	Check or Monitor	Solution: Type of Chlorination Equipment — Tablet Feeder	Solution: Type of Chlorination Equipment — Pump System
3. Coliform count fails to meet required standards for disinfection	3a. Chlorine residual too low	3a. Chlorine residual	3a. 1. Increase chlorine feed rate 2. Increase chlorine contact time	3a. 1. Increase chlorine feed rate 2. Increase chlorine contact time
	3b. Inadequate chlorine residual control	3b. Continuously record residual in effluent	3b. Use chlorine residual analyzer to monitor and control the chlorine dosage automatically	3b. Use chlorine residual analyzer to monitor and control the chlorine dosage automatically
	3c. Inadequate chlorination equipment capacity	3c. Check capacity of equipment	3c. Replace equipment as necessary to provide treatment based on maximum flow through plant	3c. Replace equipment as necessary to provide treatment based on maximum flow through plant
	3d. Solids buildup in contact chamber	3d. Visual inspection	3d. Clean contact chamber to reduce solids buildup	3d. Clean contact chamber to reduce solids buildup
	3e. Short-circuiting in contact chamber	3e. Contact time	3e. 1. Install baffling in contact chamber 2. Install mixing device in contact chamber	3e. 1. Install baffling in contact chamber 2. Install mixing device in contact chamber
	3f. Coliform regrowth in piping/sample station	3f. Effluent coliform sampling station	3f. Modify system as needed to provide adequate chlorine residual to prevent growth	3f. Modify system as needed to provide adequate chlorine residual to prevent growth
4. Chlorine residual too high in plant effluent to meet requirements	4a. Chlorine feed rate too high	4a. Chlorine residual	4a. Reduce chlorine feed rate	4a. Reduce chlorine feed rate
	4b. Malfunctioning chlorine residual control	4b. Operation of residual control system/analyzer	4b. Repair/calibrate chlorine feed/residual control loops	4b. Repair/calibrate chlorine feed/residual control loops
5. Breakout (breakaway) of Chlorine[a]	5a. Overfeeding chlorine	5a. Chlorine feed rate	5a. Decrease chlorine feed rate to minimize breakout and maximize application efficiency	5a. Decrease chlorine feed rate to minimize breakout and maximize application efficiency

[a] Breakout of Chlorine. A point at which chlorine leaves solution as a gas because the chlorine feed rate is too high. The solution is saturated and cannot dissolve any more chlorine. The maximum strength a chlorine solution can attain is approximately 3,500 mg/*L*. Beyond this concentration molecular chlorine is present which will break out of solution and cause "off-gassing" at the point of application.

TABLE 12.4 TROUBLESHOOTING GUIDE—CHLORINE SOLUTION PUMPS

Problem	Type of Pump	Check/Observe	Solution
Low discharge of chlorine solution	Finger Pump	1. Pump running, fingers activating and forcing solution through tubing but discharge is low	1a. Stop pump, disconnect power, tag, and lock out
			1b. Check tubing through pump head for cut, crimped, or plugged tubing and replace tubing if needed
			1c. Restart pump
		2. Discharge still low	2. Replace pump
	Diaphragm Pump	1. Pump running, diaphragm being activated but discharge is low	1a. Stop pump, disconnect power, tag, and lock out
			1b. Open head, check diaphragm. Replace cut, torn, or unseated diaphragm with new diaphragm
		2. Discharge still low	2. Replace pump
	Gear/Lobe Pump	1. Lobes or gears	1a. Stop pump, disconnect power, tag, and lock out
			1b. Bleed pressure from pump head
			1c. Open head and manually check operation of lobes or gears; should mesh, rotate freely
		2. Chemical buildup on gears or housing restricting operation	2a. Housing worn, permitting bypassing of gears or lobes
			2b. Restore clearances or replace pump
	Piston/Poppet Valve Pump	1. Piston or valves for damage or chemical deposits	1a. Stop pump, disconnect power, tag, and lock out
			1b. Bleed pressure from pump head
			1c. Remove head and check piston or valves for free rotation, proper meshing
		2. Pistons or valves damaged or chemical deposit blocking actions	2a. Remove deposits; pistons/valves should mesh, rotate freely
		3. Head assembly appears OK	3. Replace pump

QUESTIONS

Write your answers in a notebook and then compare your answers with those on page 208.

12.5A Why is excessive chlorine residual not required?

12.5B How does UV light disinfect?

12.5C Under what conditions is UV light not suitable?

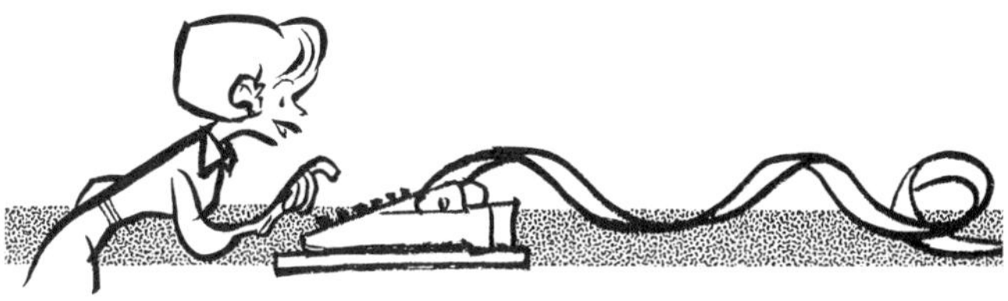

12.6 ARITHMETIC CALCULATIONS

1. A hypochlorinator is set to feed 30 pounds of chlorine per 24 hours; the wastewater flow is 0.8 MGD, and the chlorine as measured by the chlorine residual test after 30 minutes of contact time is 0.5 mg/*L*. Find the chlorine dosage and chlorine demand in mg/*L*.

Known		Unknown
Chlorine Feed, lbs/day	= 30 lbs/day	1. Chlorine Dose, mg/*L*
Flow, MGD	= 0.8 MGD	2. Chlorine Demand, mg/*L*
Chlorine Residual, mg/*L*	= 0.5 mg/*L*	

1. Calculate the chlorine dose in mg/*L*.

$$\text{Chlorine Feed or Dose, mg/}L = \frac{\text{Chlorine Feed, lbs/day}}{(\text{Flow, MGD})(8.34\text{ lbs/gal})}$$

$$= \frac{30\text{ lbs/day}}{(0.8\text{ MGD})(8.34\text{ lbs/gal})}$$

$$= 4.5\text{ mg/}L$$

Or

$$\text{Chlorine Feed or Dose, mg/}L = 30\text{ lbs/day} \div 0.8\text{ MGD} \div 8.34\text{ lbs/gal}$$

$$= 4.5\text{ mg/}L$$

2. Determine the chlorine demand in mg/*L*.

$$\text{Chlorine Demand, mg/}L = \text{Chlorine Dose, mg/}L - \text{Chlorine Residual, mg/}L$$

$$= 4.5\text{ mg/}L - 0.5\text{ mg/}L$$

$$= 4.0\text{ mg/}L$$

2. Calculate the hypochlorinator setting in pounds per 24 hours to treat a waste with a chlorine demand of 8 mg/*L*, when a chlorine residual of 1 mg/*L* is desired. The flow is 0.6 MGD.

Known		Unknown
Chlorine Demand, mg/*L*	= 8 mg/*L*	Hypochlorinator Setting, lbs/day
Chlorine Residual, mg/*L*	= 1 mg/*L*	
Flow, MGD	= 0.6 MGD	

1. Calculate the chlorine dose in mg/*L*.

$$\text{Chlorine Dose, mg/}L = \text{Chlorine Demand, mg/}L + \text{Chlorine Residual, mg/}L$$

$$= 8\text{ mg/}L + 1\text{ mg/}L$$

$$= 9\text{ mg/}L$$

2. Calculate the hypochlorinator setting in pounds per day.

$$\text{Hypochlorinator Setting, lbs/day} = (\text{Flow, MGD})(\text{Dose, mg/}L)(8.34\text{ lbs/gal})$$

$$= (0.6\text{ MGD})(9\text{ mg/}L)(8.34\text{ lbs/gal})$$

$$= 45\text{ lbs/day}$$

Or

$$\text{Hypochlorinator Setting, lbs/day} = 0.6\text{ MGD} \times 9\text{ mg/}L \times 8.34\text{ lbs/gal}$$

$$= 45\text{ lbs/day}$$

12.7 METRIC CALCULATIONS

This section contains the solutions to all problems in this chapter using metric calculations.

12.70 Conversion Factors

lb/day × 0.454	= kg/day
kg/day × 2.205	= lb/day
MGD × 3,785	= cu m/day
cu m/day × 0.000264	= MGD
1,000 *L*	= 1 cu m

12.71 Problem Solutions

1. A hypochlorinator is set to feed 25 kilograms of chlorine per 24 hours; the wastewater flow is at a rate of 3,200 cu m/day, and the chlorine as measured by the chlorine residual test after 30 minutes of contact time is 0.5 mg/*L*. Find the chlorine dosage and chlorine demand in mg/*L*.

Known		Unknown
Chlorine Feed, kg/day	= 25 kg/day	1. Chlorine Dose, mg/*L*
Flow, cu m/day	= 3,200 cu m/day	2. Chlorine Demand, mg/*L*
Chlorine Residual, mg/*L*	= 0.5 mg/*L*	

1. Calculate the chlorine dose in mg/*L*.

$$\text{Chlorine Feed or Dose, mg/}L = \frac{\text{Chlorine Feed, kg/day} \times 1{,}000\text{ gm/kg} \times 1{,}000\text{ mg/gm}}{\text{Flow, cu m/day} \times 1{,}000\text{ }L\text{/cu m}}$$

$$= \frac{25\text{ kg/day} \times 1{,}000\text{ gm/kg} \times 1{,}000\text{ mg/gm}}{3{,}200\text{ cu m/day} \times 1{,}000\text{ }L\text{/cu m}}$$

$$= 7.8\text{ mg/}L$$

Or

$$\text{Chlorine Feed or Dose, mg/}L = 25\text{ kg/day} \times 1{,}000\text{ gm/kg} \times 1{,}000\text{ mg/gm} \div 3{,}200\text{ cu m/day} \div 1{,}000\text{ }L\text{/cu m}$$

$$= 7.8\text{ mg/}L$$

2. Determine the chlorine demand in mg/*L*.

$$\text{Chlorine Demand, mg/}L = \text{Chlorine Dose, mg/}L - \text{Chlorine Residual, mg/}L$$

$$= 7.8\text{ mg/}L - 0.5\text{ mg/}L$$

$$= 7.3\text{ mg/}L$$

2. Calculate the chlorinator setting (kg per 24 hours) to treat a waste with a chlorine demand of 10 mg/*L*, when a chlorine residual of 1 mg/*L* is desired. The flow is 2,500 cu m per day.

Known		Unknown
Chlorine Demand, mg/*L*	= 10 mg/*L*	Chlorinator Setting, kg/24 hr
Chlorine Residual, mg/*L*	= 1 mg/*L*	
Flow, cu m/day	= 2,500 cu m/day	

Calculate the chlorinator setting in kg per 24 hours.

$$\text{Chlorine Dose, mg}/L = \text{Chlorine Demand, mg}/L + \text{Chlorine Residual, mg}/L$$

$$= 10\text{ mg}/L + 1\text{ mg}/L$$

$$= 11\text{ mg}/L$$

$$\text{Chlorinator Setting, kg/day} = \text{Chlorine Dose, mg}/L \times \text{Flow}\frac{\text{cu m}}{\text{day}} \times \frac{1{,}000\,L}{\text{cu m}} \times \frac{1\text{ kg}}{1{,}000{,}000\text{ mg}}$$

$$= 11\frac{\text{mg}}{L} \times 2{,}500\frac{\text{cu m}}{\text{day}} \times \frac{1{,}000\,L}{\text{cu m}} \times \frac{1\text{ kg}}{1{,}000{,}000\text{ mg}}$$

$$= 27.5\text{ kg/day}$$

$$= 27.5\text{ kg/24 hours}$$

Or

$$\text{Chlorinator Setting, kg/day} = 11\text{ mg}/L \times 2{,}500\text{ cu m/day} \times 1{,}000\ L/\text{cu m} \div 1{,}000{,}000\text{ mg/kg}$$

$$= 27.5\text{ kg/day}$$

$$= 27.5\text{ kg/24 hours}$$

12.8 ARITHMETIC ASSIGNMENT

Turn to Section A.2, *TYPICAL WASTEWATER SYSTEM PROBLEMS (ENGLISH SYSTEM)*, in the Arithmetic Appendix at the back of this manual. Work all of the problems in Section A.23, "Disinfection and Chlorination," and check the arithmetic with your calculator.

12.9 ADDITIONAL READING AND CONTACTS

1. *WATER ANALYSIS HANDBOOK*. Order No. 8374. Available from HACH Company, PO Box 389, Loveland, CO 80539-0389. No Charge.
2. *CHLORINE MANUAL*, Sixth Edition. Obtain from the Chlorine Institute, Inc., 2001 L Street, NW, Suite 506, Washington, DC 20036. Pamphlet 1. Price to members, $15.00; nonmembers, $30.00; plus 10 percent of order total for shipping and handling.
3. *LET'S TALK SAFETY—2002 SAFETY TALKS*. A series of 52 lectures on common water utility safety practices. Obtain from American Water Works Association (AWWA), Bookstore, 6666 West Quincy Avenue, Denver, CO 80235. Order No. 10123. ISBN 1-58321-145-4. Price to members, $33.50; nonmembers, $47.50; price includes cost of shipping and handling.

END OF LESSON 3 OF 3 LESSONS
on
DISINFECTION AND CHLORINATION

Please answer the discussion and review questions before continuing with the Objective Test.

DISCUSSION AND REVIEW QUESTIONS

Chapter 12. DISINFECTION AND CHLORINATION

(Lesson 3 of 3 Lessons)

Write the answers to these questions in your notebook before continuing with the Objective Test on page 209. The question numbering continues from Lesson 2.

11. How should chlorine tablets be stored?
12. What precautions would you take before starting any chlorination system?
13. What happens to hypochlorite solutions if they are not stored properly or are kept too long?
14. How can carbonate scale on hypochlorinators be removed?
15. How can fouled quartz or Teflon surfaces in UV systems be cleaned?

SUGGESTED ANSWERS

Chapter 12. DISINFECTION AND CHLORINATION

ANSWERS TO QUESTIONS IN LESSON 1

Answers to questions on page 170.

12.0A The purpose of disinfection is to destroy or inactivate pathogenic organisms. This is important to prevent the spread of waterborne diseases.

12.0B Pathogenic bacteria are destroyed or removed from water by (1) physical removal through sedimentation and filtration, (2) natural die-away in an unfavorable environment, and (3) destruction through chemical treatment.

12.0C Sterilization of wastes is impractical and unnecessary and may be detrimental to other treatment processes that are dependent on the activity of nonpathogenic saprophytes.

12.0D Chlorine is used for disinfection because it meets the general requirements of disinfection so well, and because it has been found to be the most economically useful and available chemical for disinfection.

Answers to questions on page 175.

12.0E Chlorine reacts with organic matter to form chlororganic compounds.

12.0F Chlorine Dose = Chlorine Demand + Chlorine Residual

12.0G To produce an effective disinfecting action, sufficient chlorine must be added after the chlorine demand is satisfied to produce a chlorine residual that will persist through the contact period.

Answers to questions on page 177.

12.0H Chlorine Demand is influenced by contact time, effectiveness of upstream processes, temperature, dose rate, type of chemical, pH, and numbers and types of organisms.

12. 0I

$$\text{Chlorine Dose, mg}/L = \frac{5 \text{ lbs/day}}{(0.06 \text{ MG/day})(8.34 \text{ lbs/gal})}$$

$$= 10 \frac{\text{lbs}}{\text{M lbs}}$$

$$= 10.0 \text{ mg}/L$$

Or

$$\text{Chlorine Dose, mg}/L = 5 \text{ lbs/day} \div 0.06 \text{ MG/day} \div 8.34 \text{ lbs/gal}$$

$$= 10.0 \text{ mg}/L$$

$$\text{Chlorine Demand, mg}/L = \text{Chlorine Dose} - \text{Chlorine Residual}$$

$$= 10.0 \text{ mg}/L - 0.4 \text{ mg}/L$$

$$= 9.6 \text{ mg}/L$$

12.0J The objective of disinfection is the destruction or inactivation of pathogenic bacteria, and the ultimate measure of the effectiveness is the bacteriological test result.

12.0K The ultimate measure of effectiveness is the bacteriological test result. The residual chlorine that yields satisfactory bacteriological test results in a particular plant must be determined and used as a control in that plant.

ANSWERS TO QUESTIONS IN LESSON 2

Answers to questions on page 178.

12.1A The purpose of up-sewer chlorination is to control odors and septicity, prevent deterioration of structures, and decrease BOD load.

12.1B Chlorine should be applied in sewers where odor and H_2S control is necessary. The chlorine may be applied at one or more points in the main intercepting sewer or at the upper ends of feeder lines.

12.1C Prechlorination provides partial disinfection and odor control.

12.1D Plant chlorination provides control of odors, corrosion, sludge bulking, digester foaming, filter ponding, filter flies, or sludge thickening. Be careful that chlorination does not interfere with biological treatment processes.

12.1E Postchlorination is used primarily for disinfection.

Answers to questions on page 180.

12.2A Selection of a chlorine feed- (dose-) rate control method should be based on the treatment costs and the desired or required treatment results.

12.2B Control of chlorine dosage depends on the bacterial *REDUCTION* desired; that is, the bacterial level or concentration acceptable or permissible at the point of discharge.

Answers to questions on page 182.

12.2C When hypochlorite is mixed with water, approximately half of the chlorine forms hydrochloric acid (HCl) and the other half forms hypochlorite (OCl^-) in the chlorine residual that you measured.

12.2D Find the chlorine dosage in pounds per day.

$$\text{Chlorine Dosage, lbs/day} = \text{Chlorine Dose, mg}/L \times \text{Flow, MGD} \times 8.34 \text{ lbs/gal}$$

$$= 12 \text{ mg}/L \times 0.55 \text{ MGD} \times 8.34 \text{ lbs/gal}$$

$$= 55 \text{ lbs/day}$$

$$\text{HTH Feed Rate, lbs/day} = \frac{\text{Chlorine Required, lbs/day}}{\text{Portion of Chlorine in pounds of HTH}}$$

$$= \frac{55 \text{ lbs/day}}{0.65}$$

$$= 85 \text{ lbs/day}$$

Answers to questions on page 183.

12.2E Chlorine discharge lines may be made of polyvinyl chloride (PVC) or black polyethylene flexible tubing. Rubber-lined steel pipe has been used, but rubber hose is rarely used today.

12.2F Little mixing of chlorine solution with wastewater occurs in chlorine contact basins because of the low flow velocities in a basin.

Answers to questions on page 187.

12.2G The chlorine residual must be measured as quickly as possible before chlorine concentrations change as a result of waiting too long.

12.3A Whenever hypochlorite comes in contact with your hand, immediately wash off the hypochlorite and thoroughly wash your hand. Consult a physician if the area appears burned.

Answers to questions on page 188.

12.4A The different types of chlorine feeders described in this chapter are tablet chlorinators and liquid hypochlorinators.

12.4B Flow to tablet chlorinators can be blocked by sticks, leaves, grease balls, and other matter.

Answers to questions on page 190.

12.4C Chlorine doses are adjusted to flow changes with a chlorine tablet disinfection system by allowing the amount of submergence of the tablets to fluctuate with the flows.

12.4D Chlorine is released into the wastewater by the dissolving action of the wastewater in contact with the tablets.

12.4E The tubes should always be filled with chlorine tablets to the top to ensure the longest possible periods of unattended operation.

12.4F If the feed tubes do not reach the bottom of the unit, insufficient chlorine will be released into the wastewater.

Answers to questions of page 195.

12.4G Major advantages of on-site chlorine/hypochlorite generation include (1) hypochlorite solution is generated as needed, and (2) the on-site generation system is safer.

12.4H On-site hypochlorite systems use salt, water, and electricity to produce sodium hypochlorite.

12.4I A major problem with sodium hypochlorite systems is the unreliable flow of hypochlorite solution.

Answers to questions on page 199.

12.4J Hypochlorite solutions can be prevented from freezing by the use of a strip heater or a heat lamp.

12.4K Review the spare parts inventory and then reorder as necessary to maintain adequate repair stock.

Answers to questions on page 205.

12.5A Excessive chlorine residual is not required because it is costly and can be toxic to aquatic life in the receiving streams.

12.5B UV light disinfects by damaging the DNA and RNA of organisms.

12.5C UV disinfection is not suitable for effluents with high turbidity, suspended solids, dyes, organics, or other UV-absorbing constituents.

OBJECTIVE TEST

Chapter 12. DISINFECTION AND CHLORINATION

Please write your name and mark your answers on the answer sheet, as directed at the end of Chapter 1. There may be more than one correct answer to each multiple-choice question.

True-False

1. To date there is no record of an operator becoming infected by the AIDS virus due to on-the-job duties.
   1. True
   2. False
2. The chlorine demand caused by inorganic salts must be satisfied first before any disinfection can take place.
   1. True
   2. False
3. Hypochlorination will lower the pH of the wastewater being treated.
   1. True
   2. False
4. The lowering of pH in water will increase the effectiveness of chlorination.
   1. True
   2. False
5. Due to high costs, collection system chlorination should be considered as a temporary or emergency measure in most cases.
   1. True
   2. False
6. Never use neoprene hose to carry chlorine solutions.
   1. True
   2. False
7. Chlorine solution diffusers are never constructed of the same materials used for solution lines.
   1. True
   2. False
8. Safety is everyone's problem.
   1. True
   2. False
9. Containers of hypochlorite should be stored in a cool, dry area away from other combustibles, such as gasoline, solvents, oils, greases, rags, or papers.
   1. True
   2. False
10. Hypochlorite systems can be very time consuming for the operator.
    1. True
    2. False
11. Make any necessary adjustments of the hypochlorination pump output control selector only when the pump is OFF.
    1. True
    2. False
12. UV light disinfects by damaging the DNA and RNA of organisms.
    1. True
    2. False

Best Answer (Select only the closest or best answer.)

13. What is the main purpose of disinfection?
    1. Control odors
    2. Prevent spread of waterborne diseases
    3. Reduce BOD
    4. Remove grease
14. Why is dechlorination practiced by some wastewater treatment plants?
    1. To avoid corroding materials in receiving waters
    2. To prevent the formation of trihalomethanes
    3. To protect aquatic organisms in waters where chlorinated effluents are discharged
    4. To reduce the formation of chlorine tastes and odors in drinking water
15. Why should extreme care be exercised when applying chlorine in a treatment plant?
    1. Chlorine can be very costly
    2. Chlorine can cause odor problems
    3. Chlorine can interfere with or inhibit biological treatment processes
    4. Chlorine can reduce BOD (food) to microorganisms
16. Where is the most effective place for chlorine application in a wastewater treatment plant for disinfection?
    1. After treatment and on a well-clarified effluent
    2. Ahead of the influent pumps to ensure sufficient mixing
    3. At the plant headworks to provide the longest possible contact time
    4. Before filtration to prevent interference from biological growths
17. How do manufacturers of hypochlorite compounds define available chlorine?
    1. By calculating the stoichiometric chlorine dose
    2. By weighing an equivalent amount of chlorine gas
    3. The amount of gaseous chlorine required to make the equivalent hypochlorite chlorine
    4. The amount of hypochlorite required to maintain the chlorine residual

18. Why is mixing extremely important ahead of a chlorine contact chamber or basin?
    1. Installing mixing equipment in the contact chamber is too costly
    2. Little mixing occurs after wastewater enters the contact chamber
    3. Mixing in the contact chamber will release chlorine gas
    4. Mixing prevents the deposition of solids which will cause tastes and odors

19. Where must samples be collected to determine chlorine residual?
    1. After complete mixing has occurred
    2. At the discharge end of the chlorine contact structure
    3. At the entrance to the chlorine contact chamber
    4. At the point of chlorine injection

20. How should the spill of a hypochlorite solution be cleaned up?
    1. Sweep up with a broom
    2. Vacuum up and recontainerize
    3. Wash with large volumes of water
    4. Wipe up with moist rags

21. How do tablet chlorinators work?
    1. As flow increases, larger tablets are used
    2. As flow increases, more of the tablet is submerged
    3. As flow increases, more tablets are added to the wastewater
    4. As flow increases, more tablets are ground up and added to the flow

22. How are the chlorine dose and residual in wastewater controlled by a tablet chlorinator?
    1. Adjust rotameter flow
    2. Adjust tablet dissolving rate
    3. Adjust velocity of flow
    4. Adjust weir at outlet end

23. How can operators prevent the formation of calcium carbonate coatings on the poppet valves in the hypochlorite solution feeder?
    1. Dilute solution with sulfuric acid
    2. Flush system with concentrated hypochlorite
    3. Obtain hypochlorite dilution water from a home water softener
    4. Pump hydrochloric acid through the system

24. How can water with a chlorine residual be dechlorinated?
    1. Hypochlorite
    2. Sodium hydroxide
    3. Sulfur dioxide
    4. Sulfuric acid

Multiple Choice (Select all correct answers.)

25. What types of illnesses can be caused by waterborne microorganisms found in wastewater?
    1. Bacteria-caused
    2. Cancer-caused
    3. Internal parasite-caused
    4. Lung-caused
    5. Virus-caused

26. Which beneficial uses of water need to be protected by disinfection to prevent the spread of waterborne diseases?
    1. Hydroelectric power generation
    2. Navigation
    3. Public water supplies
    4. Receiving water used for recreational uses
    5. Shellfish growing areas

27. Which major impurities in wastewater can have a serious effect on wastewater chlorination?
    1. Ammonia
    2. Hydrogen sulfide
    3. Inorganic reducing materials
    4. Pesticides
    5. Toxic heavy metals

28. Which inorganic reducing agents react with chlorine?
    1. Ferrous iron
    2. Manganese
    3. Methane
    4. Nitrite
    5. Potassium

29. Which critical factors influence disinfection?
    1. Chlorine dose
    2. Contact time
    3. Mixing
    4. pH
    5. Temperature

30. Why would chlorine be applied to wastewater collection systems?
    1. Corrosion control
    2. Fly control
    3. Foam control
    4. Odor control
    5. Slime control

31. What types of hypochlorinators are used by small wastewater systems?
    1. Acid hypochlorinators
    2. Gas/liquid hypochlorinators
    3. pH adjusted hypochlorinators
    4. Solution feed hypochlorinators
    5. Tablet hypochlorinators

32. Which factors will cause the chlorine requirement or chlorine dose to vary?
    1. Contact time
    2. pH of wastewater
    3. Temperature of wastewater
    4. Waste constituents such as hydrogen sulfide and ammonia
    5. Wastewater flow

33. What are the major causes of a large discrepancy in the chlorine requirement between laboratory and plant tests?
    1. Fluctuations in pH caused by biological treatment processes
    2. Poor mixing at the point of chlorine injection
    3. Sensitivity of analytical instruments used by different analysts
    4. Short-circuiting in the contact chamber
    5. Temperature differentials between laboratory and plant

34. What types of fittings are used to connect Schedule 80 PVC pipe?
   1. Bell and spigot
   2. Flanges
   3. Packing
   4. Solvent-welded
   5. Threads

35. Which methods and materials are approved for measuring chlorine residual?
   1. Amperometric method
   2. Color comparison strips
   3. DPD photometric method
   4. Odor panel
   5. Orthotolidine

36. Which items should be included in all safety programs?
   1. Establishment of a formal safety program
   2. Establishment of a maintenance and calibration program for safety devices and equipment
   3. Periodic hands-on training using safety equipment
   4. Provide police and fire departments with tours of facilities to locate hazardous areas
   5. Written safety rules

37. What are the advantages of tablet chlorinators?
   1. No dissolving of tablet
   2. No electrical requirements
   3. No mixing of solution
   4. No moving parts
   5. No safety hazards storing tablets

38. What maintenance procedures are required for tablet chlorine feeders?
   1. Checking and maintaining oil levels
   2. Inspecting for and correcting unusual noises and vibrations
   3. Lubricating moving parts
   4. On-stream cleaning of feeder and weir
   5. Refilling feed tubes

39. What are possible causes of the chlorine residual consistently being too low in the chlorine contact tank?
   1. Excessive dosage of disinfectant
   2. Feed tubes *NOT* reaching bottom of unit
   3. Gross hydraulic overload
   4. Higher than anticipated coliform levels
   5. Tablets jamming in stack and remaining suspended above water

40. Where should chlorine tablets be stored?
   1. Away from combustible materials
   2. In a cool area
   3. In a dry area
   4. In a moist area
   5. In open containers

41. What options are available for communities selecting a hypochlorite system?
   1. Conversion of hydrochloric acid on site to hypochlorite
   2. On-site generation of hypochlorite
   3. Purchase of dry chlorine compounds to be mixed into a solution for dispensing
   4. Purchase of gaseous hypochlorite
   5. Purchase of pre-prepared hypochlorite liquid solutions

42. What are the advantages of on-site chlorine/hypochlorite generation?
   1. Hypochlorite solution is generated as needed
   2. On-site generated hypochlorite is a higher concentration than purchased hypochlorite solutions
   3. Operators do *NOT* need to mix dry chlorine compounds
   4. Solution does *NOT* lose its strength when disinfection system is *NOT* in use
   5. System is safer for operators because they do *NOT* have to handle hypochlorite solutions

43. Hypochlorinator pumping units may develop what types of problems?
   1. Air lock
   2. Losing prime
   3. Plugging of solution tubing lines from deposits of chemicals
   4. Ruptured diaphragms
   5. Stripped gear trains

44. Under what conditions might it be necessary for an operator to adjust the dial setting on a hypochlorinator? Changes in
   1. Chlorine demand of the effluent
   2. Coliform concentrations
   3. Hypochlorite solution strength
   4. Operators on shift
   5. Pump discharge pressure

45. Normal operation of the hypochlorination process requires which tasks to be performed daily?
   1. Check chlorine residual of effluent and adjust feed rate as required
   2. Check operation of check valve
   3. Inspect the building to make sure only authorized personnel have been there
   4. Read and record level of the solution tank at the same time every day
   5. Wash the chemical storage tank

46. UV disinfection is *NOT* suitable for effluents with high levels of what types of constituents?
   1. BOD
   2. Greases
   3. Oil
   4. Suspended solids
   5. Turbidity

CHAPTER 13

ALTERNATIVE WASTEWATER TREATMENT, DISPOSAL, AND REUSE METHODS

by

John Brady

TABLE OF CONTENTS

Chapter 13. ALTERNATIVE WASTEWATER TREATMENT, DISPOSAL, AND REUSE METHODS

OBJECTIVES

Chapter 13. ALTERNATIVE WASTEWATER TREATMENT, DISPOSAL, AND REUSE METHODS

Following completion of Chapter 13, you should be able to:

1. Describe alternative wastewater treatment, disposal, and reuse methods for small systems,
2. Determine where wetlands can be used effectively,
3. Identify the components of subsurface constructed wetlands,
4. Start, operate, and maintain subsurface constructed wetlands,
5. Safely operate, maintain, and troubleshoot land treatment systems,
6. Monitor a land treatment system, and
7. Review plans and specifications for a land treatment system.

WORDS

Chapter 13. ALTERNATIVE WASTEWATER TREATMENT, DISPOSAL, AND REUSE METHODS

AEROBIC (AIR-O-bick) — AEROBIC

A condition in which atmospheric or dissolved molecular oxygen is present in the aquatic (water) environment.

ANAEROBIC (AN-air-O-bick) — ANAEROBIC

A condition in which atmospheric or dissolved molecular oxygen is *NOT* present in the aquatic (water) environment.

ANOXIC (an-OX-ick) — ANOXIC

A condition in which atmospheric or dissolved molecular oxygen is *NOT* present in the aquatic (water) environment and nitrate is present. Oxygen deficient or lacking sufficient oxygen. The term is similar to ANAEROBIC.

AVAILABLE MOISTURE CONTENT — AVAILABLE MOISTURE CONTENT

The quantity of water present that is available for use by vegetation.

CATION EXCHANGE CAPACITY — CATION EXCHANGE CAPACITY

The ability of a soil or other solid to exchange cations (positive ions such as calcium, Ca^{2+}) with a liquid.

DENITRIFICATION (dee-NYE-truh-fuh-KAY-shun) — DENITRIFICATION

(1) The anoxic biological reduction of nitrate nitrogen to nitrogen gas.

(2) The removal of some nitrogen from a system.

(3) An anoxic process that occurs when nitrite or nitrate ions are reduced to nitrogen gas and nitrogen bubbles are formed as a result of this process. The bubbles attach to the biological floc in the activated sludge process and float the floc to the surface of the secondary clarifiers. This condition is often the cause of rising sludge observed in secondary clarifiers or gravity thickeners. Also see NITRIFICATION.

DRAIN TILE SYSTEM — DRAIN TILE SYSTEM

A system of tile pipes buried under the crops that collect percolated waters and keep the groundwater table below the ground surface to prevent ponding.

DRAINAGE WELLS — DRAINAGE WELLS

Wells that can be pumped to lower the groundwater table and prevent ponding.

EVAPOTRANSPIRATION (ee-VAP-o-TRANS-purr-A-shun) — EVAPOTRANSPIRATION

(1) The process by which water vapor passes into the atmosphere from living plants. Also called TRANSPIRATION.

(2) The total water removed from an area by transpiration (plants) and by evaporation from soil, snow and water surfaces.

FREEBOARD — FREEBOARD

The vertical distance from the normal water surface to the top of the confining wall.

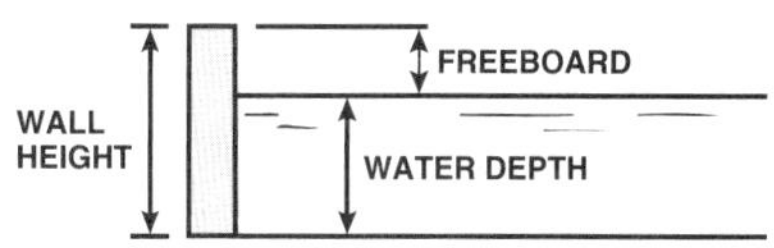

GATES — GATES

A descriptive term used on irrigation distribution piping systems instead of the word valves. Gates cover outlet ports in the pipe segments. Water flows are regulated or distributed by opening the gates by either sliding the gate up or down or by swinging the gate to one side and uncovering an individual port to permit water flow to be discharged or regulated from the pipe at that particular point.

HYDROLOGIC (HI-dro-LOJ-ick) CYCLE HYDROLOGIC CYCLE

The process of evaporation of water into the air and its return to earth by precipitation (rain or snow). This process also includes transpiration from plants, groundwater movement, and runoff into rivers, streams and the ocean. Also called the WATER CYCLE.

NITRIFICATION (NYE-truh-fuh-KAY-shun) NITRIFICATION

An aerobic process in which bacteria change the ammonia and organic nitrogen in wastewater into oxidized nitrogen (usually nitrate). The second-stage BOD is sometimes referred to as the "nitrogenous BOD" (first-stage BOD is called the "carbonaceous BOD"). Also see DENITRIFICATION.

PLUG FLOW PLUG FLOW

A type of flow that occurs in tanks, basins or reactors when a slug of wastewater moves through a tank without ever dispersing or mixing with the rest of the wastewater flowing through the tank.

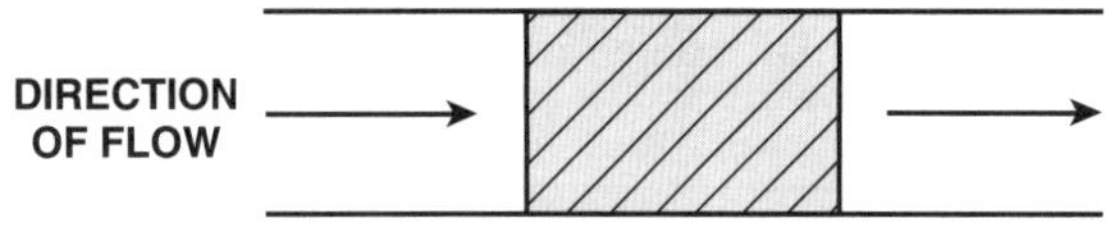

PLUG FLOW

SIDESTREAM SIDESTREAM

Wastewater flows that develop from other storage or treatment facilities. This wastewater may or may not need additional treatment.

TRANSPIRATION (TRAN-spur-RAY-shun) TRANSPIRATION

The process by which water vapor is released to the atmosphere by living plants. This process is similar to people sweating. Also see EVAPOTRANSPIRATION.

VECTOR VECTOR

An insect or other organism capable of transmitting germs or other agents of disease.

WATER CYCLE WATER CYCLE

The process of evaporation of water into the air and its return to earth by precipitation (rain or snow). This process also includes transpiration from plants, groundwater movement, and runoff into rivers, streams and the ocean. Also called the HYDROLOGIC CYCLE.

CHAPTER 13. ALTERNATIVE WASTEWATER TREATMENT, DISPOSAL, AND REUSE METHODS

(Lesson 1 of 2 Lessons)

13.0 ALTERNATIVE WASTEWATER TREATMENT METHODS FOR SMALL SYSTEMS

New wastewater treatment methods have come into use to provide high-quality effluents for disposal or reuse by small systems as an alternative to the more expensive and complex conventional secondary wastewater treatment plants. Primary treatment of the wastewater is required and, in most instances, is accomplished by septic tanks prior to the wastewater being discharged to these effluent disposal or reuse systems. The alternative treatment systems discussed in this chapter include constructed wetlands and land application of effluents.

13.1 WETLANDS

Wetlands, though in existence for millennia, have just in recent years become a real interest to society as a wastewater treatment system. In the 1970s, land application of wastewater over various grasses and feed crop vegetation was just being introduced and pilot projects reported encouraging results. At that time, NASA was looking for a reasonable, economical alternative to the conventional wastewater treatment systems at one of their facilities, which consisted of a package plant and ponds that were having difficulty meeting discharge requirements. Shallow ponds with water hyacinths growing in them were found to be an adequate treatment method in the warmer climates. Interest then turned to marshes where reeds, cattails, bulrushes, and other water-tolerant plants grew. Engineered systems using vegetation that could function in some of the colder climates such as the Dakotas and Canada were then developed. Today, many communities are using wetlands for wastewater treatment. WETLANDS systems, predominantly subsurface constructed wetlands, will be discussed in this section.

13.10 The Two Types of Wetlands

Wetlands can be either natural or constructed. Natural wetlands exist in nature as bogs, marshes, swamps, and cypress domes. Early treatment projects were associated with facilities next to existing natural wetlands. Two well-known systems are Cannon Beach, Oregon, and Arcata, California (Figure 13.1).

Constructed wetlands can be built almost anywhere and they resemble natural wetlands. Wastewater engineers describe two types of wetlands: free water surface (FWS) and subsurface flow system (SFS).

Free Water Surface (FWS) (Figures 13.2 and 13.3)

A free water surface (FWS) system consists of basins or channels with a natural or constructed subsurface barrier of clay or impervious material that prevents seepage from the wetland. Soil, sand, gravel, rock, or another suitable material is used to support the emergent vegetation and water is maintained at a relatively shallow depth flowing over the surface. In the FWS wetland, water is visible throughout the bed and maintained at a depth of 18 inches to 2 feet (45 to 60 cm) with various plants growing in the water and the bed media. Long, narrow channels are used to create *PLUG FLOW*[1] conditions, but wider, shorter basins are also in use. FWS systems are predominantly used in larger communities as a secondary or polishing wastewater treatment method. Systems in use today treat wastewater flows as high as 20,000,000 gallons per day (20 MGD or 75 M*L*D). FWS systems are not recommended for individual home sites even though a water garden appearance may be attractive.

The associated potential problems of *VECTOR*[2] and disease control must be considered and managed. Short-circuiting may occur in free water surface systems due to clumps of grasses and plants that could hinder the free flow of the water. Solids deposition can occur when trees lose their leaves and plants die. Over time, these deposits may create anaerobic zones that could release obnoxious odors.

[1] *Plug Flow. A type of flow that occurs in tanks, basins or reactors when a slug of wastewater moves through a tank without ever dispersing or mixing with the rest of the wastewater flowing through the tank.*

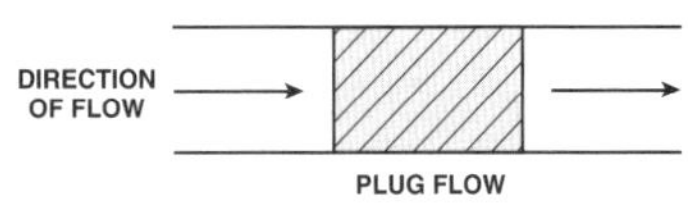

[2] *Vector. An insect or other organism capable of transmitting germs or other agents of disease.*

Inlet to No. 1 wetland cell from oxidation pond, Cannon Beach, Oregon

Wetland marsh, Arcata, California

Fig. 13.1 Natural wetlands

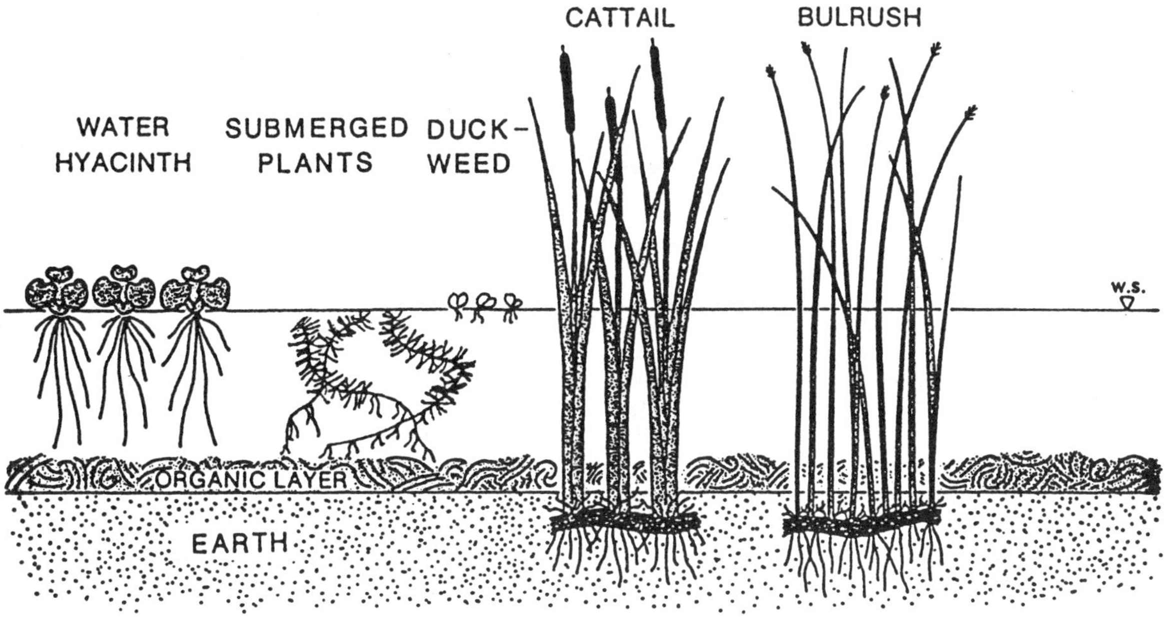

NOTE: Care must be taken to select plants that can survive local weather conditions.

Fig. 13.2 Common aquatic plants growing in a free water surface (FWS) wetland (natural wetland)

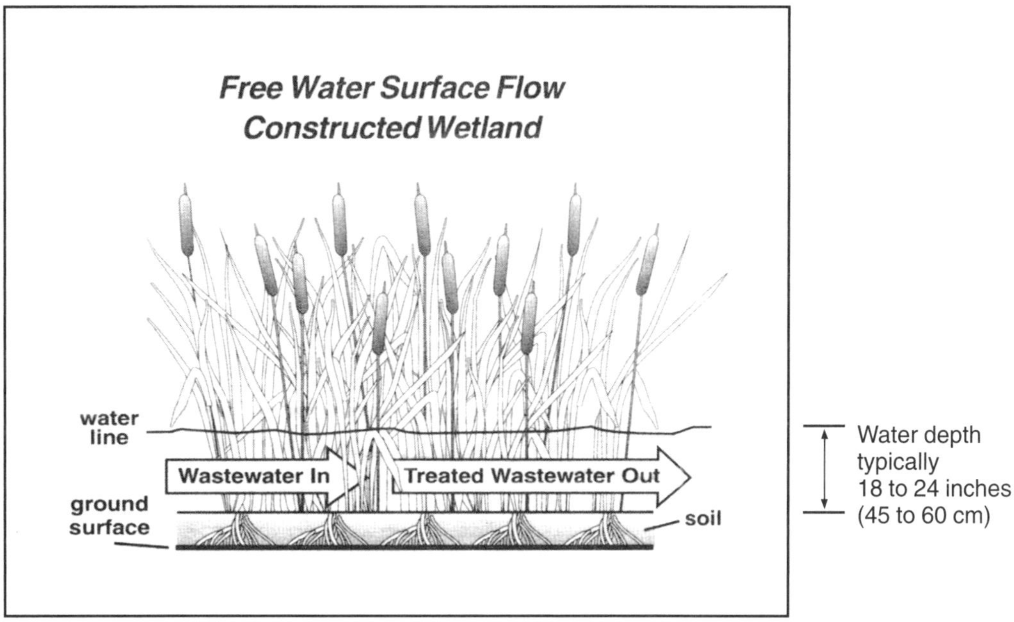

Fig. 13.3 Free water surface flow constructed wetlands

(Source: *SMALL FLOWS*)

Subsurface Flow System (SFS) (Figures 13.4 and 13.5)

A subsurface flow wetland is constructed as a trench or bed with an impermeable layer of clay or a synthetic liner. The bed contains media of rock, gravel, or soil that will support the growth of emergent vegetation. The wetland bed is constructed with a slight slope between the inlet and outlet end. Influent is applied to the wetland at the high end and permitted to flow under the media surface through the media and plant roots to the outlet end where it is collected and discharged. Water standing or pooling on the surface is not desirable; you should be able to walk on the dry surface. SFSs are used at single-family residences to provide a high degree of wastewater treatment without creating nuisance conditions or requiring complex operation and maintenance skills. The beds often look like a flower bed. Figure 13.4 shows cross-sectional views of small SFS constructed wetlands. Figure 13.5 shows SFS constructed wetlands in use at two different homes. These systems are becoming more prominent in small communities treating flows up to several million gallons per day.

The advantages of the wetlands include their lower initial construction cost if land is relatively cheap and available, reduced operation and maintenance (O & M) costs compared to conventional mechanical plants, and a greatly reduced level of operational skill needed to maintain the system and yet achieve a very high and reliable degree of wastewater treatment for the community. Over time, subsurface flow systems can become root bound, which could lead to short-circuiting and channeling. Care must be taken to ensure that the organic loading and seasonal die back of vegetation do not cause the system to operate in an anaerobic condition. If anaerobic conditions develop, obnoxious odors could be released.

13.11 Where Can Wetlands Be Used?

A wetland system can be constructed almost anywhere it is needed. The major controlling factor for the community and individual home sites is land availability. Other controlling factors include zoning and buffer distances. Constructed wetlands that discharge to surface waters require four to ten times more land area than a conventional wastewater treatment plant; for a zero discharge wetland, the land area is ten to one hundred times the area that a conventional treatment plant would need. The advantage of the wetland is that it can be built and used in places where other disposal systems (such as leach fields, seepage pits, and mounds) have failed due to tight clay soils that bind, or loose sandy soils that percolate water too fast, or in locations where rock outcroppings or a high groundwater table prevent safe wastewater treatment and disposal. The cost to construct an SFS wetland for an individual home is no more expensive than the long-used septic tank/leach field wastewater disposal system. Subsurface flow wetlands are predominantly used for homes and small communities. There are possibly two reasons for their acceptance:

1. The wastewater is not exposed to the surface until it has been treated and stabilized at the time of discharge, if a discharge does exist. This prevents contact of wastewater with children or pets, and the SFS minimizes the potential odor, insect vector, and public health problems. Studies have indicated that the possibility of contracting a disease or virus from a SFS wetland is very, very remote. The SFS further reduces the possibility as compared to the FWS wetland system.
2. The SFS bed is constructed to fit the available land area. Some designers prefer long, narrow beds; others prefer a shorter, wider bed. Wetland beds have even been constructed on the side of a hill as multiple beds or terraces; therefore, the configuration of the wetland system depends on the site. Once constructed, the wetland is planted with plants that can grow and survive in the climate of the area (Figure 13.6). There is a wide variety of plants that do well in a wetland habitat and some are very attractive, creating colorful flower beds or a different type of planting area that would not normally have been available at that site. Wastewater treatment systems now are becoming an asset and enjoyment to homeowners rather than a necessary liability.

13.12 Components of Subsurface Constructed Wetlands

These systems can be constructed as a single bed or multiple beds. The first bed or cell is made watertight by the use of clay or an impermeable liner. Liners are normally covered at the top of the berms to protect the plastic or rubber liner from ultraviolet sunlight, which will deteriorate those materials over time. The bed is constructed with a berm or wall around the perimeter (outer edge) to contain the wastewater and control hydraulic loading by preventing the entry of surface water runoff from adjoining land. The walls or berms are constructed with at least 6 inches (15 cm) *FREEBOARD*[3] above adjoining land and at least 9 to 12 inches (22.5 to 30 cm) freeboard inside the bed above the top surface of the bed media. Bed media depth is normally 12 inches (30 cm), which makes the total inside bed depth 21 to 24 inches (52 to 60 cm).

Inlet distribution piping and outlet collection pipes are placed at opposite ends of the cell with a minimum of 2 feet (60 cm) of round stone, 2 to 4 inches (5 to 10 cm) in diameter, to form a barrier between the piping and the smaller gravel media filling the bed (Figure 13.7). The stone helps distribute flows evenly and prevents plugging or clogging of the distribution and collection piping. The inlet piping to the first cell should have a cleanout or small chamber to enable visual inspection for solids accumulation that may be carried over from the primary treatment unit. Algae and other biological growths can coat the walls of the pipes and eventually could plug the pipes. Accessible cleanouts must be provided so the pipes can be cleaned.

Round ¾- to 1-inch (2- to 2.5-cm) gravel was originally used as media. Later studies indicated that using smaller pea gravel as media provided for better plant growth. In some instances, coarse sand has been used. In systems using two cells, media in the second (percolation) cell is coarse sand with a layer of mulch on the surface. The sand provides a good media for the plants, ensuring their survival in fluctuating water levels in the cell during dry periods (Figure 13.8). Cells with depths up to 30 inches (75 cm) have been constructed in colder climates to facilitate root growth for plants predominant in those areas (Figure 13.9).

[3] *Freeboard. The vertical distance from the normal water surface to the top of the confining wall.*

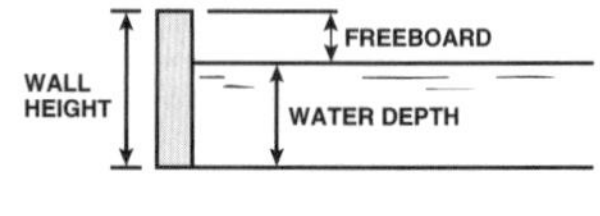

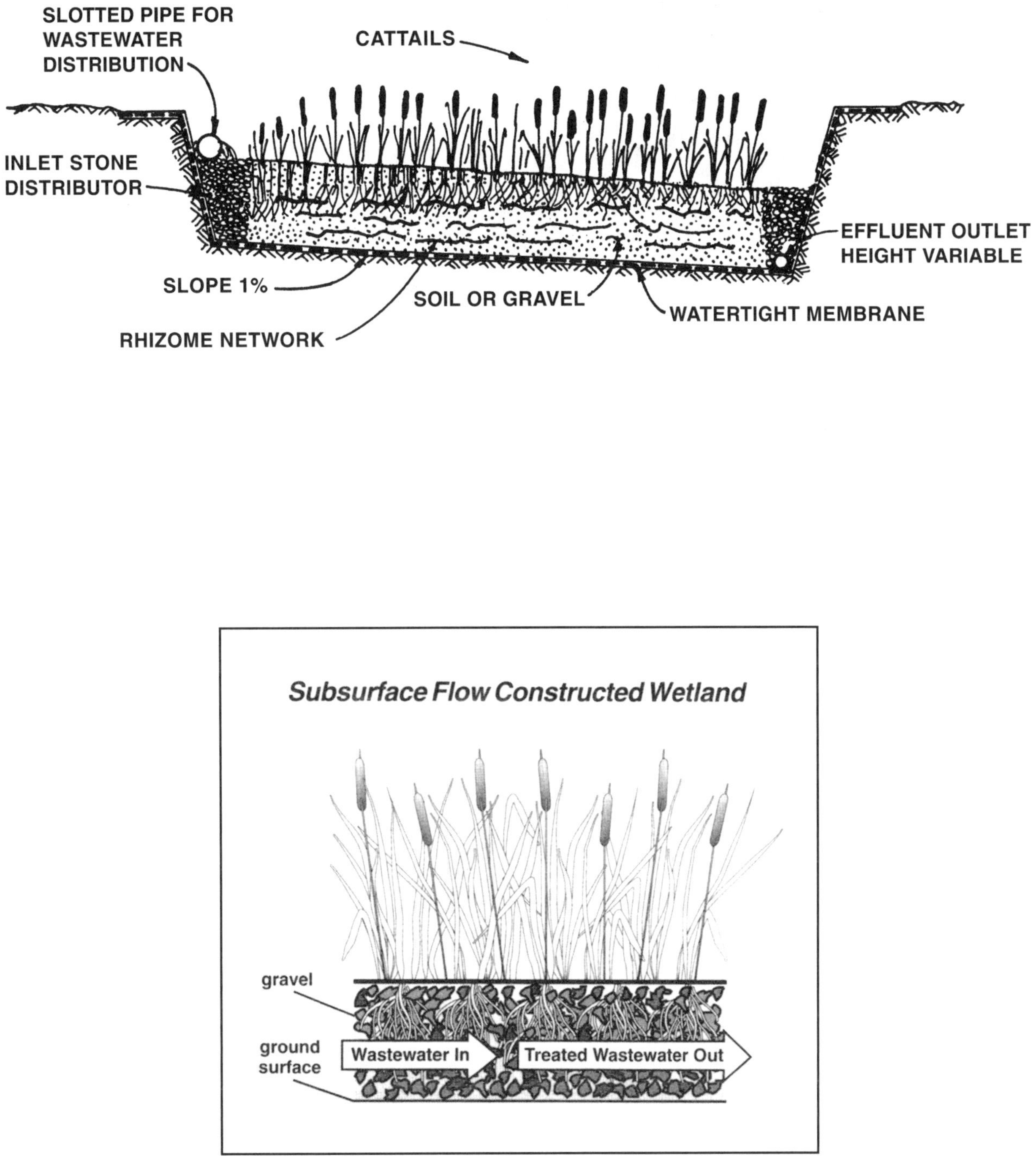

NOTE: Excessive root growth could cause these systems to become root bound.

Fig. 13.4 Subsurface flow constructed wetland

(Source of lower image: *SMALL FLOWS*)

Inspecting late winter plant growth on an on-site constructed wetland in Alexandria, Louisiana.

An on-site constructed wetland blends into the landscape near Alexandria, Louisiana.

Fig. 13.5 On-site constructed wetlands
(Source: *GENERAL DESIGN, CONSTRUCTION AND OPERATION GUIDELINES: CONSTRUCTED WETLANDS*)

Cold weather poses no problem for this constructed wetland operated by the American Crystal Sugar Company near Hillsboro, North Dakota. Although the system is frozen over and is covered with patches of snow, it continues to treat wastewater from the plant in the background.

—Photo courtesy of Donald Hammer, Ph.D., Tennessee Valley Authority.

Fig. 13.6 Cold weather constructed wetland

(Source: *SMALL FLOWS*, January 1993)

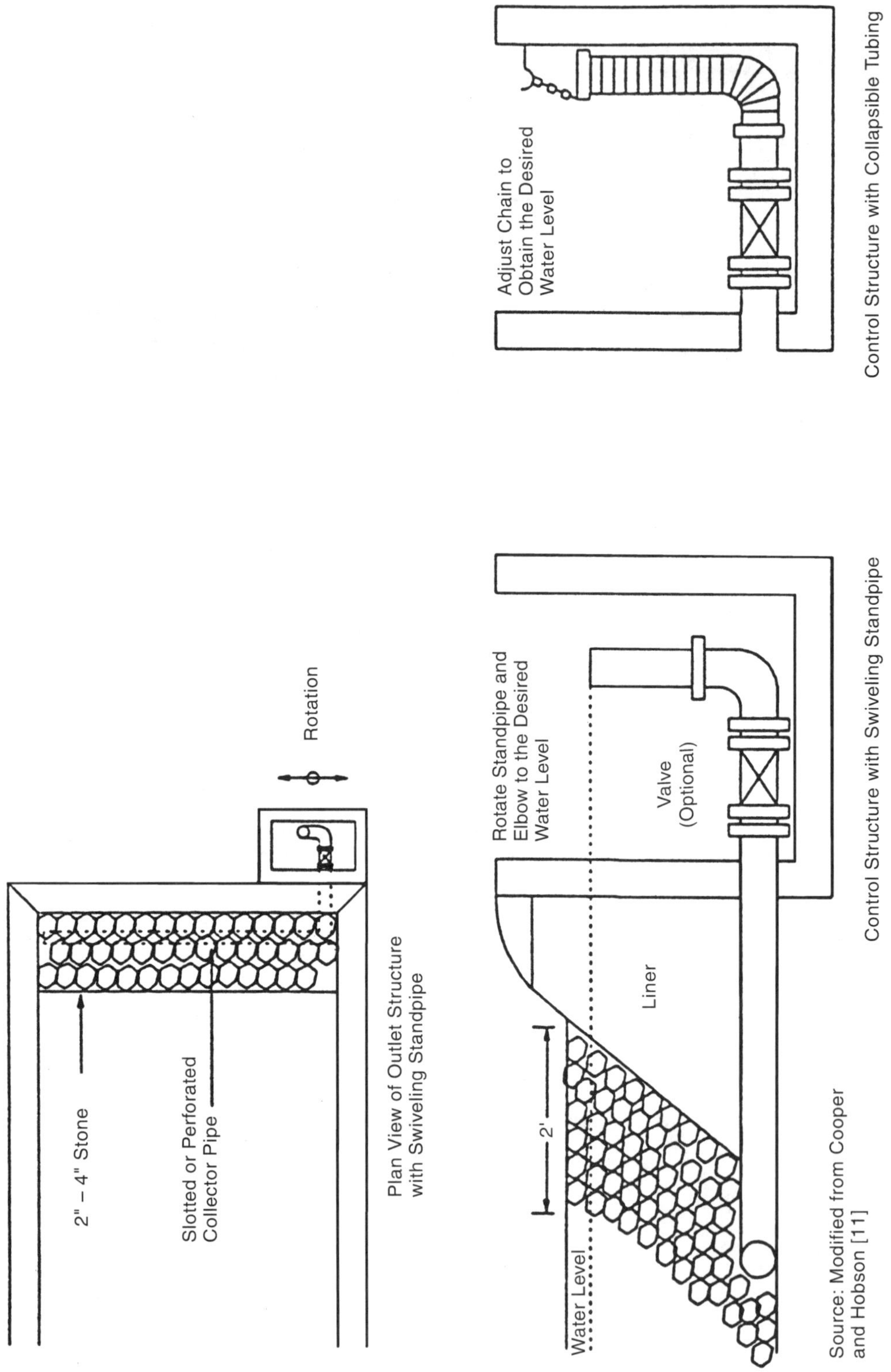

Fig. 13.7 Water level control structures
(Source: *GENERAL DESIGN CONSTRUCTION AND OPERATION GUIDELINES: CONSTRUCTED WETLANDS*)

Constructed wetlands like this one are being built throughout the nation to handle wastewater from mostly small rural communities and homes where traditional treatment systems are a problem.

Wastewater flows into the constructed wetland from a septic tank or other type of primary treatment system. Here the wastewater is evenly distributed among the plants where microorganisms and chemical reactions break down organic materials and pollutants.

Constructed wetlands provide simple, effective, and low-cost wastewater treatment when compared with conventional systems.

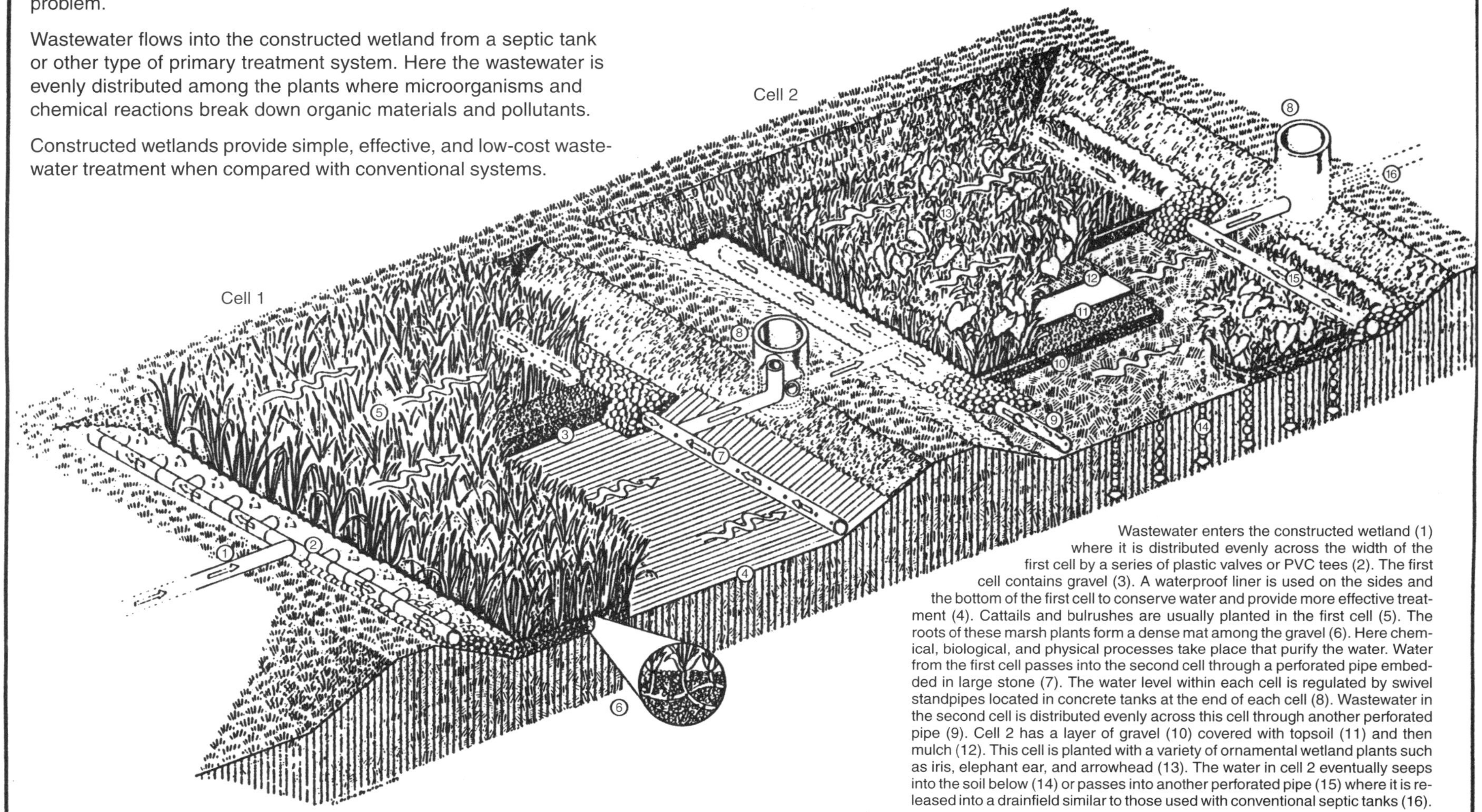

Wastewater enters the constructed wetland (1) where it is distributed evenly across the width of the first cell by a series of plastic valves or PVC tees (2). The first cell contains gravel (3). A waterproof liner is used on the sides and the bottom of the first cell to conserve water and provide more effective treatment (4). Cattails and bulrushes are usually planted in the first cell (5). The roots of these marsh plants form a dense mat among the gravel (6). Here chemical, biological, and physical processes take place that purify the water. Water from the first cell passes into the second cell through a perforated pipe embedded in large stone (7). The water level within each cell is regulated by swivel standpipes located in concrete tanks at the end of each cell (8). Wastewater in the second cell is distributed evenly across this cell through another perforated pipe (9). Cell 2 has a layer of gravel (10) covered with topsoil (11) and then mulch (12). This cell is planted with a variety of ornamental wetland plants such as iris, elephant ear, and arrowhead (13). The water in cell 2 eventually seeps into the soil below (14) or passes into another perforated pipe (15) where it is released into a drainfield similar to those used with conventional septic tanks (16).

Fig. 13.8 Cut-away perspective of a constructed wetlands system

(Source: *GENERAL DESIGN, CONSTRUCTION AND OPERATION GUIDELINES: CONSTRUCTED WETLANDS*)

The inlet distribution piping at American Crystal Sugar Company's constructed wetlands system continues to operate in cold weather.

—Photo courtesy of Donald Hammer, Ph.D., Tennessee Valley Authority

Fig. 13.9 Cells constructed in colder climates facilitate root growth for plants predominant in those areas

(Source: *SMALL FLOWS*, January 1993)

The media surface is covered with a mulch (litter) material approximately 3 inches (7.5 cm) deep. The mulch on top of the gravel prevents sun scalding of the plants, provides visual aesthetics, and helps control potential odors. In colder climates, the mulch insulates the plant roots from freezing temperatures. Reports have been made of wetlands functioning at temperatures as low as –30°F (–34°C).

Constructed wetlands are designed to be dosed by gravity flow but, where necessary, low-head pump systems are installed for communities or individual residences. Water level control is important for plant survival and wetlands health (see Figure 13.7). Roots of emergent plants must be kept wet, but plants will not survive if completely covered with water. In addition, if water is permitted to stand above the mulch layer, surface odors may develop. The water level in a subsurface flow wetland is controlled through the outlet structure regulating device by lowering or raising the overflow discharge level.

13.13 How a Subsurface Wetland Functions

Primary treated wastewater (septic tank effluent) is applied at the bottom, across the inlet end of the wetland bed. The wastewater flows through the gravel or sand media and around the plant roots, just below the surface of the gravel media in the wetland bed. Bacteria growing in the voids of the gravel use the nutrients in the wastewater for cell growth and reproduction. The major role of the plants is to transfer oxygen to the root zone. The plant stalks, roots, and rhizomes penetrate the soil or support media and transport oxygen deeper than it would naturally travel by diffusion alone. The plants also take up nutrients in the media for plant growth and remove some of the water by *TRANSPIRATION*.[4]

In areas of high groundwater or where the soil has no ability to absorb water, the effluent from the wetland is collected and can be discharged to a small percolation field or a small sand filter trench for irrigation. In areas of soil conditions that have some percolation capacity, a wetland of two-cell or bed configuration is constructed (Figure 13.10). The first cell is lined or sealed to prevent water loss into the soil to accommodate waste stabilization without contaminating groundwater aquifers. The second cell is not lined or sealed; its purpose is to dispose of the effluent from the first cell through percolation into the surrounding soil and to permit water losses due to evaporation and transpiration (review Figure 13.10).

Wetlands have been shown to significantly reduce BOD, suspended solids, nitrogen, and coliform organisms from the wastewater. Wetlands have not been shown to be efficient in removal of phosphate.

[4] *Transpiration (TRAN-spur-RAY-shun). The process by which water vapor is released to the atmosphere by living plants. This process is similar to people sweating.*

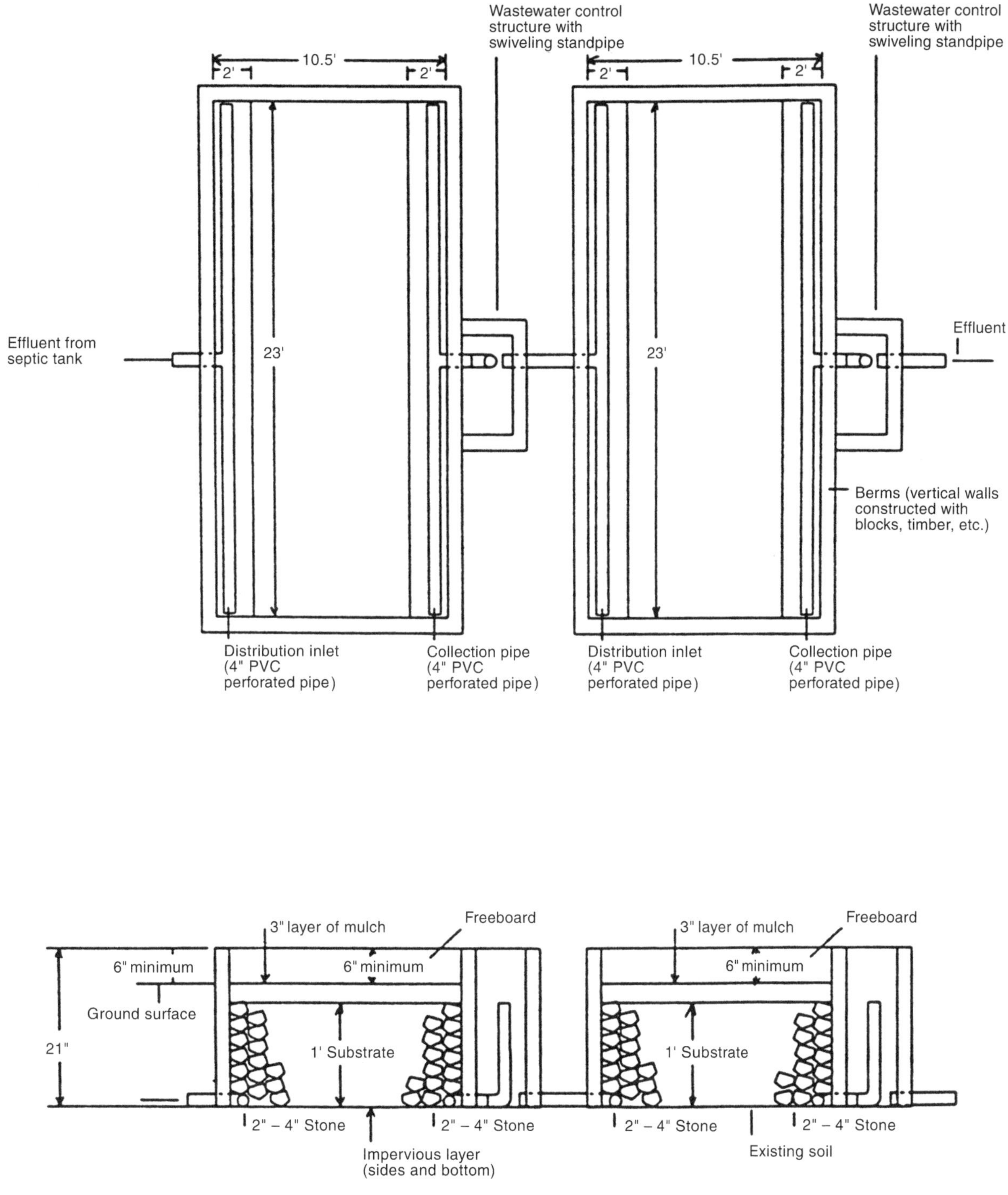

Fig. 13.10 Two-cell system

(Source: *GENERAL DESIGN, CONSTRUCTION AND OPERATION GUIDELINES: CONSTRUCTED WETLANDS*)

13.14 Constructed Wetlands Loading Criteria (FWS and SFS)

TABLE 13.1 CONSTRUCTED WETLANDS LOADING CRITERIA (FWS and SFS)

Pretreatment Required	Minimum of Primary Sedimentation
Hydraulic Loading	0.67 – 0.77 gal/day/sq ft
Hydraulic Detention Time	7 days+
Organic Loading	0.05 lb BOD/square foot/day

13.15 Starting a New SFS Wetland

The wetland bed is built at the same time the septic tank is installed. If the types of vegetation to be planted are not specified by the design, the local agriculture extension agent or a local nursery may be consulted to recommend suitable plants adapted to the local climate. Most plants require six hours of full sunlight for good growth. For shady locations, ferns or other shade-tolerant species should be selected. Decorative species such as water iris, canna lily, elephant's ear, and sweet flag can be used in warmer climates. All will grow over the entire United States (except in parts of Alaska) but some will die out in the colder regions during winter months; these areas must rely on bulrushes, cattails, and other marsh plants.

Place the plants in no less than one foot (0.3 m) apart in a grid pattern; this spacing usually will produce a uniform cover of vegetation over the entire surface of the cell in one to two years. Irrigate the new plants with domestic water for the first six to eight weeks, if possible, before applying wastewater to the bed. After the initial six weeks, add makeup water to control the water level, along with liquid fertilizer to stimulate plant growth (domestic wastewater may not have all of the nutrients required for good plant growth). In facilities that have two cells, the second cell should be given a light application of liquid fertilizer to stimulate growth.

QUESTIONS

Write your answers in a notebook and then compare your answers with those on page 253.

13.1A List the two types of wetlands.

13.1B Why are free water surface (FWS) wetlands not recommended for individual home sites?

13.1C An advantage of wetlands is that they can be built where what other types of wastewater treatment systems have failed?

13.16 Operation and Maintenance

Constructed wetlands are "natural" systems. As a result, operation is mostly passive and requires little operator intervention. Operation involves simple procedures similar to the requirements for operation of a facultative lagoon. The operator must be observant, take appropriate actions when problems develop, and conduct required monitoring and operational monitoring as necessary. The most critical items in which operator intervention is necessary are:

- Adjustment of water levels,
- Maintenance of flow uniformity (inlet and outlet structures),
- Management of vegetation,
- Odor control,
- Control of nuisance pests and insects,
- Maintenance of berms and dikes,
- Maturation of wetlands,
- Mechanical equipment maintenance, and
- Health and safety.

13.160 Water Level and Flow Control

Water level and flow control are usually the only operational variables that have a significant impact on a well-designed constructed wetland's performance. Changes in water levels affect the hydraulic residence time, atmospheric oxygen diffusion into the water phase, and plant cover. Significant changes in water levels should be investigated immediately, as they may be due to leaks, clogged outlets, breached berms, storm water drainage, or other causes.

Seasonal water level adjustment helps to prevent freezing in the winter. In cold climates, water levels should be raised approximately 18 inches (45 cm) in late fall until an ice sheet develops. Once the water surface is completely frozen, the water levels can be lowered to create an insulating air pocket under the ice and snow cover to maintain higher water temperatures in the wetland.

1. The main operational control is proper water level. In subsurface flow wetlands, water level in the first cell is maintained at one inch below the top of the gravel media at the inlet end. Water levels can be monitored easily by installing four-inch diameter pipes standing vertically at random locations throughout the cell. The pipes will act as stilling wells and provide an instant view of the water level. Immediately after an application of wastewater to a small bed, an increase in water level may result from surging. In larger community beds this is less noticeable. Water level control through the outlet structure regulating device occurs by lowering or raising the overflow discharge level. (Review Figure 13.7.)
2. Low water level. In periods of low flows or hot weather, water in the first cell may evaporate or be used by the plants during severe cold periods, it could freeze solid, damaging the roots and tubers. Makeup water may have to be added to the cell during those periods, as needed.
3. Surface pooling or ponding in the wetland bed that cannot be controlled by water level adjustment is probably caused by either excessive flows above the design hydraulic loading rate or clogging of the media by excessive solids in the influent being applied. Check the cleanout in front of the wetland. Solids accumulated in the pipe indicate excessive solids from the pretreatment unit (septic tank or tanks from the community), which must be located and pumped. If there are no solids in the cleanout, ponding is probably caused by excessive water flow that exceeds the hydraulic capacity of the media. Attempt water conservation practices to decrease flows. If ponding is not corrected, a portion of the media may have to be replaced, starting with the area near the inlet end.

The City of Davis, California, regulates water levels in its FWS wetlands seasonally. In March and April, water levels are drawn down for the summer to expose the benches along the shorelines. This creates a foraging habitat for shorebirds. Also, by keeping water off the benches, mosquitoes are discouraged. In the fall and winter months, increased stormwater flows are directed to the wetlands. The higher water levels inundate the benches, creating shallow habitat for waterfowl and wading birds. In summary, operators regulate water levels to prepare for seasonal changes and to serve the needs of local waterfowl and wildlife as appropriate.

13.161 Maintenance of Flow Uniformity

Maintaining uniform flow across the wetland through inlet and outlet adjustments is extremely important to achieve the expected treatment performance. The inlet and outlet manifolds should be inspected routinely and regularly adjusted and cleaned of debris that may clog the inlets and outlets. Debris removal and removal of bacterial slimes from weir and screen surfaces will be necessary. Submerged inlet and outlet manifolds should be flushed periodically. Additional cleaning with a high-pressure water spray or by mechanical means also may become necessary.

Wastewater suspended solids will settle and accumulate near the inlets to the wetland. These accumulations can decrease hydraulic detention times. Over time, these accumulations of solids will require removal. Subsurface flow systems cannot be desludged easily without draining and removing the media. Therefore, SFSs should not be considered for treating wastewaters with high suspended solids loads, such as facultative pond effluents, which have high algal concentrations. Also, suspended solids from septic tanks should not be allowed to enter SFSs.

13.162 Vegetation Management

Routine maintenance of wetland vegetation is not required for systems operating within their design parameters and with precise bottom-depth control of vegetation. Wetland plant communities are self-maintaining and will grow, die, and regrow each year. Plants will naturally spread to unvegetated areas with suitable environments, within the plants' depth range, and be displaced from areas that are environmentally stressful. Operators must control plants in FWS systems by harvesting to prevent them from spreading into open water areas that are intended by design to be aerobic zones through harvesting.

The primary objective in vegetation management is to maintain the desired plant communities where they are intended to be within the wetland. This is achieved through consistent pretreatment process operation, small, infrequent changes in water levels, and harvesting plants when and where necessary. Where plant cover is deficient, management activities to improve cover may include water level adjustment, reduced loadings, pesticide application, and replanting.

Harvesting and litter removal may be necessary depending on the design of the system. Plant removal from some free water surface wetlands may be required to meet the treatment goals, but a well-designed and well-operated subsurface flow system should not require routine harvesting. Harvesting of vegetation at the height of the growing season and just before the end of the growing season does help to remove some nitrogen from the system, but phosphorus removal is limited. Winter burning of vegetation, where allowed, can be used to control pests.

1. Periodically, check vegetation for signs of disease or other stress: yellowing, spots, withering or browning, and insects. Natural seasonal changes will occur with temperature and day length changes. First, check the water level in the bed. If water level is proper, obtain guidance from a local agricultural extension agent or a knowledgeable garden center for plant enhancement or correct chemical use and application rate of chemicals for insect control.
2. If vegetation does not appear healthy and water levels are correctly maintained, add a balanced liquid fertilizer periodically (three times a growing season) to the wastewater. Pour the fertilizer into an upstream toilet or into the cleanout inspection pipe just prior to the bed inlet.
3. Replace dead plants as necessary to fill open spaces in the bed.
4. Dig up weeds, trees, and shrubs from the beds and growth adjacent to the wetlands because these species will shade and crowd the desirable plants.
5. Remove mature wetland vegetation after the plants have browned in the fall, if desirable for visual aesthetics. Only cut approximately two-thirds of the height of the plants. The removed material may be left on the bed as mulch.
6. Divide and replant decorative flowering species such as irises to enhance the system's attractiveness.
7. Encourage deep root growth by slowly lowering the water level over several weeks during the dormant period. Do not drop the water level too quickly, leaving the roots stranded without water.
8. Mulch—maintain a three-inch layer on top of the media, either with litter from the wetland vegetation, pine straw or bark, or other suitable material on SFSs.
9. Native, noninvasive vegetation species are preferred to invasive species that can spread from the treatment wetlands to local wetlands, replacing native species. In addition, the plant life selected should not create a habitat for unwanted animals and invertebrates, such as mosquitoes and rats, or threatened or endangered species that could limit or eliminate the use of the treatment wetland.

13.163 Odor Control

Odors are seldom a nuisance problem in properly loaded wetlands. Odorous compounds emitted from open water areas are typically associated with anaerobic conditions, which can be created by excessive BOD and ammonia loadings. Therefore, reducing the organic and nitrogen loadings by pretreatment can control odors. If odors persist, try draining each cell, one cell at a time, and then refilling it to restore aerobic conditions in the cell. Alternatively, aerobic open water zones interspersed in areas between fully vegetated zones introduce oxygen to the system. Turbulent flow structures such as cascading outfall structures and channels with hydraulic jumps, which are used to introduce oxygen into the system effluent, can generate serious odor problems through stripping of volatile compounds such as hydrogen sulfide if the constructed wetland has failed to remove these constituents.

Standing water on SFS media surface (mulch) is the probable cause of objectionable odors. Level any low and high spots that create small standing pools on the media surface by raking or filling with additional gravel and mulch. If the water level is too high, causing standing water on most of the media surface, lower the water level to about one inch below the surface of the gravel in the bed.

13.164 Nitrogen Control

To remove ammonia by converting ammonia to nitrate, wetlands can be an effective treatment process. Wetlands require aerobic conditions to convert ammonia to nitrate (*NITRIFICATION*[5]). Aerobic conditions can be achieved if the water being treated is near the surface of submerged flow systems. After nitrification occurs, wetlands require *ANOXIC*[6] conditions for *DENITRIFICATION*[7] to occur.

13.165 Control of Nuisance Pests and Insects

Wetland plant life should NOT create a habitat for unwanted animals or invertebrates, such as mosquitoes and rats, or threatened or endangered species that could limit or eliminate the use of the treatment wetland.

Potential nuisances and vectors that may occur in FWS wetlands include burrowing animals, dangerous reptiles, mosquitoes, and odors. An infestation of burrowing animals such as muskrats and nutria can seriously damage vegetation and berms in a system. These animals use both cattails and bulrushes as food and nesting materials. These animals can be controlled during the design phase by decreasing the slope on berms to 5:1 or using a coarse riprap. Temporarily raising the operating water level may also discourage the animals. Live trapping and release may be successful, but in most cases it has been necessary to eliminate the animals. Fencing has had little success.

Dangerous reptiles are common in the southeastern states. The most common are snakes, particularly the water moccasin, and alligators. It is difficult to control these animals directly. Warning signs, fencing, raised boardwalks, and mowed hiking trails can be used to minimize human contact with the animals. Operators should be made aware of the dangers and preventive actions that can be taken to avoid dangerous situations.

Mosquito control is a critical issue in FWS wetlands. In warm climates, wetlands have been seeded with mosquito fish (gambusia) and dragonfly larvae to control mosquitoes. Mosquito fish also can be used in northern climates but they need to be restocked each year. However, mosquito fish have difficulty reaching all parts of the wetland when the accumulation of litter is too dense, particularly if cattails are grown. Other natural control methods have included erecting bat and bird houses. Desirable birds include purple martins and swallows. Bacterial larvicides, BTI (*Bacillus thuringiensis israelensis*), and BS (*Bacillus sphaericus*) have been used successfully in a number of wetlands.

13.166 Maintenance of Berms and Dikes

Berms and dikes require mowing, erosion control, and prevention of animal burrows and tree growth. When the wetland is operated at a shallow depth, periodic removal of tree seedlings from the wetland bed may be necessary. If the trees are allowed to reach maturity, they may shade out the emergent vegetation and with it the necessary conditions to enhance flocculation, sedimentation, and denitrification.

13.167 Maturation of Wetlands

Wetlands may require up to three years to "mature" and achieve expected performance. Time is required for plant roots and aquatic organisms to develop and become effective in treating wastewater.

13.168 General Maintenance

1. Pumps, if used in the system, must be lubricated and cleaned and control systems must be checked on a routine frequency.
2. Berms must be repaired immediately if erosion is noted; leaks from the beds should not be permitted. Earthen berms or dikes should be mowed to maintain an attractive site.
3. Liners of polyethylene, PVC, neoprene, or rubber that extend above the media or water level should be covered with mulch or soil to prevent ultraviolet degradation.
4. Surface drainage should be routed away from the wetland beds so as not to affect water level and hydraulic loading.

13.169 Health and Safety

Prevent children and pets from playing or digging in the wetland to avoid contact with potentially infectious microorganisms, destroying vegetation, or moving media and mulch.

13.17 Monitoring

Routine monitoring is essential in managing a wetland system. In addition to regulatory requirements, inflow and outflow rates, water quality, water levels, and indicators of biological conditions should be regularly monitored and evaluated. Monitoring of biological conditions includes measurement of microbial populations and monitoring changes in water quality, percent cover of dominant plant species, and benthic macro-

[5] *Nitrification (NYE-truh-fuh-KAY-shun). An aerobic process in which bacteria change the ammonia and organic nitrogen in wastewater into oxidized nitrogen (usually nitrate). The second-stage BOD is sometimes referred to as the "nitrogenous BOD" (first-stage BOD is called the "carbonaceous BOD"). Also see DENITRIFICATION.*

[6] *Anoxic (an-OX-ick). A condition in which atmospheric or dissolved molecular oxygen is NOT present in the aquatic (water) environment and nitrate is present. Oxygen deficient or lacking sufficient oxygen. The term is similar to anaerobic.*

[7] *Denitrification (dee-NYE-truh-fuh-KAY-shun). (1) The anoxic biological reduction of nitrate nitrogen to nitrogen gas. (2) The removal of some nitrogen from a system. (3) An anoxic process that occurs when nitrite or nitrate ions are reduced to nitrogen gas and nitrogen bubbles are formed as a result of this process. The bubbles attach to the biological floc in the activated sludge process and float the floc to the surface of the secondary clarifiers. This condition is often the cause of rising sludge observed in secondary clarifiers or gravity thickeners.*

invertebrate and fish populations at representative stations. Over time, these data help the operator to predict potential problems and select appropriate corrective actions.

13.18 Acknowledgment

The author wishes to acknowledge that a major portion of the material in this lesson on wetlands was adapted from EPA manuals, *WASTEWATER COLLECTION, TREATMENT AND DISPOSAL OPTIONS FOR SMALL COMMUNITIES* and *CONSTRUCTED WETLANDS TREATMENT OF MUNICIPAL WASTEWATERS*.

QUESTIONS

Write your answers in a notebook and then compare your answers with those on page 253.

13.1D Why should wetland vegetation be checked periodically?

13.1E What is the probable cause for odors from a subsurface flow system (SFS) wetland?

13.1F Which operational variables can have a significant impact on the performance of a well-designed constructed wetland?

13.1G What maintenance should be performed on the inlets and outlets to wetlands?

13.1H Why is vegetation harvested from wetlands?

13.1I What type of habitat should *NOT* be created by wetland plant life?

END OF LESSON 1 OF 2 LESSONS
on
ALTERNATIVE WASTEWATER TREATMENT, DISPOSAL, AND REUSE METHODS

Please answer the discussion and review questions next.

DISCUSSION AND REVIEW QUESTIONS

Chapter 13. ALTERNATIVE WASTEWATER TREATMENT, DISPOSAL, AND REUSE METHODS

(Lesson 1 of 2 Lessons)

At the end of each lesson in this chapter you will find some discussion and review questions. The purpose of these questions is to indicate to you how well you understand the material in the lesson. Write the answers to these questions in your notebook before continuing.

1. How is emergent vegetation supported in a free water surface (FWS) wetland?
2. How is the water level controlled in a subsurface flow wetland?
3. What basic tasks must an operator perform when operating and maintaining a treatment wetland?
4. What is the primary objective of vegetation management in a wetland?

CHAPTER 13. ALTERNATIVE WASTEWATER TREATMENT, DISPOSAL, AND REUSE METHODS

(Lesson 2 of 2 Lessons)

13.2 LAND TREATMENT SYSTEMS

13.20 Description of Treatment Systems

When a high-quality effluent or no discharge is required, land treatment offers a means of wastewater reclamation or ultimate disposal. Land treatment systems use soil, plants, and bacteria to treat and reclaim wastewaters. They can be designed and operated for the sole purpose of wastewater disposal, for crop production, or for both purposes. In land treatment, effluent is pretreated and applied to land by conventional irrigation methods. When systems are designed for crop production, the wastewater and its nutrients (nitrogen and phosphorus) are used as a resource. Treatment is provided by natural processes (physical, chemical, and biological) as the effluent flows through the soil and plants. Part of the wastewater returns to the *HYDROLOGIC CYCLE*[8] by *EVAPOTRANSPIRATION*[9] and part of it returns by surface runoff and/or percolation through the soil to groundwater (Figure 13.11). Land disposal of wastewater may be done by one of the following methods shown in Figure 13.12: irrigation, infiltration-percolation, and overland flow.

The method of irrigation depends on the type of crop being grown. Irrigation methods include traveling sprinklers, fixed sprinklers, furrows, and flooding. Infiltration-percolation systems are not suited for crop growth because the water is not held in the soil for crop uptake. Overland flow systems are similar to other treatment processes and have runoff that must either be discharged or recycled in the system. The other systems usually do not have a significant surface runoff. Typical loading rates for these systems are shown in Table 13.2.

Land application systems include the following components:

1. Treatment before application,
2. Transmission to the land treatment site,
3. Storage,
4. Distribution over site,
5. Runoff recovery system (if needed), and
6. Crop systems.

EXAMPLE 1

A piece of land 1,000 feet long by 800 feet wide is used for a land disposal system. If 0.4 million gallons of water is applied to the land during a 24-hour period, calculate the hydraulic loading in:

1. MGD per acre, and
2. Inches per day.

Known	Unknown
Length, ft = 1,000 ft	Hydraulic Loading
Width, ft = 800 ft	1. MGD/acre
Flow, MGD = 0.4 MGD	2. inches/day

1. Determine surface area in acres.

$$\text{Area, acres} = \frac{\text{Length, ft} \times \text{Width, ft}}{43{,}560 \text{ sq ft/acre}}$$

$$= \frac{1{,}000 \text{ ft} \times 800 \text{ ft}}{43{,}560 \text{ sq ft/acre}}$$

$$= 18.4 \text{ acres}$$

Or

$$= 1{,}000 \text{ ft} \times 800 \text{ ft} \div 43{,}560 \text{ sq ft/acre}$$

$$= 18.4 \text{ acres}$$

2. Determine the hydraulic loading, MGD/acre.

$$\text{Loading, MGD/ac} = \frac{\text{Flow, MGD}}{\text{Area, acres}}$$

$$= \frac{0.4 \text{ MGD}}{18.4 \text{ acres}}$$

$$= 0.022 \text{ MGD/acre}$$

3. Determine the hydraulic loading, inches/day.

$$\text{Loading, in/day} = \frac{\text{Flow, MGD} \times 1{,}000{,}000/\text{M} \times 12 \text{ in/ft}}{\text{Length, ft} \times \text{Width, ft} \times 7.48 \text{ gal/cu ft}}$$

$$= \frac{0.4 \text{ MGD} \times 1{,}000{,}000/\text{M} \times 12 \text{ in/ft}}{1{,}000 \text{ ft} \times 800 \text{ ft} \times 7.48 \text{ gal/cu ft}}$$

$$= 0.8 \text{ in/day}$$

Or

$$= 0.4 \text{ MGD} \times 1{,}000{,}000/\text{M} \times 12 \text{ in/ft} \div 1{,}000 \text{ ft} \div 800 \text{ ft} \div 7.48 \text{ gal/cu ft}$$

$$= 0.8 \text{ in/day}$$

[8] *Hydrologic (HI-dro-LOJ-ick) Cycle. The process of evaporation of water into the air and its return to earth by precipitation (rain or snow). This process also includes transpiration from plants, groundwater movement, and runoff into rivers, streams and the ocean. Also called the water cycle.*

[9] *Evapotranspiration (ee-VAP-o-TRANS-purr-A-shun). (1) The process by which water vapor passes into the atmosphere from living plants. Also called transpiration. (2) The total water removed from an area by transpiration (plants) and by evaporation from soil, snow and water surfaces.*

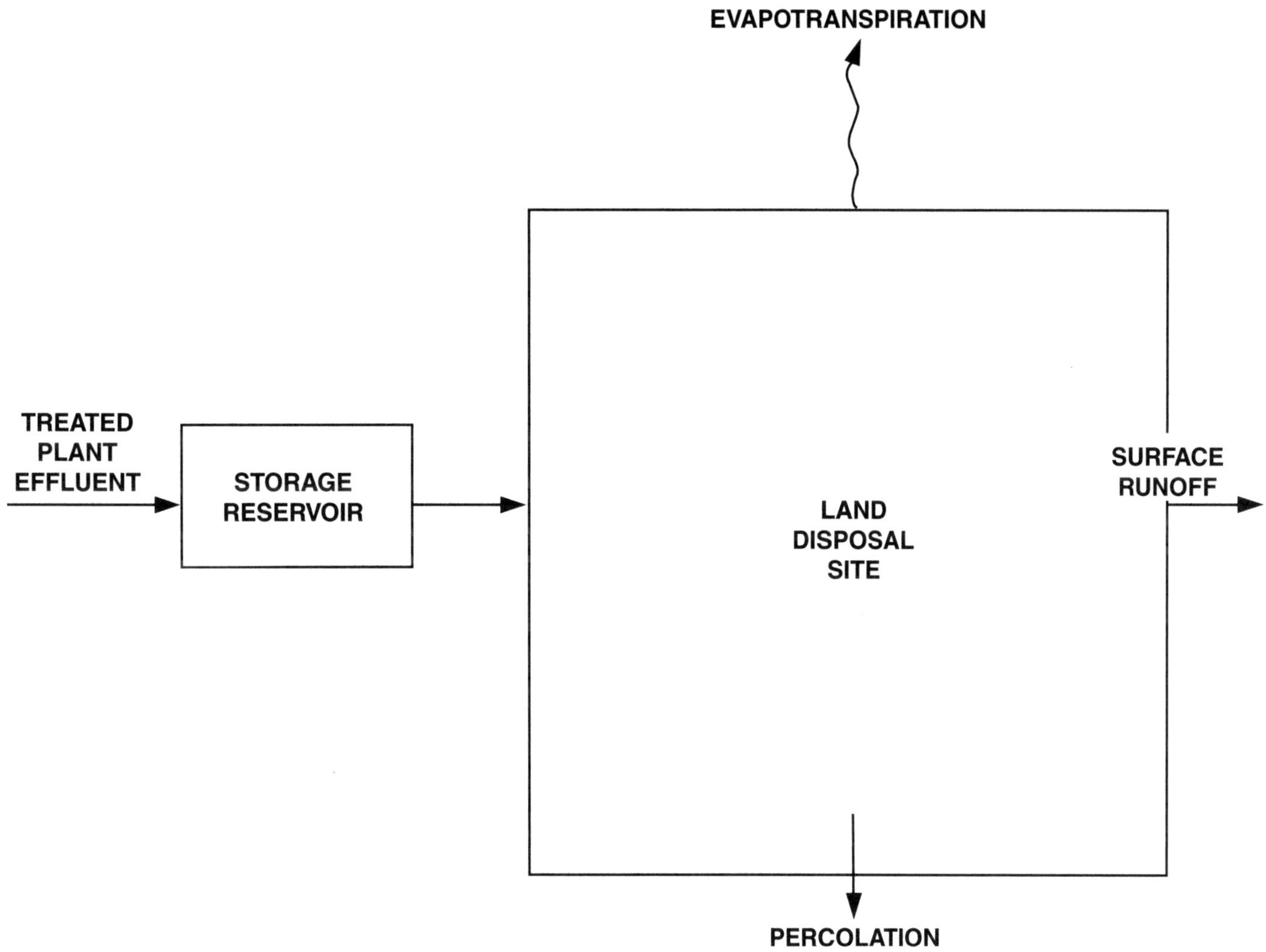

Fig. 13.11 Land disposal system schematic

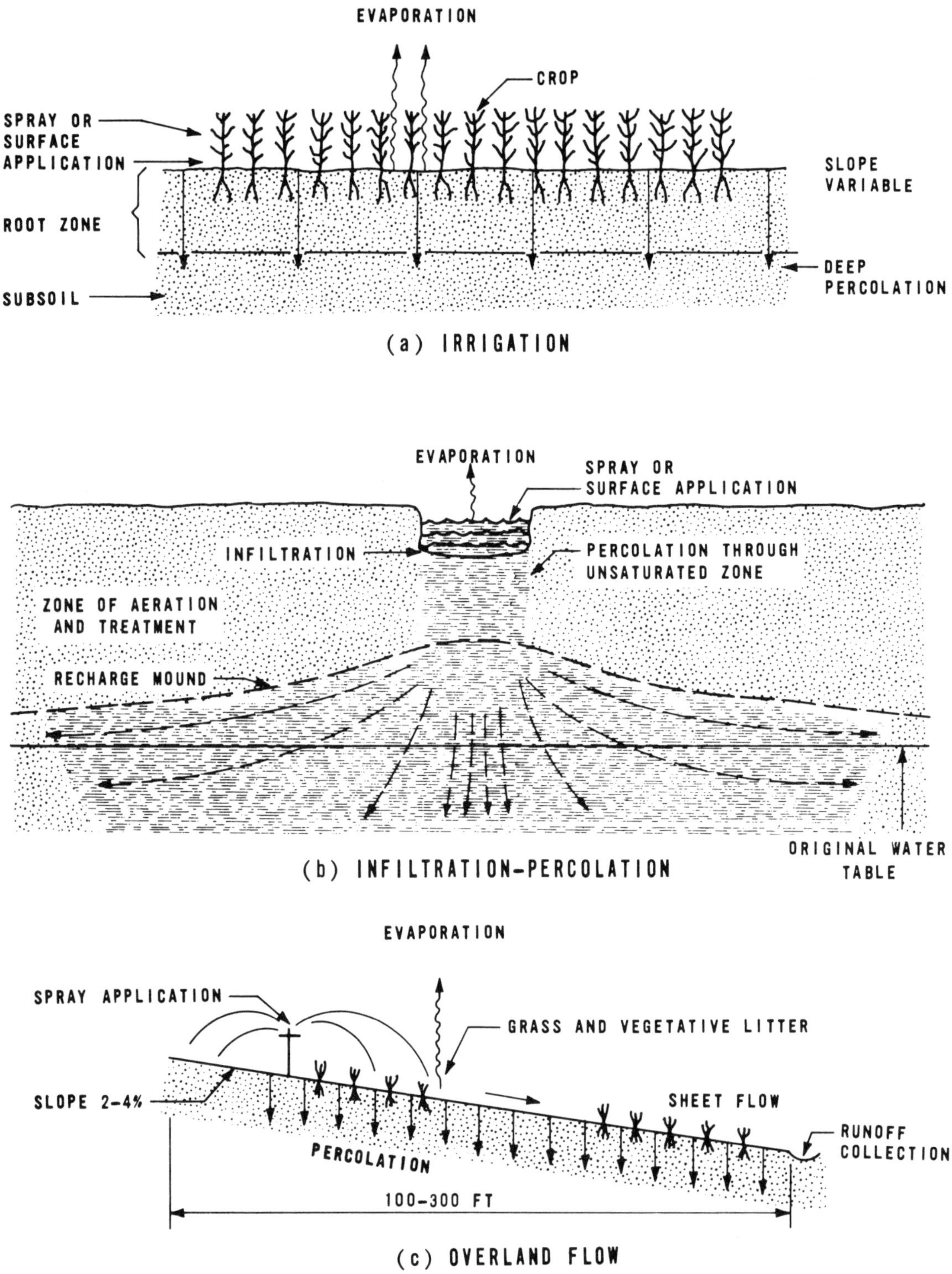

Fig. 13.12 Methods of land application

(From *COSTS OF WASTEWATER TREATMENT BY LAND APPLICATION* by Pound, C.E., et al., U.S. Environmental Protection Agency, Washington, DC 20460, EPA-430/9-75-003)

TABLE 13.2 TYPICAL LOADINGS FOR IRRIGATION, INFILTRATION-PERCOLATION, AND OVERLAND FLOW SYSTEMS[a]

Factor	Irrigation Low-Rate	Irrigation High-Rate	Infiltration-Percolation	Overland Flow
Liquid loading rate, in/wk[b]	0.5 to 1.5	1.5 to 4.0	4 to 120	2 to 9
Annual application, ft/yr[c]	2 to 4	4 to 18	18 to 500	8 to 40
Land required for 1 MGD flow rate, acres[d, e]	280 to 560	62 to 280	6 to 62	28 to 140
Application techniques	Spray or surface		Usually surface	Usually surface
Vegetation required	Yes	Yes	No	Yes
Crop production	Excellent	Good/fair	Poor/none	Fair/poor
Soils	Moderately permeable soils with good productivity when irrigated		Rapidly permeable soils, such as sands, loamy sands, and sandy loams	Slowly permeable soils such as clay loams and clays
Climatic constraints	Storage often needed		Reduce loadings in freezing weather	Storage often needed
Wastewater lost to:	Evaporation and percolation		Percolation	Surface runoff and evaporation with some percolation
Expected treatment performance				
BOD and SS removal	98+%		85 to 99%	92+%
Nitrogen removal	85+%[d]		0 to 50%	70 to 90%
Phosphorus removal	80 to 99%		60 to 95%	40 to 80%

[a] *COSTS OF WASTEWATER TREATMENT BY LAND APPLICATION* by Pound, C.E., et al., U.S. Environmental Protection Agency, Washington, DC 20460, EPA-430/9-75-003, June 1975.
[b] in/wk x 2.54 = cm/wk
[c] ft/yr x 0.3 = m/yr
[d] Dependent on crop uptake
[e] acres x 0.00107 = hectares for 1 cu m/day
or
acres x 9.24 = hectares for 1 cu m/sec

13.200 Equipment Requirements

Irrigation systems apply water by sprinkling or by surface spreading (Figure 13.13). Sprinkler systems may be fixed or movable. Fixed systems are permanently installed on the ground or buried with sprinklers set on risers that are spaced along pipelines. These systems have been used in all types of terrain. Movable systems include center pivot, side wheel roll, and traveling gun sprinklers.

Surface irrigation systems include flooding, border-check, and ridge and furrow systems. Flooding systems are very similar to overland flow systems except the slopes are nearly level. Border-check systems are simply a controlled flooding system. Ridge and furrow systems are used for row crops such as corn where the water flows through furrows between the rows and seeps into the root zone of the crop.

An overland flow system consists of effluent being sprayed or diverted over sloping terraces where it flows downslope through the vegetation. The vegetation provides a filtering action, removing suspended solids and insoluble BOD. This system can be operated as a treatment system or, with a recycle system included, as a disposal system. When operated as a disposal system, the overland flow process is nearly the same as flood irrigation so it will be included with the flood systems in the following paragraphs.

13.201 Sidestreams and Their Treatment

There are two possible sidestreams with land disposal systems, percolation from storage reservoirs and runoff from irrigated fields. Unlined storage reservoirs will result in wastewater percolating down to groundwater. If the water stored in reservoirs is the final effluent from a treatment plant, percolation down to the groundwater probably is acceptable. However, if the water is untreated or partially treated (primary effluent), the reservoir should be lined or an underground collection system should be installed to collect any percolation or seepage from the reservoir.

In some areas, percolation may cause a rise in the area's groundwater level. To prevent groundwater problems, a seepage ditch may be built around the outside of the reservoir. This ditch should be located somewhat lower than the bottom of the reservoir. Wastewater that percolates through the reservoir bottom is collected in the ditches and can be pumped back into the reservoir. The groundwater table also can be lowered by a series of shallow wells with water being pumped out as necessary. Lowering of the groundwater can result in increased percolation rates.

The other major sidestream is surface runoff. Runoff quantities vary depending on the type of irrigation system used. In all systems, provisions should be made for collecting runoff and returning it for reapplication. In some locations, runoff water can be discharged to surface water. Discharge is the preferred approach due to cost savings and minimizing operational problems. For information on discharge to surface waters, see Chapter 13, "Effluent Disposal," in *OPERATION OF WASTEWATER TREATMENT PLANTS*, Volume II, in this series of operator training manuals.

13.202 Limitations of Land Treatment

Problems encountered using land treatment systems usually involve soil problems and weather conditions. If proper care is not taken, soils can lose their ability to percolate applied water. During the cold winter season, plants and crops will not grow. Under these conditions, no water will be treated by transpiration processes. Also, precipitation can soak the soil so no wastewater can be treated. Provisions must be made to store wastewater during cold and wet weather.

One of the most common land treatment problems is the sealing (water won't percolate) of the soil by suspended solids in the final effluent. These solids are deposited on the surface of the soil and form a mat that prevents the percolation of water down through the soil. There are three possible solutions to this problem:

1. Remove the suspended solids from the effluent,
2. Apply water intermittently and allow a long enough drying period for the solids mat to dry and crack, or
3. Disc or plow the field to break the mat of solids.

Another serious problem is the buildup of salts in the soil. If the effluent has a high chloride content, there can be enough salts in the soil within one year to create a toxic condition to most grasses and plants. To overcome salinity problems:

1. Leach out the salts by applying fresh water (not effluent), or
2. Rip up the field and turn it over to a depth of 4 to 5 feet (1.2 to 1.5 meters).

The severity of both soil sealing due to suspended solids and salinity problems due to dissolved solids depends on the type of soil in the disposal area. These problems are more common and more difficult to correct in clay soils than in sandy soils.

QUESTIONS

Write your answers in a notebook and then compare your answers with those on pages 253 and 254.

13.2A Define the terms evapotranspiration and hydrologic cycle.

13.2B List the three methods by which land disposal of wastewater is accomplished.

13.2C What are the major components of a land application system?

13.2D A plot of land 2,000 feet long by 1,000 feet wide is used for a land disposal system. If one million gallons of water is applied to the land during a 24-hour period, calculate the hydraulic loading in:
1. MGD per acre, and
2. Inches per day.

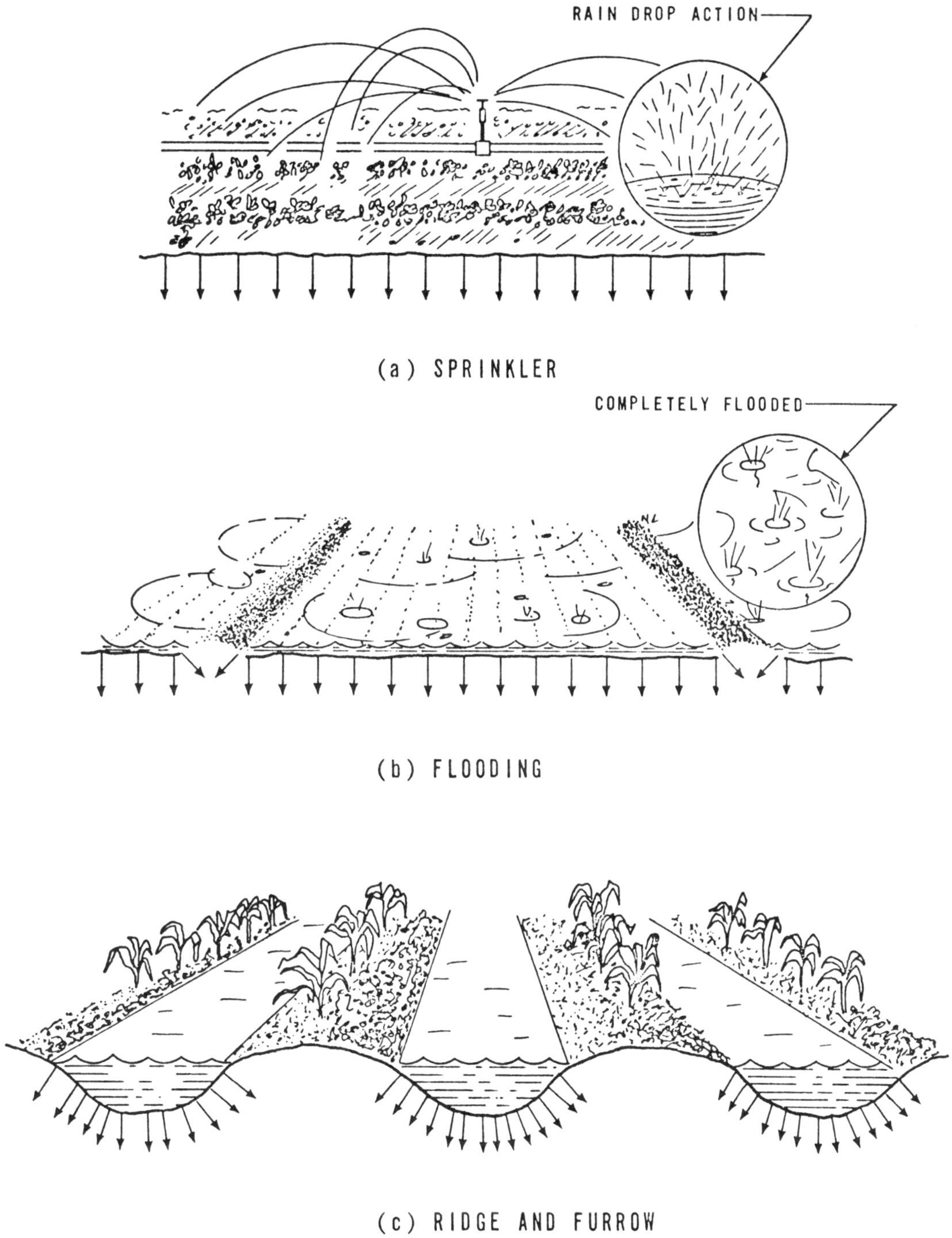

Fig. 13.13 Irrigation techniques

(From *EVALUATION OF LAND APPLICATION SYSTEMS* by Pound, C.E., et al., U.S. Environmental Protection Agency, Washington, DC 20460, EPA 430/9-74-015, September 1974)

13.21 Operating Procedures for Crops

The operating procedures described below apply to a spray irrigation system for an area where crops are to be grown. The operating procedures for this type of system are more complex than for other systems. The procedures are explained first and then an example is presented to show how to use the procedures.

13.210 Pre-Start Checklist

Table 13.3 consists of a list of items that should be checked before starting a land irrigation system. The table contains a checklist for aluminum pipe. A similar checklist can be developed for other pipe materials.

13.211 Start-Up

1. Determine the need to irrigate. The amount of water that can be applied depends on the type of crop. Some crops require a lot of water while other crops will not tolerate any excess water. The procedures in this section are prepared to help you determine when you should irrigate and how much water should be applied. *THESE PROCEDURES ARE BASED ON SOIL CONDITIONS AND MAY REQUIRE ADJUSTMENT BASED ON THE TYPE OF CROP BEING IRRIGATED.* The chart shown on Table 13.4 will aid in determining the need to irrigate based on soil conditions. To use this table you must first determine the type of soil you are irrigating. Walk around the field and try to identify the different soil types if more than one type of soil is present. Pick up a handful of soil and examine the grains or particles. Very small, hard particles indicate a sandy soil. Very fine, soft, smooth particles indicate a clay soil.

   Once you've identified the type of soil, the *AVAILABLE MOISTURE CONTENT*[10] can be estimated by squeezing the soil together in your hand. The wetter the soil, the more likely the soil will stick or cling together. By the way the soil sticks together and the type of soil, you can estimate when the available moisture content drops to 50 percent or less. Irrigate when the moisture content is 50 percent or less.

   Soils vary throughout an area of land and some areas need irrigating before others. Many farmers have specific areas in their fields which they call "hot spots." Due to soil characteristics or other reasons, these areas dry out faster than the rest of the field. Whenever the soil will not form a ball when squeezed together or crops start to show signs of stress due to lack of water, this is the time to irrigate.

2. Determine the amount of moisture to apply using Table 13.5. To use this table, determine the soil type by examining a handful of soil as described in Step 1. The root zone depth is determined by the type of crop. If necessary, dig down to determine how far down the roots are growing. By the use of Table 13.4, you can estimate the percent of available moisture remaining in the soil before irrigation. Knowing the percent available moisture in the soil before irrigation, you can use Table 13.5 to determine the net inches of water to apply when you irrigate.

3. Determine the time required to apply one inch (2.5 cm) of water from Figure 13.14 or Figure 13.15. Figure 13.14 is used for small areas (up to 60 acres) and Figure 13.15 is used for larger areas (over 60 acres). To use the tables determine (1) the area of land you wish to irrigate, and (2) the capacity of your irrigation system in gallons per minute. By starting at the bottom of the figure with the known area, draw a line vertically upward. Next draw a line from the system capacity on the left, horizontally to the right. Where these two lines intersect is the time required to apply one inch of water to the area being irrigated.

4. Multiply time required to apply one inch (2.5 cm) of water by the inches of moisture to apply to determine the total run time for the irrigation system. After you have determined the net inches of water you wish to apply from Table 13.5 (Step 2), multiply the net inches times the time required (Figures 13.14 and 13.15) to determine the total irrigation time.

   If you need help determining soil types, irrigation requirements, soil moisture content, types of crops to plant, salt tolerance of plants, depth of root zone, and any fertilizer needs, contact your local farm adviser. In many areas, an expert adviser is available free of charge though an agency of the federal, state, or local government.

5. Check the pump discharge check valve and suction screen for foreign matter.

6. Check pipelines to see that all couplings are still fastened, blocks or pipe supports haven't fallen over, and gates and valves are open.

7. Start pump.

8. Inspect the irrigation system to be sure everything is working properly.

EXAMPLE

This example shows how to use the procedures outlined in this section to determine the total time to irrigate.

1. Determine the need to irrigate. A loamy soil formed a weak ball when squeezed. Table 13.4 indicates the moisture condition is low (50 percent or less) and that it is time to irrigate.

2. The loamy (light sandy) soil has a crop with a root-zone depth of 3 feet (0.7 m) and 50 percent of the available moisture is retained in the soil at irrigation. Table 13.5 indicates that 1.5 inches (3.8 cm) of water should be applied to the soil.

3. The land to be irrigated is a 40-acre plot and the pump has a capacity of 1,000 GPM. From Figure 13.14, find 40 acres across the bottom and draw a line vertically upward to the top. Find the system capacity of 1,000 GPM along the left side and draw a line horizontally to the right. These lines intersect between the diagonal lines labeled 15 hours and 20 hours at approximately 18 hours. Therefore, we should irrigate for 18 hours to apply one inch of water.

4. Determine the total time to irrigate.

Time, hours = Time (hr), to irrigate one inch x Amount to Apply, in

= 18 hours/inch x 1.5 inches

= 27 hours

[10] *Available Moisture Content. The quantity of water present that is available for use by vegetation.*

TABLE 13.3 SPRAY IRRIGATION PRE-START CHECKLIST[a]

Check Every Time Pump Is Started	Check at the Beginning of the Season	
		Electric Motors
	______	Replace winter lubricant.
	______	Oil bath bearings. Drain oil and replace with proper weight of clean oil.
	______	Grease lube bearings. If grease gun is used, be sure old grease is purged through outlet hole.
	______	Before turning on switch, have power company check voltage.
	______	Check for proper rotation of motor and pump.
______	______	Check fuses to make sure they are still good.
	______	Check electrical contact points for excessive corrosion.
______	______	Physically inspect for rodent and insect invasion.
		Pumps
	______	Replace oil or grease with proper weight bearing lubricant.
______	______	Tighten packing gland to proper setting.
______	______	Check discharge head, discharge check valve, and suction screen thoroughly for foreign matter.
______	______	Pump shaft should turn freely without noticeable dragging.
		Aluminum Pipe
	______	If you didn't properly handle and store aluminum pipe or tubing last fall, make a mental note to do that at the end of this growing season. Always carefully drain aluminum tubing or pipe when you are finished using it. Aluminum pipe bends very easily. If you pick up a length of pipe that is half-full of water and has one end plugged, you will bend the pipe. Flush out the pipe to clean it, drain the pipe, place the pipe on a long-bed trailer for transport, and then store the pipe on racks until you are ready to use the pipe next season.
	______	Inspect pipe ends to make certain that no damage has occurred. Ends should be round for best operation. A slightly tapered wooden plug of the proper diameter can be used to round out the ends. The diameter of aluminum pipe varies from 2 to 12 inches (5 to 30 cm).
		Aluminum Pipe (continued)
	______	Check pipe for pit corrosion. If "spots" are in evidence, contact the aluminum pipe supplier for advice.
	______	Inspect pipe gaskets, couplers (irrigation pipe couplings), and gates to find those in need of replacement.
______	______	Check pipeline to see that all couplers are still fastened, pipe supports haven't fallen over, and gates and valves are still open.
______	______	Pipe makes an excellent nesting area for small animals. Flush out the pipeline before installing the end plug. Make sure you are away from power lines when you raise the pipe to drain water from the pipe.
		Sprinkler Systems
	______	Sprinkler bearing washers should be replaced if there is indication of wear.
______	______	Visually check all moving parts, seals, bearings, and flexible hose for replacement or repair.
______	______	Check to see that hose is laid out straight or on a long radius for turns. Be sure there are no kinks in the hose. There should be sufficient hose at the end of the sprinkler to act as a brake or to hold back the sprinkler system initially as it drags the hose through the field. Also check earth anchors.
______	______	If possible, operate the system to check speed adjustment, alignment, and safety switch mechanisms.
______	______	Check sprinkler oscillating arm for proper adjustment. If damage has occurred to the sprinkler oscillating arm, the arm should be replaced or bent back to the correct angle. Your dealer can help in correcting a damaged arm. The angle of water-contact surface, if not correct, will change the turning characteristics of the sprinkler. Excessive wear of sprinkler nozzles can be checked with proper size drill bit.
	REMEMBER,	inspection and corrective maintenance now may save considerable time and money later.

[a] *WATER RESOURCES MANUAL*, Norum, Edward (editor), permission of The Irrigation Association, Silver Spring, MD 20906.

TABLE 13.4 FEEL AND APPEARANCE GUIDE FOR DETERMINING SOIL MOISTURE[a]

The chart below is very useful in estimating how much available moisture is in your soil. Although the plant's daily moisture use may range from 0.1 inch to 0.4 inch per day (0.25 to 1.0 cm), it will average about 0.20 inch (0.5 cm) per day, and 0.25 inch (0.6 cm) per day during hot days.

Moisture Condition	Percent of Available Moisture Remaining in Soil %	SOIL TEXTURE		
		Sands-Sandy Loams	Loams-Silt Loams	Clay Loams-Clay
Dry	Wilting point	Dry, loose, flows through fingers.	Powdery, sometimes slightly crusted but easily broken down into powdery condition.	Hard, baked, cracked; difficult to break down into powdery condition.
Low	50% or less		Will form a weak ball when squeezed but won't stick to tools.	Pliable, but not slick; will ball under pressure—sticks to tools.
TIME TO IRRIGATE WHEN AVAILABLE MOISTURE IS 50 PERCENT OR LESS				
Fair	50 to 75%	Tends to ball under pressure but seldom will hold together when bounced in the hand.	Forms a ball somewhat plastic, will stick slightly with pressure. Doesn't stick to tools.	Forms a ball, will ribbon out between thumb and forefinger, has a slick feeling.
Good	75 to 100%	Forms a weak ball, breaks easily when bounced in the hand, can feel moistness in soil.	Forms a ball, very pliable, sticks readily, clings slightly to tools.	Easily ribbons out between thumb and forefinger, has a slick feeling, very sticky.
Ideal	Field capacity 100%	Soil mass will cling together. Upon squeezing, outline of ball is left on hand.	Wet outline of ball is left on hand when soil is squeezed. Sticks to tools.	Wet outline of ball is left on hand when soil is squeezed. Sticky enough to cling to fingers.

[a] *WATER RESOURCES MANUAL*, Norum, Edward (editor), permission of Sprinkler Irrigation Association, Silver Spring, MD 20906.

TABLE 13.5 AMOUNT OF MOISTURE TO APPLY TO VARIOUS SOILS UNDER DIFFERENT MOISTURE RETENTION CONDITIONS[a]

Soil Type	Root Zone Depth	Available Moisture Plants Will Use	Net Inches to Apply Per Irrigation With Various Percents Available Moisture Retained In the Soil Before Irrigation		
			Percent Available Moisture Before Irrigation		
	Feet	Inches	67%	50%	33%
Light Sandy	1	1.00	0.33	0.50	0.67
	1½	1.50	0.50	0.75	1.00
	2	2.00	0.56	1.00	1.33
	2½	2.50	0.83	1.25	1.67
	3	3.00	0.99	1.50	2.00
Medium	1	1.69	0.57	0.85	1.13
	1½	2.53	0.84	1.26	1.70
	2	3.38	1.11	1.69	2.26
	2½	4.21	1.39	2.11	2.82
	3	5.06	1.67	2.53	3.38
Heavy	1	2.39	0.79	1.20	1.59
	1½	3.58	1.18	1.79	2.38
	2	4.78	1.58	2.39	3.25
	2½	5.97	1.97	2.98	3.97
	3	7.17	2.36	3.58	4.77

[a] *WATER RESOURCES MANUAL*, Norum, Edward (editor), permission of Sprinkler Irrigation Association, Silver Spring, MD 20906.

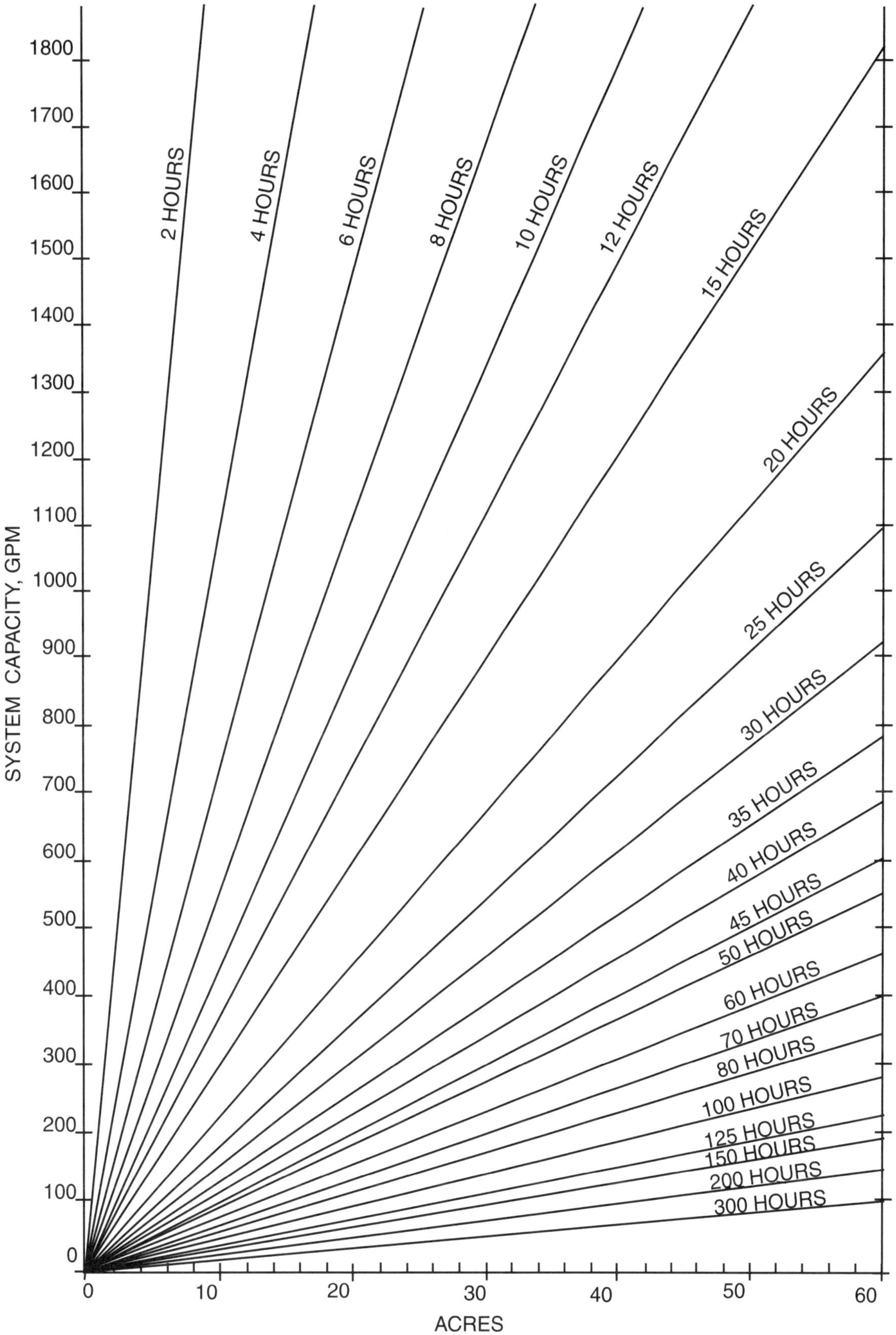

Fig. 13.14 Time required to apply one inch (2.5 cm) of water on small acreages

(From *WATER RESOURCES MANUAL*, Norum, Edward (editor), permission of Sprinkler Irrigation Association, Silver Spring, MD)

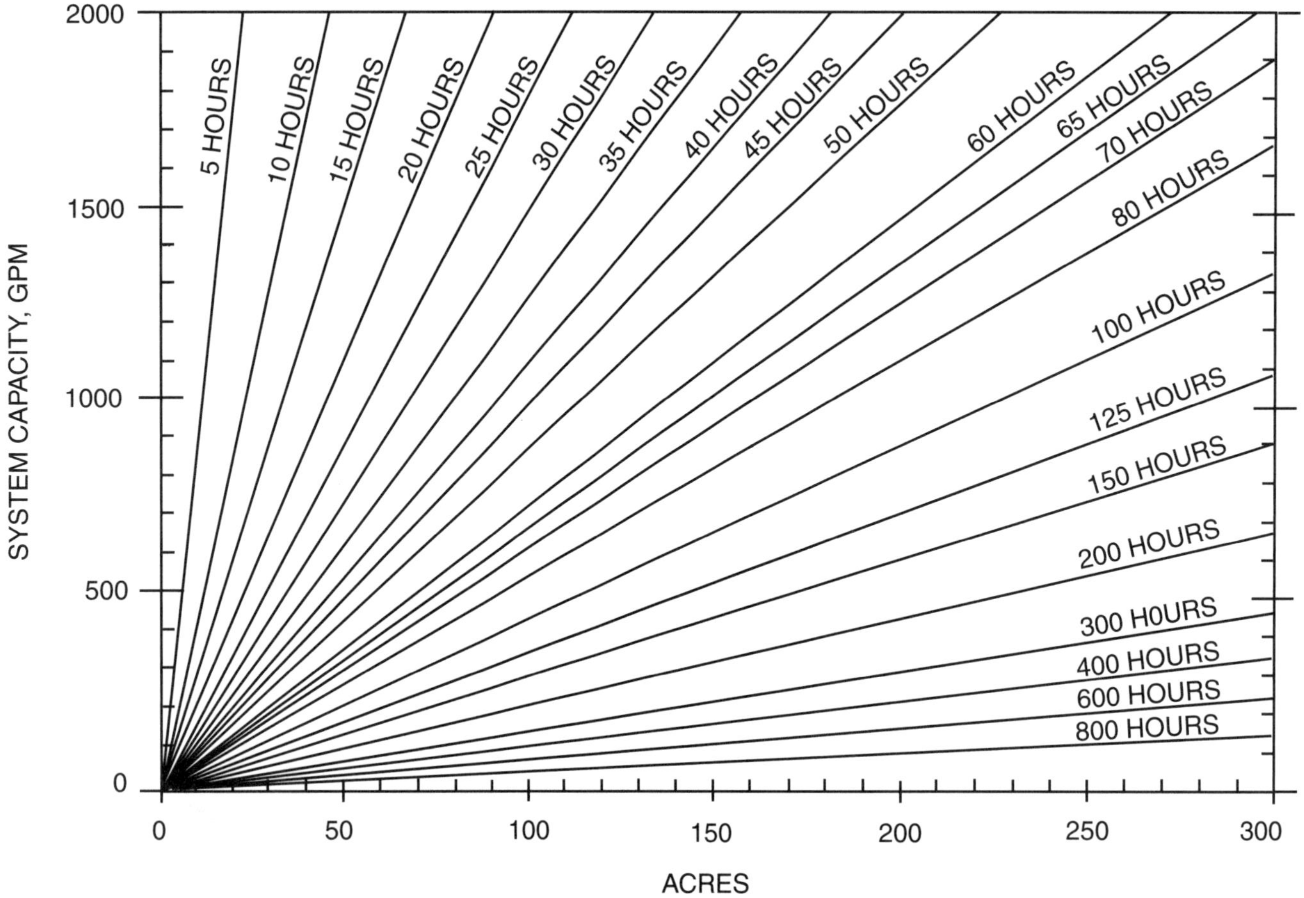

Fig. 13.15 Time required to apply one inch (2.5 cm) of water on large acreages

(From *WATER RESOURCES MANUAL*, Norum, Edward (editor), permission of Sprinkler Irrigation Association, Silver Spring, MD)

13.212 Normal Operation

1. Run the pump or pumps the time determined by Step 4 of the start-up procedure and as shown in the example.
2. Turn the pump off earlier if water begins to pond on the fields.

13.213 Shutdown

This shutdown procedure applies to the end of season shutdown.

1. Drain all lines.
2. Plug open ends of pipelines.
3. Lubricate motors and pumps for winter.
4. Store small, movable materials and equipment.
5. Store aluminum tubing or piping.

QUESTIONS

Write your answers in a notebook and then compare your answers with those on page 254.

13.2E List the major items of equipment that should be inspected before starting a spray irrigation system.

13.2F Determine the time required to irrigate 30 acres of a medium-type soil where the root-zone depth is 2 feet and 50 percent of the available moisture is retained in the soil at irrigation. Pump capacity is 1,200 GPM. Use the figures and tables in this lesson to answer this question.

13.214 Operational Strategy

Physical Control. The main objective of a land disposal system is to dispose of effluent without harming surface waters or creating nuisance conditions. An irrigation system can be designed and operated to produce a crop. The sale of this crop then helps reduce treatment costs. Physical control is then used to dispose of effluent at the highest rate possible without damaging the crop. For all types of land disposal systems, physical controls consist of valves and/or gates which are used to direct treated effluent to different disposal areas.

Process Control. There are three areas of process control: storage reservoirs, runoff and seepage water recycle systems, and soil systems where crops are grown. The first is the storage reservoir. The reservoir usually will have been equipped with aeration or mixing devices. These devices may be operated full-time or for a limited time each day by using timers. The correct length of time to operate is determined by measuring the dissolved oxygen (DO) content at several points in the reservoir. A minimum of four points should be sampled. An example showing six sampling locations is illustrated in Figure 13.16.

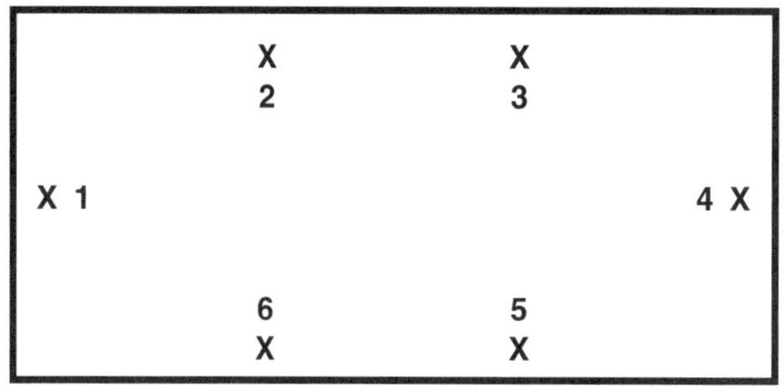

Fig. 13.16 Possible reservoir sampling locations

Each sample should have at least 0.4 mg/*L* of DO and the average of all samples should be at least 0.8 mg/*L*. Two sample sets of test results are shown below:

Set 1	Sample No.	mg/*L* DO
	1	1.2
	2	1.8
	3	2.0
	4	1.6
	5	0.2
	6	0.4

$$\text{Average} = \frac{7.2}{6} = 1.2 \text{ mg}/L$$

Set 2	Sample No.	mg/*L* DO
	1	1.0
	2	0.4
	3	0.8
	4	1.4
	5	0.6
	6	1.2

$$\text{Average} = \frac{5.4}{6} = 0.90 \text{ mg}/L$$

Set 1 does not meet the requirements. The average DO is 1.2 mg/*L* which is good, but one sample (#5) is 0.2 mg/*L*, which does not meet the minimum requirement. This result indicates that there is adequate aeration but either a portion of the system is not operating properly or the reservoir is not being adequately mixed.

Even though Set 2 has a lower average DO than Set 1, the average is greater than 0.8 and all samples are 0.4 mg/*L* or greater. Therefore, Set 2 is acceptable.

The second area of process control is the runoff and seepage water recycle systems. These recycle systems may not be necessary, depending on the particular application. For example, the storage reservoir may be lined so there would be no seepage water to recycle. Sprinkler systems that are carefully controlled will have no significant runoff. Your goal is to dispose of as much water as possible without causing runoff and seepage from the disposal area. This is done to reduce recycle pumping costs. The reduction in seepage and runoff is accomplished by taking more care in applying effluent and turning off sprinklers or closing gates when water begins to stand in the field.

The third area of process control applies to those systems where crops are grown. In dry climates such as those found in the southwestern states and western mountain states, farmers who irrigate are concerned with saline and alkaline soils. Some minerals found in the effluent may cause a decrease in crop production in soils of this type. Analyses and irrigation practices for these areas are described in detail in the U.S. Department of Agriculture's Handbook No. 60, *DIAGNOSIS AND IMPROVEMENT OF SALINE AND ALKALINE SOILS.* A diagram for the classification of irrigation waters (taken from Handbook No. 60) is shown on Figure 13.17.

A simplified version of this table with other critical constituents (substances) is shown on Table 13.6.

TABLE 13.6 CLASSIFICATION OF IRRIGATION WATERS[a]

Chemical Properties	Class I Excellent to Good "Suitable under most conditions"	Class II Good to Injurious "Suitability dependent on soil crop, climate, and other factors"	Class III Injurious to Unsatisfactory "Unsuitable under most conditions"
Total dissolved solids (mg/*L*)	Less than 700	700 – 2,000	More than 2,000
Chloride (mg/*L*)	Less than 175	175 – 350	More than 350
Sodium (percent of base constituents)	Less than 60	60 – 75	More than 75
Boron (mg/*L*)	Less than 0.5	0.5 – 2.0	More than 2.0

[a] *DIAGNOSIS AND IMPROVEMENT OF SALINE AND ALKALINE SOILS*, Richards, L.A. (editor), Agricultural Handbook No. 60, U.S. Department of Agriculture, Washington, DC.

The local U.S. Department of Agriculture, Soil Conservation Service office can provide a list of crops that can be irrigated with Class II and Class III water.

There is nothing that can be done that is economically feasible to control the concentration of these constituents. If one of the constituents exceeds the Class I limit, the crop should be changed to one that is more tolerant. Special agricultural practices can be used to minimize the effects of these constituents. These practices vary from one local area to another. The local soil conservation service office and farm adviser can assist by providing information appropriate to the area.

Important observations and interpretations were discussed earlier for determining when to irrigate. The most important observations in a system where crops are grown and surface runoff is not allowed are observing ponding or runoff. When this occurs, either the application amount was excessive or the rate of application was greater than the soil infiltration rate.

Visual appearance of the crop being grown is extremely important. Discoloration in plant leaves can indicate excess water (poor drainage) or a nutrient or mineral deficiency. Your local farm adviser can assist in diagnosing the problems.

Observations are critical in storage reservoirs. Odors can result if effluent treatment was inadequate and/or if insufficient aeration is provided to the reservoir.

13.215 Emergency Operating Procedures

Loss of power will disrupt sprinkler systems (assuming electric motors are used for pumps). If power is lost, the effluent will be retained in the storage reservoir. Gravity-flow flood irrigation systems won't be affected during power outages.

Loss of other treatment units is generally not a problem for a few days. Longer downtimes may result in an overloaded and an odorous storage reservoir and possible odors at the disposal area.

13.216 Troubleshooting Guide

See Table 13.7, page 248.

QUESTIONS

Write your answers in a notebook and then compare your answers with those on page 254.

13.2G What is the main objective of a land disposal system?

13.2H List the three main areas of process control in a land disposal system.

13.2I How many points in a storage reservoir should be sampled for DO?

13.2J What are the minimum recommended DO requirements for a storage reservoir?

13.2K What are the probable causes of water ponding in an irrigated area where ponding normally has not been observed?

13.22 Monitoring Program for an Irrigation System

13.220 Monitoring Schedule

The four monitoring areas for an irrigation system where crops are grown are: effluent, vegetation, soils, and groundwater (or collected seepage). This is reduced to the two areas of effluent and groundwater for systems where crops are not grown. Wells should be monitored to identify any adverse effects on groundwater. Testing requirements and frequencies are shown in Table 13.8.

TABLE 13.8 TESTING REQUIREMENTS

Area	Test	Frequency
Effluent and groundwater or seepage	BOD	Two times per week
	Fecal coliform	Weekly
	Total coliform	Weekly
	Flow	Continuous
	Nitrogen	Weekly
	Phosphorus	Weekly
	Suspended solids	Two times per week
	pH	Daily
	Total dissolved solids (TDS)	Monthly
	Boron	Monthly
	Chloride	Monthly
Vegetation	—variable depending on crop—	
Soils	Conductivity	Two times per month
	pH	Two times per month
	Cation exchange capacity[a]	Two times per month

[a] Cation Exchange Capacity. The ability of a soil or other solid to exchange cations (positive ions such as calcium, Ca^{2+}) with a liquid.

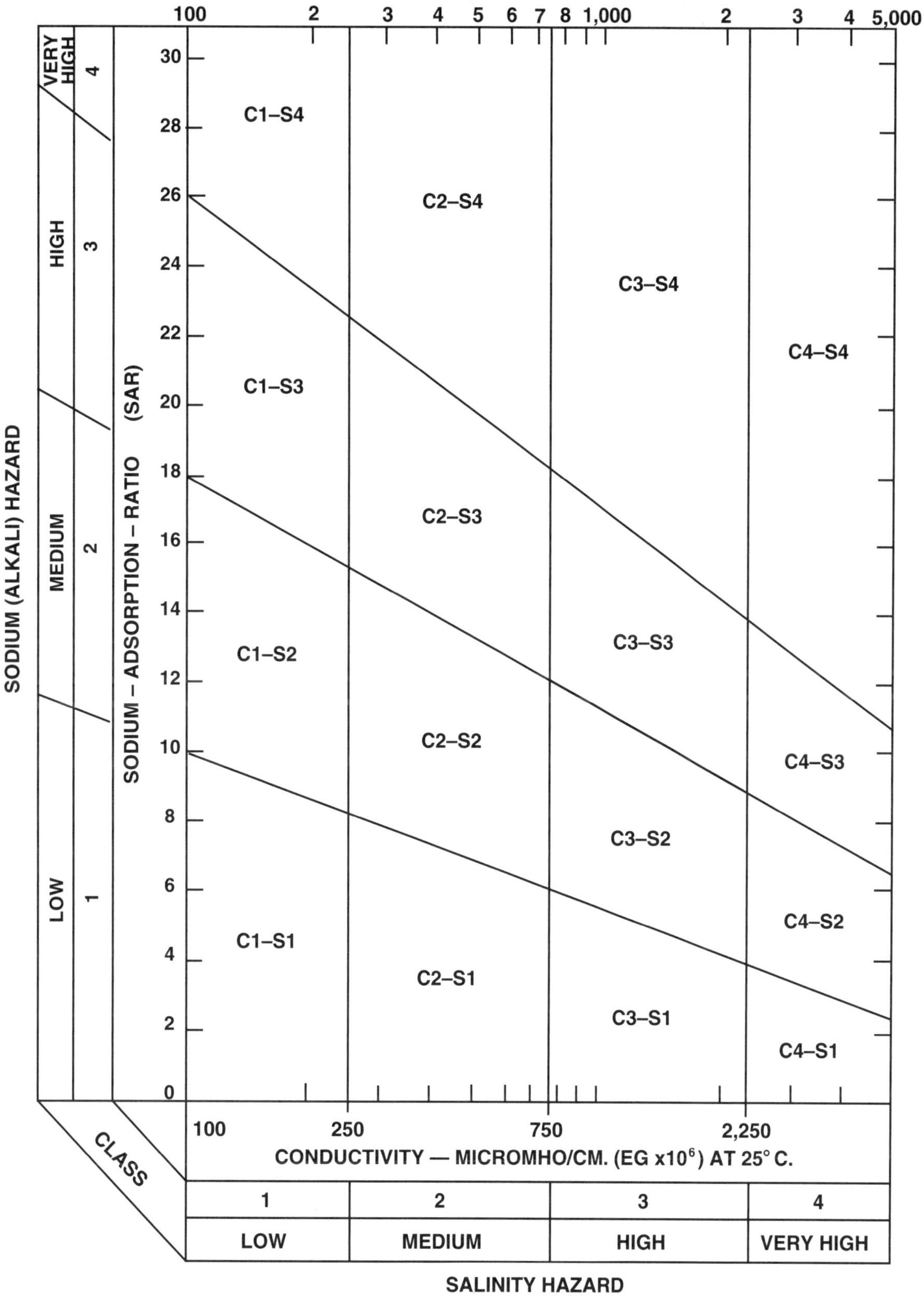

Fig. 13.17 Diagram for the classification of irrigation waters

(From *DIAGNOSIS AND IMPROVEMENT OF SALINE AND ALKALINE SOILS*, Richards, L.A. (editor), Agricultural Handbook No. 60, U.S. Department of Agriculture, Washington, DC)

TABLE 13.7 TROUBLESHOOTING GUIDE

Indicator/Observation	Probable Cause	Check or Monitor	Solution
1. Water ponding in irrigated area where ponding normally has not been observed.	1a. Application rate is excessive.	1a. Application rate.	1a. Reduce rate to normal value.
	1b. If application rate is normal, drainage rate may be inadequate.	1b(1) Seasonal variation in groundwater level.	1b(1) Irrigate portions of the site where groundwater is not a problem or store wastewater until level has dropped.
		1b(2) Operability of any drainage wells.	1b(2) Repair drainage wells or increase pumping rate.
		1b(3) Condition of drain tiles.	1b(3) Repair drain tiles.
	1c. Broken pipe in distribution system.	1c. Leaks in system.	1c. Repair pipe.
2. Lateral aluminum distribution piping deteriorating.	2a. Effluent permitted to remain in aluminum pipe too long causing electrochemical corrosion.	2a. Operating techniques.	2a. Drain aluminum lateral lines except when in use.
	2b. Dissimilar metals (steel valves and aluminum pipe).	2b. Pipe and valve specifications.	2b. Coat steel valves or install cathodic or anodic protection.
3. No flow from some sprinkler nozzles.	3. Nozzle clogged with particles from wastewater due to lack of screening at inlet side of irrigation pumps.	3. Screen may have developed hole due to partial plugging of screen.	3. Repair or replace screen.
4. Wastewater is running off of irrigated area.	4a. Sodium adsorption ratio of wastewater is too high and has caused clay soil to become impermeable.	4a. Sodium adsorption ratio (SAR) should be less than 9.	4a. Feed calcium and magnesium to adjust SAR.
	4b. Soil surface sealed by solids.	4b. Soil surface.	4b. Strip crop area.
	4c. Application rate exceeds infiltration rate of soil.	4c. Application rate.	4c. Reduce application rate until compatible with infiltration rate.
	4d. Break in distribution piping.	4d. Leaks in distribution piping.	4d. Repair breaks.
	4e. Soil permeability has decreased due to continuous application of wastewater.	4e. Duration of the continuous operation on the given area.	4e. Each area should be allowed to rest (2 – 3 days) between applications of wastewater to allow soil to drain.
	4f. Rain has saturated soil.	4f. Rainfall records.	4f. Store wastewater until soil has drained.
5. Irrigated crop is dead.	5a. Too much (or not enough) water has been applied.	5a. Water needs of specific crop versus application rate.	5a. Reduce (or increase) application rate.
	5b. Wastewater contains excessive amount of toxic elements.	5b. Analyze wastewater and consult with county agricultural agent.	5b. Eliminate industrial discharges of toxic materials.
	5c. Too much insecticide or weed killer applied.	5c. Application of insecticide or weed killer.	5c. Proper control of application of insecticide or weed killer.
	5d. Inadequate drainage has flooded root zone of crop.	5d. Water ponding.	5d. (See Item 1)
6. Growth of irrigated crop is poor.	6a. Too little nitrogen (N) or phosphorus (P) applied.	6a. N and P quantities applied—check with county agricultural agent.	6a. If increased wastewater application rates are not practical, supplement wastewater N or P with commercial fertilizer.
	6b. Timing of nutrient application not consistent with crop needs. (Also, see 5a – 5c)	6b. Consult with county agricultural agent.	6b. Adjust application schedule to meet crop needs.
7. Irrigation pumping station shows normal pressure but above-normal flow.	7a. Broken main, lateral, riser, or gasket.	7a. Inspect distribution system for leaks.	7a. Repair leak.
	7b. Missing sprinkler head or end plug.	7b. Inspect distribution system for leaks.	7b. Repair leak.

TABLE 13.7 TROUBLESHOOTING GUIDE (continued)

Indicator/Observation	Probable Cause	Check or Monitor	Solution
7. Irrigation pumping station shows normal pressure but above-normal flow. (continued)	7c. Too many laterals on at one time.	7c. Number of laterals in service.	7c. Make appropriate valving changes.
8. Irrigation pumping station shows above-average pressure but below-average flow.	8. Blockage in distribution system due to plugging sprinklers, valves, screens, or frozen water.		8. Locate blockage and eliminate.
9. Irrigation pumping station shows below-normal flow and pressure.	9a. Pump impeller is worn.	9a. Pump impeller.	9a. Replace impeller (See Section 13.28, "Review of Plans and Specifications," No. 6.)
	9b. Partially clogged inlet screen.	9b. Screen.	9b. Clean screen.
10. Excessive erosion occurring.	10a. Excessive application rate.	10a. Application rate.	10a. Reduce application rate.
	10b. Inadequate crop cover.	10b. Condition of crop cover.	10b. (See Items 5 and 6)
11. Odor complaints.	11a. Wastewater turning septic during transmission to irrigated site and odors being released as it is discharged to pretreatment.	11a. Sample wastewater as it leaves transmission system.	11a. Contain and treat off-gases from discharge point of transmission system by covering inlet with building and passing off-gases through deodorizing system.
	11b. Odors from storage reservoirs.	11b. DO in storage reservoirs.	11b. Improve pretreatment or aerate reservoirs.
12. Center pivot irrigation rigs stuck in mud.	12a. Excessive application rates.		12a. Reduce application rates.
	12b. Improper tires or rigs.		12b. Install tires with higher flotation capabilities.
	12c. Poor drainage.		12c. Improve drainage. (See Item 1b)
13. Nitrate concentration of groundwater in vicinity of irrigation site is increasing.	13a. Application of nitrogen is not in balance with crop needs.	13a. Check lbs/acre/yr of nitrogen being applied with needs of crops.	13a. Change crop to one with higher nitrogen needs.
	13b. Nitrogen being applied during period when crops are dormant.	13b. Application schedules.	13b. Apply wastewater only during periods of active crop growth.
	13c. Crop is not being harvested and removed.	13c. Farming management.	13c. Harvest and remove crop.

Sampling wells should be located within the irrigation site as well as on all sides of the site to identify any changes or trends in water quality. Typical sampling tests and frequencies are listed in Table 13.9.

13.221 Interpretation of Test Results and Follow-Up Actions

Excessive levels and concentrations greater than desired for effluent BOD, fecal and total coliforms, nitrogen, phosphorus, and suspended solids typically are not a concern for crop-growing operations. The total dissolved solids (TDS), boron, chloride, and pH are important during long periods of land treatment, but not for times less than two to three weeks. Excessive nitrogen is a potential problem in spreading basins since nitrate in water supplies can be harmful to infants. If TDS, boron, or chloride levels increase and do not return to previous levels, a change in farming practices may be necessary.

Increased levels in any of the constituents in the groundwater are unacceptable. Most likely, the only constituent that will increase is nitrate-nitrogen. If this occurs, then a nitrogen removal system (partial or complete) should be added to the treatment plant.

TABLE 13.9 SAMPLING WELL MONITORING PROGRAM

Area	Test	Frequency
Wells	Salinity	
	Conductivity	Monthly
	Chloride	Quarterly
	TDS	Quarterly
	Chemical Buildup	
	Nitrate	Monthly
	Calcium	Semiannually
	Magnesium	Semiannually
	Toxicity (Heavy Metals)	
	Cadmium	Monthly
	Lead	Annually
	Zinc	Annually
	Mercury	Annually
	Molybdenum	Annually
	Selenium	Annually
	Organics	
	Trihalomethanes	Quarterly
	Pesticides	Quarterly
	(depends on local application)	

13.23 Safety

Safe operating procedures should be practiced in all undertakings. The operation of a sprinkler irrigation system has caused fatalities among operating personnel. Many of the fatalities have resulted from contact with electricity used either to power the pumping plant or to transmit electricity associated with the area being irrigated.

Moving of portable sprinkler lateral pipelines is an extremely hazardous activity. Raising a pipeline into the air to dislodge a small animal or debris and contacting overhead electrical transmission lines has resulted in severe electric shock or death to the person holding the pipe.

A sprinkler throwing a stream of water into a power line has shorted the power to ground through the sprinkler system and resulted in severe electrical injuries to anyone touching the sprinkler system parts.

Always have the electric motor well bonded to a good ground with suitable-size conductors. Injuries have occurred from touching an ungrounded motor or pump frame having shorted electrical windings in electrically powered pumping plants.

Electric shocks have occurred from faulty starting equipment and from working on energized circuits. Always pull the line disconnect switch; lock out and tag it when making repairs or checks on electrical equipment of any kind.

Surface spreading systems can be hazardous due to wet surfaces and muddy areas. Look over each sprinkler system and mark the potential safety hazards, then avoid the hazards.

13.24 Operation and Maintenance

Maintenance of land treatment systems requires keeping the wastewater distribution piping, pumps, valves, and sprinklers in good working condition. Pump and valve maintenance is discussed in Chapter 15, "Maintenance," *OPERATION OF WASTEWATER TREATMENT PLANTS*, Volume II. Storage reservoir maintenance is similar to pond maintenance outlined in Chapter 9 of this manual, and in Chapter 9, Volume I, *OPERATION OF WASTEWATER TREATMENT PLANTS.*

The timing and application rate of treated wastewater for land treatment must be designed so the plants benefit as much as possible from the nutrients and other constituents in the wastewater without being overwhelmed by them. In addition, there is the potential that certain wastewater constituents may accumulate in the soil and plants over time and become toxic to the plants, clog the soil, or alter the soil structure. For example, too much nitrogen can result in nitrate accumulation in crops, but too little can result in reduced yields. If evaporation regularly exceeds precipitation, too much salt may remain in the soil, which can damage roots. In general, rapid infiltration of the applied water will not increase salts in the soil or the groundwater.

The particular characteristics of the wastewater must be considered in relation to such factors as climate and the individual nutrient requirements of the crops, grass, and landscape plants selected. In addition, the need to dispose of the wastewater has to be balanced with the needs of the plants during the various stages of growth and the hydraulic capacity of the soil and its ability to effectively provide treatment.

Proper operation of a land treatment system requires management of the applied wastewater, the crop, and the soil profile. Applied wastewater must be rotated around the site through the application cycle, allowing time for drying, maintenance, cultivation, and crop harvest. The soil profile must be managed to maintain infiltration rates, avoid soil compaction, and maintain soil chemical balance.

Compaction and surface sealing can reduce the soil infiltration and runoff. The compaction, solids accumulation, and crusting of surface soils may be broken up by cultivation, plowing, or discing when the soil surface is dry. A check of the soil chemical balance is required periodically to determine if the soil pH and percent exchangeable sodium are in the acceptable range. Soil pH can be adjusted by adding lime (to increase pH) or gypsum (to decrease pH). Exchangeable sodium can be reduced by adding sulfur or gypsum followed by leaching to remove the displaced sodium.

The spray headers in the spray irrigation system require regular maintenance. Open pipes and spray headers can become damaged, plugged, or frozen. Any changes in pressure in the system can alter spray patterns in the field. Therefore, spray patterns should be tested to ensure that the system still complies with all setback requirements and also covers the entire area as evenly as possible. To test whether the entire area is being covered evenly, place some pans (like an oil drain pan) in various locations and, after a specific time period, measure the volume of water in each pan using a graduated cylinder or similar device.

Erosion control is an important maintenance activity. Inspect the area to ensure that there are no bare spots that are eroding. Erosion is usually not a problem on slopes less than two percent and erosion problems can be expected on slopes greater than six to eight percent. Also, erosion can be a problem if the spray headers do not distribute the spray uniformly over the area. If sprayed water concentrates in one area, high runoff flows can result and cause erosion. Land treatment systems that do not use spray irrigation must ensure that the water enters the area and is applied uniformly over the area or erosion can be a serious problem.

Use of native grasses in land treatment systems is recommended because native grasses can thrive under local variables such as soils, moisture, and seasons. Competition from non-native grasses and broadleaf weeds can reduce expected treatment levels. Weed control consists of pulling weeds or the use of selective herbicide spraying.

13.25 Monitoring for Land Treatment

Monitoring requirements vary depending on state and local regulations, public access to the site, and system size. In some systems, regular daily or weekly monitoring is needed to check influent and effluent quality, system storage capacity, wind speed and direction, signs of ponding or runoff in the spray field, and depth to water table. Cumulative levels of nutrients, heavy metals, fecal coliforms, and wastewater constituents must be monitored in the soils (and groundwater) at some sites once or twice per year.

13.26 Drip and Subsurface Drip Irrigation Systems

Drip irrigation systems are another type of land treatment. They are also known as "trickle" systems and are an efficient and proven technology used by many small communities to recycle and dispose of wastewater. Subsurface drip systems deliver treated effluent to plant roots.

Drip systems apply treated wastewater to soil slowly and uniformly from a network of narrow tubing, usually plastic or polyethylene, placed either on the ground surface or below the ground surface at shallow depths of 6 to 12 inches (15 to 30 centimeters) in the plant root zone. Wastewater is pumped through tubes under pressure and drips out slowly from a series of evenly spaced openings.

Because subsurface drip systems release wastewater below ground directly to plant roots, they irrigate more efficiently and have advantages over surface irrigation systems. For example, the soil surface tends to stay dry, which means less water is lost to evaporation and there is almost no opportunity for the wastewater to come in contact with plant foliage, humans, or animals. Also, percolation losses are reduced because the wastewater is applied directly to the plant roots over a wide area of soil at a slow rate. Also, in drip systems the wastewater is delivered to the most biologically active part of the soil, which enhances treatment and minimizes the possibility of groundwater contamination. The constant moisture in the root zone also may increase the availability of nutrients to plants, reducing the delivery of nitrogen to groundwater.

13.27 Drip System Operation, Maintenance, and Scheduling

As with spray systems, drip irrigation must be scheduled so the plants benefit from the nutrients and other constituents in the wastewater without being overwhelmed by them, and the needs of the plants must be balanced with the capacity of the soil to treat the most restrictive components in the wastewater. These concerns must be balanced, in turn, with climate and other site factors. Less labor is usually required for operating and maintaining fixed subsurface drip system components compared to spray systems and subsurface drip systems with movable components. For small and individual home systems, the pattern of flow may be fixed or adjusted manually or automatically by the homeowner or operator, depending on the system design and sophistication. In general, the best care for subsurface drip systems is provided by following the individual manufacturer's recommendations.

13.28 Review of Plans and Specifications

Many operational and maintenance problems can be avoided by a careful review of the plans and specifications for a land treatment system. Be sure to look for the items listed in this section.

1. Ponding. Ponding problems can be avoided if the proper site is selected and provided with proper drainage. Soils at the site must be suitable for percolation and for planned crops. Adequate drainage (no ponding) can be provided by leveling or sloping of the land surface so the water will flow evenly over all of the land. *DRAINAGE WELLS*[11] or *DRAIN TILE SYSTEMS*[12] may be necessary to remove excess water and prevent ponding.
2. Plastic pipe laterals. Plastic pipe laterals installed above ground may break because of cold weather or deteriorate due to sunlight. Install plastic pipe laterals below ground.
3. Screens. Install screens on the inlet side of irrigation pumps to prevent spray nozzles from becoming plugged.
4. Buffer area. Be sure sufficient buffer area is provided around spray areas to prevent mist from drifting onto nearby homes and yards. If necessary, do not schedule spraying during days when the wind is blowing toward neighbors.
5. Potential for odor problems. If odors may be a problem, consider furrow or flood irrigation rather than spraying. Spraying can cause odor problems by releasing odors to the atmosphere.
6. Protection of pumps. Excessive wear on pumps can result from sand in the water being pumped. If sand is a problem, improve pretreatment or install a sand trap ahead of the pumps. Remember to drain out-of-service pumps before freezing weather occurs in the fall or winter.
7. Alternate place to pump effluent. An alternate location to pump or dispose of effluent is very important in case of system failure.
8. Use specially designed effluent spray irrigation heads that will not plug rather than conventional sprinkler heads.

13.29 References and Additional Reading

13.290 References

1. Pound, C.E., et al., *COSTS OF WASTEWATER TREATMENT BY LAND APPLICATION*, Environmental Protection Agency, EPA-430/9-75-003, June 1975.
2. Pound, C.E., et al., *EVALUATION OF LAND APPLICATION SYSTEMS*, Environmental Protection Agency, EPA 430/9-74-015, September 1974.
3. Norum, Edward (editor), *WATER RESOURCES MANUAL*, The Irrigation Association, 6540 Arlington Boulevard, Falls Church, VA 22042-6638. No longer available.
4. Richard, L.A. (editor), *DIAGNOSIS AND IMPROVEMENT OF SALINE AND ALKALINE SOILS*, Agricultural Handbook No. 60, U.S. Department of Agriculture, August 1969.

---

[11] *Drainage Wells. Wells that can be pumped to lower the groundwater table and prevent ponding.*

[12] *Drain Tile System. A system of tile pipes buried under the crops that collect percolated waters and keep the groundwater table below the ground surface to prevent ponding.*

13.291 Additional Reading

1. *WASTEWATER RECLAMATION AND REUSE*, Water Quality Management Library, Volume 10, edited by Takashi Asano. Obtain from CRC Press LLC, Attn: Order Entry, 2000 NW Corporate Boulevard, Boca Raton, FL 33431-9868. Price, $169.95.
2. *MANUAL OF WASTEWATER TREATMENT, "TEXAS MANUAL*," Chapters 3 and 22.* Obtain from Texas Water Utilities Association, 1106 Clayton Lane, Suite 101 East, Austin, TX 78723-1093. Price to members, $25.00; nonmembers, $35.00; plus $3.50 shipping and handling.
3. *WESTERN FERTILIZER HANDBOOK*, California Plant Health Association. Obtain from Pearson Education, Order Processing Center, PO Box 11071, Des Moines, IA 50336. ISBN 0-8134-3210-3. Price, $34.95, plus $6.99 shipping and handling.
4. *LAND TREATMENT SYSTEMS FOR MUNICIPAL AND INDUSTRIAL WASTES*. Obtain from Water Environment Federation (WEF), Publications Order Department, 601 Wythe Street, Alexandria, VA 22314-1994. Order No. R00202. Price to members, $64.00; nonmembers, $79.00; plus shipping and handling.

* Depends on edition.

QUESTIONS

Write your answers in a notebook and then compare your answers with those on page 254.

13.2L What are the four monitoring areas for an irrigation system where crops are grown?

13.2M What is the major cause of accidents to operators while working with sprinkler irrigation systems?

13.2N What equipment needs to be maintained in a land treatment system?

13.2O List the items that should be examined when reviewing plans and specifications for a land disposal system.

OF LESSON 2 OF 2 LESSONS
on
ALTERNATIVE WASTEWATER TREATMENT, DISPOSAL, AND REUSE METHODS

Please answer the discussion and review questions before continuing with the Objective Test.

DISCUSSION AND REVIEW QUESTIONS

Chapter 13. ALTERNATIVE WASTEWATER TREATMENT, DISPOSAL, AND REUSE METHODS

(Lesson 2 of 2 Lessons)

Write the answers to these questions in your notebook before continuing with the Objective Test on page 255. The question numbering continues from Lesson 1.

5. How does land treatment work?
6. What should be done with wastewater that seeps out of storage reservoirs and runs off from a land treatment system?
7. Describe the limitations of land treatment systems.
8. How long should the pumps be run while irrigating a plot of land?
9. What water quality indicators should be monitored to ensure that a land disposal system does not adversely affect a groundwater supply?
10. How can safety hazards be avoided while operating a sprinkler irrigation system?

SUGGESTED ANSWERS

Chapter 13. ALTERNATIVE WASTEWATER TREATMENT, DISPOSAL, AND REUSE METHODS

ANSWERS TO QUESTIONS IN LESSON 1

Answers to questions on page 230.

13.1A The two types of wetlands are (1) free water surface (FWS), and (2) subsurface flow system (SFS).

13.1B Free water surface (FWS) wetland systems are not recommended for individual home sites because of the associated problems of vector and disease control.

13.1C An advantage of wetlands is that they can be built where other wastewater disposal systems, such as leach fields, seepage pits, and mounds, have failed due to tight clay soils that bind, or loose sandy soils that percolate water too fast, or in locations where rock outcroppings or a high groundwater table prevent safe wastewater treatment and disposal.

Answers to questions on page 233.

13.1D Wetland vegetation should be periodically checked for signs of disease or other stress such as yellowing, spots, withering or browning, and insects.

13.1E Standing water on the media surface of a SFS wetland is the probable cause of odors.

13.1F Water level and flow control are usually the only operational variables that have a significant impact on a well-designed constructed wetland.

13.1G Inlet and outlet maintenance includes cleaning of debris that may clog the inlets and outlets. Debris removal and removal of bacterial slimes from weir and screen surfaces are necessary.

13.1H Vegetation may be harvested from wetlands to remove nitrogen.

13.1I Wetland plant life should *NOT* create a habitat for unwanted animals or invertebrates, such as mosquitoes and rats, or threatened or endangered species that could limit or eliminate the use of the treatment wetland.

ANSWERS TO QUESTIONS IN LESSON 2

Answers to questions on page 238.

13.2A EVAPOTRANSPIRATION. (1) The process by which water passes into the atmosphere from living plants. (2) The total water removed from an area by transpiration (plants) and by evaporation from soil, snow and water surfaces. Also called "transpiration."

HYDROLOGIC CYCLE. The process of evaporation of water into the air and its return to earth by precipitation (rain or snow). This process also includes transpiration from plants, groundwater movement, and runoff into rivers, streams and the ocean. Also called the "water cycle."

13.2B Land disposal of wastewater is accomplished by irrigation, overland flow, and infiltration-percolation.

13.2C The major components of land application systems include:

1. Treatment before application,
2. Transmission to the land treatment site,
3. Storage,
4. Distribution over site,
5. Runoff recovery system (if needed), and
6. Crop systems.

13.2D A plot of land 2,000 feet long by 1,000 feet wide is used for a land disposal system. If one million gallons of water is applied to the land during a 24-hour period, calculate the hydraulic loading in: (1) MGD per acre, and (2) inches per day.

Known	Unknown
Length, ft = 2,000 ft	Hydraulic Loading, 1. MGD/acre 2. inches/day
Width, ft = 1,000 ft	
Flow, MGD = 1 MGD	

1. Determine surface area in acres.

$$\text{Area, acres} = \frac{\text{Length, ft x Width, ft}}{43{,}560 \text{ sq ft/acre}}$$

$$= \frac{2{,}000 \text{ ft x } 1{,}000 \text{ ft}}{43{,}560 \text{ sq ft/acre}}$$

$$= 45.9 \text{ acres}$$

Or

$$= 2{,}000 \text{ ft x } 1{,}000 \text{ ft} \div 43{,}560 \text{ sq ft/acre}$$

$$= 45.9 \text{ acres}$$

2. Determine the hydraulic loading, MGD/acre.

$$\text{Loading, MGD/ac} = \frac{\text{Flow, MGD}}{\text{Area, acre}}$$

$$= \frac{1 \text{ MGD}}{45.9 \text{ acres}}$$

$$= 0.02 \text{ MGD/ac}$$

3. Determine the hydraulic loading, inches/day.

$$\text{Loading, in/day} = \frac{\text{Flow, MGD x 1,000,000/M x 12 in/ft}}{\text{Length, ft x Width, ft x 7.48 gal/cu ft}}$$

$$= \frac{\text{1 M Gal/day x 1,000,000/M x 12 in/ft}}{\text{2,000 ft x 1,000 ft x 7.48 gal/cu ft}}$$

$$= \text{0.8 in/day}$$

Or

$$= \text{1 MGD x 1,000,000/M x 12 in/ft} \div \text{2,000 ft} \div \text{1,000 ft} \div \text{7.48 gal/cu ft}$$

$$= \text{0.8 in/day}$$

Answers to questions on page 245.

13.2E The major items of equipment that should be inspected before starting a spray irrigation system include electric motors, pumps, aluminum tubing or piping, and sprinkler systems.

13.2F Determine the time required to irrigate 30 acres of a medium-type soil where the root-zone depth is 2 feet and 50 percent of the available moisture is retained in the soil at irrigation. Pump capacity is 1,200 GPM.

Known		Unknown
Area, ac	= 30 acres	Time to Irrigate, hr
Soil type	= Medium	
Root zone, ft	= 2 ft deep	
Moisture	= 50% retention	
Pump, GPM	= 1,200 GPM	

1. Determine inches of water to be applied.

   From Table 13.5
   Application, in = 1.69 inches

2. Determine time to irrigate 30 acres to apply 1 inch with a 1,200 GPM pumping system capacity.

   From Figure 13.15
   Time to irrigate 1 inch = 11 hours

3. Determine total time to irrigate in hours.

$$\text{Time, hr} = \text{Time (hr) to Irrigate x Amount to Apply, in}$$

$$= \text{11 hours/inch x 1.69 inches}$$

$$= \text{18.6 hours}$$

Answers to questions on page 246.

13.2G The main objective of a land disposal system is to dispose of effluent without harming surface waters or creating nuisance conditions.

13.2H The three main areas of process control in a land disposal system are:

1. Storage reservoirs,
2. Runoff and seepage water recycle systems, and
3. Soil systems where crops are grown.

13.2I A minimum of four points in a storage reservoir should be sampled for DO.

13.2J The minimum DO requirements for a storage reservoir are a minimum DO of 0.4 mg/*L* for all samples and the average of all samples should be at least 0.8 mg/*L*.

13.2K Probable causes of ponding include:

1. Application rate is excessive (too many gates open in the segments of pipe in the area),
2. If application rate is normal, drainage may be inadequate, and
3. A broken pipe in the distribution system.

Answers to questions on page 252.

13.2L The four monitoring areas for an irrigation system where crops are grown are effluent, vegetation, soils, and groundwater (or collected seepage).

13.2M The major cause of accidents to operators while working with sprinkler irrigation systems is contact with electricity used either to power the pumping plant or to transmit electricity associated with the area being irrigated.

13.2N Equipment requiring maintenance in a land treatment system includes distribution piping, pumps, valves, and sprinklers.

13.2O Items to be examined when reviewing the plans and specifications for a land disposal system include:

1. Ponding,
2. Plastic pipe laterals,
3. Screens,
4. Buffer area,
5. Potential for odor problems,
6. Protection of pumps, and
7. Alternate disposal site for emergencies.

OBJECTIVE TEST

Chapter 13. ALTERNATIVE WASTEWATER TREATMENT, DISPOSAL, AND REUSE METHODS

Please write your name and mark the correct answers on the answer sheet, as directed at the end of Chapter 1, Volume I. There may be more than one correct answer to each multiple-choice question.

True-False

1. A free water surface wetland system has a natural or constructed subsurface layer of pervious material to allow seepage from the wetland.
   1. True
   2. False
2. Wetland systems can be constructed almost anywhere they are needed.
   1. True
   2. False
3. The cost to construct a subsurface flow system wetland for an individual home is much more expensive than a septic tank/leach field wastewater disposal system.
   1. True
   2. False
4. Water level in wetlands is critical to the survival of the plants.
   1. True
   2. False
5. Wetland plant communities are self-maintaining and will grow, die, and regrow each year.
   1. True
   2. False
6. Land treatment systems using infiltration-percolation are suited for crop growth.
   1. True
   2. False
7. Flooding systems are very similar to overland flow systems except the slopes are nearly level.
   1. True
   2. False
8. Sampling wells should be monitored near land treatment systems for any adverse effects on groundwater.
   1. True
   2. False
9. Plastic pipe laterals installed above ground may break because of cold weather or deteriorate due to sunlight.
   1. True
   2. False
10. Install screens on the inlet side of irrigation pumps to prevent spray nozzles from becoming plugged.
    1. True
    2. False

Best Answer (Select only the closest or best answer.)

11. How is primary treatment of wastewater usually accomplished in small systems prior to the wastewater being discharged to alternative treatment, disposal, and reuse methods?
    1. Comminutors
    2. Grit chambers
    3. Screens
    4. Septic tanks
12. Why are media depths in subsurface flow system wetlands deeper in colder climates?
    1. To facilitate root growth for plants predominant in those areas
    2. To minimize evaporation from the system
    3. To preserve heat in the system
    4. To prevent freezing in the entire column of water in the media
13. Under what conditions might surface odors develop in a subsurface flow system wetland?
    1. When aerobic conditions develop in the media
    2. When groundwater high in sulfate is used to maintain water levels
    3. When the wind is strong enough to blow away the upper layer of mulch
    4. When water is permitted to stand above the mulch layer
14. What is the major role of plants in subsurface flow system wetlands?
    1. Providing vegetation habitat for wildlife
    2. Removing carbon dioxide from the atmosphere
    3. Serving as a source of food for grazing animals
    4. Transferring oxygen to the root zone
15. What is the primary objective in wetland vegetation management?
    1. To keep the wetland vegetation alive
    2. To maintain the desired plant communities where they are intended to be within the wetland
    3. To prevent the generation of odors and production of mosquitoes
    4. To properly regulate the water level

16. Why is vegetation burned during the winter in wetlands?
    1. To allow new plantings in the spring
    2. To control pests
    3. To take advantage of favorable weather conditions
    4. To warm the wetlands
17. When irrigating a crop using spray irrigation, when should the pump be turned off?
    1. When erosion starts
    2. When mildew begins to appear on the soil
    3. When vegetation turns yellow
    4. When water begins to pond on the fields
18. Which effluent constituent could cause a decrease in crop production in saline and alkaline soils?
    1. Biochemical oxygen demand
    2. Dissolved oxygen
    3. Minerals
    4. Nutrients
19. Why are increased levels of nitrate in groundwater unacceptable?
    1. Harmful to aquatic life
    2. Harmful to fisheries
    3. Harmful to infants
    4. Harmful to vegetation
20. How has the operation of a sprinkler irrigation system caused fatalities among operating personnel?
    1. Contact with careless drivers
    2. Contact with deep water (drowning)
    3. Contact with electricity
    4. Contact with pathogenic organisms

Multiple Choice (Select all correct answers.)

21. What are the major types of wetlands?
    1. Free water surface
    2. Marshes
    3. Subsurface flow systems
    4. Swamps
    5. Tidelands
22. What are the advantages of wetlands for small, rural communities?
    1. Can achieve a very high and reliable degree of wastewater treatment
    2. High energy requirements for treatment processes
    3. Lower construction costs
    4. Reduced cost in operation and maintenance
    5. Reduced operational skill to maintain the system
23. Where can constructed wetlands be used even though other disposal systems, such as leach fields, seepage pits, and mounds, have failed or are not acceptable?
    1. Extremely cold weather
    2. High groundwater table
    3. Locations within rock outcroppings
    4. Loose, sandy soils that percolate water too fast
    5. Tight clay soils that bind
24. Which items in a constructed wetland are affected by changes in water levels?
    1. Atmospheric oxygen diffusion into the water phase
    2. Hydraulic residence time
    3. Monitoring requirements
    4. Plant cover
    5. Recordkeeping/reporting
25. What could cause significant changes in water levels in constructed wetlands?
    1. Breached berms
    2. Clogged outlets
    3. Evapotranspiration
    4. Leaks
    5. Storm water drainage
26. What could be the cause of surface pooling or ponding in the wetland bed?
    1. Clogging of the media
    2. Excessive flows
    3. High evaporation rates
    4. Improper water level adjustments
    5. Insufficient solids in the influent
27. Potential nuisances and vectors that may occur in free water surface wetlands include which items?
    1. Burrowing animals
    2. Dangerous reptiles
    3. Mosquitoes
    4. Odors
    5. Water fowl
28. What types of maintenance are required on berms and dikes?
    1. Encouraging grazing animals
    2. Erosion control
    3. Mowing
    4. Prevention of animal burrows
    5. Prevention of tree growth
29. Monitoring of biological conditions in a wetland includes which items?
    1. Benthic macroinvertebrate populations
    2. Changes in water quality
    3. Measurement of fish populations
    4. Measurement of microbial populations
    5. Percent cover of dominant plant species
30. How do land treatment systems treat and reclaim wastewater? By using
    1. Bacteria
    2. Chemical precipitation
    3. Electrowinning
    4. Plants
    5. Soil
31. Which methods are used for land disposal of wastewater?
    1. Infiltration-percolation
    2. Inflow/infiltration
    3. Irrigation
    4. Overland flow
    5. Rotating biological contactors
32. Which are possible sidestreams with land disposal systems?
    1. Condensation
    2. Evapotranspiration
    3. Supernatant
    4. Surface runoff
    5. Wastewater percolating down to groundwater

33. What types of problems may be encountered using land treatment systems?
    1. Buildup of salts in the soil
    2. During the cold weather season, no water will be treated by the transpiration process
    3. Nutrients are taken up by vegetation
    4. Precipitation can soak the soil so no wastewater will be treated
    5. Soils can lose their ability to percolate applied water

34. What are the main areas of process control for a land treatment system?
    1. Erosion and sediment control
    2. Runoff and seepage water recycle systems
    3. Soil systems where crops are grown
    4. Solids balance and inventory
    5. Storage reservoirs

35. Which items are indicators of a problem with an irrigated land treatment system?
    1. Growth of irrigated crop is poor
    2. No flow from some sprinkler nozzles
    3. Odor complaints
    4. Wastewater running off the irrigated area
    5. Water ponding in the irrigated area

36. What could be the cause of wastewater running off an irrigated area?
    1. Application rate exceeds infiltration rate of soil
    2. Break in distribution piping
    3. Rain has saturated soil
    4. Sodium adsorption ratio of wastewater is too high
    5. Soil surface sealed by solids

37. What are the monitoring areas for an irrigation system where crops are grown?
    1. Distribution system
    2. Effluent
    3. Groundwater (or collected seepage)
    4. Soils
    5. Vegetation

38. Which types of tests should be included in a sampling well monitoring program?
    1. Chemical buildup
    2. Dissolved oxygen/biochemical oxygen demand
    3. Organics
    4. Salinity
    5. Toxicity (heavy metals)

39. When operating and maintaining a land treatment system, how must the soil profile be managed?
    1. Avoid soil compaction
    2. Backwash when necessary to prevent clogging
    3. Control distribution of particle sizes
    4. Maintain infiltration rates
    5. Maintain soil chemical balance

40. What are the advantages of a subsurface drip irrigation system?
    1. Easy to clean and maintain drip nozzles
    2. Minimum opportunity for wastewater to come in contact with plant foliage, humans, or animals
    3. Percolation losses are reduced because of application directly to plant roots
    4. Release wastewater directly to plant roots
    5. Soil surface tends to stay dry, less water lost to evaporation

END OF OBJECTIVE TEST

CHAPTER 14

LABORATORY PROCEDURES

by

John Brady

and

Ken Kerri

Much of this text is taken from Chapter 16, "Laboratory Procedures and Chemistry," *OPERATION OF WASTEWATER TREATMENT PLANTS*, Volume II, by James Paterson, Joe Nagano, and James Sequeira.

TABLE OF CONTENTS

Chapter 14. LABORATORY PROCEDURES

OBJECTIVES

Chapter 14. LABORATORY PROCEDURES

Following completion of Chapter 14, you should be able to:

1. Work safely in a laboratory,
2. Operate laboratory equipment,
3. Collect representative samples of influents to and effluents from a treatment process as well as sample the process,
4. Prepare samples for analysis in the plant lab,
5. Perform plant control tests,
6. Record laboratory test results, and
7. Collect and prepare samples of plant effluent for analysis in accordance with NPDES permit requirements.

CHAPTER 14. LABORATORY PROCEDURES

(Lesson 1 of 7 Lessons)

14.0 IMPORTANCE OF LABORATORY PROCEDURES

Laboratory control tests are the means by which we collect the data we use to control the efficiency of our wastewater treatment processes. By relating laboratory results to operations, the plant operator can select the most effective operational procedures, determine the efficiency of treatment unit processes, and identify developing treatment problems before they seriously affect effluent quality. For these reasons, a clear understanding of laboratory procedures is a must to any operator.

14.00 Material in This Chapter

The lab procedures outlined in this chapter are those simple and easy procedures used by many operators to control wastewater treatment processes. See Section 14.4, "Laboratory Procedures for Plant Control," for procedures for successful plant control and operation. When lab data must be submitted to regulatory agencies for monitoring and enforcement purposes, you must use approved test procedures. See Section 14.5, "Laboratory Procedures for *NPDES*[1] Monitoring."

Most small wastewater system operators collect samples for NPDES monitoring and submit these samples to an approved laboratory for analysis. For details on laboratory procedures used by approved laboratories that are not included in this chapter, operators may refer to the following references: (1) Chapter 16, "Laboratory Procedures and Chemistry," (1)[2]; (2) "CWEA Lab Procedures" (2); (3) "Water Analysis Handbook" (3); and (4) "Standard Methods" (4). The "Water Analysis Handbook," by HACH Company, is an excellent lab procedure reference for small wastewater system operators.

Each test section contains the following information.

1. Discussion of test.
2. What is tested?
3. Apparatus.
4. Reagents.
5. Procedures.
6. Precautions.
7. Examples.
8. Calculations.

14.01 References

1. *OPERATION OF WASTEWATER TREATMENT PLANTS,* Volume II, Fifth Edition, 2001, Office of Water Programs, California State University, Sacramento, 6000 J Street, Sacramento, CA 95819-6025. Price, $27.00, includes shipping and handling.
2. *LABORATORY PROCEDURES AND CHEMISTRY FOR OPERATORS OF WATER POLLUTION CONTROL PLANTS.* This publication is a reproduction of *OPERATION OF WASTEWATER TREATMENT PLANTS,* Volume II, Chapter 16, "Laboratory Procedures and Chemistry." Obtain from California Water Environment Association (CWEA), 7677 Oakport Street, Suite 525, Oakland, CA 94621. Price to members, $10.00; nonmembers, $14.60; plus $3.50 shipping and handling.
3. *WATER ANALYSIS HANDBOOK.* Order No. 8374. Available from HACH Company, PO Box 389, Loveland, CO 80539-0389. No charge.
4. *STANDARD METHODS FOR THE EXAMINATION OF WATER AND WASTEWATER,* 20th Edition, 1998. Obtain from Water Environment Federation (WEF), Publications Order Department, 601 Wythe Street, Alexandria, VA 22314-1994. Order No. S82010. Price to members, $155.00; nonmembers, $200.00; plus shipping and handling.

14.02 Acknowledgments

Many of the illustrated laboratory procedures were provided by Joe Nagano, Laboratory Director, Hyperion Treatment Plant, City of Los Angeles, California. These procedures originally appeared in *LABORATORY PROCEDURES FOR OPERATORS OF WATER POLLUTION CONTROL PLANTS,* prepared by Mr. Nagano and published by the California Water Pollution Control Association. Information in this chapter is a condensed version of Chapter 16, "Laboratory Procedures and Chemistry," by James Paterson and revised by James Sequeira, in *OPERATION OF WASTEWATER TREATMENT PLANTS,* Volume II.

---

[1] *NPDES Permit. **N**ational **P**ollutant **D**ischarge **E**limination **S**ystem permit is the regulatory agency document issued by either a federal or state agency which is designed to control all discharges of potential pollutants from point sources and storm water runoff into U.S. waterways. NPDES permits regulate discharges into navigable waters from all point sources of pollution, including industries, municipal wastewater treatment plants, sanitary landfills, large agricultural feedlots and return irrigation flows.*

[2] *Numbers in parentheses refer to references in Section 14.01.*

14.1 BASIC LABORATORY WORDS, EQUIPMENT, AND TECHNIQUES

14.10 Laboratory Words

>GREATER THAN

DO >5 mg/*L* would be read as DO GREATER THAN 5 mg/*L*.

<LESS THAN

DO <5 mg/*L* would be read as DO LESS THAN 5 mg/*L*.

ALIQUOT (AL-li-kwot)

Representative portion of a sample. Often an equally divided portion of a sample.

AMBIENT (AM-bee-ent) TEMPERATURE

Temperature of the surroundings.

AMPEROMETRIC (am-PURR-o-MET-rick) AMPEROMETRIC

A method of measurement that records electric current flowing or generated, rather than recording voltage. Amperometric titration is a means of measuring concentrations of certain substances in water.

ANAEROBIC (AN-air-O-bick) ANAEROBIC

A condition in which atmospheric or dissolved molecular oxygen is *NOT* present in the aquatic (water) environment.

ASEPTIC (a-SEP-tick) ASEPTIC

Free from the living germs of disease, fermentation, or putrefaction. Sterile.

BIOASSAY (BUY-o-ass-SAY) BIOASSAY

(1) A way of showing or measuring the effect of biological treatment on a particular substance or waste, or

(2) A method of determining the relative toxicity of a test sample of industrial wastes or other wastes by using live test organisms such as fish.

BIOMONITORING BIOMONITORING

A term used to describe methods of evaluating or measuring the effects of toxic substances in effluents on aquatic organisms in receiving waters. There are two types of biomonitoring, the BIOSURVEY and the BIOASSAY.

BIOSURVEY BIOSURVEY

A survey of the types and numbers of organisms naturally present in the receiving waters upstream and downstream from plant effluents. Comparisons are made between the aquatic organisms upstream and those organisms downstream of the discharge.

BLANK BLANK

A bottle containing only dilution water or distilled water; the sample being tested is not added. Tests are frequently run on a *SAMPLE* and a *BLANK* and the differences are compared. This procedure helps to eliminate or reduce test result errors that could be caused when the dilution water or distilled water used is contaminated.

BUFFER BUFFER

A solution or liquid whose chemical makeup neutralizes acids or bases without a great change in pH.

BUFFER CAPACITY BUFFER CAPACITY

A measure of the capacity of a solution or liquid to neutralize acids or bases. This is a measure of the capacity of water or wastewater for offering a resistance to changes in pH.

CHAIN OF CUSTODY CHAIN OF CUSTODY

A record of each person involved in the handling and possession of a sample from the person who collected the sample to the person who analyzed the sample in the laboratory and to the person who witnessed disposal of the sample.

COLORIMETRIC MEASUREMENT COLORIMETRIC MEASUREMENT

A means of measuring unknown chemical concentrations in water by *MEASURING A SAMPLE'S COLOR INTENSITY.* The specific color of the sample, developed by addition of chemical reagents, is measured with a photoelectric colorimeter or is compared with "color standards" using, or corresponding with, known concentrations of the chemical.

COMPOSITE (come-PAH-zit) (PROPORTIONAL) SAMPLE | COMPOSITE (PROPORTIONAL) SAMPLE

A composite sample is a collection of individual samples obtained at regular intervals, usually every one or two hours during a 24-hour time span. Each individual sample is combined with the others in proportion to the rate of flow when the sample was collected. The resulting mixture (composite sample) forms a representative sample and is analyzed to determine the average conditions during the sampling period.

COMPOUND | COMPOUND

A pure substance composed of two or more elements whose composition is constant. For example, table salt (sodium chloride, NaCl) is a compound.

DESICCATOR (DESS-uh-KAY-tor) | DESICCATOR

A closed container into which heated weighing or drying dishes are placed to cool in a dry environment in preparation for weighing. The dishes may be empty or they may contain a sample. Desiccators contain a substance, such as anhydrous calcium chloride, which absorbs moisture and keeps the relative humidity near zero so that the dish or sample will not gain weight from absorbed moisture.

DISTILLATE (DIS-tuh-late) | DISTILLATE

In the distillation of a sample, a portion is collected by evaporation and recondensation; the part that is recondensed is the distillate.

ELEMENT | ELEMENT

A substance which cannot be separated into its constituent parts and still retain its chemical identity. For example, sodium (Na) is an element.

END POINT | END POINT

Samples are titrated to the end point. This means that a chemical is added, drop by drop, to a sample until a certain color change (blue to clear, for example) occurs. This is called the *END POINT* of the titration. In addition to a color change, an end point may be reached by the formation of a precipitate or the reaching of a specified pH. An end point may be detected by the use of an electronic device such as a pH meter. The completion of a desired chemical reaction.

FACULTATIVE (FACK-ul-TAY-tive) | FACULTATIVE

Facultative bacteria can use either dissolved molecular oxygen or oxygen obtained from food materials such as sulfate or nitrate ions. In other words, facultative bacteria can live under aerobic, anoxic, or anaerobic conditions.

FLAME POLISHED | FLAME POLISHED

Melted by a flame to smooth out irregularities. Sharp or broken edges of glass (such as the end of a glass tube) are rotated in a flame until the edge melts slightly and becomes smooth.

GRAB SAMPLE | GRAB SAMPLE

A single sample of water collected at a particular time and place which represents the composition of the water only at that time and place.

GRAVIMETRIC | GRAVIMETRIC

A means of measuring unknown concentrations of water quality indicators in a sample by *WEIGHING* a precipitate or residue of the sample.

INDICATOR (CHEMICAL) | INDICATOR (CHEMICAL)

A substance that gives a visible change, usually of color, at a desired point in a chemical reaction, generally at a specified end point.

M or MOLAR | ***M*** or MOLAR

A molar solution consists of one gram molecular weight of a compound dissolved in enough water to make one liter of solution. A gram molecular weight is the molecular weight of a compound in grams. For example, the molecular weight of sulfuric acid (H_2SO_4) is 98. A one *M* solution of sulfuric acid would consist of 98 grams of H_2SO_4 dissolved in enough distilled water to make one liter of solution.

MPN (pronounce as separate letters) | MPN

MPN is the **M**ost **P**robable **N**umber of coliform-group organisms per unit volume of sample water. Expressed as a density or population of organisms per 100 m*L* of sample water.

MSDS MSDS

Material **S**afety **D**ata **S**heet. A document which provides pertinent information and a profile of a particular hazardous substance or mixture. An MSDS is normally developed by the manufacturer or formulator of the hazardous substance or mixture. The MSDS is required to be made available to employees and operators whenever there is the likelihood of the hazardous substance or mixture being introduced into the workplace. Some manufacturers are preparing MSDSs for products that are not considered to be hazardous to show that the product or substance is *NOT* hazardous.

MENISCUS (meh-NIS-cuss) MENISCUS

The curved surface of a column of liquid (water, oil, mercury) in a small tube. When the liquid wets the sides of the container (as with water), the curve forms a valley. When the confining sides are not wetted (as with mercury), the curve forms a hill or upward bulge. When a meniscus forms in a measuring device, the top of the liquid level of the sample is determined by the bottom of the meniscus.

MOLECULAR WEIGHT MOLECULAR WEIGHT

The molecular weight of a compound in grams is the sum of the atomic weights of the elements in the compound. The molecular weight of sulfuric acid (H_2SO_4) in grams is 98.

Element	Atomic Weight	Number of Atoms	Molecular Weight
H	1	2	2
S	32	1	32
O	16	4	64
			98

MOLECULE (MOLL-uh-KULE) MOLECULE

The smallest division of a compound that still retains or exhibits all the properties of the substance.

N or NORMAL ***N*** or NORMAL

A normal solution contains one gram equivalent weight of reactant (compound) per liter of solution. The equivalent weight of an acid is that weight which contains one gram atom of ionizable hydrogen or its chemical equivalent. For example, the equivalent weight of sulfuric acid (H_2SO_4) is 49 (98 divided by 2 because there are two replaceable hydrogen ions). A one *N* solution of sulfuric acid would consist of 49 grams of H_2SO_4 dissolved in enough water to make one liter of solution.

NPDES PERMIT NPDES PERMIT

National **P**ollutant **D**ischarge **E**limination **S**ystem permit is the regulatory agency document issued by either a federal or state agency which is designed to control all discharges of potential pollutants from point sources and storm water runoff into U.S. waterways. NPDES permits regulate discharges into navigable waters from all point sources of pollution, including industries, municipal wastewater treatment plants, sanitary landfills, large agricultural feedlots and return irrigation flows.

NONVOLATILE MATTER NONVOLATILE MATTER

Material such as sand, salt, iron, calcium, and other mineral materials which are only slightly affected by the actions of organisms and are not lost on ignition of the dry solids at 550°C. Volatile materials are chemical substances usually of animal or plant origin. Also see INORGANIC WASTE and VOLATILE MATTER or VOLATILE SOLIDS.

OXIDATION (ox-uh-DAY-shun) OXIDATION

Oxidation is the addition of oxygen, removal of hydrogen, or the removal of electrons from an element or compound. In wastewater treatment, organic matter is oxidized to more stable substances. The opposite of REDUCTION.

OXIDATION STATE/OXIDATION NUMBER OXIDATION STATE/OXIDATION NUMBER

In a chemical formula, a number accompanied by a polarity indication (+ or –) that together indicate the charge of an ion as well as the extent to which the ion has been oxidized or reduced in a REDOX REACTION.

Due to the loss of electrons, the charge of an ion that has been oxidized would go from negative toward or to neutral, from neutral to positive, or from positive to more positive. As an example, an oxidation number of 2+ would indicate that an ion has lost two electrons and that its charge has become positive (that it now has an excess of two protons).

Due to the gain of electrons, the charge of the ion that has been reduced would go from positive toward or to neutral, from neutral to negative, or from negative to more negative. As an example, an oxidation number of 2– would indicate that an ion has gained two electrons and that its charge has become negative (that it now has an excess of two electrons). As an ion gains electrons, its oxidation state (or the extent to which it is oxidized) lowers; that is, its oxidation state is reduced.

OXIDATION-REDUCTION POTENTIAL (ORP) OXIDATION-REDUCTION POTENTIAL (ORP)

The electrical potential required to transfer electrons from one compound or element (the oxidant) to another compound or element (the reductant); used as a qualitative measure of the state of oxidation in wastewater treatment systems. ORP is measured in millivolts, with negative values indicating a tendency to reduce compounds or elements and positive values indicating a tendency to oxidize compounds or elements.

OXIDATION-REDUCTION (REDOX) REACTION — OXIDATION-REDUCTION (REDOX) REACTION

A two-part reaction between two substances involving a transfer of electrons from one substance to the other. Oxidation is the loss of electrons by one substance, and reduction is the acceptance of electrons by the other substance. Reduction refers to the lowering of the OXIDATION STATE or OXIDATION NUMBER of the substance accepting the electrons.

In a redox reaction, the substance that gives up the electrons (that is oxidized) is called the reductant because it causes a reduction in the oxidation state or number of the substance that accepts the transferred electrons. The substance that receives the electrons (that is reduced) is called the oxidant because it causes oxidation of the other substance. Oxidation and reduction always occur simultaneously.

PERCENT SATURATION — PERCENT SATURATION

The amount of a substance that is dissolved in a solution compared with the amount dissolved in the solution at saturation, expressed as a percent.

$$\text{Percent Saturation, \%} = \frac{\text{Amount of Substance That Is Dissolved x 100\%}}{\text{Amount Dissolved in Solution at Saturation}}$$

pH (pronounce as separate letters) — pH

pH is an expression of the intensity of the basic or acidic condition of a liquid. Mathematically, pH is the logarithm (base 10) of the reciprocal of the hydrogen ion activity.

$$\text{pH} = \text{Log}\,\frac{1}{[H^+]}$$

The pH may range from 0 to 14, where 0 is most acidic, 14 most basic, and 7 neutral. Natural waters usually have a pH between 6.5 and 8.5.

PILLOWS — PILLOWS

Plastic tubes shaped like pillows that contain exact amounts of chemicals or reagents. Cut open the pillow, pour the reagents into the sample being tested, mix thoroughly and follow test procedures.

PLUG FLOW — PLUG FLOW

A type of flow that occurs in tanks, basins or reactors when a slug of wastewater moves through a tank without ever dispersing or mixing with the rest of the wastewater flowing through the tank.

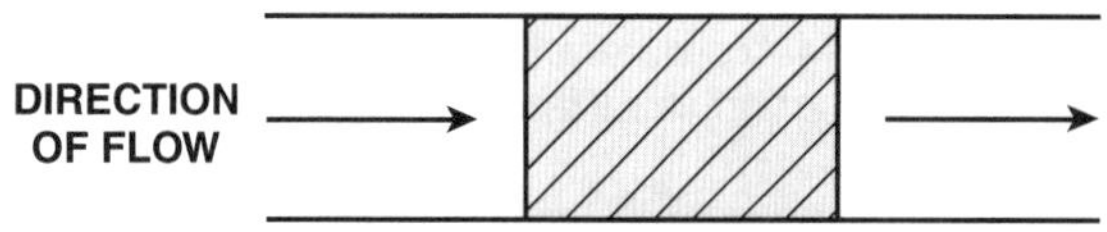

PLUG FLOW

PRECIPITATE (pre-SIP-uh-TATE) — PRECIPITATE

(1) An insoluble, finely divided substance which is a product of a chemical reaction within a liquid.

(2) The separation from solution of an insoluble substance.

REAGENT (re-A-gent) — REAGENT

A pure chemical substance that is used to make new products or is used in chemical tests to measure, detect, or examine other substances.

REDUCTION (re-DUCK-shun) — REDUCTION

Reduction is the addition of hydrogen, removal of oxygen, or the addition of electrons to an element or compound. Under anaerobic conditions (no dissolved oxygen present), sulfur compounds are reduced to odor-producing hydrogen sulfide (H_2S) and other compounds. The opposite of OXIDATION.

REFLUX — REFLUX

Flow back. A sample is heated, evaporates, cools, condenses, and flows back to the flask.

REPRESENTATIVE SAMPLE — REPRESENTATIVE SAMPLE

A sample portion of material or wastestream that is as nearly identical in content and consistency as possible to that in the larger body of material or wastestream being sampled.

SOLUTION — SOLUTION

A liquid mixture of dissolved substances. In a solution it is impossible to see all the separate parts.

SPECIFIC GRAVITY — SPECIFIC GRAVITY

(1) Weight of a particle, substance or chemical solution in relation to the weight of an equal volume of water. Water has a specific gravity of 1.000 at 4°C (39°F). Wastewater particles or substances usually have a specific gravity of 0.5 to 2.5.

(2) Weight of a particular gas in relation to the weight of an equal volume of air at the same temperature and pressure (air has a specific gravity of 1.0). Chlorine has a specific gravity of 2.5 as a gas.

STANDARD SOLUTION — STANDARD SOLUTION

A solution in which the exact concentration of a chemical or compound is known.

STANDARDIZE — STANDARDIZE

To compare with a standard.

(1) In wet chemistry, to find out the exact strength of a solution by comparing it with a standard of known strength. This information is used to adjust the strength by adding more water or more of the substance dissolved.

(2) To set up an instrument or device to read a standard. This allows you to adjust the instrument so that it reads accurately, or enables you to apply a correction factor to the readings.

SURFACTANT (sir-FAC-tent) — SURFACTANT

Abbreviation for surface-active agent. The active agent in detergents that possesses a high cleaning ability.

THIEF HOLE — THIEF HOLE

A digester sampling well which allows sampling of the digester contents without venting digester gas.

TITRATE (TIE-trate) — TITRATE

To *TITRATE* a sample, a chemical solution of known strength is added drop by drop until a certain color change, precipitate, or pH change in the sample is observed (end point). Titration is the process of adding the chemical reagent in increments until completion of the reaction, as signaled by the end point.

TOTALIZER — TOTALIZER

A device or meter that continuously measures and calculates (adds) a process rate variable in cumulative fashion; for example, total flows displayed in gallons, million gallons, cubic feet, or some other unit of volume measurement. Also called an integrator.

TURBIDITY UNITS (TU) — TURBIDITY UNITS (TU)

Turbidity units are a measure of the cloudiness of water. If measured by a nephelometric (deflected light) instrumental procedure, turbidity units are expressed in nephelometric turbidity units (NTU) or simply TU. Those turbidity units obtained by visual methods are expressed in Jackson Turbidity Units (JTU) which are a measure of the cloudiness of water; they are used to indicate the clarity of water. There is no real connection between NTUs and JTUs. The Jackson turbidimeter is a visual method and the nephelometer is an instrumental method based on deflected light.

VOLATILE (VOL-uh-tull) — VOLATILE

(1) A volatile substance is one that is capable of being evaporated or changed to a vapor at relatively low temperatures. Volatile substances also can be partially removed from water or wastewater by the air stripping process.

(2) In terms of solids analysis, volatile refers to materials lost (including most organic matter) upon ignition in a muffle furnace for 60 minutes at 550°C. Natural volatile materials are chemical substances usually of animal or plant origin. Manufactured or synthetic volatile materials such as plastics, ether, acetone, and carbon tetrachloride are highly volatile and not of plant or animal origin. Also see NONVOLATILE MATTER.

VOLATILE ACIDS — VOLATILE ACIDS

Fatty acids produced during digestion which are soluble in water and can be steam-distilled at atmospheric pressure. Also called organic acids. Volatile acids are commonly reported as equivalent to acetic acid.

VOLATILE LIQUIDS — VOLATILE LIQUIDS

Liquids which easily vaporize or evaporate at room temperature.

VOLATILE SOLIDS — VOLATILE SOLIDS

Those solids in water, wastewater, or other liquids that are lost on ignition of the dry solids at 550°C.

VOLUMETRIC — VOLUMETRIC

A measurement based on the volume of some factor. Volumetric titration is a means of measuring unknown concentrations of water quality indicators in a sample *BY DETERMINING THE VOLUME* of titrant or liquid reagent needed to complete particular reactions.

14.11 The Metric System

The metric system is based on the decimal system. All units of length, volume, and weight (mass) use factors of 10 to express larger or smaller quantities of these units. The metric system is used exclusively in the wastewater plant laboratory. Below is a summary of metric and English unit names and abbreviations.

Type of Measurement	English System	Metric Name	Metric Abbreviation
Length	inch feet yard	meter	m
Volume	quart gallon	liter	*L*
Weight	ounce pound	gram	gm
Temperature	Fahrenheit	Celsius	°C
Time	second	second	sec

In the laboratory, we sometimes use smaller amounts than a meter, a liter, or a gram. To express these smaller amounts, prefixes are added to the names of the metric units. There are many prefixes in use; however, we commonly use two prefixes more than any others in the laboratory.

Prefix	Abbreviation	Meaning
milli-	m	1/1,000 of, *OR* 0.001 times
centi-	c	1/100 of, *OR* 0.01 times

One milliliter (m*L*) is 1/1,000 of a liter and likewise one centimeter (cm) is 1/100 of a meter.

EXAMPLE:

(1) Convert 2 grams into milligrams.

1 milligram = 1 mg = 1/1,000 gm

therefore, 1 gram = 1,000 milligrams

2 grams x 1,000 mg/gram = 2,000 mg

(2) Convert 500 m*L* to liters.

1 m*L* = 1/1,000 liter

therefore, 1 liter = 1,000 m*L*

500 m*L* x 1 liter/1,000 m*L* = 0.500 liters

Or

500 m*L* ÷ 1,000 m*L*/*L* = 0.050 liters

The Celsius (or centigrade) temperature scale is used in the laboratory rather than the more familiar Fahrenheit scale.

	Fahrenheit (°F)	Celsius (°C)
Boiling point of water	212	100
Freezing point of water	32	0

To convert Fahrenheit to Celsius you can use the following formula:

$$\text{Temperature, °C} = \frac{5}{9}(\text{°F} - 32°)$$

EXAMPLE: Convert 68°F to °C.

$$\begin{aligned}\text{Temperature, °C} &= \frac{5}{9}(\text{°F} - 32°)\\ &= \frac{5}{9}(68\text{°F} - 32°)\\ &= \frac{5}{9}(36)\\ &= 20\text{°C}\end{aligned}$$

Or

$$\begin{aligned}\text{Temp, °C} &= 5 \div 9 \times 36\\ &= 20\text{°C}\end{aligned}$$

To convert Celsius to Fahrenheit you can use the following formula:

$$\text{Temperature, °F} = \left(\text{°C} \times \frac{9}{5}\right) + 32°$$

EXAMPLE: Convert 35°C to °F.

$$\begin{aligned}\text{Temperature, °F} &= \left(\text{°C} \times \frac{9}{5}\right) + 32°\\ &= \left(35\text{°C} \times \frac{9}{5}\right) + 32°\\ &= 63 + 32\\ &= 95\text{°F}\end{aligned}$$

Or

$$\begin{aligned}\text{Temp, °F} &= 35\text{°C} \times 9 \div 5 + 32\\ &= 95\text{°F}\end{aligned}$$

14.12 Chemical Names and Formulas

Chemical symbols are "shorthand" for the names of the elements. Names and symbols for some of these elements are listed below.

Chemical Name	Symbol
Calcium	Ca
Carbon	C
Chlorine	Cl
Hydrogen	H
Iron	Fe
Oxygen	O
Potassium	K
Sodium	Na
Sulfur	S

Many different compounds can be made from the same two or three elements, therefore, you must read the formula and name carefully to prevent errors and accidents. A chemical formula is a "shorthand" or abbreviated way to write the name of a compound. For example, the name sodium chloride (table salt) can be written "NaCl." Table 14.1 lists commonly used chemicals found in the wastewater treatment plant laboratory.

TABLE 14.1 NAMES AND FORMULAS OF CHEMICALS COMMONLY USED IN WASTEWATER ANALYSES

Chemical Name	Chemical Formula
Ammonium chloride	NH_4Cl
Calcium chloride (heptahydrate)*	$CaCl_2 \cdot 7H_2O$
Dipotassium hydrogen phosphate	K_2HPO_4
Disodium hydrogen phosphate (heptahydrate)*	$Na_2HPO_4 \cdot 7H_2O$
Ferric chloride	$FeCl_3$
Magnesium sulfate (heptahydrate)*	$MgSO_4 \cdot 7H_2O$
Manganous sulfate (tetrahydrate)*	$MnSO_4 \cdot 4H_2O$
Phenylarsine oxide	C_6H_5AsO
Potassium iodide	KI
Sodium azide	NaN_3
Sodium chloride	NaCl
Sodium hydroxide	NaOH
Sodium iodide	NaI
Sodium thiosulfate (pentahydrate)*	$Na_2S_2O_3 \cdot 5H_2O$
Sulfuric acid	H_2SO_4

* Note that: tetra = 4, penta = 5, and hepta = 7, thus heptahydrate = $7H_2O$

The following procedures are given to show the use of some of the chemicals whose names and formulas you will see in wastewater tests:

Dissolved Oxygen Procedure

". . . To 300 milliliter (m*L*) sample placed in a BOD bottle, add 1 m*L* of manganous sulfate solution. Now, add 1 m*L* of alkaline iodine-sodium azide solution. Shake well and add 1 m*L* of concentrated sulfuric acid. Titrate with sodium thiosulfate solution until . . ."

Preparation of Manganous Sulfate Solution

". . . Weigh 480 grams (gm) $MnSO_4 \cdot H_2O$ and dissolve in distilled water. Dilute to 1 liter . . ."

Poor results and safety hazards are often caused by using a chemical from the shelf that is *NOT* the same chemical called for in the procedure. This mistake usually occurs when the chemicals have similar names or formulas. This problem can be eliminated if you use *BOTH* the chemical name and formula as a double check. As you can see in the table of chemical names, the spellings of many chemicals are very similar. These slight differences are critical because the chemicals do not behave alike. For example, the chemicals potassium nitr*A*te (KNO_3) and potassium nitr*I*te (KNO_2) are just as different in meaning chemically as the words f*A*t and f*I*t are to your doctor.

14.13 Laboratory Equipment

The equipment in a wastewater laboratory is the lab analyst's "tools of the trade." In any laboratory, there are certain pieces of equipment that are used routinely to perform chemical tests such as those in wastewater analyses. The pictures and names of these are given with a statement concerning the use of each piece.

Volumetric glassware (graduated cylinders and pipets) is calibrated either "to contain" (TC) or "to deliver" (TD). Glassware designed "to deliver" will do so accurately only when the inner surface is so scrupulously clean that water wets the surface immediately and forms a uniform film on the surface upon emptying.

BEAKERS. Beakers are the most common pieces of laboratory equipment. They come in sizes from 1 m*L* to 4,000 m*L*. They are used mainly for mixing chemicals and to measure approximate volumes.

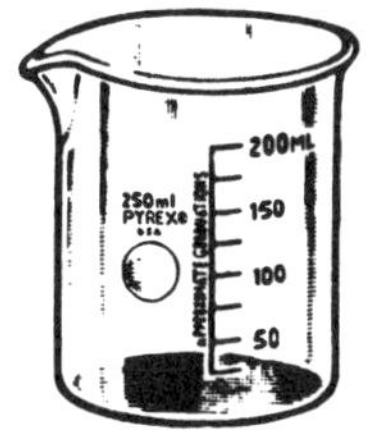

Beaker

GRADUATED CYLINDERS. Graduated cylinders also are basic to any laboratory and come in sizes from 5 m*L* to 4,000 m*L*. They are used to measure volumes more accurately than beakers.

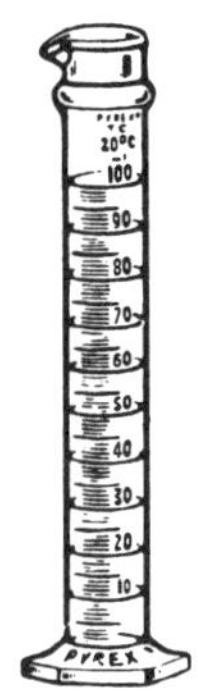

Cylinder, Graduated

PIPETS. Pipets are used to deliver accurate volumes and range in size from 0.1 m*L* to 100 m*L*.

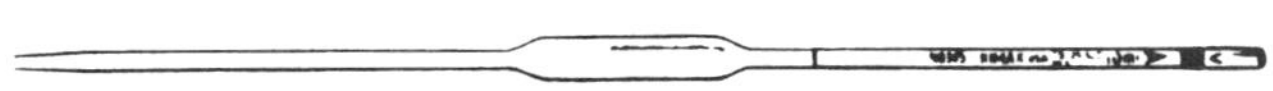

Pipet (pie-PET), Volumetric

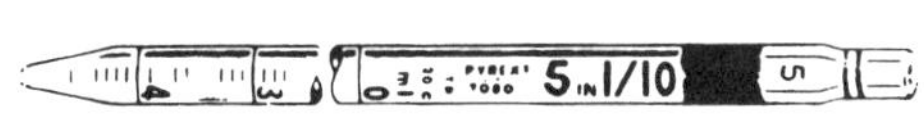

Pipet, Serological

BURETS. Burets are also used to deliver accurate volumes. They are especially useful in a procedure called "titration." Burets come in sizes from 10 to 1,000 m*L*.

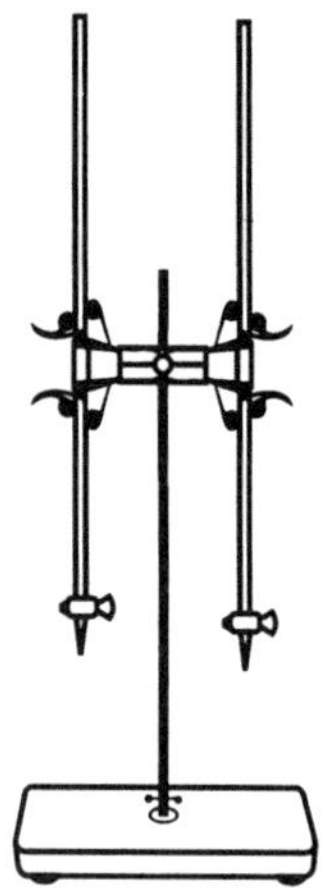

Support, Buret, and Buret Clamp

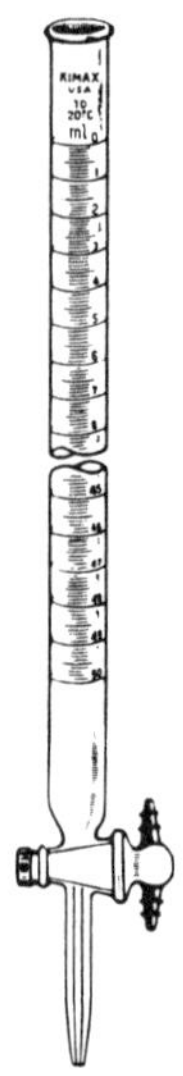

Buret (bur-RET)

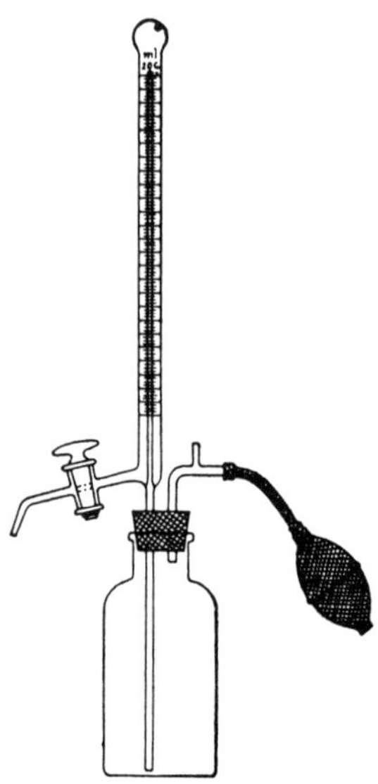

Automatic Buret

FLASKS. Flasks are used for containing and mixing chemicals. There are many different sizes and shapes.

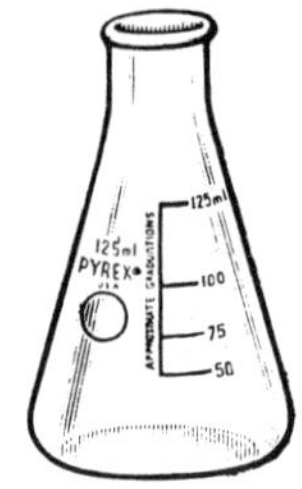

Flask, Erlenmeyer (ER-len-MY-er) Wide Mouth

Flask, Boiling Flat Bottom

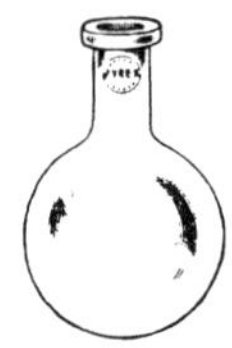

Flask, Boiling Round Bottom Short Neck

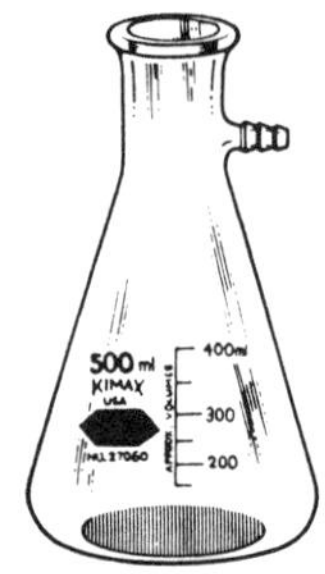

Flask, Filtering

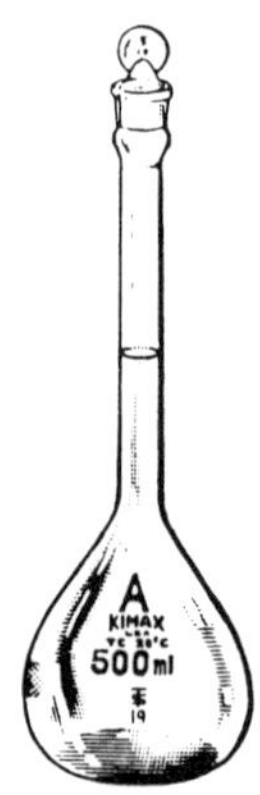

Flask, Volumetric

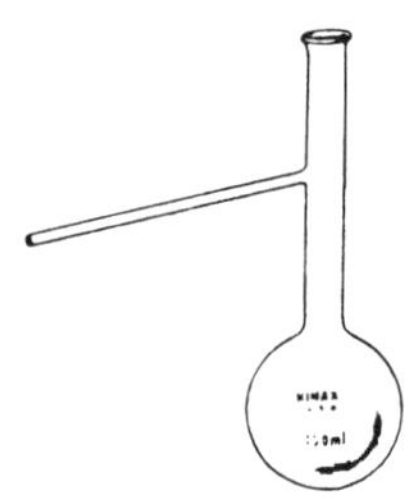

Flask, Distilling

Kjeldahl Flask (KELL-doll)

BOTTLES. Bottles are used to store chemicals, to collect samples for testing purposes, and to dispense liquids.

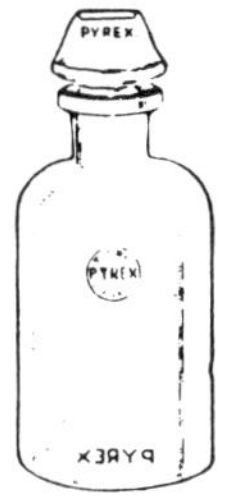

Bottle, Reagent

Bottle, BOD

Separatory funnels are used to separate one chemical mixture from another. The separated chemical usually is dissolved in one or two layers of liquid.

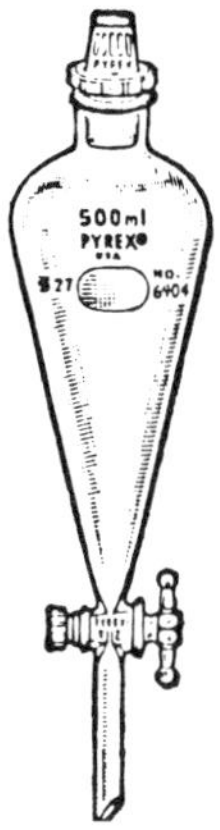

Separatory Funnel

FUNNELS. A funnel is used for pouring solutions or transferring solid chemicals. This funnel can also be used with filter paper to remove solids from a solution.

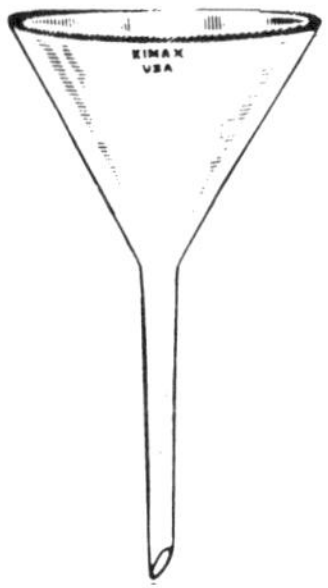

Funnel

A Buchner funnel is used to separate solids from a mixture. It is used with a filter flask and a vacuum.

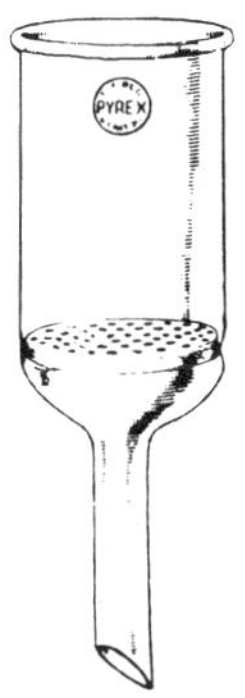

Funnel, Buchner With Perforated Plate

TUBES. Test tubes are used for mixing small quantities of chemicals. They are also used as containers for bacterial testing (culture tubes).

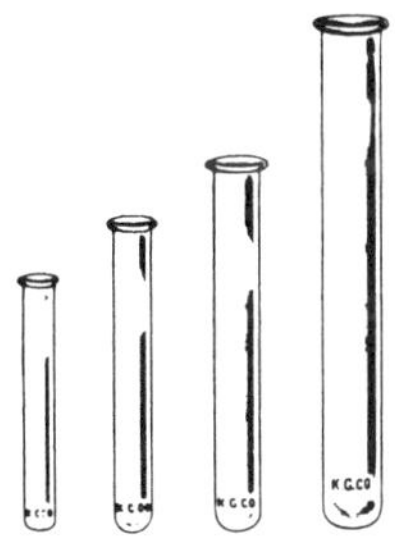

Test Tube

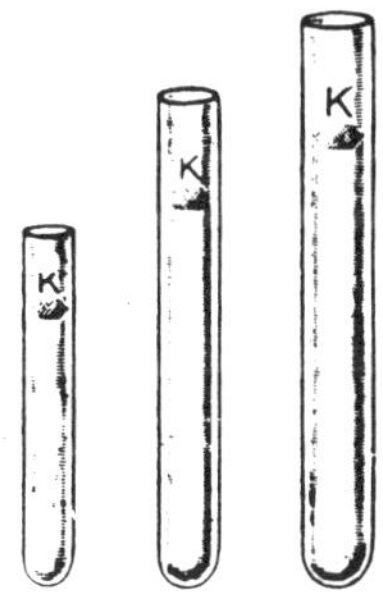

Culture Tube Without Lip

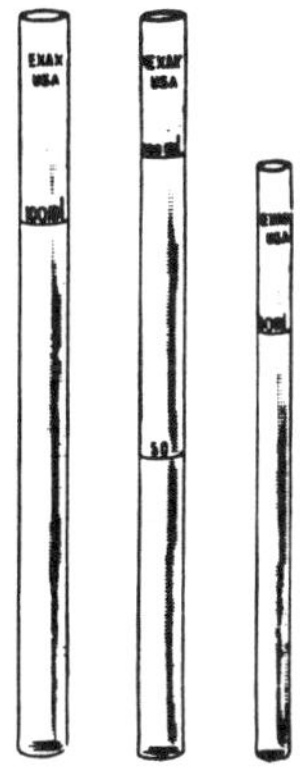

Color Comparison Tubes, Nessler

OTHER LABWARE AND EQUIPMENT.

Cone, Imhoff (IM-hoff)

Cone Support

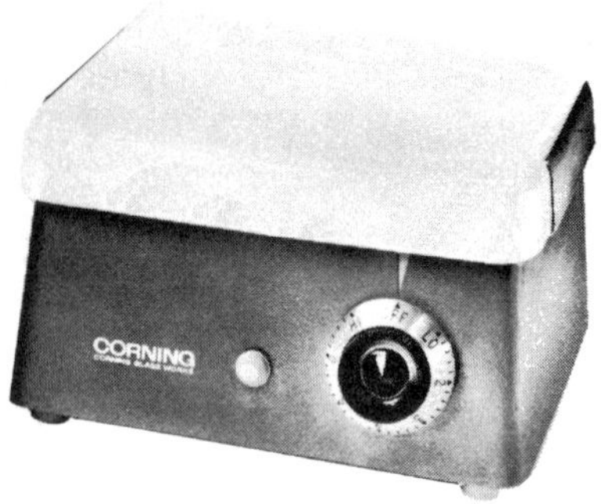

Hot Plate

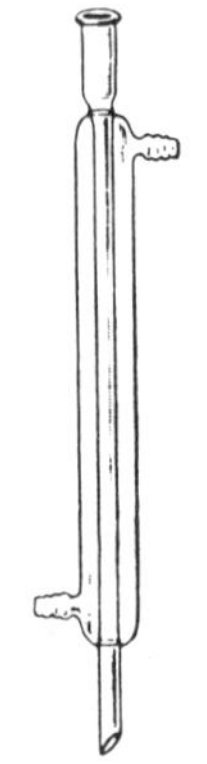

Condenser

Dish, Petri

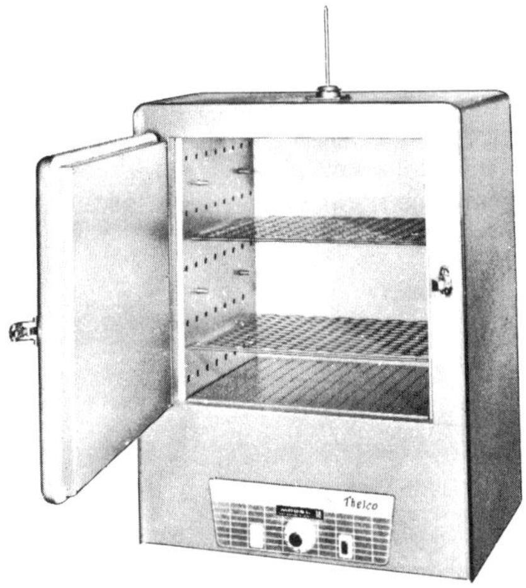

Oven, Mechanical Convection

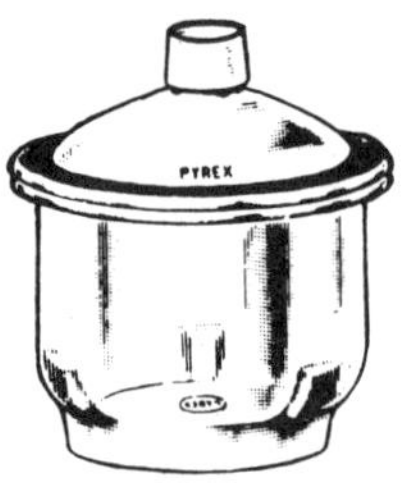

Desiccator (DES-uh-KAY-tor)

Thermometer, Dial

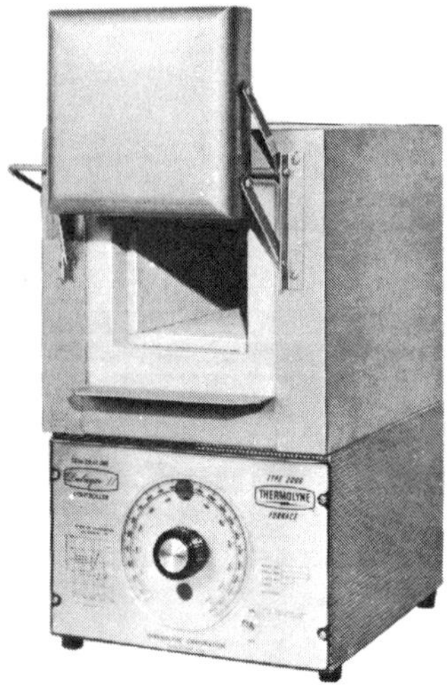

Muffle Furnace, Electric

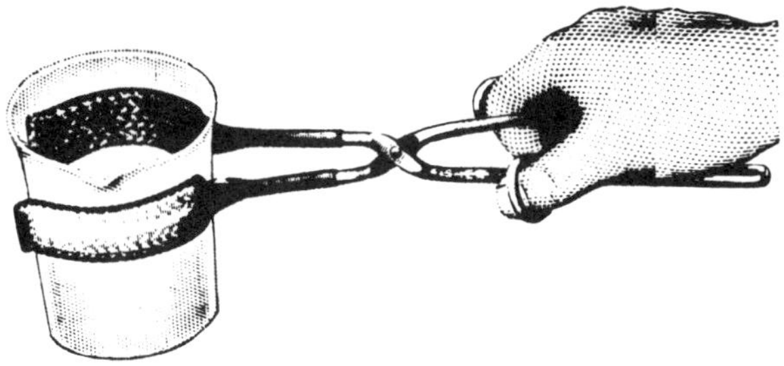

Clamp, Beaker,
Safety Tongs

Clamp, Dish,
Safety Tongs

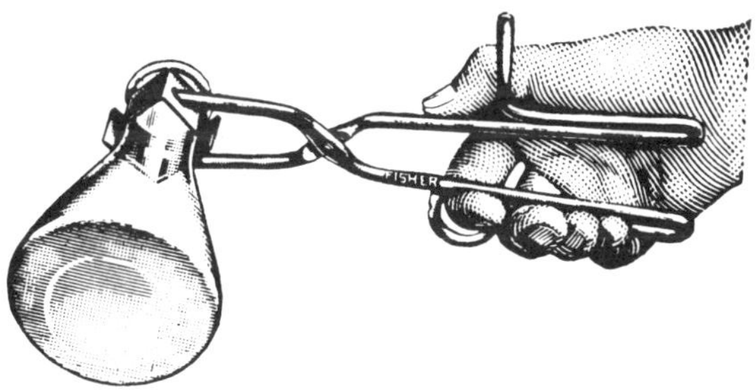

Clamp, Flask,
Safety Tongs

Clamp, Test Tube

Clamp Holder

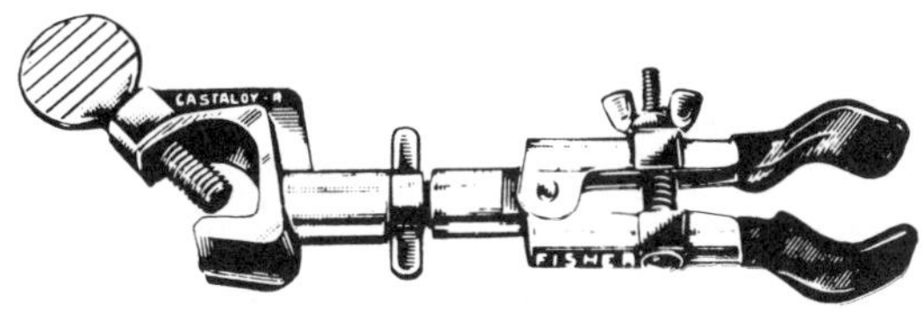

Clamp, Utility

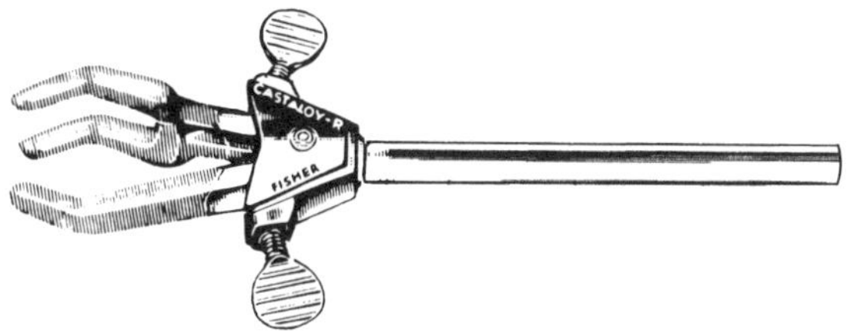

Clamp

Tripod, Concentric
Ring

Burner, Bunsen

Triangle,
Fused

Fume Hood

Portable Dissolved Oxygen Meter
(with computer docking station)
(Courtesy of HACH Company)

Portable pH Meter
(Courtesy of HACH Company)

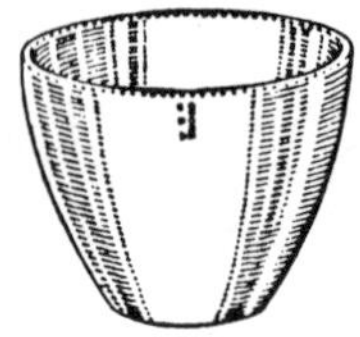

Crucible,
Porcelain

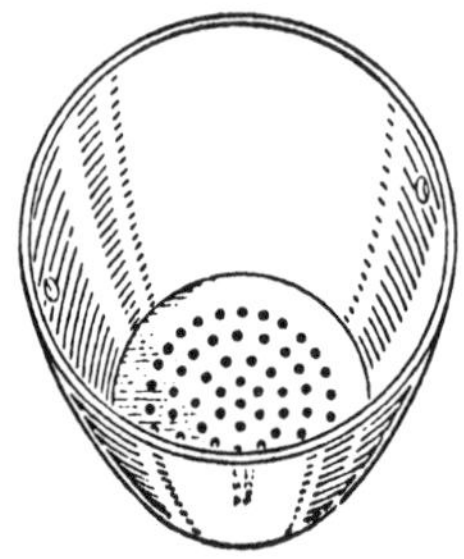

Crucible,
Gooch
(GOO-ch)
Porcelain

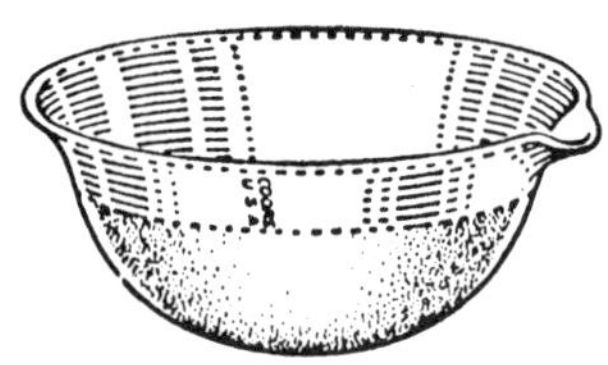

Dish,
Evaporating

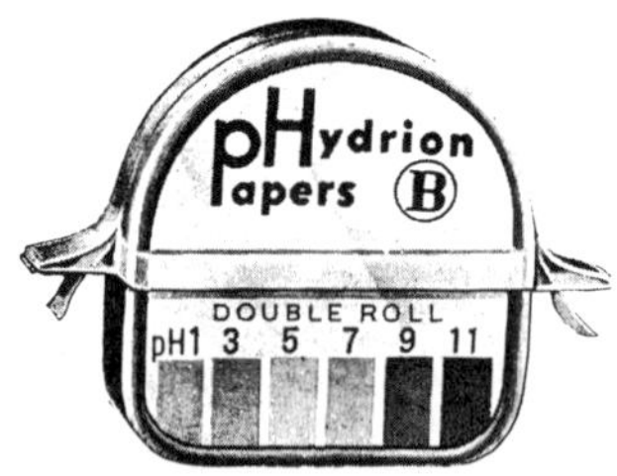

Test Paper, pH 1 – 11

Pump, Air Pressure and Vacuum

BOD Incubator

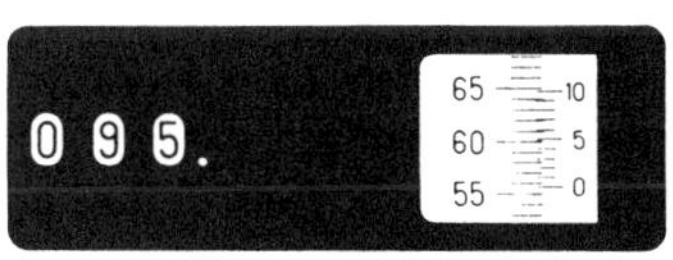

Weight = 95.5580 gm.

Balance, Analytical

(Permission of Mettler)

14.14 Use of Laboratory Glassware

BURETS

A buret is used to give accurate measurements of liquid volumes. The stopcock controls the amount of liquid that will flow from the buret. A glass stopcock must be lubricated (stopcock grease) and should not be used with alkaline solutions. A Teflon stopcock never needs to be lubricated.

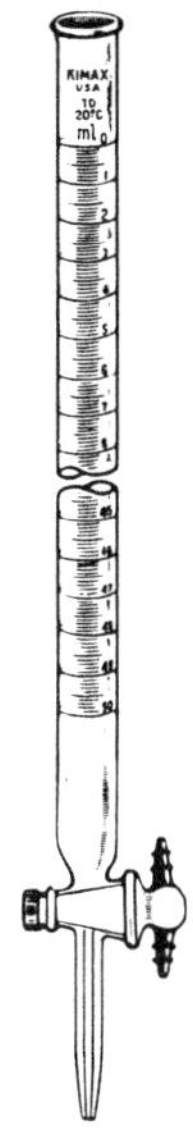

Buret

Burets come in several sizes, with those holding 10 or 25 milliliters used most frequently.

When a buret is filled with liquid, the surface of the liquid is curved. This curve of the surface is called the meniscus (meh-NIS-cuss). Depending on the liquid, the curve forms a valley, as with water, or forms a hill, as with mercury. Since most solutions used in the laboratory are water-based, always read the bottom of the meniscus with your eye at the same level (Figure 14.1). If you have the meniscus at eye level, the closest marks that go all the way around the buret will appear as straight lines, not circles.

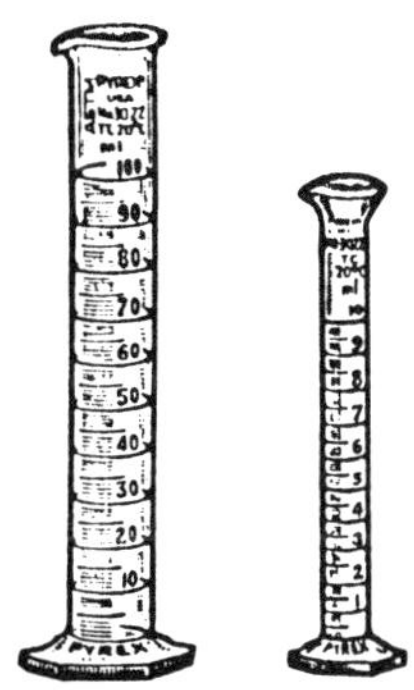

GRADUATED CYLINDERS

The graduated cylinder or "graduate" is one of the most often used pieces of laboratory equipment. This cylinder is made either of glass or of plastic and ranges in size from 10 m*L* to 4 liters. The graduate is used to measure volumes of liquid with an accuracy *LESS* than burets but *GREATER* than beakers or flasks. Graduated cylinders should never be heated in an open flame because they will break.

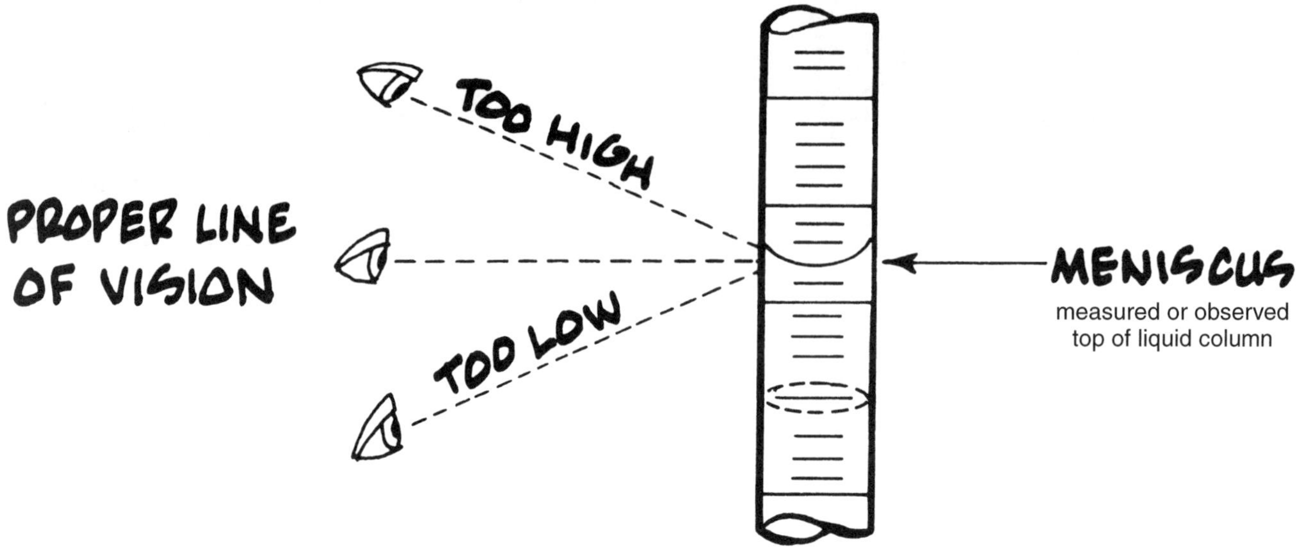

Fig. 14.1 How to read meniscus

FLASKS AND BEAKERS

Beakers and flasks are used for mixing, heating, and weighing chemicals. Most beakers and flasks are *NOT* calibrated with exact volume lines; however, they are sometimes marked with approximate volumes and can be used to estimate volumes.

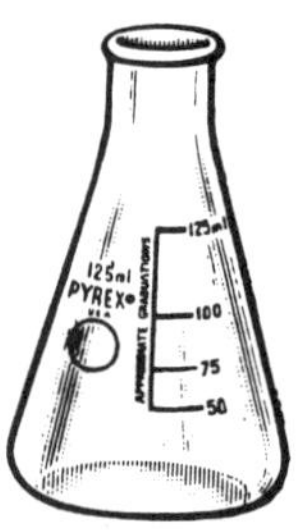

Flask

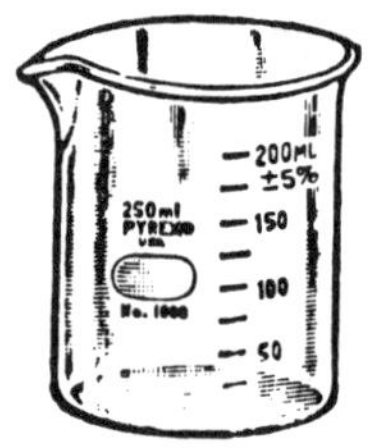

Beaker

VOLUMETRIC FLASKS

Volumetric flasks are used to prepare solutions and come in sizes from 10 to 2,000 m*L*. Volumetric flasks should *NEVER* be heated. Rather than store liquid chemicals in volumetric flasks, the chemicals should be transferred to a storage bottle. Volumetric flasks are more accurate than graduated cylinders.

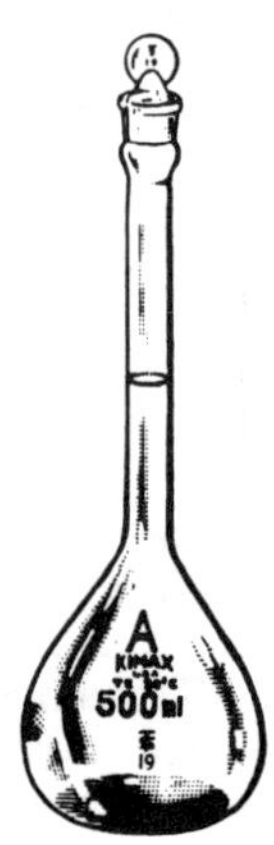

PIPETS

Pipets are used for accurate volume measurements and transfer. There are three types of pipets commonly used in the laboratory—volumetric pipets, graduated (measuring) or Mohr pipets, and serological pipets.

Volumetric Pipet

Graduated or Measuring Pipet

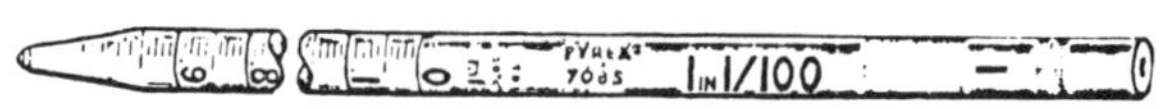

Serological Pipet

Volumetric pipets are available in sizes such as 1, 10, 25, 50, and 100 m*L*. They are used to deliver a single volume. Measuring (graduated) and serological pipets, however, will deliver fractions of the total volume indicated on the pipet.

Volumetric pipets should be held in a vertical position when emptying and the outflow should be unrestricted. The tip should be touched to the wet surface of the receiving vessel and kept in contact with it until the emptying is complete. Under no circumstance should the small amount remaining in the tip be blown out.

Measuring (graduated) and serological pipets should be held in the vertical position. After outflow has stopped, the tip should be touched to the wet surface of the receiving vessel. No drainage period is allowed. Where the small amount remaining in the tip is to be blown out and added, indication is made by a frosted band near the top of the pipet.

14.15 Solutions

Many laboratory procedures do not give the concentrations of standard solutions in grams/liter or milligrams/liter. Instead, the concentrations are usually given as normality (*N*), which is the standard designation for solution strengths in chemistry.

EXAMPLES:

0.025 $N\ H_2SO_4$ means a 0.025 normal solution of sulfuric acid

2 *N* NaOH means that the normality of the sodium hydroxide solution is 2

The *LARGER* the number in front of the *N*, the *MORE* concentrated the solution. For example, 1 *N* NaOH solution is more concentrated than a 0.2 *N* NaOH solution.

When the exact concentration of a chemical or compound in a solution is known, it is referred to as a "standard solution." Many times, standard solutions can be ordered already prepared. Once a standard has been prepared, it can then be used to standardize other solutions. To standardize a solution means to determine its concentration accurately, thereby making it a standard solution. "Standardization" is the process of using one solution of known concentration to determine the concentration of another solution. This action often involves a procedure called a "titration."

14.16 Titrations

A titration involves the addition of one solution, which is usually in a buret, to another solution in a flask or beaker. The solution in the buret is referred to as the "titrant" and is added to the other solution until there is a measurable change in the solution in the flask or beaker. This change is frequently a color change as a result of the addition of a special chemical called an "indicator" to the solution in the flask before the titration begins. The solution in the buret is added slowly to the flask until the change, which is called the "end point," is reached. The entire process is a "titration."

14.17 Use of a Spectrophotometer

In the field of wastewater analysis, many determinations such as phosphorus, nitrite, and nitrate are based on measuring the intensity of color at a particular wavelength. The color is formed in the sample by adding a specific developing reagent to it. The intensity of the color formed is directly related to the amount of material (such as phosphorus) in the sample. For the analysis of phosphorus present in wastewater, for example, ammonium molybdate reagent is added as the developing reagent; if phosphorus is present, a blue color develops. The more phosphorus there is, the deeper and darker the blue color.

The human eye can detect some differences in color intensity; however, for very precise measurements an instrument called a spectrophotometer is used.

THE SPECTROPHOTOMETER. A spectrophotometer is an instrument generally used to measure the color intensity of a chemical solution. A spectrophotometer in its simplest form consists of a light source that is focused on a prism or other suitable light dispersion device to separate the light into its separate bands of energy. Each different wavelength or color may be selectively focused through a narrow slit. This beam of light then passes through the sample to be measured. The sample is usually contained in a glass tube called a cuvette (que-VET). Most cuvettes are standardized to have a 1.0-cm light path length, however, many other sizes are available.

After the selected beam of light has passed through the sample, it emerges and strikes a photoelectric cell. If the solution in the sample cell has absorbed any of the light, the total energy content will be reduced. If the solution in the sample cell does not absorb the light, then there will be no change in energy. When the transmitted light beam strikes the photoelectric tube, it generates an electric current that is proportional to the intensity of light energy striking it. By connecting the photoelectric tube to a galvanometer (a device for measuring electric current) with a graduated scale, a means of measuring the intensity of the transmitted beam is achieved.

The diagram below illustrates the working parts of a spectrophotometer.

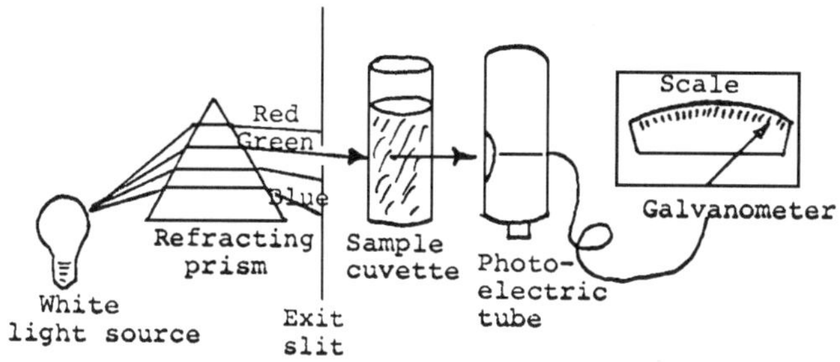

The operator should always follow the working instructions provided with the instrument.

UNITS OF SPECTROSCOPIC MEASUREMENT. The scale on spectrophotometers is generally graduated in two ways:

1. In units of percent transmittance (%T), an arithmetic scale with units graded from 0 to 100%; and
2. In units of absorbance (A), a logarithmic scale of nonequal divisions graduated from 0.0 to 2.0.

Some specialized spectrophotometers also contain a scale that directly reads the concentration of one chemical constituent in milligrams per liter.

Both the units percent transmittance and absorbance are associated with the words color intensity. That is, a sample that has a low color intensity will have a high percent transmittance but a low absorbance.

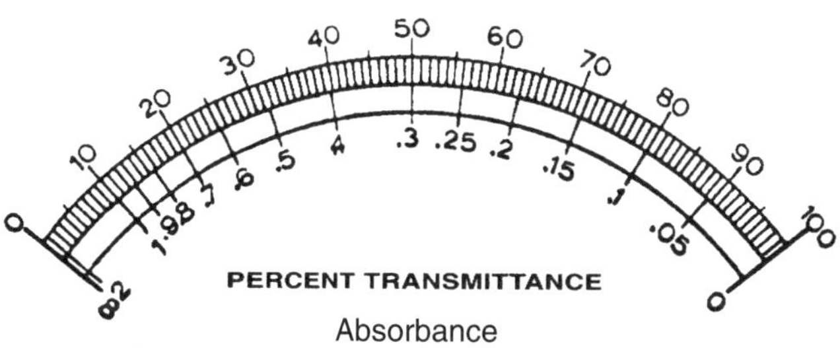

As illustrated above, the absorbance scale is ordinarily calibrated on the same scale as percent transmittance on spectrophotometers. The chief usefulness of absorbance lies in the fact that it is a logarithmic function rather than linear (arithmetic) and a law known as Beer's Law states that the concentration of a light absorbing colored solution is directly proportional to absorbance over a given range of concentrations. If one were to plot a graph showing %T or percent transmittance versus concentration on straight graph or line paper and another showing absorbance versus concentration on the same paper, the following curves (graphs) would result:

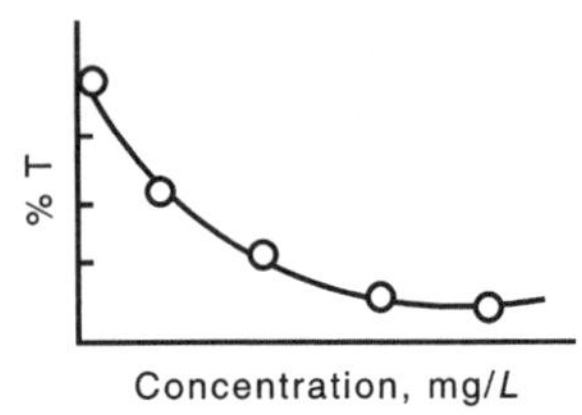

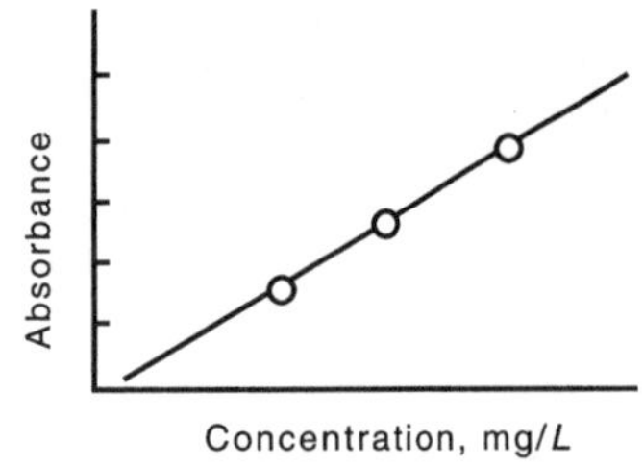

CALIBRATION CURVES. The calibration curve is used to determine the concentration of the water quality indicator (phosphorus, nitrite) contained in a sample. Three steps must be completed in order to prepare a calibration graph.

First, a series of standards must be prepared. A standard is a solution that contains a known amount of the same chemical constituent that is being determined in the sample.

Second, these standard solutions and a sample containing none of the constituent being tested for (usually distilled water and generally referred to as a blank) must be treated with the developing reagent in the same manner as the sample would be treated.

Third, using a spectrophotometer the absorbance or transmittance at the specified wavelength of the standards and blank must be determined. From the values obtained, a calibration curve of absorbance versus concentration can be plotted. Once these several points have been plotted, you can then extend the plotted points by connecting the known points with a straight line. For example, with the data given below one could construct the following calibration curve.

Absorbance	Concentration, mg/*L*
0.0	0.0
0.30	0.25
0.55	0.50
0.80	0.75

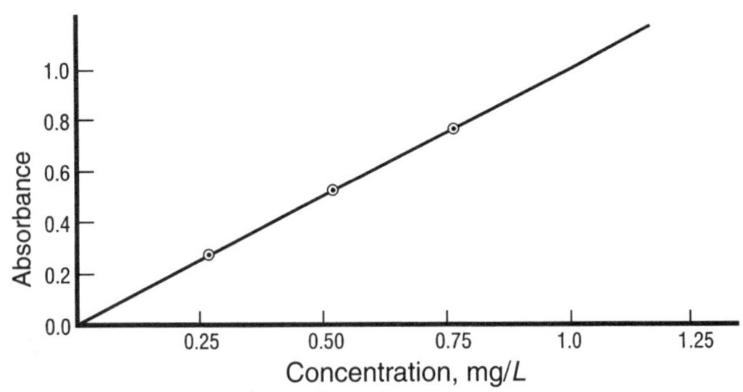

Once you have established a calibration curve for the water quality indicator in question, you can easily determine the amount of that substance contained in a solution of unknown concentration. You merely take an absorbance reading on the color developed by the unknown and locate it on the vertical axis. Then a straight line is drawn to the right on the graph until it intersects with the experimental standard curve. A line is then dropped to the horizontal axis and this value identifies the concentration of your unknown water quality indicator.

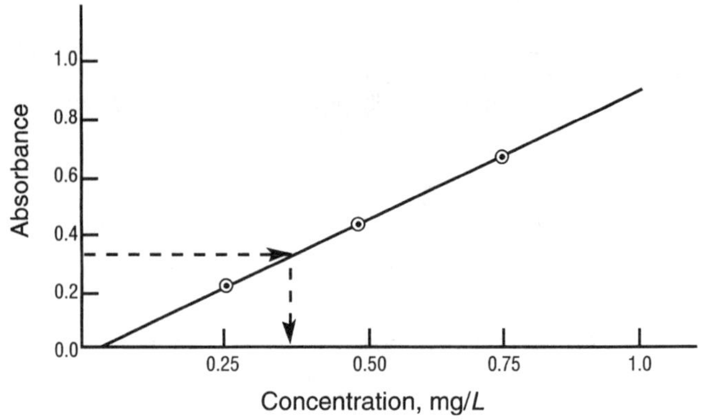

In this example, an absorbance reading of 0.32 was read on the unknown solution or sample, which indicates a concentration of about 0.37 mg/*L*.

14.18 Indicators

Indicators used in water testing are chemicals that will change color in the presence of specific chemical ions. Even though the indicator changes color as the test is run, it does not get involved or interfere with the reaction that takes place during the test.

There are many different indicators that are used in the testing of water samples. The person who sets up the test procedure will select the indicator that will give the most recognizable and reproducible results. Generally, the test kits and procedures that are used in small plants will specify or include the best indicator for each test.

Indicators are used in three different levels of testing.

1. Qualitative Analysis. In this type of test, the indicator is added to an unmeasured amount of sample. The only purpose of the test is to find out if a specific chemical, one that will cause the color change, is present in the sample.
2. Quantitative Titration. In this test a reagent solution that has a very accurately known concentration is added to an accurately measured volume of a test sample that contains the selected indicator. The reagent can be added by using a standardized dropper bottle or a laboratory apparatus called a buret. The purpose of the test is to measure how much reagent it takes to react with all of the test chemical that is in the sample. When the indicator changes color, it signals that the reaction is complete. Then, by using the magic of chemistry-mathematics, the amount of reagent

used in the test is converted to the amount of test chemical present in the sample. Charts provided in the test kits or laboratory manuals show the equivalent concentrations of the chemical being tested in milligrams per liter.

3. Quantitative Colorimetric Analysis. In this test, an accurately measured volume of sample is placed in a glass tube called a cuvette and the indicator is added to the sample to develop the color. The intensity of the color indicates the concentration of the test chemical that is in the sample. The color intensity is measured in one of two ways:

   a. Color comparator: This method enables the tester to match the color of the test sample to the color of one of the standards provided with the test kit.

   b. Photoelectric cell: The prepared test sample is placed in a device or meter that directs a light source through the sample. A photocell then measures the amount of light that passes through the sample and, using the magic of photo electronics, converts the amount of light to the concentration of the test chemical in the sample. The test devices either report the concentration directly on a dial on the meter, or they report a value that can be used to calculate the concentration in milligrams per liter.

14.19 Laboratory Work Sheets and Notebooks

The plant operator has two goals in using a laboratory notebook and work sheets: (1) to record data, and (2) to arrange data in an orderly manner. Often days of work can be wasted if data are written down on a scrap of paper, which is usually misplaced or thrown away. Work sheets and notebooks help prevent error and provide a record of the work. The routine use of work sheets and notebooks is the only way an operator can be sure that all important information for a test is properly recorded.

There is no standard laboratory form. Most treatment plants usually develop their own data sheets for recording laboratory results and other important plant data. The data sheets are prepared in a manner that makes it easy for you to record the data, review it, and recover it when necessary. Each plant may have different needs for collecting and recording data and many plants may use from five to eight different data sheets. Figures 14.2 and 14.3 illustrate typical laboratory work sheets (sometimes called bench sheets).

Laboratory work sheets provide an organized method for recording data. They are used to review effluent quality, identify problems, and search for the cause of problems. These sheets provide the information needed to complete NPDES permit reporting forms.

Acknowledgment

Pictures of laboratory glassware and equipment in this manual are reproduced with the permission of VWR Scientific, San Francisco, California.

QUESTIONS

Write your answers in a notebook and then compare your answers with those on page 337.

14.1A For each of the items listed below, describe the item and its use or purpose.

1. Beakers
2. Graduated Cylinders
3. Pipets
4. Burets

14.1B Why should graduated cylinders never be heated in an open flame?

14.1C What is a bench sheet?

Please answer the discussion and review questions next.

DISCUSSION AND REVIEW QUESTIONS

Chapter 14. LABORATORY PROCEDURES

(Lesson 1 of 7 Lessons)

At the end of each lesson in this chapter you will find some discussion and review questions. The purpose of these questions is to indicate to you how well you understand the material in the lesson. Write the answers to these questions in your notebook before continuing.

1. Why must chemicals be properly labeled?
2. How are pipets emptied or drained? Why is this procedure important?
3. How would you titrate a solution?

PLANT ______________________

DATE ______________________

SUSPENDED SOLIDS AND DISSOLVED SOLIDS

SAMPLE							
Crucible							
Sample, m*L*							
Wt Dry + Dish, gm Wt Dish, gm							
Wt Dry, gm							
$\text{mg}/L = \dfrac{\text{Wt Dry, gm} \times 1{,}000{,}000}{\text{Sample, m}L}$							
Wt Dish + Dry, gm Wt Dish + Ash, gm							
Wt Volatile, gm							
$\% = \dfrac{\text{Wt Vol}}{\text{Wt Dry}} \times 100\%$							

BOD

\# Blank ______________

SAMPLE										
DO Sample										
Bottle #										
% Sample										
Blank or adjusted blank DO after incubation										
Depletion, 5 days										
Depletion, %										

SETTLEABLE SOLIDS

Sample, m*L* ______________ ______________

Direct m*L*/*L* ______________ ______________

COD

Sample ______________ ______________

Blank Titration ______________ ______________

Sample Titration ______________ ______________

Depletion ______________ ______________

$\text{mg}/L = \dfrac{\text{Dep} \times N\ \text{FAS} \times 8{,}000}{\text{Sample, m}L}$ ______________ ______________

Fig. 14.2 Typical laboratory work sheet

TOTAL SOLIDS

SAMPLE					
Dish No.					
Wt Dish + Wet, gm Wt Dish, gm					
Wt Wet, gm					
Wt Dish + Dry, gm Wt Dish, gm					
Wt Dry, gm					
% Solids = $\frac{\text{Wt Dry}}{\text{Wt Wet}}$ x 100%					
Wt Dish + Dry, gm Wt Dish + Ash, gm					
Wt Volatile, gm					
% Volatile = $\frac{\text{Wt Vol x 100\%}}{\text{Wt Dry}}$					
pH					
Volatile Acid, mg/*L*					
Alkalinity as $CaCO_3$, mg/*L*					

Grease

Sample ____________________

Sample, m*L* ____________________

Wt Flask + Grease, mg ____________________

Wt Flask, mg ____________________

Wt Grease, mg ____________________

mg/*L* = $\frac{\text{Wt Grease, mg x 1,000}}{\text{Sample, m}L}$ ____________________

Fig. 14.3 Typical laboratory work sheet

CHAPTER 14. LABORATORY PROCEDURES

(Lesson 2 of 7 Lessons)

14.2 SAFETY AND HYGIENE IN THE LABORATORY

Safety is just as important in the laboratory as in the rest of the treatment plant. State laws and the Occupational Safety and Health Act (OSHA) demand proper safety procedures to be exercised in the laboratory at all times. OSHA specifically deals with "safety at the place of work." The Act requires that "each employer has the general duty to furnish all employees, employment free from recognized hazards causing, or likely to cause death or serious physical harm."

Personnel working in a wastewater treatment plant laboratory must realize that a number of hazardous materials and conditions exist. Be alert and careful. Be aware of potential dangers at all times. Safe practice in the laboratory requires hardly any more effort than unsafe practice, and the important results are prevention of injury or bodily damage.

Each laboratory must develop and implement a written hazard communication program to ensure that personnel dealing with chemicals are appropriately informed of any health and safety considerations and that they are effectively trained in methods to safely deal with the chemicals. As a minimum, the written program must describe how criteria are met concerning labels and other forms of warnings, **M**aterial **S**afety **D**ata **S**heets (MSDSs), and employee information and training. Plants that have a laboratory must also develop and implement a written Chemical Hygiene Plan (CHP).

14.20 Laboratory Hazards

Working with chemicals and other materials in the wastewater treatment plant laboratory can be dangerous. Dangers include:

Infectious Materials	Electric Shock
Poisons	Toxic Fumes
Explosions	Fire
Cuts and Bruises	Burns

The above dangers to yourself and others can be minimized, however, by using proper techniques and equipment.

14.200 Infectious Materials

Wastewater and sludge contain millions of biological organisms. Some of these are infectious and dangerous and can cause diseases such as tetanus, typhoid, dysentery, and hepatitis. Personnel handling these materials should thoroughly wash their hands with soap and water, particularly before handling food or smoking.

Do not pipet *ANYTHING* by mouth. Pipet all samples and reagents by mechanical means (rubber pipet bulbs) to avoid taking a chance on severe illness or death

Never drink from a beaker or other laboratory glassware. A beaker left "specifically" for drinking is a menace to the laboratory.

DON'T PIPET HAZARDOUS LIQUIDS BY MOUTH.

Though not mandatory, inoculations by your local Health Department are recommended for each employee to reduce the possibility of contracting diseases.

14.201 Corrosive Chemicals

Acids

1. Examples: Sulfuric (H_2SO_4), hydrochloric or muriatic (HCl), nitric (NHO_3), glacial acetic ($H_4C_2O_2$), and chromic acid.
2. Acids are extremely corrosive to human tissue, metals, clothing, wood, cement, stone, and concrete.
3. Commercially available spill cleanup materials should be kept on hand to neutralize acid in the event of an accidental spill.

Bases

1. Examples: Sodium hydroxide (caustic soda or lye—NaOH), potassium hydroxide (KOH), alkaline iodine—sodium azide solution (used in the dissolved oxygen test).
2. Bases are extremely corrosive to skin, clothing, and leather.
3. Commercially available spill cleanup materials should be kept on hand for use in the event of an accidental spill.

Miscellaneous

1. Examples: Chlorine, ferric salts (ferric chloride), and other strong oxidants.

14.202 Toxic Materials

1. Solids: Cyanide, chromium, orthotolidine, cadmium, and other heavy metals.
2. Liquids: Carbon tetrachloride, ammonium hydroxide, nitric acid, bromine, chlorine water, aniline dyes, formaldehyde, and carbon disulfide. Carbon tetrachloride is absorbed into the skin on contact; its vapors will damage the lungs; and it will build up in your body to a dangerous level.
3. Gases: Hydrogen sulfide, chlorine, ammonia, and sulfur dioxide.

14.203 Explosive or Flammable Materials

1. Liquids: Carbon disulfide, benzene, ethyl ether, petroleum ether, acetone, and gasoline.
2. Gases: Acetylene and hydrogen.

14.21 Personal Safety and Hygiene

14.210 Laboratory Safety

Laboratory work can be dangerous if proper precautions and techniques are not taken. *ALWAYS* follow these basic rules:

1. *NEVER* work alone in the laboratory. Someone should always be available to help in case you should have an accident that blinds you, leaves you unconscious, or starts a fire you can't handle. If necessary, have someone check on you regularly to be sure you are OK.
2. Wear protective goggles or safety glasses at all times in the laboratory. Contact lenses should not be worn, even under safety goggles, because fumes can seep between the lens and the eyeball and irritate the eye.

Safety Glasses

3. Wear a face shield if there is any danger of a hot liquid erupting from a container or flying pieces of glassware from an exploded apparatus. If in doubt as to its need—wear it!
4. Wear protective or insulated gloves when handling hot equipment or very cold objects, or when handling liquids or solids that are skin irritants.
5. Always wear a lab coat or apron in the laboratory to protect your skin and clothes.
6. Never pipet hazardous materials by your mouth.
7. Never eat or smoke in the laboratory. Never use laboratory glassware for serving food or drinks.
8. Don't keep food in a refrigerator that is used for chemical and/or sample storage.
9. Use ventilated laboratory fume hoods whenever handling toxic chemicals.
10. Maintain clear access to emergency eye wash stations and deluge showers.
11. Practice good housekeeping as an effective way to prevent accidents.

14.211 Personal Hygiene

Although it is highly unlikely that personnel will contract diseases by working in wastewater treatment plants, such a possibility does exist with certain diseases.

1. Some diseases are contracted through breaks in skin, cuts, or puncture wounds. In such cases the bacteria causing the disease may be covered over and trapped by flesh, creating a suitable *ANAEROBIC*[3] environment in which the bacteria may thrive and produce toxins that may spread throughout the body.

   For protection against diseases contracted through breaks in the skin, cuts, or puncture wounds, everyone working in or around wastewater should receive immunization from tetanus. Immunization must be received *BEFORE* the infection occurs. To prevent diseases from entering open wounds, care must be taken to keep wounds protected either with band-aids or, if necessary, with rubber gloves or waterproof protective clothing.
2. Diseases that may be contracted through the gastrointestinal system or through the mouth are typhoid, cholera, dysentery, amebiasis, worms, salmonella, infectious hepatitis, and polio virus. These diseases are transmitted when the infected wastewater materials are ingested or swallowed by

[3] *Anaerobic (AN-air-O-bick). A condition in which atmospheric or dissolved molecular oxygen is NOT present in the aquatic (water) environment.*

careless persons. The best protection against those diseases is furnished by *THOROUGH CLEANSING.* Hands, face, and body should be thoroughly washed with soap and water, particularly the hands, in order to prevent the transfer of any unsanitary materials or germs to the mouth while eating.

A change of working clothes into street clothes before leaving work is highly recommended to prevent carrying unsanitary materials to your home. Personal hygiene, thorough cleansing, and washing of the hands are effective means of protection. Store your work clothes and street clothing in separate lockers.

Immunization is provided for typhoid, polio, and hepatitis B. Little is known about infectious hepatitis A except that it can be transmitted by wastewater and/or dried sludge solids. Hepatitis is frequently associated with gross wastewater pollution.

3. Diseases that may be contracted by breathing contaminated air include tuberculosis, infectious hepatitis, and San Joaquin fever. There has been no past evidence to indicate the transmission of tuberculosis through the air at wastewater treatment plants. However, there was one case of tuberculosis being contracted by an employee who fell into wastewater and, while swimming, inhaled wastewater into his lungs. San Joaquin fever is caused by a fungus that may be present in wastewater. However, there is no record of operators contracting the disease while on the job.

The best insurance against these diseases is proper personal hygiene and immunization. Your plant should have an immunization program against (1) tetanus, (2) polio, and (3) hepatitis B. Immunization against typhoid should be considered if the possibility exists that typhoid could be found in your area. The immunizations should be provided to protect you. Check with your local or state health department for recommendations regarding immunization.

In the washing of hands, the kind of soap is less important than the thorough use of the soap. (Special disinfectant soaps are not essential.)

The use of protective clothing is very important, particularly gloves and boots. The protection of wounds and cuts is also important. Report injuries and take care of them.

The responsibility rests upon *YOU.*

There is no absolute insurance against contraction of disease in a wastewater treatment plant. However, the likelihood of transmission is practically negligible. There appears to be no special risk in working at treatment plants. In fact, operators may develop a natural immunity by working in this environment.

The possibility that Acquired Immune Deficiency Syndrome (AIDS), which is caused by a virus, can be contracted from exposure to raw wastewater has been discounted by researchers. Researchers have found that, although the AIDS virus is present in the wastes from AIDS victims, the raw wastewater environment is hostile to the virus itself and has not been identified as a mode of transmission to date. Needle sticks from potentially contaminated syringes should remain a concern to operators, maintenance personnel, and laboratory analysts. Fluids in or on syringes may provide a less severe environment for pathogens than raw wastewater where dilution and chlorination significantly reduce infection potential.

14.22 Prevention of Laboratory Accidents

14.220 Chemical Storage

An adequate storeroom is essential for safety in the wastewater laboratory. The storeroom should be properly ventilated and lighted and be laid out to segregate incompatible chemicals. Flammable liquids, acids, bases, and oxidizing agents should be separated from each other by distances, partitions, or other means so as to prevent accidental contact between them. Order and cleanliness must be maintained. All chemicals and bottles or reagents must be clearly labeled and dated. Never handle chemicals with bare hands. Use a spoon, spatula, or tongs.

Heavy items should be stored on or as near to the floor as possible. *VOLATILE LIQUIDS*[4] that may escape as a gas, such as ether, must be kept away from heat sources, sunlight, and electric switches. Do not store ether in the refrigerator. Several laboratory explosions have occurred because the ether gas will ignite when there is a small area and when the refrigerator light comes on.

CLAMPS, RAISED SHELF EDGES, AND PROPER ARRANGEMENT PREVENT STOCKROOM FALLOUT.

Cylinders of gas in storage should be capped and secured to prevent rolling or tipping. They should also be placed away from any possible sources of heat or open flames.

Flammable gases must be stored separately. The storage room should be fitted with explosion-proof wiring and lighting fixtures. Appropriate warning signs prohibiting sources of ignition must be posted in conspicuous locations.

The usual common sense rules of storage should be followed. Good housekeeping is a most significant contribution toward an active safety campaign.

[4] *Volatile Liquids. Liquids which easily vaporize or evaporate at room temperature.*

14.221 Movement of Chemicals

The next area of concern is the transfer of chemicals, apparatus, gases, or other hazardous materials from the storeroom to the laboratory for use. Use cradles or tilters to facilitate handling carboys or other larger chemical vessels.

Drum Tilter

In transporting cylinders of compressed gases, use a trussed hand truck. Never move or transport a cylinder without the valve protection hood in place. Never roll a cylinder by its valve. Immediately after they are positioned for use, cylinders should be clamped securely into place to prevent shifting or toppling.

Carry flammable liquids in safety cans or, in the case of reagent-grade chemicals, protect the bottle with a carrier. Always wear protective gloves, safety shoes, and rubber aprons in case of accidental spilling of chemical containers.

14.222 Proper Laboratory Techniques

Faulty technique is one of the chief causes of accidents and, because it involves the human element, is one of the most difficult to correct.

Because of their nature and prevalence in the laboratory, acids and other corrosive materials constitute a series of hazards ranging from poisoning, burning, and gassing through explosion. Always flush the outsides of acid bottles before opening them. Don't lay the stopper down on the countertop where a person might lay a hand or rest an arm on it. Keep all acids tightly stoppered when not in use and make sure no spilled acid remains on the floor, table, or bottle after use. To avoid splashing, don't pour water into acid; *ALWAYS POUR ACID INTO WATER*, like you would when adding acid to a swimming pool.

Mercury requires special care. Even a small amount in the bottom of a drawer can poison the atmosphere in a room. After an accident involving mercury, go over the entire area carefully until there are no globules remaining. Follow all of the safety requirements specified on the MSDS. Keep all mercury containers tightly stoppered.

14.223 Accident Prevention

ELECTRIC SHOCK. Whenever there are electrical outlets, plugs, and wiring connections, there is a danger of electric shock. The usual "do's" and "don'ts" of protection against shock in the home are equally applicable in the laboratory. Don't use worn or frayed wires. Replace connections when there is any sign of thinning insulation. Ground all apparatus using plugs or pigtail adapters. Ground-fault circuit interrupters (GFIs) should be installed on all electric circuits near laboratory sinks or liquid operations or activities. Don't continue to run a motor after liquid has spilled on it. Turn it off immediately and clean and dry the inside thoroughly before attempting to use it again.

Electrical units that are operated in an area exposed to flammable vapors should be explosion-proof. All permanent wiring should be installed by an electrician with proper conduit or BX cable to eliminate any danger of circuit overloading.

CUTS. Some of the pieces of glass used in the laboratory, such as glass tubing, thermometers, and funnels, must be inserted through rubber stoppers. If the glass is forced through the hole in the stopper by applying a lot of pressure, the glass usually breaks. This is one of the most common sources of cuts in the laboratory.

Use care in making rubber-to-glass connections. Lengths of glass tubing should be supported while they are being inserted into rubber. The ends of the glass should be *FLAME POLISHED*[5] and either wetted or covered with a lubricating jelly for ease in joining connections. Never use oil or grease. Gloves should be worn when making such connections, and the tubing should be held as close to the end being inserted as possible to prevent bending or breaking. Never try to force rubber tubing or stoppers from glassware. Cut off the rubber or material.

A FIRST-AID kit must be available in the laboratory.

BURNS. All glassware and porcelain look cold after the red from heating has disappeared. The red is gone in seconds but the glass is hot enough to burn for several minutes. After heating a piece of glass, put it out of the way until cool.

Spattering from acids, caustic materials, and strong oxidizing solutions should be washed off immediately with large quantities of water. Neutralize an acid with sodium carbonate or bicarbonate. Every worker in the wastewater laboratory should have access to a sink and an emergency deluge shower and eye wash station.

Many safeguards against burns are available. Special gloves, safety tongs, and aprons are but a few examples. Never decide it is too much trouble to put on a pair of gloves or use a pair of tongs to handle a dish or flask that has been heated.

USE TONGS—DON'T JUGGLE HOT CONTAINERS.

[5] *Flame Polished. Melted by a flame to smooth out irregularities. Sharp or broken edges of glass (such as the end of a glass tube) are rotated in a flame until the edge melts slightly and becomes smooth.*

Perhaps the most harmful and painful chemical burn occurs when small objects, chemicals, or fumes get into your eyes. Immediately flood your eyes with water or a special "eye-wash" solution from a safety kit or from an eye wash station or fountain. Washing with large amounts of water for at least 15 minutes is recommended.

TOXIC FUMES. Certain chemicals are dangerous to breathe, therefore, use a ventilated laboratory fume hood for routine reagent preparation or whenever handling substances that may generate harmful atmospheric contaminants. Select a hood that has adequate air displacement and expels harmful vapors and gases at their source. The face velocity of laboratory fume hoods must average from 100 to 150 FPM (feet per minute) (30 to 45 meters per minute), depending on the substance handled. A method to continuously indicate that air is flowing should also be provided. Check with your local safety regulatory agency for specific requirements for your facility. An annual check should be made of the entire laboratory building. Sometimes noxious fumes are spread by the heating and cooling system of the building.

WASTE DISPOSAL. A good safety program requires constant care in disposal of laboratory waste. Hazardous chemicals/substances must be disposed of by methods that comply with local environmental regulations. Confirm the local requirements *BEFORE* disposal.

To protect maintenance personnel, use separate, clearly marked, covered containers to dispose of broken glass.

DON'T POUR VOLATILE LIQUIDS INTO THE SINK.

FIRE. The laboratory should be equipped with a fire blanket. The fire blanket is used to smother clothing fires. Small fires that occur in an evaporating dish or beaker may be put out by covering the container with a glass plate, wet towel, or wet blanket. For larger fires, or ones that may spread rapidly, promptly use a fire extinguisher. Do not use a fire extinguisher on small beaker fires because the force of the spray will knock over the beaker and spread the fire. Become familiar with the operation and use of your fire extinguisher *BEFORE* an emergency arises.

The use of the proper type extinguisher for each class of fire will give the best control of the situation and avoid compounding the problem. The class of fire is based on the type of material being consumed, and the class of extinguisher corresponds with the class of fire it will extinguish.

CHOOSE AN EXTINGUISHER BY CLASS OF FIRE—DON'T GUESS.

Fire classifications are important for determining the type of fire extinguisher needed to control the fire. Classifications also aid in recordkeeping. Fires are classified as A, B, C, or D fires based on the type of material being consumed: A, ordinary combustibles; B, flammable liquids and vapors; C, energized electrical equipment; and D, combustible metals. Fire extinguishers are also classified as A, B, C, or D to correspond with the class of fire each will extinguish.

Class A fires: ordinary combustibles such as wood, paper, cloth, rubber, many plastics, dried grass, hay, and stubble. Use foam, water, soda-acid, carbon dioxide gas, or almost any type of extinguisher.

Class B fires: flammable and combustible liquids such as gasoline, oil, grease, tar, oil-based paint, lacquer, and solvents, and also flammable gases. Use foam, carbon dioxide, or dry chemical extinguishers.

Class C fires: energized electrical equipment such as starters, breakers, and motors. Use carbon dioxide or dry chemical extinguishers to smother the fire; both types are nonconductors of electricity.

Class D fires: combustible metals such as magnesium, sodium, zinc, and potassium. Operators rarely encounter this type of fire. Use a Class D extinguisher or use fine dry soda ash, sand, or graphite to smother the fire. Consult with your local fire department about the best methods to use for specific hazards that exist at your facility.

Multipurpose extinguishers are also available, such as a Class BC carbon dioxide extinguisher that can be used to smother Class B and Class C fires. A multipurpose ABC carbon dioxide extinguisher will handle most laboratory fire situations. (When using carbon dioxide extinguishers, remember that the carbon dioxide can displace oxygen—take appropriate precautions.)

There is no single type of fire extinguisher that is effective for all fires so it is important that you understand the class of fire you are trying to control. You must be trained in the use of the different types of extinguishers, and the proper type should be located near the area where that class of fire may occur.

14.23 Acknowledgments

Portions of this section were taken from material written by A. E. Greenberg, "Safety and Hygiene," which appeared in the California Water Pollution Control Association's *OPERATORS' LABORATORY MANUAL*. Some of the ideas and material also came from the *FISHER SAFETY MANUAL*.

14.24 Additional Reading

1. *FISHER SCIENTIFIC SAFETY PRODUCTS REFERENCE MANUAL*. Obtain from Fisher Scientific Company, Safety Division, 9999 Veterans Memorial Drive, Houston, TX 77038.
2. *GENERAL INDUSTRY, OSHA. SAFETY AND HEALTH STANDARDS* (CFR, Title 29, Labor Pt. 1900-1910 (most recent edition)). Obtain from the U.S. Government Printing Office, Superintendent of Documents, PO Box 371954, Pittsburgh, PA 15250-7954. Order No. 869-048-00104-2. Price, $58.00.

QUESTIONS

Write your answers in a notebook and then compare your answers with those on page 337.

14.2A Why should you always use a rubber bulb to pipet wastewater or polluted water?

14.2B Why are inoculations against disease recommended for people working around wastewater?

14.2C True or False? You may *ADD ACID TO WATER*, but never water to acid.

14.2D What would you do if you spilled a concentrated acid on your hand?

Please answer the discussion and review questions next.

DISCUSSION AND REVIEW QUESTIONS

Chapter 14. LABORATORY PROCEDURES

(Lesson 2 of 7 Lessons)

Write the answers to these questions in your notebook before continuing. The question numbering continues from Lesson 1.

4. What precautions should you take to protect yourself from diseases when working in a wastewater treatment plant?
5. Discuss the basic rules for working in a laboratory.
6. Why should work with certain chemicals be conducted under a ventilated laboratory fume hood?

CHAPTER 14. LABORATORY PROCEDURES

(Lesson 3 of 7 Lessons)

14.3 SAMPLING

14.30 Importance

The basis for any plant monitoring program rests on information obtained by sampling. Decisions based on incorrect data may be made if sampling is performed in a careless and thoughtless manner. Obtaining good results will depend to a great extent on the following factors:

1. Ensuring that the sample taken is truly representative of the wastestream,
2. Using proper sampling techniques, and
3. Protecting and preserving the samples until they are analyzed.
4. Following a proper *CHAIN OF CUSTODY*[6] procedure if samples are sent off site for analysis.

THE GREATEST ERRORS PRODUCED IN LABORATORY TESTS ARE USUALLY CAUSED BY IMPROPER SAMPLING, POOR PRESERVATION, OR LACK OF ENOUGH MIXING DURING COMPOSITING[7] *AND TESTING.*

14.31 Representative Sampling

You must always remember that wastewater flows can vary widely in quantity and composition over a 24-hour period. Also, composition can vary within a given stream at any single time due to partial settling of solids or floating of light materials. Samples should therefore be taken from the wastestream where it is well mixed. Obtaining a representative sample should be of major concern in any sampling and monitoring program.

Laboratory equipment, in itself, is generally quite accurate. Analytical balances weigh to 0.1 milligram. Graduated cylinders, pipets, and burets usually measure to 1 percent accuracy, so that the errors introduced by these items should total less than 5 percent, and under the worst possible conditions only 10 percent. Under ideal conditions, let us assume that a test of raw wastewater for suspended solids should run about 300 mg/*L*. Because of the previously mentioned equipment or apparatus variables, the value may actually range from 270 to 330 mg/*L*. Results in this range are reasonable for operation. Other less obvious factors are usually present that make it quite possible to obtain results that are 25, 50, or even 100 percent in error, unless certain precautions are taken. The following example will illustrate how these errors are produced.

The Dumpmore Wastewater Treatment Plant is a secondary treatment facility with a flow of 8 million gallons per day (30 M*L*/day). The plant has an aerated grit basin, two circular primary clarifiers of 750,000-gallon (2.8-M*L*) capacity, two digesters, two aeration basins, two secondary clarifiers, four chlorinators, and two chlorine contact basins.

Monthly summary calculations based on the suspended solids test showed that about 8,000 pounds (3,640 kg) of suspended solids were being captured per day during primary sedimentation, assuming 200 mg/*L* for the influent and 100 mg/*L* for the effluent. However, it also appeared that 12,000 pounds (5,450 kg) per day of raw sludge solids were being pumped out of the primary clarifiers and to the digester. Obviously, if sampling and analyses had been perfect, these weights would have been balanced, provided the waste activated sludge was not returned to the primary clarifiers. The capture should equal the removal of solids. A study was made to determine why the variance in these values was so great. It would seem logical to expect that the problem could be due to (1) incorrect testing procedures, (2) poor sampling, (3) incorrect metering of the wastewater or sludge flow, or (4) any combination of the three or all of them.

In the first case, the equipment was in excellent condition. The operator was a conscientious and able person who was found to have carried out the laboratory procedures carefully and who had previously run successful tests on comparative samples. It was concluded that the equipment and test procedures were completely satisfactory.

A survey was then made to determine if sampling stations were in need of relocation. By using Imhoff cones and running settleable solids tests along the influent channel and the aerated grit chamber, one could quickly recognize that the best mixed and most representative samples were to be taken from the aerated grit chamber rather than the influent channel.

The settleable solids ran 13 m*L*/*L* in the aerated grit chamber against 10 m*L*/*L* in the channel. By the simple process of determining the best sampling station, the suspended solids value in the influent was corrected from 200 mg/*L* to the more representative 300 mg/*L*. Calculations, using the correct figures, changed the solids capture from 8,000 pounds to 12,000 pounds per day (3,640 kg to 5,450 kg/day) and a balance was obtained.

This example clearly illustrates the importance of selecting a good sampling point in securing a truly representative sample. It emphasizes the point that *EVEN THOUGH A TEST IS*

---

[6] *Chain of Custody. A record of each person involved in the handling and possession of a sample from the person who collected the sample to the person who analyzed the sample in the laboratory and to the person who witnessed disposal of the sample.*

[7] *Composite (come-PAH-zit) (Proportional) Sample. A composite sample is a collection of individual samples obtained at regular intervals, usually every one or two hours during a 24-hour time span. Each individual sample is combined with the others in proportion to the rate of flow when the sample was collected. The resulting mixture (composite sample) forms a representative sample and is analyzed to determine the average conditions during the sampling period.*

ACCURATELY PERFORMED, THE RESULTS MAY BE ENTIRELY ERRONEOUS AND MEANINGLESS INSOFAR AS USE FOR PROCESS CONTROL IS CONCERNED, UNLESS A GOOD REPRESENTATIVE SAMPLE IS TAKEN. FURTHERMORE, A GOOD SAMPLE IS HIGHLY DEPENDENT ON THE SAMPLING STATION. Whenever possible, select a place where mixing is thorough and the wastewater quality is uniform. As the solids concentration increases, above about 200 mg/*L*, mixing becomes even more significant because the wastewater solids will tend to separate rapidly with the heavier solids settling toward the bottom, the lighter solids in the middle, and the floatables rising toward the surface. If, as is usual, a one-gallon portion is taken as representative of a million gallon flow, the job of sample location and sampling must be taken seriously.

14.32 Time of Sampling

Let us consider next the time and frequency of sampling. In carrying out a testing program, particularly where personnel and time are limited due to the press of operational responsibilities, testing may necessarily be restricted to about one test day per week. If you decide to start your tests early in the week by taking samples early on Monday morning, you may wind up with some very odd results.

One such incident will be cited. During a test for *SURFACTANTS*,[8] samples were taken early on Monday morning and rushed into the laboratory for testing. Due to the detention time in the sewers, these wastewater samples actually represented Sunday flow on the graveyard shift, the weakest wastewater obtainable. The surfactant content was only 1 mg/*L*, whereas it would usually run 8 to 10 mg/*L*. So, the time and day of sampling is quite important. Samples should be taken to represent typical weekdays or even varied from day to day within the week for a good indication of the characteristics of the wastewater.

14.33 Types of Samples

The two types of samples collected in treatment plants are known as (1) grab samples, and (2) composite samples, and either may be obtained manually or automatically.

GRAB SAMPLES

A grab sample is a single sample of wastewater taken at neither a set time nor flow. The grab samples show the waste characteristics only at the time the sample is taken. A grab sample may be preferred over a composite sample when:

1. The wastewater to be sampled does not flow on a continuous basis,
2. The wastewater characteristics are relatively constant,
3. You wish to determine whether or not a composite sample obscures extreme conditions of the waste, and
4. The wastewater is to be analyzed for dissolved gases (DO), coliforms, residual chlorine, temperature, and pH. The analyses for these water quality indicators must be performed immediately. (*NOTE:* Grab samples for these water quality indicators may be collected at set times or specific time intervals.)

COMPOSITE SAMPLES

Since the wastewater quality changes from moment to moment and hour to hour, the best results would be obtained by using some sort of continuous sampler-analyzer. However, since operators are usually the sampler-analyzer, continuous analysis would leave little time for anything but sampling and testing. Except for tests that cannot wait due to rapid chemical or biological change of the sample, such as tests for dissolved oxygen and sulfide, a fair compromise may be reached by taking samples throughout the day at hourly or two-hour intervals.

When the samples are taken, they should be refrigerated immediately to preserve them from continued bacterial decomposition. When all of the samples have been collected for a 24-hour period, the samples from a specific location should be combined or composited together according to flow to form a single 24-hour composite sample.

If an equal volume of sample was collected each hour and mixed, this would simply be a composite sample. To prepare a proportional composite sample, (1) the rate of wastewater flow must be known, and (2) each grab sample must then be taken and measured out in direct proportion to the volume of flow at that time. For example, Table 14.2 illustrates the hourly flow and sample volume to be measured out for a 12-hour proportional composite sample.

Large wastewater solids should be excluded from a sample, particularly those greater than one-quarter inch (6 mm) in diameter.

A very important point should be emphasized. *DURING COMPOSITING AND AT THE EXACT MOMENT OF TESTING, THE SAMPLES MUST BE VIGOROUSLY REMIXED*[9] *SO THAT THEY WILL BE OF THE SAME COMPOSITION*

TABLE 14.2 DATA COLLECTED TO PREPARE PROPORTIONAL COMPOSITE SAMPLE

Time	Flow, MGD	Factor	Sample Vol, m*L*	Time	Flow, MGD	Factor	Sample Vol, m*L*
6 AM	0.2	100	20	12 N	1.5	100	150
7 AM	0.4	100	40	1 PM	1.2	100	120
8 AM	0.6	100	60	2 PM	1.0	100	100
9 AM	1.0	100	100	3 PM	1.0	100	100
10 AM	1.2	100	120	4 PM	1.0	100	100
11 AM	1.4	100	140	5 PM	0.9	100	90
							1,140

A sample composited in this manner would total 1,140 m*L*.

[8] *Surfactant (sir-FAC-tent). Abbreviation for surface-active agent. The active agent in detergents that possesses a high cleaning ability.*
[9] *NOTE: If the sample has a low buffer capacity and the real pH is 6.5 or less, vigorous shaking can cause a significant change in pH level.*

AND AS WELL MIXED AS WHEN THEY WERE ORIGINALLY SAMPLED. Sometimes such remixing may become lax, so that all the solids are not uniformly suspended. Lack of mixing can cause low results in samples of solids that settle out rapidly, such as those in activated sludge or raw wastewater. Samples must therefore be mixed thoroughly and poured quickly before any settling occurs. If this is not done, errors of 25 to 50 percent may easily occur. For example, on the same mixed liquor sample, one person may find 3,000 mg/*L* suspended solids while another person may determine that there are only 2,000 mg/*L* due to poor mixing. When a well-mixed, proportional composite sample is tested, a reasonably accurate measurement of the quality of the day's flow can be made.

If a 24-hour sampling program is not possible, perhaps due to insufficient personnel or the absence of a night shift, single representative samples should be taken at a time when typical characteristic qualities are present in the wastewater. The samples should be taken in accordance with the detention time required for treatment. For example, this period may exist between 10 AM and 5 PM for the sampling of raw influent. If a sample is taken at 12 Noon, other samples should be taken in accordance with the detention periods of the serial processes of treatment in order to follow this slug of wastewater or *PLUG FLOW.*[10] In primary settling, if the detention time in the primaries is two hours, the primary effluent should be sampled at 2 PM. If the detention time in the succeeding secondary treatment process required three hours, this sample should be taken at 5 PM.

14.34 Sludge Sampling

In sampling raw sludge and feeding a digester, a few important points should be kept in mind as shown in the following illustrative table.

For raw sludge pumped from a primary clarifier, the sludge solids vary considerably with pumping time as shown by samples withdrawn every one-half minute in Table 14.3.

TABLE 14.3 DECREASE IN PERCENT TOTAL SOLIDS DURING PUMPING

Pumping Time In Minutes	Total Solids Percent	Cumulative Solids Average
0.5	7.0	7.0
1.0	7.1	7.1
1.5	7.4	7.2
2.0	7.3	7.2
2.5	6.7	7.1
3.0	5.3	6.8
3.5	4.0	6.4
4.0	2.3	5.9
4.5	2.0	5.5
5.0	1.5	5.1

1. Table 14.3 shows that the solids were heavy during the first 2.5 minutes, and thereafter rapidly became thinner and watery. Since sludge solids should be fed to a digester with solids as heavy as possible and a minimum of water, the pumping should probably have been stopped at about 3 minutes. After 3 minutes, the water content did become greater than desirable.
2. In sampling this sludge, the sample should be taken as a composite by mixing small equal portions taken every 0.5 minute during pumping. If only a single portion of sludge is taken for the sample, there is a chance that the sludge sample may be too thick or too thin, depending on the moment the sample is taken. A composite sample will prevent this possibility.
3. Remember that as a sludge sample stands, the solids and liquid separate due to gasification and flotation or settling of the solids, and that it is absolutely necessary to thoroughly remix the sample back into its original form as a mixture before pouring it for a test. Do not store sludge samples in sealed containers. Sludge digestion can occur even under refrigeration. The gases produced can cause sludge to spray out of the container when the container is opened.
4. When individual samples are taken at regular intervals in this manner, they should be carefully preserved to prevent sample deterioration by bacterial action. Refrigeration is an excellent method of preservation and is generally preferable to chemicals. Chemicals may interfere with tests such as biochemical oxygen demand (BOD) and chemical oxygen demand (COD).

14.35 Sampling Devices

Automatic sampling devices are wonderful timesavers and should be used where possible. However, as with anything automatic, problems do arise and the operator should be aware of potential difficulties. Sample lines to auto-samplers may build up growths that may periodically slough off and contaminate the sample with a high solids content. Very regular cleanout of the intake line is required.

Manual sampling equipment includes dippers, weighted bottles, hand-operated pumps, and cross-section samplers. Dippers consist of wide-mouth, corrosion-resistant containers (such as cans or jars) on long handles that collect a sample for testing. A weighted bottle is a collection container that is lowered to a desired depth. At this location a cord or wire removes the bottle stopper so the bottle can be filled. Sampling pumps allow the inlet to the suction hose to be lowered to the sampling depth. Cross-sectional samplers are used to sample where the wastewater and sludge may be in layers, such as in a digester or clarifier. The sampler consists of a tube, open at both ends, that is lowered at the sampling location. When the tube is at the proper depth, the ends of the tube are closed and a sample is obtained from different layers.

Many operators build their own sampler (Figure 14.4) using the material described below:

1. *SAMPLING BUCKET.* A coffee can attached to an 8-foot (2.5-m) length of ½-inch (1.2-cm) electrical conduit or a wooden broom handle with a ¼-inch (0.6-cm) diameter spring in a 4-inch (10-cm) loop. A section of thin-walled PVC pipe (Schedule 40) of appropriate diameter with a cap on one end will be a more durable sampling bucket than a coffee can.

   Surface samples also can be collected using a plastic bucket or a gallon milk jug on a rope with an opening cut

[10] *Plug Flow. A type of flow that occurs in tanks, basins or reactors when a slug of wastewater moves through a tank without ever dispersing or mixing with the rest of the wastewater flowing through the tank.*

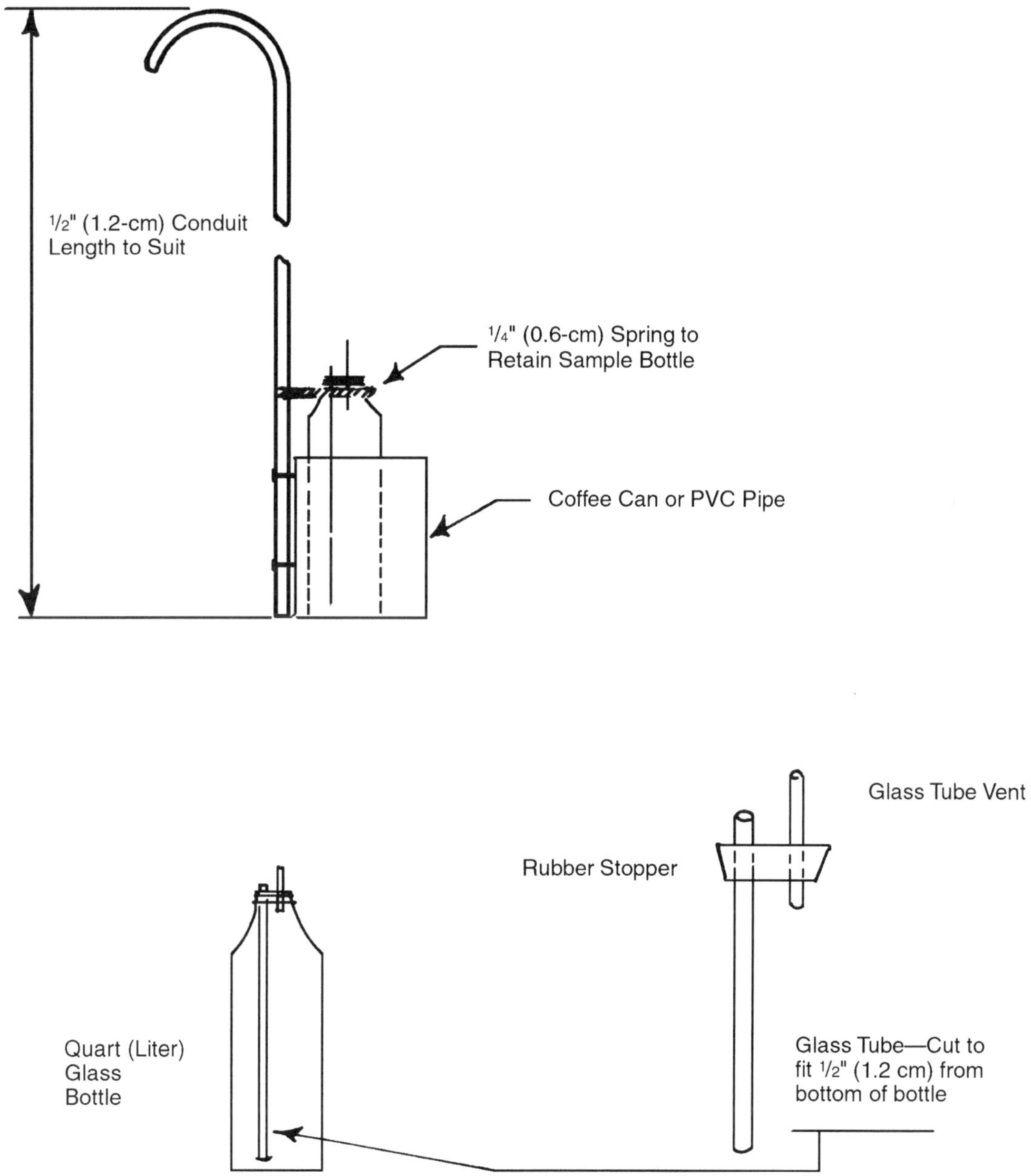

Fig. 14.4 Dissolved oxygen sampling bottle

out opposite from the handle. Placing a large bolt and nut near one edge of the sampler will cause the bucket to tip so that the sample will fill the bucket, rather than the bucket floating on the water surface.

2. *SAMPLING BOTTLE.* Glass bottle with rubber stopper equipped with two ³⁄₈-inch (1-cm) glass tubes, one ending near the bottom of bottle to allow sample to enter and the other ending at the bottom of the stopper to allow the air in the bottle to escape while the sample is filling the bottle.

For sample containers, wide-mouth glass bottles are recommended. Glass bottles, though somewhat expensive initially, greatly reduce the problem of metal contamination. The wide-mouth bottles ease the washing problem. For regular samples, sets of glass bottles bearing identification labels should be used.

14.36 Preservation of Samples

Sample deterioration starts immediately after collection for most wastewaters. The shorter the time that elapses between collection and analysis, the more reliable will be the analytical results.

In many instances, however, laboratory analysis cannot be started immediately due to the remoteness of the laboratory or the operator's workload. A summary of acceptable EPA (U.S. Environmental Protection Agency) methods of preservation appears in Table 14.4. Note that some constituents, such as temperature, pH, and DO, must be analyzed immediately on site because they cannot be preserved.

14.37 Quality Control in the Wastewater Laboratory

Having good equipment and using the correct methods are not enough to ensure correct analytical results. Each operator must be constantly alert to factors in the plant that can cause poor data quality. Such factors include: sloppy laboratory technique, deteriorated reagents, poorly operating instruments, and calculation mistakes.

14.38 Summary

1. Representative samples must be taken before any tests are made.
2. Select a good sampling location.
3. Collect samples and, if necessary, properly preserve them.
4. Mix samples thoroughly before compositing and at time of test.

TABLE 14.4 US EPA RECOMMENDED PRESERVATION METHODS FOR WATER AND WASTEWATER SAMPLES[a]

Test	Container[b]	Preservation Method	Maximum Recommended Holding Time
Acidity/Alkalinity	P,G	Store at 4°C	14 days
Ammonia	P,G	Add H_2SO_4 to pH <2 Store at 4°C	28 days
BOD	P,G	Store at 4°C	48 hours
COD	P,G	Add H_2SO_4 to pH <2 Store at 4°C	28 days
Chloride	P,G	None required	28 days
Chlorine, residual	P,G	Det. on site	No holding
Cyanide	P,G	Add NaOH to pH >12 Store at 4°C	14 days
Dissolved Oxygen	G	Det. on site	No holding
Fluoride	P	None required	28 days
Mercury	P,G	Add HNO_3 to pH <2	28 days
Metals	P,G	Add HNO_3 to pH <2	6 months
Nitrate	P,G	Store at 4°C	48 hours
Nitrate-nitrite	P,G	Add H_2SO_4 to pH <2 Store at 4°C	28 days
Nitrite	P,G	Store at 4°C	48 hours
Oil & Grease	G	Add HCl or H_2SO_4 to pH <2 Store at 4°C	28 days
Organic Carbon	P,G	Add HCl or H_2SO_4 or H_3PO_4 to pH <2 Store at 4°C	28 days
pH	P,G	Det. on site	No holding
Phenols	G	H_2SO_4 to pH <2 Store at 4°C	28 days
Phosphorus, ortho	P,G	Filter on site Store at 4°C	48 hours
Phosphorus, total	P,G	Add H_2SO_4 to pH <2 Store at 4°C	28 days
Solids	P,G	Store at 4°C	7 days
Specific Conductance	P,G	Store at 4°C	28 days
Sulfate	P,G	Store at 4°C	28 days
Sulfide	P,G	Add zinc acetate & NaOH to pH >9 Store at 4°C	7 days
Temperature	P,G	Det. on site	No holding
T. Kjeldahl Nitrogen	P,G	H_2SO_4 to pH <2 Store at 4°C	28 days
Turbidity	P,G	Store at 4°C	48 hours

[a] *40 CFR 136.3*, July 1, 2001, "Required Containers, Preservation Techniques, and Holding Times." U.S. Government Printing Office, Superintendent of Documents, PO Box 371954, Pittsburgh, PA 15250-7954.
[b] Polyethylene (P) or glass (G).

14.39 Additional Reading

1. *HANDBOOK FOR MONITORING INDUSTRIAL WASTEWATER*, U.S. Environmental Protection Agency. Obtain from National Technical Information Service (NTIS), 5285 Port Royal Road, Springfield, VA 22161. Order No. PB-259146/LL. EPA No. 625-6-73-002. Price, $51.00, plus $5.00 shipping and handling per order.

2. *HANDBOOK FOR ANALYTICAL QUALITY CONTROL IN WATER AND WASTEWATER LABORATORIES*, U.S. Environmental Protection Agency. Obtain from National Technical Information Service (NTIS), 5285 Port Royal Road, Springfield, VA 22161. Order No. PB-297451/7. EPA No. 600-4-79-019. Price, $45.00, plus $5.00 shipping and handling per order.

QUESTIONS

Write your answers in a notebook and then compare your answers with those on page 337.

14.3A What are the largest sources of errors found in laboratory results?

14.3B Why must a representative sample be collected?

14.3C How would you prepare a proportional composite sample?

Please answer the discussion and review questions next.

DISCUSSION AND REVIEW QUESTIONS

Chapter 14. LABORATORY PROCEDURES

(Lesson 3 of 7 Lessons)

Write the answers to these questions in your notebook before continuing. The question numbering continues from Lesson 2.

7. What is meant by representative sample?

8. Where would you obtain a representative sample?

9. Under what conditions and why would you preserve a sample?

CHAPTER 14. LABORATORY PROCEDURES

(Lesson 4 of 7 Lessons)

14.4 LABORATORY PROCEDURES FOR PLANT CONTROL

NOTICE: Lesson 4 outlines procedures used by operators to analyze samples for plant process control. These tests are frequently performed by operators to quickly obtain the results and make any necessary process adjustments. Lessons 6 and 7 are laboratory procedures for NPDES monitoring. These test procedures may be performed by operators, but are also performed by laboratory analysts, especially in larger plants.

14.40 Clarity

A. *DISCUSSION*

All plant effluents should be examined daily to observe the clarity of the effluent. Clarity is an indication of the quality of the effluent with respect to color, solids, and turbidity. The purpose of this test is to determine whether the plant effluent is staying the same, improving, or deteriorating.

The clarity test can be performed in the lab by looking down through the effluent in a graduated cylinder, or it can be performed in the field by looking down through the effluent in a clarifier or chlorine contact basin. The effluent should be examined at the same time each day and under the same conditions; otherwise, the results will not be comparable from day to day. By noting how far down you can see into the cylinder or basin and comparing each day's test results with the previous day's results, you will be able to see whether the general quality of the effluent is changing.

Sometimes, this test is referred to as a turbidity measurement; however, for the purposes described here, you are interested in the *clarity of the effluent* rather than the *quantity of solids* in the effluent (also see Section 14.5, Procedure 6, "Turbidity").

B. *WHAT IS TESTED?*

Sample	Common Range (Field Test) Poor	Good
Secondary Clarifiers:		
Trickling Filter	1 ft (0.3 m)	3 ft (0.9 m)
Activated Sludge	3 ft (0.9 m)	6 ft (1.8 m)
Activated Sludge Blanket in Secondary Clarifier	1 ft (0.3 m)	4 ft (1.2 m)
Chlorine Contact Basins	3 ft (0.9 m)	6 ft (1.8 m)

C. *APPARATUS*

1. One clarity unit (Secchi (SECK-key) Disc) and attached rope marked in one-foot units
2. One 1,000-m*L* graduated cylinder

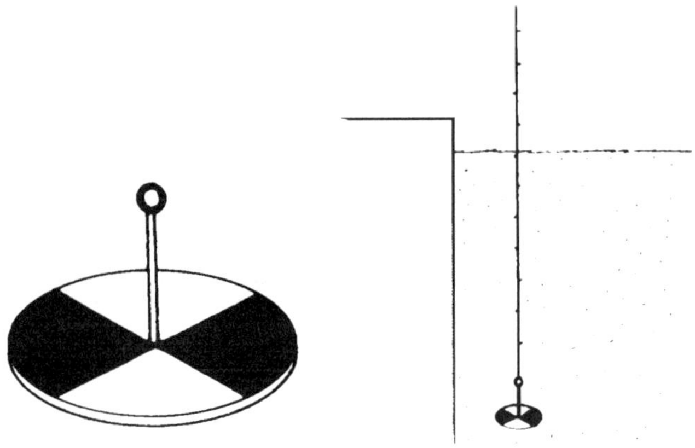

Secchi Disk

D. *REAGENTS*

None

E. *PROCEDURES*

1. *FIELD TEST.* Tie end of marked rope to handrail where tests will be run, for example, in final sedimentation unit. Always take tests at the same time each day for comparable results. Lower disc slowly until you just lose sight of it. Stop. Bring up slowly until just visible. Stop. Look at the marks on the rope to see the depth of water that you can see the disc through. Bring up disc, clean, and store. *RECORD RESULTS.*
2. *LAB TEST.* Use a clean 1,000-m*L* graduate. Fill with a well-mixed sample up to the 1,000-m*L* mark. During every test the same lighting conditions in the lab should be maintained. Look down through the liquid in the cylinder and read the last visible number etched on the side of the graduate and *RECORD RESULTS.*

Whether you use one or both of these tests, you should run each test at the same time every day and under similar conditions for comparable results.

F. *TEST RESULTS*

1. Each foot of depth is better clarity with Secchi disc.
2. Each 100-m*L* mark seen in depth is better clarity.

QUESTIONS

Write your answers in a notebook and then compare your answers with those on page 337.

14.4A What does the clarity test indicate?

14.4B What happens when you attempt to measure clarity under different conditions, such as lighting and clarifier loadings?

14.41 Hydrogen Sulfide (H_2S)

Hydrogen sulfide is measured because it causes corrosion and odors that should be controlled. The major portion of the sulfide content in wastewater is produced in the conversion of sulfate (SO_4^{2-}) to sulfide (S^{2-}) by bacteria found in the wastewater. Oxygen-reducing bacteria will use any available sulfur-containing compound as food. This process can produce odorous, reduced-sulfur compounds, including hydrogen sulfide (H_2S).

H_2S production in wastewater can be controlled by up-sewer maintenance, which reduces H_2S formation and, in some cases, by the application of chemicals such as chlorine, oxygen, hydrogen peroxide, iron salts, or bases to increase the pH.

14.410 H_2S in the Atmosphere

A. *DISCUSSION*

The rate of corrosion in the sewer collection system and treatment plant is often directly related to the rate of H_2S production or the amount of H_2S in the atmosphere. In addition, hydrogen sulfide gas is toxic to your respiratory system and is both flammable and explosive under certain conditions. The explosive limits of hydrogen sulfide range from 4.3% (LEL or Lower Explosive Limit) to 46% (UEL or Upper Explosive Limit). The parts per million (ppm) concentration of hydrogen sulfide in the sewer atmosphere is quite different from that *IN* wastewater. A concentration of 1 mg/*L* (ppm) in turbulent wastewater can quickly produce a concentration of 300 ppm in an unventilated, enclosed atmospheric space. The minimum concentration of H_2S in the atmosphere known to cause death is 300 ppm.

B. *METHODS*

Two examples of the methods available for testing H_2S in the atmosphere are listed below. The first method, using paper tape or tiles with lead acetate, will give a rough estimate of H_2S. The second method is a faster and more accurate instrumental measurement of a hazardous level or of the actual concentration of sulfide present.

C. *WHAT IS TESTED?*

Sample	Common Range
Atmosphere in sewers, outlets from force mains, wet pits, pumping stations, and influent areas to treatment plant.	1. Lead Acetate Method Not black in 24 hours = Good, 24+ hr Black in less than 1 hour = Bad, <1 hour 2. H_2S Detector Method <3 ppm = good >10 ppm = toxic

D. *APPARATUS*

Method 1: Lead acetate paper or unglazed tile soaked in lead acetate solution

Method 2: H_2S Detectors or H_2S Detector Tubes

These devices are available in three types:

(1) Personal. Attach to your belt or use a shoulder strap. These have an alarm signal; however, some units also have a readout for the level of hydrogen sulfide. You should rely on alarm signals rather than readouts.

(2) Portable. Carry by hand and place near work site. These have an alarm signal and meter indicating level of H_2S.

(3) Stationary. Install permanently in lift stations, gas compressor rooms, and other potentially dangerous areas. These have an alarm signal and may have a meter indicating level of H_2S.

Contact your local safety equipment supplier for information on atmospheric monitors.

E. *REAGENTS*

Method 1: Lead Acetate Solution, saturated. Dissolve 50 grams lead acetate in 80 m*L* distilled water.

Method 2: No reagents are required.

F. *PROCEDURE*

Method 1:

1. Obtain pieces of unglazed tile or use lead acetate paper. Cut tile with hacksaw into 1/2-inch (1.2-cm) strips.
2. Soak strips of tile or paper in lead acetate solution.
3. Dry tile in drying oven or air dry.
4. An open manhole or any point where wastewater is exposed to the atmosphere is a good test site. Drive a nail between metal crown ring of manhole, concrete, or other convenient place. Tie paper or tile with cotton string to nail.

Replace manhole cover and return in half an hour or less. If tile is not black or substantially colored, return periodically until black. If H_2S is present as indicated by a color change, then measure flow, temperature, pH, and BOD for further evaluation of problem.

Method 2: The instructions are included with instrument.

G. *CALCULATIONS*

Method 1: None required.

Method 2: None required.

14.411 H_2S in Wastewater

A. *DISCUSSION*

In sewers, when there is no longer any dissolved oxygen, H_2S tests are run to determine the rate of H_2S increase as the wastewater travels to a pumping station or treatment plant. If the wastewater is exposed to the atmosphere, H_2S will be released and a typical rotten egg odor will be detected. Anaerobic bacteria found in wastewater can liberate H_2S from the solids. When the gas leaves the wastewater stream, some of it

(Settleable Solids)

dissolves in the condensed moisture on the concrete. Sulfur-oxidizing bacteria convert the hydrogen sulfide to sulfuric acid, which is very corrosive to concrete. Only the concrete exposed to the atmosphere is corroded and the concrete below the low water line is not attacked.

Not all odors in wastewater are from H_2S, and there is no correlation between H_2S and other odors. The total H_2S procedure is good up to 18 mg/*L*, and higher concentrations must be diluted before testing. H_2S production can be controlled by up-sewer maintenance, which reduces H_2S formation in the wastewater and protects the collection system. In some severe cases chemicals are applied to flows in the collection system for H_2S control. Chemicals used include chlorine, oxygen, hydrogen peroxide, iron salts, or bases to increase the pH.

B. *WHAT IS TESTED?*

Sample Wastewater From the Following Locations	H_2S Concentrations Possible Results, mg/*L*	
	Good	**Bad**
Sewers	0.1	1
Outlets from force mains	0.1	1
Wet pits, pumping stations	0.1	0.5
Influents to treatment plants	0	0.5

All of the above locations should be sampled, if pertinent, when using upstream H_2S control in the collection system.

C. *APPARATUS*

1. LaMotte-Pomeroy Sulfide Testing Kit to test:
   a. Total Sulfide
   b. Dissolved Sulfide
   c. Hydrogen Sulfide in Solution

   Obtain from LaMotte Company. Order by Code #4630, $108.15, PO Box 329, Chestertown, MD 21620.

2. *OR* Hach Hydrogen Sulfide Test Kit

   Obtain from HACH Company. Order by Catalog No. 2238-01, $120.00, PO Box 389, Loveland, CO 80539-0389.

D. *REAGENTS*

The reagents are included in the kits.

E. *PROCEDURE*

The instructions are in the kits.

F. *EXAMPLE*

The instructions are in the kits.

G. *CALCULATIONS*

The instructions are in the kits.

H. *ADDITIONAL READING*

1. *STANDARD METHODS FOR THE EXAMINATION OF WATER AND WASTEWATER*, 20th Edition, 1998. Obtain from Water Environment Federation (WEF), Publications Order Department, 601 Wythe Street, Alexandria, VA 22314-1994. Order No. S82010. Price to members, $155.00; nonmembers, $200.00; plus shipping and handling.
2. *OPERATION AND MAINTENANCE OF WASTEWATER COLLECTION SYSTEMS*, Volumes I and II, prepared by the Office of Water Programs, California State University, Sacramento, 6000 J Street, Sacramento, CA 95819-6025. Price: $22.00 for each volume.

QUESTION

Write your answer in a notebook and then compare your answer with the one on page 337.

14.4C Why would you measure the H_2S concentration in:

1. The atmosphere?
2. Wastewater?

14.42 Settleable Solids

A. *DISCUSSION*

The settleable solids test measures the volume of settleable solids in one liter of sample that will settle to the bottom of an Imhoff cone during one hour. The test is an indication of the volume of solids removed by sedimentation in sedimentation tanks, clarifiers, or ponds. The results are read directly in milliliters from the Imhoff cone. The actual volume of solids pumped may disagree with the calculated volume from the settleable solids test due to compacting of the sludge in the clarifier, the sludge may settle differently in Imhoff cones than in the clarifier, and the clarifier may not capture all of the solids that should settle.

B. *WHAT IS TESTED?*

Sample	Common Ranges Found
Influent	6 m*L*/*L* weak wastewater 12 m*L*/*L* medium wastewater 20 m*L*/*L* strong wastewater
Primary Effluent	0.1 m*L*/*L* to 3 m*L*/*L* Over 3 m*L*/*L* poor
Secondary Effluent	Trace to 0.5 m*L*/*L* Over 0.5 m*L*/*L* poor

C. *APPARATUS*

1. Imhoff cones
2. Rack for holding Imhoff cones
3. Glass stirring rod, or wire
4. Time clock or watch

D. *PROCEDURE*

OUTLINE OF PROCEDURE FOR SETTLEABLE SOLIDS

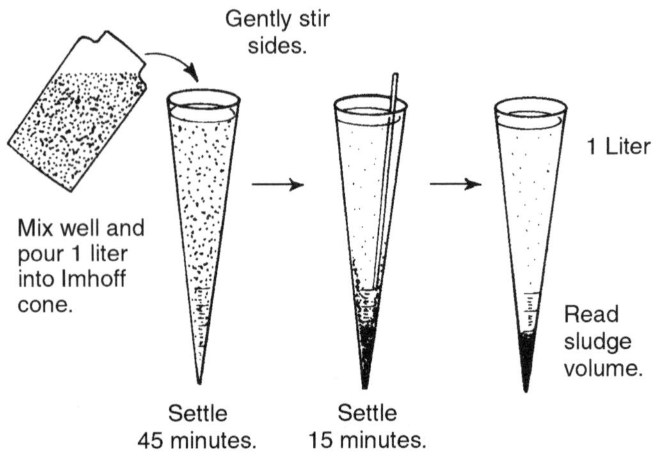

1. Thoroughly mix the wastewater sample by shaking and immediately fill an Imhoff cone to the liter mark.
2. Record the time of day that the cone was filled. T = _____.
3. Allow the wastewater sample to settle for 45 minutes.
4. Gently stir sides of the cone to facilitate settling of material adhering to the side of the cone.
5. After one hour, record the number of milliliters of settleable solids in the Imhoff cone. Make allowance for voids among the settled material. For example, if you read a sludge volume of 3.0 m*L* and voids or spaces in the sludge occupy approximately 0.2 m*L*, record a sludge volume of 2.8 m*L*.
6. Record the settleable solids as m*L*/*L* or milliliters per liter.

   Settleable Solids, Influent = ______________ m*L*/*L*

   Settleable Solids, Effluent = ______________ m*L*/*L*

   Settleable Solids, Removal = ______________ m*L*/*L*

E. *EXAMPLE*

Samples were collected from the influent and effluent of a primary clarifier. After one hour, the following results were recorded:

	Settleable Solids, m*L*/*L*
Influent	12.0
Effluent	0.2

F. *CALCULATIONS*

1. Calculate the efficiency or percent removal of the above primary clarifier in removing settleable solids.

$$\begin{aligned}\text{\% Removal of Set Sol} &= \frac{(\text{Infl Set Sol, m}L/L - \text{Effl Set Sol, m}L/L)}{\text{Influent Set Sol, m}L/L} \times 100\% \\ &= \frac{(12 \text{ m}L/L - 0.2 \text{ m}L/L)}{12 \text{ m}L/L} \times 100\% \\ &= \frac{11.8}{12} \times 100\% \\ &= 98\%\end{aligned}$$

Or

$$\begin{aligned}\text{\% Removal of Set Sol} &= (12 \text{ m}L/L - 0.2 \text{ m}L/L) \times 100\% \div 12 \text{ m}L/L \\ &= 98.3\%\end{aligned}$$

2. Estimate the gallons per day of sludge that should be pumped from the above primary clarifier if the flow is 1 MGD (1 million gallons per day). In your plant, the Imhoff cone may not measure or indicate the exact performance of your clarifier or sedimentation tank, but with some experience you should be able to relate or compare your lab tests with actual performance.

   Sludge Removed by Clarifier, m*L*/*L*

$$\begin{aligned}&= \text{Influent Set Sol, m}L/L - \text{Effluent Set Sol, m}L/L \\ &= 12 \text{ m}L/L - 0.2 \text{ m}L/L \\ &= 11.8 \text{ m}L/L\end{aligned}$$

To estimate the GPD (gallons per day) of sludge pumped to a digester, use the following formula:

Sludge to Digester, GPD

$$\begin{aligned}&= \text{Total Set Sol Rem, m}L/L \times 1{,}000 \text{ mg/m}L \times \text{Flow, MGD} \\ &= 11.8 \frac{\text{m}L}{\text{M mg}} \times \frac{1{,}000 \text{ mg}}{\text{m}L} \times \frac{1 \text{ M gal}}{\text{day}} \\ &= 11{,}800 \text{ GPD}\end{aligned}$$

This value may be reduced by 30 to 75 percent due to compaction of the sludge in the clarifier.

If you figure sludge removed as a percentage (1.18%), the sludge pumped to the digester would be calculated as follows:

$$\frac{1.18\%}{100\%} = \frac{\text{Sludge to Digester, GPD}}{\text{Flow of 1,000,000 GPD}}$$

$$\begin{aligned}\text{Sludge to Digester, GPD} &= \frac{(1.18\% \times 1{,}000{,}000 \text{ GPD})}{100\%} \\ &= 11{,}800 \text{ GPD}\end{aligned}$$

Or

$$\begin{aligned}\text{Sludge to Digester, GPD} &= 1.18\% \times 1{,}000{,}000 \text{ GPD} \div 100\% \\ &= 11{,}800 \text{ GPD}\end{aligned}$$

G. *CLINICAL CENTRIFUGE*

Settleable solids may also be measured by a small clinical centrifuge. This method, however, should be used for plant control only and not for NPDES monitoring. A mixed sample is placed in 15-m*L* graduated centrifuge tubes and spun for 15 minutes. The solid deposition in the tip of the tube is compared with a curve prepared by plotting settleable solids versus centrifuge solid deposition. This test provides a quick estimate of the settleable solids.

QUESTION

Write your answer in a notebook and then compare your answer with the one on page 337.

14.4D Estimate the volume of solids pumped to a digester in gallons per day (GPD) if the flow is 1 MGD, the influent settleable solids are 10 m*L*/*L*, and the effluent settleable solids are 0.4 m*L*/*L* for a primary clarifier.

14.43 Suspended Solids

A. *DISCUSSION*

One of the tests run on wastewater is used to determine the amount of material suspended within the sample. The result obtained from the suspended solids test does not mean that all of the suspended solids settle out in the primary clarifier or, for that matter, in the final clarifier. Some of the particles are of such size and weight (*SPECIFIC GRAVITY*[11]) that they will not float or settle without additional treatment. Therefore, suspended solids are a combination of settleable solids and those solids that remain in suspension.

[11] *Specific Gravity. (1) Weight of a particle, substance or chemical solution in relation to the weight of an equal volume of water. Water has a specific gravity of 1.000 at 4°C (39°F). Wastewater particles or substances usually have a specific gravity of 0.5 to 2.5. (2) Weight of a particular gas in relation to the weight of an equal volume of air at the same temperature and pressure (air has a specific gravity of 1.0). Chlorine has a specific gravity of 2.5 as a gas.*

(Suspended Solids)

B. *WHAT IS TESTED?*

Sample	Common Ranges, mg/*L*
Influent	Weak 150 to 400+ Strong
Primary Effluent	Weak 60 to 150+ Strong
Secondary Effluent	Good 10 to 60+ Bad
Activated Sludge Tests:	Depending on Type of Process
Mixed Liquor	1,000 to <5,000
Return or Waste Sludge	2,000 to <12,000
Digester Tests:	
Supernatant	3,000 to <10,000

When supernatant suspended solids are greater than 10,000 mg/*L*, the total solids test is usually performed.

C. *APPARATUS*

1. Glass-fiber filter disks without organic binder (Whatman grade 934H; Gelman type A/E; Millipore type AP40; or equivalent).
2. Filtration apparatus. One of the following, which must be suitable for the filter disk selected above:
   a. Membrane filter funnel.
   b. Gooch crucible, 25-m*L* to 40-m*L* capacity with Gooch crucible adapter, or
   c. Filtration apparatus with reservoir and coarse 40- to 60-µm/fritted (high density porcelain) disk as filter support.
3. Flask, suction.
4. Vacuum source.
5. Drying oven, 103° to 105°C.
6. Desiccator.
7. Analytical balance.

D. *PROCEDURE*

Preparation of Gooch Crucible

1. Insert a 2.2-cm glass-fiber filter into 25-m*L* Gooch crucible and center it.
2. Apply suction. Wash filter with 20 m*L* of distilled or deionized water three times to seat filter properly.
3. Dry at 103°C for one hour. If volatile suspended solids are to be determined, ignite crucible in muffle furnace for one hour at 550°C.
4. Cool in desiccator.
5. Weigh crucible and record weight. (This is known as "tare weight.")

Sample Analysis

6. Depending on the suspended solids content, measure out a 25-, 50-, or 100-m*L* portion of a *WELL-MIXED* sample into a graduated cylinder. Use 25 m*L* if sample filters slowly. Use larger volumes of sample if sample filters easily, such as secondary effluent. Try to limit filtration time to about 15 minutes or less. Wet prepared Gooch crucible with distilled water and apply suction.
7. Filter sample through the Gooch crucible.
8. Wash out dissolved solids on the filter with about 20 m*L* of distilled water. (Use two 10-m*L* portions.)
9. Dry crucible at 103°C for at least one hour. Some samples may require up to three hours to dry if the residue is thick.
10. Cool in desiccator for 20 to 30 minutes.
11. Weigh crucible plus suspended solids.
12. Repeat drying cycle until constant weight is attained or until weight loss is less than 0.5 mg.
13. Record weight:

    Total Weight = ____________________ gm

    Tare Weight = ____________________ gm

    Solids Weight = ____________________ gm

E. *PRECAUTIONS*

1. Check and regulate the oven temperature at 103° to 105°C.
2. Observe crucible and glass fiber for any possible leaks. A leak will cause solids to pass through and give low results. The glass-fiber filter may become unseated and leaky when the crucible is placed on the filter flask. The filter should be reseated by adding distilled water to the filter in the crucible and applying vacuum before filtering the sample.
3. Mix the sample thoroughly so that it is completely uniform in suspended solids when measured into a graduated cylinder *BEFORE SAMPLE CAN SETTLE OUT.* This is especially true of samples heavy in suspended solids, such as raw wastewater and mixed liquor in activated sludge, which settle rapidly. The test can be no better than the mix.
4. A good practice is to prepare a number of extra Gooch crucibles for additional tests if the need arises. If a test result appears faulty or questionable, the test should be repeated. Check filtration rate and clarity of water passing through the filter.

F. *EXAMPLE AND CALCULATIONS*

This section is provided to show you the detailed calculations. After some practice, most operators use the lab work sheet as shown at the end of the calculations.

CALCULATIONS FOR SUSPENDED SOLIDS TEST

(Or use lab work sheet at end of calculations)

Always record the crucible number for the crucible used for the test. Different crucibles will have slightly different weights.

EXAMPLE: Assume the following data.

Crucible #015

Volume of Sample = 50 m*L*

	Recorded Weights
Crucible Weight	21.6329 gm
Crucible Plus Dry Solids	21.6531 gm
Crucible Plus Ash[12]	21.6360 gm

[12] *Obtained by placing the crucible plus dry solids in a muffle furnace at 550°C for one hour. The crucible plus remaining ash are cooled and weighed.*

(Suspended Solids)

OUTLINE OF PROCEDURE FOR SUSPENDED SOLIDS

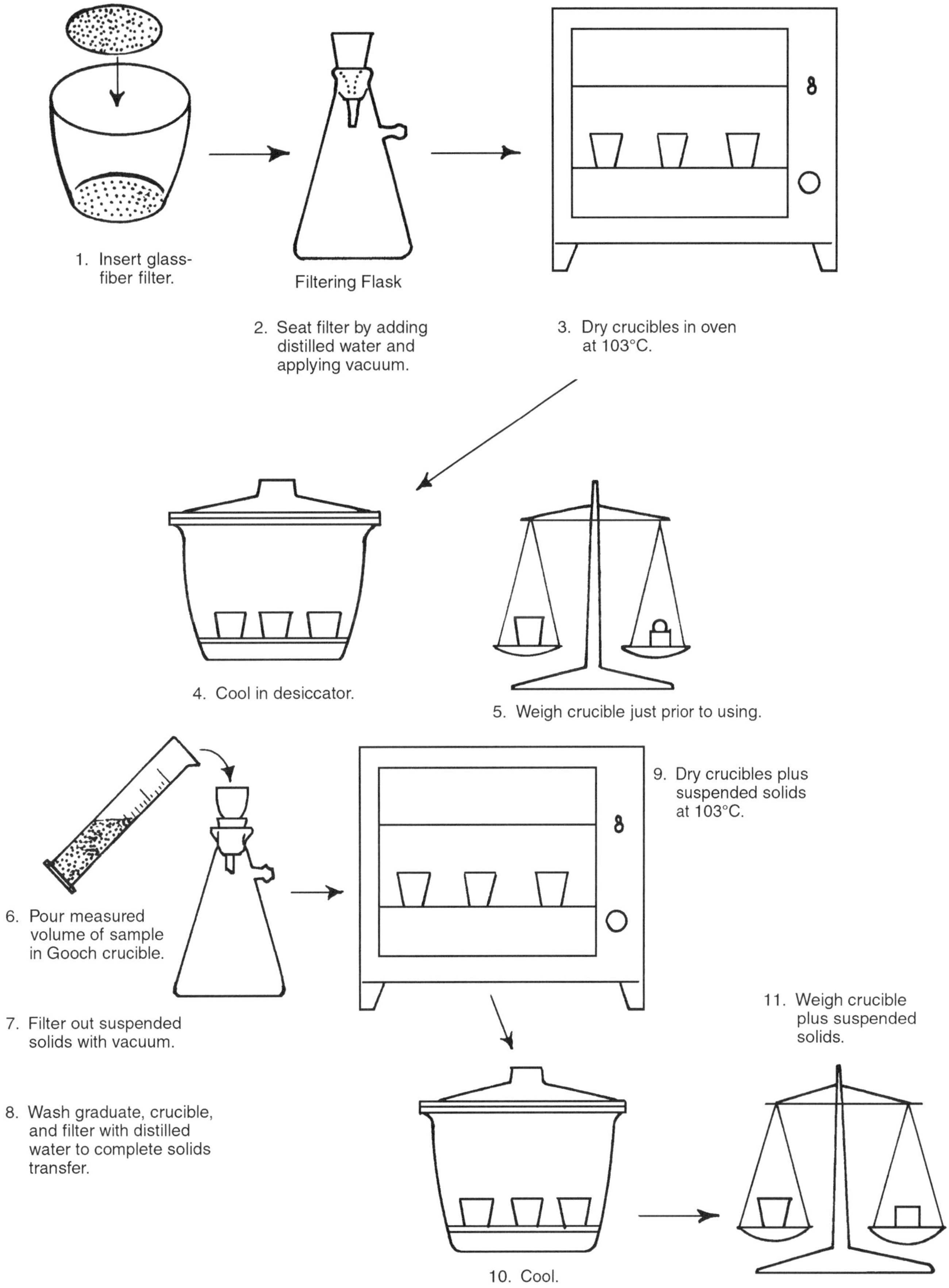

(Suspended Solids)

1. Compute total suspended solids.

21.6531 gm	Weight of Crucible Plus Dry Solids, grams
− 21.6329 gm	− Weight of Crucible, grams
= 0.0202 gm	= Weight of Dry Solids, grams

Or

= 20.2 mg

1,000 milligrams (mg) = 1 gram (gm)

Or

20.2 mg = 0.0202 gm

$$\text{Total Suspended Solids, mg}/L = \frac{\text{Weight of Solids, mg x 1,000 m}L/L}{\text{Sample Volume, m}L}$$

$$= \frac{\text{20.2 mg x 1,000 m}L/L}{\text{50 m}L}$$

$$= 404 \text{ mg}/L$$

Or

$$= \text{20.2 mg x 1,000 m}L/L \div \text{50 m}L$$

$$= 404 \text{ mg}/L$$

2. Compute volatile or organic suspended solids.

21.6531 gm	Weight of Crucible Plus Dry Solids, gm
− 21.6360 gm	− Weight of Crucible Plus Ash, gm
= 0.0171 gm	= Weight of Volatile Solids, gm

Or

= 17.1 mg

$$\text{Volatile Suspended Solids, mg}/L = \frac{\text{Weight of Volatile Solids, mg x 1,000 m}L/L}{\text{Sample Volume, m}L}$$

$$= \frac{\text{17.1 mg x 1,000 m}L/L}{\text{50 m}L}$$

$$= 342 \text{ mg}/L$$

3. Compute the percent volatile solids.

$$\text{Volatile Solids, \%} = \frac{\text{Weight Volatile, mg x 100\%}}{\text{Weight Total Dry Solids, mg}}$$

$$= \frac{\text{17.1 mg}}{\text{20.2 mg}} \times 100\%$$

$$= 84.7\%$$

The previous calculations are also performed on a Laboratory Work Sheet (Figure 14.5) to illustrate the use of the work sheet.

4. Compute fixed or inorganic suspended solids.

21.6360 gm	Weight of Crucible Plus Ash, gm
− 21.6329 gm	− Weight of Crucible, gm
= 0.0031 gm	= Weight of Fixed Solids, gm

Or

= 3.1 mg

$$\text{Fixed Suspended Solids, mg}/L = \frac{\text{Weight of Fixed Solids, mg x 1,000 m}L/L}{\text{Sample Volume, m}L}$$

$$= \frac{\text{3.1 mg x 1,000 m}L/L}{\text{50 m}L}$$

$$= 62 \text{ mg}/L$$

To check your work:

$$\text{Fixed Susp Solids, mg}/L = \text{Total Susp Solids, mg}/L - \text{Volatile Susp Solids, mg}/L$$

$$= 404 \text{ mg}/L - 342 \text{ mg}/L$$

$$= 62 \text{ mg}/L \text{ (check)}$$

5. Compute the percent fixed solids.

$$\text{Fixed Solids, \%} = \frac{\text{Weight Fixed, mg}}{\text{Weight Total, mg}} \times 100\%$$

$$= \frac{\text{3.1 mg}}{\text{20.2 mg}} \times 100\%$$

$$= 15.3\%$$

CALCULATIONS FOR OVERALL PLANT REMOVAL OF SUSPENDED SOLIDS IN PERCENT

EXAMPLE: Assume the following data.

Influent Suspended Solids	202 mg/*L*
Primary Effluent Suspended Solids	110 mg/*L*
Secondary Effluent Suspended Solids	52 mg/*L*
Final Effluent Suspended Solids	12 mg/*L*

To calculate the percent removal or treatment efficiency for a particular process or the overall plant, use the following formula:

$$\text{Removal, \%} = \frac{(\text{In} - \text{Out})}{\text{In}} \times 100\%$$

(Suspended Solids)

PLANT CLEAN WATER

DATE ____________________

SUSPENDED SOLIDS AND DISSOLVED SOLIDS

SAMPLE	INFL.					
Crucible	#015					
Sample, m*L*	50					
Wt Dry + Dish, gm Wt Dish, gm	21.6531 21.6329					
Wt Dry, gm	0.0202					
$mg/L = \frac{\text{Wt Dry, gm} \times 1{,}000{,}000}{\text{Sample, mL}}$	404 mg/*L*					
Wt Dish + Dry, gm Wt Dish + Ash, gm	21.6531 21.6360					
Wt Volatile, gm	0.0171					
$\% = \frac{\text{Wt Vol}}{\text{Wt Dry}} \times 100\%$	84.7%					

BOD

Blank ____________

SAMPLE					
DO Sample					
Bottle #					
% Sample					
Blank or adjusted blank DO after incubation					
Depletion, 5 days					
Depletion, %					

SETTLEABLE SOLIDS

Sample, m*L* ____________ ____________

Direct m*L*/*L* ____________ ____________

COD

Sample ____________ ____________

Blank Titration ____________ ____________

Sample Titration ____________ ____________

Depletion ____________ ____________

$mg/L = \frac{\text{Dep} \times N \text{ FAS} \times 8{,}000}{\text{Sample, mL}}$ ____________ ____________

Fig. 14.5 Calculation of solids content on laboratory work sheet

(Total Sludge Solids—Volatile and Fixed)

Compute percentage removed between influent and primary effluent:

$$\text{Removal, \%} = \frac{(\text{In} - \text{Out})}{\text{In}} \times 100\%$$

$$= \frac{(202 \text{ mg}/L - 110 \text{ mg}/L)}{202 \text{ mg}/L} \times 100\%$$

$$= \frac{92}{202} \times 100\%$$

$$= 45.5\%$$

Or

$$= (202 \text{ mg}/L - 110 \text{ mg}/L) \times 100\% \div 202 \text{ mg}/L$$

$$= 45.5\%$$

Compute percentage removed between influent and secondary effluent:

$$\text{Removal, \%} = \frac{(\text{In} - \text{Out})}{\text{In}} \times 100\%$$

$$= \frac{(202 \text{ mg}/L - 52 \text{ mg}/L)}{202 \text{ mg}/L} \times 100\%$$

$$= \frac{150}{202} \times 100\%$$

$$= 74.3\%$$

Compute percentage removed between influent and final effluent (overall plant percentage removed):

$$\text{Removal, \%} = \frac{(\text{In} - \text{Out})}{\text{In}} \times 100\%$$

$$= \frac{(202 \text{ mg}/L - 12 \text{ mg}/L)}{202 \text{ mg}/L} \times 100\%$$

$$= \frac{190}{202} \times 100\%$$

= 94.1% Suspended Solids Removal for the Plant

CALCULATIONS FOR POUNDS SUSPENDED SOLIDS REMOVED PER DAY

EXAMPLE: Assume the following data.

Influent Suspended Solids = 200 mg/*L*
Effluent Suspended Solids = 10 mg/*L*
Flow in million gallons/day = 2 MGD
1 gallon of water weighs = 8.34 lbs

Compute pounds suspended solids removed:

The general formula for computing pounds removed is:

Material Removed, lbs/day = (Concentration In, mg/*L* – Concentration Out, mg/*L*) x Flow, MGD x 8.34 lbs/gal

= (200 mg/*L* – 10 mg/*L*) x 2 MGD x 8.34 lbs/gal

= 190 x 2 x 8.34

= 3,169 lbs/day of Suspended Solids Removed by Plant

Or

= (200 mg/*L* – 10 mg/*L*) x 2 MGD x 8.34 lbs/gal

= 3,169 lbs/day of SS Removed

DERIVATION

This section is not essential to efficient plant operation, but is provided to furnish you with a better understanding of the calculation if you are interested. For practical purposes,

1 mg/*L* = 1 ppm or 1 part per million

Or = 1 mg/million mg, because 1 liter = 1,000,000 mg

Therefore:

$$\frac{\text{lbs}}{\text{day}} = \frac{\text{mg}}{\text{M mg}} \times \frac{\text{M gal}}{\text{day}} \times \frac{\text{lbs}}{\text{gal}}$$

= lbs/day

QUESTIONS

Write your answers in a notebook and then compare your answers with those on pages 337, 338, and 339.

14.4E Why does some of the suspended material in wastewater fail to be removed by settling or flotation?

14.4F Given the following data:

Volume of Sample = 100 m*L*
Crucible Weight = 19.3241 gm
Crucible Plus Dry Solids = 19.3902 gm
Crucible Plus Ash = 19.3469 gm

Compute:

1. Total Suspended Solids, mg/*L*
2. Volatile Suspended Solids, mg/*L*
3. Volatile Solids, %
4. Fixed Suspended Solids, mg/*L*
5. Fixed Solids, %

14.4G Suspended solids data from a wastewater treatment plant are given below:

Influent Suspended Solids = 221 mg/*L*
Primary Effluent SS = 159 mg/*L*
Final Effluent SS = 33 mg/*L*

Compute the percent removal of suspended solids by the:

1. Primary clarifier,
2. Secondary process (removal between primary effluent and secondary effluent), and
3. Overall plant.

14.4H If the data in problem 14.4G are from a 1.5 MGD plant, calculate the pounds of suspended solids removed by the:

1. Primary unit,
2. Secondary unit, and
3. Overall plant.

14.44 Total Sludge Solids (Volatile and Fixed)

A. *DISCUSSION*

Total solids are the combined amounts of suspended and dissolved materials in the sample.

(Total Sludge Solids—Volatile and Fixed)

This test is used for wastewater sludges or where the solids can be expressed in percentages by weight. The weight can be measured on an inexpensive beam balance to the nearest 0.01 of a gram. The total solids are composed of two components, volatile and fixed solids. Volatile solids are composed of organic compounds that are of either plant or animal origin. Volatile solids in a treatment plant indicate the waste material that may be treated by biological processes. Fixed solids are inorganic compounds such as sand, gravel, minerals, or salts.

B. *WHAT IS TESTED?*

	Common Range, % by Weight		
Sample	**Total**	**Volatile**	**Fixed**
Raw Sludge	6% to 9%	75%	25% ± 6%
Raw Sludge Plus Waste Activated Sludge	2% to 5%	80%	20% ± 5%
Recirculated Sludge	1.5% to 3%	75%	25% ± 5%
Supernatant:			
Good Quality, has Suspended Solids	<1%	50%	50% ± 10%
Poor Quality	>1%	50%	50% ± 10%
Digested Sludge to Air Dry	3% too Thin to <8%	50%	50% ± 10%

C. *APPARATUS*

1. Evaporating dish
2. Analytical balance
3. Drying oven, 103° to 105°C
4. Measuring device—graduated cylinder
5. Muffle furnace, 550°C

D. *PROCEDURE*

OUTLINE OF PROCEDURE FOR TOTAL SOLIDS

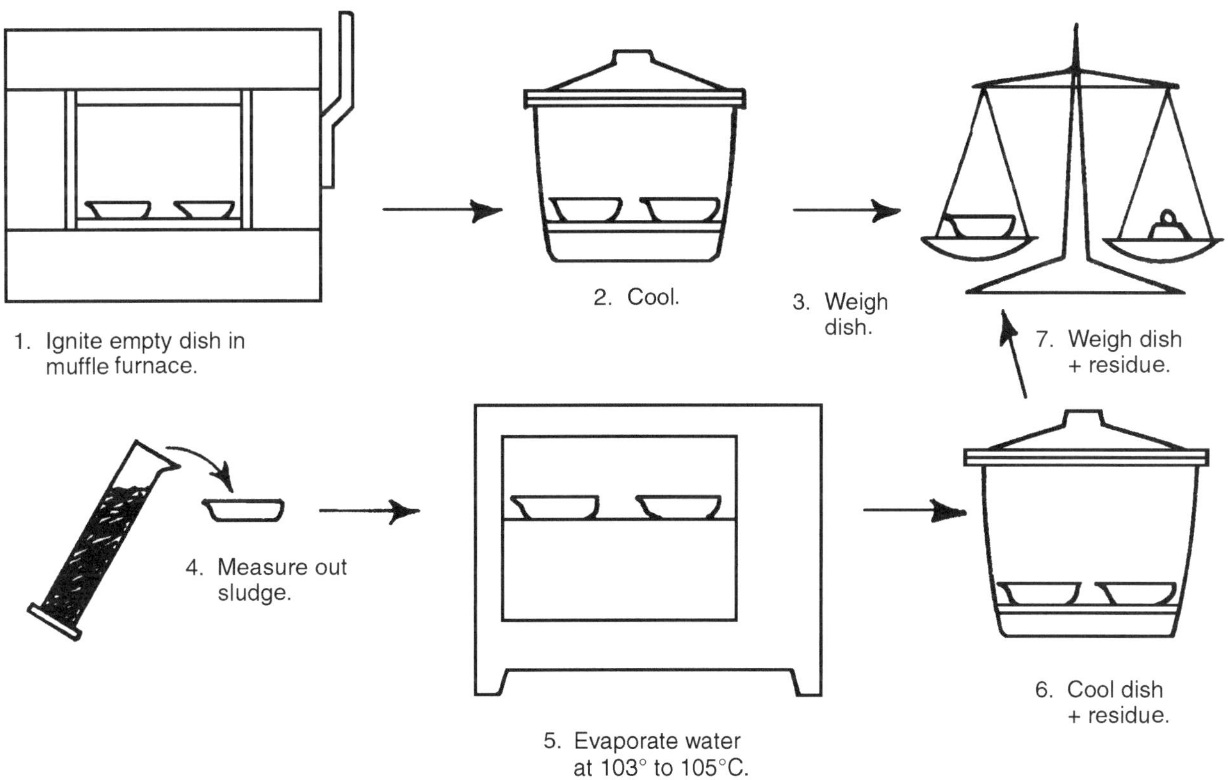

1. Dry the dish by ignition in a muffle furnace at 550°C for one hour. Cool dish in desiccator.
2. Tare the evaporating dish to the nearest 10 milligrams, or 0.01 gm on a single pan balance. Record the weight as Tare Weight = _______ gm.
3. Weigh dish plus 50 to 100 m*L* of WELL-MIXED sludge sample. Record total weight to nearest 0.01 gram as Gross Weight = _______ gm.
4. Evaporate the sludge sample to dryness in the 103°C drying oven.
5. Weigh the dried residue in the evaporating dish to the nearest 10 milligrams, or 0.01 gm. Record the weight as

   Dry Sample and Dish = _______ gm.
6. Compute the net weight of the residue by subtracting the tare weight of the dish from the dry sample and dish.

(Total Sludge Solids—Volatile and Fixed)

E. *PRECAUTIONS*

1. *BE SURE THAT THE SAMPLE IS THOROUGHLY MIXED* and is representative of the sludge being pumped. Generally, where sludge pumping is intermittent, sludge is much heavier at the beginning and is less dense toward the end of pumping. Take several equal portions of sludge at regular intervals and mix for a good sample.

2. Take a large enough sample. Measuring a 50- or 100-m*L* sample, which is closely equal to 50 or 100 grams, is recommended. Since this material is so heterogeneous (non-uniform), it is difficult to obtain a good representative sample with less volume. Smaller volumes will show greater variations in answers, due to the uneven and lumpy nature of the material.

3. Control oven temperature closely at 103° to 105°C. Some solids are lost at any drying temperature. Close control of oven temperature is necessary because higher temperatures increase the losses of volatile solids in addition to the evaporated water.

4. Heat dish long enough to ensure evaporation of water, usually about 3 to 4 hours. If heat drying and weighing are repeated, stop when the weight change becomes less than 4 percent of the previous weight, or 0.5 mg, whichever is less. The oxidation, dehydration, and degradation of the volatile fraction won't completely stabilize until it is carbonized or becomes ash.

5. Since sludge is so non-uniform, weighing on the analytical balance should probably be made only to the nearest 0.01 gram or 10 milligrams.

F. *PROCEDURE*

(continued from total solids procedure)

OUTLINE OF PROCEDURE FOR VOLATILE SOLIDS

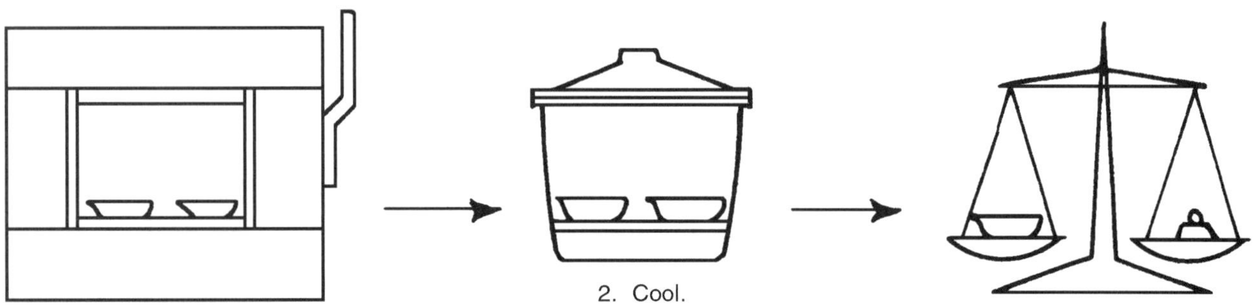

1. Ignite dried solids at 550°C.

2. Cool.

3. Weigh fixed solids.

1. Determine the total solids as previously described in Section D.
2. Ignite the dish and residue from total solids test at 550°C for one hour or until a white ash remains.
3. Cool in desiccator for about 30 minutes.
4. Weigh and record weight of Dish Plus Ash = _______ gm.

G. *EXAMPLE*

Dish No. 7

Weight of Dish (Tare)	= 20.31 gm
Weight of Dish Plus Wet Solids (Gross)	= 70.31 gm
Weight of Dish Plus Dry Solids	= 22.81 gm
Weight of Dish Plus Ash	= 20.93 gm

H. *CALCULATIONS*

See Laboratory Work Sheet (Figure 14.6) or calculations shown below.

1. Find weight of sample.

Weight of Dish Plus Wet Solids (Gross)	= 70.31 gm
Weight of Dish (Tare)	= 20.31 gm
Weight of Sample	= 50.00 gm

2. Find weight of total solids.

Weight of Dish Plus Dry Solids	= 22.81 gm
Weight of Dish (Tare)	= 20.31 gm
Weight of Total Solids	2.50 gm

3. Find percent sludge.

$$\text{Solids, \%} = \frac{(\text{Weight of Solids, gm})100\%}{\text{Weight of Sample, gm}}$$

$$= \frac{(2.50\text{ gm})100\%}{50.00\text{ gm}}$$

$$= 5\%$$

Or

$$= 2.50\text{ gm} \times 100\% \div 50.00\text{ gm}$$

$$= 5\%$$

4. Find weight of volatile solids.

Weight of Dish Plus Dry Solids	= 22.81 gm
Weight of Dish Plus Ash	= 20.93 gm
Weight of Volatile Solids	= 1.88 gm

TOTAL SOLIDS

SAMPLE	RAW				
Dish No.	7				
Wt Dish + Wet, gm Wt Dish, gm	70.31 20.31				
Wt Wet, gm	50.00				
Wt Dish + Dry, gm Wt Dish, gm	22.81 20.31				
Wt Dry, gm	2.50				
% Solids = $\frac{\text{Wt Dry}}{\text{Wt Wet}}$ x 100%	5.0%				
Wt Dish + Dry, gm Wt Dish + Ash, gm	22.81 20.93				
Wt Volatile, gm	1.88				
% Volatile = $\frac{\text{Wt Vol x 100\%}}{\text{Wt Dry}}$	75%				
pH					
Volatile Acid, mg/*L*					
Alkalinity as $CaCO_3$, mg/*L*					

Grease

Sample ________________

Sample, m*L* ________________

Wt Flask + Grease, mg ________________

Wt Flask, mg ________________

Wt Grease, mg ________________

$$mg/L = \frac{\text{Wt Grease, mg x 1,000}}{\text{Sample, m}L}$$ ________________

Fig. 14.6 Calculation of total solids on laboratory work sheet

5. Find percent volatile solids.

$$\text{Volatile Solids, \%} = \frac{(\text{Weight of Volatile Solids, gm})100\%}{\text{Weight of Total Solids, gm}}$$

$$= \frac{(1.88\text{ gm})100\%}{2.50\text{ gm}}$$

$$= 75\%$$

Or

$$= 1.88\text{ gm} \times 100\% \div 2.50\text{ gm}$$

$$= 75\%$$

QUESTIONS

Write your answers in a notebook and then compare your answers with those on page 339.

14.4I What is the origin of the volatile solids found in a digester?

14.4J What is the significance of volatile solids in a treatment plant?

Please answer the discussion and review questions next.

DISCUSSION AND REVIEW QUESTIONS

Chapter 14. LABORATORY PROCEDURES

(Lesson 4 of 7 Lessons)

Write the answers to these questions in your notebook before continuing. The question numbering continues from Lesson 3.

10. Why must the clarity test always be run under the same conditions?

11. Hydrogen sulfide is measured because it causes ______ ____ .

12. What causes H_2S formation in sewers?

13. Why does the actual volume of solids pumped from a clarifier not agree exactly with calculations based on the settleable solids test?

14. Calculate the efficiency or percent removal by a primary clarifier when the influent settleable solids are 10 m*L*/*L* and the effluent settleable solids are 0.3 m*L*/*L*.

15. Given the following data:

Volume of Sample	= 100 m*L*
Crucible Weight	= 19.9850 gm
Crucible Plus Dry Solids	= 20.0503 gm
Crucible Plus Ash	= 20.0068 gm

Compute:

1. Total Suspended Solids in mg and mg/*L*
2. Volatile Suspended Solids in mg and mg/*L*
3. Volatile Solids %

16. Estimate the pounds of solids removed per day by a primary clarifier if the influent suspended solids are 220 mg/*L* and the effluent suspended solids are 120 mg/*L* when the flow is 1.5 MGD.

17. Why are solids only weighed to the nearest 0.01 gram when determining the total volatile solids content of digesters?

CHAPTER 14. LABORATORY PROCEDURES

(Lesson 5 of 7 Lessons)

14.45 Tests for Activated Sludge Control

14.450 Settleability

A. *DISCUSSION*

This test is run on mixed liquor or return sludge and the results are plotted on a graph (Figure 14.7). All pertinent information is filled in for process control of aerators.

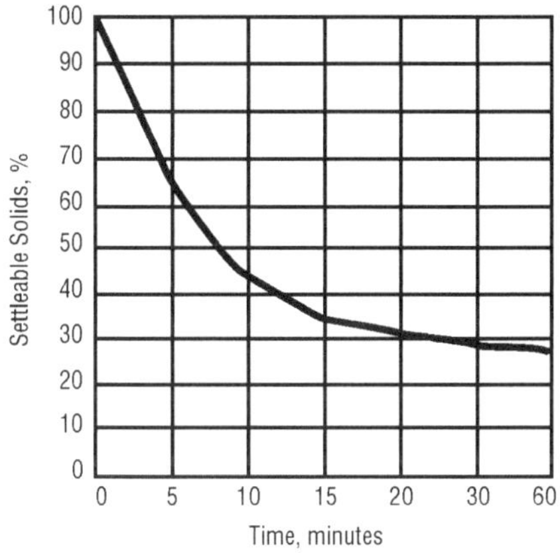

Fig. 14.7 Settleability of activated sludge solids

Settleability is important in determining the ability of the solids to separate from the liquid in the final clarifier. The activated sludge solids should be returned to the aeration tank, and the quality of the effluent is dependent on the absence of solids flowing over the effluent weir.

The suspended solids test should be run on the same sample of mixed liquor that is used for the settleability test. This will allow you to calculate the Sludge Volume Index (SVI) or the Sludge Density Index (SDI), which are explained in the following sections.

The 2,000 m*L* graduate that is filled with mixed liquor in the settleability test is supposed to indicate what will happen to the mixed liquor in the final clarifier—the rate of sludge settling, turbidity, color, and volume of sludge at the end of 60 minutes.

B. *WHAT IS TESTED?*

Sample	Working Range
Mixed Liquor or Return Sludge	Depends on desirable mixed liquor concentration

C. *APPARATUS*

2,000 m*L* graduated cylinder[13]

D. *REAGENTS*

None

E. *PROCEDURE* *(see illustration on page 310)*

1. Collect a sample of mixed liquor or return sludge.
2. Carefully mix sample and pour into 2,000-m*L* graduate. Vigorous shaking or mixing tends to break up floc and produces slower settling or poorer separation.
3. Record settleable solids, %, at regular intervals.

NOTE: 1. If a 1,000-m*L* graduate is used, the percent settleable solids is easier to record.

2. Some plants are using a video camera in the lab to observe and record settleability.

F. *EXAMPLE AND CALCULATION*

The percent settling rate can be compared for the various days of the week and with other measurements—suspended solids, SVI, percent sludge solids returned, aeration rate, and plant inflow. A very slow-settling mixed liquor usually requires air and solids adjustment to encourage increased stabilization during aeration. Both very rapidly settling and very slowly settling mixed liquors can give poor effluent clarification.

14.451 Sludge Volume Index (SVI)

A. *DISCUSSION*

The Sludge Volume Index (SVI) is used to indicate the condition of sludge (aeration solids or suspended solids) for settleability in a secondary or final clarifier. The SVI is the VOLUME in m*L* occupied by one gram of mixed liquor suspended solids after 30 minutes of settling. It is a useful test to indicate changes in sludge characteristics. The proper SVI range for your plant is determined at the time your final effluent is in the *BEST* condition regarding solids and BOD removals and clarity.

B. *WHAT IS TESTED?*

Sample	Preferable Range, SVI
Aeration Tank Suspended Solids	50 m*L*/gm to 150 m*L*/gm

C. *APPARATUS*

See Section 14.45, "Tests for Activated Sludge Control," 14.450, "Settleability," and Section 14.43, "Suspended Solids."

D. *REAGENTS*

None

[13] *Mallory Direct Reading Settlometer (a 2-liter graduated cylinder approximately 5 inches (12.5 cm) in diameter and 7 inches (17.5 cm) high). Obtain from Customer Service, Wilmad Glass/Lab Glass, PO Box 610, Vineland, NJ 08360. Catalog No. LG5601-100. Price, $143.43 each.*

(Settleability—SDI)

E. *PROCEDURE*

See Sections on "Settleability" and "Suspended Solids."

F. *EXAMPLE*

30-minute Settleable Solids Test = 360 m*L* in 2,000 m*L* graduate or 18%

Mixed Liquor Suspended Solids = 1,500 mg/*L*

G. *CALCULATIONS*

$$\text{Sludge Volume Index, SVI, m}L\text{/gm} = \frac{\text{\% Settleable Solids x 1,000 mg/gm x 1,000 m}L/L}{\text{Mixed Liquor Suspended Solids, mg/}L \text{ x 100\%}}$$

$$= \frac{\text{\% Settleable Solids x 10,000}}{\text{Mixed Liquor Suspended Solids, mg/}L}$$

$$= \frac{18 \times 10{,}000}{1{,}500}$$

$$= \frac{1{,}800}{15}$$

$$= 120 \text{ m}L\text{/gm}$$

Or

$$= 18\% \times 1{,}000 \text{ mg/gm} \times 1{,}000 \text{ m}L/L \div 1{,}500 \text{ mg/}L \div 100\%$$

$$= 120 \text{ m}L\text{/gm}$$

14.452 Sludge Density Index (SDI)

A. *DISCUSSION*

The Sludge Density Index (SDI) is used in a way similar to the SVI to indicate the settleability of a sludge in a secondary clarifier or effluent. The SDI is the weight in mg of one milliliter of mixed liquor suspended solids after 30 minutes of settling. The calculation of the SDI requires the same information as the SVI test.

$$\text{SDI} = \frac{\text{mg/}L \text{ of Suspended Solids in Mixed Liquor}}{\text{m}L/L \text{ of Settled Mixed Liquor Solids x 10}}$$

Or

SDI = 100/SVI

B. *WHAT IS TESTED?*

Sample	Preferable Range, SDI
Aeration Tank Suspended Solids	0.6 gm/100 m*L* to 2.0 gm/100 m*L*

C. *C THROUGH G OF SECTION 14.451.*

These items are not included because of their similarity to the SVI test.

QUESTIONS

Write your answers in a notebook and then compare your answers with those on page 339.

14.4K Why should you run settleability tests on mixed liquor?

14.4L What is the Sludge Volume Index (SVI)?

14.4M Why is the SVI test run?

14.4N What is the relationship between the Sludge Density Index (SDI) and SVI?

OUTLINE OF PROCEDURE FOR SETTLEABLE SOLIDS

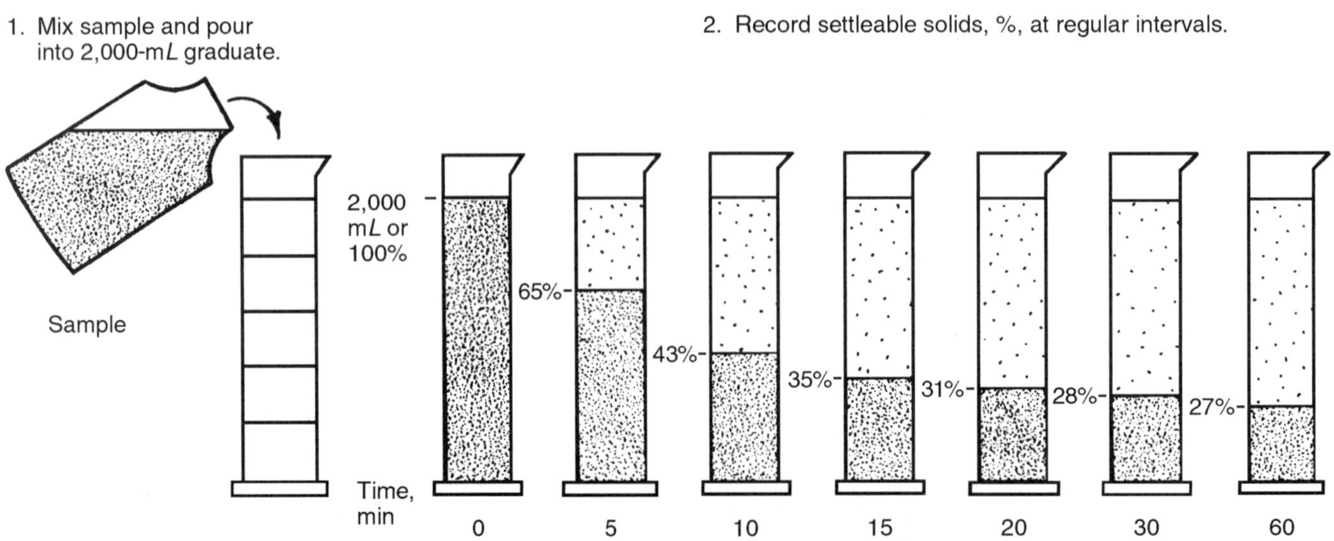

14.453 *Sludge Age*

A. *DISCUSSION*

"Sludge age" is a control guide that is widely used and is a rough indicator of the length of time a pound of solids is maintained under aeration in the system. The basis for calculating the sludge age is weight of suspended solids in the mixed liquor in the aeration tank divided by weight of suspended solids added per day to the aerator.

$$\text{Sludge Age, days} = \frac{\text{Suspended Solids in Mixed Liquor, mg/}L \text{ x Aerator Volume in MG x 8.34 lbs/gal}}{\text{SS in Primary Effluent, mg/}L^{*} \text{ x Daily Flow, MGD x 8.34 lbs/gal}}$$

---

* *NOTE:* Sludge age is calculated by three different methods:

1. Suspended solids in primary effluent, mg/*L*
2. Suspended solids removed from primary effluent, mg/*L*, or primary effluent suspended solids, mg/*L* – final effluent suspended solids, mg/*L*
3. BOD or COD in primary effluent, mg/*L*

Any significant additional loading placed on the aerator by the digester supernatant liquor must be added to the above loadings by considering the additional flow (MGD) and concentration (mg/*L*) if the supernatant is returned following the primary clarifiers. If supernatant is returned ahead of the primary clarifiers, only additional flow must be considered because the solids concentration will be represented in the primary effluent analysis. The selection of the method of determining sludge age is discussed in Chapter 10, "Activated Sludge." Since this method is based on suspended solids, it should not be used if the soluble organic (BOD) portion of the wastewater is more than 50 percent of the total organic (BOD) component.

B. *WHAT IS TESTED?*

Sample	Common Range
Suspended Solids in Aerator and Suspended Solids in Primary Effluent	mg/*L*, Depends on process
Sludge Age	Conventional process, 2.5 to 6 days

C. *APPARATUS*

See Section 14.43, "Suspended Solids."

D. *REAGENTS*

None

E. *PROCEDURE*

See Section 14.43, "Suspended Solids."

F. *EXAMPLE*

Suspended Solids in Mixed Liquor	=	1,500 mg/*L*
Aeration Tank Volume	=	0.50 MG
Suspended Solids in Primary Effl	=	100 mg/*L*
Daily Flow	=	2.0 MGD

G. *CALCULATIONS*

$$\text{Sludge Age, days} = \frac{\text{Susp Solids in Mixed Liquor, mg/}L \text{ x Aerator Vol, MG x 8.34 lbs/gal}}{\text{Susp Solids in Primary Effl, mg/}L \text{ x Flow, MGD x 8.34 lbs/gal}}$$

$$= \frac{\text{Mixed Liquor Susp Solids, lbs}}{\text{Primary Effluent SS, lbs/day}}$$

$$= \frac{1{,}500 \text{ mg/}L \text{ x 0.50 MG x 8.34 lbs/gal}}{100 \text{ mg/}L \text{ x 2.0 MGD x 8.34 lbs/gal}}$$

$$= \frac{1{,}500 \times 0.50}{100 \times 2.0}$$

$$= \frac{7.5}{2.0}$$

$$= 3.75 \text{ days}$$

Or

$$= 1{,}500 \text{ mg/}L \text{ x 0.50 MG x 8.34 lbs/gal} \div 100 \text{ mg/}L \div 2.0 \text{ MGD} \div 8.34 \text{ lbs/gal}$$

$$= 3.75 \text{ days}$$

QUESTION

Write your answer in a notebook and then compare your answer with the one on page 339.

14.4O Determine the sludge age in an activated sludge process if the volume of the aeration tank is 200,000 gallons and the suspended solids in the mixed liquor equals 2,000 mg/*L*. The primary effluent SS are 115 mg/*L*, and the average daily flow is 1.8 MGD.

14.454 *Dissolved Oxygen in Aerator*

A. *DISCUSSION*

This modification is used for biological flocs that have high oxygen utilization rates in the activated sludge process, and when a DO probe is not available. It is very important that some oxygen be present in aeration tanks at all times to maintain aerobic conditions.

This test is similar to the regular DO test (Section 14.5, Procedure 2) except that copper sulfate is added to kill oxygen-consuming organisms, and sulfamic acid is added to combat nitrite before the regular DO test is run.

(DO in Aerator)

NOTE: *IF THE RESULTS INDICATE A DO OF LESS THAN 1 MG/L, IT IS POSSIBLE THAT THE DO IN THE AERATION TANK IS ZERO!* When the DO in the aeration tank is near zero, considerable DO from the surrounding atmosphere can mix with the sample when it is collected, when the inhibitor is added, while the solids are settling, and when the sample is transferred to a BOD bottle for the DO test. If you use this test, use a deep container and avoid stirring.

B. *WHAT IS TESTED?*

Sample	Common DO Range, mg/*L*
Aerator Mixed Liquor	0.1 to 3.0

C. *APPARATUS*

1. One tall bottle, approximately 1,000 m*L*
2. Regular DO apparatus

D. *REAGENTS*

1. Copper sulfate-sulfamic acid inhibitor solution. Dissolve 32 gm technical grade sulfamic acid (NH_2SO_2OH) without heat in 475 m*L* distilled water. Dissolve 50 gm copper sulfate, $CuSO_4 \cdot 5H_2O$, in 500 m*L* water. Mix the two solutions together and add 25 m*L* concentrated acetic acid.
2. Regular DO reagents (See Section 14.5, Procedure 2).

E. *PROCEDURE*

OUTLINE OF PROCEDURE FOR DISSOLVED OXYGEN

1. Add 10 m*L* of inhibitor to DO sampling bottle (Figure 14.4, page 293).
2. Dip into mixed liquor and let sampling bottle fill. Stopper bottle.
3. Remove glass tube and stopper and allow floc to settle.
4. Siphon over 300 m*L* of sample into BOD bottle.

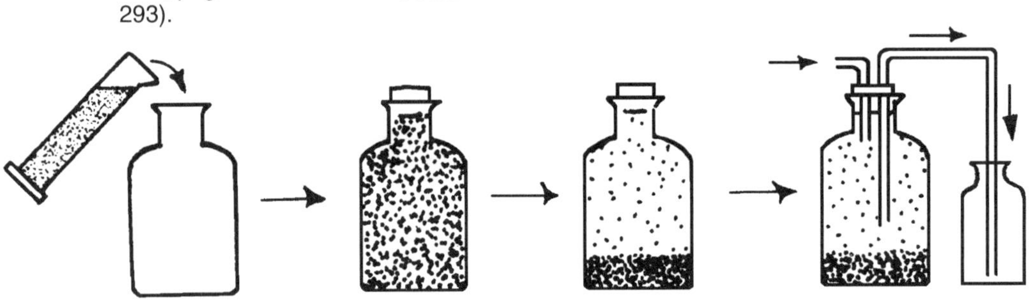

5. Test for DO.

1. Add at least 10 m*L* of inhibitor (5 m*L* copper sulfate and 5 m*L* sulfamic acid) to any *TALL* bottle (1-quart or 1-liter bottle) with an approximate volume of 1,000 m*L*. Place filling tube near the bottom. An emptying tube is placed approximately 1/4 inch (6 mm) from the top of the bottle cork. Attach bottle to rod or aluminum conduit and lower into aeration tank. See Figure 14.4 on page 293.
2. Submerge bottle 1.5 to 2.0 feet (0.45 to 0.60 m) below the surface of the aerator and allow bottle to fill with mixed liquor. Remove bottle from aeration tank.
3. Remove glass tube and stopper from bottle. Insert lid in bottle. Allow bottle to stand until solids (floc) settle and leave a clear supernatant liquor. Siphon the supernatant liquor into a 300-m*L* BOD bottle. Keep the outlet of the siphon below the water surface. Do not aerate in transfer.
4. Perform regular DO test as outlined on page 322.

F. and G. *EXAMPLE AND CALCULATIONS*

Same as regular DO test (see Procedure 2, page 322).

QUESTION

Write your answer in a notebook and then compare your answer with the one on page 339.

14.4P What are the limitations of the copper sulfate-sulfamic acid procedure for measuring DO in the aeration tank when the DO in the tank is very low?

14.455 Suspended Solids in Aerator

Centrifuge Method

A. *DISCUSSION*

This procedure is frequently used in plants as a quick and easy method to estimate the suspended solids concentration of the mixed liquor in the aeration tank instead of the regular suspended solids test. Many operators control the solids in their aerator on the basis of centrifuge readings. Others prefer to control solids using Figure 14.8 to estimate the suspended solids. In either case, the operator should periodically com-

(Suspended Solids—Centrifuge)

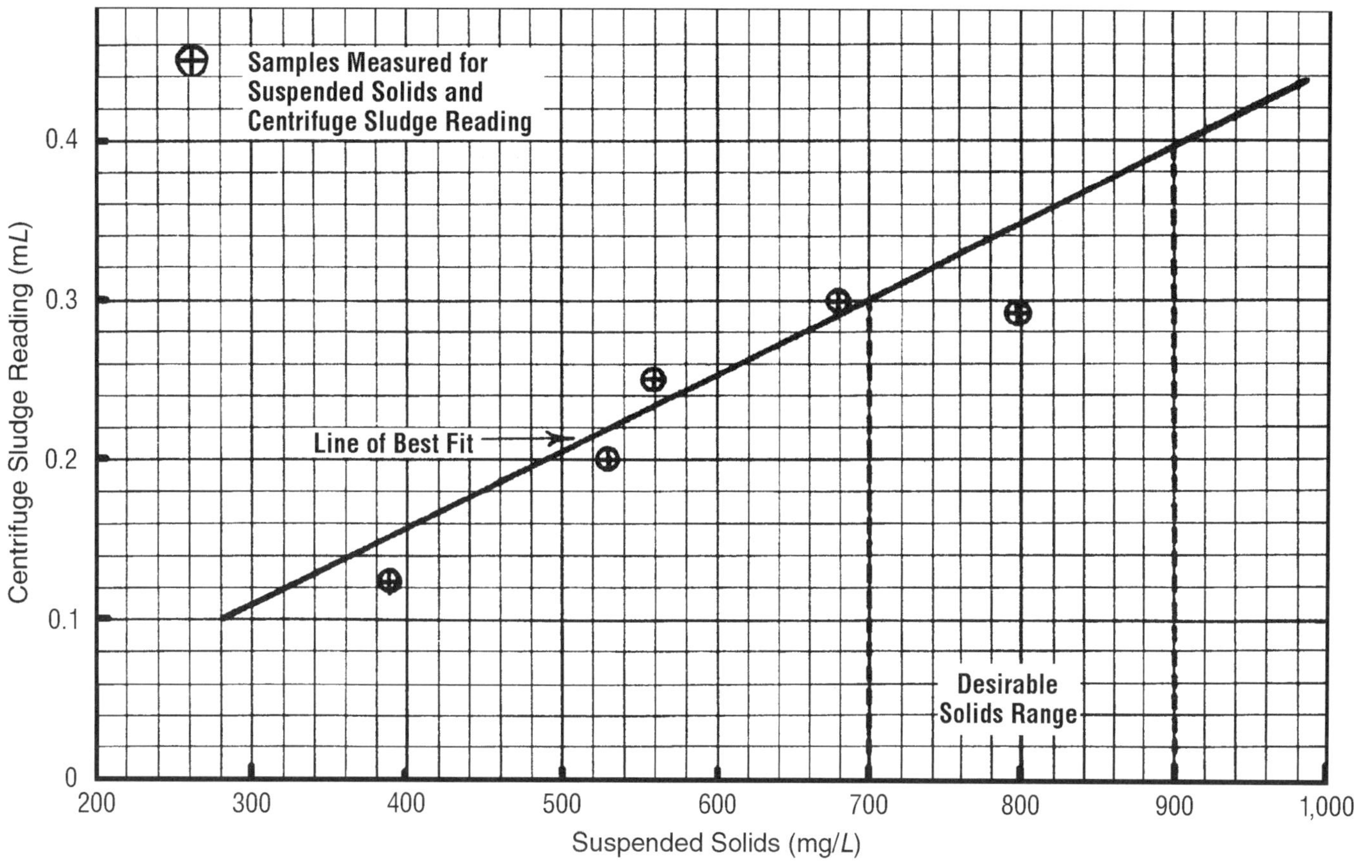

Fig. 14.8 Plant control by centrifuge solids in aeration tank, centrifuge speed, 1,750 RPM

pare centrifuge readings with values obtained from suspended solids tests. If the solids are in a good settling condition, a one percent centrifuge solids reading could have a suspended solids concentration of 1,000 mg/*L*. However, if the sludge is feathery, a one percent centrifuge solids reading could have a suspended solids concentration of 600 mg/*L*.

The centrifuge reading versus mg/*L* suspended solids chart (Figure 14.8) must be developed for each plant by comparing centrifuge readings with suspended solids determined by the regular Gooch crucible method. The points are plotted and a line of best fit is drawn as shown in Figure 14.8. This line must be periodically checked by comparing centrifuge readings with regular suspended solids tests because of the large number of variables influencing the relationship, such as characteristics of influent waste, mixing in aerator, and organisms in aerator. If you don't have a centrifuge or if your solids content is over 1,500 mg/*L*, determine suspended solids by the regular method.

B. *WHAT IS TESTED?*

Sample	Common Range
Suspended Solids in Aeration Tanks	800 to 5,000 mg/*L*

C. *APPARATUS*

1. Centrifuge
2. Graduated centrifuge tubes, 15 m*L*

D. *REAGENTS*

None

E. *PROCEDURE*

1. Collect sample.
2. Mix sample well and fill each centrifuge tube to the 15-m*L* line with sample.
3. Place filled sample tubes in centrifuge holders.
4. Crank centrifuge at fast speed as you count slowly to 60. Be sure to count and crank at the same speed for all tests. It is extremely important to perform each step exactly the same every time.
5. Remove one tube and read the amount (in tenths of a milliliter) of suspended solids concentrated in the bottom of the tube. Results in other tubes should be compared.
6. Refer to the conversion curve to determine suspended solids in mg/*L*.

NOTE: The reason for filling tubes to the 15-m*L* mark is that the curve (Figure 14.8) is computed for samples of this size. This curve was developed for a specific mechanically aerated activated sludge plant. Seven hundred to nine hundred mg/*L* MLSS is the best range for this plant. Each plant must develop its own curve based on actual data.

(Mean Cell Residence Time (MCRT))

F. *EXAMPLE*

Suspended solids concentration on bottom of centrifuge tube is 0.4 m*L*.

G. *CALCULATIONS*

From Figure 14.8, find 0.4 m*L* on the centrifuge reading side (vertical axis). Follow the horizontal line across the chart from 0.4 m*L* until it intersects the line of best fit. Drop a line downward to the horizontal axis and read the result—900 mg/*L* suspended solids.

If the suspended solids concentration is above or below the desired range, then you should make the proper changes in the pumping rate of the waste and return sludge. For details on controlling the solids concentration, refer to Chapter 10, "Activated Sludge."

H. *DEVELOPMENT OF FIGURE 14.8*

To develop Figure 14.8, take a sample from the aeration tank and measure suspended solids and also centrifuge a portion of the sample to obtain the centrifuge sludge reading in m*L* of sludge at the bottom of the tube. Obtain other samples of different solids concentrations to obtain the points on the graph. Draw a line of best fit through the points. Periodically, the points should be checked because the influent characteristics and conditions in the aeration tank change.

I. *PRECAUTIONS*

This test works best for low mixed liquor suspended solids (MLSS) concentrations below 1,500 mg/*L*. Above 1,500 mg/*L*, the centrifuge results might not allow an accurate estimate of the MLSS.

QUESTION

Write your answer in a notebook and then compare your answer with the one on page 339.

14.4Q What is the advantage of the centrifuge test for determining suspended solids in an aeration tank in comparison with other methods of measuring suspended solids?

14.456 Mean Cell Residence Time (MCRT)

A. *DISCUSSION*

The mean cell residence time (MCRT) is in a sense a much more precise "sludge age" calculation. MCRT describes the mean (average) residence time of an activated sludge particle in the activated sludge system and is a true measure of the age of the activated sludge. MCRT considers the solids removed from the process by wasting and the solids removed in the effluent. The MCRT calculation also considers the solids in the system. To use MCRT for the control of an activated sludge process, several measurements are required. Representative composite samples of mixed liquor suspended solids, effluent suspended solids, and waste sludge suspended solids are essential. Also, measure influent and waste sludge flows.

B. *WHAT IS TESTED?*

Sample	Common Range, days
Mean Cell Residence Time (MCRT)	5 to 15 days

C. *APPARATUS AND REAGENTS*

See Section 14.43 for volatile suspended solids test procedures.

D. *PROCEDURE*

To calculate MCRT the following data are required:

1. Influent flow, MGD
2. Waste sludge flow, MGD
3. Volume of aeration basins in million gallons
4. Mixed liquor volatile suspended solids concentration in mg/*L*
5. Effluent volatile suspended solids in mg/*L*
6. Waste sludge volatile suspended solids in mg/*L*

E. *EXAMPLE*

Influent Flow, Q = 3 MGD

Waste Sludge Flow, W = 0.040 MGD

Volume of Aeration Basins, V = 1.0 MG

Mixed Liquor Suspended Solids Concentration, X_1 = 1,600 mg/*L*

Effluent Suspended Solids, X_2 = 8 mg/*L*

Waste Sludge Suspended Solids, X_w = 4,700 mg/*L*

F. *CALCULATIONS*

$$\text{Mean Cell Residence Time, days} = \frac{(V, MG)(X_1, mg/L)}{(Q, MGD)(X_2, mg/L) + (W, MGD)(X_w, mg/L)}$$

$$= \frac{(1.0\ MG)(1{,}600\ mg/L)}{(3\ MGD)(8\ mg/L) + (0.040\ MGD)(4{,}700\ mg/L)}$$

$$= \frac{1{,}600}{24 + 188}$$

$$= 7.5 \text{ days}$$

NOTE: The mixed liquor, effluent and waste sludge *VOLATILE* suspended solids may be used instead of suspended solids to calculate the MCRT. See Chapter 11, "Activated Sludge," Section 11.552, "Mean Cell Residence Time (MCRT)," *OPERATION OF WASTEWATER TREATMENT PLANTS*, Volume II, for a discussion of ways to calculate MCRT.

QUESTION

Write your answer in a notebook and then compare your answer with the one on page 339.

14.4R What measurements are required in order to calculate the Mean Cell Residence Time (MCRT)?

DISCUSSION AND REVIEW QUESTIONS

Chapter 14. LABORATORY PROCEDURES

(Lesson 5 of 7 Lessons)

Write the answers to these questions in your notebook before continuing. The question numbering continues from Lesson 4.

18. Calculate the SVI if the mixed liquor suspended solids are 2,000 mg/*L* and the 30-minute settleable solids test is 500 m*L* in 2 liters or 25 percent.

19. Calculate the SDI if the SVI is 125.

20. What does sludge age measure?

21. What is the difference between sludge age and mean cell residence time (MCRT)?

CHAPTER 14. LABORATORY PROCEDURES

(Lesson 6 of 7 Lessons)

14.5 LABORATORY PROCEDURES FOR NPDES MONITORING

14.50 Need for Approved Procedures

Tests in this section are designed for the treatment plant operator or laboratory analyst who is required to monitor effluent discharges under a National Pollutant Discharge Elimination System (NPDES) permit. Principal tests include Dissolved Oxygen (DO), Biochemical Oxygen Demand (BOD), pH, Residual Chlorine, and Suspended Solids. These and other tests included in this section are listed below. (Test procedures for ammonia nitrogen are described in *OPERATION OF WASTEWATER TREATMENT PLANTS*, Volume II, Chapter 16, "Laboratory Procedures and Chemistry.")

Test

Biochemical Oxygen Demand (BOD) (See Dissolved Oxygen)

1. Chlorine, Total Residual
2. Dissolved Oxygen (DO) and Biochemical Oxygen Demand (BOD)
3. Hydrogen Ion (pH)

   pH (See Hydrogen Ion)
4. Solids, Total, Dissolved, and Suspended
5. Temperature
6. Turbidity

Remember that monitoring data required by an NPDES permit *MUST* be obtained by using APPROVED test procedures. A list of approved test procedures can be found in *40 CFR 136*, July 1, 2002, "Guidelines Establishing Test Procedures for the Analysis of Pollutants." U.S. Government Printing Office, Superintendent of Documents, PO Box 371954, Pittsburgh, PA 15250-7954.

14.51 Test Procedures

1. Chlorine Residual (Total)

A. *DISCUSSION*

The many uses of chlorination in wastewater treatment include disinfecting, reducing BOD, controlling odor, improving scum and grease removal, controlling activated sludge bulking, controlling foam, and aiding in chemical coagulation. The most important use of chlorine in the treatment of wastewater is for disinfection. The amount of residual chlorine remaining in the treated wastewater after passing through a contact basin or chamber may be related to the numbers of coliform bacteria allowed in the effluent by regulatory agencies. In many cases, the residual chlorine must be reduced (for most practical purposes removed) before final discharge to a receiving stream for protection of fish and other aquatic life.

Chlorine reacts quickly (within minutes) and completely with ammonia in wastewater to produce monochloramine and dichloramine. Residual chlorine in the monochloramine or dichloramine state is termed "combined residual chlorine." With the amount of ammonia usually found in wastewater, the chlorine residual will contain all combined chlorine and no free chlorine. Because chlorine residual in wastewater is in a combined state, the determination of residual chlorine presents special problems.

B. *WHAT IS TESTED?*

Sample	Common Range, mg/*L* (After 30 minutes)
Effluent	0.5 to 2.0 mg/*L*

C. *METHODS*

The Iodometric Method for measuring residual chlorine is used for samples containing wastewater, such as plant effluents or receiving waters. The DPD Titrimetric Method is applicable to wastewaters that do not contain iodine-reducing substances. This method also has the advantage that it can be modified to determine free residual chlorine, monochloramine, and dichloramine. Colorimetric tests for residual chlorine have special limitations and should generally be avoided in wastewater. The *AMPEROMETRIC*[14] Titration Method gives the best results, but the titrator instrument is expensive. The Amperometric Titration Method can detect and measure chlorine residual below 0.2 mg/*L* but it is not possible to distinguish between free and combined chlorine forms at this low level. However, such differentiation may not be necessary for wastewater.

D. *APPARATUS REQUIRED*

Iodometric Method

1. Graduated cylinder, 250 m*L*
2. Pipets, 5 and 10 m*L*
3. Erlenmeyer flask, 500 m*L*
4. Buret, 5 m*L*
5. Magnetic stirrer

DPD Titrimetric Method

1. Graduated cylinder, 100 m*L*
2. Pipets, 1 and 10 m*L*

[14] *Amperometric (am-PURR-o-MET-rick). A method of measurement that records electric current flowing or generated, rather than recording voltage. Amperometric titration is a means of measuring concentrations of certain substances in water.*

(Chlorine Residual)

3. Erlenmeyer flask, 250 m*L*
4. Buret, 10 m*L*
5. Magnetic stirrer

Amperometric Titration Method (All Residual Levels)

1. Amperometric titrator
2. End point detection apparatus
3. Agitator
4. Buret

See page 4-58, *STANDARD METHODS*, 20th Edition, and amperometric titrator manufacturer's instruction manual.

E. *REAGENTS*

NOTE: Standardized solutions may be purchased from chemical suppliers.

F. *VOLUME OF SAMPLE*

For residual chlorine concentrations of 10 mg/*L* or less, take a 200-m*L* sample for titration. For residuals greater than 10 mg/*L*, use proportionately less sample.

G. *PROCEDURE*

Iodometric Method

1. Pipet 5.00 m*L* 0.00564 *N* PAO solution into an Erlenmeyer flask.
2. Add excess KI (approximately 1 gm).
3. Add 4 m*L* acetate buffer solution, or enough to lower the pH to between 3.5 and 4.2.
4. Pour in 200 m*L* of sample.
5. Mix with magnetic stirrer.
6. Add 1 m*L* starch solution just before titration.
7. Titrate with 0.0282 *N* Iodine to the first appearance of blue color that remains after complete mixing.

DPD Titrimetric Method

1. Place 5 m*L* each of buffer reagent and DPD indicator in a 250-m*L* flask and mix.
2. Add 2 drops (0.1 m*L*) KI solution and mix.
3. Add 100 m*L* of sample and mix.
4. Let stand for 2 minutes and then titrate with FAS titrant.

Amperometric Titration Method

Free chlorine

1. Unless sample pH is known to be between 6.5 and 7.5, add 1 m*L* pH 7 phosphate buffer solution to produce a pH of 6.5 to 7.5.
2. Titrate with standard phenylarsine oxide (PAO), observing current changes on microammeter. Add titrant (PAO) in smaller amounts until all needle movement stops. Record buret readings when needle action becomes sluggish, signaling approach of end point. Subtract last small increment of titrant that causes no needle movement because of overtitration.

Combined chlorine

3. To sample remaining from free-chlorine titration, add 1.00 m*L* KI solution and 1 m*L* acetate buffer solution, in that order.
4. Titrate with phenylarsine oxide (PAO) titrant to the end point, as in Step 2, above. Do not refill buret but simply continue titration after recording buret reading for free chlorine. Again, subtract last small increment of titrant to give amount of titrant actually used in reaction with chlorine. If titrant was continued without refilling buret, this buret reading represents total chlorine. Subtracting free chlorine from total gives combined chlorine.
5. Wash apparatus and sample cell thoroughly to remove iodide ion to avoid inaccuracies when the titrator is used subsequently for a free chlorine determination.

Monochloramine

6. After titrating for free chlorine, add 0.2 m*L* KI solution to same sample and, without refilling buret, continue to titrate with phenylarsine oxide (PAO) to end point. Subtract last increment of PAO to obtain net volume of PAO consumed by monochloramine.

Dichloramine

7. Add 1 m*L* acetate buffer solution and 1 m*L* KI solution to same sample and titrate final dichloramine fraction by titrating with PAO to end point. Subtract last increment of PAO to obtain net volume of PAO consumed by dichloramine.

Amperometric Titration Method (Low Residual Levels)

1. Select a sample volume requiring no more than 2 m*L* PAO titrant. A 200-m*L* sample will be adequate for sample containing less than 0.2 mg total chlorine/*L* (see Section I, Note 11).
2. Add 200 m*L* sample to sample container and add approximately 1.5 gm KI crystal. Dissolve using a stirrer or mixer.
3. Add 1.0 m*L* acetate buffer and place container in end point detection apparatus.
4. When the current (amperage) signal stabilizes, record the reading. Initially, adjust meter to a near full-scale deflection.
5. Titrate by adding small, known volume of titrant. After each addition, record cumulative volume added and current reading when the signal stabilizes (see Section I, Note 12).
6. Continue adding titrant until no further meter deflection occurs.

(Chlorine Residual)

OUTLINE OF PROCEDURE FOR CHLORINE RESIDUAL
IODOMETRIC METHOD

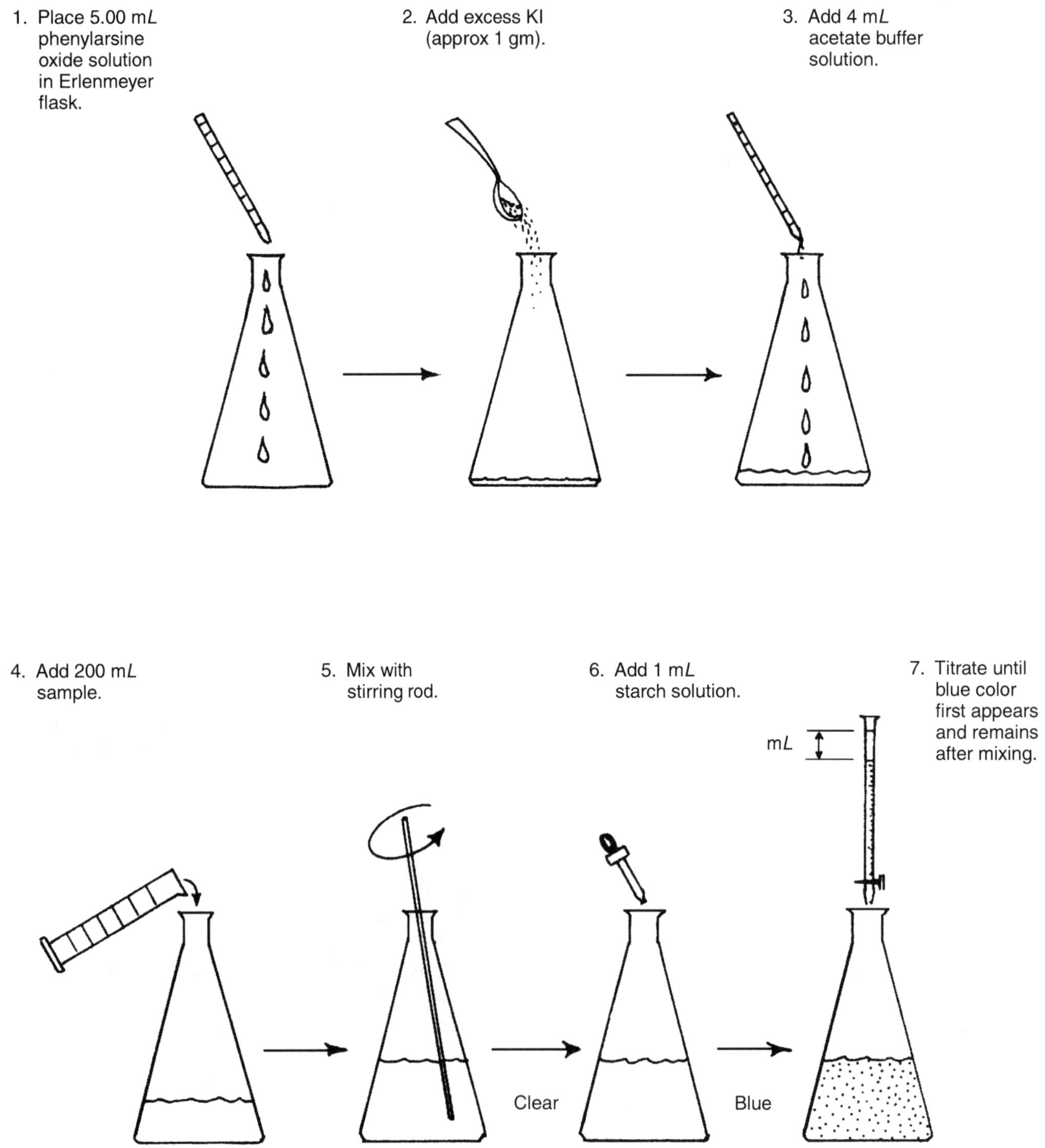

7. Determine equivalence point by plotting total meter deflection against titrant volume added. Draw straight line through the first several points in the plot and a second, horizontal line corresponding to the final total deflection in the meter (Figure 14.9).
8. Read equivalence point as the volume of titrant added at the intersection of these two lines.

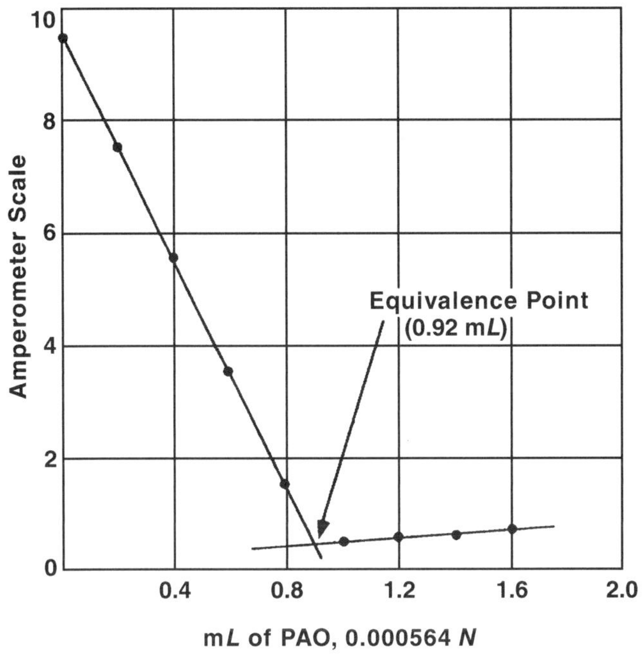

Fig. 14.9 Determination of low-level total chlorine
(Source: Leonard Ashack)

H. *CALCULATIONS*

Iodometric Method

$$\text{Total Residual Chlorine, mg}/L = \frac{(\text{A} - 5\text{B}) \times 200}{\text{C}}$$

A = m*L* 0.00564 *N* PAO

B = m*L* 0.0282 *N* I_2

C = m*L* of Sample Used

EXAMPLE:

Titration of a 200-m*L* sample requires 0.3 m*L* I_2 solution and 5 m*L* of PAO.

$$\text{Total Residual Chlorine, mg}/L = \frac{(\text{A} - 5\text{B}) \times 200}{\text{C}}$$

$$= \frac{[5 - (5 \times 0.3)](200)}{200 \text{ m}L}$$

$$= (5 - 1.5)$$

$$= 3.5 \text{ mg}/L$$

DPD Titrimetric Method

For a 100-m*L* sample, 1.00 m*L* standard FAS titrant = 1.0 mg/*L* residual chlorine.

EXAMPLE:

100 m*L* of sample required 3.4 m*L* standard FAS titrant, therefore

Total Residual Chlorine, mg/*L* = 3.4 mg/*L*

Amperometric Titration Method

$$\text{Chlorine Residual, mg}/L = \frac{\text{A} \times 200}{\text{m}L \text{ Sample}}$$

Where:

A = m*L* phenylarsine oxide (PAO) titration

EXAMPLE:

100 m*L* of sample required 2.4 m*L* PAO for titration of free and combined chlorine.

$$\text{Total Residual Chlorine, mg}/L = \frac{\text{A} \times 200}{\text{m}L \text{ Sample}}$$

$$= \frac{2.4 \times 200}{100 \text{ m}L}$$

$$= 4.8 \text{ mg}/L$$

Amperometric Titration Method (Low Residual Levels)

$$\text{mg Cl as } Cl_2/L = \frac{\text{A} \times 200 \times N}{\text{B} \times 0.000564\ N}$$

where: A = m*L* Titrant at Equivalence Point

B = Sample Volume, m*L*

N = Phenylarsine (PAO) Normality

EXAMPLE:

The results of PAO titration on a 200-m*L* sample of chlorinated and dechlorinated effluent sample are as follows:

Volume of PAO, m*L*	**Current Reading on Amperometer**
0	9.5
0.2	7.5
0.4	5.6
0.6	3.6
0.8	1.6
1.0	0.5
1.2	0.6
1.4	0.6
1.6	0.7

(*DO* and BOD)

From Figure 14.9, the equivalence point was determined at 0.92 m*L*.

$$\text{mg}/L \text{ of Cl} = \frac{0.92 \times 200 \times 0.000564}{200 \times 0.000564}$$

$$= 0.92 \text{ mg}/L$$

Or

$$= 0.92 \times 200 \times 0.000564 \div 200 \div 0.000564$$

$$= 0.92$$

I. *NOTES*

1. All chlorine residuals are reported as milligrams of chlorine as Cl_2 per liter (mg Cl as Cl_2/L).
2. Some organic chloramines can interfere with the test. Monochloramine can intrude into the free chlorine fraction and dichloramine can interfere in the monochloramine fraction, especially at high temperatures and prolonged titration times.
3. Free halogens (iodine, bromine) other than chlorine also will titrate as free chlorine.
4. Combined chlorine reacts with iodide ions to produce iodine.
5. When titration for free chlorine follows a combined chlorine titration that requires addition of KI, erroneous results may occur unless the measuring cell is rinsed thoroughly with distilled water between titrations.
6. Interference from copper has been observed in samples taken from copper pipe or after heavy copper sulfate treatment of reservoirs.
7. Silver ions poison the electrode.
8. Interference occurs in some highly colored waters and in waters containing surface-active agents.
9. Violent stirring can lower chlorine values by volatilization.
10. Very low temperatures slow the response of the measuring cell and longer time is required for the titration, but precision is not affected.
11. It is possible to use the same method noted above and measure effluent chlorine concentration greater than 0.2 mg/*L* by using higher strength of PAO. According to the formula in Section H above, varying the concentration of PAO would allow flexibility as shown below.

Strength of PAO	Dilution from 0.00564 *N*	Chlorine Conc, in mg/*L* per m*L* Titrant	Maximum* Concentration Measured
0.000564	1 m*L* into 10 m*L*	0.1 mg/*L*	0.2 mg/*L*
0.001128	1 m*L* into 5 m*L*	0.2 mg/*L*	0.4 mg/*L*
0.00282	1 m*L* into 2 m*L*	1.0 mg/*L*	1.0 mg/*L*
0.00564	No dilution	1.0 mg/*L*	2.0 mg/*L*

* Assumed the maximum titrant volume is 2.0 m*L*.

12. By keeping the tip of titration needle touching the surface of sample being titrated, drop formation can be avoided. This will hasten the meter response.
13. Operators using the DPD colorimetric method to test water for a free chlorine residual need to be aware of a potential error that may occur. If the DPD test is run on water containing a combined chlorine residual, a precipitate may form during the test. The particles of precipitated material will give the sample a turbid appearance or the appearance of having color. This turbidity can produce a positive test result for free chlorine residual when there is actually no chlorine present. Operators call this error a "false positive" chlorine residual reading.
14. Home swimming pool chlorine residual analyzers are not approved, but could be used in an emergency for a short period of time. The results are not accurate.

J. *REFERENCES*

Iodometric Method: See page 4-55, *STANDARD METHODS*, 20th Edition.

DPD Titrimetric Method: See page 4-61, *STANDARD METHODS*, 20th Edition.

Amperometric Titration Method: See page 4-58, *STANDARD METHODS*, 20th Edition.

Low-Level Amperometric Titration Method: See page 4-60, *STANDARD METHODS*, 20th Edition.

QUESTIONS

Write your answers in a notebook and then compare your answers with those on page 339.

14.5A Why should plant effluents be chlorinated?

14.5B Discuss the important advantages and disadvantages between the iodometric titration and amperometric titration methods of measuring chlorine residual.

2. Dissolved Oxygen (DO) and Biochemical Oxygen Demand (BOD)

A. *DISCUSSION*

The dissolved oxygen (DO) test is, as the name implies, the testing procedure to determine the amount of oxygen dissolved in samples of water or wastewater. The analysis for DO is a key test in water pollution control activities and waste treatment process control. There are various types of tests that can be run to obtain the amount of dissolved oxygen. This procedure is the Sodium Azide Modification of the Winkler Method and is best suited for relatively clean waters. Interfering substances include color, organics, suspended solids, sulfide, chlorine, and ferrous and ferric iron. Nitrite will not interfere with the test if fresh azide is used.

The generalized principle is that iodine will be released in proportion to the amount of dissolved oxygen present in the sample. By using sodium thiosulfate with starch as the indicator, one can titrate the sample and determine the amount of dissolved oxygen.

B. *WHAT IS TESTED?*

Sample	Common Range, mg/*L*
Influent	Usually 0, >1 is very good.
Primary Clarifier Effluent	Usually 0, recirculated from filters >2 is good.
Secondary Effluent	50% to 95% Saturation, 3 to >8 is good.
Oxidation Ponds	1 to 25+*
Activated Sludge—Aeration Tank Outlet	>2 desirable

> means greater than
* supersaturated with oxygen

C. *APPARATUS*

METHOD A (Sodium Azide Modification of Winkler Method)

1. Buret, graduated to 0.1 m*L*
2. Three 300-m*L* glass-stoppered BOD bottles
3. Wide-mouth Erlenmeyer flask, 500 m*L*
4. One 10-m*L* measuring pipet
5. One 1-liter reagent bottle to collect activated sludge

METHOD B (DO Probe)

Follow manufacturer's instructions. See Section H, "DO Probe," on page 322.

D. *REAGENTS*

Standardized solutions may be purchased from chemical suppliers.

E. *PROCEDURE (page 322)*

SODIUM AZIDE MODIFICATION OF THE WINKLER METHOD

NOTE: The sodium azide destroys nitrate, which will interfere with this test.

The reagents are to be added in the quantities, order, and methods as follows:

1. Collect a sample to be tested in 300-m*L* (BOD) bottle taking special care to avoid aeration of the liquid being collected. Fill bottle completely and add cap.
2. Remove cap and add 1 m*L* of manganous sulfate solution below surface of the liquid.
3. Add 1 m*L* of alkaline-iodide-sodium azide solution below the surface of the liquid.
4. Replace the stopper, avoid trapping air bubbles, and mix well by inverting the bottle several times. Repeat this mixing after the floc has settled halfway. Allow the floc to settle halfway a second time.
5. Acidify with 1 m*L* of concentrated sulfuric acid by allowing the acid to run down the neck of the bottle above the surface of the liquid.
6. Restopper and mix well until the precipitate has dissolved. The solution will then be ready to titrate. Handle the bottle carefully to avoid acid burns.
7. Pour 200 m*L* of original sample from bottle into an Erlenmeyer flask.
8. If the solution is brown in color, titrate with 0.025 *N* PAO until the solution is pale yellow color. Add a small quantity of starch indicator and proceed to Step 10. (Note: Either PAO or 0.025 *N* sodium thiosulfate can be used.)
9. If the solution has no brown color, or is only slightly colored, add a small quantity of starch indicator. If no blue color develops, there is zero Dissolved Oxygen. If a blue color does develop, proceed to Step 10.

(*DO* and BOD)

10. Titrate to the first disappearance of the blue color. Record the number of m*L* of PAO used. If the blue color returns, this indicates that nitrate ions are present.
11. The amount of oxygen dissolved in the original solution will be equal to the total number of m*L* of PAO used in the titration provided significant interfering substances are not present.

    mg/*L* DO = m*L* PAO

F. *EXAMPLE*

The DO titration of a 200-m*L* sample requires 5.0 m*L* of 0.025 *N* PAO. Therefore, the dissolved oxygen concentration in the sample is 5 mg/*L*.

G. *CALCULATION*

You will want to find the percent saturation of DO in the effluent of your secondary plant. The DO is 5.0 mg/*L* and the temperature is 20°C. At 20°C, 100 percent DO saturation is 9.2 mg/*L*.

The dissolved oxygen saturation values are given in Table 14.5. Note that as the temperature of water increases, the DO saturation value (100% Saturation column) decreases. Table 14.5 gives 100 percent DO saturation values for temperatures in °C and °F.

TABLE 14.5 EFFECT OF TEMPERATURE ON OXYGEN SATURATION FOR A CHLORIDE CONCENTRATION OF ZERO mg/*L*

°C	°F	mg/*L* DO at 100% Saturation
0	32.0	14.6
1	33.8	14.2
2	35.6	13.8
3	37.4	13.5
4	39.2	13.1
5	41.0	12.8
6	42.8	12.5
7	44.6	12.2
8	46.4	11.9
9	48.2	11.6
10	50.0	11.3
11	51.8	11.1
12	53.6	10.8
13	55.4	10.6
14	57.2	10.4
15	60.0	10.2
16	61.8	10.0
17	63.6	9.7
18	65.4	9.5
19	67.2	9.4
20	68.0	9.2
21	69.8	9.0
22	71.6	8.8
23	73.4	8.7
24	75.2	8.5
25	77.0	8.4

(*DO* and BOD)

OUTLINE OF PROCEDURE FOR PREPARING REAGENTS

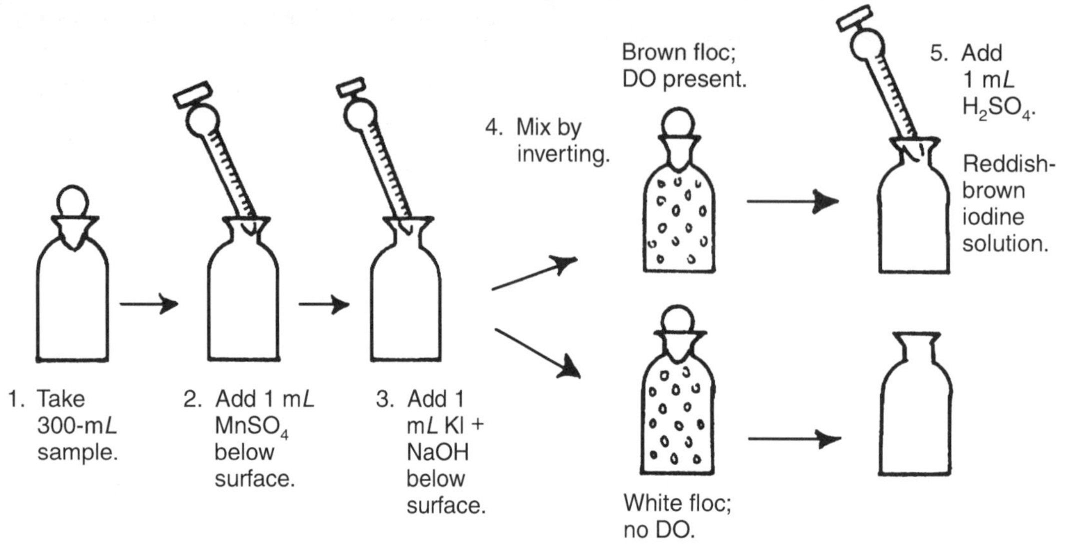

Titration of Iodine Solution:

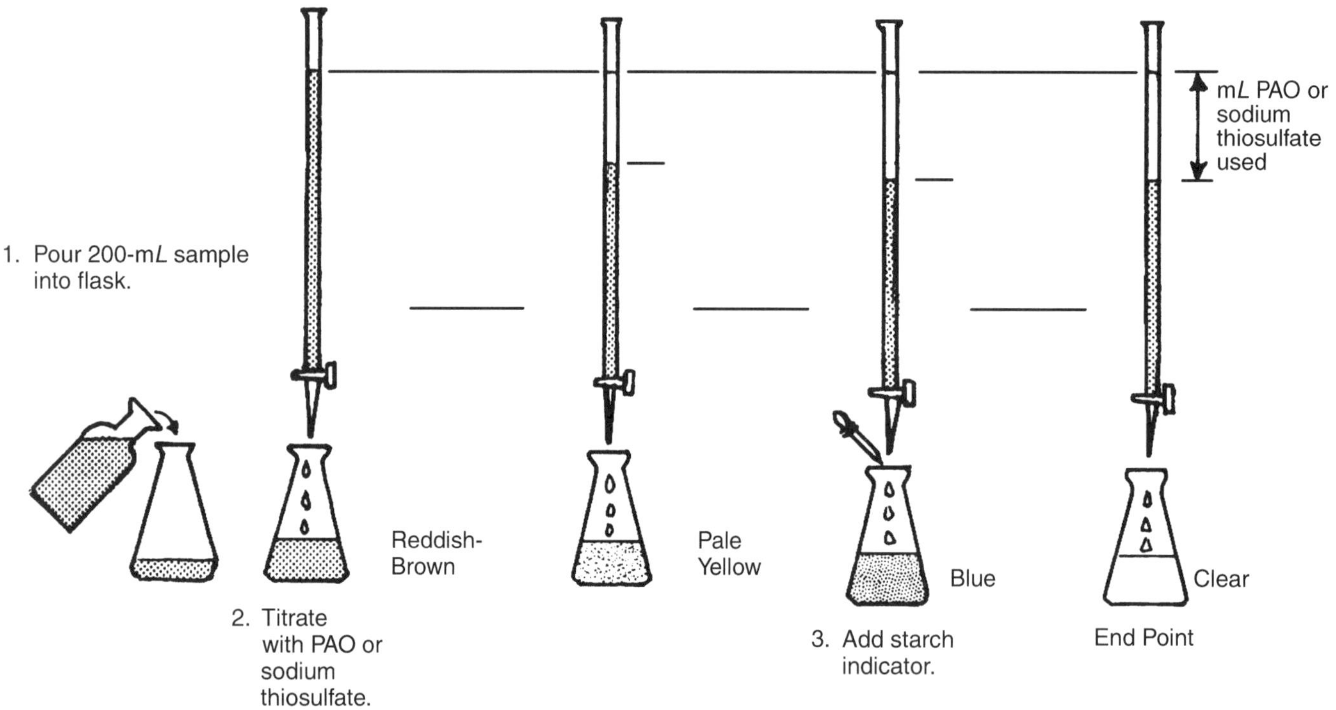

$$\text{DO Saturation, \%} = \frac{\text{DO of Sample, mg}/L \times 100\%}{\text{DO at 100\% Saturation, mg}/L}$$

$$= \frac{5.0 \text{ mg}/L}{9.2 \text{ mg}/L} \times 100\%$$

$$= .54 \times 100\%$$

$$= 54\%$$

H. *DO PROBE*

1. *DISCUSSION*

Measurement of the dissolved oxygen (DO) concentration with a probe and electronic readout meter is a satisfactory substitute for the Sodium Azide Modification of the Winkler Method under many circumstances. The probe is recommended when samples contain substances that interfere with the Modified Winkler procedure, such as sulfite, thiosulfate, polythionate, mercaptans, free chlorine or hypochlorite, organic substances readily hydrolyzed in alkaline solutions, free iodine, intense color or turbidity, and biological flocs. A continuous record of the dissolved oxygen content of aeration tanks and receiving waters may be obtained using a probe. In determining the BOD of samples, a probe may be used to determine the DO initially and after the five-day incubation period of the blanks and sample dilutions.

2. *PROCEDURE*

Follow manufacturer's instructions.

3. *CALIBRATION*

To ensure that the DO probe reading provides the dissolved oxygen content of the sample, the probe must be calibrated. Take a sample that does not contain substances that interfere with either the probe reading or the Modified Winkler procedure. Split the sample. Measure the DO in one portion of the sample using the Modified Winkler procedure and compare this result with the DO probe reading on the other portion of the sample. Adjust the probe reading to agree with the results from the Modified Winkler procedure.

When calibrating the probe in an aeration tank of the activated sludge process, do not attempt to measure the dissolved oxygen in the aerator and then adjust the probe. The biological flocs in the aerator will interfere with the Modified Winkler procedure, and the copper sulfate-sulfamic acid procedure is not sufficiently accurate to calibrate the probe. An aeration tank probe may be calibrated by splitting an effluent sample, measuring the DO by the Modified Winkler procedure, and comparing results with the probe readings. Always keep the membrane in the tip of the probe from drying because the probe can lose its accuracy until reconditioned.

4. *PRECAUTIONS*

a. Periodically check the calibration of the probe.

b. Keep the membrane in the tip of the probe from drying out.

c. Dissolved inorganic salts, such as found in seawater, can influence the readings from a probe.

d. Reactive compounds, such as reactive gases and sulfur compounds, can interfere with the output of a probe.

e. Don't place the probe directly over a diffuser because you want to measure the dissolved oxygen in the water being treated, not the oxygen in the air supply to the aerator.

QUESTIONS

Write your answers in a notebook and then compare your answers with those on page 340.

14.5C Calculate the percent dissolved oxygen saturation if the receiving water DO is 7.9 mg/*L* and the temperature is 10°C.

14.5D How would you calibrate the DO probe in an aeration tank?

BIOCHEMICAL OXYGEN DEMAND OR BOD

A. *DISCUSSION*

The BOD test measures the amount of oxygen used by microorganisms as they use the substrate (food) in wastewater when placed in a controlled temperature for five days. The DO (dissolved oxygen) is measured at the beginning and recorded. During the five-day period, microorganisms in the sample break down complex organic matter in the sample, using up oxygen in the process. After the five-day dark incubation period, the DO is again determined. The BOD is then calculated on the basis of the reduction of DO and the size of sample. This test is an estimate of the availability of food in the sample (food or organisms that take up oxygen) expressed in terms of oxygen use. Results of a BOD test indicate the rate of oxidation and provide an indirect estimate of the availability to organisms or concentration of the waste.

Actual environmental conditions of temperature, organism population, water movement, sunlight, and oxygen concentration cannot be accurately reproduced in the laboratory. Results obtained from this test must take into account these factors when relating BOD results to receiving water oxygen demands.

Samples for the BOD test should be collected before chlorination because chlorine interferes with the organisms in the test. It is difficult to obtain accurate results with dechlorinated samples.

Samples are incubated for a standard period of five days because a fraction of the total BOD will be exerted during this period. The ultimate or total BOD is normally never run for plant control. A limitation of the BOD test is that the results are not available until five days after the sample was collected.

B. *WHAT IS TESTED?*

Sample	Common Range, mg/*L*
Influent	150 to 400
Primary Effluent	60 to 160
Secondary Effluent	5 to 30
Digester Supernatant	1,000 to 4,000+
Industrial Wastes	100 to 3,000+

C. *APPARATUS*

1. 300 m*L* BOD bottles with ground glass stoppers
2. Incubator, 20°C ± 1°C
3. Pipets, 10 m*L* graduated, 1/32- to 1/16-inch (0.8- to 1.6-mm) diameter tip
4. Buret and stand
5. Erlenmeyer flask, 500 m*L*

D. *REAGENTS*

Standardized solutions may be purchased from chemical suppliers.

1. Distilled water. Water used for solutions and for preparation of the solution water must be of the highest quality. It must contain less than 0.1 mg/*L* copper and be free of chlorine, chloramines, caustic, organic material, and acids. Ordinary distilled water for your car's battery is not good enough.
2. Dilution water. Add 1 m*L* each of phosphate buffer, magnesium sulfate, calcium chloride, and ferric chloride solutions for each liter of distilled water. Saturate with DO by shaking in a partially filled bottle or by aerating with filtered air. Store at a temperature as close to 20°C as possible for at least 24 hours to allow the water to become stabilized. This water should not show a drop in DO of more than 0.2 mg/*L* on incubation for five days.

Many plants do not prepare reagents. Small plants and plants that do not run many tests find it quicker and easier to purchase commercially prepared reagents. These reagents may be available in the desired strength or they may consist of dry *PILLOWS*[15] that are added to the sample, rather than the

[15] *Pillows. Plastic tubes shaped like pillows that contain exact amounts of chemicals or reagents. Cut open the pillow, pour the reagents into the sample being tested, mix thoroughly and follow test procedures.*

(DO and *BOD*)

liquid reagent. Check with your chemical supplier for these reagents.

E. *PROCEDURE FOR UNCHLORINATED SAMPLES*

The test is made by measuring the oxygen used or depleted during a five-day period at 20°C by a measured quantity of wastewater sample seeded into a reservoir of dilution water saturated with oxygen. This is compared to an unseeded or blank reservoir of dilution water by subtracting the difference and multiplying by a factor for dilution.

PROCEDURE

1. BOD bottles should be of 300-m*L* capacity with graduations and ground-glass stoppers. To clean the bottles, carefully rinse with tap water followed by distilled water. Each BOD bottle and fitted stopper are numbered. Always insert each numbered stopper in the BOD bottle with the same number. If the numbers do not match, errors could occur in the test results.
2. Fill two bottles completely with dilution water and insert the stopper tightly so that no air is trapped beneath the stopper. Siphon dilution water from its container when filling BOD bottles.
3. Set up one or more dilutions of the sample to cover the estimated range of BOD values. From the estimated BOD, calculate the volume of raw sample to be added to the BOD bottle based on the fact that:

   The most valid DO depletion is 4 mg/*L*. Therefore,

$$\text{m}L \text{ of Sample Added per 300 m}L = \frac{(4 \text{ mg}/L)(300 \text{ m}L)}{\text{Estimated BOD, mg}/L} = \frac{1{,}200}{\text{Estimated BOD, mg}/L}$$

EXAMPLES:

a. Estimated BOD = 400 mg/*L*

$$\text{m}L \text{ of Sample Added to BOD Bottle} = \frac{1{,}200}{400} = 3 \text{ m}L$$

b. Estimated BOD = 200 mg/*L*: use 6 m*L*

100 mg/*L*: use 12 m*L*

20 mg/*L*: use 60 m*L*

When the BOD is unknown, select more than one sample size. For example, place several samples—1 m*L*, 3 m*L*, 6 m*L*, and 12 m*L*—into four BOD bottles.

For samples with very high BOD values, it may be difficult to accurately measure small volumes or to get a truly representative sample. In such a case, initial dilution should first be made on the sample. A dilution of 1:10 is convenient.

4. To perform the BOD test, first fill two BOD bottles with BOD dilution water (blanks). Nos. (1) and (2) in illustration, page 325.
5. Next, for each sample to be tested, carefully measure out the two portions of sample and place them into two new BOD bottles, Nos. (3) and (4). Add dilution water until the bottles are completely filled. Insert the stoppers. Avoid entrapping air bubbles. Be sure that there are water seals on the stoppers. A water seal is obtained by water in the small indentation between the stopper and the rim of the bottle.
6. On bottles (2) and (4), immediately determine the initial dissolved oxygen.
7. Incubate the remaining dilution water blank and diluted sample in the dark at 20°C for five days. These are bottles (1) and (3).
8. At the end of exactly five days (±3 hours), test bottles (1) and (3) for their *DISSOLVED OXYGEN* by using the sodium azide modification of the Winkler method or a DO probe. At the end of five days, the oxygen content should be at least 1 mg/*L*. Also, a depletion of 2 mg/*L* or more is desirable. Bottles (1) and (2) (blanks) are only used to check the dilution water quality. Their difference should be less than 0.2 mg/*L* after five days if the quality is good and free of impurities. The difference in blank readings is not used as a blank correction, but merely as a check on the quality of dilution water. Differences of greater than 0.2 mg/*L* could possibly be due to contamination and/or dirty BOD bottles.

F. *PRECAUTIONS*

1. The temperature of the incubator must be at 20°C. Other temperatures will change the rate of oxygen used.
2. The dilution water must be made according to *STANDARD METHODS* (as outlined on page 323, Section D, 2) for the most favorable growth rate of the bacteria. This water must be free of copper, which is often present when copper stills are used by commercial dealers. Use all glass or stainless-steel stills or demineralized water.
3. The wastewater must also be free of toxic wastes, such as hexavalent chromium.
4. If you use a cleaning solution to wash BOD bottles, be sure to rinse the bottles several times. Cleaning agents are toxic and if any residue remains in a BOD bottle, a BOD test could be ruined.
5. Unchlorinated wastewater normally contains an ample supply of seed bacteria; therefore seeding is usually not necessary.
6. Since this is a bioassay (BUY-o-ass-SAY), that is, living organisms are used for the test, environmental conditions must be quite exact.

G. *EXAMPLE*

BOD Bottle Volume	= 300 m*L*
Sample Volume	= 15 m*L*
Initial DO of Diluted Sample	= 8.0 mg/*L*
DO of Sample and Dilution After 5-Day Incubation	= 4.0 mg/*L*

H. *CALCULATIONS*

$$\text{BOD, mg}/L = \left[\begin{array}{l}\text{Initial DO of}\\ \text{Diluted Sam-}\\ \text{ple, mg}/L\end{array} - \begin{array}{l}\text{DO of Diluted}\\ \text{Sample After}\\ \text{5-Day Incuba-}\\ \text{tion, mg}/L\end{array}\right]\left[\frac{\text{BOD Bottle Vol, m}L}{\text{Sample Volume, m}L}\right]$$

$$= (8.0 \text{ mg}/L - 4.0 \text{ mg}/L)\left[\frac{300 \text{ m}L}{15 \text{ m}L}\right]$$

$$= \frac{(4.0)(300)}{15}$$

$$= 80 \text{ mg}/L$$

(DO and *BOD*)

OUTLINE OF PROCEDURE FOR BOD

(Unchlorinated Samples)

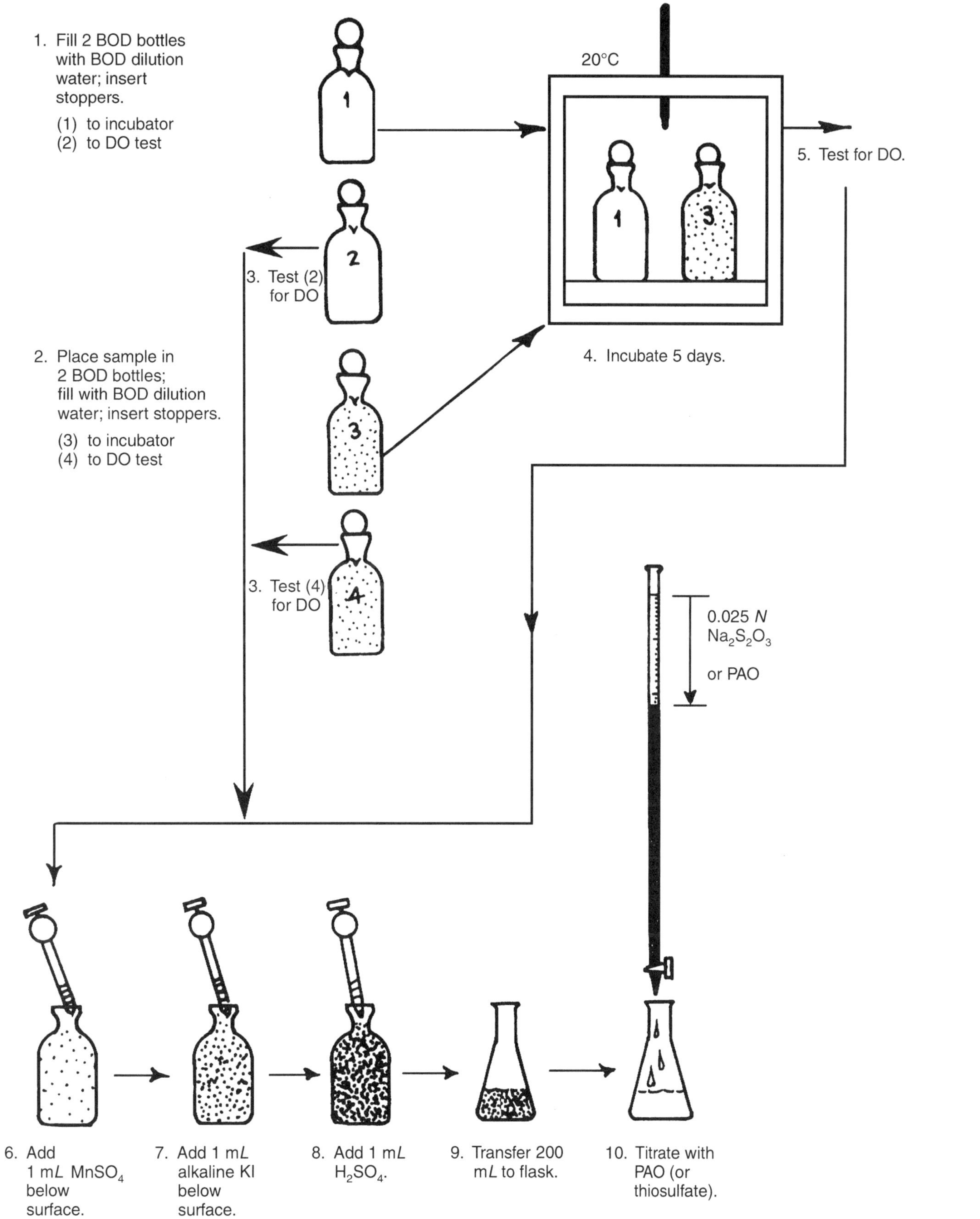

(DO and *BOD*)

For acceptable results, the percent depletion of oxygen in the BOD test should range from 30 percent to 80 percent depletion. Also, if a residual DO of at least 1 mg/*L* remains or a DO uptake of at least 2 mg/*L* results after five days of incubation, the results also are considered reliable.

$$\% \text{ Depletion} = \frac{(\text{DO of Diluted Sample, mg}/L - \text{DO After 5 Days, mg}/L)}{\text{DO of Diluted Sample, mg}/L} \times 100\%$$

$$= \frac{(8.0 \text{ mg}/L - 4.0 \text{ mg}/L)}{8.0 \text{ mg}/L} \times 100\%$$

$$= \frac{4}{8} \times 100\%$$

$$= 50\%$$

I. *PROCEDURE FOR CHLORINATED SAMPLES*

Dechlorination of Samples

Whenever chlorinated wastewater samples are collected for BOD analysis, sufficient dechlorinating agent must be added to destroy the residual chlorine. The following procedure should be followed:

1. Using the procedure given for residual chlorine analysis on page 316, test the sample for chlorine.
2. If there is no residual chlorine, proceed with test procedures below. If residual chlorine is present, add sufficient 0.025 *N* sodium sulfite until residual chlorine is absent. If a chlorine residual is present, the BOD results will be low. If too much sodium sulfite is added, the BOD results will be high. Prepare 0.025 *N* sodium sulfite by dissolving 1.575 grams of anhydrous Na_2SO_3 in 1,000 m*L* of distilled water. This solution is not stable and must be prepared daily.

Seeding of Sample

When a sample contains very few microorganisms as a result, for example, of chlorination or extreme pH, microorganisms must be added to the sample.

1. Collect about one liter of unchlorinated raw wastewater or primary effluent about 24 hours prior to the time you wish to start the BOD test. Let sample settle at 20°C for 24 to 36 hours. Filter through glass wool to remove large particles and grease clumps. Use the filtered sample for "seed."
2. Fill two 300-m*L* BOD bottles with dilution water (see page 327) and insert stopper tightly so no air bubbles are trapped. These are called blanks.
3. Set up one or more dilutions of samples in duplicate as shown on page 327.
4. Add 1 m*L* seed from Step 1 to each BOD bottle containing a dechlorinated sample.
5. Set up blanks of seed material to determine the amount of oxygen depletion that will be due to the added seed material. Use 3, 6, and 9 m*L* of seed and determine the five-day oxygen depletion due to 1 m*L* of seed. Seed material should produce a correction of at least 0.6 mg/*L* per m*L* of seed. Make these samples also in duplicate.
6. Determine initial DO on one set of duplicate bottles.
7. Incubate dilution water blank, diluted samples, and seeded blanks at 20°C for five days.
8. At the end of five days, test bottles for dissolved oxygen.
9. Calculate five-day BOD.

EXAMPLE:

	Bottle 2	**Bottle 4**	**Bottle 6**
Sample	blank	effluent	seed blank
Sample volume	—	10 m*L*	0
Seed volume	—	1 m*L*	3 m*L*
Initial DO	9.2 mg/*L*	9.2 mg/*L*	9.2 mg/*L*
DO after 5 days	9.2 mg/*L*	5.1 mg/*L*	7.1 mg/*L*
Depletion	0.0 mg/*L*	4.1 mg/*L*	2.1 mg/*L*

CALCULATION:

For seed correction:

$$\text{mg}/L \text{ DO Depletion Caused by 1 m}L \text{ of Seed} = \frac{\text{5-day Depletion of Seed Sample}}{\text{m}L \text{ of Seed}}$$

$$= \frac{9.2 \text{ mg}/L - 7.1 \text{ mg}/L}{3 \text{ m}L}$$

$$= 0.7 \text{ mg}/L/\text{m}L$$

5-day BOD:

$$\text{BOD, mg}/L = \frac{[(\text{Initial DO} - \text{DO after 5 days}) - (\text{Seed Correction})] \times 300 \text{ m}L}{\text{m}L \text{ of Sample Volume}}$$

$$= \frac{[(9.2 \text{ mg}/L - 5.1 \text{ mg}/L) - (0.7 \text{ mg}/L/\text{m}L)(1 \text{ m}L)] \times 300 \text{ m}L}{10 \text{ m}L}$$

$$= \frac{(4.1 \text{ mg}/L - 0.7 \text{ mg}/L) 300 \text{ m}L}{10 \text{ m}L}$$

$$= 102 \text{ mg}/L$$

J. *PRECAUTIONS*

1. On effluent samples where the DO is run on the sample and the blue bounces back on the end point titration, this indicates nitrite interference and can cause the BOD test result to be higher than actual by as much as 10 to 15 percent. This fact should be considered in interpreting your results. The end point also may waver because of decomposition of azide in an old reagent or resuspension of sample solids. To correct a wavering end point, try preparing a new alkaline-azide solution or more of the old solution should be used because it may be decomposing.
2. A minimum of two dilutions per sample should be used. Use only analyses with oxygen depletions of greater than 2 mg/*L* and residual DOs of greater than 1 mg/*L* after five days of incubation at 20°C.
3. Samples should be *WELL MIXED* before dilutions are made. A wide-tip pipet should be used for making dilutions. The wide tip will not clog with suspended solids.
4. Wastewaters that have been partially nitrified may produce high BOD results. This increased oxygen demand results from the oxidation of ammonia to nitrate. The use of chemicals such as allythiourea or other commercially available nitrification inhibitors in the dilution water will inhibit the nitrifiers and alleviate this problem.
5. For effluent samples with a low DO and BOD, saturate with DO by shaking in a partially filled bottle.

K. *PROCEDURE FOR INDUSTRIAL WASTE SAMPLES*

Some industrial waste samples may require special seeding because of a low microbial population or because the wastes contain organic compounds that are not readily oxidized by domestic wastewater seed. To obtain the necessary special-

OUTLINE OF PROCEDURE FOR BOD

(Chlorinated Samples)

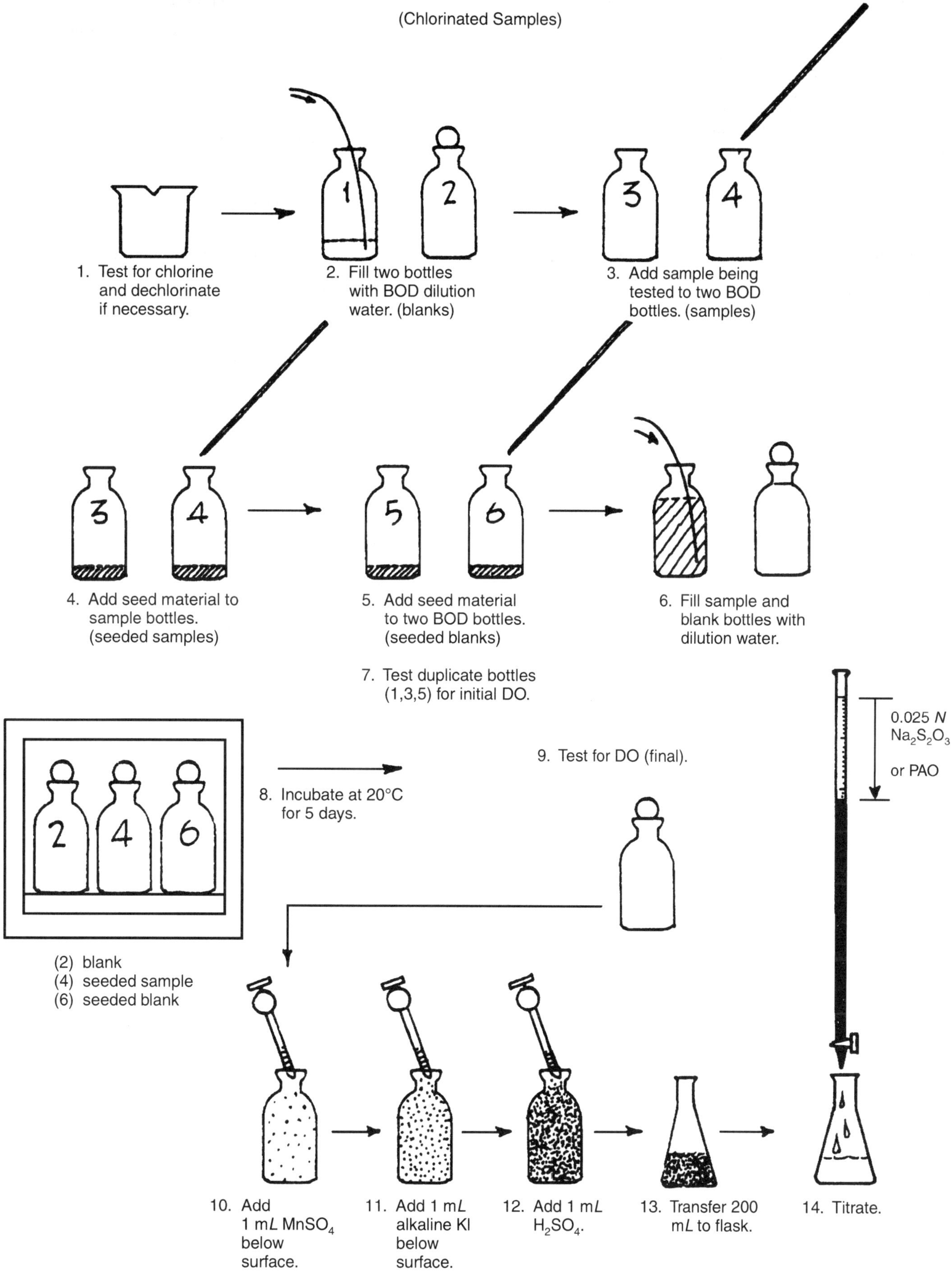

ized seed material (microorganisms) adapted or acclimated to the industrial organic compounds, collect a sample of adapted seed from the effluent of a biological treatment process (activated sludge aeration tank) treating the industrial waste. When this source of adapted seed is not available, develop the adapted seed in the laboratory by continuously aerating a large sample of water and feeding it with small daily portions of the particular waste, together with soil or settled domestic wastewater, until a satisfactory microbial population has developed.

Once a satisfactory adapted seed is available, follow the procedures in the previous Section I, "Procedure for Chlorinated Samples, Seeding of Sample." Start with Step 2, "Fill two 300-m*L* BOD bottles with dilution water," and follow the steps except use the industrial waste sample instead of the dechlorinated sample.

L. *REFERENCE*

See pages 4-129 and 5-2, *STANDARD METHODS*, 20th Edition.

QUESTIONS

Write your answers in a notebook and then compare your answers with those on page 340.

14.5E How would you determine the amount of organic material in wastewater?

14.5F Why should samples for the BOD test be collected before chlorination?

14.5G Why should opened bottles of sodium thiosulfate be used or re-standardized within two weeks?

14.5H How would you prepare dilutions to measure the BOD of cannery waste having an expected BOD of 2,000 mg/*L*?

14.5I What is the BOD of a sample of wastewater if a 2-m*L* sample in a 300-m*L* BOD bottle had an initial DO of 7.5 mg/*L* and a final DO of 3.9 mg/*L*?

END OF LESSON 6 OF 7 LESSONS
ON
LABORATORY PROCEDURES

Please answer the discussion and review questions next.

DISCUSSION AND REVIEW QUESTIONS

Chapter 14. LABORATORY PROCEDURES

(Lesson 6 of 7 Lessons)

Write the answers to these questions in your notebook before continuing. The question numbering continues from Lesson 5.

22. Why is the effluent from a treatment plant chlorinated?
23. What is the formula for calculating the percent saturation of DO?
24. What precautions should be exercised when using a DO probe?
25. What is a limitation of the BOD test?
26. What is a blank, as referred to in laboratory procedures?
27. What precautions should be taken when running a BOD test on an unchlorinated sample?
28. Calculate the BOD of a 5-m*L* sample if the initial DO of the diluted sample was 7.5 mg/*L* and the DO of the diluted sample after five-day incubation was 3.0 mg/*L*.

CHAPTER 14. LABORATORY PROCEDURES

(Lesson 7 of 7 Lessons)

3. Hydrogen Ion (pH)

A. *DISCUSSION*

The intensity of the basic or acidic strength of water is expressed by its pH.

Mathematically, pH is the logarithm of the reciprocal of the hydrogen ion activity, or the negative logarithm of the hydrogen ion activity.

$$pH = Log \frac{1}{[H^+]} = -Log\ [H^+]$$

EXAMPLE

If a wastewater has a pH of 1, then the hydrogen ion activity $[H^+] = 10^{-1} = 0.1$.

If pH = 7, then $[H^+] = 10^{-7} = 0.0000001$.

pH Scale

0	increasing acidity — 7 — increasing alkalinity	14
1 ← 2 ← 3 ← 4 ← 5 ← 6	∧	8 → 9 → 10 → 11 → 12 → 13
	Neutral 6 through 8	

In a solution, both hydrogen ions $[H^+]$ and the hydroxyl ions $[OH^-]$ are always present. At a pH of 7, the activity of both hydrogen and hydroxyl ions equals 10^{-7} moles per liter. When the pH is less than 7, the activity of hydrogen ions is greater than the hydroxyl ions. The hydroxyl ion activity is greater than the hydrogen ions in solutions with a pH greater than 7.

The pH test indicates whether a treatment process may continue to function properly at the pH measured. Each process in the plant has its own favorable range of pH, which must be checked routinely. Generally, a pH value from 6 to 8 is acceptable for best organism activity. Most wastewater contains many dissolved solids and buffers that tend to minimize pH changes.

B. *WHAT IS TESTED?*

Wastewater	Common Range
Influent or Raw Wastewater (domestic)	6.8 to 8.0
Raw Sludge (domestic)	5.6 to 7.0
Digester Recirculated Sludge or Supernatant	6.8 to 7.2
Plant Effluent Depending on Type of Treatment	6.8 to 8.0

C. *MINIMUM APPARATUS LIST*

1. pH meter
2. Glass electrode
3. Reference electrode
4. Magnetic stirrer (optional)
5. Beaker for sample

D. *REAGENTS*

1. Buffer tablets of various pH values
2. Distilled water

E. *PROCEDURE*

1. Due to the differences between the various makes and models of pH meters commercially available, specific instructions cannot be provided for the correct operation of all instruments. In each case, follow the manufacturer's instructions for preparing the electrodes and operating the instrument.
2. Standardize the instrument against a buffer solution with a pH approaching that of the sample.
3. Rinse electrodes thoroughly with distilled water after removal from buffer solution.
4. Place electrodes in sample and measure pH.
5. Remove electrodes from sample, rinse thoroughly with distilled water.
6. Immerse electrode ends in beaker of pH 7 buffer solution.
7. Shut off meter.

F. *PRECAUTIONS*

1. To avoid faulty instrument calibration, prepare fresh buffer solutions as needed, once per week, from commercially available buffer tablets.
2. pH meter, buffer solution, and samples should all be at the same temperature (constant) because temperature variations will give erroneous results. Allow a few minutes for the probe to adjust to the buffers before calibrating a pH meter to ensure accurate pH readings.
3. Watch for erratic results arising from electrodes, faulty connections, or fouling of electrodes with oily or precipitated matter. Films may be removed from electrodes by placing isopropanol on a tissue or Q-tip and cleaning the probe.
4. If you're measuring pH in colored samples or samples with high solids content, or if you're taking measurements that need to be reported to the USEPA, you should use a pH electrode and meter instead of a colorimetric method or test papers. pH meters are capable of providing ± 0.1 pH accuracy in most applications. In contrast, colorimetric tests provide ± 0.1 pH accuracy only in a limited range. pH papers provide even less accuracy.

(Total Solids)

QUESTION

Write your answer in a notebook and then compare your answer with the one on page 340.

14.5J What precautions should be exercised when using a pH meter?

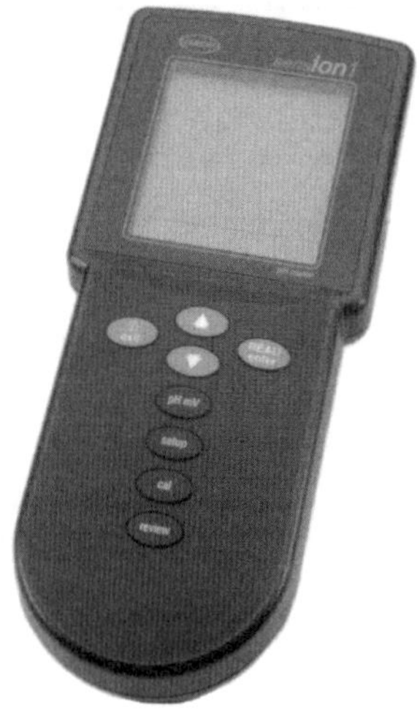

4. Total Solids (Residue)

The term "solids" or "residue" refers to the matter suspended and dissolved in wastewater or water.

I. Total Solids

A. *DISCUSSION*

"Total solids" is the term applied to material left in the container after evaporation of a sample in an oven at 103° to 105°C.

B. *WHAT IS TESTED?*

Sample	Common Range, mg/*L*
Influent and Effluent Sludges	300 to 1,200

C. *APPARATUS REQUIRED*

1. Evaporating dishes (100 m*L* capacity)
2. Drying oven
3. Desiccator
4. Analytical balance
5. Muffle furnace

D. *PROCEDURE*

Preparation of evaporating dish.

1. Ignite a clean, porcelain evaporating dish at 550° ± 50°C for one hour in a muffle furnace.
2. Cool in desiccator, weigh, and record weight. Store in desiccator until ready for use.

NOTE: Aluminum evaporating dishes can be used directly from container.

Sample analysis.

1. Shake sample vigorously and transfer a measured amount of sample to the pre-weighed dish. This can be done by using a graduated cylinder that is graduated to deliver the desired volume. Due to the difficulty in accurately measuring the volume of some sludge samples, an unknown quantity of sludge can be poured into the dish and then weighed to determine the initial sample quantity. This is possible because the specific gravity of sludge is only slightly higher than 1.0. Choose a sample volume that will yield a maximum residue of 25 to 250 milligrams. If necessary, add additional portions of sample to dish.
2. Place the sample in an oven and evaporate to dryness. The oven temperature should be 2°C lower than boiling (that is 98°C) to prevent spattering. Once the water has evaporated, raise the temperature to 103°C and maintain this temperature for at least one hour. Repeat drying cycle until a constant weight is obtained or until weight loss is less than 0.5 milligrams. The exact time is usually determined by experience, but at 103°C some sludge samples will need to be dried overnight.
3. Remove dried sample to desiccator and allow to cool completely. Weigh dish plus total solids and save if volatile portion is to be determined.

E. *PRECAUTIONS*

1. Wastewater and sludge may contain infectious organisms and must be handled with care.
2. Use tongs or gloves when handling hot items.
3. Be sure the balance is properly leveled, zeroed, and maintained.
4. Do not weigh items when they are hot because the heat will cause errors by creating convection currents in the balance.

F. *EXAMPLE*

Results from an effluent sample were:

Weight of Clean Dish	=	80.1526 grams
Weight of Residue and Dish	=	80.1732 grams
Sample Volume	=	100 m*L*

G. *CALCULATIONS*

1. $$\text{Total Solids, mg}/L = \frac{(A - B)(1{,}000\ \text{m}L/L)}{\text{m}L\ \text{of Sample}}$$

where, A = Weight of Dish and Residue in milligrams

B = Weight of Dish in milligrams

2. From example,

$$\text{Total Solids, mg}/L = \frac{(A - B)(1{,}000\ \text{m}L/L)}{\text{m}L\ \text{of Sample}}$$

$$= \frac{(80{,}173.2\ \text{mg} - 80{,}152.6\ \text{mg})(1{,}000\ \text{m}L/L)}{100\ \text{m}L}$$

$$= 206\ \text{mg}/L$$

Or

$$= (80{,}173.2\ \text{mg} - 80{,}152.6\ \text{mg}) \times 1{,}000\ \text{m}L/L \div 100\ \text{m}L$$

$$= 206\ \text{mg}/L$$

H. *REFERENCE*

See page 2-54, *STANDARD METHODS*, 20th Edition.

OUTLINE OF PROCEDURE FOR TOTAL SOLIDS

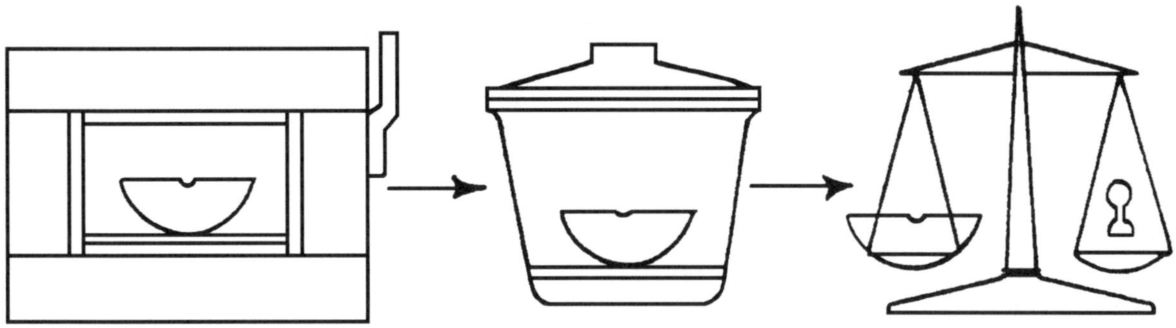

1. Ignite dish at 550°C.
2. Cool in desiccator.
3. Weigh dish.

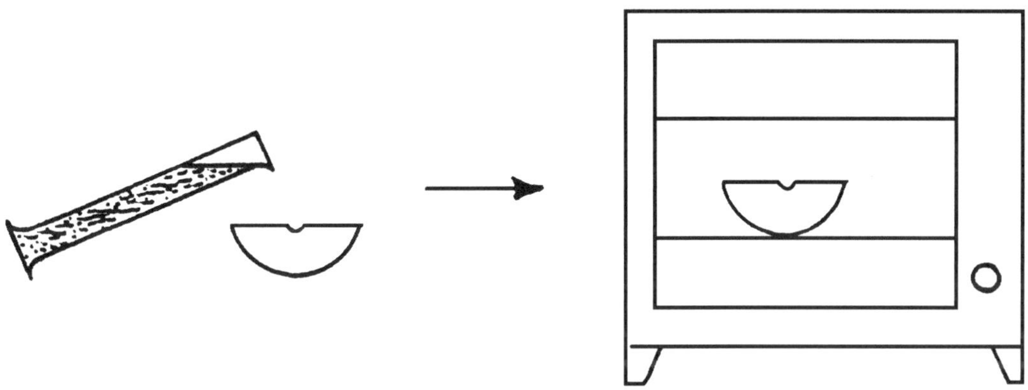

4. Pour measured volume in dish.
5. Dry dish plus sample at 103°C to dryness.

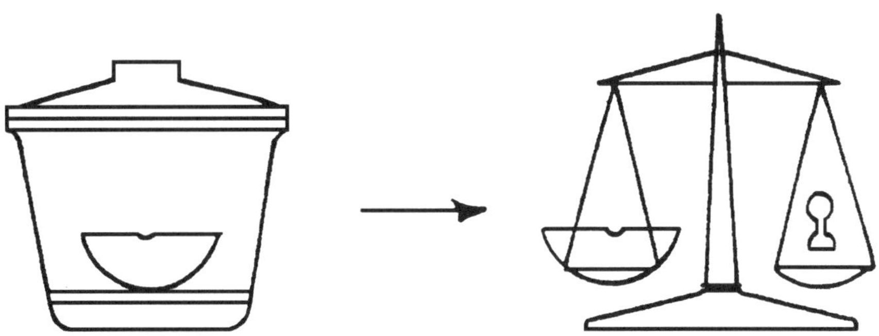

6. Cool in desiccator.
7. Weigh dish plus total solids.

(Total Solids)

II. Total Dissolved (Filterable) Solids

A. *DISCUSSION*

"Total dissolved solids" (TDS) refers to material that passes through a standard glass-fiber filter disc and remains after evaporation at 180°C.

B. *WHAT IS TESTED?*

Sample	Common Range, mg/*L*
Influent and Effluent	150 to 600

C. *APPARATUS REQUIRED*

1. Glass-fiber filter discs (Reeve Angel Type 934A, 984H; or Gelman Type A/E)
2. Flask, suction 500 m*L*
3. Filter holder or Gooch crucible adapter
4. Gooch crucibles (25 m*L* if 2.2 cm filter used)
5. Evaporating dishes, 100 m*L*
6. Drying oven, 180°C
7. Steam bath
8. Vacuum source
9. Desiccator
10. Analytical balance
11. Muffle furnace, 550°C

D. *PROCEDURE*

Preparation of Dish

1. Ignite a clean evaporating dish at 550 ± 50°C for one hour in muffle furnace.
2. Cool in desiccator then weigh and record weight. Store in desiccator until needed.

Preparation of Glass-Fiber Filter Disc

1. Place the disc on the filter apparatus or insert into the bottom of a suitable Gooch crucible. While vacuum is applied, wash the filter disc with three successive 20-m*L* volumes of distilled water. Continue the suction to remove all traces of water from the disc and discard the washings.

Sample Analysis

1. Shake the sample vigorously and transfer 125 to 150 m*L* to the funnel or Gooch crucible by means of a 150-m*L* graduated cylinder.
2. Filter the sample through the glass-fiber filter and continue to apply vacuum for about three minutes after filtration is complete to remove as much water as possible.
3. Transfer 100 m*L* of the filtrate to the weighed evaporating dish and evaporate to dryness on a steam bath.
4. Dry the evaporated sample for at least one hour at 180°C. Cool in desiccator and weigh. Repeat drying cycle until constant weight is obtained or until weight loss is less than 0.5 mg.

E. *EXAMPLE*

Results from weighings were:

Clean Dish	= 47.0028 grams (47,002.8 mg)
Dissolved Residue + Dish	= 47.0453 grams (47,045.3 mg)
Sample Volume	= 100 m*L*

F. *CALCULATIONS*

1. $$\text{Total Dissolved Solids, mg}/L = \frac{(A - B)(1{,}000\ \text{m}L/L)}{\text{m}L\ \text{Sample Volume}}$$

Where:

A = Weight of Dish and Dissolved Material in milligrams (mg)

B = Weight of Clean Dish in milligrams (mg)

2. From example,

$$\text{Total Dissolved Solids, mg}/L = \frac{(A - B)(1{,}000\ \text{m}L/L)}{\text{m}L\ \text{Sample Volume}}$$

$$= \frac{(47{,}045.3\ \text{mg} - 47{,}002.8\ \text{mg})(1{,}000\ \text{m}L/L)}{100\ \text{m}L}$$

$$= 425\ \text{mg}/L$$

Or

$$= (47{,}045.3\ \text{mg} - 47{,}002.8\ \text{mg}) \times 1{,}000\ \text{m}L/L \div 100\ \text{m}L$$

$$= 425\ \text{mg}/L$$

G. *COMMENTS*

Because excessive residue in the evaporating dish may form a water-entrapping crust, use a sample that yields no more than 200 mg of residue.

H. *REFERENCE*

See page 2-55, *STANDARD METHODS*, 20th Edition.

III. Suspended Solids

See Section 14.43 for procedure to measure Suspended Solids.

QUESTION

Write your answer in a notebook and then compare your answer with the one on page 340.

14.5K Determine the total solids (residue) and total dissolved (filterable) solids from the following lab test results.

Weight of Empty Dish	= 64,328.9 mg
Weight of Dish and Residue	= 64,381.2 mg
Weight of Dish and Dissolved Material	= 64,351.2 mg
Sample Volume	= 100 m*L*

(Total Solids)

OUTLINE OF PROCEDURE FOR TOTAL SOLIDS

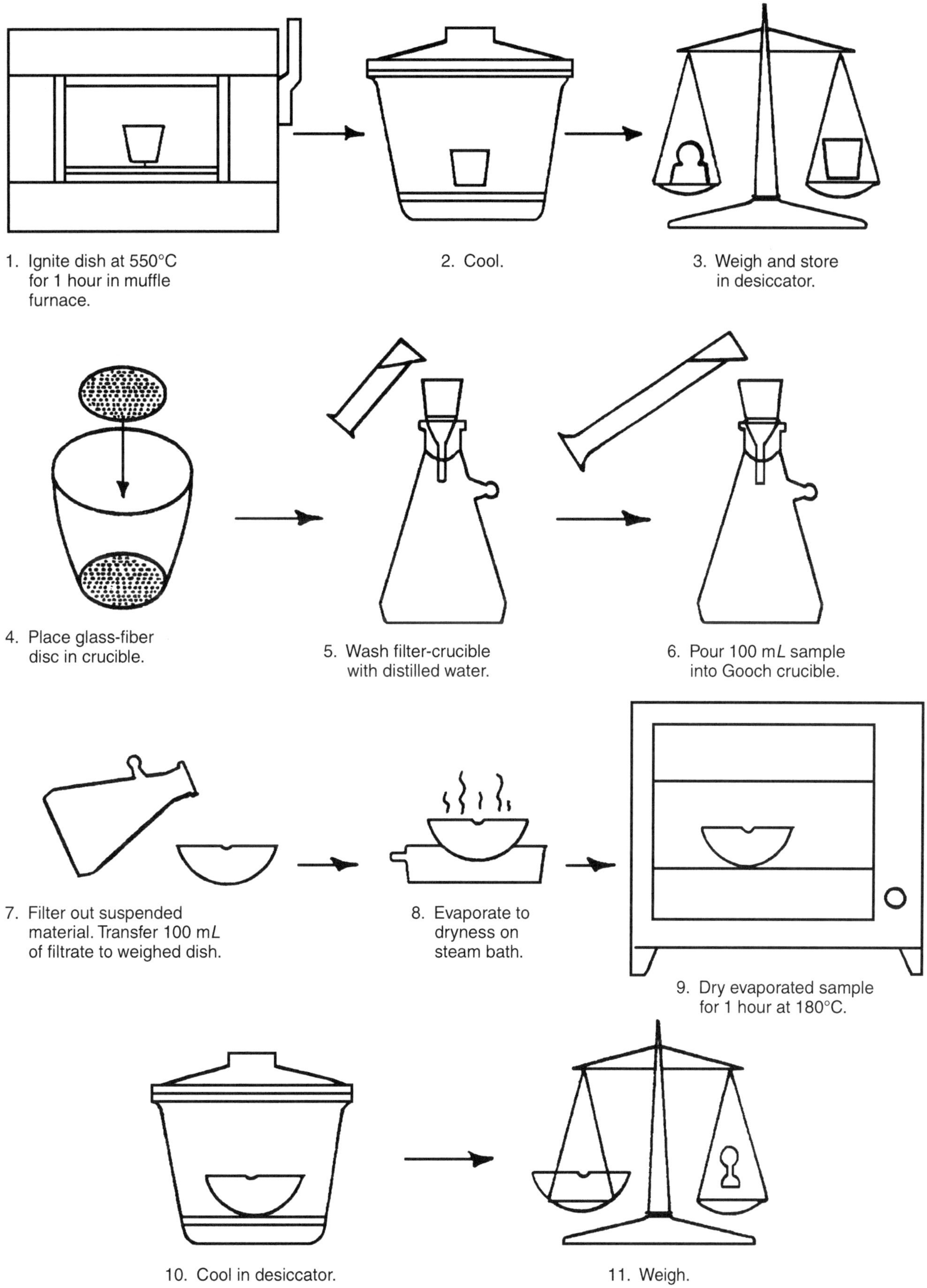

(Temperature)

5. Temperature

A. *DISCUSSION*

Temperature is one of the most important factors affecting biological growth. Temperature measurements can be helpful in detecting changes in raw wastewater quality. For example, an influent temperature drop may indicate large volumes of cold water from infiltration. An increase in temperature may indicate that hot water discharges by industry are reaching your plant.

Temperature is one of the most frequently taken tests. One of the many uses is to calculate the percent saturation of dissolved oxygen in the DO test.

A temperature measurement should be taken where samples are collected for other tests. This test is always immediately performed on a grab sample because temperature changes so rapidly. Always leave the thermometer in the liquid while reading the temperature because the reading will change as soon as you remove the thermometer from the liquid. Record temperature on a suitable work sheet, including time, location, and sampler's name.

Thermometers are calibrated for either total immersion or partial immersion. A thermometer calibrated for total immersion must be completely immersed in the wastewater sample to give a correct reading, while a partial immersion thermometer must be immersed in the sample to the depth of the etched circle around the stem for a correct reading.

B. *WHAT IS TESTED?*

Sample	Common Range
Influent[16]	65°F to 85°F[17] (18°C to 29°C)
Effluent[16]	60°F to 95°F (16°C to 35°C) or higher from ponds
Receiving Water	60°F (16°C) to ambient temperature[18]
Digester (Recirculated Sludge before Heat Exchanger and Supernatant)	60°F to 100°F (16°C to 38°C)

C. *APPARATUS*

1. One NBS (National Bureau of Standards) thermometer for calibration of the other thermometers
2. One Fahrenheit mercury-filled, 1° subdivided thermometer
3. One Celsius (formerly called Centigrade) mercury-filled, 1° subdivided thermometer
4. One metal case to fit each thermometer

There are three types of thermometers and two scales.

SCALES

1. Fahrenheit, marked °F
2. Celsius (formerly Centigrade), marked °C

THERMOMETERS

1. Total immersion. This type of thermometer must be totally immersed when read. This will change most rapidly when removed from the liquid to be recorded.
2. Partial immersion. This type thermometer will have a solid line around the stem below the point where the scale starts.
3. Dial. This type has a dial that can be easily read while the thermometer is still immersed. Dial thermometer readings should be checked (calibrated) against the NBS thermometer. Some dial thermometers can be recalibrated (adjusted) to read the correct temperature of the NBS thermometer.

D. *REAGENTS*

None

E. *PROCEDURES*

Use a large volume of sample, preferably at least a 2-pound coffee can or equivalent volume. The temperature will have less chance to change in a large volume than in a small container. Collect sample in container and immediately measure and record temperature. Do not touch the bottom or sides of the sample container with the thermometer. To avoid breaking or damaging a glass thermometer, store it in a shielded metal case. Check your thermometer accuracy against the NBS certified thermometer by measuring the temperature of a sample with both thermometers simultaneously. Some of the poorer quality thermometers are substantially inaccurate (off as much as 6°F or 3°C).

F. *EXAMPLE*

To measure influent temperature, obtain sample in large coffee can, immediately immerse thermometer in can, and record temperature when reading becomes constant. For example, 72°F or 22°C.

G. *CALCULATIONS*

Normally, we measure and record temperatures using a thermometer with the proper scale. However, we could measure a temperature in °F and convert to °C, or we might measure a temperature in °C and convert to °F. The following formulas are used to convert temperatures from one scale to the other.

1. Measure in °F, want °C: $°C = 5/9(°F - 32°)$
2. Measure in °C, want °F: $°F = 9/5(°C) + 32°$
3. Example Calculation:

The measured influent temperature was 77°F. What was the temperature in °C?

$$
\begin{aligned}
°C &= 5/9(°F - 32°) \\
&= 5/9(77° - 32°) \\
&= 5/9(45°) \\
&= 25°C
\end{aligned}
$$

Or

$$
\begin{aligned}
&= (77° - 32°) \times 5 \div 9 \\
&= 25°C
\end{aligned}
$$

---

[16] *If dissolved oxygen (DO) measurements are performed on any samples, the temperature should be measured and recorded.*

[17] *Depends on season, location, and temperature of water supply.*

[18] *Ambient (AM-bee-ent) Temperature. Temperature of the surroundings.*

(Turbidity)

QUESTIONS

Write your answers in a notebook and then compare your answers with those on page 340.

14.5L What could a change in influent temperature indicate?

14.5M Why should the thermometer remain immersed in the liquid while being read?

14.5N Why should thermometers be calibrated against an accurate National Bureau of Standards (NBS) certified thermometer?

6. Turbidity

A. *DISCUSSION*

The term turbidity is simply an expression of the physical clarity of water or wastewater. Turbidity can be caused by the presence of suspended matter such as mud, finely divided organic and inorganic matter, and microscopic organisms such as algae.

The turbidity measurement is useful for plant effluent monitoring; however problems can be encountered when one instrument's readings are compared with those of another. Commercial turbidimeters come in many shapes and sizes. They each can read different turbidity values on the same sample even though they have been calibrated using the procedure given later in this section. The operator should simply be aware of this shortcoming.

The accepted method used to measure turbidity is called nephelometric. The nephelometric turbidimeter is designed to measure particle-reflected light at an angle of 90 degrees. The greater the intensity of the scattered light, the higher the turbidity.

The turbidity unit (NTU) is an empirical quantity that is based on the amount of light that is scattered by particles of a polymer reference standard called formazin, which produces particles that scatter light in a reproducible manner. Formazin, the primary turbidity standard, is an aqueous suspension of an insoluble polymer formed by the condensation reaction between hydrazine sulfate and hexamethylenetetramine.

Secondary turbidity standards are suspensions of various materials formulated to match the primary formazin solutions. These secondary standards are generally used because of their convenience and the instability of dilute formazin primary standard solutions. Examples of these secondary standards include "standards" that are supplied by the turbidimeter manufacturer with the instrument. Periodic checks of these secondary standards against the primary formazin standard are a must and will provide assurance of measurement accuracy.

B. *WHAT IS TESTED?*

Sample	Common Range, NTU
Effluent	10 to 50

C. *APPARATUS*

1. Turbidimeter: The turbidimeter should consist of a nephelometer with a light source illuminating the samples and one or more photoelectric detectors with a readout device to indicate the intensity of scattered light. Turbidimeters used to test plant effluents should be approved by the U.S. Environmental Protection Agency.
2. Sample tubes

D. *REAGENTS*

1. Turbidity-free water: Pass distilled water through a membrane filter having a pore size no greater than 100 microns. Discard the first 200 m*L* collected. If filtration does not reduce turbidity, use distilled water.
2. Stock Formazin turbidity suspension:[19]
   a. Solution I. Dissolve 1.000 gm hydrazine sulfate in distilled water and dilute to 100 m*L* in a volumetric flask.
   b. Solution II. Dissolve 10.00 gm hexamethylenetetramine in distilled water and dilute to 100 m*L* in a volumetric flask.
   c. In a 100-m*L* volumetric flask, add (using 5 m*L* volumetric pipets) 5.0 m*L* Solution I and 5.0 m*L* of Solution II. Mix and allow to stand 24 hours at 25°C. Then dilute to the mark and mix. The turbidity of this suspension is 400 NTU.
   d. Prepare solutions and suspensions monthly.
3. Standard turbidity suspensions. Dilute 10.00 m*L* stock turbidity suspension to 100 m*L* with turbidity-free water. Prepare weekly. The turbidity of this suspension is defined as 40 NTU.
4. Dilute turbidity standards. Dilute portions of the standard turbidity suspension with turbidity-free water as required. Prepare weekly.

E. *PROCEDURE*

1. Turbidimeter calibration. The manufacturer's operating instructions should be followed. Measure standards on the turbidimeter covering the range of interest. If the instrument is already calibrated in standard *TURBIDITY UNITS*,[20] this procedure will check the accuracy of the calibration scales. At least one standard should be run in each instrument range to be used. Some instruments permit adjustments of sensitivity so that scale values will correspond to turbidities. Reliance on a manufacturer's solid scattering standard for setting overall instrument sensitivity for all ranges is not an acceptable practice unless the turbidimeter has been shown to be free of drift on all ranges. If a pre-calibrated scale is not supplied, then calibration curves should be prepared for each range of the instrument.
2. Turbidities less than 40 units: Shake the sample to thoroughly disperse the solids. Wait until air bubbles disappear, then pour the sample into the turbidimeter tube. Read the turbidity directly from the instrument scale or from the appropriate calibration curve.

---

[19] *Stock secondary standard turbidity suspensions that require no preparation are available from commercial suppliers.*

[20] *Turbidity Units (TU). Turbidity units are a measure of the cloudiness of water. If measured by a nephelometric (deflected light) instrumental procedure, turbidity units are expressed in nephelometric turbidity units (NTU) or simply TU. Those turbidity units obtained by visual methods are expressed in Jackson Turbidity Units (JTU) which are a measure of the cloudiness of water; they are used to indicate the clarity of water. There is no real connection between NTUs and JTUs. The Jackson turbidimeter is a visual method and the nephelometer is an instrumental method based on deflected light.*

(Turbidity)

3. Turbidities exceeding 40 units: Dilute the sample with one or more volumes of turbidity-free water until the turbidity falls below 40 units. The turbidity of the original sample is then computed from the turbidity of the diluted sample and the dilution factor.

F. *EXAMPLE*

If 5 volumes of turbidity-free water were added to 1 volume of sample, and the diluted sample showed a turbidity of 20 units, then the turbidity of the original sample was 120 units.

G. *CALCULATIONS*

Sample Reading x Dilution = Actual Turbidity

Report results as follows:

NTU	Record to Nearest:
0.0 to 1.0	0.05
1 to 10	0.01
10 to 40	1
40 to 100	5
100 to 400	10
400 to 1,000	50
>1,000	100

H. *NOTES AND PRECAUTIONS*

1. A commercially available polymer standard that requires no preparation is also approved for use. This standard is identified as AMCO Clear, available from APS Analytical Standards, Inc., 123 Saginaw Drive, Redwood City, CA 94063.
2. Sample tubes must be kept scrupulously clean, both inside and out. Discard them when they become scratched or etched. Never handle the area where the light strikes the tube during the test.
3. Fill the tubes with samples and standards that have been agitated thoroughly, and allow sufficient time for bubbles to escape.

I. *REFERENCE*

See page 2-8, *STANDARD METHODS*, 20th Edition.

QUESTION

Write your answer in a notebook and then compare your answer with the one on page 340.

14.5O What causes turbidity in wastewater?

14.6 ARITHMETIC ASSIGNMENT

Turn to the Appendix, "How to Solve Wastewater Treatment Plant Arithmetic Problems," at the back of this manual and read Section A.25, "Laboratory." Also, work the example problems and check the arithmetic using your calculator. You should be able to get the same answers.

Please answer the discussion and review questions before continuing with the Objective Test.

DISCUSSION AND REVIEW QUESTIONS

Chapter 14. LABORATORY PROCEDURES

(Lesson 7 of 7 Lessons)

Write the answers to these questions in your notebook before continuing with the Objective Test on page 341. The question numbering continues from Lesson 6.

29. What does the total solids (residue) test measure?
30. What is the ambient temperature?
31. Convert a temperature reading of 50°F to degrees Celsius.

SUGGESTED ANSWERS

Chapter 14. LABORATORY PROCEDURES

ANSWERS TO QUESTIONS IN LESSON 1

Answers to questions on page 281.

14.1A Descriptions of laboratory items and their use or purpose.

Item	Description	Use or Purpose
1. Beakers	Short, wide cylinders in sizes from 1 m*L* to 4,000 m*L*.	Mixing chemicals.
2. Graduated Cylinders	Long, narrow cylinders in sizes from 5 m*L* to 4,000 m*L*.	Measuring volumes.
3. Pipets	Very thin tubes with a pointed tip in sizes from 0.1 m*L* to 100 m*L*.	Delivering accurate volumes.
4. Burets	Long tubes with graduated walls and a stopcock in sizes from 10 to 1,000 m*L*.	Delivering and measuring accurate volumes used in "titrations."

14.1B Never heat graduated cylinders in an open flame because they will break.

14.1C *A BENCH SHEET* is used to record data and arrange data in an orderly manner. Bench sheets also are called *LABORATORY WORK SHEETS.*

ANSWERS TO QUESTIONS IN LESSON 2

Answers to questions on page 289.

14.2A A bulb should always be used to pipet wastewater or polluted water to prevent infectious materials from entering your mouth.

14.2B Inoculations are recommended to reduce the possibility of contracting diseases.

14.2C True. You may add acid to water like in a swimming pool, but never add water to acid.

14.2D Immediately wash area where acid spilled with water and neutralize the acid with sodium carbonate or bicarbonate.

ANSWERS TO QUESTIONS IN LESSON 3

Answers to questions on page 295.

14.3A The largest sources of errors found in laboratory results are usually caused by improper sampling, poor preservation, and lack of sufficient mixing during compositing and testing.

14.3B A representative sample must be collected or the test results will not have any significant meaning. To efficiently operate a wastewater treatment plant, the operator must rely on test results to indicate what is happening in the treatment process.

14.3C A proportional composite sample may be prepared by collecting a sample every hour. The size of this sample is proportional to the rate of flow when the sample is collected. All of these proportional samples are mixed together to produce a proportional composite sample. If an equal volume of sample was collected each hour and mixed, this would simply be a composite sample.

ANSWERS TO QUESTIONS IN LESSON 4

Answers to questions on page 296.

14.4A The clarity test indicates the quality of the effluent with respect to color, solids, and turbidity.

14.4B When clarity is measured under different conditions the results cannot be compared. You won't be able to tell whether your plant performance is improving, staying the same, or deteriorating.

Answer to question on page 298.

14.4C
1. H_2S in the atmosphere produces a rotten egg odor. It is indicative of anaerobic decomposition of organics in wastewater, which occurs in the absence of oxygen. High levels of H_2S are toxic to your respiratory system and can create flammable and explosive conditions.
2. You would measure the H_2S in the wastewater to know the strength of H_2S and to provide an indication of the corrosion taking place on the concrete.

Answer to question on page 299.

14.4D

$$\text{Sludge to Digester, GPD} = (\text{Total SS Removed, m}L/L)(1{,}000\text{ mg/m}L)(\text{Flow, MGD})$$

$$= (10\text{ m}L/L - 0.4\text{ m}L/L)(1{,}000\text{ mg/m}L)(1\text{ M gal/day})$$

$$= \frac{9.6\text{ m}L}{\text{M mg}} \times \frac{1{,}000\text{ mg}}{\text{m}L} \times \frac{1\text{ M gal}}{\text{day}}$$

$$= 9{,}600\text{ GPD}$$

Or

$$= (10\text{ m}L/L - 0.4\text{ m}L/L) \times 1{,}000\text{ mg/m}L \times 1\text{ MGD}$$

$$= 9{,}600\text{ GPD}$$

This value may be reduced by 30 to 75 percent due to compaction of the sludge in the clarifier.

Answers to questions on page 304.

14.4E Some suspended material in wastewater will not be removed because the specific gravity is very near that of water and the material is not light enough to float or heavy enough to settle.

14.4F Solids calculations will be shown in detail here to illustrate the computational approach and the units involved. After you understand this approach, use of the laboratory work sheet is more convenient.

1. Total Suspended Solids, mg/*L*

Volume of Sample, m*L* = 100 m*L*

Weight of Dried Sample + Dish, grams = 19.3902 gm

Weight of Dish (Tare Weight), grams = 19.3241 gm

Dry Weight = 0.0661 gm

or = 66.1 mg

$$\text{Total Suspended Solids, mg/}L = \frac{\text{Weight of Solids, mg x 1,000 m}L/L}{\text{Volume of Sample, m}L}$$

$$= \frac{66.1 \text{ mg x } 1{,}000 \text{ m}L/L}{100 \text{ m}L}$$

$$= 661 \text{ mg/}L$$

Or

$$= 66.1 \text{ mg x } 1{,}000 \text{ m}L/L \div 100 \text{ m}L$$

$$= 661 \text{ mg/}L$$

2. Volatile Suspended Solids, mg/*L*

Weight of Dried Sample + Dish, grams = 19.3902 gm

Weight of Ash + Dish, grams = 19.3469 gm

Weight Volatile Solids, grams = 0.0433 gm

or = 43.3 mg

$$\text{Volatile Suspended Solids, mg/}L = \frac{\text{Weight of Vol Sol, mg x 1,000 m}L/L}{\text{Volume of Sample, m}L}$$

$$= \frac{(43.3 \text{ mg})(1{,}000 \text{ m}L/L)}{100 \text{ m}L}$$

$$= 433 \text{ mg/}L$$

3. Volatile Solids, %

$$\text{\% Volatile Solids} = \frac{\text{Weight Volatile, mg x 100\%}}{\text{Weight Dry, mg}}$$

$$= \frac{43.3 \text{ mg}}{66.1 \text{ mg}} \text{ x } 100\%$$

$$= 65.5\%$$

4. Fixed Suspended Solids, mg/*L*

Total Suspended Solids, mg/*L* = 661 mg/*L*

Volatile Suspended Solids, mg/*L* = 433 mg/*L*

Fixed Suspended Solids, mg/*L* = 228 mg/*L*

5. Fixed Solids, %

Total Solids, % = 100.00%

Volatile Solids, % = 65.50%

Fixed Solids, % = 34.5%

Or

$$\text{\% Fixed} = \frac{\text{Fixed, mg}}{\text{Total, mg}} \text{ x } 100\%$$

$$= \frac{22.8 \text{ mg}}{66.1 \text{ mg}} \text{ x } 100\%$$

$$= 34.5\% \text{ (Check)}$$

14.4G 1. Calculate the percent removal of suspended solids by the primary clarifier.

In = Suspended solids entering the plant or unit

Out = Suspended solids leaving the plant or unit

$$\text{\% Removal} = \frac{(\text{In} - \text{Out})}{\text{In}} \text{ x } 100\%$$

$$= \frac{(221 \text{ mg/}L - 159 \text{ mg/}L)}{221 \text{ mg/}L} \text{ x } 100\%$$

$$= \frac{62}{221} \text{ x } 100\%$$

= 28% Removal by Primary Clarifier

Or

$$= (221 \text{ mg/}L - 159 \text{ mg/}L) \text{ x } 100\% \div 221 \text{ mg/}L$$

= 28% Removal by Primary Clarifier

2. Calculate the percent removal by secondary process.

In = 159 mg/*L* SS in Primary Effluent

Out = 33 mg/*L* SS in Final Effluent

$$\text{\% Removal} = \frac{(\text{In} - \text{Out})}{\text{In}} \text{ x } 100\%$$

$$= \frac{(159 \text{ mg/}L - 33 \text{ mg/}L)}{159 \text{ mg/}L} \text{ x } 100\%$$

$$= \frac{126}{159} \text{ x } 100\%$$

= 79.2% Removal From Primary Effluent to Final Effluent

3. Calculate the percent removal by the plant overall or overall plant efficiency.

In = 221 mg/*L* SS in Plant Influent

Out = 33 mg/*L* SS in Plant Effluent

$$\text{\% Removal} = \frac{(\text{In} - \text{Out})}{\text{In}} \text{ x } 100\%$$

$$= \frac{(221 \text{ mg/}L - 33 \text{ mg/}L)}{221 \text{ mg/}L} \text{ x } 100\%$$

$$= \frac{188}{221} \text{ x } 100\%$$

= 85% Overall Plant Removal

14.4H Calculate the pounds of solids removed per day by each unit:

Amount Removed, lbs/day = Conc Rem, mg/L x Flow, MGD x 8.34 lbs/gal

where MGD = million gallons per day

1. Influent, mg/L = 221 mg/L

   Primary Effluent, mg/L = 159 mg/L

   Primary Removal, mg/L = 62 mg/L

Amount Removed, lbs/day (Primary) = (62 mg/L)(1.5 MGD)(8.34 lbs/gal)

= 775.6 lbs/day Removed by Primary

Or

= (221 mg/L – 159 mg/L) x 1.5 MGD x 8.34 lbs/gal

= 775.6 lbs/day Removed by Primary

2. Primary Effluent, mg/L = 159 mg/L

   Final Effluent, mg/L = 33 mg/L

   Secondary Removal, mg/L = 126 mg/L

Amount Removed, lbs/day (Secondary) = (126 mg/L)(1.5 MGD)(8.34 lbs/gal)

= 1,576 lbs/day Removed by Secondary

3. Influent, mg/L = 221 mg/L

   Final Effluent, mg/L = 33 mg/L

   Overall Removal, mg/L = 188 mg/L

Amount Removed, lbs/day = (188 mg/L)(1.5 MGD)(8.34 lbs/gal)

= 2,352 lbs/day Removed by Plant

= Primary Removal, lbs/day + Secondary, lbs/day

Or = 776 + 1,576

= 2,352 (Check)

Answers to questions on page 308.

14.4I Volatile solids found in a digester are organic compounds of either plant or animal origin.

14.4J Volatile solids in a treatment plant indicate the portion of material that may be treated by biological processes.

ANSWERS TO QUESTIONS IN LESSON 5

Answers to questions on page 310.

14.4K Mixed liquor settleability tests are run to get an indication of what will happen to the mixed liquor in the final clarifier—the rate of sludge settling, turbidity, color, and volume of sludge at the end of 60 minutes.

14.4L The SVI is the *VOLUME* in mL occupied by one gram of mixed liquor suspended solids after 30 minutes of settling.

14.4M The SVI test is used to indicate changes in sludge characteristics.

14.4N Sludge Density Index (SDI) = 100/SVI.

Answer to question on page 311.

14.4O The sludge age of 200,000 gallon aeration tank that has 2,000 mg/L mixed liquor suspended solids, a primary effluent of 115 mg/L SS, and an average flow of 1.8 MGD:

$$\text{Sludge Age, days} = \frac{\text{Vol of Aeration Tank, MG x Sus Solids, mg/}L}{\text{Flow, MGD x Primary Effluent, mg/}L}$$

$$= \frac{0.2 \text{ MG x } 2{,}000 \text{ mg/}L}{1.8 \text{ MGD x } 115 \text{ mg/}L}$$

= 1.93 days

Or

= 0.2 MG x 2,000 mg/L ÷ 1.8 MGD ÷ 115 mg/L

= 1.93 days

Answer to question on page 312.

14.4P When the DO in the aeration tank is very low, the copper sulfate-sulfamic acid test procedure can give DO results higher than actual DO in the aerator. The results may be high because oxygen may enter the sample from the air when the sample is collected, when the copper sulfate-sulfamic acid inhibitor is added, while the solids are settling, and when the sample is transferred to a BOD bottle for the DO test.

Answer to question on page 314.

14.4Q The advantages of the centrifuge over the regular suspended solids test are:

1. Speed of answer! Not as accurate as other methods, but results are sufficiently close.
2. Answers very acceptable if suspended solids concentration is below 1,500 mg/L.

Answer to question on page 314.

14.4R In order to calculate the Mean Cell Residence Time (MCRT), measure representative composite samples of mixed liquor suspended solids, effluent suspended solids, and waste sludge suspended solids. Also, measure influent and waste sludge flow.

ANSWERS TO QUESTIONS IN LESSON 6

Answers to questions on page 320.

14.5A Plant effluents should be chlorinated for disinfection purposes to protect the bacteriological quality of the receiving waters.

14.5B The iodometric titration method gives good results with samples containing wastewater, such as plant effluent or receiving waters. Amperometric titration gives the best results, but the equipment is expensive.

Answers to questions on page 323.

14.5C DO Saturation, % $= \frac{\text{DO of Sample, mg}/L \times 100\%}{\text{DO at 100\% Saturation, mg}/L}$

$= \frac{(7.9 \text{ mg}/L)100\%}{11.3 \text{ mg}/L}$

$= 70\%$

Or

$= 7.9 \text{ mg}/L \times 100\% \div 11.3 \text{ mg}/L$

$= 70\%$

14.5D To calibrate the DO probe in an aeration tank, a sample of the effluent can be collected and split. The DO of the effluent is measured by the Modified Winkler procedure, and the probe DO reading is adjusted to agree with the Winkler results.

Answers to questions on page 328.

14.5E The amount of organic material in wastewater can be estimated indirectly by measuring oxygen use with the BOD test.

14.5F Samples for the BOD test should be collected before chlorination because chlorine interferes with the organisms in the test. It is difficult to obtain accurate results with dechlorinated samples.

14.5G A solution of sodium thiosulfate at 0.025 *N* is very weak and unstable and will not remain accurate over two weeks.

14.5H To prepare dilutions for a cannery waste with an expected BOD of 2,000 mg/*L*, take 10 m*L* of sample and add 90 m*L* of dilution water to obtain a new sample with an estimated BOD of 200 mg/*L* (10 to 1 dilution):

BOD Dilution, m*L* $= \frac{1{,}200}{\text{Estimated BOD, mg}/L}$

$= \frac{1{,}200}{200}$

$= 6 \text{ m}L$

14.5I

$$\text{BOD, mg}/L = \left[\begin{matrix}\text{Initial DO of} \\ \text{Diluted Sample, mg}/L\end{matrix} - \begin{matrix}\text{DO of Diluted} \\ \text{Sample After} \\ \text{5-Day Incubation, mg}/L\end{matrix}\right]\left[\frac{\text{BOD Bottle Vol, m}L}{\text{Sample Volume, m}L}\right]$$

$= (7.5 \text{ mg}/L - 3.9 \text{ mg}/L)\left[\frac{300 \text{ m}L}{2 \text{ m}L}\right]$

$= (3.6 \text{ mg}/L)(150)$

$= 540 \text{ mg}/L$

Or

$= (7.5 \text{ mg}/L - 3.9 \text{ mg}/L) \times 300 \text{ m}L \div 2 \text{ m}L$

$= 540 \text{ mg}/L$

ANSWERS TO QUESTIONS IN LESSON 7

Answer to question on page 330.

14.5J Precautions to be exercised when using a pH meter include:

1. Prepare fresh buffer solution weekly for calibration purposes.
2. pH meter, samples, and buffer solutions should all be at the same temperature.
3. Watch for erratic results arising from faulty operation of pH meter or fouling of electrodes with interfering matter.

Answer to question on page 332.

14.5K

Known		Unknown
Empty Dish, mg	= 64,328.9 mg	1. Total Solids, mg/*L*
Dish and Residue, mg	= 64,381.2 mg	2. Total Dissolved Solids, mg/*L*
Dish and Dissolved Solids, mg	= 64,351.2 mg	
Sample Volume, m*L*	= 100 m*L*	

1. Calculate Total Solids (Residue)

Total Solids, mg/*L* $= \frac{(A - B)(1{,}000 \text{ m}L/L)}{\text{m}L \text{ of Sample}}$

$= \frac{(64{,}381.2 \text{ mg} - 64{,}328.9 \text{ mg})(1{,}000 \text{ m}L/L)}{100 \text{ m}L}$

$= 523 \text{ mg}/L$

Or

$= (64{,}381.2 \text{ mg} - 64{,}328.9 \text{ mg}) \times 1{,}000 \text{ m}L/L \div 100 \text{ m}L$

$= 523 \text{ mg}/L$

2. Calculate Total Dissolved Solids (Filterable)

Total Dissolved Solids, mg/*L* $= \frac{(A - B)(1{,}000 \text{ m}L/L)}{\text{m}L \text{ of Sample}}$

$= \frac{(64{,}351.2 \text{ mg} - 64{,}328.9 \text{ mg})(1{,}000 \text{ m}L/L)}{100 \text{ m}L}$

$= 223 \text{ mg}/L$

Answers to questions on page 335.

14.5L Changes in influent temperature could indicate a new influent source. A drop in temperature could be caused by cold water from infiltration, and an increase in temperature could be caused by an industrial waste discharge.

14.5M The thermometer should remain immersed in the liquid while being read for accurate results. When removed from the liquid, the reading will change.

14.5N All thermometers should be calibrated against an accurate National Bureau of Standards thermometer because some thermometers that are substantially inaccurate (off as much as 6°F) can be purchased.

Answer to question on page 336.

14.5O Turbidity in wastewater is caused by the presence of suspended matter such as clay, silt, finely divided organic and inorganic matter, and microscopic organisms.

OBJECTIVE TEST

Chapter 14. LABORATORY PROCEDURES

Please write your name and mark the correct answers on the answer sheet, as directed at the end of Chapter 1, Volume I. There may be more than one correct answer to each multiple-choice question.

True-False

1. A Teflon stopcock always needs to be lubricated.
   1. True
   2. False
2. Volumetric flasks should never be heated.
   1. True
   2. False
3. Safe practice in the laboratory requires hardly any more effort than unsafe practice.
   1. True
   2. False
4. Never drink from a beaker or other laboratory glassware.
   1. True
   2. False
5. Storage of perishable foods in a laboratory refrigerator is an acceptable practice.
   1. True
   2. False
6. The greatest errors in laboratory tests are usually caused by improper sampling, poor preservation, or lack of enough mixing during compositing and testing.
   1. True
   2. False
7. Sample deterioration starts immediately after collection for most wastewaters.
   1. True
   2. False
8. Samples should be mixed thoroughly before compositing and at time of test.
   1. True
   2. False
9. All odors in wastewater are from hydrogen sulfide.
   1. True
   2. False
10. Each process in a wastewater treatment plant has its own favorable range of pH, which must be checked routinely.
    1. True
    2. False

Best Answer (Select only the closest or best answer.)

11. What is the metric name for a volume measurement?
    1. Gram
    2. Liter
    3. Meter
    4. Second
12. Which piece of laboratory glassware provides the most accurate measurement of liquid volumes?
    1. Beaker
    2. Buret
    3. Flask
    4. Graduated cylinder
13. Why should operators change from working clothes into street clothes before leaving work?
    1. To avoid bringing odors from wastewater home
    2. To avoid having odors permeate the interior of your car
    3. To prevent carrying unsanitary materials home
    4. To prevent overloading the family washing machine
14. How can maintenance personnel be protected from broken glass?
    1. Dispose of broken glass in separate, clearly marked, covered containers
    2. Flame polish all sharp edges of glass before disposal
    3. Grind broken glass into small pieces before disposal
    4. Personally dispose of all broken glass
15. What is a grab sample?
    1. A sample composed of a mixture of samples
    2. A sample grabbed from a flowing wastestream
    3. A sample prepared on the basis of wastestream flows
    4. A single sample of wastewater taken at neither a set time nor flow
16. Why should samples be refrigerated immediately?
    1. To lower the temperature as quickly as possible
    2. To maintain the sample in a uniform mix status
    3. To minimize transport time to the laboratory
    4. To preserve samples from continued bacterial decomposition
17. Why should sample lines to automatic samplers be cleaned regularly?
    1. To comply with the vendor's warranty requirements
    2. To maintain scouring velocities in the sample lines
    3. To prevent sloughing growths, which will contaminate the sample with high solids content
    4. To save on pumping costs for energy necessary to overcome excessive friction losses

18. What does the settleable solids test indicate?
    1. Need to pump sludge from the clarifiers
    2. Percent efficiency of sedimentation basins
    3. Rate of BOD removed by sedimentation
    4. Volume of solids removed by sedimentation
19. What does the suspended solids test represent?
    1. Combination of settleable solids and those solids that remain in suspension
    2. Combination of solids and BOD in sample
    3. Those solids that will remain in suspension
    4. Those solids that will settle
20. Why is the settleability test run on activated sludge?
    1. To determine the ability of the solids to separate from the liquid in the final clarifier
    2. To determine the activity of the microorganisms in the activated sludge
    3. To determine the respiration rate of the microorganisms in the final clarifier
    4. To determine the settling velocity of the solids in the final clarifier
21. Why should a suspended solids test be run on the same sample of mixed liquor that is used for the settleability test?
    1. To calculate the BOD
    2. To calculate the MCRT
    3. To calculate the OUR
    4. To calculate the SVI
22. What is the purpose of the Sludge Volume Index (SVI) test?
    1. To indicate the condition of sludge for settleability
    2. To indicate the index for the microorganisms
    3. To indicate the need to adjust the sludge wasting rate
    4. To indicate the volume of sludge that will settle
23. Why is it very important that some dissolved oxygen be present in aeration tanks at all times?
    1. To maintain aerobic conditions
    2. To maintain algal growth potential
    3. To maintain anaerobic conditions
    4. To maintain anoxic conditions
24. Why should residual chlorine be removed before final discharge to a receiving stream in many locations?
    1. For inactivation of pathogenic organisms
    2. For protection of fish and other aquatic life
    3. For protection of water contact sport activities
    4. For ultimate BOD reduction
25. Which chlorine residual test method gives the best results?
    1. Amperometric titration
    2. Colorimetric
    3. DPD titrimetric
    4. Iodometric
26. Why should samples for the BOD test be collected before chlorination?
    1. Chlorine can discolor the test sample
    2. Chlorine gas can be released from the sample
    3. Chlorine interferes with the organisms in the test
    4. Chlorine produces disinfection by-products
27. What is a limitation of the BOD test?
    1. Results are *NOT* available until five days after sample was collected
    2. Results are *NOT* repeatable
    3. Results depend on DO measurements
    4. Results estimate the availability of food in the sample
28. What are total dissolved solids (TDS)?
    1. Material consisting entirely of fixed or inorganic salts
    2. Material consisting of inorganic and organic salts
    3. Material that completely dissolves in a solution and leaves no particle residues
    4. Material that passes through a standard glass-fiber filter disc and remains after evaporation at 180°C
29. Why should the thermometer be left in the liquid while reading the temperature?
    1. Because reading will change as soon as thermometer is removed from liquid
    2. Because the thermometer is easy to read while immersed in the liquid
    3. Because this approach reduces the probability of breaking thermometer
    4. Because this is quickest way to measure temperature
30. What does turbidity measure?
    1. Dissolved solids in wastewater
    2. Hydrogen ion activity of wastewater
    3. Physical clarity of wastewater
    4. Suspended solids in wastewater

Multiple Choice (Select all correct answers.)

31. Why must operators understand laboratory procedures and results?
    1. To complete laboratory data sheets
    2. To control treatment processes
    3. To identify developing treatment problems
    4. To purchase laboratory reagents
    5. To select most effective operational procedures
32. What are flasks and beakers used for in the laboratory?
    1. Calibrating
    2. Heating
    3. Measuring
    4. Mixing
    5. Weighing
33. How does a plant operator use a laboratory notebook and work sheets?
    1. To arrange data in an orderly manner
    2. To fill in the blank spaces
    3. To forecast the weather
    4. To order more reagents
    5. To record data
34. What types of hazards are people exposed to when working in a wastewater laboratory?
    1. Burns
    2. Electric shock
    3. Infectious materials
    4. Reckless drivers
    5. Toxic fumes

35. What types of hazards can be caused by acids?
    1. Burning
    2. Gassing through explosion
    3. Infecting
    4. Neutralizing
    5. Poisoning

36. What types of safeguards against burns are available for persons working in a laboratory?
    1. Aprons
    2. Emergency deluge showers
    3. Fire extinguishers
    4. Safety tongs
    5. Special gloves

37. Obtaining good lab test results depends on which factors?
    1. Ensuring that the sample taken is truly representative of the wastestream
    2. Protecting and preserving the samples until they are analyzed
    3. Reporting results in a timely fashion to regulatory agencies
    4. Using proper sampling techniques
    5. Writing legibly on lab bench sheet

38. Which items should be considered when developing a sampling program?
    1. Preservation of samples
    2. Representative sampling
    3. Sampling devices
    4. Time of sampling
    5. Types of samples

39. When may a grab sample be preferred over a composite sample?
    1. The wastewater characteristics are relatively constant
    2. The wastewater is to be analyzed for coliforms
    3. The wastewater is to be analyzed for residual chlorine
    4. The wastewater to be sampled does not flow on a continuous basis
    5. You wish to determine whether or not a composite sample obscures extreme conditions of the waste

40. Which wastewater constituents or characteristics must be analyzed immediately on site because they cannot be preserved?
    1. Biochemical oxygen demand (BOD)
    2. Chlorine residual
    3. Dissolved oxygen
    4. pH
    5. Temperature

41. Which factors can cause poor laboratory data quality?
    1. Calculation mistakes
    2. Calibrated instruments
    3. Deteriorated reagents
    4. Poorly operating instruments
    5. Sloppy laboratory technique

42. What is the purpose of the effluent clarity test? To obtain an indication of the quality of the effluent with respect to
    1. BOD
    2. Coliforms
    3. Color
    4. Solids
    5. Turbidity

43. Which chemicals can be used to control hydrogen sulfide in wastewater?
    1. Bases to increase the pH
    2. Chlorine
    3. Hydrochloric acid
    4. Hydrogen peroxide
    5. Oxygen

44. Why should hydrogen sulfide be controlled? Because hydrogen sulfide is
    1. Corrosive
    2. Explosive
    3. Flammable
    4. Odorous
    5. Toxic to your respiratory system

45. What are the uses of chlorination in wastewater treatment?
    1. Controlling activated sludge bulking
    2. Controlling foam
    3. Controlling odors
    4. Disinfecting
    5. Improving scum and grease removal

46. What substances can interfere with the Sodium Azide Modification of the Winkler Method for measuring dissolved oxygen?
    1. Color
    2. Nitrite
    3. Organics
    4. Sulfide
    5. Suspended Solids

47. What precautions should be observed when using a DO probe?
    1. Dissolved inorganic salts can influence readings from a probe
    2. Don't place the probe directly over a diffuser
    3. Keep the membrane in the tip of the probe from drying out
    4. Periodically check the calibration of the probe
    5. Reactive compounds, such as reactive gases and sulfur compounds, can interfere with the output of a probe

48. What precautions should be observed when measuring total solids?
    1. Adjust pH to optimum range
    2. Be sure the balance is properly leveled, zeroed, and maintained
    3. Do *NOT* weigh items when they are hot
    4. Use tongs or gloves when handling hot items
    5. Wastewater and sludge may contain infectious organisms and must be handled with care

49. Why is temperature an important measurement?
    1. Affects biological growth
    2. Detects changes in raw wastewater quality
    3. Determines mixed liquor suspended solids in aeration tank
    4. Estimates sludge volume pumped from primary clarifier
    5. Influences percent saturation of dissolved oxygen

50. What causes turbidity in wastewater?
    1. Dissolved salts
    2. Finely divided inorganic matter
    3. Finely divided organic matter
    4. Microscopic organisms
    5. Mud

PLANT ______________________

DATE ______________________

SUSPENDED SOLIDS AND DISSOLVED SOLIDS

SAMPLE							
Crucible							
Sample, mL							
Wt Dry + Dish, gm Wt Dish, gm							
Wt Dry, gm							
mg/L = $\frac{\text{Wt Dry, gm x 1,000,000}}{\text{Sample, m}L}$							
Wt Dish + Dry, gm Wt Dish + Ash, gm							
Wt Volatile, gm							
% = $\frac{\text{Wt Vol}}{\text{Wt Dry}}$ x 100%							

BOD

\# Blank ______________

SAMPLE										
DO Sample										
Bottle #										
% Sample										
Blank or adjusted blank DO after incubation										
Depletion, 5 days										
Depletion, %										

SETTLEABLE SOLIDS

Sample, mL ______________ ______________

Direct mL/L ______________ ______________

COD

Sample ______________ ______________

Blank Titration ______________ ______________

Sample Titration ______________ ______________

Depletion ______________ ______________

mg/L = $\frac{\text{Dep x } N \text{ FAS x 8,000}}{\text{Sample, m}L}$ ______________ ______________

Typical laboratory work sheet

TOTAL SOLIDS

SAMPLE					
Dish No.					
Wt Dish + Wet, gm Wt Dish, gm					
Wt Wet, gm					
Wt Dish + Dry, gm Wt Dish, gm					
Wt Dry, gm					
% Solids = $\frac{\text{Wt Dry}}{\text{Wt Wet}}$ x 100%					
Wt Dish + Dry, gm Wt Dish + Ash, gm					
Wt Volatile, gm					
% Volatile = $\frac{\text{Wt Vol x 100\%}}{\text{Wt Dry}}$					
pH					
Volatile Acid, mg/*L*					
Alkalinity as $CaCO_3$, mg/*L*					

Grease

Sample ____________________

Sample, m*L* ____________________

Wt Flask + Grease, mg ____________________

Wt Flask, mg ____________________

Wt Grease, mg ____________________

$$\text{mg}/L = \frac{\text{Wt Grease, mg x 1,000}}{\text{Sample, m}L}$$ ____________________

Typical laboratory work sheet (continued)

CHAPTER 15

MANAGEMENT

by

John Brady

and

Ken Kerri

Adapted from *UTILITY MANAGEMENT*

by Lorene Lindsay

TABLE OF CONTENTS

Chapter 15. MANAGEMENT

OBJECTIVES

Chapter 15. MANAGEMENT

Following completion of Chapter 15, you should be able to:

1. Identify the functions of a manager,
2. Describe the benefits of short-term, long-term, and emergency planning,
3. Define the following terms:

   a. Authority,
   b. Responsibility,
   c. Delegation,
   d. Accountability, and
   e. Unity of command,

4. Read and construct an organizational chart identifying lines of authority and responsibility,
5. Prepare a written or oral report on the utility's operations,
6. Communicate effectively within the organization, with media representatives, and with the community,
7. Describe the financial strength of your utility,
8. Calculate your utility's operating ratio, coverage ratio, and simple payback,
9. Develop and implement capital improvement plans for your utility,
10. List and describe the types of funding sources available to the utility, and
11. Collect, organize, file, retrieve, use, and dispose of utility records.

WORDS

Chapter 15. MANAGEMENT

ACCOUNTABILITY | ACCOUNTABILITY

When a manager gives power/responsibility to an employee, the employee ensures that the manager is informed of results or events.

AUTHORITY | AUTHORITY

The power and resources to do a specific job or to get that job done.

BOND | BOND

(1) A written promise to pay a specified sum of money (called the face value) at a fixed time in the future (called the date of maturity). A bond also carries interest at a fixed rate, payable periodically. The difference between a note and a bond is that a bond usually runs for a longer period of time and requires greater formality. Utility agencies use bonds as a means of obtaining large amounts of money for capital improvements.

(2) A warranty by an underwriting organization, such as an insurance company, guaranteeing honesty, performance, or payment by a contractor.

CALL DATE | CALL DATE

First date a bond can be paid off.

CERTIFICATION EXAMINATION | CERTIFICATION EXAMINATION

Certification Examination. An examination administered by a state agency or professional association that small wastewater system operators take to indicate a level of professional competence. In many states and provinces, certification of small wastewater system operators may be voluntary. Current trends indicate that more states, provinces, and employers will require small wastewater system operators to be "certified" in the future.

CERTIFIED OPERATOR | CERTIFIED OPERATOR

A person who has the education and experience required to operate a specific class of treatment facility as indicated by possessing a certificate of professional competence given by a state agency or professional association.

CODE OF FEDERAL REGULATIONS (CFR) | CODE OF FEDERAL REGULATIONS (CFR)

A publication of the United States Government which contains all of the proposed and finalized federal regulations, including environmental regulations.

CONFINED SPACE | CONFINED SPACE

Confined space means a space that:

A. Is large enough and so configured that an employee can bodily enter and perform assigned work; and

B. Has limited or restricted means for entry or exit (for example, manholes, tanks, vessels, silos, storage bins, hoppers, vaults, and pits are spaces that may have limited means of entry); and

C. Is not designed for continuous employee occupancy.

(Definition from the Code of Federal Regulations (CFR) Title 29 Part 1910.146.)

CONFINED SPACE, PERMIT-REQUIRED (PERMIT SPACE) | CONFINED SPACE, PERMIT-REQUIRED (PERMIT SPACE)

A confined space that has one or more of the following characteristics:

- Contains or has a potential to contain a hazardous atmosphere,
- Contains a material that has the potential for engulfing an entrant,
- Has an internal configuration such that an entrant could be trapped or asphyxiated by inwardly converging walls or by a floor which slopes downward and tapers to a smaller cross section, or
- Contains any other recognized serious safety or health hazard.

(Definition from the Code of Federal Regulations (CFR) Title 29 Part 1910.146.)

COVERAGE RATIO | COVERAGE RATIO

The coverage ratio is a measure of the ability of the utility to pay the principal and interest on loans and bonds (this is known as "debt service") in addition to any unexpected expenses.

DEBT SERVICE | DEBT SERVICE

The amount of money required annually to pay the (1) interest on outstanding debts; or (2) funds due on a maturing bonded debt or the redemption of bonds.

DELEGATION | DELEGATION

The act in which power is given to another person in the organization to accomplish a specific job.

MATERIAL SAFETY DATA SHEET (MSDS) | MATERIAL SAFETY DATA SHEET (MSDS)

A document which provides pertinent information and a profile of a particular hazardous substance or mixture. An MSDS is normally developed by the manufacturer or formulator of the hazardous substance or mixture. The MSDS is required to be made available to employees and operators whenever there is the likelihood of the hazardous substance or mixture being introduced into the workplace. Some manufacturers are preparing MSDSs for products that are not considered to be hazardous to show that the product or substance is *NOT* hazardous.

OSHA (O-shuh) | OSHA

The Williams-Steiger **O**ccupational **S**afety and **H**ealth **A**ct of 1970 (OSHA) is a federal law designed to protect the health and safety of workers, including collection system and treatment plant operators. It regulates the design, construction, operation and maintenance of industrial plants and wastewater collection systems and treatment plants. The Act does not apply directly to municipalities, *EXCEPT* in those states that have approved plans and have asserted jurisdiction under Section 18 of the OSHA Act. *HOWEVER, CONTRACT OPERATORS AND PRIVATE FACILITIES DO HAVE TO COMPLY WITH OSHA REQUIREMENTS.* Wastewater treatment plants have come under stricter regulation in all phases of activity as a result of OSHA standards. OSHA also refers to the federal and state agencies which administer the OSHA regulations.

OPERATING RATIO | OPERATING RATIO

The operating ratio is a measure of the total revenues divided by the total operating expenses.

ORGANIZING | ORGANIZING

Deciding who does what work and delegating authority to the appropriate persons. A utility should have a written organizational plan and written policies.

PLANNING | PLANNING

Management of utilities to build the resources and financial capability to provide for future needs.

PRESENT WORTH | PRESENT WORTH

The value of a long-term project expressed in today's dollars. Present worth is calculated by converting (discounting) all future benefits and costs over the life of the project to a single economic value at the start of the project. Calculating the present worth of alternative projects makes it possible to compare them and select the one with the largest positive (beneficial) present worth or minimum present cost.

RESPONSIBILITY | RESPONSIBILITY

Answering to those above in the chain of command to explain how and why you have used your authority.

CHAPTER 15. MANAGEMENT

15.0 NEED FOR MANAGEMENT

The management of a public or private utility, large or small, is a complex and challenging job. Communities are concerned about their drinking water and their wastewater as well as a demand for service to be provided without interruption. They are aware of past environmental disasters and they want to protect their communities, but they want this uninterrupted service and environmental protection with a minimum investment of money. In addition to the local community demands, the utility manager must also keep up with increasingly stringent regulations and monitoring from regulatory agencies. While meeting these "external" (outside the utility) concerns, the manager faces the normal challenges from within the organization: personnel, resources, equipment, and preparing for the future. An important responsibility for managers is to deal with councils, boards, or commissions that control the administrative policies of the utility. For the successful manager, all of these responsibilities combine to create an exciting and rewarding job.

A brief quiz is given in Table 15.1 that asks some basic management questions. This quiz can be used as a guide to management areas that may need some attention in your utility. You should be able to answer yes to most of the questions; however, all utilities have areas that can be improved. In the environmental field, as well as other fields, the workforce itself is changing. Minorities, women, and people with disabilities provide new opportunities for growth in the utility. For the employee, however, overcoming employment barriers can be difficult, especially when the workload is demanding and physically challenging. The utility manager must provide adequate support services for these workers.

Changes in the environmental workplace also are created by advances in technology. The environmental field has exploded with new technologies, such as video monitoring, robotics in the collection system, and computer-controlled treatment processes. The utility manager must keep up with these changes and provide the leadership to keep everyone at the utility up to speed on new ways of doing things. In addition, the utility manager must provide a safer, cleaner work environment while constantly training and retraining employees to understand new technologies.

QUESTIONS

Write your answers in a notebook and then compare your answers with those on page 382.

15.0A What are the local community demands on a utility manager?

15.0B What has created changes in the environmental workplace?

15.1 FUNCTIONS OF A MANAGER

The functions of a utility manager are the same as for the CEO (Chief Executive Officer) of any big company: planning, organizing, staffing, directing, and controlling. In many communities, the utility manager may be the only one who has these responsibilities and the community depends on the manager to handle everything.

Planning (see Section 15.2) consists of determining the goals, policies, procedures, and other elements to achieve the goals and objectives of the agency. Planning requires the manager to collect and analyze data, consider alternatives, and then make decisions. Planning must be done before the other managing functions. Planning may be the most difficult in smaller communities, where the future may involve a decline in population instead of growth.

Organizing (see Section 15.3) means that the manager decides who does what work and delegates authority to the appropriate operators. The organizational function in some utilities may be fairly loose while some communities are very tightly controlled.

Staffing (see Section 15.4) is the recruiting of new operators and staff and determining if there are enough qualified operators and staff to fill available positions. The utility manager's staffing responsibilities include selecting and training employees, evaluating their performance, and providing opportunities for advancement for operators and staff in the agency.

Directing includes guiding, teaching, motivating, and supervising operators and utility staff members. Direction includes issuing orders and instructions so that activities at the facilities or in the field are performed safely and are properly completed.

Controlling involves taking the steps necessary to ensure that essential activities are performed so that objectives will be achieved as planned. Controlling means being sure that progress is being made toward objectives and taking corrective action as necessary. The utility manager is directly involved in controlling the treatment process to ensure that water or wastewater is being properly treated and to make sure that the utility is meeting its short- and long-term goals.

TABLE 15.1 HOW WELL DOES YOUR SYSTEM MANAGE?

The following self-test is designed for small water or wastewater treatment facilities to provide a guide for identifying areas of concern and for improving small system management.

1. Is the treatment system budget separate from other accounts so that the true cost of treatment can be determined?
2. Are the funds adequate to cover operating costs, debt service, and future capital improvements?
3. Do operational personnel have input into the budget process?
4. Is there a monthly or quarterly review of the actual operating costs compared to the budgeted costs?
5. Does the user charge system adequately reflect the cost of treatment?
6. Are all users properly metered and does the unaccounted for water not exceed 20 percent of the total flow?
7. Are plant discharges and monitoring tests representative of plant performance?
8. Are operational control decisions based on process control testing within the plant?
9. Are provisions made for continued training for plant personnel?
10. Are qualified personnel available to fill job vacancies and is job turnover relatively low?
11. Are the energy costs for the system not more than 20 to 30 percent of the total operating costs?
12. Is the ratio of corrective (reactive) maintenance to preventive (proactive) maintenance remaining stable and is it less than 1.0?
13. Are maintenance records available for review?
14. Is the spare parts inventory adequate to prevent long delays in equipment repairs?
15. Are old or outdated pieces of equipment replaced as necessary to prevent excessive equipment downtime, inefficient process performance, or unreliability?
16. Are technical resources and tools available for repairing, maintaining, and installing equipment?
17. Is the utility's equipment providing the expected design performance?
18. Are standby units for key equipment available to maintain process performance during breakdowns or during preventive maintenance activities?
19. Are the plant processes adequate to meet the demand for treatment?
20. Does the facility have an adequate emergency response plan including a wastewater bypass storage procedure or an alternate water source?

QUESTIONS

Write your answers in a notebook and then compare your answers with those on page 382.

15.1A What are the functions of a utility manager?

15.1B In small communities, what does the community depend on the utility manager to do?

15.2 PLANNING[1]

A very large portion of any manager's typical work day will be spent on activities that can be described as planning activities since nearly every area of a manager's responsibilities require some type of planning.

Planning is one of the most important functions of utility management and one of the most difficult. Communities must have good, clean water and the management of water or wastewater utilities must include building the resources and financial capability to provide for future needs. The utility must plan for future growth, including commercial and industrial development, and be ready to provide the water and waste treatment systems that will be needed as the community grows. The most difficult problem for some small communities is recognizing and planning for a decline in population. The utility manager must develop reliable information to plan for growth or decline. Decisions must be made about goals, both short- and long-term. The manager must prepare plans for the next two years and the next 10 to 20 years. Remember that utility planning should include operational personnel, local officials (decision makers), and the public. Everyone must understand the importance of planning and be willing to contribute to the process.

Operation and maintenance of a utility also involves planning by the utility manager. A preventive maintenance program should be established to keep the system performing as intended and to protect the community's investment in water or wastewater facilities. (Section 15.9 describes the various types of maintenance and the benefits of establishing maintenance programs.)

The utility also must have an emergency response plan to deal with natural or human disasters. Without adequate planning, your utility will be facing system failures, inability to meet compliance regulations, and inadequate service capacity to meet community needs. Plan today and avoid disaster tomorrow. (Section 15.10, "Emergency Response Plan," describes the basic elements of an emergency operations plan.)

15.3 ORGANIZING[2]

A utility should have a written organizational plan and written policies. In some communities, the organizational plan and policies are part of the overall community plan. In either case, the utility manager and all plant personnel should have a copy of the organizational plan and written policies of the utility.

The purpose of the organizational plan is to show who reports to whom and to identify the lines of authority. The organizational plan should show each person or job position in the organization with a direct line showing to whom each person reports in the organization. Remember, an employee can serve only one supervisor (unity of command) and each supervisor should ideally manage only six or seven employees. The organizational plan should include a job description for each of the positions on the organizational chart. When the organizational plan is in place, employees know who is their immediate boss and confusion about job tasks is eliminated.

To understand organization and its role in management, we need to understand some other terms including authority, responsibility, delegation, and accountability. AUTHORITY means the power and resources to do a specific job or to get that job done. Authority may be given to an employee due to their position in the organization (this is formal authority) or authority may be given to the employee informally by their co-workers when the employee has earned their respect. RESPONSIBILITY may be described as answering to those above in the chain of command to explain how and why you have used your authority. DELEGATION is the act in which power is given to another person in the organization to accomplish a specific job. Finally, when a manager gives power/responsibility to an employee, then the employee is held *ACCOUNTABLE*[3] for the results.

Organization and effective delegation are very important to keep any utility operating well. Effective delegation is uncomfortable for many managers since it requires giving up power and responsibility. Many managers believe that they can do the job better than others, they believe that other employees are not well trained or experienced, and they are afraid of mistakes. The utility manager retains some responsibility even after delegating to another employee and, therefore, the man-

[1] *Planning. Management of utilities to build the resources and financial capability to provide for future needs.*

[2] *Organizing. Deciding who does what work and delegating authority to the appropriate persons. A utility should have a written organizational plan and written policies.*

[3] *Accountability. When a manager gives power/responsibility to an employee, the employee ensures that the manager is informed of results or events.*

ager is often reluctant to delegate or may delegate the responsibility but not the authority to get the job done. For the utility manager, good organization means that employees are ready to accept responsibility for their own actions and have the power and resources to make sure that the job gets done.

Employees should not be asked to accept responsibilities for job tasks that are beyond their level of authority in the organizational structure. For example, an operator or lead utility worker should not be asked to accept responsibility for additional lab testing unless qualified. The responsibility for additional lab testing must be delegated to the lab supervisor. Authority and responsibility must be delegated properly to be effective. When these three components—proper job assignments, authority, and responsibility—are all present, the supervisor has successfully delegated. The success of delegation is dependent upon all three components.

An important and often overlooked part of delegation is follow-up by the supervisor. A good manager will delegate and follow up on progress to make sure that the employee has the necessary resources to get the job done. Well-organized managers can delegate effectively and do not try to do all the work themselves, but are responsible for getting good results.

QUESTIONS

Write your answers in a notebook and then compare your answers with those on page 382.

15.2A Who should be included in utility planning?

15.3A What is the purpose of an organizational plan?

15.3B Why is it sometimes difficult or uncomfortable for supervisors or managers to delegate effectively?

15.3C What is an important and often overlooked part of delegation?

15.4 STAFFING

15.40 The Utility Manager's Responsibilities

The utility manager is also responsible for staffing, which includes hiring new employees, training employees, and evaluating job performance. The utility should have established procedures for job hiring that include requirements for advertising the position, application procedures, and the procedures for conducting interviews.

In the area of staffing, more than any other area of responsibility, a manager must be extremely cautious and consider the consequences before taking action. Personnel management practices have changed dramatically in the past few years and continue to be redefined almost daily by the courts. A manager who violates an employee's or job applicant's rights can be held both personally and professionally liable in court. Throughout this section on staffing you will repeatedly find references to two terms: **job-related** and **documentation**. These are key concepts in personnel management today. Any personnel action you take must be job-related, from the questions you ask during interviews to disciplinary actions or promotions. And, while almost no one wants more paperwork, documentation of personnel actions detailing what you did, when you did it, and why you did it (the reasons will be job-related, of course) is absolutely essential. There is no way to predict when you might be called upon to defend your actions in court. Good records not only serve to refresh your memory about past events but can also be used to demonstrate your pattern of lawful behavior over time.

NOTICE

The information provided in this section on staffing should ***not*** be viewed as ***legal advice***. The purpose of this section is simply to identify and describe in general terms the major components of a utility manager's responsibilities in the area of staffing. One issue, harassment, is discussed in somewhat greater detail to illustrate the broad scope of a manager's responsibilities within a single policy area. Personnel administration is affected by many federal and state regulations. Legal requirements of legislation such as the Americans With Disabilities Act (ADA), Equal Employment Opportunity (EEO) Act, Family and Medical Leave Act, and wages and hours laws are complex and beyond the scope of this manual. If your utility does not have established personnel policies and procedures, consider getting help from a labor law attorney to develop appropriate policies. At the very least, you should get help from a recruitment specialist to develop and document hiring procedures that meet the federal guidelines for Equal Employment Opportunity.

15.41 How Many Employees Are Needed?

There is a common tendency for organizations to add personnel in response to changing conditions without first examining how the existing workforce might be reorganized to achieve greater efficiency and meet the new work demands. In water and wastewater utilities, aging of the system, changes in use, and expansion of the system often mean changes in the operation and maintenance tasks being performed. The manager of a utility should periodically review the agency's work requirements and staffing to ensure that the utility is operating as efficiently as possible. A good time to conduct such a review is during the annual budgeting process or when you are considering hiring a new employee because the workload seems to be greater than the current staff can adequately handle. Another option that a manager might consider is overtime. When you consider the lower overhead costs associated with overtime, it can be an inexpensive alternative for meeting temporary needs.

The staffing analysis procedure outlined in this section illustrates how to conduct a comprehensive analysis of the type needed for a complete reorganization of the agency. In practice, however, a complete reorganization may not be desirable or even possible. Frequent organizational changes can make employees anxious about their jobs and may interfere with their work performance. Some employees show strong resistance to any change in job responsibilities. Nonetheless, by thoroughly examining the functions and staffing of the utility on a periodic basis, the manager may spot trends (such as an increase in the amount of time spent maintaining certain equipment or portions of the system) or discover inefficiencies that could be corrected over a period of time.

The first step in analyzing the utility's staffing needs is to prepare a detailed list of all the tasks to be performed to operate and maintain the utility. Next, estimate the number of staff hours per year required to perform each task. Be sure to include the time required for supervision and training.

When you have completed the task analysis, prepare a list of the utility's current employees. Assign tasks to each employee based on the person's skills and abilities. To the extent possible, try to minimize the number of different work activities assigned to each person but also keep in mind the need to provide opportunities for career advancement. One full-time staff year equals 260 days: (52 wk/yr)(5 days/wk) = 260 days/yr.

Consideration must be given to the amount and causes of employee overtime, and to identify possible ways to correct the problem. If excessive overtime and the related costs are due to lack of staff, this information can be used in a request for increased staffing.

You can expect to find that this "ideal" staffing arrangement does not exactly match up with your current employees' job assignments. Most likely, you will also find that the number of staff hours required does not exactly equal the number of staff hours available. Your responsibility as a manager is to create the best possible fit between the work to be done and the personnel/skills available to do it. In addition to shifting work assignments between employees, other options you might consider are contracting out some types of work, hiring part-time or seasonal staff, or setting up a second shift (to make fuller use of existing equipment). Of course, you may find that it is time to hire another full- or part-time operator.

15.42 Qualifications Profile

Hiring new employees requires careful planning before the personal interview process. In an effort to limit discriminatory hiring practices, the law and administrative policy have carefully defined the hiring methods and guidelines employers may use. The selection method and examination process used to evaluate applicants must be limited to the applicant's knowledge, skills, and abilities to perform relevant job-related activities. In all but rare cases, factors such as age and level of education may not be used to screen candidates in place of performance testing. A description of the duties and qualifications for the job must be clearly defined in writing. The job description may be used to develop a qualifications profile. This qualifications profile clearly and precisely identifies the required job qualifications. All job qualifications must be relevant to the actual job duties that will be performed in that position. The following list of typical job qualifications may be used to help you develop your own qualifications profiles with advice from a recruitment specialist.

1. General Requirements:
   a. Knowledge of methods, tools, equipment, and materials used in water/wastewater utilities,
   b. Knowledge of work hazards and applicable safety precautions,
   c. Ability to establish and maintain effective working relations with employees and the general public, and
   d. Possession of a valid state driver's license for the class of equipment the employee is expected to drive.
2. General Educational Development:
   a. Reasoning: Apply common sense understanding to carry out instructions furnished in oral, written, or diagrammatical form,
   b. Mathematical: Use a pocket calculator to make arithmetic calculations relevant to the utility's operation and maintenance processes, and
   c. Language: Communicate with fellow employees and train subordinates in work methods. Fill out maintenance report forms.
3. Specific Vocational Preparation: Three years of experience in water/wastewater utility operation and maintenance.
4. Interests: May or may not be relevant to knowledge, skills, and ability; for example, an interest in activities concerned with objects and machines, ecology, or business management.
5. Temperament: Must adjust to a variety of tasks requiring frequent change and must routinely use established standards and procedures.
6. Physical Demands: Medium to heavy work involving lifting, climbing, kneeling, crouching, crawling, reaching, hearing, and seeing. Must be able to lift and carry _____ number of pounds for a distance of _____ feet.
7. Working Conditions: The work is outdoors and involves wet conditions, cramped and awkward spaces, noise, risks of bodily injury, and exposure to weather.

For more information and details regarding staffing, we recommend reading Sections 5.3, "Applications and the Selection Process," and 5.5, "Employment Policies and Procedures," in *UTILITY MANAGEMENT* in this series of operator training manuals.

15.43 New Employee Orientation

During the first day of work, a new employee should be given all the information available in written and verbal form on the policies and practices of the utility including compensation, benefits, attendance expectations, alcohol and drug testing (if the utility does this), and employer/employee relations. Answer any questions from the new employee at this time and try to explain the overall structure of the utility as well as identify who can answer employee questions when they arise. Introduce the new employee to co-workers and tour the work area. Every utility should have a safety training session for all new employees and specific safety training for some job categories. Provide safety training (see Section 15.11, "Safety Program") for new employees on the first day of employment or as soon thereafter as possible. Establishing safe work practices is a very important function of management. The utility agency should have a checklist of all the topics or tasks that are a part of the orientation. Each item should be checked off as delivered and then signed by the employee.

QUESTIONS

Write your answers in a notebook and then compare your answers with those on page 382.

15.4A What do staffing responsibilities include?

15.4B What are two key personnel management concepts a manager should always keep in mind?

15.4C List the steps involved in a staffing analysis.

15.4D What is a qualifications profile?

15.4E When should a new employee's safety training begin?

15.5 COMMUNICATION

Good communication is an essential part of good management skills. Both written and oral communication skills are needed to effectively organize and direct the operation of a treatment facility. Remember that communication is a two-part process; information must be given and it must be understood. Good listening skills are as important in communication as the information you need to communicate. As the manager of a wastewater utility, you will need to communicate with employees, with your governing body, and with the public. Your communication style will be slightly different with each of these groups but you should be able to adjust easily to your audience.

15.50 Oral Communication

Oral communication may be informal, such as talking with employees, or it may be formal, such as giving a technical presentation. In both cases your words should be appropriate to the audience, for example, avoid technical jargon when talking with nontechnical audiences. As you talk, you should be observing your audience to be sure that what you are saying is getting across. If you are talking with an employee, it is a good idea to ask for feedback from the employee, especially if you are giving instructions. When the employee is talking, watch and listen carefully and clarify areas that seem unclear. Likewise, in a more formal presentation, watching your audience will give you feedback about how well your message is being received. Some tips for preparing a formal speech are given in Table 15.2.

15.51 Written Communication

Written communication is more demanding than oral communication and requires more careful preparation. Again, keep your audience in mind and use language that will be understood. Written communication requires more organization since you cannot clarify and explain ideas in response to your audience. Before you begin, you should have a clear idea of exactly what you wish to communicate, then keep your language as concise as possible. Extra words and phrases tend to confuse and clutter your message. Good writing skills develop slowly, but you should be able to find good writing classes in your community if you need help improving your skills. In addition, many publications and computer software programs are available that can assist you in writing memos, letters, press releases, resumes, monitoring reports, and the annual report.

Before you can write a report you must first organize your thoughts. Ask yourself, what is the objective of this report? Am I trying to persuade someone of something? What information is important to communicate in this report? For whom is the report being written? How can I make it interesting? What does the reader want to learn from this report?

After you have answered the above questions, the next step is to prepare a general outline of how you intend to proceed with the preparation of the report. List not only key topics, but try to list all of the related topics. Then arrange the key topics in sequence so there is a workable, smooth flow from one topic to the next. Do not attempt to make your outline perfect. It is just a guide. It should be flexible. As you write you will find that you need to remove nonessential points and expand on more important points.

You might, for example, outline the following points in preparation for writing a report on a polymer testing program.

- A problem condition of high turbidity was discovered
- Polymer testing offered the best means of reducing turbidity
- Funds, equipment, and material were acquired
- Operators were trained
- Tests were conducted
- Results and conclusions were reached
- Corrective actions were planned and taken
- Conclusion, the tests did or did not produce the anticipated results or correct the problem

Once you are fairly sure you have included all the major topics you will want to discuss, go through the outline and write down facts you want to include on each topic. As you work through it, you may decide to move material from one topic to another. The outline will help you organize your ideas and facts.

When your outline is complete, you will have the essentials of your report. Now you need to tailor it to the audience that will be reading it. Take a few minutes to think about your audi-

TABLE 15.2 TIPS FOR GIVING AN ORAL PRESENTATION

1. Arrive early. Give yourself plenty of time to become familiar with the room, practice using your audiovisuals, and make any necessary changes in room setup.
2. Be ready to avoid mistakes. Number the pages in your presentation and your audiovisuals. Check the order of the pages before the meeting begins.
3. Pace yourself. Don't speak too quickly, speak slowly and carefully. Keep a careful eye on audience reaction to be sure that you are speaking at a pace that can be understood.
4. Project yourself. Speak loudly and look at the audience. Do not talk with your back to the audience. Check that those in the back can hear you.
5. Be natural. Try not to read your presentation. Practice ahead of time so that you can speak normally and keep eye contact with your audience.
6. Connect with the audience. Try to smile and make eye contact with audience.
7. Involve the audience. Allow for audience questions and invite their comments.
8. Repeat audience questions. Always repeat the question so everyone can hear and to be sure that you hear the question correctly.
9. Know when to stop. Keep your remarks within the time allocated for your presentation and be aware that long, rambling speeches create a negative impression on the audience.
10. Use readable audiovisuals. Audiovisuals should enhance and reinforce your words. Be sure that all members of the audience can see and read your audiovisuals. Normal typewritten text is not readable on overheads; use large type so everyone can see. Use no more than five to seven key ideas per overhead.
11. Organize your presentation. Prepare an introduction, body of the speech, and conclusion. "Tell them what you're going to say, say it, and then tell them what you said." The presentation should have three to five main points presented in some logical order, for example, chronologically or from simple to complex.

ence. What information do they want? What aspects of the topics will they be most interested in reading? Each of the following groups may be interested in specific topics in the report. Consider these interests as you write.

1. Management

   Management will have specific interests that relate to the cost effectiveness of the program. A report to management should include a summary that presents the essential information, procedures used, an analysis of the data (including trends), and conclusions. Be sure to include complete cost information. Did the benefits warrant the costs? As a result of the tests, can future expenses be reduced? Backup information and field data can be included in an appendix for those who want more information.

2. Other Utilities

   Other utilities will be interested in costs but will also want more detailed information about how the program was performed. They will also be interested in the results and benefits of the program. Explain how the tests were done, the procedures, size of the crew, equipment used, source and availability of materials, difficulties encountered and how they were overcome.

3. Citizen Groups

   Citizens' interest will be more general. What is a polymer test and why is it needed? Is the polymer harmless? Will it injure fish or birds? How does the polymer test work? Who pays for the test? How much will it cost to implement results and will they be effective?

Your report may be written to include all of these groups. Adjust the outline to include the topics of interest to each group and identify the topics so readers can find the information most interesting to them. Keep the following information in mind as you write your report:

- Drafts. Good reports are not perfect the first time. Re-read and improve your report several times.
- Facts. Confine your writing to the facts and events that occurred. Include figures and statistics only when they make the report more effective. Include only the relevant facts. Large amounts of data should be put in the appendix. Do not clutter the report with unimportant data. Do not make subjective comments or offer personal opinions.
- Continuity. To be interesting and understood, a report must have continuity. It must make sense to the reader and be organized logically. In the report on the polymer testing, the report should be organized to show you had a problem, you had to find a way to identify where the problem existed, you did the testing, you identified the problems and the corrective actions.
- Effective. To be effective, a report should achieve the objective for which it was written. In this example we wanted to justify the costs for the program to management, help

other utilities in conducting a similar program, and help citizens to understand what we were doing and why.

- Candid. A good report should be frank and straightforward. Keep the language appropriate for the audience. Do not try to impress your readers with technical terms they do not understand. Your purpose is to communicate information. Keep the information accurate and easy to understand.

The annual report is an important part of the management of the utility. It is one of the most involved writing projects that the utility must put together. The annual report should be a review of what and how the utility operated during the past year and it should also include the goals for the next year. In many small communities the annual report may be presented orally to the city council rather than written. If this report is well written, it can be used to highlight accomplishments and provide support for future planning.

The first step to organizing the report is to make a list of three or four major accomplishments of the last year, then make a list of the top three goals for next year. These accomplishments and goals should be the focus of the report. The annual report should be a summary of the expenses, treatment services provided, and revenues generated over the last year. As you organize this information, keep those accomplishments in mind and let the data tell the story of how the utility accomplished last year's tasks. The data by itself may seem boring but as you organize the data it becomes a meaningful description of the year's accomplishments. Conclude with projections for next year. The facts and figures should tell the audience how you plan to accomplish your goals for the next year. The annual report may be simple or complex depending on your community needs. A sample Table of Contents for a medium-sized utility is given in Table 15.3. When you are finished, the annual report will be a valuable planning tool for the utility and can be used to build support for new projects.

TABLE 15.3 EXAMPLE OF TABLE OF CONTENTS FOR THE ANNUAL REPORT OF A UTILITY

Table of Contents	
Executive Summary	
Summary of the Treatment Process Including Flows and Costs	
Review of Goals and Objectives for the Year	
Special Projects Completed	
Professional Awards or Recognition for the Utility or Its Staff	
General Operating Conditions Including Regulatory Requirements	
Expectations for the Next Year—Goals and Objectives	
Recommended Changes for the Utility in Organizing, Staffing, Equipment, or Resources Summary	
System deficiencies including regulatory notifications and causes of problems	
Appendixes:	Operating Data
	Budget
	Information on Special Projects

QUESTIONS

Write your answers in a notebook and then compare your answers with those on page 382.

15.5A What kinds of communication skills are needed by a manager?

15.5B What are the most common written documents that a utility manager must write?

15.5C What should be included in the annual report?

15.6 CONDUCTING MEETINGS

As a utility manager, you will be asked to conduct meetings. These meetings may be with employees, your governing board, the public, or with other professionals in your field. Many new managers fail to prepare for these meetings and the meetings end up as a terrible waste of time. As a manager, you need to learn to conduct meetings in a way that is productive and guides the participants into an active role. The following steps should be taken to conduct a productive meeting.

Before the meeting:

- Prepare an agenda and distribute it to all participants.
- Find an adequate meeting room.
- Set a beginning and ending time for the meeting.

During the meeting:

- Start the meeting on time.
- Clearly state the purpose and objectives of the meeting.
- Involve all the participants.
- Do not let one or two individuals dominate the meeting.
- Keep the discussion on track and on time with the agenda.
- When the group makes a decision or reaches consensus, restate your understanding of the results.
- Make clear assignments for participants and review them with everyone during the meeting.
- Establish a time schedule for progress reports and/or completion of assignments.

After the meeting:

- Send out minutes of the meeting.
- Send out reminders, when appropriate, about any assignments made for participants, and the next meeting time.

QUESTIONS

Write your answers in a notebook and then compare your answers with those on page 382.

15.6A With whom may a utility manager be asked to conduct meetings?

15.6B What should be done before a meeting?

15.7 PUBLIC RELATIONS

15.70 Establish Objectives

The first step in organizing an effective public relations campaign is to establish objectives. The only way to know whether your program is a success is to have a clear idea of what you expect to achieve—for example, better customer relations, greater water conservation, and enhanced organizational credibility. Each objective must be specific, achievable, and measurable. It is also important to know your audience and tailor various elements of your public relations effort to specific groups you wish to reach, such as community leaders, school children, or the average customer. Your objective may be the same in each case, but what you say and how you say it will depend upon your target audience.

15.71 Utility Operations

Good public relations begin at home. Dedicated, service-oriented employees provide for better public relations than paid advertising or complicated public relations campaigns. For most people, contact with an agency employee establishes their first impression of the competence of the organization, and those initial opinions are difficult to change.

In addition to ensuring that employees are adequately trained to do their jobs and knowledgeable about the utility's operations, management has the responsibility to keep employees informed about the organization's plans, practices, and goals. Newsletters, bulletin boards, and regular, open communication between supervisors and subordinates will help build understanding and contribute to a team spirit.

Despite the old adage to the contrary, the customer is not always right. Management should try to instill among its employees the attitude that while the customer may be confused or unclear about the situation, everyone is entitled to courteous treatment and a factual explanation. Whenever possible, employees should phrase responses as positively, or neutrally, as possible, avoiding negative language. For example, "Your complaint" is better stated as "Your question." "You should have..." is likely to make the customer defensive, while "Will you please..." is courteous and respectful. "You made a mistake" emphasizes the negative, "What we'll do..." is a positive, problem-solving approach.

15.72 The Mass Media

We live in the age of communications, and one of the most effective and least expensive ways to reach people is through the mass media—radio, television, and newspapers. Each medium has different needs and deadlines, and obtaining coverage for your issue or event is easier if you are aware of these constraints. Television must have strong visuals, for example. When scheduling a press conference, provide an interesting setting and be prepared to suggest good shots to the reporter. Radio's main advantage over television and newspapers is immediacy, so have a spokesperson available and prepared to give the interview over the telephone, if necessary. Newspapers give more thorough, in-depth coverage to stories than do the broadcast media, so be prepared to spend extra time with print reporters and provide written backup information and additional contacts.

It is not difficult to get press coverage for your event or press conference if a few simple guidelines are followed:

1. Demonstrate that your story is newsworthy, that it involves something unusual or interesting.
2. Make sure your story will fit the targeted format (television, radio, or newspaper).
3. Provide a spokesperson who is interesting, articulate, and well prepared.

Whenever possible, all media contact should be restricted to the manager, a designated spokesperson, or, in some cases, an appointed attorney. An agency policy should clearly state that "employees must refer all mass media people to the proper spokesperson." If an employee wishes to contact the media, the employee should be accompanied by the manager.

15.73 Being Interviewed

Whether you are preparing for a scheduled interview or are simply contacted by the press on a breaking news story, here are some key hints to keep in mind when being interviewed.

1. Speak in personal terms, free of institutional jargon.
2. Do not argue or show anger if the reporter appears to be rude or overly aggressive.
3. If you don't know an answer, say so and offer to find out. Don't bluff.
4. If you say you will call back by a certain time, do so. Reporters face tight deadlines.
5. State your key points early in the interview, concisely and clearly. If the reporter wants more information, he or she will ask for it.
6. If a question contains language or concepts with which you disagree, don't repeat them, even to deny them.
7. Know your facts.
8. Never ask to see a story before it is printed or broadcast. Doing so indicates that you doubt the reporter's ability and professionalism.

15.74 Public Speaking

Direct contact with people in your community is another effective tool in promoting your utility. Though the audiences tend to be small, a personal, face-to-face presentation generally leaves a strong and long-lasting impact on the listener.

Depending upon the size of the organization, your utility may wish to establish a speaker's bureau and send a list of topics to service clubs in the area. Visits to high schools and college campuses can also be beneficial, and educators are often looking for new and interesting topics to supplement their curriculum.

Effective public speaking takes practice. It is important to be well prepared while retaining a personal, informal style. Find out how long your talk is expected to be, and don't exceed that time frame. Have a definite beginning, middle, and end to your presentation. Visual aids such as charts, slides, or models can assist in conveying your message. The use of humor and anecdotes can help to warm up the audience and build rapport between the speaker and the listener. Just be sure the humor is natural, not forced, and that the point of your story is accessible to the particular audience. Try to keep in mind that audiences only expect you to do your best. They are interested in learning about their utility and will appreciate that you are making a sincere effort to inform them about an important subject.

15.75 Telephone Contacts

First impressions are extremely important and, frequently, a person's first contact with your utility is over the telephone. A person who answers the phone in a courteous, pleasant, and helpful manner goes a long way toward establishing a friendly, cooperative atmosphere. Be sure anyone answering telephone inquiries receives appropriate training and conveys a positive image for the utility.

Following a few simple guidelines will help to start your utility off on the right note with your customers:

1. ANSWER CALLS PROMPTLY. Your conversation will get off to a better start if the phone is answered by the third or fourth ring.
2. IDENTIFY YOURSELF. This adds a personal note and lets the caller know whom he or she is talking to.
3. PAY ATTENTION. Don't conduct side conversations. Minimize distractions so you can give the caller your full attention, avoiding repetitions of names, addresses, and other pertinent facts.
4. MINIMIZE TRANSFERS. Nobody likes to get the runaround. Few things are more frustrating to a caller than being transferred from office to office, repeating the situation, problem, or concern over and over again. Transfer only those calls that must be transferred, and make certain you are referring the caller to the right person. Then, explain why you are transferring the call. This lets the caller know you are referring him or her to a co-worker for a reason and reassures the customer that the problem or question will be dealt with. In some cases, it may be better to take a message and have someone return the call than to keep transferring the customer's call.

15.76 Customer Inquiries

No single set of rules can possibly apply to all types of customer questions or complaints about utility service. There are, however, basic principles to follow in responding to inquiries and concerns.

1. BE PREPARED. Your employees should be familiar enough with your utility's organization, services, and policies to either respond to the question or complaint or locate the person who can.
2. LISTEN. Ask the customer to describe the problem and listen carefully to the explanation. Take written notes of the facts and addresses.
3. DON'T ARGUE. Callers often express a great deal of pent-up frustration in their contacts with a utility. Give the caller your full attention. Once you've heard them out, most people will calm down and state their problems in more reasonable terms.
4. AVOID JARGON. The average consumer lacks the technical knowledge to understand the complexities of water quality or wastewater treatment. Use plain, nontechnical language and avoid telling the consumer more than he or she needs to know.
5. SUMMARIZE THE PROBLEM. Repeat your understanding of the situation back to the caller. This will assure the customer that you understand the problem and offer the opportunity to clear up any confusion or missed communication.
6. PROMISE SPECIFIC ACTION. Make an effort to give the customer an immediate, clear, and accurate answer to the problem. Be as specific as possible *WITHOUT PROMISING SOMETHING YOU CAN'T DELIVER*.

In some cases, you may wish to have a representative of the utility visit the customer and observe the problem first hand. If the complaint involves water quality, take samples, if necessary, and report back to the customer to be sure the problem has been resolved.

Complaints can be a valuable asset in determining consumer acceptance and pinpointing problems. Customer calls are frequently your first indication that something may be wrong. Responding to complaints and inquiries promptly can save the utility money and staff resources, and minimize the number of customers who are inconvenienced. Keep a record or log of phone contacts indicating who called, when, why, problem, and promises or action taken. Still, education can greatly reduce complaints about utility services. Information brochures, utility bill inserts, and other educational tools help to inform customers and avoid future complaints.

15.77 Plant Tours

Tours of water and wastewater treatment plants can be an excellent way to inform the public about your utility's efforts. Political leaders, such as the City Council and members of the Board of Supervisors, should be invited and encouraged to tour the facilities, as should school groups and service clubs.

A brochure describing your utility's goals, accomplishments, operations, and processes can be a good supplement to the tour and should be handed out at the end of the visit. The more visually interesting the brochure is, the more likely that it will be read, and the use of color, photographs, graphics, or other design features is encouraged. If you have access to the necessary equipment, production of a videotape program about the utility can also add interest to the facility tour.

The tour itself should be conducted by an employee who is very familiar with plant operations and can answer the types of questions that are likely to arise. Consider including:

1. A description of the sources of water supply (if appropriate),
2. History of the plant, the years of operation, modifications, and innovations over the years,
3. Major plant design features, including plant capacity and safety features,

4. Observation of the treatment processes,
5. A visit to the laboratory, and
6. Anticipated improvements, expansions, and long-range plans for meeting future service needs.

Plant tours can contribute to a utility's overall program to gain financing for capital improvements. If the City Council or other governing board has seen the treatment process first hand, it is more likely to understand the need for enhancement and support future funding.

QUESTIONS

Write your answers in a notebook and then compare your answers with those on page 382.

15.7A What is the first step in organizing a public relations campaign?

15.7B How can employees be kept informed of the utility's plans, practices, and goals?

15.7C Which news medium is more likely to give a story thorough, in-depth coverage?

15.7D What is the key to effective public speaking?

15.7E How do customer complaints help a utility?

15.8 FINANCIAL MANAGEMENT

Financial management for a utility should include providing financial stability for the utility, careful budgeting, and providing capital improvement funds for future utility expansion. These three areas must be examined on a routine basis to ensure the continued operation of the utility. They may be formally reviewed on an annual basis or more frequently when the utility is changing rapidly. The utility manager should understand what is required for each of the three areas and be able to develop record systems that keep the utility on track and financially prepared for the future. Also, the manager must identify areas of financial concern that are or will place the utility agency in peril. This information must then be presented to the administration for action.

15.80 Financial Stability

How do you measure financial stability for a utility? Two very simple calculations can be used to help you determine how healthy and stable the finances are for the utility. These two calculations are the OPERATING RATIO and the COVERAGE RATIO. The operating ratio is a measure of the total revenues divided by the total operating expenses. The coverage ratio is a measure of the ability of the utility to pay the principal and interest on loans and bonds (this is known as *DEBT SERVICE*[4]) in addition to any unexpected expenses. A utility that is in good financial shape will have an operating ratio and coverage ratio above 1.0. In fact, most bonds and loans require the utility to have a coverage ratio of at least 1.25. As state and federal funds for utility improvements have become much more difficult to obtain, these financial indicators have become more important for utilities. Being able to show and document the financial stability of the utility is an important part of getting funding for more capital improvements.

The operating ratio is perhaps the simplest measure of a utility's financial stability. In essence, the utility must be generating enough revenue to pay its operating expenses. The actual ratio is usually computed on a yearly basis, since many utilities may have monthly variations that do not reflect the overall performance. The total revenue is calculated by adding up all revenue generated by user fees, hook-up charges, taxes or assessments, interest income, and special income. Next, determine the total operating expenses by adding up the expenses of the utility, including administrative costs, salaries, benefits, energy costs, chemicals, supplies, fuel, equipment costs, equipment replacement fund, principal and interest payments, and other miscellaneous expenses.

EXAMPLE 1

The total revenues for a utility are $1,686,000 and the operating expenses for the utility are $1,278,899. The debt service expenses are $560,000. What is the operating ratio? What is the coverage ratio?

Known		Unknown
Total Revenue, $	= $1,686,000	Operating Ratio
Operating Expenses, $	= $1,278,899	Coverage Ratio
Debt Service Expenses, $	= $560,000	

1. Calculate operating ratio.

$$\text{Operating Ratio} = \frac{\text{Total Revenue, \$}}{\text{Operating Expenses, \$}}$$

$$= \frac{\$1{,}686{,}000}{\$1{,}278{,}899}$$

$$= 1.32$$

2. Calculate non-debt expenses.

$$\begin{aligned}\text{Non-Debt Expenses, \$} &= \text{Operating Exp, \$} - \text{Debt Service Exp, \$}\\ &= \$1{,}278{,}899 - \$560{,}000\\ &= \$718{,}899\end{aligned}$$

3. Calculate coverage ratio.

$$\text{Coverage Ratio} = \frac{\text{Total Revenue, \$} - \text{Non-Debt Expenses, \$}}{\text{Debt Service Expenses, \$}}$$

$$= \frac{\$1{,}686{,}000 - \$718{,}899}{\$560{,}000}$$

$$= 1.73$$

These calculations provide a good starting point for looking at the financial strength of the utility. Both of these calculations use the total revenue for the utility, which is an important component for any utility budgeting. As managers, we often focus on the expense side and forget to look carefully at the revenue side of utility management. The fees collected by the utility, including hook-up fees and user fees, must accurately reflect the cost of providing service. These fees must be reviewed annually and they must be increased as expenses rise to maintain financial stability. Some other areas to examine on the

[4] *Debt Service. The amount of money required annually to pay the (1) interest on outstanding debts; or (2) funds due on a maturing bonded debt or the redemption of bonds.*

revenue side include how often and how well user fees are collected, the number of delinquent accounts, and the accuracy of meters (for drinking water utilities). Some small communities have found they can cut their administrative costs significantly by switching to a quarterly billing cycle. The utility must have the support of the community to determine and collect user fees, and the utility must keep track of revenue generation as carefully as resource spending.

15.81 Budgeting

Budgeting for the utility is perhaps the most challenging task of the year for many managers. The list of needs usually is much larger than the possible revenue for the utility. The only way for the manager to prepare a good budget is to have good records from the year before. A system of recording or filing purchase orders or a requisition records system (see Section 15.124, "Procurement Records") must be in place to keep track of expenses and prevent spending money that is not in the budget.

To budget effectively, a manager needs to understand how the money has been spent over the last year, the needs of the utility, and how the needs should be prioritized. The manager also must take into account cost increases that cannot be controlled while trying to minimize the expenses as much as possible. The following problem is an example of the types of decisions a manager must make to keep the budget in line while also improving service from the utility.

EXAMPLE 2

A wastewater pump that has been in operation for 25 years pumps a constant 600 GPM through 47 feet of dynamic head. The pump uses 6,071 kilowatt-hours of electricity per month, at a cost of $0.085 per kilowatt-hr. The old pump efficiency has dropped to 63 percent. Assuming a new pump that operates at 86 percent efficiency is available for $9,730.00, how long would it take to pay for replacing the old pump?

Known		Unknown
Electricity, kW-hr/mo	= 6,071 kW-hr/mo	New Pump Payback Time, yr
Electricity Cost, $/kW-hr	= $0.085/kW-hr	
Old Pump Efficiency, %	= 63%	
New Pump Efficiency, %	= 86%	
New Pump Cost, $	= $9,730	

1. Calculate old pump operating costs in dollars per month.

$$\text{Old Pump Operating Costs, \$/mo} = (\text{Electricity, kW-hr/mo})(\text{Electricity Cost, \$/kW-hr})$$
$$= (6{,}071\ \text{kW-hr/mo})(\$0.085/\text{kW-hr})$$
$$= \$516.04/\text{mo}$$

2. Calculate new pump operating electricity requirements.

$$\text{New Pump Electricity, kW-hr/mo} = (\text{Old Pump Electricity, kW-hr/mo})\frac{(\text{Old Pump Eff, \%})}{(\text{New Pump Eff, \%})}$$
$$= (6{,}071\ \text{kW-hr/mo})\frac{(63\%)}{(86\%)}$$
$$= 4{,}447\ \text{kW-hr/mo}$$

3. Calculate new pump operating costs in dollars per month.

$$\text{New Pump Operating Costs, \$/mo} = (\text{Electricity, kW-hr/mo})(\text{Electricity Cost, \$/kW-hr})$$
$$= (4{,}447\ \text{kW-hr/mo})(\$0.085/\text{kW-hr})$$
$$= \$378.03/\text{mo}$$

4. Calculate annual cost savings of new pump.

$$\text{Cost Savings, \$/yr} = (\text{Old Costs, \$/mo} - \text{New Costs, \$/mo})(12\ \text{mo/yr})$$
$$= (\$516.04/\text{mo} - \$378.03/\text{mo})(12\ \text{mo/yr})$$
$$= \$1{,}656.12/\text{yr}$$

5. Calculate the new pump payback time in years.

$$\text{Payback Time, yr} = \frac{\text{Initial Cost, \$}}{\text{Savings, \$/yr}}$$
$$= \frac{\$9{,}730.00}{\$1{,}656.12/\text{yr}}$$
$$= 5.9\ \text{years}$$

In this example, a payback time of 5.9 years is acceptable and would probably justify the expense for a new pump. This calculation was a simple payback calculation, which did not take into account the maintenance on each pump, depreciation, and inflation. Many excellent references are available from EPA to help utility managers make more complex decisions about purchasing new equipment.

The annual report should be used to help develop the budget so that long-term planning will have its place in the budgeting process. The utility manager must track revenue generation and expenses with adequate records to budget effectively. The manager must also get input from other personnel in the utility as well as community leaders as the budgeting process proceeds. This input from others is invaluable to gain support for the budget and to keep the budget on track once adopted.

15.82 Equipment Repair/Replacement Fund

To adequately plan for the future, every utility must have a repair/ replacement fund. The purpose of this fund is to generate additional revenue to pay for the repair and replacement of capital equipment as the equipment wears out. To prepare adequately for this repair/replacement, the manager should make a list of all capital equipment (this is called an asset inventory) and estimate the replacement cost for each item. The expected life span of the equipment must be used to determine how much money should be collected over time. When a treatment plant is new, the balance in the repair/replacement fund should be increasing each year. As the plant gets older, the funds will have to be used and the balance may get dangerously low as equipment breakdowns occur. Perhaps the hardest job for the utility manager is to maintain a positive balance in this account with the understanding that this account is not meant to generate a "profit" for the utility but rather to plan for future equipment needs. In wastewater facilities constructed with federal funds under the construction grant program, providing an adequate repair/replacement fund was one of the grant conditions, but if this repair/replacement fund hasn't been reviewed annually, it must be updated.

To set up a repair/replacement fund for your utility, you should first put together a list of the equipment required for each process in your utility. Once you have this list, you need to estimate the life expectancy of the equipment and the replacement cost. From this list you can predict the amount of money you should set aside each year so that when each piece of equipment wears out, you will have enough money to replace that piece of equipment. Several EPA publications listed in Section 15.14, "Additional Reading," at the end of this chapter are excellent references for utility planning.

15.83 Capital Improvements and Funding in the Future

A capital improvements fund must be a part of the utility budget and included in the operating ratio. Your responsibility as the utility manager is to be sure that everyone, your governing body and the public, understands the capital improvement fund is not a profit for the utility but a replacement fund to keep the utility operating in the future.

Capital planning starts with a look at changes in the community. Where are the areas of growth in the community, where are the areas of decline, and what are the anticipated changes in industry within the community? After identifying the changing needs in the community, you should examine the existing utility structure. Identify your weak spots (in the collection or distribution system, or with in-plant processes). Make a list of the areas that will be experiencing growth, weak spots in the system, and anticipated new regulatory requirements. The list should include expected capital improvements that will need to be made over the next year, two years, five years, and ten years. You can use the information in your annual reports and other operational logs to help compile the list.

Once you have compiled this information, prioritize the list and make a timetable for improving each of the areas. Starting at the top of the priority list, estimate the costs for improvements and incorporate these costs into your capital improvement budget. The calculations you have made previously, including operating ratio, coverage ratio, and payback time, will all be useful in prioritizing and streamlining your list of needs. Another useful ratio is the corrective to preventive maintenance ratio.

You may find that some of your capital improvement needs could be met in more than one way. How do you decide which of several options is the most cost-effective? How do you compare fundamentally different solutions? For example, assume your community's population is growing rapidly and you will need to increase treated water production by 20 percent by the end of the next ten years. Your existing wells cannot provide that amount of additional water and your filtration equipment is operating at 90 percent of design capacity. Possible solutions might include a combination of the following options, some of which might be implemented immediately while others might be brought on line in five or ten years:

- Rehabilitate some declining wells,
- Drill additional new wells,
- Develop an available surface water source,
- Install another filtration unit, and/or
- Install additional distribution system storage reservoirs.

To compare alternative plans, you will need to calculate the present value (or *PRESENT WORTH*[5]) of each plan; that is, the costs and benefits of each plan in today's dollars. This is done by identifying all the costs and benefits of each alternative plan over the same time period or time horizon. Costs should include not only the initial purchase price or construction costs, but also financing costs over the life of the loans or bonds and all operation and maintenance costs. Benefits include all of the revenue that would be produced by this facility or equipment, including connection and user fees. With the help of an experienced accountant, apply standard inflation, depreciation, and other economic discount factors to calculate the present value of all the benefits and costs of each plan during the same planning period. This will give you the cost of each plan in the equivalent of today's dollars.

Remember to involve all of your local officials and the public in this capital improvement budget so they understand what will be needed.

Long-term capital improvements such as a new plant or a new treatment process are usually anticipated in your 10-year or 20-year projection. These long-term capital improvements usually require some additional financing. The basic ways for a utility to finance capital improvements are through general obligation bonds, revenue bonds, or loan funding programs.

General obligation bonds or *ad valorem* (based on value) taxes are assessed based on property taxes. These bonds usually have a lower interest rate and longer payback time, but the total bond limit is determined for the entire community. This means that the water or wastewater utility will have available only a portion of the total bond capacity of the community. These bonds are not often used for funding water and wastewater utility improvements today.

The second type of bond, the revenue bond, is commonly used to fund utility improvements. This bond has no limit on the amount of funds available and the user charges provide repayment on the bond. To qualify for these bonds, the utility

[5] *Present Worth. The value of a long-term project expressed in today's dollars. Present worth is calculated by converting (discounting) all future benefits and costs over the life of the project to a single economic value at the start of the project. Calculating the present worth of alternative projects makes it possible to compare them and select the one with the largest positive (beneficial) present worth or minimum present cost.*

must show sound financial management and the ability to repay the bond. As the utility manager, you should be aware of the provisions of the bond. Be sure the bond has a call date, which is the first date when you can pay off the bond. The common practice is for a 20-year bond to have a 10-year call date and for a 15-year bond to have an 8-year call date. The bond will also have a call premium, which is the amount of extra funds needed to pay off the debt on the call date. You should try to get your bonds a call premium of no more than 102 percent par. This means that for a debt of $200,000 on the call date, the total payoff would be $204,000, which includes the extra two percent for the call premium. You will need to get help from a financial advisor to prepare for and issue the bonds. These advisors will help you negotiate the best bond structure for your community.

Special assessment bonds may be used to extend services into specific areas. The direct users pay the capital costs and the assessment is usually based on frontage or area of real estate. These special assessments carry a greater risk to investors but may be the best way to extend service to some areas.

The most common way to finance water and wastewater improvements in the past has been federal and state grant programs. The Block Grants from HUD are still available for some projects and Rural Utilities Service (RUS) loans may also be used as a funding source. In addition, state revolving fund (SRF) programs provide loans (but not direct grants) for improvements. The SRF program has already been implemented with wastewater improvements and the new Safe Drinking Water Regulations include an SRF program for funding water treatment improvements. These SRF programs are very competitive and utilities must provide evidence of sound financial management to qualify for these loans. You should contact your state regulatory agency to find out more about the SRF program in your state.

15.84 Financial Assistance

Many small utility systems need additional funds to repair and upgrade their systems. Potential funding sources include loans and grants from federal and state agencies, banks, foundations, and other sources. Some of the federal funding programs for small public utility systems include:

- Appalachian Regional Commission (ARC),
- Department of Housing and Urban Development (HUD) (provides Community Development Block Grants),
- Economic Development Administration (EDA),
- Indian Health Service (IHS), and
- Rural Utilities Service (RUS) (formerly Farmer's Home Administration (FmHA) and Rural Development Administration (RDA)).

For additional information, see "Financing Assistance Available for Small Public Water Systems," by Susan Campbell, Benjamin W. Lykins, Jr., and James A. Goodrich, *Journal American Water Works Association*, June 1993, pages 47-53.

Another valuable contact is the Environmental Financing Information Network (EFIN), which provides information on financing alternatives for state and local environmental programs and projects in the form of abstracts of publications, case studies, and contacts. Contact Environmental Financing Information Network, U.S. Environmental Protection Agency (EPA), EFIN (mail code 2731R), Ariel Rios Building, 1200 Pennsylvania Avenue, NW, Washington, DC 20460. Phone (202) 564-4994 and FAX (202) 565-2587.

Also many states have one or more special financing mechanisms for small public utility systems. These funds may be in the form of grants, loans, bonds, or revolving loan funds. Contact your state for more information.

QUESTIONS

Write your answers in a notebook and then compare your answers with those on page 383.

15.8A List the three main areas of financial management for a utility.

15.8B How is a utility's operating ratio calculated?

15.8C Why is it important for a manager to consult with other utility personnel and with community leaders during the budget process?

15.8D How can long-term capital improvements be financed?

15.8E What is a revenue bond?

15.9 OPERATIONS AND MAINTENANCE

15.90 The Manager's Responsibilities

A utility manager's specific operation and maintenance (O & M) responsibilities vary depending on the size of the utility. At a small utility, the manager may oversee all utility operations while also serving as chief operator and supervising a small staff of operations and maintenance personnel. In larger utility agencies, the manager may have no direct, day-to-day responsibility for operations and maintenance but is ultimately responsible for efficient, cost-effective operation of the entire utility. Whether large or small, every utility needs an effective operations and maintenance program.

15.91 Purpose of O & M Programs

The purpose of O & M programs is to maintain design functionality (capacity) and/or to restore the system components to their original condition and thus functionality. Stated another way, does the system perform as designed and intended? The ability to effectively operate and maintain a water or wastewater utility so it performs as intended depends greatly on proper design (including selection of appropriate materials and equipment), construction and inspection, acceptance, and system start-up. Permanent system deficiencies that affect O & M of the system are frequently the result of these phases. O & M staff should be involved at the beginning of each project, including planning, design, construction, acceptance, and start-up. When a utility system is designed with future O & M considerations in mind, the result is a more effective O & M program in terms of O & M cost and performance.

Effective O & M programs are based on knowing what components make up the system, where they are located, and the condition of the components. With that information, proactive maintenance can be planned and scheduled, rehabilitation needs identified, and long-term Capital Improvement Programs (CIPs) planned and budgeted. High-performing agencies have all developed performance measurements of their O & M program and track the information necessary to evaluate performance.

15.92 Types of Maintenance

Water or wastewater system maintenance can be either a proactive or a reactive activity. Commonly accepted types of maintenance include three classifications: corrective maintenance, preventive maintenance, and predictive maintenance.

Corrective maintenance, including emergency maintenance, is reactive. For example, a piece of equipment or a system is allowed to operate until it fails, with little or no scheduled maintenance occurring prior to the failure. Only when the equipment or system fails is maintenance performed. Reliance on reactive maintenance will always result in poor system performance, especially as the system ages. Utility agencies taking a corrective maintenance approach are characterized by:

- The inability to plan and schedule work,
- The inability to budget adequately,
- Poor use of resources,
- A high incidence of equipment and system failures, and
- High corrective and/or repair costs with related overtime pay.

Emergency maintenance involves two types of emergencies: normal emergencies and extraordinary emergencies. Public utilities are faced with normal emergencies on a daily basis, whether it is a water main break or a blockage in a sewer. Normal emergencies can be reduced by an effective maintenance program. Extraordinary emergencies, such as high-intensity rainstorms, hurricanes, floods, and earthquakes, will always be unpredictable occurrences. However, the effects of extraordinary emergencies on the utility's performance can be minimized by implementation of a planned maintenance program and development of a comprehensive emergency response plan (see Section 15.10).

Preventive maintenance is proactive and is defined as a programmed, systematic approach to maintenance activities. This type of maintenance will always result in improved system performance except in the case where major chronic problems are the result of design and/or construction flaws that cannot be corrected by O & M activities. Proactive maintenance is performed on a periodic (preventive) basis or an as-needed (predictive) basis. Preventive maintenance can be scheduled on the basis of specific criteria such as equipment operating time since the last maintenance was performed, or passage of a certain amount of time (calendar period). For example, performing television inspection of a gravity sewer system on a five-year cycle (calendar period) would require that 20 percent of the system be televised each year. At the end of the five-year period, the cycle would start over again. Similarly, lubrication of motors is frequently based on running time.

The major elements of a good preventive maintenance program include the following:

1. Planning and scheduling,
2. Records management,
3. Spare parts management,
4. Cost and budget control,
5. Emergency repair procedures, and
6. Training program.

Some benefits of taking a preventive maintenance approach are:

1. Maintenance can be planned and scheduled,
2. Work backlog can be identified,
3. Adequate resources necessary to support the maintenance program can be budgeted,
4. Capital Improvement Program (CIP) items can be identified and budgeted for, and
5. Human and material resources can be used effectively.

Predictive maintenance, which is also proactive, is a method of establishing baseline performance data, monitoring performance criteria over a period of time, and observing changes in performance so that failure can be predicted and maintenance can be performed on a planned, scheduled basis. Knowing the condition of the system makes it possible to plan and schedule maintenance on an "as-required" basis and thus avoid unnecessary maintenance. An example of predictive maintenance in a collection system would be visually inspecting manholes or monitoring flows; when changes in flow conditions are observed, cleaning is scheduled accordingly.

In reality, every agency operates their system with corrective and emergency maintenance, preventive maintenance, and predictive maintenance methods. The goal, however, is to reduce the corrective and emergency maintenance efforts by performing preventive maintenance, which will minimize system failures that result in stoppages and overflows.

System performance is frequently a reliable indicator of how the system is operated and maintained. Agencies that rely primarily on corrective maintenance as their method of operating and maintaining the system are never able to focus on preventive and predictive maintenance since most of their resources are directed at corrective maintenance activities and it is difficult to free up these resources to begin developing preventive maintenance programs. For an agency to develop an effective proactive maintenance program, it must add initial resources over and above those currently existing in order to establish preventive and predictive maintenance programs.

15.93 Benefits of Managing Maintenance

The goal of managing maintenance is to minimize investments of labor, materials, money, and equipment. In other words, we want to manage our human and material resources as effectively as possible, while delivering a high level of service to our customers. The benefits of an effective operation and maintenance program are as follows:

- Ensuring the availability of facilities and equipment as intended.
- Maintaining the reliability of the equipment and facilities as designed. Utility systems are required to operate 24 hours per day, 7 days per week, 365 days per year. Reliability is a critical component of the operation and maintenance program. If equipment and facilities are not reliable, then the ability of the system to perform as designed is impaired.
- Maintaining the value of the investment. Water and wastewater systems represent major capital investments for communities and are major capital assets of the community. If maintenance of the system is not managed, equipment and facilities will deteriorate through normal use and age. Maintaining the value of the capital asset is one of the utility manager's major responsibilities. Accomplishing this goal requires ongoing investment to maintain existing facilities and equipment and extend the life of the system, and establishing a comprehensive O & M program.
- Obtaining full use of the system throughout its design life.
- Collecting accurate information and data on which to base the operation and maintenance of the system and justify requests for the financial resources necessary to support it.

QUESTIONS

Write your answers in a notebook and then compare your answers with those on page 383.

15.9A What is the purpose of an operation and maintenance (O & M) program?

15.9B What are the three common types of maintenance?

15.9C List the major elements of a good preventive maintenance program.

15.10 EMERGENCY RESPONSE PLAN

Contingency planning is an essential facet of utility management and one that is often overlooked. Natural disasters such as floods, earthquakes, hurricanes, forest fires, avalanches, and blizzards are a more or less routine occurrence for some utilities. Although utilities in various locations will be vulnerable to somewhat different kinds of natural disasters, the effects of these disasters in many cases will be quite similar. When such catastrophic emergencies occur, the utility must be prepared to minimize the effects of the event and have a plan for rapid recovery. Such preparation should be a specific obligation of every utility manager.

In addition to emergencies associated with natural disasters, all utilities suffer from common problems such as equipment breakdowns and leaking pipes. During the past few years there has also been an increasing amount of vandalism, civil disorder, toxic spills, and employee strikes, which have threatened to disrupt utility operations. In observing today's international tension and the potential for nuclear war or the effects of terrorist-induced chemical or biological warfare, water and wastewater utilities must seriously consider how to respond.

As a first step toward developing an effective emergency response plan, each utility should make an assessment of its own vulnerability during various types of emergency situations. The following steps should be taken in assessing the vulnerability of a system:

1. Identify and describe the system components,
2. List assumed disaster characteristics,
3. Estimate disaster effects on system components,
4. Estimate customer demand for service following a potential disaster, and
5. Identify key system components that would be primarily responsible for system failure.

If the assessment shows a system is unable to meet estimated requirements because of the failure of one or more critical components, the vulnerable elements have been identified. Repeating this procedure using several "typical" disasters will usually point out system weaknesses. Frequently, the same vulnerable element appears for a variety of assumed disaster events. Eliminate as many system weaknesses as possible. For the remaining weaknesses identified in the vulnerability assessment, estimate the probable extent of damage that would be caused by each of the emergency events most likely to occur in the area and develop and implement a comprehensive plan of action for responding to emergencies.

Although all system elements are important for the utility to function at peak efficiency, experience with disasters points out elements that are most subject to disruption. These elements are:

1. The absence of trained personnel to make critical decisions and carry out orders,
2. The loss of power to the utility's facilities,
3. An inadequate amount of supplies and materials, and
4. Inadequate communication equipment.

An emergency operations plan need not be too detailed, since all types of emergencies cannot be anticipated and a complex response program can be more confusing than helpful. Supervisory personnel must have a detailed description of their responsibilities during emergencies. They will need information, supplies, equipment, and the assistance of trained personnel. All these can be provided through a properly constructed emergency operations plan that is not extremely detailed.

The following outline can be used as the basis for developing an emergency operations plan:

1. Make a vulnerability assessment,
2. Inventory organizational personnel,
3. Provide for a recovery operation (plan),
4. Provide training programs for operators in carrying out the plan,
5. Coordinate with local and regional agencies such as the health, police, and fire departments to develop procedures for carrying out the plan,
6. Establish a communications procedure, and
7. Provide protection for personnel, plant equipment, records, and maps.

By following these steps, an emergency plan can be developed and maintained even though changes in personnel may occur. "Emergency Simulation" training sessions, including the use of standby power, equipment, and field test equipment will ensure that equipment and personnel are ready at times of emergency.

A list of phone numbers for operators to call in an emergency should be prepared and posted by a phone for emergency use. The list should include:

1. Plant supervisor,
2. Director of public works or head of utility agency,
3. Police,
4. Fire,
5. Doctor (2 or more),
6. Ambulance (2 or more),
7. Hospital (2 or more),

If appropriate for your utility, also include the following phone numbers on the emergency list:

8. Chlorine supplier and manufacturer,
9. CHEMTREC (800-424-9300) for the hazardous chemical spills; sponsored by the Manufacturing Chemists Association,
10. U.S. Coast Guard's National Response Center (800-424-8802),
11. Local and state poison control centers, and
12. Local hazardous materials spill response team.

You should prepare a list for your plant NOW, if you have not already done so.

QUESTIONS

Write your answers in a notebook and then compare your answers with those on page 383.

15.10A What is the first step toward an effective contingency plan for emergencies?

15.10B Why is too detailed an emergency operations plan not needed or even desirable?

15.11 SAFETY PROGRAM

The utility manager is responsible for the safety of the agency's personnel and the public exposed to the water or wastewater utility's operations. Therefore, the manager must develop and administer an effective safety program and must provide new employee safety training as well as ongoing training for all employees. The basic elements of a safety program include a safety policy statement, safety training and promotion, and accident investigation and reporting.

15.110 Policy Statement

A safety policy statement should be prepared by the top management of the utility. The purpose of the statement is to let employees know that the safety program has the full support of the agency and its management. The statement should:

1. Define the goals and objectives of the program,
2. Identify the persons responsible for each element of the program,
3. Affirm management's intent to enforce safety regulations, and
4. Describe the disciplinary actions that will be taken to enforce safe work practices.

Give a copy of the safety policy statement to every current employee and each new employee during orientation. Figure 15.1 is an example of a safety policy statement for a water supply utility.

15.111 Responsibilities

The following list of responsibilities for safety is from the *PLANT MANAGER'S HANDBOOK.*[6] These responsibilities represent a typical list but may be incomplete if your agency is subject to stricter local, state, and/or federal regulations than what is shown here. Check with your safety professional.

Management has the responsibility to:

1. Formulate a written safety policy,
2. Provide a safe workplace,
3. Set achievable safety goals,
4. Provide adequate training, and
5. Delegate authority to ensure that the program is properly implemented.

[6] *PLANT MANAGER'S HANDBOOK (MOP SM-4), Water Environment Federation (WEF), no longer in print.*

SAFETY POLICY STATEMENT

It is the policy of the Las Vegas Valley Water District that every employee shall have a safe and healthy place to work. It is the District's responsibility; its greatest asset, the employees and their safety.

When a person enters the employ of the District, he or she has a right to expect to be provided a proper work environment, as well as proper equipment and tools, so that they will be able to devote their energies to the work without undue danger. Only under such circumstances can the association between employer and employee be mutually profitable and harmonious. It is the District's desire and intention to provide a safe workplace, safe equipment, proper materials, and to establish and insist on safe work methods and practices at all times. It is a basic responsibility of all District employees to make the SAFETY of human beings a matter for their daily and hourly concern. This responsibility must be accepted by everyone who works at the District, regardless of whether he or she functions in a management, supervisory, staff, or the operative capacity. Employees must use the SAFETY equipment provided; Rules of Conduct and SAFETY shall be observed; and SAFETY equipment must not be destroyed or abused. Further, it is the policy of the Water District to be concerned with the safety of the general public. Accordingly, District employees have the responsibility of performing their duties in such a manner that the public's safety will not be jeopardized.

The joint cooperation of employees and management in the implementation and continuing observance of this policy will provide safe working conditions and relatively accident-free performance to the mutual benefit of all involved. The Water District considers the SAFETY of its personnel to be of primary importance, and asks each employee's full cooperation in making this policy effective.

Fig. 15.1 Example safety policy statement
(Permission of Las Vegas Valley Water District)

The manager is the key to any safety program. Implementation and enforcement of the program is the responsibility of the manager. The manager also has the responsibility to:

1. Ensure that all employees are trained and periodically retrained in proper safe work practices,
2. Ensure that proper safety practices are implemented and continued as long as the policy is in effect,
3. Investigate all accidents and injuries to determine their cause; prepare and implement a written procedure for preventing reoccurrence of the accident,
4. Institute corrective measures where unsafe conditions or work methods exist,
5. Ensure that equipment, tools, and the work are maintained to comply with established safety standards, and
6. Respond to and correct all identified unsafe conditions and activities.

The utility operators are the direct beneficiaries of a safety program. They share the responsibility to:

1. Observe prescribed work procedures with respect to personal safety and that of their co-workers,
2. Report any detected hazard to a manager immediately,
3. Report any accident, including a minor accident that causes minor injuries,
4. Report near-miss accidents so that hazards can be removed or procedures changed to avoid problems in the future,
5. Use all protective devices and safety equipment supplied to reduce the possibility of injury, and
6. An employee has the right, privilege, and responsibility to refuse to perform unsafe orders, but must be aware of the need to avoid retribution for nonperformance.

15.112 Hazard Communication Program and Worker Right-To-Know (RTK) Laws

In the past few years there has been an increased emphasis nationally on hazardous materials and wastes. Much of this attention has focused on hazardous and toxic waste dumps and the efforts to clean them up after the long-term effects on human health were recognized. Each year thousands of new chemical compounds are produced for industrial, commercial, and household use. Frequently, the long-term effects of these chemicals are unknown. As a result, federal and state laws have been enacted to control all aspects of hazardous materials handling and use. These laws are more commonly known as Worker Right-To-Know (RTK) laws. The law that has had the greatest impact is the *OCCUPATIONAL SAFETY AND HEALTH ACT OF 1970 (OSHA)*,[7] Public Law 91-596, which took effect on December 29, 1970.

---

[7] *OSHA (O-shuh). The Williams-Steiger **O**ccupational **S**afety and **H**ealth **A**ct of 1970 (OSHA) is a federal law designed to protect the health and safety of workers, including collection system and treatment plant operators. It regulates the design, construction, operation and maintenance of industrial plants and wastewater collection systems and treatment plants. The Act does not apply directly to municipalities, EXCEPT in those states that have approved plans and have asserted jurisdiction under Section 18 of the OSHA Act. HOWEVER, CONTRACT OPERATORS AND PRIVATE FACILITIES DO HAVE TO COMPLY WITH OSHA REQUIREMENTS. Wastewater treatment plants have come under stricter regulation in all phases of activity as a result of OSHA standards. OSHA also refers to the federal and state agencies which administer the OSHA regulations.*

The intent of the OSHA regulations is to create a place of employment that is free from recognized hazards that could cause serious physical harm or death to an operator (or other employee). In many cases, the individual states have the authority under the OSHA Standard to develop their own state RTK laws and most states have adopted their own laws. The Federal OSHA Standard 29 *CFR*[8] 1910.1200—Hazard Communication forms the basis of most of these state RTK laws. Unfortunately, state laws vary significantly from state to state. The state laws that have been passed are at least as stringent as the federal standard and, in most cases, are even more stringent. State laws are also under continuous revision and, because a strong emphasis is being placed on hazardous materials and worker exposure, state laws can be expected to be amended in the future to apply to virtually everybody in the workplace. Managers should become familiar with both the state and federal OSHA regulations that apply to their organizations.

The basic elements of a hazard communication program are described in the following paragraphs.

1. Identify Hazardous Materials—While there are thousands of chemical compounds that could be considered hazardous, focus on the ones to which operators and other personnel in your utility are most likely to be exposed.

2. Obtain Chemical Information and Define Hazardous Conditions—Once the inventory of hazardous chemicals is complete, the next step is to obtain specific information on each of the chemicals. This information is generally incorporated into a standard format form called the *MATERIAL SAFETY DATA SHEET (MSDS)*.[9] This information is commonly available from manufacturers. Many agencies request an MSDS when the purchase order is generated and will refuse to accept delivery of the shipment if the MSDS is not included. Figure 15.2 shows OSHA's standard MSDS form, but other forms are also acceptable provided they contain all of the required information.

   The purpose of the MSDS is to have a readily available reference document that includes complete information on common names, safe exposure level, effects of exposure, symptoms of exposure, flammability rating, type of first-aid procedures, and other information about each hazardous substance. Operators should be trained to read and understand the MSDS forms. The forms themselves should be stored in a convenient location where they are readily available for reference.

3. Properly Label Hazards—Once the physical, chemical, and health hazards have been identified and listed, a labeling and training program must be implemented. To meet labeling requirements on hazardous materials, specialized labeling is available from a number of sources, including commercial label manufacturers. Exemptions to labeling requirements do exist, so consult your local safety regulatory agency for specific details.

4. Train Operators—The last element in the hazard communication program is training and making information available to utility personnel. A common-sense approach eliminates the confusing issue of which of the thousands of substances operators should be trained for, and concentrates on those that they will be exposed to or use in everyday maintenance routines.

The Hazard Communication Standard and the individual state requirements are obviously a very complex set of regulations. Remember, however, the ultimate goal of these regulations is to provide additional operator protection. These standards and regulations, once the intent is understood, are relatively easy to implement.

15.113 Lockout/Tagout Procedures
(OSHA 29 CFR 1910.147, "Control of Hazardous Energy")

Wastewater treatment plant operators are frequently required to work on mechanical, hydraulic, and electrical systems. All of these have the potential of causing serious injury unless some action is taken to prevent the accidental start-up of a system while maintenance work is being performed.

Many forms of energy are present in a typical wastewater collection and treatment system: electrical energy in a pump station, mechanical energy in a surge-relief valve, thermal energy from high temperatures, and energy associated with hydraulic pressure such as static head pressure in a pipeline. Typical equipment systems where these hazards exist are motor control centers, pumps, belts, and spring-loaded devices such as check valves. It is essential for the operator to take appropriate precautions to prevent the accidental discharge of energy from these systems because it could cause serious injury or death. Before performing any maintenance, be sure you thoroughly understand the systems you are working on and the precautions you need to take.

The following is a general lockout/tagout procedure that you can modify for your specific needs when performing maintenance where potential hazards exist.

Whenever it is necessary to work on a piece of equipment or machinery, the following procedure shall be adhered to:

1. The main power source shall be locked out, in the OFF position, with a multiple lockout device and padlock. (*NOTE:* the pulling of fuses shall not be considered a substitute for locking out.)

2. All pneumatic, hydraulic, and other fluid lines shall be bled, drained, purged, or blanked off to prevent pressure and/or contents from causing movement and mechanisms under spring tension or compression shall be blocked, clamped, or chained in position. Blocks may also be needed on some machinery or equipment to prevent gravitational movement.

3. Never place locks where the disconnect can be bypassed at other locations.

4. Multiple lockout devices shall be used in all cases. Each employee who is engaged in working on machinery or equipment shall install a lock on the multiple lockout device. Each lock shall be individually keyed to prevent unautho-

---

[8] *Code of Federal Regulations (CFR). A publication of the United States Government which contains all of the proposed and finalized federal regulations, including environmental regulations.*

[9] *Material Safety Data Sheet (MSDS). A document which provides pertinent information and a profile of a particular hazardous substance or mixture. An MSDS is normally developed by the manufacturer or formulator of the hazardous substance or mixture. The MSDS is required to be made available to employees and operators whenever there is the likelihood of the hazardous substance or mixture being introduced into the workplace. Some manufacturers are preparing MSDSs for products that are not considered to be hazardous to show that the product or substance is NOT hazardous.*

Material Safety Data Sheet May be used to comply with OSHA's Hazard Communication Standard, 29 CFR 1910.1200 Standard must be consulted for specific requirements.	**U.S. Department of Labor** Occupational Safety and Health Administration (Non-Mandatory Form) Form Approved OMB No. 1218-0072
IDENTITY *(As Used on Label and List)*	*Note: Blank spaces are not permitted. If any item is not applicable, or no information is available, the space must be marked to indicate that.*

Section I

Manufacturer's Name	Emergency Telephone Number
Address *(Number, Street, City, State, and ZIP Code)*	Telephone Number for Information
	Date Prepared
	Signature of Preparer *(optional)*

Section II — Hazardous Ingredients/Identity Information

Hazardous Components (Specific Chemical Identity: Common Name(s))	OSHA PEL	ACGIH TLV	Other Limits Recommended	%*(optional)*

Section III — Physical/Chemical Characteristics

Boiling Point		Specific Gravity (H_2O = 1)	
Vapor Pressure (mm Hg.)		Melting Point	
Vapor Density (Air = 1)		Evaporation Rate (Butyl Acetate = 1)	
Solubility in Water			
Appearance and Odor			

Section IV — Fire and Explosion Hazard Data

Flash Point (Method Used)	Flammable Limits	LEL	UEL
Extinguishing Media			
Special Fire Fighting Procedures			
Unusual Fire and Explosion Hazards			

(Reproduce locally) OSHA 174, Sept. 1985

Fig 15.2 OSHA's standard Material Safety Data Sheet

Section V — Reactivity Data

Stability	Unstable		Conditions to Avoid
	Stable		

Incompatibility *(Materials to Avoid)*

Hazardous Decomposition or Byproducts

Hazardous Polymerization	May Occur		Condition to Avoid
	Will Not Occur		

Section VI — Health Hazard Data

Route(s) of Entry:	Inhalation?	Skin?	Ingestion?

Health Hazards *(Acute and Chronic)*

Carcinogenicity	NTP?	IARC Monographs?	OSHA Regulated?

Signs and Symptoms of Exposure

Medical Conditions
Generally Aggravated by Exposure

Emergency and First Aid Procedures

Section VII — Precautions for Safe Handling and Use

Steps to Be Taken in Case Material is Released or Spilled

Waste Disposal Method

Precautions to be Taken in Handling and Storing

Other Precautions

Section VIII — Control Measures

Respiratory Protection (Specify Type)

Ventilation	Local Exhaust	Special
	Mechanical *(General)*	Other

Protective Gloves	Eye Protection

Other Protective Clothing or Equipment

Work/Hygienic Practices

Page 2 * USGPO 1986-491-529/45775

Fig 15.2 OSHA's standard Material Safety Data Sheet (continued) rized removal from the lockout device. The operator using the padlock is the only person normally authorized to remove it (see step 9 below).

5. A tagging system shall be used that will advise other employees of the work being performed.
6. *CAUTION:* It is mandatory that the disconnect or valve be tried to make sure it cannot be moved to ON. In addition, the machine controls themselves shall be tried to make certain that the energy is OFF.
7. This procedure shall be followed at the start of each shift or workday.
8. The immediate supervisor shall check to determine that all requirements of this procedure have been complied with and shall give the personnel involved the clearance to proceed with the required task
9. In case of emergency, supervisory personnel may authorize removal of the lock after all precautions have been taken. An attempt shall be made to contact the person who initially locked out the equipment (see step 5). A check shall be made to ensure that all personnel, tools, and other items will not be exposed to harm or injury before the lock is removed. The operator signing the tag shall be notified of the removal of the lock at the start of the operator's next scheduled workday.

Your agency should develop a site-specific written lockout/tagout procedure and provide training to all of the operators who may use it. Each operator who must use lockout/tagout to safely do the job must be aware of applicable energy sources and the methods necessary for their effective isolation and control.

If you hire a contractor you must inform each other of your respective lockout/tagout procedures. You must ensure that you and your employees or co-workers comply with the restrictions and prohibitions of the contractor's program.

Periodic inspections of your lockout/tagout program are also required (at least annually) to ensure that the requirements of the program are being followed. The inspection(s) must be done by someone other than the one(s) using the procedure.

15.114 Confined Space Entry Procedures

CONFINED SPACES[10,11] pose significant risks for a large number of workers, including many utility operators. OSHA has therefore defined very specific procedures to protect the health and safety of operators whose jobs require them to enter or work in a confined space. The regulations (which can be found in the Code of Federal Regulations at 29 CFR 1910.146) require conditions in the confined space to be tested and evaluated before anyone enters the space. If conditions exceed OSHA's limits for safe exposure, additional safety precautions must be taken and a confined space entry permit (Figure 15.3) must be approved by the appropriate authorities prior to anyone entering the space.

The managers of water and wastewater utilities may or may not be involved in the day-to-day details of enforcing the agency's confined space policy and procedures. However, every utility manager should be aware of the current OSHA requirements and should ensure that the utility's policies not only comply with current regulations, but that the agency's policies are vigorously enforced for the safety of all operators.

Managers should provide confined space entry training for all employees who will be working in or near confined spaces. It is a criminal offense to order an employee into a confined space without complying with all of the applicable regulations.

15.115 Reporting

Regardless of the size of the utility, recordkeeping is an important part of an effective safety program. All injuries should be reported, even if they are minor in nature, so as to establish a record in case the injury develops into a serious injury. It may be difficult at a later date to prove whether the accident occurred on or off the job and this information may determine who is responsible for the costs. The responsibility for reporting accidents affects several levels of personnel. First, of

[10] *Confined Space. Confined space means a space that:*
A. Is large enough and so configured that an employee can bodily enter and perform assigned work; and
B. Has limited or restricted means for entry or exit (for example, manholes, tanks, vessels, silos, storage bins, hoppers, vaults, and pits are spaces that may have limited means of entry); and
C. Is not designed for continuous employee occupancy.
(Definition from the Code of Federal Regulations (CFR) Title 29 Part 1910.146.)

[11] *Confined Space, Permit-Required (Permit Space). A confined space that has one or more of the following characteristics:*
- *Contains or has a potential to contain a hazardous atmosphere,*
- *Contains a material that has the potential for engulfing an entrant,*
- *Has an internal configuration such that an entrant could be trapped or asphyxiated by inwardly converging walls or by a floor which slopes downward and tapers to a smaller cross section, or*
- *Contains any other recognized serious safety or health hazard.*

(Definition from the Code of Federal Regulations (CFR) Title 29 Part 1910.146.)

Confined Space Pre-Entry Checklist/Confined Space Entry Permit

Date and Time Issued: ________ Date and Time Expires: ________ Job Site/Space I.D.: ________

Job Supervisor: ________ Equipment to be worked on: ________ Work to be performed: ________

Standby personnel: ________ ________ ________

1. Atmospheric Checks: Time ________ Oxygen ________ % Toxic ________ ppm
   Explosive ________ % LEL Carbon Monoxide ________ ppm

2. Tester's signature: ________
3. Source isolation: (No Entry) N/A Yes No
   Pumps or lines blinded, disconnected, or blocked ( ) ( ) ( )
4. Ventilation Modification: N/A Yes No
   Mechanical ( ) ( ) ( )
   Natural ventilation only ( ) ( ) ( )
5. Atmospheric check after isolation and ventilation: Time ________
   Oxygen ________ % > 19.5% < 23.5% Toxic ________ ppm < 10 ppm H_2S
   Explosive ________ % LEL < 10% Carbon Monoxide ________ ppm < 35 ppm CO

Tester's signature: ________

6. Communication procedures: ________
7. Rescue procedures: ________
   ____

8. Entry, standby, and backup persons Yes No
   Successfully completed required training? ( ) ( )
   Is training current? ( ) ( )
9. Equipment: N/A Yes No
   Direct reading gas monitor tested ( ) ( ) ( )
   Safety harnesses and lifelines for entry and standby persons ( ) ( ) ( )
   Hoisting equipment ( ) ( ) ( )
   Powered communications ( ) ( ) ( )
   SCBAs for entry and standby persons ( ) ( ) ( )
   Protective clothing ( ) ( ) ( )
   All electric equipment listed for Class I, Division I, Groups A, B, C, and D, and nonsparking tools ( ) ( ) ( )
10. Periodic atmospheric tests:
    Oxygen: ____% Time ____; ____% Time ____; ____% Time ____; ____% Time ____;
    Explosive: ____% Time ____; ____% Time ____; ____% Time ____; ____% Time ____;
    Toxic: ____ppm Time ____; ____ppm Time ____; ____ppm Time ____; ____ppm Time ____;
    Carbon Monoxide: ____ppm Time ____; ____ppm Time ____; ____ppm Time ____; ____ppm Time ____;

We have reviewed the work authorized by this permit and the information contained herein. Written instructions and safety procedures have been received and are understood. Entry cannot be approved if any brackets ( ) are marked in the "No" column. This permit is not valid unless all appropriate items are completed.

Permit Prepared By: (Supervisor) ________ Approved By: (Unit Supervisor) ________

Reviewed By: (CS Operations Personnel) ________
(Entrant) (Attendant) (Entry Supervisor)

This permit to be kept at job site. Return job site copy to Safety Office following job completion.

Fig. 15.3 Confined space pre-entry checklist/confined space entry permit course, is the injured person. Next, it is the responsibility of the supervisor, and finally, it is the

Responsibility of Management to review the causes and take steps to prevent such accidents from happening in the future.

Accident report forms may be very simple. However, they must record all details required by law and all data needed for statistical purposes. The forms shown in Figures 15.4 and 15.5 are examples for you to consider for use in your utility.

In addition to reports needed by the utility, other reports may be required by state or federal agencies. For example, vehicle accident reports must be submitted to local police departments. If a member of the public is injured, additional forms are needed because of possible subsequent claims for damages. If the accident is one of occupational injury, causing lost time, other reports may be required. Follow-up investigations to identify causes and responsibility may require the development of other specific types of record forms.

Emphasis on the prevention of future accidents cannot be overstressed. We must identify the causes of accidents and implement whatever measures are necessary to protect operators from becoming injured.

S SAFETY FIRST
A ACCIDENTS COST LIVES
F FASTER IS NOT ALWAYS BETTER
E EXPECT THE UNEXPECTED
T THINK BEFORE YOU ACT
Y YOU CAN MAKE THE DIFFERENCE

ACCIDENTS DON'T JUST HAPPEN... THEY ARE CAUSED!

QUESTIONS

Write your answers in a notebook and then compare your answers with those on page 383.

15.11A Who is responsible for the implementation of a safety program?

15.11B What should be included in a utility's policy statement on safety?

15.11C List the basic elements of a hazard communication program.

15.11D What are a manager's responsibilities for ensuring the safety of operators entering or working in confined spaces?

15.11E Why should a report be prepared whenever an injury occurs?

15.12 RECORDKEEPING

15.120 Purpose of Records

Accurate records are essential for effective utility management and to satisfy legal requirements. Records are also a valuable source of information. They can save time when trouble develops and provide proof that problems were identified and solved. Pertinent and complete records should be used as a basis for plant operation, interpreting results of water or wastewater treatment, preparing preventive maintenance programs, and preparing budget requests. When accurately kept, records provide a sound basis for design of future changes or expansions of the treatment plant. If legal questions or problems occur in connection with the treatment processes or the operation of the plant, accurate and complete records will provide evidence of what actually occurred and what procedures were followed.

15.121 Types of Records

Many different types of records are required for effective management and operation of water supply, treatment, and distribution system facilities or wastewater collection, treatment, and disposal facilities. The following Sections, 15.122 through 15.126, describe some of the most important types of records that should be kept and Section 15.127 discusses how long records need to be kept.

15.122 Equipment and Maintenance Records

A good plant maintenance effort depends heavily on good recordkeeping. You will need to keep accurate records to monitor the operation and maintenance of each piece of plant equipment. Equipment control cards and work orders can be used to:

- Record important equipment data such as make, model, serial number, and date purchased,
- Record maintenance and repair work performed to date,
- Anticipate preventive maintenance needs, and
- Schedule future maintenance work.

Whenever a piece of equipment is changed, repaired, or tested, the work performed should be recorded on an equipment history card of some type. Complete, up-to-date equipment records will enable the plant operators to evaluate the reliability of equipment and will provide the basis for a realistic preventive maintenance program.

15.123 Plant Operations Data

Plant operations logs can be as different as the treatment plants whose information they record. The differences in amount, nature, and format of data are so significant that any attempt to prepare a "typical" log would be very difficult. For detailed information and example recordkeeping forms for a water supply utility, see *WATER TREATMENT PLANT OPERATION*, Volume I, Chapter 10, "Plant Operation," Section 10.6, "Operating Records and Reports." For further information regarding the types of daily and monthly operating records required for a wastewater treatment plant, see *OPERATION OF WASTEWATER TREATMENT PLANTS*, Volume II, Chapter 19, "Records and Report Writing," Section 19.11, "Type of Records."

15.124 Procurement Records

Ordering repair parts and supplies usually is done when the on-hand quantity of a stocked part or chemical falls below the reorder point, a new item is added to stock, or an item has

Date ____________

Name of injured employee ____________ Employee # ____________ Area ____________

Date of accident ____________ Time ________ Employee's Occupation ____________

Location of accident ____________ Nature of injury ____________

Name of doctor ____________ Address ____________

Name of hospital ____________ Address ____________

Witnesses (name & address) ____________

PHYSICAL CAUSES

Indicate below by an "X" whether, in your opinion, the accident was caused by:

- _____ Improper guarding
- _____ Defective substances or equipment
- _____ Hazardous arrangement
- _____ Improper illumination
- _____ Improper dress or apparel
- _____ Not listed (describe briefly) ____________
- _____ No mechanical cause
- _____ Working methods
- _____ Lack of knowledge or skill
- _____ Wrong attitude
- _____ Physical defect

UNSAFE ACTS

Sometimes the injured person is not directly associated with the causes of an accident. Using an "X" to represent the injured worker and an "O" to represent any other person involved, indicate whether, in your opinion, the accident was caused by:

- _____ Operating without authority
- _____ Failure to secure or warn
- _____ Working at unsafe speed
- _____ Made safety device inoperative
- _____ Unsafe equipment or hands instead of equipment
- _____ No unsafe act
- _____ Not listed (describe briefly) ____________
- _____ Unsafe loading, placement & etc.
- _____ Took unsafe position
- _____ Worked on moving equipment
- _____ Teased, abused, distracted & etc.
- _____ Did not use safe clothing or personal protective equipment

What job was the employee doing? ____________

What specific action caused the accident? ____________

What steps will be taken to prevent recurrence? ____________

Date of Report ____________ Immediate Supervisor ____________

REVIEWING AUTHORITY

Comments:	Comments:
Safety Officer Date	Department Director Date

Fig. 15.4 Supervisor's accident report

INJURED: COMPLETE THIS SECTION

Name ______________________________ Age_____ Sex ______

Address ______________________________

Title ____________________ Dept. Assigned ____________________

Place of Accident ______________________________

Street or Intersection ______________________________

Date ____________________ Hour______ A.M. ______ P.M. ______

Type of Job You Were Doing When Injured

____

Object Which Directly Injured You Part of Body Injured

____

How Did Accident Happen? (Be specific and give details; use back of sheet if necessary.)

____

Did You Report Accident or Exposure at Once? (Explain "No") Yes ☐ No ☐

Did You Report Accident or Exposure to Supervisor? Yes ☐ No ☐
Give Name

Were There Witnesses to Accident or Exposure? Yes ☐ No ☐
Give Names

Did You See a Doctor? (If Yes, Give Name) Yes ☐ No ☐

Are You Going to See a Doctor? (Give Name) Yes ☐ No ☐

____________________ ____________________
Date Signature

SUPERVISOR: COMPLETE THIS SECTION — (Return to Personnel as Soon as Possible)

Was an Investigation of Unsafe Conditions and/or
Unsafe Acts Made? If Yes, Please Submit Copy. Yes ☐ No ☐

Was Injured Intoxicated or Behaving Inappropriately
at Time of Accident? (Explain "Yes") Yes ☐ No ☐

Date Disability Commenced ______________ Last Day Wages Earned ______________ Date Back on Job ______________

Date Report Completed ______________ 19____ Signed By ______________

Title ______________

Distribution: Canary - Department Head, Pink - Supervisor, White - Personnel

Fig. 15.5 Accident report

been requested that is not stocked. Most organizations require employees to submit a requisition (Figure 15.6) when they need to purchase equipment or supplies. When the requisition has been approved by the authorized person (a supervisor or purchasing agent, in most cases) the items are ordered using a form called a purchase order. A purchase order contains a number of important items, including: (1) the date, (2) a complete description of each item and quantity needed, (3) prices, (4) the name of the vendor, and (5) a purchase order number.

A copy of the purchase order should be retained in a suspense file or on a clipboard until the ordered items arrive. This procedure helps keep track of the items that have been ordered but have not yet been received. The copy of the purchase order should be attached to the item delivery slip and then sent to an accounts payable procedure. This will ensure that items received will be paid for.

All supplies should be processed through the storeroom immediately upon arrival. When an item is received, it should be so recorded on an inventory card. The inventory card will keep track of the numbers of an item in stock, when last ordered, cost, and other information. Furthermore, by always logging in supplies immediately upon receipt, you are in a position to reject defective or damaged shipments and control shortages or errors in billing. Many utilities now use personal computers to keep track of orders and deliveries.

15.125 Inventory Records

An inventory consists of the supplies the treatment plant needs to keep on hand to operate the facility. These maintenance supplies may include repair parts, spare valves, electrical supplies, tools, and lubricants. The purpose of maintaining an inventory is to provide needed parts and supplies quickly, thereby reducing equipment downtime and work delays.

In deciding what supplies to stock, keep in mind the economics involved in buying and stocking an item as opposed to depending on outside availability to provide needed supplies. Is the item critical to continued plant or process operation? Should certain frequently used repair parts be kept on hand? Does the item have a shelf life?

Inventory costs can be held to a minimum by keeping on hand only those parts and supplies for which a definite need exists or that would take too long to obtain from an outside vendor. A "definite need" for an item is usually demonstrated by a history of regular use. Some items may be infrequently used but may be vital in the event of an emergency; these items should also be stocked. Take care to exclude any parts and supplies that may become obsolete, and do not stock parts for equipment scheduled for replacement.

15.126 Personnel Records

Documentation of all aspects of personnel management provides an important measure of legal protection for the utility. If an employee files a lawsuit alleging discriminatory hiring practices or treatment, harassment, breach of contract, or other grievance, the utility's ability to defend its practices and procedures will depend almost entirely on complete and accurate records. Similarly, if an employee is injured on the job, written records can help establish whether the utility was responsible for the accident.

Each personnel action should be fully documented in writing and filed. Even verbal discussions of a supervisor with an employee about job performance should be summarized in writing upon completion of the conversation. Also, file copies of all written warnings and disciplinary actions.

An employee's personnel file should also contain a complete record of accomplishments, certificates earned, commendations received, and formal performance reviews.

As previously mentioned, personnel records often contain sensitive, confidential information; therefore, access to these records should be closely controlled.

15.127 Disposition of Plant Records

An important question is how long records should be kept. As a general rule, records should be kept as long as they may be useful or as long as legally required. Some information will become useless after a short time, while other data may be valuable for many years. Data that might be used for future design or expansion should be kept indefinitely. Laboratory data will always be useful and should be kept indefinitely. Regulatory agencies may require you to keep certain water quality analyses (bacteriological test results) and customer complaint records on file for specified time periods (10 years for chemical analyses and bacteriological tests).

Even if old records are not consulted every day, this does not lessen their potential value. For orderly records handling and storage, set up a schedule to periodically review old records and to dispose of those records that are no longer needed. A decision can be made when a record is established regarding the time period for which it must be retained.

QUESTIONS

Write your answers in a notebook and then compare your answers with those on page 383.

15.12A What are some of the benefits of keeping complete, up-to-date records?

15.12B List the important items usually contained on a purchase order.

15.12C What is the purpose of maintaining an inventory?

15.12D As a general rule, how long should utility records be kept?

15.13 ACKNOWLEDGMENTS

During the writing of this manual, Lynne Scarpa, Phil Scott, Chris Smith, and Rich von Langen, all members of California Water Environment Association (CWEA), provided many excellent materials and suggestions for improvement. Their generous contributions are greatly appreciated.

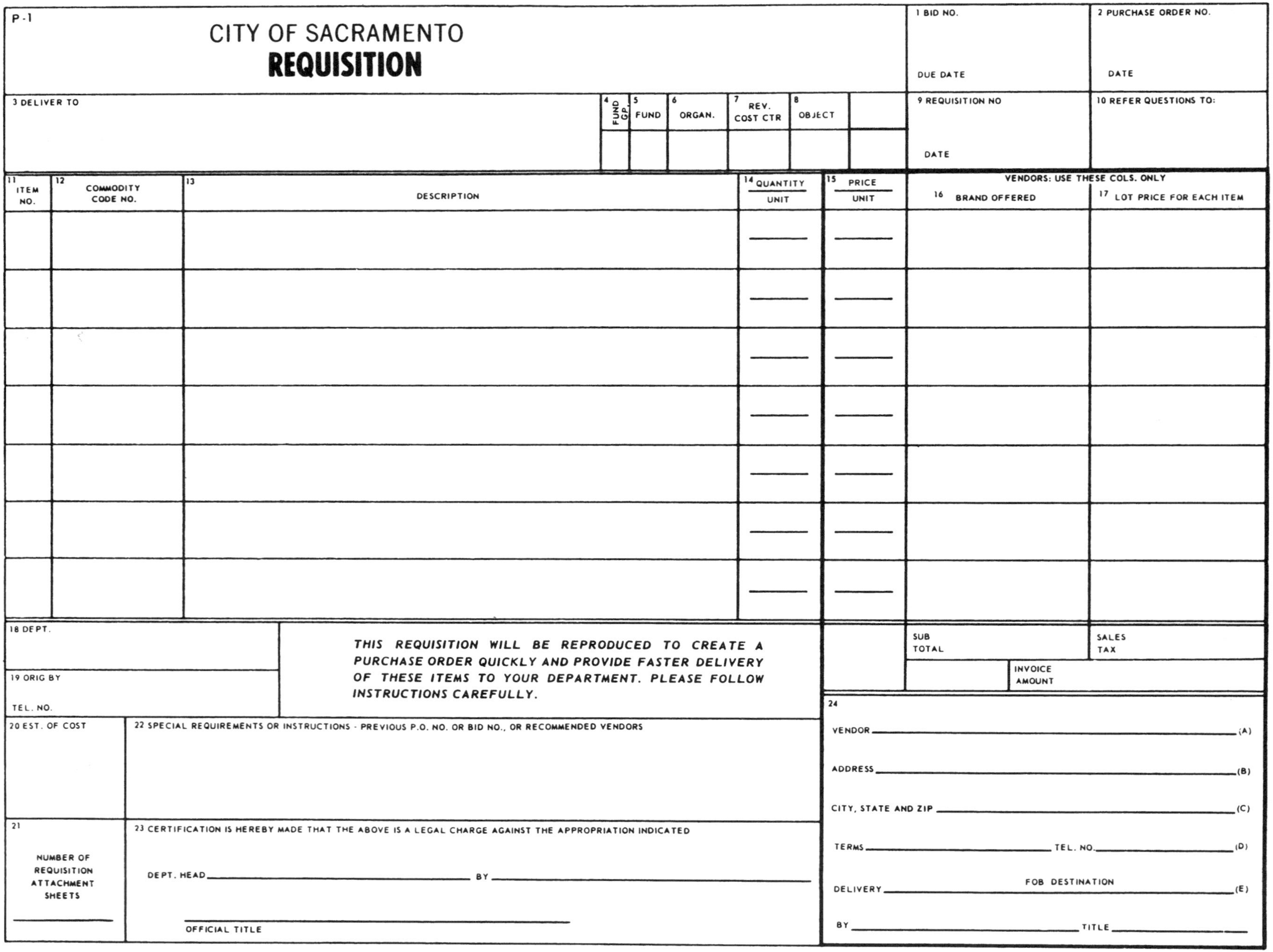

P-1

CITY OF SACRAMENTO
REQUISITION

1 BID NO.
DUE DATE

2 PURCHASE ORDER NO.
DATE

3 DELIVER TO

4 FUND GP.	5 FUND	6 ORGAN.	7 REV. COST CTR	8 OBJECT	

9 REQUISITION NO
DATE

10 REFER QUESTIONS TO:

11 ITEM NO.	12 COMMODITY CODE NO.	13 DESCRIPTION	14 QUANTITY UNIT	15 PRICE UNIT	16 BRAND OFFERED	17 LOT PRICE FOR EACH ITEM

VENDORS: USE THESE COLS. ONLY (16, 17)

18 DEPT.

19 ORIG BY

TEL. NO.

THIS REQUISITION WILL BE REPRODUCED TO CREATE A PURCHASE ORDER QUICKLY AND PROVIDE FASTER DELIVERY OF THESE ITEMS TO YOUR DEPARTMENT. PLEASE FOLLOW INSTRUCTIONS CAREFULLY.

SUB TOTAL

SALES TAX

INVOICE AMOUNT

20 EST. OF COST

22 SPECIAL REQUIREMENTS OR INSTRUCTIONS - PREVIOUS P.O. NO. OR BID NO., OR RECOMMENDED VENDORS

21

NUMBER OF REQUISITION ATTACHMENT SHEETS

23 CERTIFICATION IS HEREBY MADE THAT THE ABOVE IS A LEGAL CHARGE AGAINST THE APPROPRIATION INDICATED

DEPT. HEAD ________ BY ________

OFFICIAL TITLE

24

VENDOR ________ (A)

ADDRESS ________ (B)

CITY, STATE AND ZIP ________ (C)

TERMS ________ TEL. NO. ________ (D)

DELIVERY ____ FOB DESTINATION ____ (E)

BY ________ TITLE ________

Fig. 15.6 Example requisition/purchase order form

15.14 ADDITIONAL READING

1. *WATER UTILITY MANAGEMENT* (M5). Obtain from American Water Works Association (AWWA), Bookstore, 6666 West Quincy Avenue, Denver, CO 80235. Order No. 30005. ISBN 0-89867-063-2. Price to members, $62.50; nonmembers, $92.50; price includes cost of shipping and handling.
2. *A WATER AND WASTEWATER MANAGER'S GUIDE FOR STAYING FINANCIALLY HEALTHY*, U.S. Environmental Protection Agency. Obtain from National Technical Information Service (NTIS), 5285 Port Royal Road, Springfield, VA 22161. Order No. PB90-114455. EPA No. 430-09-89-004. Price, $28.50, plus $5.00 shipping and handling per order.
3. *WASTEWATER UTILITY RECORDKEEPING, REPORTING AND MANAGEMENT INFORMATION SYSTEMS*, U.S. Environmental Protection Agency. Obtain from National Technical Information Service (NTIS), 5285 Port Royal Road, Springfield, VA 22161. Order No. PB83-109348. EPA No. 430-9-82-006. Price, $34.00, plus $5.00 shipping and handling per order.
4. *SUPERVISION: CONCEPTS AND PRACTICES OF MANAGEMENT*, Eighth Edition, 2001, Hilgert, Raymond L., and Edwin Leonard, Jr. Obtain from Thomson Learning, Attn: Order Fulfillment, PO Box 6904, Florence, KY 41022. ISBN 0-324-01389-2. Price, $89.95, plus shipping and handling.

Please answer the discussion and review questions before continuing with the Objective Test.

DISCUSSION AND REVIEW QUESTIONS

Chapter 15. MANAGEMENT

Write the answers to these questions in your notebook before continuing with the Objective Test on page 384. The purpose of these questions is to indicate to you how well you understand the material in the chapter.

1. What are the different types of demands on a utility manager?
2. List the basic functions of a manager.
3. What can happen without adequate utility planning?
4. What is the purpose of an organizational plan?
5. Define the following terms:
   1. Authority,
   2. Responsibility,
   3. Delegation, and
   4. Accountability.
6. When has a supervisor successfully delegated?
7. What information should be provided to a new employee during orientation?
8. With whom do managers need to communicate?
9. What information should be included in the utility's annual report?
10. List four steps that can be taken during a meeting to make sure it is a productive meeting.
11. What happens any time you or a member of your utility comes in contact with the public?
12. What attitude should management try to develop among its employees regarding the consumer?
13. What is the value of consumer complaints?
14. How do you measure financial stability for a utility?
15. How can a manager prepare a good budget?
16. What can happen when agencies rely primarily on corrective maintenance to keep the system running?
17. What types of information should be included in an emergency operations plan?
18. What is the intent of the OSHA regulations?
19. List four major types of records a utility must maintain.
20. Why is it important to document all aspects of personnel management?

SUGGESTED ANSWERS

Chapter 15. MANAGEMENT

Answers to questions on page 353.

15.0A Local utility demands on a utility manager include protection from environmental disasters with a minimum investment of money.

15.0B Changes in the environmental workplace are created by changes in the workforce and advances in technology.

Answers to questions on page 355.

15.1A The functions of a utility manager include planning, organizing, staffing, directing, and controlling.

15.1B In small communities, the community depends on the manager to handle everything.

Answers to questions on page 356.

15.2A Utility planning should include operational personnel, local officials (decision makers), and the public.

15.3A The purpose of an organizational plan is to show who reports to whom and to identify the lines of authority.

15.3B Effective delegation is uncomfortable for many managers since it requires giving up power and responsibility. Many managers believe that they can do the job better than others, they believe that other employees are not well trained or experienced, and they are afraid of mistakes. The utility manager retains some responsibility even after delegating to another employee and, therefore, the manager is often reluctant to delegate or may delegate the responsibility but not the authority to get the job done.

15.3C An important and often overlooked part of delegation is follow-up by the supervisor.

Answers to questions on page 358.

15.4A Staffing responsibilities include hiring new employees, training employees, and evaluating job performance.

15.4B The two personnel management concepts a manager should always keep in mind are "job-related" and "documentation."

15.4C The steps involved in a staffing analysis include:

1. List the tasks to be performed,
2. Estimate the number of staff hours per year required to perform each task,
3. List the utility's current employees,
4. Assign tasks based on each employee's skills and abilities, and
5. Adjust the work assignments as necessary to achieve the best possible fit between the work to be done and the personnel/skills available to do it.

15.4D A qualifications profile is a clear statement of the knowledge, skills, and abilities a person must possess to perform the essential job duties of a particular position.

15.4E A new employee's safety training should begin on the first day of employment or as soon thereafter as possible.

Answers to questions on page 360.

15.5A A manager needs both written and oral communication skills.

15.5B The most common written documents that a utility manager must write include memos, business letters, press releases, monitoring reports, and the annual report.

15.5C The annual report should be a review of what and how the utility operated during the past year and also the goals for the next year.

Answers to questions on page 360.

15.6A A utility manager may be asked to conduct meetings with employees, the governing board, the public, and with other professionals in your field.

15.6B Before a meeting (1) prepare an agenda and distribute, (2) find an adequate meeting room, and (3) set a beginning and ending time.

Answers to questions on page 363.

15.7A The first step in organizing a public relations campaign is to establish objectives so you will have a clear idea of what you expect to achieve.

15.7B Employees can be informed about the utility's plans, practices, and goals through newsletters, bulletin boards, and regular, open communication between supervisors and subordinates.

15.7C Newspapers give more thorough, in-depth coverage to stories than do the broadcast media.

15.7D Practice is the key to effective public speaking.

15.7E Complaints can be a valuable asset in determining consumer acceptance and pinpointing problems. Customer calls are frequently the first indication that something may be wrong. Responding to complaints and inquiries promptly can save the utility money and staff resources, and minimize customer inconvenience.

Answers to questions on page 366.

15.8A The three main areas of financial management for a utility include providing financial stability for the utility, careful budgeting, and providing capital improvement funds for future utility expansion.

15.8B The operating ratio for a utility is calculated by dividing total revenues by total operating expenses.

15.8C It is important for a manager to get input from other personnel in the utility as well as community leaders as the budgeting process proceeds in order to gain support for the budget and to keep the budget on track once adopted.

15.8D The basic ways for a utility to finance capital improvements are through general obligation bonds, revenue bonds, or loan funding programs.

15.8E A revenue bond is a debt incurred by the community, often to finance utility improvements. User charges provide repayment on the bond.

Answers to questions on page 368.

15.9A The purpose of an O & M program is to maintain design functionality and/or to restore the system components to their original condition and thus functionality. That is, to ensure that the system performs as designed and intended.

15.9B Commonly accepted types of maintenance include corrective maintenance, preventive maintenance, and predictive maintenance.

15.9C The major elements of a good preventive maintenance program include the following:

1. Planning and scheduling,
2. Records management,
3. Spare parts management,
4. Cost and budget control,
5. Emergency repair procedures, and
6. Training program.

Answers to questions on page 369.

15.10A The first step toward an effective contingency plan for emergencies is to make an assessment of vulnerability. Then, a comprehensive plan of action can be developed and implemented.

15.10B A detailed emergency operations plan is not needed since all types of emergencies cannot be anticipated and a complex response program can be more confusing than helpful.

Answers to questions on page 376.

15.11A The utility manager is responsible for the safety of the agency's personnel and the public exposed to the water or wastewater utility's operations. Therefore, the manager must develop and administer an effective safety program and must provide new employee safety training as well as ongoing training for all employees.

15.11B A utility's policy statement on safety should include:

1. A definition of the goals and objectives of the program,
2. Identification of the persons responsible for each element of the program,
3. A statement affirming management's intent to enforce safety regulations, and
4. A description of the disciplinary actions that will be taken to enforce safe work practices.

15.11C The basic elements of a hazard communication program include the following:

1. Identify hazardous materials,
2. Obtain chemical information and define hazardous conditions,
3. Properly label hazards, and
4. Train operators.

15.11D A utility manager may or may not be involved in the day-to-day details of enforcing the agency's confined space policy and procedures. However, every utility manager should be aware of the current OSHA requirements and should ensure that the utility's policies not only comply with current regulations, but that the agency's policies are vigorously enforced for the safety of all operators.

15.11E All injuries should be reported, even if they are minor in nature, so as to establish a record in case the injury develops into a serious injury. It may be difficult at a later date to prove whether the accident occurred on or off the job and this information may determine who is responsible for the costs.

Answers to questions on page 379.

15.12A Keeping complete, up-to-date records contributes to more effective utility management, helps to satisfy legal requirements, provides valuable operations and maintenance information, and assists in preparing budget requests. Accurate records provide a sound basis for design of future changes or expansions of the treatment plant. If legal questions or problems occur in connection with the treatment processes or the operation of the plant, accurate and complete records will provide evidence of what actually occurred and what procedures were followed.

15.12B A purchase order usually contains: (1) the date, (2) a complete description of each item and quantity needed, (3) prices, (4) the name of the vendor, and (5) a purchase order number.

15.12C The purpose of maintaining an inventory is to provide needed parts and supplies quickly, thereby reducing equipment downtime and work delays.

15.12D As a general rule, utility records should be kept for as long as they may be useful or as long as legally required.

OBJECTIVE TEST

Chapter 15. MANAGEMENT

Please write your name and mark the correct answers on the answer sheet, as directed at the end of Chapter 1, Volume I. There may be more than one correct answer to each multiple-choice question.

True-False

1. Planning must be done before the other managing functions.
   1. True
   2. False
2. Planning is one of the most important functions of utility management and one of the easiest.
   1. True
   2. False
3. An employee can serve only one supervisor.
   1. True
   2. False
4. Good records about past events can be used to demonstrate your pattern of lawful behavior over time.
   1. True
   2. False
5. Good communication is an essential part of good management skills.
   1. True
   2. False
6. Oral communication is more demanding than written communication and requires more careful preparation.
   1. True
   2. False
7. Each objective in a public relations campaign must be specific, achievable, and measurable.
   1. True
   2. False
8. The customer is always right.
   1. True
   2. False
9. Complaints can be a valuable asset in determining consumer acceptance and pinpointing problems.
   1. True
   2. False
10. Education can greatly increase complaints about utility service.
    1. True
    2. False
11. A utility must keep track of revenue generation as carefully as resource spending.
    1. True
    2. False
12. For a utility agency to develop an effective proactive maintenance program, the agency must add initial resources over and above those currently existing in order to establish preventive and predictive maintenance programs.
    1. True
    2. False
13. The utility manager is responsible for the safety of the agency's personnel and the public exposed to the utility's operations.
    1. True
    2. False
14. Federal and state laws that have been enacted to control all aspects of hazardous materials handling and use are known as Worker Right-To-Know (RTK) Laws.
    1. True
    2. False
15. A copy of the purchase order should be retained in a suspense file or on a clipboard until the ordered items arrive.
    1. True
    2. False

Best Answer (Select only the closest or best answer.)

16. Planning consists of which tasks performed by a manager?
    1. Deciding who does what work and delegating authority
    2. Determining the goals, policies, procedures, and other elements to achieve the goals and objectives of the agency
    3. Guiding, teaching, motivating, and supervising
    4. Taking the steps necessary to ensure that essential activities are performed so that objectives will be achieved as planned
17. What is the purpose of a utility's organizational plan?
    1. To show how the public can participate in management activities
    2. To show the utility's organizational policy
    3. To show the utility's plan for future expansion
    4. To show who reports to whom and to identify the lines of authority

18. What does accountability mean?
    1. When manager gives power/responsibility to an employee, employee ensures manager is informed
    2. When manager gives resources to get a job done, employee minimizes use of resources
    3. When manager receives information on how or why something happened
    4. When manager receives results after an activity is completed
19. How can a manager start to determine how many employees are needed?
    1. Compare needs with other similar utilities
    2. Estimate budget allocation and funds available for staff
    3. Identify equipment available and staff requirements for equipment
    4. Prepare a detailed list of all the tasks to be performed
20. What is the first step in writing a report?
    1. Make a list of all topics
    2. Organize thoughts
    3. Prepare a general outline
    4. Write the abstract
21. How can a report be effective?
    1. By achieving the objective for which it was written
    2. By being frank and straightforward
    3. By covering the facts and events that occurred
    4. By making sense to the reader and being organized logically
22. What is the first step in organizing an effective public relations campaign?
    1. Contact media
    2. Establish objectives
    3. Request interviews
    4. Talk to employees
23. How do you measure the financial stability for a utility?
    1. Coverage ratio
    2. Profit
    3. Rate of return
    4. Revenue
24. How is the operating ratio measured?
    1. Total revenues divided by debt service expenses
    2. Total revenues divided by total operating expenses
    3. Total revenues minus non-debt expenses
    4. Total revenues minus total operating expenses
25. What does a manager need in order to prepare a good budget?
    1. Good bookkeeper
    2. Good records from the year before
    3. Good relationship with budget approval people
    4. Good revenue stream forecast
26. What is the purpose of an equipment repair/replacement fund? To generate additional revenue for
    1. Extra maintenance costs
    2. Increased operation costs
    3. Profits
    4. Repair/replacement as equipment wears out
27. What is the purpose of O & M programs?
    1. To ensure continued operation of facilities
    2. To ensure system performs as designed and intended
    3. To keep equipment lubricated
    4. To maintain compliance with permits and regulations
28. What is the first step in developing an emergency response plan?
    1. Assess vulnerability
    2. Develop a plan of action
    3. Implement a plan of action
    4. Prepare a budget
29. What is the purpose of a safety policy statement?
    1. To indicate that the utility has the appropriate policy statements
    2. To inform the public that the utility is safety conscious
    3. To let employees know that the safety program has the full support of the agency and its management
    4. To reduce insurance premiums by documenting a viable safety program
30. Why should all aspects of personnel management be documented?
    1. To allow statistical surveys of staff demographics
    2. To ensure that the best people get promoted
    3. To help with budget preparation
    4. To provide an important measure of legal protection for the utility

Multiple Choice (Select all correct answers.)

31. Communities expect managers to be aware of what kinds of concerns?
    1. Effective communication
    2. Minimize investment of money
    3. Protect the environment
    4. Provide safe drinking water
    5. Public relations
32. What are the functions of a utility manager?
    1. Controlling
    2. Directing
    3. Organizing
    4. Planning
    5. Staffing
33. Why should a utility agency have adequate emergency planning? To avoid
    1. Enhanced public relations program
    2. Inability to meet compliance regulations
    3. Inadequate service capacity to meet community needs
    4. Running out of money
    5. System failures
34. What are the key concepts in personnel management today?
    1. Benefits
    2. Documentation
    3. Equity
    4. Essential
    5. Job-related
35. What information may be collected to evaluate an applicant for a job?
    1. Age
    2. Knowledge, skills, and abilities to perform job
    3. Level of education
    4. Number of dependents
    5. Residence status

36. During the first day of work, what kind of information should be given a new employee?
    1. Attendance expectations
    2. Compensation and benefits
    3. Information available on policies and practices of the utility
    4. Safety training
    5. Seat assignment in utility vehicles

37. What types or groups of people will you need to communicate with as a utility manager?
    1. Employees
    2. Governing body
    3. Public
    4. Regulators
    5. School children touring facilities

38. Which items could be included in the annual report of a utility?
    1. Professional awards or recognition for the utility or its staff
    2. Recommended changes for the utility
    3. Review of goals and objectives for the year
    4. Special projects completed
    5. Summary of costs

39. Which steps should be taken to conduct a productive meeting?
    1. Allow one or two individuals to dominate the meeting
    2. Clearly state the purpose and objectives of the meeting
    3. Keep the discussion on track and on time with the agenda
    4. Prepare an agenda and distribute it to all participants
    5. Send out minutes of the meeting

40. What guidelines should be followed to obtain press coverage for an event or press conference?
    1. Be prepared to argue with reporters who are rude or overly aggressive
    2. Be sure to review the story to ensure accuracy
    3. Demonstrate that your story is newsworthy
    4. Make sure your story will fit the targeted format
    5. Provide a spokesperson who is interesting, articulate, and well prepared

41. What guidelines should be followed with telephone contacts?
    1. Answer calls promptly
    2. Be courteous, pleasant, and helpful
    3. Identify yourself
    4. Minimize transfers
    5. Pay attention

42. Which of the following items are basic principles to follow in responding to customer inquiries and concerns?
    1. Argue
    2. Be prepared
    3. Listen
    4. Promise specific action
    5. Use jargon

43. Financial management for a utility should include which items?
    1. Analyzing municipal bonds issued to other utilities
    2. Careful budgeting
    3. Maintaining the petty cash fund at a minimum level
    4. Providing capital improvement funds
    5. Providing financial stability

44. What does a manager need in order to budget effectively?
    1. Consider account cost increases that cannot be controlled
    2. How needs should be prioritized
    3. Needs of the utility
    4. Try to minimize expenses as much as possible
    5. Understand how money has been spent over past year

45. What are the basic ways for a utility to finance capital improvements?
    1. Bonded indebtedness
    2. General obligation bonds
    3. Loan funding programs
    4. Revenue bonds
    5. Savings bonds

46. What types of maintenance are considered reactive maintenance?
    1. Corrective maintenance
    2. Emergency maintenance
    3. Predictive maintenance
    4. Preventive maintenance
    5. Proactive maintenance

47. What are the elements of a good preventive maintenance program?
    1. Cost and budget control
    2. Planning and scheduling
    3. Records management
    4. Spare parts management
    5. Training program

48. What is the goal of managing maintenance? To minimize investments of
    1. Equipment
    2. Labor
    3. Materials
    4. Money
    5. Training

49. Which elements important for a utility to function during a disaster are most subject to disruption?
    1. Absence of trained personnel to make critical decisions and carry out orders
    2. Inadequate amount of supplies and materials
    3. Inadequate communication equipment
    4. Insufficient emergency funds
    5. Loss of power to the utility's facilities

50. What are the basic elements of a safety program?
    1. Accident insurance premiums
    2. Accident investigation and reporting
    3. Lost time injury statistics
    4. Safety policy statement
    5. Safety training and promotion

END OF OBJECTIVE TEST

APPENDIX

SMALL WASTEWATER SYSTEM OPERATION AND MAINTENANCE

(VOLUME II)

Final Examination and Suggested Answers

How To Solve Wastewater System Arithmetic Problems

Wastewater Words

Subject Index

FINAL EXAMINATION

VOLUME II

This final examination was prepared *TO HELP YOU REVIEW* the material in this manual. The questions are divided into four types:

1. True-False,
2. Best Answer,
3. Multiple Choice, and
4. Short Answer.

To work this examination:

1. Write the answer to each question in your notebook,
2. After you have worked a group of questions (you decide how many), check your answers with the suggested answers at the end of this exam, and
3. If you missed a question and don't understand why, reread the material in the manual.

You may wish to use this examination for review purposes when preparing for civil service and certification examinations.

Since you have already completed this course, you do not have to send your answers to California State University, Sacramento.

True-False

1. Plants or grasses with long roots should be allowed to grow on levees to prevent erosion.
   1. True
   2. False
2. The treatment plant operator has significant control over the makeup or amount of influent coming into the wastewater treatment plant.
   1. True
   2. False
3. The oxidation ditch is a modified form of oxidation ponds.
   1. True
   2. False
4. Usually, it is necessary for the operator to vary the amount of MLSS in the oxidation ditch as seasons change.
   1. True
   2. False
5. As the flows to a rotating biological contactor (RBC) decrease, the percent of BOD removal decreases.
   1. True
   2. False
6. RBC process efficiency can be increased by washing the slimes off the media.
   1. True
   2. False
7. To date there is no record of an operator becoming infected by the AIDS virus due to on-the-job duties.
   1. True
   2. False
8. The chlorine demand caused by inorganic salts must be satisfied first before any disinfection can take place.
   1. True
   2. False
9. The lowering of pH water will increase the disinfection effectiveness of chlorination.
   1. True
   2. False
10. Wetland systems can be constructed anywhere they are needed.
    1. True
    2. False
11. Water level in wetlands is critical to the survival of the plants.
    1. True
    2. False
12. Wells should be monitored near land treatment systems for any adverse effects on groundwater.
    1. True
    2. False
13. A Teflon stopcock always needs to be lubricated.
    1. True
    2. False

14 Never drink from a beaker or other laboratory glassware.
   1. True
   2. False

15. Sample deterioration starts immediately after collection for most wastewaters.
    1. True
    2. False
16. Planning must be done before the other managing functions.
    1. True
    2. False
17. Good records about past events can be used to demonstrate your pattern of lawful behavior over time.
    1. True
    2. False
18. The customer is always right.
    1. True
    2. False

19. A utility must keep track of revenue generation as carefully as resource spending.
    1. True
    2. False

Best Answer (Select only the closest or best answer.)

1. What happens to oxygen in a pond at night?
    1. Algae produce additional oxygen at night
    2. Algae use up the oxygen by respiration
    3. Oxygen levels increase due to lack of algal activity
    4. Oxygen levels remain constant
2. Why is duckweed undesirable in waste treatment ponds?
    1. Attracts ducks
    2. Causes supersaturation
    3. Presents control problems
    4. Stops sunlight penetration
3. What is a common cause of sludge bulking in the activated sludge process?
    1. Filamentous bacteria
    2. High dissolved oxygen (DO in the aeration tank)
    3. Pin floc
    4. Properly settling solids
4. How can problems in the activated sludge process caused by toxic substances be corrected?
    1. Add detoxifying chemical
    2. Change to more tolerant microorganisms
    3. Locate source and prevent future discharges
    4. Provide pretreatment
5. What should an operator do if the surface of an oxidation ditch has a thick, dark foam?
    1. Decrease the brush rotor depth
    2. Decrease the wasting rate
    3. Increase the brush rotor depth
    4. Increase the wasting rate
6. Why is staging used in the rotating biological contactor process?
    1. To keep the wastewater flowing uniformly
    2. To maximize the effectiveness of a given amount of media
    3. To minimize the settling of solids in each tank
    4. To prevent short circuiting
7. How can snails be controlled on RBC media?
    1. Chlorination
    2. Decrease the pH to 4
    3. Scrape snails off media
    4. Wash snails off media
8. What is the main purpose of disinfection?
    1. Control odors
    2. Prevent spread of waterborne diseases
    3. Reduce BOD
    4. Remove grease
9. How do manufacturers of hypochlorite compounds define available chlorine?
    1. By calculating the stoichiometric chlorine dose
    2. By weighing an equivalent amount of chlorine gas
    3. The amount of gaseous chlorine required to make the equivalent hypochlorite chlorine
    4. The amount of hypochlorite required to maintain the chlorine residual
10. How should the spill of a hypochlorite solution be cleaned up?
    1. Sweep up with a broom
    2. Vacuum up and recontainerize
    3. Wash with large volumes of water
    4. Wipe up with moist rags
11. Under what conditions might surface odors develop in a subsurface flow system wetland?
    1. When aerobic conditions develop in the media
    2. When groundwater high in sulfate is used to maintain water levels
    3. When the wind is strong enough to blow away the upper layer of mulch
    4. When water is permitted to stand above the mulch layer
12. What is the primary objective in wetland vegetation management?
    1. To keep the wetland vegetation alive
    2. To maintain the desired plant communities where they are intended to be within the wetland
    3. To prevent the generation of odors and production of mosquitoes
    4. To properly regulate the water level
13. When irrigating a crop using spray irrigation, when should the pump be turned off?
    1. When erosion starts
    2. When mildew begins to appear on the soil
    3. When vegetation turns yellow
    4. When water begins to pond on the fields
14. Which piece of laboratory glassware provides the most accurate measurement of liquid volumes?
    1. Beaker
    2. Buret
    3. Flask
    4. Graduated cylinder
15. What is a grab sample?
    1. A sample composed of a mixture of samples
    2. A sample grabbed from a flowing wastestream
    3. A sample prepared on the basis of wastestream flow
    4. A single sample of wastewater taken at neither a set time nor flow
16. Why should residual chlorine be removed before final discharge to a receiving stream in many locations?
    1. For inactivation of pathogenic organisms
    2. For protection of fish and other aquatic life
    3. For protection of water contact sport activities
    4. For ultimate BOD reduction
17. What is the purpose of a utility's organizational plan?
    1. To show how the public can participate in management activities
    2. To show the utility's organizational policy
    3. To show the utility's plan for future expansion
    4. To show who reports to whom and to identify the lines of authority
18. How can a report be effective?
    1. By achieving the objective for which it was written
    2. By being frank and straightforward
    3. By covering the facts and events that occurred
    4. By making sense to the reader and being organized logically

19. How do you measure the financial stability for a utility?
    1. Coverage ratio
    2. Profit
    3. Rate of return
    4. Revenue

20. Why should all aspects of personnel management be documented?
    1. To allow statistical surveys of staff demographics
    2. To ensure that the best people get promoted
    3. To help with budget preparation
    4. To provide an important measure of legal protection for the utility

Multiple Choice (Select all correct answers.)

1. How can oxygen be supplied to ponds?
    1. By adding sodium hypochlorite to the pond influent
    2. By algae during daylight hours
    3. By compressors bubbling air through the pond
    4. By mechanical agitation of the surface
    5. By promoting pond overturn

2. Why are weeds around the edge of a pond objectionable?
    1. Allow scum accumulation
    2. Allow sheltered area for mosquito breeding
    3. Contribute to evapotranspiration
    4. Provide habitat for water fowl
    5. Provide source of food for mosquito fish

3. Which items could be probable causes of poor quality pond effluent?
    1. Blockage of light by excessive plant growth near dikes
    2. Excessive turbidity from scum mat
    3. Low temperature
    4. Organic overload
    5. Toxic material in influent

4. What must the operator control for the activated sludge process to work properly?
    1. Dissolved oxygen in the aeration tanks
    2. Influent organic loading
    3. Number of organisms in the aeration tank
    4. Temperature of influent wastewater
    5. Treatment time

5. How can an operator confirm that enough air is being supplied by the aeration equipment to the aeration tank?
    1. Aeration compartment has a rotten egg odor
    2. Measure DO with a probe
    3. Water is clear in settling compartment
    4. Water is cloudy
    5. Water is murky

6. How can rusting of metal pipes be prevented in a package activated sludge plant?
    1. Apply protective coatings
    2. Keep debris from flowing in pipes
    3. Maintain sacrificial anodes
    4. Prevent excessive flow velocities
    5. Use iron-free water

7. For an oxidation ditch to obtain maximum performance efficiency, which process control methods must be maintained?
    1. Proper DO levels in the oxidation ditch
    2. Proper ditch detention time to treat the wastes by control of the adjustable weir
    3. Proper ditch environment (no toxicants and sufficient microorganisms to treat the wastes)
    4. Proper food supply (measured as BOD or COD and no toxicants) for the microorganisms
    5. Proper water/solids separation in the clarifier

8. Why are rotating biological contactors covered?
    1. Avoid exposure of media to sunlight, which may cause the media to become brittle
    2. Prevent intense rains from washing off some of the slime growths
    3. Protect biological slime growths from freezing
    4. Provide protection for operators from sun, rain, snow, or wind while maintaining equipment
    5. Stop exposure of media to direct sunlight to prevent growth of algae

9. What could cause excess sloughing or increased loss of biomass from a rotating biological contactor media?
    1. Changes in slime growth reproduction dynamics
    2. Influent dissolved oxygen fluctuations
    3. Seasonal weather patterns
    4. Toxic or inhibitory substances
    5. Unusual variation in flow and/or organic loading

10. What could be the probable cause of rapid wear on a roller chain drive?
    1. Faulty lubrication
    2. Loose parts
    3. Misaligned parts
    4. Rust or corrosion
    5. Worn parts

11. Which beneficial uses of water need to be protected by disinfection to prevent the spread of waterborne diseases?
    1. Hydroelectric power generation
    2. Navigation
    3. Public water supplies
    4. Receiving waters used for recreational uses
    5. Shellfish growing areas

12. Which critical factors influence effectiveness of disinfection?
    1. Chlorine dose
    2. Contact time
    3. Mixing
    4. pH
    5. Temperature

13. Which factors will cause the chlorine requirement or chlorine dose to vary?
    1. Contact time
    2. pH of wastewater
    3. Temperature of wastewater
    4. Wastewater constituents such as hydrogen sulfide and ammonia
    5. Wastewater flow

14. Where should chlorine tablets be stored?
    1. Away from combustible materials
    2. In a cool area
    3. In a dry area
    4. In a moist area
    5. In open containers

15. What are the advantages of wetlands for small, rural communities?
    1. Can achieve a very high and reliable degree of wastewater treatment
    2. Higher energy requirements for treatment processes
    3. Lower construction costs
    4. Reduced cost in operation and maintenance
    5. Reduced operational skills to maintain the system

16. What could be the cause of surface pooling or ponding in the wetland bed?
    1. Clogging of the media
    2. Excessive flows
    3. High evaporation rates
    4. Improper water level adjustments
    5. Insufficient solids in the influent

17. How do land treatment systems treat and reclaim wastewater? By using
    1. Bacteria
    2. Chemical precipitation
    3. Electrowinning
    4. Plants
    5. Soil

18. What are the main areas of process control for a land treatment system?
    1. Erosion and sediment control
    2. Runoff and seepage water recycle systems
    3. Soil systems where crops are grown
    4. Solids balance and inventory
    5. Storage reservoirs

19. Why must operators understand laboratory procedures and results?
    1. To complete laboratory data sheets
    2. To control treatment processes
    3. To identify developing treatment problems
    4. To purchase laboratory reagents
    5. To select most effective operational procedures

20. What types of hazards are people exposed to when working in a wastewater laboratory?
    1. Burns
    2. Electric shock
    3. Infectious materials
    4. Reckless drivers
    5. Toxic fumes

21. Obtaining good test results depends on which factors?
    1. Ensuring that the sample taken is truly representative of the wastestream
    2. Protecting and preserving the samples until they are analyzed
    3. Reporting results in a timely fashion to regulatory agencies
    4. Using proper sampling techniques
    5. Writing legibly on lab bench sheet

22. When may a grab sample be preferred over a composite sample?
    1. The wastewater characteristics are relatively constant
    2. The wastewater is to be analyzed for coliforms
    3. The wastewater is to be analyzed for residual chlorine
    4. The wastewater to be sampled does not flow on a continuous basis
    5. You wish to determine whether or not a composite sample obscures extreme conditions of the waste

23. Why should a utility agency have adequate emergency planning? To avoid
    1. Enhanced public relations program
    2. Inability to meet compliance regulations
    3. Inadequate service capacity to meet community needs
    4. Running out of money
    5. System failures

24. What information may be collected to evaluate an applicant for a job?
    1. Age
    2. Knowledge, skills, and abilities to perform job
    3. Level of education
    4. Number of dependents
    5. Residence status

25. What guidelines should be followed with telephone contacts?
    1. Answer calls promptly
    2. Be courteous, pleasant, and helpful
    3. Identify yourself
    4. Minimize transfers
    5. Pay attention

26. What does a manager need in order to budget effectively?
    1. Consider account cost increases that cannot be controlled
    2. How needs should be prioritized
    3. Needs of the utility
    4. Try to minimize expenses as much as possible
    5. Understand how money has been spent over past years

27. What is the goal of managing maintenance? To minimize investments of
    1. Equipment
    2. Labor
    3. Materials
    4. Money
    5. Training

Short Answer

1. If a pond is giving off objectionable odors, are the wastes being effectively treated? Explain your answer.
2. Why should the inlet to a pond be submerged?
3. Why is good housekeeping an important factor in operating a properly functioning pond?
4. Why is air added to the aeration tank in the activated sludge process?
5. What should you do if you take a sample in a jar from the aeration compartment and after 30 minutes solids do not settle to the bottom half of the jar?

6. When an oxidation ditch is operating properly, how should the activated sludge solids appear?
7. How can an operator determine if the sludge wasting should be increased or decreased from an oxidation ditch?
8. How does a rotating biological contactor (RBC) treat wastewater?
9. Why should debris, grit, and suspended solids be removed before the wastewater being treated reaches the RBC unit?
10. How are pathogenic bacteria destroyed or removed from water?
11. How much chlorine must be added to wastewater to produce disinfecting action?
12. How are chlorine doses adjusted to flow changes with a chlorine tablet disinfection system?
13. What are the major advantages of on-site chlorine/hypochlorite generation?
14. Why are free water surface (FWS) wetlands not recommended for individual homesites?
15. What maintenance should be performed on the inlets and outlets to wetlands?
16. What are the major parts of a land application system?
17. What is the main objective of a land disposal system?
18. What is a bench sheet?
19. What are the largest sources of errors found in laboratory results?
20. What happens when you attempt to measure clarity under different conditions, such as lighting and clarifier loadings?
21. Why should plant effluents be chlorinated?
22. What are the local community demands on a utility manager?
23. Why is it sometimes difficult or uncomfortable for supervisors or managers to delegate effectively?
24. When should a new employee's safety training begin?
25. What kinds of communication skills are needed by a manager?
26. What is the key to effective public speaking?
27. Who is responsible for the implementation of a safety program?

SUGGESTED ANSWERS FOR FINAL EXAMINATION

True-False

1. False — Plants or grasses with long roots should *NOT* be allowed to grow on levees.
2. False — The operator has little (*NOT* significant) control over the makeup or amount of influent coming into the wastewater treatment plant.
3. False — The oxidation ditch is a modified form of the activated sludge process, *NOT* oxidation ponds.
4. True — Usually, it is necessary for the operator to vary the amount of MLSS in the oxidation ditch as seasons change.
5. False — As flows to an RBC decrease, percent BOD removal increases.
6. False — Washing slimes off media can decrease process efficiency.
7. True — To date there is no record of an operator becoming infected by the AIDS virus due to on-the-job duties.
8. True — Chlorine demand caused by inorganic salts must be satisfied first before any disinfection can take place.
9. True — The lowering of pH will increase the disinfection effectiveness of chlorination.
10. True — Wetland systems can be constructed anywhere they are needed.
11. True — Water level is critical to the survival of plants in wetlands.
12. True — Monitor wells near land treatment systems for any adverse effects on groundwater.
13. False — A Teflon stopcock never needs to be lubricated.
14. True — Never drink from a beaker or laboratory glassware.
15. True — Sample deterioration starts immediately after collection.
16. True — Planning must be done before the other managing functions.
17. True — Good records about past events are a good documentation of lawful behavior over time.
18. False — The customer is *NOT* always right.
19. True — A utility must keep track of revenue generation as carefully as resource spending.

Best Answer

1. 2 — Algae use up oxygen in a pond at night by respiration.
2. 4 — Duckweed is undesirable in waste treatment ponds because it stops sunlight penetration.
3. 1 — Filamentous bacteria are a common cause of sludge bulking in the activated sludge process.
4. 3 — Problems caused by toxic substances can be corrected by locating the source and preventing future discharges.
5. 4 — To correct a thick, dark foam on the surface of an oxidation ditch, increase the wasting rate.
6. 2 — Staging is used in the rotating biological contactor process to maximize the effectiveness of a given amount of media.
7. 1 — Snails can be controlled on RBC media by the use of chlorination.
8. 2 — The main purpose of disinfection is to prevent the spread of waterborne diseases.
9. 3 — Manufacturers of hypochlorite compounds define available chlorine as the amount of gaseous chlorine required to make the equivalent hypochlorite.
10. 3 — Clean up a spill of hypochlorite solution by washing with large volumes of water.
11. 4 — Surface odors may be produced when water is permitted to stand above the mulch layer.
12. 2 — The primary objective of wetland vegetation management is to maintain the desired plant communities where they are intended to be within the wetland.
13. 4 — Turn the spray irrigation pump off when water begins to pond on the fields.
14. 1 — Burets provide the most accurate measurement of liquid volumes.
15. 4 — A grab sample is a single sample taken at neither a set time nor flow.
16. 2 — Residual chlorine should be removed before final discharge to a recovery stream for the protection of fish and other aquatic life.
17. 4 — The purpose of a utility's organizational plan is to show who reports to whom and to identify the lines of authority.

18. 1 — A report can be effective by achieving the objective for which it was written.

19. 1 — Financial stability for a utility is measured by its coverage ratio and operating ratio.

20. 4 — All aspects of personnel management should be documented to provide a measure of legal protection for the utility.

Multiple Choice

1. 2, 3, 4 — Oxygen can be supplied to ponds by algae during daylight hours, by compressors bubbling air through the pond, and by mechanical agitation of the surface.

2. 1, 2 — Weeds are objectionable around the edge of a pond because they allow scum accumulation and a sheltered area for mosquito breeding.

3. 1, 2, 3, 4, 5 — Probable causes of poor quality pond effluent include blockage of light by excessive plant growth near dikes, excessive turbidity caused by a scum mat, low temperature, organic overload, and toxic material in influent.

4. 1, 3, 5 — For the activated sludge process to work properly, the operator must control dissolved oxygen in the aeration tanks, number of organisms in the aeration tanks, and treatment time.

5. 2, 3 — Operators can determine if enough air is being supplied by aeration equipment by using a DO probe and observing clear water in the settling compartment.

6. 1, 3 — Rusting of metal pipes in package activated sludge plants can be prevented by applying protective coatings and maintaining sacrificial anodes.

7. 1, 2, 3, 4, 5 — For an oxidation ditch to obtain maximum performance efficiency, operators must maintain proper DO levels in the oxidation ditch, proper ditch detention time to treat the wastes, proper ditch environment (no toxicants and sufficient microorganisms), proper food supply (measured as BOD or COD), and proper water/solids separation in the clarifier.

8. 1, 2, 3, 4, 5 — Rotating biological contactors are covered to avoid exposure of media to sunlight, to prevent intense rains from washing off some of the slime growths, to protect biological slime growths from freezing, to provide protection for operators from adverse weather, and to stop exposure of media to direct sunlight to prevent growth of algae.

9. 4, 5 — Excess sloughing or increased loss of biomass could be caused by toxic or inhibitory substances and also unusual variation in flow and/or organic loading.

10. 1, 2, 3 — Probable causes of rapid wear on a roller chain drive include faulty lubrication, loose parts, and misaligned parts.

11. 3, 4, 5 — Beneficial uses of water that need to be protected by disinfection to prevent the spread of waterborne diseases include public water supplies, receiving waters used for recreational purposes, and shellfish growing areas.

12. 1, 2, 3, 4, 5 — Critical factors that influence the effectiveness of disinfection include chlorine dose, contact time, mixing, pH, and temperature.

13. 1, 2, 3, 4, 5 — Factors that will cause the chlorine requirement or chlorine dose to vary include contact time, pH of wastewater, temperature of wastewater, wastewater constituents, and wastewater flow.

14. 1, 2, 3 — Chlorine tablets should be stored away from combustible materials and in a cool, dry area.

15. 1, 3, 4, 5 — Advantages of wetlands for small communities include high and reliable degree of treatment, lower construction costs, reduced cost in operation and maintenance, and reduced operational skills needed to maintain the system.

16. 1, 2, 4 — Causes of surface pooling or ponding include clogging of media, excessive flows, and improper water level adjustments.

17. 1, 4, 5 — Land treatment systems treat and reclaim wastewater by using bacteria, plants, and soil.

18. 2, 3, 5 — The main areas of process control for a land treatment system are runoff and seepage water recycle systems, soil systems where crops are grown, and storage reservoirs.

19. 2, 3, 5 — Operators must understand laboratory procedures and results to control treatment processes, to identify developing treatment problems, and to select most effective operational procedures.

20. 1, 2, 3, 5 — People working in a wastewater laboratory are exposed to many hazards, including burns, electric shock, infectious materials, and toxic fumes.

21. 1, 2, 4 — Obtaining good lab test results depends on ensuring that the sample taken is truly representative, on protecting and preserving the samples until they are analyzed, and on using proper sampling techniques.

22. 1, 2, 3, 4, 5 — Grab samples may be preferred over a composite sample when wastewater characteristics are relatively constant, wastewater is to be analyzed for coliforms or residual chlorine, wastewater to be sampled does not flow on a continuous basis, and you wish to determine whether or not a composite sample obscures extreme conditions of the waste.

23. 2, 3, 5 — Adequate emergency planning helps avoid inability to comply with regulations, inadequate service capacity to meet community needs, and system failures.

24. 2 Knowledge, skills, and abilities are the information that may be collected to evaluate an applicant for a job.

25. 1, 2, 3, 4, 5 Guidelines that should be followed with telephone contacts include answering calls promptly, being courteous, pleasant, and helpful, identifying yourself, minimizing transfers, and paying attention.

26. 1, 2, 3, 4, 5 In order to budget effectively, a manager needs to consider account cost increases that cannot be controlled, how needs should be prioritized, needs of the utility, trying to minimize expenses as much as possible, and understanding how money has been spent over past years.

27. 1, 2, 3, 4 The goal of managing maintenance is to minimize investments of equipment, labor, materials, and money.

Short Answer

1. Probably. When the greatest organic load per unit of area (high treatment efficiency) is destroyed, objectionable odors may develop.
2. The inlet to a pond should be submerged to distribute the heat of the influent as much as possible and to minimize the occurrence of floating material.
3. Good housekeeping is an important factor in operating a properly functioning pond because many complaints about ponds originate from poor housekeeping practices.
4. Air is added to the aeration tank in the activated sludge process to provide oxygen to sustain the living organisms as they oxidize wastes to obtain energy for growth. The application of air also encourages mixing in the aerator.
5. If solids do not settle in the jar, the aeration rate should be increased and the MLSS concentration checked for proper level.
6. When an oxidation ditch is operating properly, the solids should look like particles, golden to rich dark brown in color, with sharply defined edges.
7. The appearance of the surface of the oxidation ditch can be a helpful indication of whether the sludge wasting rate should be increased or decreased. If the foam on the surface is white and crisp, reduce the wasting rate. A thick, dark foam indicates that the wasting rate should be increased.
8. A rotating biological contactor treats wastewater by allowing the slime growths to develop on the surface of plastic media. The slime growths contain organisms that remove the organic materials from the wastewater.
9. Debris, grit, and suspended solids should be removed to prevent sludge deposits forming beneath the media. These sludge deposits can reduce the effective tank volume, produce septic conditions, scrape the slimes from the media, and possibly stall the unit.
10. Pathogenic bacteria are destroyed or removed from water by (1) physical removal through sedimentation or filtration, (2) natural die-away in an unfavorable environment by storage, and (3) destruction through chemical treatment.
11. To produce an effective disinfecting action, sufficient chlorine must be added after the chlorine demand is satisfied to produce a chlorine residual likely to persist through the contact period.
12. Chlorine doses are adjusted to flow changes with a chlorine tablet disinfection system by allowing the amount of submergence of the tablets to fluctuate with the flows.
13. Major advantages of on-site chlorine/hypochlorite generation include (1) hypochlorite solution is generated as needed, and (2) the on-site generation system is safer.
14. Free water surface (FWS) wetland systems are *NOT* recommended for individual homesites because of the associated problems of vector and disease control.
15. Keep inlets and outlets clear of vegetation and be sure they can be easily raised and lowered.
16. The major parts of land application systems include:
    1. Treatment before application,
    2. Transmission to the land treatment site,
    3. Storage,
    4. Distribution over site, and
    5. Crop systems.
17. The main objective of a land disposal system is to dispose of effluent without harming surface waters or creating nuisance conditions.
18. A bench sheet is used to record data and arrange data in an orderly manner. Bench sheets are also called *LABORATORY WORK SHEETS*.
19. The largest sources of errors found in laboratory results are usually caused by improper sampling, poor preservation, and lack of sufficient mixing during compositing and testing.
20. When clarity is measured under difficult conditions, the results cannot be compared. You won't be able to tell whether your plant performance is improving, staying the same, or deteriorating.
21. Plant effluents should be chlorinated for disinfection purposes to protect the bacteriological quality of the receiving waters.
22. Local utility demands on a utility manager include protection from environmental disasters with a minimum investment of money.
23. Effective delegation is uncomfortable for many managers since it requires giving up power and responsibility. Many managers believe that they can do the job better than others, they believe that other employees are not well trained or experienced, and they are afraid of mistakes. The utility manager retains some responsibility even after delegating to another employee and, therefore, the manager is often reluctant to delegate or may delegate the responsibility but not the authority to get the job done.
24. A new employee's safety training should begin on the first day of employment or as soon thereafter as possible.
25. A manager needs both written and oral communication skills.
26. Practice is the key to effective public speaking.
27. The utility manager is responsible for the safety of the agency's personnel and the public exposed to the water or wastewater utility's operations. Therefore, the manager must develop and administer an effective safety program and must provide new employee safety training as well as ongoing training for all employees.

APPENDIX

HOW TO SOLVE WASTEWATER SYSTEM ARITHMETIC PROBLEMS

(VOLUME II)

TABLE OF CONTENTS

HOW TO SOLVE WASTEWATER SYSTEM ARITHMETIC PROBLEMS

HOW TO SOLVE WASTEWATER SYSTEM ARITHMETIC PROBLEMS

(VOLUME II)

A.0 BASIC CONVERSION FACTORS (ENGLISH SYSTEM)

UNITS		
1,000,000	= 1 Million	1,000,000/1 Million
LENGTH		
12 in	= 1 ft	12 in/ft
3 ft	= 1 yd	3 ft/yd
5,280 ft	= 1 mi	5,280 ft/mi
AREA		
144 sq in	= 1 sq ft	144 sq in/sq ft
43,560 sq ft	= 1 acre	43,560 sq ft/ac
VOLUME		
7.48 gal	= 1 cu ft	7.48 gal/cu ft
1,000 m*L*	= 1 liter	1,000 m*L*/*L*
3.785 *L*	= 1 gal	3.785 *L*/gal
231 cu in	= 1 gal	231 cu in/gal
WEIGHT		
1,000 mg	= 1 gm	1,000 mg/gm
1,000 gm	= 1 kg	1,000 gm/kg
454 gm	= 1 lb	454 gm/lb
2.2 lbs	= 1 kg	2.2 lbs/kg
POWER		
0.746 kW	= 1 HP	0.746 kW/HP

DENSITY		
8.34 lbs	= 1 gal	8.34 lbs/gal
62.4 lbs	= 1 cu ft	62.4 lbs/cu ft
DOSAGE		
17.1 mg/*L*	= 1 grain/gal	17.1 mg/*L*/gpg
64.7 mg	= 1 grain	64.7 mg/grain
PRESSURE		
2.31 ft water	= 1 psi	2.31 ft water/psi
0.433 psi	= 1 ft water	0.433 psi/ft water
1.133 ft water	= 1 in Mercury	1.133 ft water/in Mercury
FLOW		
694 GPM	= 1 MGD	694 GPM/MGD
1.55 CFS	= 1 MGD	1.55 CFS/MGD
TIME		
60 sec	= 1 min	60 sec/min
60 min	= 1 hr	60 min/hr
24 hr	= 1 day	24 hr/day

NOTE: In our equations the values in the right-hand column may be written either as 24 hr/day or 1 day/24 hours depending on which units we wish to convert to obtain our desired results.

A.1 BASIC FORMULAS

WASTEWATER STABILIZATION PONDS

1. $$\text{BOD Removal, \%} = \frac{(\text{In} - \text{Out})}{\text{In}} \times 100\%$$

2. $$\text{Detention Time, days} = \frac{\text{Pond Volume, ac-ft}}{\text{Flow Rate, ac-ft/day}}$$

   where

   $$\text{Pond Area, acres} = \frac{(\text{Average Width, ft})(\text{Average Length, ft})}{43{,}560\text{ sq ft/acre}}$$

   $$\text{Pond Volume, ac-ft} = (\text{Avg Area, ac})(\text{Depth, ft})$$

   $$\text{Flow Rate, ac-ft/day} = \frac{\text{Flow, gal/day}}{(7.48\text{ gal/cu ft})(43{,}560\text{ sq ft/acre})}$$

3. $$\text{Population Loading, persons/ac} = \frac{\text{Population Served, persons}}{\text{Pond Surface Area, acres}}$$

4. $$\text{Hydraulic Loading, inches/day} = \frac{\text{Depth of Pond, inches}}{\text{Detention Time, days}}$$

5. Organic Loading, lbs BOD/day/acre $= \frac{\text{(Flow, MGD)(BOD, mg/}L\text{)(8.34 lbs/gal)*}}{\text{Area, acres}}$

---

* Remember that 1 liter = 1,000,000 mg and that

$$\frac{\text{lbs}}{\text{day}} = \frac{\text{Mil Gal}}{\text{day}} \times \frac{\text{mg}}{\text{Mil mg}} \times \frac{\text{lbs}}{\text{gal}} = \frac{\text{lbs}}{\text{day}}$$

ACTIVATED SLUDGE (OXIDATION DITCHES)

6. BOD Loading, lbs BOD/day = (Flow, MGD)(BOD, mg/*L*)(8.34 lbs/gal)

7. BOD Loading, lbs/day/1,000 cu ft $= \frac{\text{BOD, lbs/day}}{\text{Ditch Volume, 1,000 cu ft}}$

8. $\frac{\text{Food}}{\text{Microorganism}} = \frac{\text{BOD, lbs/day}}{\text{MLVSS, lbs}}$

   where

   MLVSS, lbs = (Volume, M Gal)(MLSS, mg/*L*)(VM)(8.34 lbs/gal)

9. Sludge Age, days $= \frac{\text{Aeration Solids, lbs}}{\text{Solids Added, lbs/day}}$

   where

   Aeration Solids, lbs = (Volume, M Gal)(MLSS, mg/*L*)(8.34 lbs/gal)

   and

   Solids Added, lbs/day = (Flow, MGD)(Infl SS, mg/*L*)(8.34 lbs/gal)

10. Detention Time, hours $= \frac{\text{(Ditch Volume, M Gal)(24 hr/day)}}{\text{Flow, MGD}}$

ROTATING BIOLOGICAL CONTACTORS (RBCs)

11. Hydraulic Loading, GPD/sq ft $= \frac{\text{Flow, gal/day}}{\text{Surface Area, sq ft}}$

12. Organic (BOD) Loading, lbs BOD/day/1,000 sq ft $= \frac{\text{Soluble BOD Applied, lbs BOD/day}}{\text{Surface Area of Media (in 1,000 sq ft)}}$

    where

12a. Soluble BOD, mg/*L* = Total BOD, mg/*L* − [(K)(Suspended Solids, mg/*L*)]

12b. Soluble BOD, lbs/day = (Flow, MGD)(Soluble BOD, mg/*L*)(8.34 lbs/gal)

DISINFECTION AND CHLORINATION

13. Chlorinated Lime Feed Rate, lbs/day $= \frac{\text{Chlorine Required, lbs/day}}{\text{Portion of Chlorine in lb of Hypochlorite}}$

14. Chlorine Feed or Dose, mg/*L* $= \frac{\text{Chlorine Feed, lbs/day}}{\text{(Flow, MGD)(8.34 lbs/gal)}}$

15. Chlorine Demand, mg/*L* = Chlorine Dose, mg/*L* − Chlorine Residual, mg/*L*

16. Hypochlorinator Setting, lbs/day = (Flow, MGD)(Dose, mg/*L*)(8.34 lbs/gal)

LAND TREATMENT

17. Area, acres $= \frac{\text{(Length, ft)(Width, ft)}}{\text{43,560 sq ft/acre}}$

18. Hydraulic Loading, MGD/ac $= \frac{\text{Flow, MGD}}{\text{Area, acres}}$

19. Hydraulic Loading, inches/day $= \frac{\text{(Flow, MGD)(1,000,000/M)(12 in/ft)}}{\text{(Length, ft)(Width, ft)(7.48 gal/cu ft)}}$

LABORATORY

20a. $°F = (9/5)(°C) + 32°$

20b. $°C = (5/9)(°F - 32°)$

MANAGEMENT

21. $\text{Operating Ratio} = \dfrac{\text{Total Revenue, \$}}{\text{Operating Expenses, \$}}$

22. $\text{Coverage Ratio} = \dfrac{\text{Total Revenue, \$} - \text{Non-Debt Expenses, \$}}{\text{Debt Service Expenses, \$}}$

A.2 TYPICAL WASTEWATER SYSTEM PROBLEMS (ENGLISH SYSTEM)

A.20 Wastewater Stabilization Ponds

EXAMPLE 1

The influent BOD to a series of ponds is 240 mg/*L*, and the effluent BOD is 50 mg/*L*. What is the pond efficiency in removing BOD?

Known		**Unknown**
BOD In, mg/*L*	= 240 mg/*L*	BOD Removal, %
BOD Out, mg/*L*	= 50 mg/*L*	

Calculate the pond efficiency in removing BOD.

$$\text{BOD Removal, \%} = \frac{(\text{In} - \text{Out}) \times 100\%}{\text{In}}$$

$$= \frac{(240\text{ mg}/L - 50\text{ mg}/L)}{240\text{ mg}/L} \times 100\%$$

$$= 79\%$$

EXAMPLES 2, 3, 4, and 5

To calculate the different loadings on a pond, the information listed in the known column must be available.

Known

Avg Depth, ft	= 4 ft
Avg Width, ft	= 100 ft
Avg Length, ft	= 150 ft
Flow, GPD	= 30,000 GPD
Flow, MGD	= 0.030 MGD
BOD In, mg/*L*	= 160 mg/*L*
Population, persons	= 320 persons

Unknown

2. Detention Time, days
3. Population Loading, persons/acre
4. Hydraulic Load, inches/day
5. Organic Load, lbs BOD/day/acre

2. *DETENTION TIME*

$$\text{Detention Time, days} = \frac{\text{Pond Volume, ac-ft}}{\text{Flow Rate, ac-ft/day}}$$

$$\text{Pond Area, acres} = \frac{(\text{Avg Width, ft})(\text{Avg Length, ft})}{43{,}560\text{ sq ft/acre}}$$

$$= \frac{(100\text{ ft})(150\text{ ft})}{43{,}560\text{ sq ft/acre}}$$

$$= 0.344\text{ acre}$$

$$\text{Pond Volume, ac-ft} = (\text{Avg Area, ac})(\text{Depth, ft})$$

$$= (0.344\text{ acre})(4\text{ ft})$$

$$= 1.38\text{ ac-ft}$$

$$\text{Flow Rate, ac-ft/day} = \frac{\text{Flow, GPD}}{(7.48\text{ gal/cu ft})(43{,}560\text{ sq ft/ac})}$$

$$= \frac{30{,}000\text{ gal/day}}{(7.48\text{ gal/cu ft})(43{,}560\text{ sq ft/ac})}$$

$$= 0.092\text{ ac-ft/day}$$

$$\text{Detention Time, days} = \frac{\text{Pond Volume, ac-ft}}{\text{Flow Rate, ac-ft/day}}$$

$$= \frac{1.38\text{ ac-ft}}{0.092\text{ ac-ft/day}}$$

$$= 15\text{ days}$$

3. *POPULATION LOADING*

$$\text{Population Loading, persons/acre} = \frac{\text{Population Served, persons}}{\text{Pond Surface Area, acres*}}$$

$$= \frac{320\text{ persons}}{0.344\text{ acre}}$$

$$= 930\text{ persons/acre}$$

* *NOTE:* We used average width and length rather than surface width and length to calculate surface area to simplify our calculations.

4. *HYDRAULIC LOADING*

$$\text{Hydraulic Loading, inches/day} = \frac{\text{Depth of Pond, inches}}{\text{Detention Time, days}}$$

$$= \frac{(4\text{ ft})(12\text{ in/ft})}{15\text{ days}}$$

$$= 3.2\text{ inches/day}$$

5. *ORGANIC LOADING*

$$\text{Organic Loading, lbs BOD/day/ac} = \frac{(\text{Flow, MGD})(\text{BOD, mg}/L)(8.34\text{ lbs/gal})}{\text{Area, acres}}$$

$$= \frac{(0.030\text{ MGD})(160\text{ mg}/L)(8.34\text{ lbs/gal})}{0.344\text{ acres}}$$

$$= 116\text{ lbs BOD/day/acre}$$

A.21 Activated Sludge (Oxidation Ditches)

EXAMPLES 6, 7, 8, 9, and 10

To calculate the different loadings on an oxidation ditch, the information listed in the known column must be available.

Known

Ditch Volume, cu ft	= 20,000 cu ft
	= 20.0 • (1,000 cu ft)
Ditch Volume, MG	= 0.15 M Gal
Flow, MGD	= 0.2 MGD
BOD, mg/*L*	= 150 mg/*L*
Infl SS, mg/*L*	= 170 mg/*L*
MLSS, mg/*L*	= 3,500 mg/*L*
Volatile Matter in MLSS	= 70%

Unknown

6. BOD Loading, lbs BOD/day
7. BOD Loading, lbs BOD/day/1,000 cu ft
8. $\frac{\text{Food}}{\text{Microorganism}}, \frac{\text{BOD, lbs/day}}{\text{MLVSS, lbs}}$
9. Sludge Age, days
10. Detention Time, hours

6. *BOD LOADING*

$$\text{BOD Loading, lbs/day} = (\text{Flow, MGD})(\text{BOD, mg}/L)(8.34\text{ lbs/gal})$$
$$= (0.2\text{ MGD})(150\text{ mg}/L)(8.34\text{ lbs/gal})$$
$$= 250\text{ lbs BOD/day}$$

7. *BOD LOADING*

$$\text{BOD Loading, lbs/day/1,000 cu ft} = \frac{\text{BOD, lbs/day}}{\text{Ditch Volume, 1,000 cu ft}}$$
$$= \frac{250\text{ lbs BOD/day}}{20 \cdot (1{,}000\text{ cu ft})}$$
$$= 12.5\text{ lbs BOD/day/1,000 cu ft}$$

8. *FOOD/MICROORGANISM RATIO*

$$\frac{F}{M} = \frac{\text{Food, lbs BOD/day}}{\text{Microorganisms, lbs MLVSS}}$$

$$\text{MLVSS, lbs} = (\text{Vol, MG})(\text{MLSS, mg}/L)(\text{VM})(8.34\text{ lbs/gal})$$
$$= (0.15\text{ MG})(3{,}500\text{ mg}/L)(0.70)(8.34\text{ lbs/gal})$$
$$= 3{,}065\text{ lbs MLVSS}$$

$$\frac{F}{M} = \frac{\text{Food, lbs BOD/day}}{\text{Microorganisms, lbs MLVSS}}$$
$$= \frac{250\text{ lbs BOD/day}}{3{,}065\text{ lbs MLVSS}}$$
$$= 0.08\text{ lb BOD/day/lb MLVSS}$$

9. *SLUDGE AGE*

$$\text{Sludge Age, days} = \frac{\text{Aeration Solids, lbs}}{\text{Solids Added, lbs/day}}$$

a. Calculate the pounds of solids under aeration.

$$\text{Aeration Solids, lbs} = (\text{Vol, MG})(\text{MLSS, mg}/L)(8.34\text{ lbs/gal})$$
$$= (0.15\text{ M Gal})(3{,}500\text{ mg}/L)(8.34\text{ lbs/gal})$$
$$= 4{,}379\text{ lbs}$$

b. Calculate solids fed to ditch in pounds per day.

$$\text{Solids Added, lbs/day} = (\text{Flow, MGD})(\text{Infl SS, mg}/L)(8.34\text{ lbs/gal})$$
$$= (0.2\text{ MGD})(170\text{ mg}/L)(8.34\text{ lbs/gal})$$
$$= 284\text{ lbs/day}$$

c. Calculate the sludge age in days.

$$\text{Sludge Age, days} = \frac{\text{Solids Under Aeration, lbs}}{\text{Solids Added, lbs/day}}$$
$$= \frac{4{,}379\text{ lbs}}{284\text{ lbs/day}}$$
$$= 15.4\text{ days}$$

10. *DETENTION TIME*

$$\text{Detention Time, hours} = \frac{(\text{Ditch Volume, MG})(24\text{ hrs/day})}{\text{Flow, MGD}}$$
$$= \frac{(0.15\text{ MG})(24\text{ hrs/day})}{0.2\text{ MGD}}$$
$$= 18\text{ hours}$$

A.22 Rotating Biological Contactors (RBCs)

EXAMPLE 11

A rotating biological contactor treats a flow of 1.5 MGD. The surface area of the media is 500,000 square feet (provided by manufacturer). What is the hydraulic loading in GPD/sq ft?

Known		Unknown
Flow, MGD	= 1.5 MGD	Hydraulic Loading, GPD/sq ft
Surface Area, sq ft	= 500,000 sq ft	

Calculate the hydraulic loading in gallons per day of wastewater per square foot of surface media.

$$\text{Hydraulic Loading, GPD/sq ft} = \frac{\text{Flow, gal/day}}{\text{Surface Area, sq ft}}$$
$$= \frac{1{,}500{,}000\text{ gal/day}}{500{,}000\text{ sq ft}}$$
$$= 3.0\text{ GPD/sq ft}$$

EXAMPLE 12

The rotating biological contactor in Example 11 treats an industrial influent with a total BOD of 220 mg/*L* and suspended solids of 300 mg/*L*. Assume a K value of 0.5 to calculate the soluble BOD. What is the organic loading in pounds of soluble BOD per day per 1,000 square feet of media surface?

Known		Unknown
Flow, MGD	= 1.5 MGD	Organic Loading, lbs BOD/day/1,000 sq ft
Surface Area, sq ft	= 500,000 sq ft	
Total BOD, mg/*L*	= 220 mg/*L*	
SS, mg/*L*	= 300 mg/*L*	
K	= 0.5	

a. Estimate the soluble BOD treated by the rotating biological contactor.

$$\text{Soluble BOD, mg}/L = \text{Total BOD, mg}/L - (\text{K} \times \text{Suspended Solids, mg}/L)$$
$$= 220\text{ mg}/L - (0.5 \times 300\text{ mg}/L)$$
$$= 220\text{ mg}/L - 150\text{ mg}/L$$
$$= 70\text{ mg}/L$$

b. Determine the soluble BOD applied to the rotating biological contactor in pounds of soluble BOD per day.

$$\text{BOD Applied, lbs/day} = (\text{Flow, MGD})(\text{Soluble BOD, mg}/L)(8.34\text{ lbs/gal})$$

$$= (1.5\text{ MGD})(70\text{ mg}/L)(8.34\text{ lbs/gal})$$

$$= 876\text{ lbs Soluble BOD/day}$$

c. Calculate the organic loading in pounds of soluble BOD per day per 1,000 square feet of media surface.

$$\text{Organic BOD Loading, lbs BOD/day/1,000 sq ft} = \frac{\text{Soluble BOD Applied, lbs BOD/day}}{\text{Surface Area of Media (in 1,000 sq ft)}}$$

$$= \frac{876\text{ lbs Soluble BOD/day}}{500 \cdot 1{,}000\text{ sq ft}}$$

$$= 1.8\text{ lbs BOD/day/1,000 sq ft}$$

A.23 Disinfection and Chlorination

EXAMPLE 13

A wastewater requires a chlorine feed rate of 12 pounds chlorine per day. How many pounds of chlorinated lime will be required to provide the needed chlorine? Assume each pound of chlorinated lime contains 0.34 lb of available chlorine.

Known		Unknown
Chlorine Required, lbs/day	= 12 lbs/day	Chlorinated Lime Feed Rate, lbs/day
Portion of Chlorine in lb of Hypochlorite	= 0.34	

Calculate pounds of chlorinated lime required per day.

$$\text{Chlorinated Lime Feed Rate, lbs/day} = \frac{\text{Chlorine Required, lbs/day}}{\text{Portion of Chlorine in lb of Hypochlorite}}$$

$$= \frac{12\text{ lbs Chlorine/day}}{0.34}$$

$$= 35.3\text{ lbs Chlorinated Lime/day}$$

EXAMPLES 14 and 15

A hypochlorinator is set to feed 15 pounds of chlorine per 24 hours; the wastewater flow is 0.4 MGD and the chlorine as measured by the chlorine residual test after 30 minutes of contact time is 0.4 mg/*L*. Find the chlorine dosage and chlorine demand in mg/*L*.

Known		Unknown
Chlorine Feed, lbs/day	= 15 lbs/day	14. Chlorine Dose, mg/*L*
Flow, MGD	= 0.4 MGD	15. Chlorine Demand, mg/*L*
Chlorine Residual, mg/*L*	= 0.4 mg/*L*	

14. Calculate the chlorine dose in mg/*L*.

$$\text{Chlorine Feed or Dose, mg}/L = \frac{\text{Chlorine Feed, lbs/day}}{(\text{Flow, MGD})(8.34\text{ lbs/gal})}$$

$$= \frac{15\text{ lbs/day}}{(0.4\text{ MGD})(8.34\text{ lbs/gal})}$$

$$= 4.5\text{ mg}/L$$

15. Determine the chlorine demand in mg/*L*.

$$\text{Chlorine Demand, mg}/L = \text{Chlorine Dose, mg}/L - \text{Chlorine Residual, mg}/L$$

$$= 4.5\text{ mg}/L - 0.4\text{ mg}/L$$

$$= 4.1\text{ mg}/L$$

EXAMPLE 16

Calculate the hypochlorinator setting in pounds per 24 hours to treat a wastewater with a chlorine demand of 10 mg/*L*, when a chlorine residual of 2 mg/*L* is desired. The flow is 0.4 MGD.

Known		Unknown
Chlorine Demand, mg/*L*	= 10 mg/*L*	Hypochlorinator Setting, lbs/day
Chlorine Residual, mg/*L*	= 2 mg/*L*	
Flow, MGD	= 0.4 MGD	

a. Calculate the chlorine dose in mg/*L*.

$$\text{Chlorine Dose, mg}/L = \text{Chlorine Demand, mg}/L + \text{Chlorine Residual, mg}/L$$

$$= 10\text{ mg}/L + 2\text{ mg}/L$$

$$= 12\text{ mg}/L$$

b. Calculate the hypochlorinator setting in pounds per day

$$\text{Hypochlorinator Setting, lbs/day} = (\text{Flow, MGD})(\text{Dose, mg}/L)(8.34\text{ lbs/gal})$$

$$= (0.4\text{ MGD})(12\text{ mg}/L)(8.34\text{ lbs/gal})$$

$$= 40\text{ lbs/day}$$

A.24 Land Treatment

EXAMPLES 17, 18, and 19

A plot of land 1,600 feet long by 600 feet wide is used for a land disposal system. If 0.5 million gallons of water is applied to the land during a 24-hour period, calculate the land area in acres and the hydraulic loading in MGD per acre and also in inches per day.

Known		Unknown
Length, ft	= 1,600 ft	17. Area, acres
Width, ft	= 600 ft	18. Hydraulic Loading, MGD/ac
Flow, MGD	= 0.5 MGD	19. Hydraulic Loading, in/day

17. Determine the surface area in acres.

$$\text{Area, acres} = \frac{(\text{Length, ft})(\text{Width, ft})}{43{,}560\text{ sq ft/acre}}$$

$$= \frac{(1{,}600\text{ ft})(600\text{ ft})}{43{,}560\text{ sq ft/acre}}$$

$$= 22.0\text{ acres}$$

18. Determine the hydraulic loading in MGD per acre.

$$\text{Hydraulic Loading, MGD/ac} = \frac{\text{Flow, MGD}}{\text{Area, acres}}$$

$$= \frac{0.5\text{ MGD}}{22.0\text{ acres}}$$

$$= 0.023\text{ MGD/acre}$$

19. Determine the hydraulic loading in inches per day.

$$\text{Hydraulic Loading, inches/day} = \frac{(\text{Flow, MGD})(1{,}000{,}000/\text{M})(12\text{ in/ft})}{(\text{Length, ft})(\text{Width, ft})(7.48\text{ gal/cu ft})}$$

$$= \frac{(0.5\text{ MGD})(1{,}000{,}000/\text{M})(12\text{ in/ft})}{(1{,}600\text{ ft})(600\text{ ft})(7.48\text{ gal/cu ft})}$$

$$= 0.84\text{ in/day}$$

A.25 Laboratory

EXAMPLE 20

a. If a water temperature is 20°C, what is the temperature in °F?

b. If a water temperature is 40°F, what is the temperature in °C?

a. Convert water temperature from 20°C to °F.

$$°F = (9/5)(°C) + 32°$$

$$= (9/5)(20°C) + 32°$$

$$= 68°F$$

b. Convert water temperature from 40°F to °C.

$$°C = (5/9)(°F - 32°)$$

$$= (5/9)(40°F - 32°)$$

$$= 4.4°C$$

A.26 Management

EXAMPLES 21 and 22

The total revenues for a utility are $1,471,000 and the operating expenses for the utility are $1,108,000. The debt service expenses are $480,000. What is the operating ratio? What is the coverage ratio?

Known		Unknown
Total Revenue, $	= $1,471,000	21. Operating Ratio
Operating Expenses, $	= $1,108,000	22. Coverage Ratio
Debt Service Expenses, $	= $480,000	

21. Calculate the operating ratio.

$$\text{Operating Ratio} = \frac{\text{Total Revenue, \$}}{\text{Operating Expenses, \$}}$$

$$= \frac{\$1{,}471{,}000}{\$1{,}108{,}000}$$

$$= 1.33$$

22. Calculate the coverage ratio.

a. Calculate the non-debt expenses.

$$\text{Non-Debt Expenses, \$} = \text{Operating Exp, \$} - \text{Debt Service Exp, \$}$$

$$= \$1{,}108{,}000 - \$480{,}000$$

$$= \$628{,}000$$

b. Calculate the coverage ratio.

$$\text{Coverage Ratio} = \frac{\text{Total Revenue, \$} - \text{Non-Debt Expenses, \$}}{\text{Debt Service Expenses, \$}}$$

$$= \frac{\$1{,}471{,}000 - \$628{,}000}{\$480{,}000}$$

$$= 1.76$$

A.3 BASIC CONVERSION FACTORS (METRIC SYSTEM)

LENGTH		
100 cm	= 1 m	100 cm/m
3.281 ft	= 1 m	3.281 ft/m
AREA		
2.4711 ac	= 1 ha*	2.4711 ac/ha
10,000 sq m	= 1 ha	10,000 sq m/ha
VOLUME		
1,000 m*L*	= 1 liter	1,000 m*L*/*L*
1,000 *L*	= 1 cu m	1,000 *L*/cu m
3.785 *L*	= 1 gal	3.785 *L*/gal
WEIGHT		
1,000 mg	= 1 gm	1,000 mg/gm
1,000 gm	= 1 kg	1,000 gm/kg
DENSITY		
1 kg	= 1 liter	1 kg/*L*
PRESSURE		
10.015 m	= 1 kg/sq cm	10.015 m/kg/sq cm
1 Pascal	= 1 N/sq m	1 Pa/N/sq m
1 psi	= 6,895 Pa	1 psi/6,895 Pa
FLOW		
3,785 cu m/day	= 1 MGD	3,785 cu m/day/MGD
3.785 M*L*/day	= 1 MGD	3.785 M*L*/day/MGD

* hectare

A.4 TYPICAL WASTEWATER SYSTEM PROBLEMS (METRIC SYSTEM)

A.40 Wastewater Stabilization Ponds

EXAMPLE 1

The influent BOD to a series of ponds is 180 mg/*L*, and the effluent BOD is 40 mg/*L*. What is the pond efficiency in removing BOD?

Known		Unknown
BOD In, mg/*L*	= 180 mg/*L*	BOD Removal, %
BOD Out, mg/*L*	= 40 mg/*L*	

Calculate the pond efficiency in removing BOD.

$$\text{BOD Removal, \%} = \frac{(\text{In} - \text{Out}) \times 100\%}{\text{In}}$$

$$= \frac{(180\text{ mg}/L - 40\text{ mg}/L)}{180\text{ mg}/L} \times 100\%$$

$$= 78\%$$

EXAMPLES 2, 3, 4, and 5

To calculate the different loadings on a pond, the information listed in the known column must be available.

Known

Avg Depth, m	=	1.4 m
Avg Width, m	=	30 m
Avg Length, m	=	45 m
Flow, cu m/day	=	150 cu m/day
BOD In, mg/*L*	=	140 mg/*L*
Population, persons	=	320 persons

Unknown

2. Detention Time, days
3. Population Loading, persons/sq m
4. Hydraulic Load, cm/day
5. Organic Load, gm BOD/day/sq m

2. *DETENTION TIME*

$$\text{Detention Time, days} = \frac{\text{Pond Volume, cu m}}{\text{Flow Rate, cu m/day}}$$

$$\text{Pond Area, sq m} = (\text{Avg Width, m})(\text{Avg Length, m})$$

$$= (30\text{ m})(45\text{ m})$$

$$= 1{,}350\text{ sq m}$$

$$\text{Pond Volume, cu m} = (\text{Avg Area, sq m})(\text{Depth, m})$$

$$= (1{,}350\text{ sq m})(1.4\text{ m})$$

$$= 1{,}890\text{ cu m}$$

$$\text{Flow Rate, cu m/day} = 150\text{ cu m/day}$$

$$\text{Detention Time, days} = \frac{\text{Pond Volume, cu m}}{\text{Flow Rate, cu m/day}}$$

$$= \frac{1{,}890\text{ cu m}}{150\text{ cu m/day}}$$

$$= 12.6\text{ days}$$

3. *POPULATION LOADING*

$$\text{Population Loading, persons/sq m} = \frac{\text{Population Served, persons}}{\text{Pond Surface Area, sq m*}}$$

$$= \frac{320\text{ persons}}{1{,}350\text{ sq m}}$$

$$= 0.237\text{ persons/sq m}$$

* *NOTE:* We used average width and length rather than surface width and length to calculate surface area to simplify our calculations.

4. *HYDRAULIC LOADING*

$$\text{Hydraulic Loading, cm/day} = \frac{(\text{Depth of Pond, m})(100\text{ cm/m})}{\text{Detention Time, days}}$$

$$= \frac{(1.4\text{ m})(100\text{ cm/m})}{12.6\text{ days}}$$

$$= 11.1\text{ cm/day}$$

5. *ORGANIC LOADING*

$$\text{Organic Loading, gm BOD/day/sq m} = \frac{(\text{BOD, mg}/L)\left(\text{Flow, }\frac{\text{cu m}}{\text{day}}\right)\left(\frac{1{,}000\ L}{1\text{ cu m}}\right)\left(\frac{1\text{ gm}}{1{,}000\text{ mg}}\right)}{\text{Area, sq m}}$$

$$= \frac{(140\text{ mg}/L)\left(150\ \frac{\text{cu m}}{\text{day}}\right)\left(\frac{1{,}000\ L}{1\text{ cu m}}\right)\left(\frac{1\text{ gm}}{1{,}000\text{ mg}}\right)}{1{,}350\text{ sq m}}$$

$$= 15.6\text{ gm BOD/day/sq m}$$

A.41 Activated Sludge (Oxidation Ditches)

EXAMPLES 6, 7, 8, 9, and 10

To calculate the different loadings on an oxidation ditch, the information listed in the known column must be available.

Known

Ditch Volume, cu m	=	700 cu m
	=	0.70 • 1,000 cu m
Flow, cu m/day	=	800 cu m/day
BOD, mg/*L*	=	180 mg/*L*
Infl SS, mg/*L*	=	210 mg/*L*
MLSS, mg/*L*	=	3,800 mg/*L*
Volatile Matter in MLSS	=	70%

Unknown

6. BOD Loading, kg BOD/day
7. BOD Loading, kg BOD/day/1,000 cu m
8. $\frac{\text{Food}}{\text{Microorganism}}$, $\frac{\text{BOD, kg/day}}{\text{MLVSS, kg}}$
9. Sludge Age, days
10. Detention Time, hours

6. *BOD LOADING*

$$\text{BOD Loading, kg/day} = \left(\text{Flow, }\frac{\text{cu m}}{\text{day}}\right)\left(\text{BOD, }\frac{\text{mg}}{L}\right)\left(\frac{1\text{ kg}}{1{,}000{,}000\text{ mg}}\right)\left(\frac{1{,}000\ L}{1\text{ cu m}}\right)$$

$$= \left(800\ \frac{\text{cu m}}{\text{day}}\right)\left(180\ \frac{\text{mg}}{L}\right)\left(\frac{1\text{ kg}}{1{,}000{,}000\text{ mg}}\right)\left(\frac{1{,}000\ L}{1\text{ cu m}}\right)$$

$$= 144\text{ kg BOD/day}$$

7. *BOD LOADING*

$$\text{BOD Loading, kg/day/1,000 cu m} = \frac{\text{BOD, kg/day}}{\text{Ditch Volume, 1,000 cu m}}$$

$$= \frac{\text{144 kg BOD/day}}{0.7 \cdot \text{(1,000 cu m)}}$$

$$= \text{206 kg BOD/day/1,000 cu m}$$

8. *FOOD/MICROORGANISM*

$$\frac{F}{M} = \frac{\text{Food, kg BOD/day}}{\text{MLVSS, kg}}$$

$$\text{MLVSS, kg} = \text{(Vol, cu m)}\left(\text{MLSS, }\frac{mg}{L}\right)\text{(VM)}\left(\frac{\text{1 kg}}{\text{1,000,000 mg}}\right)\left(\frac{1{,}000\ L}{\text{1 cu m}}\right)$$

$$= \text{(700 cu m)}\left(3{,}800\,\frac{mg}{L}\right)(0.70)\left(\frac{\text{1 kg}}{\text{1,000,000 mg}}\right)\left(\frac{1{,}000\ L}{\text{1 cu m}}\right)$$

$$= \text{1,862 kg}$$

$$\frac{F}{M} = \frac{\text{Food, kg BOD/day}}{\text{Microorganisms, kg MLVSS}}$$

$$= \frac{\text{144 kg BOD/day}}{\text{1,862 kg MLVSS}}$$

$$= \text{0.07 kg BOD/day/kg MLVSS}$$

9. *SLUDGE AGE*

$$\text{Sludge Age, days} = \frac{\text{Aeration Solids, kg}}{\text{Solids Added, kg/day}}$$

a. Calculate the kilograms of solids under aeration.

$$\text{Aeration Solids, kg} = \text{(Vol, cu m)}\left(\text{MLSS,}\frac{mg}{L}\right)\left(\frac{\text{1 kg}}{\text{1,000,000 mg}}\right)\left(\frac{1{,}000\ L}{\text{1 cu m}}\right)$$

$$= \text{(700 cu m)}\left(3{,}800\,\frac{mg}{L}\right)\left(\frac{\text{1 kg}}{\text{1,000,000 mg}}\right)\left(\frac{1{,}000\ L}{\text{1 cu m}}\right)$$

$$= \text{2,660 kg}$$

b. Calculate solids fed to ditch in kilograms per day.

$$\text{Solids Added, kg/day} = \text{(Flow, cu m/day)(Infl SS, mg/}L)\left(\frac{\text{1 kg}}{\text{1,000,000 mg}}\right)\left(\frac{1{,}000\ L}{\text{1 cu m}}\right)$$

$$= \text{(700 cu m/day)(210 mg/}L)\left(\frac{\text{1 kg}}{\text{1,000,000 mg}}\right)\left(\frac{1{,}000\ L}{\text{1 cu m}}\right)$$

$$= \text{147 kg/day}$$

c. Calculate the sludge age in days.

$$\text{Sludge Age, days} = \frac{\text{Solids Under Aeration, kg}}{\text{Solids Added, kg/day}}$$

$$= \frac{\text{2,660 kg}}{\text{147 kg/day}}$$

$$= \text{18.1 days}$$

10. *DETENTION TIME*

$$\text{Detention Time, hours} = \frac{\text{(Ditch Volume, cu m)(24 hrs/day)}}{\text{Flow, cu m/day}}$$

$$= \frac{\text{(700 cu m)(24 hrs/day)}}{\text{800 cu m/day}}$$

$$= \text{21 hours}$$

A.42 Rotating Biological Contactors (RBCs)

EXAMPLE 11

A rotating biological contactor treats a flow of 5,000 cubic meters per day. The surface area of the media is 40,000 square meters (provided by manufacturer). What is the hydraulic loading in cu m/day/sq m?

Known		Unknown
Flow, cu m/day	= 5,000 cu m/day	Hydraulic Loading, cu m/day/sq m
Surface Area, sq m	= 40,000 sq m	

Calculate the hydraulic loading in cubic meters per day of wastewater per square meter of surface media.

$$\text{Hydraulic Loading, cu m/day/sq m} = \frac{\text{Flow, cu m/day}}{\text{Surface Area, sq m}}$$

$$= \frac{\text{5,000 cu m/day}}{\text{40,000 sq m}}$$

$$= \text{0.125 cu m/day/sq m}$$

EXAMPLE 12

The rotating biological contactor in Example 11 treats an industrial influent with a total BOD of 220 mg/*L* and suspended solids of 300 mg/*L*. Assume a K value of 0.5 to calculate the soluble BOD. What is the organic loading in grams of soluble BOD per day per square meter of media surface?

Known		Unknown
Flow, cu m/day	= 5,000 cu m/day	Organic Loading, gm BOD/day/sq m
Surface Area, sq m	= 40,000 sq m	
Total BOD, mg/*L*	= 220 mg/*L*	
SS, mg/*L*	= 300 mg/*L*	
K	= 0.5	

a. Estimate the soluble BOD treated by the rotating biological contactor.

$$\text{Soluble BOD, mg/}L = \text{Total BOD, mg/}L - (\text{K} \times \text{Suspended Solids, mg/}L)$$

$$= \text{220 mg/}L - (0.5 \times \text{300 mg/}L)$$

$$= \text{220 mg/}L - \text{150 mg/}L$$

$$= \text{70 mg/}L$$

b. Determine the soluble BOD applied to the rotating biological contactor in grams of soluble BOD per day.

$$\text{BOD Applied, gm/day} = \text{(Flow, cu m/day)(Soluble BOD, mg/}L\text{)(1,000 kg/cu m)(1,000 gm/kg)}$$

$$= \text{(5,000 cu m/day)}\left(\frac{\text{70 mg}}{\text{1,000,000 mg}}\right)\left(\frac{\text{1,000 kg}}{\text{cu m}}\right)\left(\frac{\text{1,000 gm}}{\text{kg}}\right)$$

$$= \text{350,000 gm Soluble BOD/day}$$

c. Calculate the organic loading in grams of soluble BOD per day per square meter of media surface.

$$\text{Organic BOD Loading, gm BOD/day/sq m} = \frac{\text{Soluble BOD Applied, gm BOD/day}}{\text{Surface Area of Media, sq m}}$$

$$= \frac{\text{350,000 gm Soluble BOD/day}}{\text{40,000 sq m}}$$

$$= \text{8.75 gm BOD/day/sq m}$$

A.43 Disinfection and Chlorination

EXAMPLE 13

A wastewater requires a chlorine feed rate of 6 kilograms chlorine per day. How many kilograms of chlorinated lime will be required to provide the needed chlorine? Assume each kilogram of chlorinated lime contains 0.34 kilogram of available chlorine.

Known		Unknown
Chlorine Required, kg/day	= 6 kg/day	Chlorinated Lime Feed Rate, kg/day
Portion of Chlorine in kg of Hypochlorite	= 0.34	

Calculate kilograms of chlorinated lime required per day.

$$\text{Chlorinated Lime Feed Rate, kg/day} = \frac{\text{Chlorine Required, kg/day}}{\text{Portion of Chlorine in kg of Hypochlorite}}$$

$$= \frac{\text{6 kg Chlorine/day}}{0.34}$$

$$= \text{17.6 kg Chlorinated Lime/day}$$

EXAMPLES 14 and 15

A hypochlorinator is set to feed 15 kilograms of chlorine per 24 hours; the wastewater flow is at a rate of 1,400 cu m/day and the chlorine as measured by the chlorine residual test after 30 minutes of contact time is 0.6 mg/*L*. Find the chlorine dosage and chlorine demand in mg/*L*.

Known		Unknown
Chlorine Feed, kg/day	= 15 kg/day	14. Chlorine Dose, mg/*L*
Flow, cu m/day	= 1,400 cu m/day	15. Chlorine Demand, mg/*L*
Chlorine Residual, mg/*L*	= 0.6 mg/*L*	

14. Calculate the chlorine dose in mg/*L*.

$$\text{Chlorine Feed or Dose, mg/}L = \frac{\text{(Chlorine Feed, kg/day)(1,000 gm/kg)(1,000 mg/gm)}}{\text{(Flow, cu m/day)(1,000 }L\text{/cu m)}}$$

$$= \frac{\text{(15 kg/day)(1,000 gm/kg)(1,000 mg/gm)}}{\text{(1,400 cu m/day)(1,000 }L\text{/cu m)}}$$

$$= \text{10.7 mg/}L$$

15. Determine the chlorine demand in mg/*L*.

$$\text{Chlorine Demand, mg/}L = \text{Chlorine Dose, mg/}L - \text{Chlorine Residual, mg/}L$$

$$= \text{10.7 mg/}L - \text{0.6 mg/}L$$

$$= \text{10.1 mg/}L$$

EXAMPLE 16

Calculate the hypochlorinator setting in kilograms per 24 hours to treat a wastewater with a chlorine demand of 12 mg/*L*, when a chlorine residual of 1.5 mg/*L* is desired. The flow is 1,100 cu m per day.

Known		Unknown
Chlorine Demand, mg/*L*	= 12 mg/*L*	Hypochlorinator Setting, kg/day
Chlorine Residual, mg/*L*	= 1.5 mg/*L*	
Flow, cu m/day	= 1,100 cu m/day	

a. Calculate the chlorine dose in mg/*L*.

$$\text{Chlorine Dose, mg/}L = \text{Chlorine Demand, mg/}L + \text{Chlorine Residual, mg/}L$$

$$= \text{12 mg/}L + \text{1.5 mg/}L$$

$$= \text{13.5 mg/}L$$

b. Calculate the hypochlorinator setting in kilograms per day.

$$\text{Hypochlorinator Setting, kg/day} = \left(\text{Flow, }\frac{\text{cu m}}{\text{day}}\right)\left(\text{Dose, }\frac{\text{mg}}{L}\right)\left(\frac{\text{1 kg}}{\text{1,000,000 mg}}\right)\left(\frac{1{,}000\ L}{\text{1 cu m}}\right)$$

$$= \left(1{,}100\ \frac{\text{cu m}}{\text{day}}\right)\left(13.5\ \frac{\text{mg}}{L}\right)\left(\frac{\text{1 kg}}{\text{1,000,000 mg}}\right)\left(\frac{1{,}000\ L}{\text{1 cu m}}\right)$$

$$= \text{14.9 kg/day}$$

$$= \text{14.9 kg/24 hours}$$

A.44 Land Treatment

EXAMPLES 17, 18, and 19

A plot of land 500 meters long by 200 meters wide is used for a land disposal system. If 2,000 cubic meters of water is applied to the land during a 24-hour period, calculate the land area in hectares and the hydraulic loading in cubic meters per day per hectare and also in centimeters per day.

Known		Unknown
Length, m	= 500 m	17. Area, hectares
Width, m	= 200 m	18. Hydraulic Loading, cu m/day/ha
Flow, cu m/day	= 2,000 cu m/day	19. Hydraulic Loading, cm/day

17. Determine the surface area in hectares.

$$\text{Area, hectares} = \frac{\text{(Length, m)(Width, m)}}{\text{10,000 sq m/ha}}$$

$$= \frac{\text{(500 m)(200 m)}}{\text{10,000 sq m/ha}}$$

$$= \text{10 hectares}$$

18. Determine the hydraulic loading in cubic meters per day per hectare.

$$\text{Hydraulic Loading, cu m/day/ha} = \frac{\text{Flow, cu m/day}}{\text{Area, ha}}$$

$$= \frac{\text{2,000 cu m/day}}{\text{10 ha}}$$

$$= \text{200 cu m/day/ha}$$

19. Determine the hydraulic loading in centimeters per day.

$$\text{Hydraulic Loading, cm/day} = \frac{\text{(Flow, cu m/day)(100 cm/m)}}{\text{(Length, m)(Width, m)}}$$

$$= \frac{\text{(2,000 cu m/day)(100 cm/m)}}{\text{(500 m)(200 m)}}$$

$$= \text{2 cm/day}$$

A.45 Laboratory

EXAMPLE 20

a. If a water temperature is 30°C, what is the temperature in °F?

b. If a water temperature is 60°F, what is the temperature in °C?

a. Convert water temperature from 30°C to °F.

$$°F = (9/5)(°C) + 32°$$
$$= (9/5)(30°C) + 32°$$
$$= 86°F$$

b. Convert water temperature from 60°F to °C.

$$°C = (5/9)(°F - 32°)$$
$$= (5/9)(60°F - 32°)$$
$$= 15.6°C$$

A.46 Management

EXAMPLES 21 and 22

The total revenues for a utility are $1,283,000 and the operating expenses for the utility are $978,000. The debt service expenses are $320,000. What is the operating ratio? What is the coverage ratio?

Known		Unknown
Total Revenue, $	= $1,283,000	21. Operating Ratio
Operating Expenses, $	= $978,000	22. Coverage Ratio
Debt Service Expenses, $	= $320,000	

21. Calculate the operating ratio.

$$\text{Operating Ratio} = \frac{\text{Total Revenue, \$}}{\text{Operating Expenses, \$}}$$
$$= \frac{\$1,283,000}{\$978,000}$$
$$= 1.31$$

22. Calculate the coverage ratio.

a. Calculate the non-debt expenses.

$$\text{Non-Debt Expenses, \$} = \text{Operating Exp, \$} - \text{Debt Service Exp, \$}$$
$$= \$978,000 - \$320,000$$
$$= \$658,000$$

b. Calculate the coverage ratio.

$$\text{Coverage Ratio} = \frac{\text{Total Revenue, \$} - \text{Non-Debt Expenses, \$}}{\text{Debt Service Expenses, \$}}$$
$$= \frac{\$1,283,000 - \$658,000}{\$320,000}$$
$$= 1.95$$

ABBREVIATIONS

ac	acre
ac-ft	acre-feet
af	acre feet
amp	ampere
°C	degrees Celsius
CFM	cubic feet per minute
CFS	cubic feet per second
cm	centimeter
cu ft	cubic feet
cu in	cubic inch
cu m	cubic meter
cu yd	cubic yard
°F	degrees Fahrenheit
ft	feet or foot
ft-lb/min	foot-pounds per minute
g	gravity
gal	gallon
gal/day	gallons per day
gm	gram
GPD	gallons per day
gpg	grains per gallon
GPM	gallons per minute
gr	grain
ha	hectare
HP	horsepower
hr	hour
in	inch
k	kilo
kg	kilogram
km	kilometer
kN	kilonewton
kW	kilowatt
kWh	kilowatt-hour
L	liter
lb	pound
lbs/sq in	pounds per square inch
m	meter
M	mega
M	million
mg	milligram
MGD	million gallons per day
mg/*L*	milligram per liter
min	minute
m*L*	milliliter
mm	millimeter
N	Newton
ohm	ohm
Pa	Pascal
ppb	parts per billion
ppm	parts per million
psf	pounds per square foot
psi	pounds per square inch
psig	pounds per square inch gage
RPM	revolutions per minute
sec	second
sq ft	square feet
sq in	square inches
W	watt

WASTEWATER WORDS

A Summary of the Words Defined

in

SMALL WASTEWATER SYSTEM OPERATION AND MAINTENANCE

(VOLUMES I and II)

PROJECT PRONUNCIATION KEY

by Warren L. Prentice

The Project Pronunciation Key is designed to aid you in the pronunciation of new words. While this key is based primarily on familiar sounds, it does not attempt to follow any particular pronunciation guide. This key is designed solely to aid operators in this program.

You may find it helpful to refer to other available sources for pronunciation help. Each current standard dictionary contains a guide to its own pronunciation key. Each key will be different from each other and from this key. Examples of the difference between the key used in this program and the *WEBSTER'S NEW WORLD COLLEGE DICTIONARY*[1] "Key" are shown below.

In using this key, you should accent (say louder) the syllable that appears in capital letters. The following chart is presented to give examples of how to pronounce words using the Project Key.

	SYLLABLE				
WORD	1st	2nd	3rd	4th	5th
acid	AS	id			
coliform	COAL	i	form		
biological	BUY	o	LODGE	ik	cull

The first word, *ACID*, has its first syllable accented. The second word, *COLIFORM*, has its first syllable accented. The third word, *BIOLOGICAL*, has its first and third syllables accented.

We hope you will find the key useful in unlocking the pronunciation of any new word.

[1] *The WEBSTER'S NEW WORLD COLLEGE DICTIONARY, Fourth Edition, 1999, was chosen rather than an unabridged dictionary because of its availability to the operator. Other editions may be slightly different.*

WORDS

>GREATER THAN >GREATER THAN

DO >5 mg/*L* would be read as DO GREATER THAN 5 mg/*L*.

<LESS THAN <LESS THAN

DO <5 mg/*L* would be read as DO LESS THAN 5 mg/*L*.

A

ACEOPS ACEOPS

See **A**LLIANCE OF **CE**RTIFIED **OP**ERATOR**S**, LAB ANALYSTS, INSPECTORS, AND SPECIALISTS (ACEOPS).

ABSORPTION (ab-SORP-shun) ABSORPTION

The taking in or soaking up of one substance into the body of another by molecular or chemical action (as tree roots absorb dissolved nutrients in the soil).

ACCOUNTABILITY ACCOUNTABILITY

When a manager gives power/responsibility to an employee, the employee ensures that the manager is informed of results or events.

ACTIVATED SLUDGE (ACK-ta-VATE-ed sluj) ACTIVATED SLUDGE

Sludge particles produced in raw or settled wastewater (primary effluent) by the growth of organisms (including zoogleal bacteria) in aeration tanks in the presence of dissolved oxygen. The term "activated" comes from the fact that the particles are teeming with bacteria, fungi, and protozoa. Activated sludge is different from primary sludge in that the sludge particles contain many living organisms which can feed on the incoming wastewater.

ACTIVATED SLUDGE (ACK-ta-VATE-ed sluj) PROCESS ACTIVATED SLUDGE PROCESS

A biological wastewater treatment process which speeds up the decomposition of wastes in the wastewater being treated. Activated sludge is added to wastewater and the mixture (mixed liquor) is aerated and agitated. After some time in the aeration tank, the activated sludge is allowed to settle out by sedimentation and is disposed of (wasted) or reused (returned to the aeration tank) as needed. The remaining wastewater then undergoes more treatment.

ACUTE HEALTH EFFECT ACUTE HEALTH EFFECT

An adverse effect on a human or animal body, with symptoms developing rapidly.

ADSORPTION (add-SORP-shun) ADSORPTION

The gathering of a gas, liquid, or dissolved substance on the surface or interface zone of another material.

ADVANCED WASTE TREATMENT ADVANCED WASTE TREATMENT

Any process of water renovation that upgrades treated wastewater to meet specific reuse requirements. May include general cleanup of water or removal of specific parts of wastes insufficiently removed by conventional treatment processes. Typical processes include chemical treatment and pressure filtration. Also called TERTIARY TREATMENT.

AERATION (air-A-shun) LIQUOR AERATION LIQUOR

Mixed liquor. The contents of the aeration tank including living organisms and material carried into the tank by either untreated wastewater or primary effluent.

AERATION (air-A-shun) TANK | AERATION TANK

The tank where raw or settled wastewater is mixed with return sludge and aerated. The same as aeration bay, aerator, or reactor.

AEROBES | AEROBES

Bacteria that must have molecular (dissolved) oxygen (DO) to survive. Aerobes are aerobic bacteria.

AEROBIC (AIR-O-bick) | AEROBIC

A condition in which atmospheric or dissolved molecular oxygen is present in the aquatic (water) environment.

AEROBIC BACTERIA (AIR-O-bick back-TEAR-e-ah) | AEROBIC BACTERIA

Bacteria which will live and reproduce only in an environment containing oxygen which is available for their respiration (breathing), namely atmospheric oxygen or oxygen dissolved in water. Oxygen combined chemically, such as in water molecules (H_2O), cannot be used for respiration by aerobic bacteria.

AEROBIC (AIR-O-bick) DIGESTION | AEROBIC DIGESTION

The breakdown of wastes by microorganisms in the presence of dissolved oxygen. This digestion process may be used to treat only waste activated sludge, or trickling filter sludge and primary (raw) sludge, or waste sludge from activated sludge treatment plants designed without primary settling. The sludge to be treated is placed in a large aerated tank where aerobic microorganisms decompose the organic matter in the sludge. This is an extension of the activated sludge process.

AESTHETIC (es-THET-ick) | AESTHETIC

Attractive or appealing.

AGGLOMERATION (a-GLOM-er-A-shun) | AGGLOMERATION

The growing or coming together of small scattered particles into larger flocs or particles which settle rapidly. Also see FLOC.

AIR BINDING | AIR BINDING

The clogging of a pipe or pump due to the presence of air released from water.

AIR LIFT PUMP | AIR LIFT PUMP

A special type of pump. This device consists of a vertical riser pipe submerged in the wastewater or sludge to be pumped. Compressed air is injected into a tail piece at the bottom of the pipe. Fine air bubbles mix with the wastewater or sludge to form a mixture lighter than the surrounding water which causes the mixture to rise in the discharge pipe to the outlet. An air-lift pump works like the center stand in a percolator coffee pot.

AIR RELIEF | AIR RELIEF

A type of valve used to allow air caught in high spots in pipes to escape.

ALGAE (AL-gee) | ALGAE

Microscopic plants which contain chlorophyll and live floating or suspended in water. They also may be attached to structures, rocks, or other submerged surfaces. Algae produce oxygen during sunlight hours and use oxygen during the night hours. Their biological activities appreciably affect the pH, alkalinity, and dissolved oxygen of the water.

ALGAL (AL-gull) BLOOM | ALGAL BLOOM

Sudden, massive growths of microscopic and macroscopic plant life, such as green or blue-green algae, which can, under the proper conditions, develop in lakes, reservoirs, and also in ponds.

ALIQUOT (AL-li-kwot) | ALIQUOT

Representative portion of a sample. Often an equally divided portion of a sample.

ALLIANCE OF CERTIFIED OPERATORS, LAB ANALYSTS, INSPECTORS, AND SPECIALISTS (ACEOPS) | ALLIANCE OF CERTIFIED OPERATORS, LAB ANALYSTS, INSPECTORS, AND SPECIALISTS (ACEOPS)

A professional organization for operators, lab analysts, inspectors, and specialists dedicated to improving professionalism; expanding training, certification, and job opportunities; increasing information exchange; and advocating the importance of certified operators, lab analysts, inspectors, and specialists. For information on membership, contact ACEOPS, 1810 Bel Air Drive, Ames, IA 50010-5125, phone (515) 663-4128 or e-mail: ACEOPS@aol.com.

AMBIENT (AM-bee-ent) TEMPERATURE | AMBIENT TEMPERATURE

Temperature of the surroundings.

AMPERAGE (AM-purr-age) AMPERAGE

The strength of an electric current measured in amperes. The amount of electric current flow, similar to the flow of water in gallons per minute.

AMPEROMETRIC (am-PURR-o-MET-rick) AMPEROMETRIC

A method of measurement that records electric current flowing or generated, rather than recording voltage. Amperometric titration is a means of measuring concentrations of certain substances in water.

AMPLITUDE AMPLITUDE

The maximum strength of an alternating current during its cycle, as distinguished from the mean or effective strength.

ANAEROBES ANAEROBES

Bacteria that do not need molecular (dissolved) oxygen (DO) to survive.

ANAEROBIC (AN-air-O-bick) ANAEROBIC

A condition in which atmospheric or dissolved molecular oxygen is *NOT* present in the aquatic (water) environment.

ANAEROBIC BACTERIA (AN-air-O-bick back-TEAR-e-ah) ANAEROBIC BACTERIA

Bacteria that live and reproduce in an environment containing no "free" or dissolved oxygen. Anaerobic bacteria obtain their oxygen supply by breaking down chemical compounds which contain oxygen, such as sulfate (SO_4^{2-}).

ANAEROBIC (AN-air-O-bick) DECOMPOSITION ANAEROBIC DECOMPOSITION

The decay or breaking down of organic material in an environment containing no "free" or dissolved oxygen.

ANAEROBIC (AN-air-O-bick) DIGESTER ANAEROBIC DIGESTER

A wastewater solids treatment device in which the solids and water (about 5 percent solids, 95 percent water) are placed in a large tank where bacteria decompose the solids in the absence of dissolved oxygen.

ANODE (an-O-d) ANODE

The positive pole or electrode of an electrolytic system, such as a battery. The anode attracts negatively charged particles or ions (anions).

ANOXIC (an-OX-ick) ANOXIC

A condition in which atmospheric or dissolved molecular oxygen is *NOT* present in the aquatic (water) environment and nitrate is present. Oxygen deficient or lacking sufficient oxygen. The term is similar to ANAEROBIC.

APPURTENANCE (uh-PURR-ten-nans) APPURTENANCE

Machinery, appliances, structures and other parts of the main structure necessary to allow it to operate as intended, but not considered part of the main structure.

AQUIFER (ACK-wi-fer) AQUIFER

A natural underground layer of porous, water-bearing materials (sand, gravel) usually capable of yielding a large amount or supply of water.

ASEPTIC (a-SEP-tick) ASEPTIC

Free from the living germs of disease, fermentation, or putrefaction. Sterile.

ASHING ASHING

Formation of an activated sludge floc in a clarifier effluent that is well oxidized and floats on the water surface (has the appearance of gray ash).

AUGER (AW-grr) AUGER

A sharp tool used to go through and break up or remove various materials that become lodged in sewers.

AUTHORITY AUTHORITY

The power and resources to do a specific job or to get that job done.

AVAILABLE MOISTURE CONTENT AVAILABLE MOISTURE CONTENT

The quantity of water present that is available for use by vegetation.

B

BOD (pronounce as separate letters) BOD

Biochemical **O**xygen **D**emand. The rate at which organisms use the oxygen in water or wastewater while stabilizing decomposable organic matter under aerobic conditions. In decomposition, organic matter serves as food for the bacteria and energy results from its oxidation. BOD measurements are used as a measure of the organic strength of wastes in water.

BOD_5 BOD_5

BOD_5 refers to the five-day biochemical oxygen demand. The total amount of oxygen used by microorganisms decomposing organic matter increases each day until the ultimate BOD is reached, usually in 50 to 70 days. BOD usually refers to the five-day BOD or BOD_5.

BACTERIA (back-TEAR-e-ah) BACTERIA

Bacteria are living organisms, microscopic in size, which usually consist of a single cell. Most bacteria use organic matter for their food and produce waste products as a result of their life processes.

BACTERIAL (back-TEAR-e-al) CULTURE BACTERIAL CULTURE

In the case of activated sludge, the bacterial culture refers to the group of bacteria classified as AEROBES, and FACULTATIVE organisms, which covers a wide range of organisms. Most treatment processes in the United States grow facultative organisms which use the carbonaceous (carbon compounds) BOD. Facultative organisms can live when oxygen resources are low. When "nitrification" is required, the nitrifying organisms are OBLIGATE AEROBES (require oxygen) and must have at least 0.5 mg/*L* of dissolved oxygen throughout the whole system to function properly.

BASE-EXTRA CAPACITY METHOD BASE-EXTRA CAPACITY METHOD

A cost allocation method used by utilities to determine rates for various groups. This method considers base costs (O & M expenses and capital costs), extra capacity costs (additional costs for maximum day and maximum hour demands), customer costs (meter maintenance and reading, billing, collection, accounting) and fire protection costs.

BATCH PROCESS BATCH PROCESS

A treatment process in which a tank or reactor is filled, the wastewater (or other solution) is treated or a chemical solution is prepared, and the tank is emptied. The tank may then be filled and the process repeated. Batch processes are also used to cleanse, stabilize or condition chemical solutions for use in industrial manufacturing and treatment processes.

BIOASSAY (BUY-o-ass-SAY) BIOASSAY

(1) A way of showing or measuring the effect of biological treatment on a particular substance or waste, or

(2) A method of determining the relative toxicity of a test sample of industrial wastes or other wastes by using live test organisms such as fish.

BIOCHEMICAL OXYGEN DEMAND (BOD) BIOCHEMICAL OXYGEN DEMAND (BOD)

The rate at which organisms use the oxygen in water or wastewater while stabilizing decomposable organic matter under aerobic conditions. In decomposition, organic matter serves as food for the bacteria and energy results from its oxidation. BOD measurements are used as a measure of the organic strength of wastes in water.

BIODEGRADABLE (BUY-o-dee-GRADE-able) BIODEGRADABLE

Organic matter that can be broken down by bacteria to more stable forms which will not create a nuisance or give off foul odors is considered biodegradable.

BIOFLOCCULATION (BUY-o-flock-u-LAY-shun) BIOFLOCCULATION

The clumping together of fine, dispersed organic particles by the action of certain bacteria and algae. This results in faster and more complete settling of the organic solids in wastewater.

BIOMASS (BUY-o-MASS) BIOMASS

A mass or clump of organic material consisting of living organisms feeding on the wastes in wastewater, dead organisms and other debris. Also see ZOOGLEAL MASS.

BIOMONITORING BIOMONITORING

A term used to describe methods of evaluating or measuring the effects of toxic substances in effluents on aquatic organisms in receiving waters. There are two types of biomonitoring, the BIOSURVEY and the BIOASSAY.

BIOSURVEY BIOSURVEY

A survey of the types and numbers of organisms naturally present in the receiving waters upstream and downstream from plant effluents. Comparisons are made between the aquatic organisms upstream and those organisms downstream of the discharge.

BLANK BLANK

A bottle containing only dilution water or distilled water; the sample being tested is not added. Tests are frequently run on a *SAMPLE* and a *BLANK* and the differences are compared. This procedure helps to eliminate or reduce test result errors that could be caused when the dilution water or distilled water used is contaminated.

BOND BOND

(1) A written promise to pay a specified sum of money (called the face value) at a fixed time in the future (called the date of **maturity**). A bond also carries interest at a fixed rate, payable periodically. The difference between a note and a bond is that a bond usually runs for a longer period of time and requires greater formality. Utility agencies use bonds as a means of obtaining large amounts of money for capital improvements.

(2) A warranty by an underwriting organization, such as an insurance company, guaranteeing honesty, performance, or payment by a contractor.

BREAKOUT OF CHLORINE BREAKOUT OF CHLORINE

A point at which chlorine leaves solution as a gas because the chlorine feed rate is too high. The solution is saturated and cannot dissolve any more chlorine. The maximum strength a chlorine solution can attain is approximately 3,500 mg/*L*. Beyond this concentration molecular chlorine is present which will break out of solution and cause "off-gassing" at the point of application.

BREAKPOINT CHLORINATION BREAKPOINT CHLORINATION

Addition of chlorine to water or wastewater until the chlorine demand has been satisfied. At this point, further additions of chlorine result in a residual that is directly proportional to the amount of chlorine added beyond the breakpoint.

BUFFER BUFFER

A solution or liquid whose chemical makeup neutralizes acids or bases without a great change in pH.

BUFFER CAPACITY BUFFER CAPACITY

A measure of the capacity of a solution or liquid to neutralize acids or bases. This is a measure of the capacity of water or wastewater for offering a resistance to changes in pH.

BULKING (BULK-ing) BULKING

Clouds of billowing sludge that occur throughout secondary clarifiers and sludge thickeners when the sludge does not settle properly. In the activated sludge process, bulking is usually caused by filamentous bacteria or bound water.

BURPING BURPING

"Billowing" solids that result when the settling tank sludge blanket becomes too deep (occupies too much volume in the bottom of the tank) and solids are swept by the water up and out over the effluent weirs in the discharged effluent.

C

COD (pronounce as separate letters) COD

Chemical **O**xygen **D**emand. A measure of the oxygen-consuming capacity of organic matter present in wastewater. COD is expressed as the amount of oxygen consumed from a chemical oxidant in mg/*L* during a specific test. Results are not necessarily related to the biochemical oxygen demand (BOD) because the chemical oxidant may react with substances that bacteria do not stabilize.

CALL DATE CALL DATE

First date a bond can be paid off.

CAPILLARY (CAP-i-larry) ACTION CAPILLARY ACTION

The movement of water through very small spaces due to molecular forces.

CATHODIC (ca-THOD-ick) PROTECTION CATHODIC PROTECTION

An electrical system for prevention of rust, corrosion, and pitting of steel and iron surfaces in contact with water, wastewater or soil. A low-voltage current is made to flow through a liquid (water) or a soil in contact with the metal in such a manner that the external electromotive force renders the metal structure cathodic. This concentrates corrosion on auxiliary anodic parts which are deliberately allowed to corrode instead of letting the structure corrode.

CATION EXCHANGE CAPACITY CATION EXCHANGE CAPACITY

The ability of a soil or other solid to exchange cations (positive ions such as calcium, Ca^{2+}) with a liquid.

CAVITATION (CAV-uh-TAY-shun) CAVITATION

The formation and collapse of a gas pocket or bubble on the blade of an impeller or the gate of a valve. The collapse of this gas pocket or bubble drives water into the impeller or gate with a terrific force that can cause pitting on the impeller or gate surface. Cavitation is accompanied by loud noises that sound like someone is pounding on the impeller or gate with a hammer.

CENTRIFUGE CENTRIFUGE

A mechanical device that uses centrifugal or rotational forces to separate solids from liquids.

CERTIFICATION EXAMINATION CERTIFICATION EXAMINATION

An examination administered by a state agency or professional association that small wastewater system operators take to indicate a level of professional competence. In many states and provinces, certification of small wastewater system operators may be voluntary. Current trends indicate that more states, provinces, and employers will require small wastewater system operators to be "certified" in the future.

CERTIFIED OPERATOR CERTIFIED OPERATOR

A person who has the education and experience required to operate a specific class of treatment facility as indicated by possessing a certificate of professional competence given by a state agency or professional association.

CHAIN OF CUSTODY CHAIN OF CUSTODY

A record of each person involved in the handling and possession of a sample from the person who collected the sample to the person who analyzed the sample in the laboratory and to the person who witnessed disposal of the sample.

CHEMICAL OXYGEN DEMAND (COD) CHEMICAL OXYGEN DEMAND (COD)

A measure of the oxygen-consuming capacity of organic matter present in wastewater. COD is expressed as the amount of oxygen consumed from a chemical oxidant in mg/*L* during a specific test. Results are not necessarily related to the biochemical oxygen demand (BOD) because the chemical oxidant may react with substances that bacteria do not stabilize.

CHLORAMINES (KLOR-uh-means) CHLORAMINES

Compounds formed by the reaction of hypochlorous acid (or aqueous chlorine) with ammonia.

CHLORINATION (KLOR-uh-NAY-shun) CHLORINATION

The application of chlorine to water or wastewater, generally for the purpose of disinfection, but frequently for accomplishing other biological or chemical results.

CHLORINE CONTACT CHAMBER CHLORINE CONTACT CHAMBER

A baffled basin that provides sufficient detention time of chlorine contact with wastewater for disinfection to occur. The minimum contact time is usually 30 minutes. (Also commonly referred to as basin or tank.)

CHLORINE DEMAND CHLORINE DEMAND

Chlorine demand is the difference between the amount of chlorine added to wastewater and the amount of residual chlorine remaining after a given contact time. Chlorine demand may change with dosage, time, temperature, pH, and nature and amount of the impurities in the water.

Chlorine Demand, mg/*L* = Chlorine Applied, mg/*L* – Chlorine Residual, mg/*L*

CHLORINE REQUIREMENT CHLORINE REQUIREMENT

The amount of chlorine which is needed for a particular purpose. Some reasons for adding chlorine are reducing the number of coliform bacteria (Most Probable Number), obtaining a particular chlorine residual, or oxidizing some substance in the water. In each case a definite dosage of chlorine will be necessary. This dosage is the chlorine requirement.

CHLORINE RESIDUAL CHLORINE RESIDUAL

The concentration of chlorine present in water after the chlorine demand has been satisfied. The concentration is expressed in terms of the total chlorine residual, which includes both the free and combined or chemically bound chlorine residuals.

CHLORINE RESIDUAL ANALYZER CHLORINE RESIDUAL ANALYZER

An instrument used to measure chlorine residual in water or wastewater. This instrument also can be used to control the chlorine dose rate.

CHLORORGANIC (klor-or-GAN-ick) CHLORORGANIC

Organic compounds combined with chlorine. These compounds generally originate from, or are associated with, living or dead organic materials.

COAGULATION (co-AGG-you-LAY-shun) COAGULATION

The clumping together of very fine particles into larger particles (floc) caused by the use of chemicals (coagulants). The chemicals neutralize the electrical charges of the fine particles, allowing them to come closer and form larger clumps. This clumping together makes it easier to separate the solids from the water by settling, skimming, draining or filtering.

CODE OF FEDERAL REGULATIONS (CFR) CODE OF FEDERAL REGULATIONS (CFR)

A publication of the United States Government which contains all of the proposed and finalized federal regulations, including environmental regulations.

COLIFORM (COAL-i-form) COLIFORM

One type of bacteria. The presence of coliform-group bacteria is an indication of possible pathogenic bacterial contamination. The human intestinal tract is one of the main habitats of coliform bacteria. They may also be found in the intestinal tracts of warm-blooded animals, and in plants, soil, air, and the aquatic environment. Fecal coliforms are those coliforms found in the feces of various warm-blooded animals; whereas the term "coliform" also includes other environmental sources.

COLORIMETRIC MEASUREMENT COLORIMETRIC MEASUREMENT

A means of measuring unknown chemical concentrations in water by *MEASURING A SAMPLE'S COLOR INTENSITY.* The specific color of the sample, developed by addition of chemical reagents, is measured with a photoelectric colorimeter or is compared with "color standards" using, or corresponding with, known concentrations of the chemical.

COMBINED AVAILABLE CHLORINE COMBINED AVAILABLE CHLORINE

The total chlorine, present as chloramine or other derivatives, that is present in a water and is still available for disinfection and for oxidation of organic matter. The combined chlorine compounds are more stable than free chlorine forms, but they are somewhat slower in disinfection action.

COMBINED AVAILABLE CHLORINE RESIDUAL COMBINED AVAILABLE CHLORINE RESIDUAL

The concentration of residual chlorine that is combined with ammonia, organic nitrogen, or both in water as a chloramine (or other chloro derivative) and yet is still available to oxidize organic matter and help kill bacteria.

COMBINED CHLORINE COMBINED CHLORINE

The sum of the chlorine species composed of free chlorine and ammonia, including monochloramine, dichloramine, and trichloramine (nitrogen trichloride). Dichloramine is the strongest disinfectant of these chlorine species, but it has less oxidative capacity than free chlorine.

COMMINUTION (com-mi-NEW-shun) COMMINUTION

Shredding. A mechanical treatment process which cuts large pieces of wastes into smaller pieces so they won't plug pipes or damage equipment. COMMINUTION and SHREDDING usually mean the same thing.

COMMINUTOR (com-mih-NEW-ter) COMMINUTOR

A device used to reduce the size of the solid chunks in wastewater by shredding (comminuting). The shredding action is like many scissors cutting or chopping to shreds all the large influent solids material in the wastewater.

COMMODITY-DEMAND METHOD COMMODITY-DEMAND METHOD

A cost allocation method used by water utilities to determine water rates for the various water user groups. This method considers the commodity costs (water, chemicals, power, amount of water use), demand costs (treatment, storage, distribution), customer costs (meter maintenance and reading, billing, collection, accounting) and fire protection costs.

COMPOSITE (come-PAH-zit) (PROPORTIONAL) SAMPLE COMPOSITE (PROPORTIONAL) SAMPLE

A composite sample is a collection of individual samples obtained at regular intervals, usually every one or two hours during a 24-hour time span. Each individual sample is combined with the others in proportion to the rate of flow when the sample was collected. The resulting mixture (composite sample) forms a representative sample and is analyzed to determine the average conditions during the sampling period.

COMPOUND COMPOUND

A pure substance composed of two or more elements whose composition is constant. For example, table salt (sodium chloride, NaCl) is a compound.

CONFINED SPACE CONFINED SPACE

Confined space means a space that:

A. Is large enough and so configured that an employee can bodily enter and perform assigned work; and

B. Has limited or restricted means for entry or exit (for example, manholes, tanks, vessels, silos, storage bins, hoppers, vaults, and pits are spaces that may have limited means of entry); and

C. Is not designed for continuous employee occupancy.

(Definition from the Code of Federal Regulations (CFR) Title 29 Part 1910.146.)

CONFINED SPACE, PERMIT-REQUIRED (PERMIT SPACE) CONFINED SPACE, PERMIT-REQUIRED (PERMIT SPACE)

A confined space that has one or more of the following characteristics:

- Contains or has a potential to contain a hazardous atmosphere,
- Contains a material that has the potential for engulfing an entrant,
- Has an internal configuration such that an entrant could be trapped or asphyxiated by inwardly converging walls or by a floor which slopes downward and tapers to a smaller cross section, or
- Contains any other recognized serious safety or health hazard.

(Definition from the Code of Federal Regulations (CFR) Title 29 Part 1910.146.)

CONING (CONE-ing) CONING

Development of a cone-shaped flow of liquid, like a whirlpool, through sludge. This can occur in a sludge hopper during sludge withdrawal when the sludge becomes too thick. Part of the sludge remains in place while liquid rather than sludge flows out of the hopper. Also called coring.

CONTACT STABILIZATION CONTACT STABILIZATION

Contact stabilization is a modification of the conventional activated sludge process. In contact stabilization, two aeration tanks are used. One tank is for separate reaeration of the return sludge for at least four hours before it is permitted to flow into the other aeration tank to be mixed with the primary effluent requiring treatment. The process may also occur in one long tank.

CORPORATION STOP CORPORATION STOP

A service shutoff valve located at a street main. This valve cannot be operated from the ground surface because it is buried and there is no valve box. Also called a corporation cock.

COVERAGE RATIO COVERAGE RATIO

The coverage ratio is a measure of the ability of the utility to pay the principal and interest on loans and bonds (this is known as "debt service") in addition to any unexpected expenses.

CURVILINEAR (KER-vuh-LYNN-e-ur) CURVILINEAR

In the shape of a curved line.

CYCLE CYCLE

A complete alternation of voltage and/or current in an alternating current (A.C.) circuit.

D

DNA (pronounce as separate letters) DNA

Deoxyribo**n**ucleic **a**cid. A chemical that encodes genetic information that is transmitted between generations of cells.

DANGEROUS AIR CONTAMINATION DANGEROUS AIR CONTAMINATION

An atmosphere presenting a threat of causing death, injury, acute illness, or disablement due to the presence of flammable and/or explosive, toxic or otherwise injurious or incapacitating substances.

A. Dangerous air contamination due to the flammability of a gas, vapor, or mist is defined as an atmosphere containing the gas, vapor, or mist at a concentration greater than 10 percent of its lower explosive (lower flammable) limit (LEL).

B. Dangerous air contamination due to a combustible particulate is defined as a concentration that meets or exceeds the particulate's lower explosive limit (LEL).

C. Dangerous air contamination due to the toxicity of a substance is defined as the atmospheric concentration which could result in employee exposure in excess of the substance's permissible exposure limit (PEL).

NOTE: A dangerous situation also occurs when the oxygen level in air drops below 19.5 percent oxygen.

DEBT SERVICE DEBT SERVICE

The amount of money required annually to pay the (1) interest on outstanding debts; or (2) funds due on a maturing bonded debt or the redemption of bonds.

DECHLORINATION (dee-KLOR-uh-NAY-shun) DECHLORINATION

The removal of chlorine from the effluent of a treatment plant. Chlorine needs to be removed because chlorine is toxic to fish and other aquatic life.

DECIBEL (DES-uh-bull) DECIBEL

A unit for expressing the relative intensity of sounds on a scale from zero for the average least perceptible sound to about 130 for the average level at which sound causes pain to humans. Abbreviated dB.

DECOMPOSITION, DECAY DECOMPOSITION, DECAY

The conversion of chemically unstable materials to more stable forms by chemical or biological action. If organic matter decays when there is no oxygen present (anaerobic conditions or putrefaction), undesirable tastes and odors are produced. Decay of organic matter when oxygen is present (aerobic conditions) tends to produce much less objectionable tastes and odors.

DECOMPOSITION, DECAY DECOMPOSITION, DECAY

Processes that convert unstable materials into more stable forms by chemical or biological action. Waste treatment encourages decay in a controlled situation so that material may be disposed of in a stable form. When organic matter decays under anaerobic conditions (putrefaction), undesirable odors are produced. The aerobic processes in common use for wastewater treatment produce much less objectionable odors.

DELEGATION DELEGATION

The act in which power is given to another person in the organization to accomplish a specific job.

DENITRIFICATION (dee-NYE-truh-fuh-KAY-shun) (ACTIVATED SLUDGE) DENITRIFICATION

(1) The anoxic biological reduction of nitrate nitrogen to nitrogen gas.

(2) The removal of some nitrogen from a system.

(3) An anoxic process that occurs when nitrite or nitrate ions are reduced to nitrogen gas and nitrogen bubbles are formed as a result of this process. The bubbles attach to the biological floc in the activated sludge process and float the floc to the surface of the secondary clarifiers. This condition is often the cause of rising sludge observed in secondary clarifiers or gravity thickeners. Also see NITRIFICATION.

DENITRIFICATION (dee-NYE-truh-fuh-KAY-shun) (RBCs) DENITRIFICATION

(1) The anoxic biological reduction of nitrate nitrogen to nitrogen gas.

(2) The removal of some nitrogen from a system.

(3) An anoxic process that occurs when nitrite or nitrate ions are reduced to nitrogen gas and nitrogen bubbles are formed as a result of this process. The bubbles attach to the biological floc and float the floc to the surface of the secondary clarifiers. This condition is often the cause of rising sludge observed in secondary clarifiers or gravity thickeners. Also see NITRIFICATION.

DEPRECIATION DEPRECIATION

The gradual loss in service value of a facility or piece of equipment due to all the factors causing the ultimate retirement of the facility or equipment. This loss can be caused by sudden physical damage, wearing out due to age, obsolescence, inadequacy or availability of a newer, more efficient facility or equipment. The value cannot be restored by maintenance.

DESICCATOR (DESS-uh-KAY-tor) DESICCATOR

A closed container into which heated weighing or drying dishes are placed to cool in a dry environment in preparation for weighing. The dishes may be empty or they may contain a sample. Desiccators contain a substance, such as anhydrous calcium chloride, which absorbs moisture and keeps the relative humidity near zero so that the dish or sample will not gain weight from absorbed moisture.

DETENTION TIME DETENTION TIME

The time required to fill a tank at a given flow or the theoretical time required for a given flow of wastewater to pass through a tank. In septic tanks this detention time will decrease as the volumes of sludge and scum increase.

DEWATER DEWATER

To remove or separate a portion of the water present in a sludge or slurry.

DIFFUSED-AIR AERATION DIFFUSED-AIR AERATION

A diffused-air activated sludge plant takes air, compresses it, and then discharges the air below the water surface of the aerator through some type of air diffusion device.

DIFFUSER DIFFUSER

A device (porous plate, tube, bag) used to break the air stream from the blower system into fine bubbles in an aeration tank or reactor.

DISCHARGE HEAD DISCHARGE HEAD

The pressure (in pounds per square inch or psi) measured at the centerline of a pump discharge and very close to the discharge flange, converted into feet. The pressure is measured from the centerline of the pump to the hydraulic grade line of the water in the discharge pipe.

$$\text{Discharge Head, ft} = (\text{Discharge Pressure, psi})(2.31\ \text{ft/psi})$$

DISINFECTION (dis-in-FECT-shun) DISINFECTION

The process designed to kill or inactivate most microorganisms in wastewater, including essentially all pathogenic (disease-causing) bacteria. There are several ways to disinfect, with chlorination being the most frequently used in water and wastewater treatment plants. Compare with STERILIZATION.

DISSOLVED OXYGEN DISSOLVED OXYGEN

Molecular (atmospheric) oxygen dissolved in water or wastewater, usually abbreviated DO.

DISTILLATE (DIS-tuh-late) DISTILLATE

In the distillation of a sample, a portion is collected by evaporation and recondensation; the part that is recondensed is the distillate.

DRAIN TILE SYSTEM DRAIN TILE SYSTEM

A system of tile pipes buried under the crops that collect percolated waters and keep the groundwater table below the ground surface to prevent ponding.

DRAINAGE WELLS DRAINAGE WELLS

Wells that can be pumped to lower the groundwater table and prevent ponding.

DRY WELL DRY WELL

A dry room or compartment in a lift station where the pumps are located, usually located next to the wet well.

DUCKWEED DUCKWEED

A small, green, cloverleaf-shaped floating plant, about one-quarter inch (6 mm) across which appears as a grainy "scum" on the surface of a pond.

DYNAMIC HEAD DYNAMIC HEAD

When a pump is operating, the vertical distance (in feet) from a point to the energy grade line. Also see TOTAL DYNAMIC HEAD, STATIC HEAD, and ENERGY GRADE LINE.

E

EGL EGL

See ENERGY GRADE LINE (EGL).

EASEMENT EASEMENT

Legal right to use the property of others for a specific purpose. For example, a utility company may have a five-foot easement along the property line of a home. This gives the utility the legal right to install and maintain a sewer line within the easement.

EDUCTOR (e-DUCK-ter) EDUCTOR

A hydraulic device used to create a negative pressure (suction) by forcing a liquid through a restriction, such as a Venturi. An eductor or aspirator (the hydraulic device) may be used in the laboratory in place of a vacuum pump. As an injector, it is used to produce vacuum for chlorinators. Sometimes used instead of a suction pump.

EFFECTIVE SOIL DEPTH EFFECTIVE SOIL DEPTH

The depth of soil in the leach field trench that provides a satisfactory percolation area for the septic tank effluent.

EFFLUENT (EF-lu-ent) EFFLUENT

Wastewater or other liquid—raw (untreated), partially or completely treated—flowing *FROM* an interceptor, reservoir, basin, treatment process, or treatment plant.

ELECTROLYSIS (ee-leck-TRAWL-uh-sis) ELECTROLYSIS

The decomposition of material by an outside electric current.

ELEMENT ELEMENT

A substance which cannot be separated into its constituent parts and still retain its chemical identity. For example, sodium (Na) is an element.

END POINT END POINT

Samples of water or wastewater are titrated to the end point. This means that a chemical is added, drop by drop, to a sample until a certain color change (blue to clear, for example) occurs. This is called the *END POINT* of the titration. In addition to a color change, an end point may be reached by the formation of a precipitate or the reaching of a specified pH. An end point may be detected by the use of an electronic device such as a pH meter. The completion of a desired chemical reaction.

ENDOGENOUS (en-DODGE-en-us) RESPIRATION ENDOGENOUS RESPIRATION

A situation where living organisms oxidize some of their own cellular mass instead of new organic matter they adsorb or absorb from their environment.

ENERGY GRADE LINE (EGL) ENERGY GRADE LINE (EGL)

A line that represents the elevation of energy head (in feet) of water flowing in a pipe, conduit or channel. The line is drawn above the hydraulic grade line (gradient) a distance equal to the velocity head ($V^2/2g$) of the water flowing at each section or point along the pipe or channel. Also see HYDRAULIC GRADE LINE.

[SEE DRAWING ON PAGE 424]

ENTRAIN ENTRAIN

To trap bubbles in water either mechanically through turbulence or chemically through a reaction.

ENZYMES (EN-zimes) ENZYMES

Organic substances (produced by living organisms) which cause or speed up chemical reactions. Organic catalysts and/or biochemical catalysts.

EQUITY EQUITY

The value of an investment in a facility.

EVAPOTRANSPIRATION (ee-VAP-o-TRANS-purr-A-shun) EVAPOTRANSPIRATION

(1) The process by which water vapor passes into the atmosphere from living plants. Also called TRANSPIRATION.

(2) The total water removed from an area by transpiration (plants) and by evaporation from soil, snow and water surfaces.

EXFILTRATION (EX-fill-TRAY-shun) EXFILTRATION

Liquid wastes and liquid-carried wastes which unintentionally leak out of a sewer pipe system and into the environment.

F

F/M RATIO F/M RATIO

See FOOD/MICROORGANISM RATIO.

FACULTATIVE (FACK-ul-TAY-tive) FACULTATIVE

Facultative bacteria can use either dissolved molecular oxygen or oxygen obtained from food materials such as sulfate or nitrate ions. In other words, facultative bacteria can live under aerobic, anoxic, or anaerobic conditions.

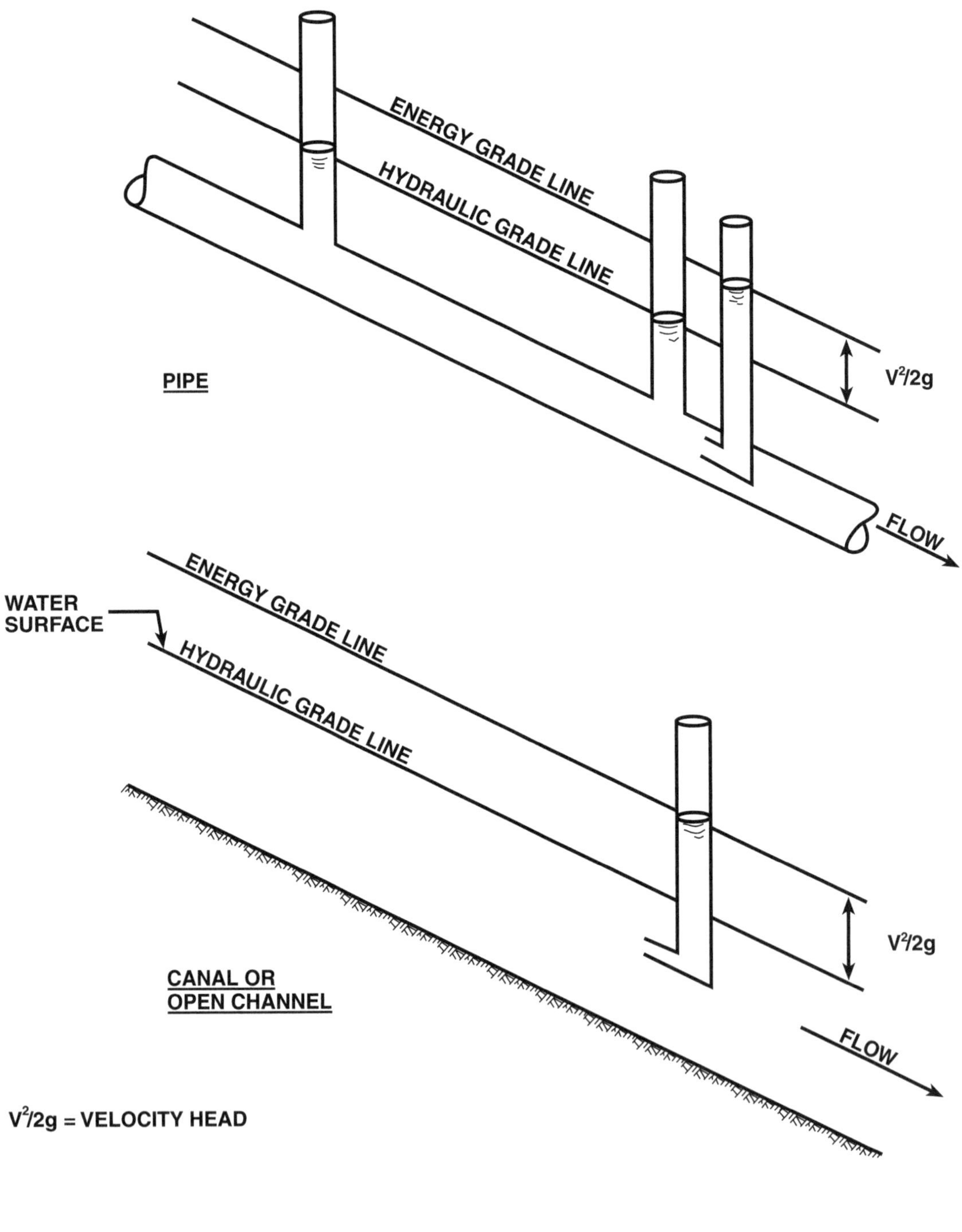

ENERGY GRADE LINE and HYDRAULIC GRADE LINE

FACULTATIVE (FACK-ul-TAY-tive) POND FACULTATIVE POND

The most common type of pond in current use. The upper portion (supernatant) is aerobic, while the bottom layer is anaerobic. Algae supply most of the oxygen to the supernatant.

FILAMENTOUS (FILL-a-MEN-tuss) ORGANISMS FILAMENTOUS ORGANISMS

Organisms that grow in a thread or filamentous form. Common types are *Thiothrix* and *Actinomycetes.* A common cause of sludge bulking in the activated sludge process.

FIXED COSTS FIXED COSTS

Costs that a utility must cover or pay even if there is no service provided. Also see VARIABLE COSTS.

FLAME POLISHED FLAME POLISHED

Melted by a flame to smooth out irregularities. Sharp or broken edges of glass (such as the end of a glass tube) are rotated in a flame until the edge melts slightly and becomes smooth.

FLIGHTS FLIGHTS

Scraper boards, made from redwood or other rot-resistant woods or plastic, used to collect and move settled sludge or floating scum.

FLOC FLOC

Clumps of bacteria and particles or coagulants and impurities that have come together and formed a cluster. Found in aeration tanks, secondary clarifiers and chemical precipitation processes.

FLOW LINE FLOW LINE

(1) The top of the wetted line, the water surface or the hydraulic grade line of water flowing in an open channel or partially full conduit.

(2) The lowest point of the channel inside a pipe or manhole. See INVERT. *NOTE:* (2) is an improper definition, although used by some contractors.

FOOD/MICROORGANISM RATIO FOOD/MICROORGANISM RATIO

Food to microorganism ratio. A measure of food provided to bacteria in an aeration tank.

$$\frac{\text{Food}}{\text{Microorganisms}} = \frac{\text{BOD, lbs/day}}{\text{MLVSS, lbs}}$$

$$= \frac{\text{Flow, MGD} \times \text{BOD, mg}/L \times 8.34\text{ lbs/gal}}{\text{Volume, MG} \times \text{MLVSS, mg}/L \times 8.34\text{ lbs/gal}}$$

or by calculator math system

$$= \text{Flow, MGD} \times \text{BOD, mg}/L \div \text{Volume, MG} \div \text{MLVSS, mg}/L$$

or metric

$$= \frac{\text{BOD, kg/day}}{\text{MLVSS, kg}}$$

$$= \frac{\text{Flow, M}L\text{/day} \times \text{BOD, mg}/L \times 1\text{ kg/M mg}}{\text{Volume, M}L \times \text{MLVSS, mg}/L \times 1\text{ kg/M mg}}$$

Commonly abbreviated F/M Ratio.

FREE AVAILABLE CHLORINE FREE AVAILABLE CHLORINE

The amount of chlorine available in water. This chlorine may be in the form of dissolved gas (Cl_2), hypochlorous acid (HOCl), or hypochlorite ion (OCl^-), but does not include chlorine combined with an amine (ammonia or nitrogen) or other organic compound.

FREE OXYGEN FREE OXYGEN

Molecular oxygen available for respiration by organisms. Molecular oxygen is the oxygen molecule, O_2, that is not combined with another element to form a compound.

FREEBOARD FREEBOARD

The vertical distance from the normal water surface to the top of the confining wall.

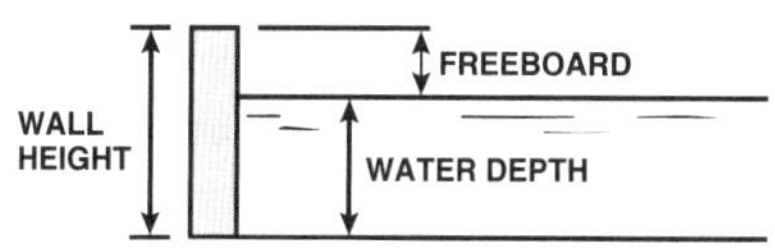

FRICTION LOSSES FRICTION LOSSES

The head, pressure or energy (they are the same) lost by water flowing in a pipe or channel as a result of turbulence caused by the velocity of the flowing water and the roughness of the pipe, channel walls, or restrictions caused by fittings. Water flowing in a pipe loses head, pressure or energy as a result of friction losses. Also see HEAD LOSS.

G

GIS GIS

Geographic **I**nformation **S**ystem. A computer program that combines mapping with detailed information about the physical locations of structures such as pipes, valves, and manholes within geographic areas. The system is used to help operators and maintenance personnel locate utility system features or structures and to assist with the scheduling and performance of maintenance activities.

GARNET GARNET

A group of hard, reddish, glassy, mineral sands made up of silicates of base metals (calcium, magnesium, iron and manganese). Garnet has a higher density than sand.

GAS/LIQUID GAS/LIQUID

Gaseous/Liquid. Gaseous/liquid chlorination refers to the fact that free chlorine is delivered to small treatment plants in containers (150-pound chlorine cylinders) that hold liquid chlorine with a free chlorine gas above the liquid in the container. The release of chlorine gas from the liquid chlorine surface depends on the temperature of the liquid and the pressure of the chlorine gas on the liquid surface. As chlorine gas is removed from the container, the gas pressure drops and more liquid chlorine becomes chlorine gas.

GATES GATES

A descriptive term used on irrigation distribution piping systems instead of the word valves. Gates cover outlet ports in the pipe segments. Water flows are regulated or distributed by opening the gates by either sliding the gate up or down or by swinging the gate to one side and uncovering an individual port to permit water flow to be discharged or regulated from the pipe at that particular point.

GEOGRAPHIC INFORMATION SYSTEM (GIS) GEOGRAPHIC INFORMATION SYSTEM (GIS)

A computer program that combines mapping with detailed information about the physical locations of structures such as pipes, valves, and manholes within geographic areas. The system is used to help operators and maintenance personnel locate utility system features or structures and to assist with the scheduling and performance of maintenance activities.

GRAB SAMPLE GRAB SAMPLE

A single sample of water collected at a particular time and place which represents the composition of the water only at that time and place.

GRADE GRADE

(1) The elevation of the invert (or bottom) of a pipeline, canal, culvert, or similar conduit.

(2) The inclination or slope of a pipeline, conduit, stream channel, or natural ground surface; usually expressed in terms of the ratio or percentage of number of units of vertical rise or fall per unit of horizontal distance. A 0.5 percent grade would be a drop of one-half foot per hundred feet of pipe.

GRAVIMETRIC GRAVIMETRIC

A means of measuring unknown concentrations of water quality indicators in a sample by *WEIGHING* a precipitate or residue of the sample.

GRINDER PUMP GRINDER PUMP

A small submersible centrifugal pump with an impeller designed to serve as a pump with grinder elements which grind solids into small pieces that pass through the pump impellers.

GROUNDWATER GROUNDWATER

Subsurface water in the saturation zone from which wells and springs are fed. In a strict sense the term applies only to water below the water table.

GROUNDWATER TABLE GROUNDWATER TABLE

The average depth or elevation of the groundwater over a selected area.

H

HGL HGL

See HYDRAULIC GRADE LINE (HGL).

HEAD HEAD

The vertical distance, height or energy of water above a point. A head of water may be measured in either height (feet or meters) or pressure (pounds per square inch or kilograms per square centimeter). Also see DISCHARGE HEAD, DYNAMIC HEAD, STATIC HEAD, SUCTION HEAD, SUCTION LIFT and VELOCITY HEAD.

HEAD LOSS HEAD LOSS

An indirect measure of loss of energy or pressure. Flowing water will lose some of its energy when it passes through a pipe, bar screen, comminutor, filter or other obstruction. The amount of energy or pressure lost is called "head loss." Head loss is measured as the difference in elevation between the upstream water surface and the downstream water surface and may be expressed in feet or meters.

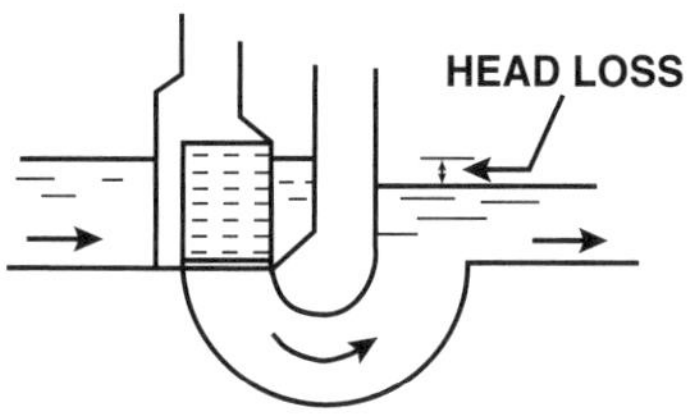

HEADER HEADER

A large pipe to which the ends of a series of smaller pipes are connected. Also called a MANIFOLD.

HEPATITIS (HEP-uh-TIE-tis) HEPATITIS

Hepatitis is an inflammation of the liver caused by an acute viral infection. Yellow jaundice is one symptom of hepatitis.

HERTZ HERTZ

The number of complete electromagnetic cycles or waves in one second of an electric or electronic circuit. Also called the frequency of the current. Abbreviated Hz.

HOT TAP HOT TAP

Tapping into a sewer line under pressure, such as a force main or a small-diameter sewer under pressure.

HYDRAULIC GRADE LINE (HGL) HYDRAULIC GRADE LINE (HGL)

The surface or profile of water flowing in an open channel or a pipe flowing partially full. If a pipe is under pressure, the hydraulic grade line is at the level water would rise to in a small tube connected to the pipe. To reduce the release of odors from wastewater, the water surface or hydraulic grade line should be kept as smooth as possible. Also see ENERGY GRADE LINE.

[SEE DRAWING ON PAGE 424]

HYDRAULIC JUMP HYDRAULIC JUMP

The sudden and usually turbulent abrupt rise in water surface in an open channel when water flowing at high velocity is suddenly retarded to a slow velocity.

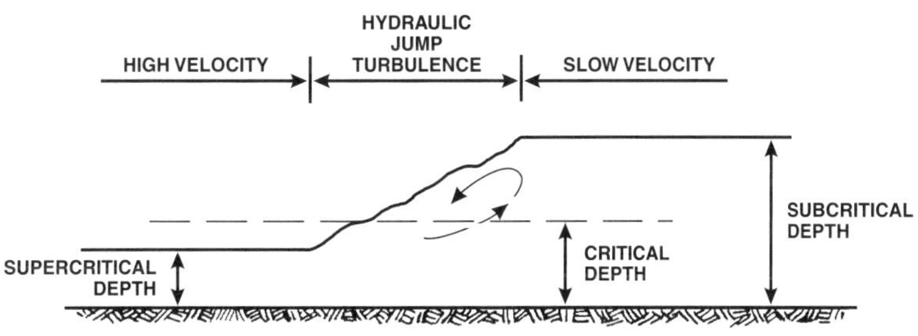

HYDROGEN SULFIDE GAS (H_2S) HYDROGEN SULFIDE GAS (H_2S)

Hydrogen sulfide is a gas with a rotten egg odor. This gas is produced under anaerobic conditions. Hydrogen sulfide gas is particularly dangerous because it dulls the sense of smell so that you don't notice it after you have been around it for a while. In high concentrations, hydrogen sulfide gas is only noticeable for a very short time before it dulls the sense of smell. The gas is very poisonous to the respiratory system, explosive, flammable, colorless and heavier than air.

HYDROLOGIC (HI-dro-LOJ-ick) CYCLE — HYDROLOGIC CYCLE

The process of evaporation of water into the air and its return to earth by precipitation (rain or snow). This process also includes transpiration from plants, groundwater movement, and runoff into rivers, streams and the ocean. Also called the WATER CYCLE.

HYDROPHILIC (HI-dro-FILL-ick) — HYDROPHILIC

Having a strong affinity (liking) for water. The opposite of HYDROPHOBIC.

HYDROPHOBIC (HI-dro-FOE-bick) — HYDROPHOBIC

Having a strong aversion (dislike) for water. The opposite of HYDROPHILIC.

HYPOCHLORINATION (HI-poe-KLOR-uh-NAY-shun) — HYPOCHLORINATION

The application of hypochlorite compounds to water or wastewater for the purpose of disinfection.

HYPOCHLORINATORS (HI-poe-KLOR-uh-NAY-tors) — HYPOCHLORINATORS

Chlorine pumps, chemical feed pumps or devices used to dispense chlorine solutions made from hypochlorites such as bleach (sodium hypochlorite) or calcium hypochlorite into the water being treated.

HYPOCHLORITE (HI-poe-KLOR-ite) — HYPOCHLORITE

Chemical compounds containing available chlorine; used for disinfection. They are available as liquids (bleach) or solids (powder, granules, and pellets) in barrels, drums, and cans. Salts of hypochlorous acid.

I

IMHOFF CONE — IMHOFF CONE

A clear, cone-shaped container marked with graduations. The cone is used to measure the volume of settleable solids in a specific volume (usually one liter) of wastewater.

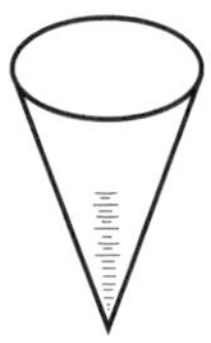

IMPELLER — IMPELLER

A rotating set of vanes in a pump or compressor designed to pump or move water or air.

INDICATOR (CHEMICAL) — INDICATOR (CHEMICAL)

A substance that gives a visible change, usually of color, at a desired point in a chemical reaction, generally at a specified end point.

INFILTRATION (IN-fill-TRAY-shun) — INFILTRATION

The seepage of groundwater into a sewer system, including service connections. Seepage frequently occurs through defective or cracked pipes, pipe joints and connections, interceptor access risers and covers, or manhole walls.

INFLOW — INFLOW

Water discharged into a sewer system and service connections from sources other than regular connections. This includes flow from yard drains, foundations and around access and manhole covers. Inflow differs from infiltration in that it is a direct discharge into the sewer rather than a leak in the sewer itself.

INFLUENT (IN-flu-ent) — INFLUENT

Wastewater or other liquid—raw (untreated) or partially treated—flowing *INTO* an interceptor, reservoir, basin, treatment process, or treatment plant.

INORGANIC — INORGANIC

Material such as sand, salt, iron, calcium salts and other mineral materials. Inorganic substances are of mineral origin, whereas organic substances are usually of animal or plant origin. Also see ORGANIC.

INORGANIC WASTE — INORGANIC WASTE

Waste material such as sand, salt, iron, calcium, and other mineral materials which are only slightly affected by the action of organisms. Inorganic wastes are chemical substances of mineral origin; whereas organic wastes are chemical substances usually of animal or plant origin. Also see NONVOLATILE MATTER.

INSECTICIDE INSECTICIDE

Any substance or chemical formulated to kill or control insects.

INTEGRATOR INTEGRATOR

A device or meter that continuously measures and calculates (adds) a process rate variable in cumulative fashion; for example, total flows displayed in gallons, million gallons, cubic feet, or some other unit of volume measurement. Also called a TOTALIZER.

INTERCEPTOR INTERCEPTOR

A septic tank or other holding tank which serves as a temporary wastewater storage reservoir for a Septic Tank Effluent Pump (STEP) system. See SEPTIC TANK.

INTERCEPTOR SEWER INTERCEPTOR SEWER

A large sewer that receives flow from a number of sewers and conducts the wastewater to a treatment plant. Often called an interceptor. The term interceptor is sometimes used in small communities to describe a septic tank or other holding tank which serves as a temporary wastewater storage reservoir for a Septic Tank Effluent Pump (STEP) system.

INTERFACE INTERFACE

The common boundary layer between two substances such as water and a solid.

INVERT (IN-vert) INVERT

The lowest point of the channel inside a pipe or manhole. See FLOW LINE.

INVERTED SIPHON INVERTED SIPHON

A pressure pipeline used to carry wastewater flowing in a gravity collection system under a depression such as a valley or roadway or under a structure such as a building. Also called a "depressed sewer."

J

(NO LISTINGS)

K

(NO LISTINGS)

L

LIFT LIFT

Vertical distance water is lifted from upstream water surface up to downstream water surface (which is at a higher elevation).

LINEAL (LIN-e-al) LINEAL

The length in one direction of a line. For example, a board 12 feet long has 12 lineal feet in its length.

LOWER EXPLOSIVE LIMIT (LEL) LOWER EXPLOSIVE LIMIT (LEL)

The lowest concentration of gas or vapor (percent by volume in air) that explodes if an ignition source is present at ambient temperature. At temperatures above 250°F the LEL decreases because explosibility increases with higher temperature.

M

M or MOLAR ***M*** or MOLAR

A molar solution consists of one gram molecular weight of a compound dissolved in enough water to make one liter of solution. A gram molecular weight is the molecular weight of a compound in grams. For example, the molecular weight of sulfuric acid (H_2SO_4) is 98. A one *M* solution of sulfuric acid would consist of 98 grams of H_2SO_4 dissolved in enough distilled water to make one liter of solution.

mg/*L* mg/*L*

See MILLIGRAMS PER LITER, mg/*L*.

MLSS MLSS

See MIXED LIQUOR SUSPENDED SOLIDS (MLSS).

MPN (pronounce as separate letters) MPN

MPN is the **M**ost **P**robable **N**umber of coliform-group organisms per unit volume of sample water. Expressed as a density or population of organisms per 100 m*L* of sample water.

MSDS MSDS

See MATERIAL SAFETY DATA SHEET (MSDS).

MANIFOLD MANIFOLD

A large pipe to which the ends of a series of smaller pipes are connected. Also called a HEADER.

MANOMETER (man-NAH-mut-ter) MANOMETER

Usually a glass tube filled with a liquid and used to measure the difference in pressure across a flow measuring device such as an orifice or a Venturi meter.

MATERIAL SAFETY DATA SHEET (MSDS) MATERIAL SAFETY DATA SHEET (MSDS)

A document which provides pertinent information and a profile of a particular hazardous substance or mixture. An MSDS is normally developed by the manufacturer or formulator of the hazardous substance or mixture. The MSDS is required to be made available to employees and operators whenever there is the likelihood of the hazardous substance or mixture being introduced into the workplace. Some manufacturers are preparing MSDSs for products that are not considered to be hazardous to show that the product or substance is *NOT* hazardous.

MEAN CELL RESIDENCE TIME (MCRT) MEAN CELL RESIDENCE TIME (MCRT)

An expression of the average time that a microorganism will spend in the activated sludge process.

$$\text{MCRT, days} = \frac{\text{Total Suspended Solids in Activated Sludge Process, lbs}}{\text{Total Suspended Solids Removed From Process, lbs/day}}$$

$$\text{or metric} = \frac{\text{Total Suspended Solids in Activated Sludge Process, kg}}{\text{Total Suspended Solids Removed From Process, kg/day}}$$

NOTE: Operators at different plants calculate the Total Suspended Solids (TSS) in the Activated Sludge Process, lbs (kg), by three different methods.

1. TSS in the Aeration Basin or Reactor Zone, lbs (kg),
2. TSS in Aeration Basin and Secondary Clarifier, lbs (kg), or
3. TSS in Aeration Basin and Secondary Clarifier Sludge Blanket, lbs (kg).

These three different methods make it difficult to compare MCRTs in days among different plants unless everyone uses the same method.

MECHANICAL AERATION MECHANICAL AERATION

The use of machinery to mix air and water so that oxygen can be absorbed into the water. Some examples are: paddle wheels, mixers, or rotating brushes to agitate the surface of an aeration tank; pumps to create fountains; and pumps to discharge water down a series of steps forming falls or cascades.

MEDIAN MEDIAN

The middle measurement or value. When several measurements are ranked by magnitude (largest to smallest), half of the measurements will be larger and half will be smaller.

MEGOHM MEGOHM

Meg means one million, so 5 megohms means 5 million ohms. A megger reads in millions of ohms.

MENISCUS (meh-NIS-cuss) MENISCUS

The curved surface of a column of liquid (water, oil, mercury) in a small tube. When the liquid wets the sides of the container (as with water), the curve forms a valley. When the confining sides are not wetted (as with mercury), the curve forms a hill or upward bulge. When a meniscus forms in a measuring device, the top of the liquid level of the sample is determined by the bottom of the meniscus.

MICRON (MY-kron) MICRON

μm, Micrometer or Micron. A unit of length. One millionth of a meter or one thousandth of a millimeter. One micron equals 0.00004 of an inch.

MICROORGANISMS (MY-crow-OR-gan-IS-zums) MICROORGANISMS

Very small organisms that can be seen only through a microscope. Some microorganisms use the wastes in wastewater for food and thus remove or alter much of the undesirable matter.

MIL MIL

A unit of length equal to 0.001 of an inch. The diameter of wires and tubing is measured in mils, as is the thickness of plastic sheeting.

MILLIGRAMS PER LITER, mg/*L* (MILL-i-GRAMS per LEET-er) MILLIGRAMS PER LITER, mg/*L*

A measure of the concentration by weight of a substance per unit volume. For practical purposes, one mg/*L* of a substance in water is equal to one part per million parts (ppm). Thus a liter of water with a specific gravity of 1.0 weighs one million milligrams. If it contains 10 milligrams of dissolved oxygen, the concentration is 10 milligrams per million milligrams, or 10 milligrams per liter (10 mg/*L*), or 10 parts of oxygen per million parts of water, or 10 parts per million (10 ppm) or 10 pounds dissolved oxygen in 1 million pounds of water (10 ppm).

MIXED LIQUOR MIXED LIQUOR

When the activated sludge in an aeration tank is mixed with primary effluent or the raw wastewater and return sludge, this mixture is then referred to as mixed liquor as long as it is in the aeration tank. Mixed liquor also may refer to the contents of mixed aerobic or anaerobic digesters.

MIXED LIQUOR SUSPENDED SOLIDS (MLSS) MIXED LIQUOR SUSPENDED SOLIDS (MLSS)

Suspended solids in the mixed liquor of an aeration tank.

MIXED LIQUOR VOLATILE SUSPENDED SOLIDS (MLVSS) MIXED LIQUOR VOLATILE SUSPENDED SOLIDS (MLVSS)

The organic or volatile suspended solids in the mixed liquor of an aeration tank. This volatile portion is used as a measure or indication of the microorganisms present.

MOLECULAR OXYGEN MOLECULAR OXYGEN

The oxygen molecule, O_2, that is not combined with another element to form a compound.

MOLECULAR WEIGHT MOLECULAR WEIGHT

The molecular weight of a compound in grams is the sum of the atomic weights of the elements in the compound. The molecular weight of sulfuric acid (H_2SO_4) in grams is 98.

Element	Atomic Weight	Number of Atoms	Molecular Weight
H	1	2	2
S	32	1	32
O	16	4	64
			98

MOLECULE (MOLL-uh-KULE) MOLECULE

The smallest division of a compound that still retains or exhibits all the properties of the substance.

N

N or NORMAL ***N*** or NORMAL

A normal solution contains one gram equivalent weight of reactant (compound) per liter of solution. The equivalent weight of an acid is that weight which contains one gram atom of ionizable hydrogen or its chemical equivalent. For example, the equivalent weight of sulfuric acid (H_2SO_4) is 49 (98 divided by 2 because there are two replaceable hydrogen ions). A one *N* solution of sulfuric acid would consist of 49 grams of H_2SO_4 dissolved in enough water to make one liter of solution.

NPDES PERMIT NPDES PERMIT

National **P**ollutant **D**ischarge **E**limination **S**ystem permit is the regulatory agency document issued by either a federal or state agency which is designed to control all discharges of potential pollutants from point sources and storm water runoff into U.S. waterways. NPDES permits regulate discharges into navigable waters from all point sources of pollution, including industries, municipal wastewater treatment plants, sanitary landfills, large agricultural feedlots and return irrigation flows.

NAMEPLATE NAMEPLATE

A durable metal plate found on equipment which lists critical operating conditions for the equipment.

NATURAL CYCLES NATURAL CYCLES

Cycles that take place in nature such as the water or hydrologic cycle where water is transformed or changed from one form to another until the water has returned to the original form, thus completing the cycle. Other natural cycles include the life cycles of aquatic organisms and plants, nutrient cycles, and cycles of self- or natural purification.

NEUTRALIZATION (new-trall-i-ZAY-shun) NEUTRALIZATION

Addition of an acid or alkali (base) to a liquid to cause the pH of the liquid to move toward a neutral pH of 7.0.

NITRIFICATION (NYE-truh-fuh-KAY-shun) NITRIFICATION

An aerobic process in which bacteria change the ammonia and organic nitrogen in wastewater into oxidized nitrogen (usually nitrate). The second-stage BOD is sometimes referred to as the "nitrogenous BOD" (first-stage BOD is called the "carbonaceous BOD"). Also see DENITRIFICATION.

NITROGENOUS (nye-TRAH-jen-us) NITROGENOUS

A term used to describe chemical compounds (usually organic) containing nitrogen in combined forms. Proteins and nitrate are nitrogenous compounds.

NONVOLATILE MATTER NONVOLATILE MATTER

Material such as sand, salt, iron, calcium, and other mineral materials which are only slightly affected by the actions of organisms and are not lost on ignition of the dry solids at 550°C. Volatile materials are chemical substances usually of animal or plant origin. Also see INORGANIC WASTE and VOLATILE MATTER or VOLATILE SOLIDS.

NUTRIENT CYCLE NUTRIENT CYCLE

The transformation or change of a nutrient from one form to another until the nutrient has returned to the original form, thus completing the cycle. The cycle may take place under either aerobic or anaerobic conditions.

NUTRIENTS NUTRIENTS

Substances which are required to support living plants and organisms. Major nutrients are carbon, hydrogen, oxygen, sulfur, nitrogen, and phosphorus. Nitrogen and phosphorus are difficult to remove from wastewater by conventional treatment processes because they are water soluble and tend to recycle. Also see NUTRIENT CYCLE.

O

O & M MANUAL O & M MANUAL

Operation and **M**aintenance Manual. A manual that describes detailed procedures for operators to follow to operate and maintain a specific wastewater treatment plant and the equipment of that plant

OSHA (O-shuh) OSHA

The Williams-Steiger **O**ccupational **S**afety and **H**ealth **A**ct of 1970 (OSHA) is a federal law designed to protect the health and safety of workers, including collection system and treatment plant operators. It regulates the design, construction, operation and maintenance of industrial plants and wastewater collection systems and treatment plants. The Act does not apply directly to municipalities, *EXCEPT* in those states that have approved plans and have asserted jurisdiction under Section 18 of the OSHA Act. *HOWEVER, CONTRACT OPERATORS AND PRIVATE FACILITIES DO HAVE TO COMPLY WITH OSHA REQUIREMENTS.* Wastewater treatment plants have come under stricter regulation in all phases of activity as a result of OSHA standards. OSHA also refers to the federal and state agencies which administer the OSHA regulations.

OBLIGATE AEROBES OBLIGATE AEROBES

Bacteria that must have atmospheric or dissolved molecular oxygen to live and reproduce.

OFFSET JOINT OFFSET JOINT

A pipe joint that is not exactly in line and centered.

OPERATING RATIO OPERATING RATIO

The operating ratio is a measure of the total revenues divided by the total operating expenses.

ORGANIC ORGANIC

Substances that come from animal or plant sources. Organic substances always contain carbon. (Inorganic materials are chemical substances of mineral origin.) Also see INORGANIC.

ORGANIC WASTE ORGANIC WASTE

Waste material which comes mainly from animal or plant sources. Organic wastes generally can be consumed by bacteria and other small organisms. Inorganic wastes are chemical substances of mineral origin.

ORGANIZING ORGANIZING

Deciding who does what work and delegating authority to the appropriate persons. A utility should have a written organizational plan and written policies.

ORIFICE (OR-uh-fiss) ORIFICE

An opening (hole) in a plate, wall, or partition. An orifice flange or plate placed in a pipe consists of a slot or a calibrated circular hole smaller than the pipe diameter. The difference in pressure in the pipe above and at the orifice may be used to determine the flow in the pipe.

OUTFALL OUTFALL

(1) The point, location or structure where wastewater or drainage discharges from a sewer, drain, or other conduit.

(2) The conduit leading to the final disposal point or area.

OVERHEAD OVERHEAD

Indirect costs necessary for a utility to function properly. These costs are not related to the actual collection, treatment, and disposal of wastewater, but include the costs of rent, lights, office supplies, management, and administration.

OVERTURN OVERTURN

The almost spontaneous mixing of all layers of water in a reservoir or lake when the water temperature becomes similar from top to bottom. This may occur in the fall/winter when the surface waters cool to the same temperature as the bottom waters and also in the spring when the surface waters warm after the ice melts. This is also called "turnover."

OXIDATION (ox-uh-DAY-shun) OXIDATION

Oxidation is the addition of oxygen, removal of hydrogen, or the removal of electrons from an element or compound. In wastewater treatment, organic matter is oxidized to more stable substances. The opposite of REDUCTION.

OXIDATION DITCH OXIDATION DITCH

The oxidation ditch is a modified form of the activated sludge process. The ditch consists of two channels placed side by side and connected at the ends to produce one continuous loop of wastewater flow and a brush rotator assembly placed across the channel to provide aeration and circulation.

OXIDATION STATE/OXIDATION NUMBER OXIDATION STATE/OXIDATION NUMBER

In a chemical formula, a number accompanied by a polarity indication (+ or –) that together indicate the charge of an ion as well as the extent to which the ion has been oxidized or reduced in a REDOX REACTION.

Due to the loss of electrons, the charge of an ion that has been oxidized would go from negative toward or to neutral, from neutral to positive, or from positive to more positive. As an example, an oxidation number of 2+ would indicate that an ion has lost two electrons and that its charge has become positive (that it now has an excess of two protons).

Due to the gain of electrons, the charge of the ion that has been reduced would go from positive toward or to neutral, from neutral to negative, or from negative to more negative. As an example, an oxidation number of 2– would indicate that an ion has gained two electrons and that its charge has become negative (that it now has an excess of two electrons). As an ion gains electrons, its oxidation state (or the extent to which it is oxidized) lowers; that is, its oxidation state is reduced.

OXIDATION-REDUCTION POTENTIAL (ORP) OXIDATION-REDUCTION POTENTIAL (ORP)

The electrical potential required to transfer electrons from one compound or element (the oxidant) to another compound or element (the reductant); used as a qualitative measure of the state of oxidation in wastewater treatment systems. ORP is measured in millivolts, with negative values indicating a tendency to reduce compounds or elements and positive values indicating a tendency to oxidize compounds or elements.

OXIDATION-REDUCTION (REDOX) REACTION OXIDATION-REDUCTION (REDOX) REACTION

A two-part reaction between two substances involving a transfer of electrons from one substance to the other. Oxidation is the loss of electrons by one substance, and reduction is the acceptance of electrons by the other substance. Reduction refers to the lowering of the OXIDATION STATE or OXIDATION NUMBER of the substance accepting the electrons.

In a redox reaction, the substance that gives up the electrons (that is oxidized) is called the reductant because it causes a reduction in the oxidation state or number of the substance that accepts the transferred electrons. The substance that receives the electrons (that is reduced) is called the oxidant because it causes oxidation of the other substance. Oxidation and reduction always occur simultaneously.

OXIDIZING AGENT OXIDIZING AGENT

Any substance, such as oxygen (O_2) or chlorine (Cl_2), that will readily add (take on) electrons. When oxygen or chlorine is added to wastewater, organic substances are oxidized. These oxidized organic substances are more stable and less likely to give off odors or to contain disease-causing bacteria. The opposite is a REDUCING AGENT.

OZONATION (O-zoe-NAY-shun) OZONATION

The application of ozone to water, wastewater, or air, generally for the purposes of disinfection or odor control.

P

PEL PEL

See PERMISSIBLE EXPOSURE LIMIT (PEL).

PARALLEL OPERATION PARALLEL OPERATION

Wastewater being treated is split and a portion flows to one treatment unit while the remainder flows to another similar treatment unit. Also see SERIES OPERATION.

PARASITIC (PAIR-a-SIT-tick) BACTERIA PARASITIC BACTERIA

Parasitic bacteria are those bacteria which normally live off another living organism, known as the "host."

PATHOGENIC (PATH-o-JEN-ick) ORGANISMS PATHOGENIC ORGANISMS

Bacteria, viruses, cysts, or protozoa which can cause disease (giardiasis, cryptosporidiosis, typhoid, cholera, dysentery) in a host (such as a person). There are many types of organisms which do *NOT* cause disease and which are *NOT* called pathogenic. Many beneficial bacteria are found in wastewater treatment processes actively cleaning up organic wastes.

PEAKING FACTOR PEAKING FACTOR

Ratio of a maximum flow to the average flow, such as maximum hourly flow or maximum daily flow to the average daily flow.

PERCENT SATURATION PERCENT SATURATION

The amount of a substance that is dissolved in a solution compared with the amount dissolved in the solution at saturation, expressed as a percent.

$$\text{Percent Saturation, \%} = \frac{\text{Amount of Substance That Is Dissolved} \times 100\%}{\text{Amount Dissolved in Solution at Saturation}}$$

PERCOLATION (PURR-co-LAY-shun) PERCOLATION

The movement or flow of water through soil or rocks.

PERMEABILITY (PURR-me-uh-BILL-uh-tee) PERMEABILITY

The property of a material or soil that permits considerable movement of water through it when it is saturated.

PERMISSIBLE EXPOSURE LIMIT (PEL) PERMISSIBLE EXPOSURE LIMIT (PEL)

The maximum 8-hour time weighted average of any airborne contaminant (such as dust, mist, vapor, gas, noise) to which an operator may be exposed. At no time may the exposure level exceed the ceiling concentration for that contaminant. Ceiling levels of regulated contaminants are listed in the Code of Federal Regulations (CFR) Title 29 Part 1910, Subparts G and Z. Also see TIME WEIGHTED AVERAGE (TWA).

PERMIT-REQUIRED CONFINED SPACE (PERMIT SPACE) PERMIT-REQUIRED CONFINED SPACE (PERMIT SPACE)

See CONFINED SPACE, PERMIT-REQUIRED (PERMIT SPACE).

pH (pronounce as separate letters) pH

pH is an expression of the intensity of the basic or acidic condition of a liquid. Mathematically, pH is the logarithm (base 10) of the reciprocal of the hydrogen ion activity.

$$pH = \text{Log}\ \frac{1}{[H^+]}$$

The pH may range from 0 to 14, where 0 is most acidic, 14 most basic, and 7 neutral. Natural waters usually have a pH between 6.5 and 8.5.

PHOTOSYNTHESIS (foe-toe-SIN-thuh-sis) PHOTOSYNTHESIS

A process in which organisms, with the aid of chlorophyll (green plant enzyme), convert carbon dioxide and inorganic substances into oxygen and additional plant material, using sunlight for energy. All green plants grow by this process.

PIG PIG

Refers to a poly pig which is a bullet-shaped device made of hard rubber or similar material. This device is used to clean pipes. It is inserted in one end of a pipe, moves through the pipe under pressure, and is removed from the other end of the pipe.

PILLOWS PILLOWS

Plastic tubes shaped like pillows that contain exact amounts of chemicals or reagents. Cut open the pillow, pour the reagents into the sample being tested, mix thoroughly and follow test procedures.

PLAN PLAN

A drawing showing the *TOP* view of sewers, manholes and streets.

PLANNING PLANNING

Management of utilities to build the resources and financial capability to provide for future needs.

PLUG FLOW PLUG FLOW

A type of flow that occurs in tanks, basins or reactors when a slug of wastewater moves through a tank without ever dispersing or mixing with the rest of the wastewater flowing through the tank.

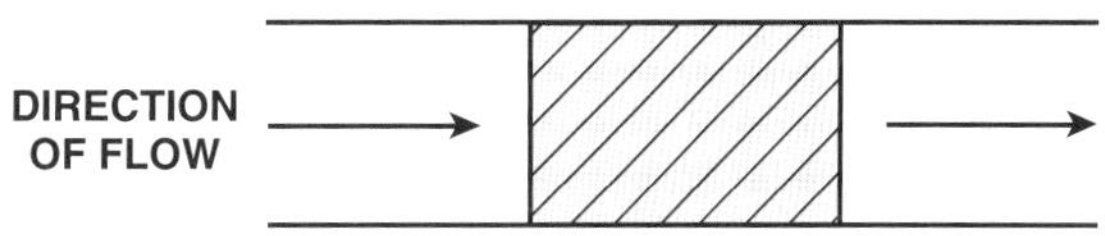

PLUG FLOW

POINT SOURCE POINT SOURCE

A discharge that comes out the end of a pipe. A nonpoint source refers to runoff or a "sheet flow" discharge from a field or similar source.

POLYELECTROLYTE (POLY-ee-LECK-tro-lite) POLYELECTROLYTE

A high-molecular-weight substance that is formed by either a natural or synthetic process. Natural polyelectrolytes may be of biological origin or derived from starch products, cellulose derivatives, and alignates. Synthetic polyelectrolytes consist of simple substances that have been made into complex, high-molecular-weight substances. Often called a POLYMER.

POLYMER (POLY-mer) POLYMER

A long chain molecule formed by the union of many monomers (molecules of lower molecular weight). Polymers are used with other chemical coagulants to aid in binding small suspended particles to larger chemical flocs for their removal from water.

POPULATION EQUIVALENT POPULATION EQUIVALENT

A means of expressing the strength of organic material in wastewater. In a domestic wastewater system, microorganisms use up about 0.2 pound of oxygen per day for each person using the system (as measured by the standard BOD test). May also be expressed as flow (100 gallons per day per person) or suspended solids (0.2 lb SS/day/person).

$$\text{Population Equivalent, persons} = \frac{\text{Flow, MGD} \times \text{BOD, mg}/L \times 8.34\text{ lbs/gal}}{0.2\text{ lb BOD/day/person}}$$

PORE PORE

A very small open space in a rock or granular material. Also called an interstice, VOID, or void space. Also see VOID.

POSTCHLORINATION POSTCHLORINATION

The addition of chlorine to the plant discharge or effluent, *FOLLOWING* plant treatment, for disinfection purposes.

POTABLE (POE-tuh-bull) WATER POTABLE WATER

Water that does not contain objectionable pollution, contamination, minerals, or infective agents and is considered satisfactory for drinking.

PRECHLORINATION PRECHLORINATION

The addition of chlorine in the collection system serving the plant or at the headworks of the plant *PRIOR TO* other treatment processes mainly for odor and corrosion control. Also applied to aid disinfection, to reduce plant BOD load, to aid in settling, to control foaming in Imhoff units and to help remove oil.

PRECIPITATE (pre-SIP-uh-TATE) PRECIPITATE

(1) An insoluble, finely divided substance which is a product of a chemical reaction within a liquid.

(2) The separation from solution of an insoluble substance.

PRESENT WORTH PRESENT WORTH

The value of a long-term project expressed in today's dollars. Present worth is calculated by converting (discounting) all future benefits and costs over the life of the project to a single economic value at the start of the project. Calculating the present worth of alternative projects makes it possible to compare them and select the one with the largest positive (beneficial) present worth or minimum present cost.

PRIMARY CLARIFIER PRIMARY CLARIFIER

A wastewater treatment device which consists of a rectangular or circular tank that allows those substances in wastewater that readily settle or float to be separated from the wastewater being treated.

PRIMARY TREATMENT PRIMARY TREATMENT

A wastewater treatment process that takes place in a rectangular or circular tank and allows those substances in wastewater that readily settle or float to be separated from the wastewater being treated. A septic tank is considered primary treatment.

PRIME PRIME

The action of filling a pump casing with water to remove the air. Most pumps must be primed before start-up or they will not pump any water.

PROFILE PROFILE

A drawing showing the *SIDE* view of sewers and manholes.

PROTOZOA (pro-toe-ZOE-ah) PROTOZOA

A group of motile microscopic organisms (usually single-celled and aerobic) that sometimes cluster into colonies and generally consume bacteria as an energy source.

PUTREFACTION (PEW-truh-FACK-shun) PUTREFACTION

Biological decomposition of organic matter with the production of foul-smelling products associated with anaerobic conditions.

PYROMETER (pie-ROM-uh-ter) PYROMETER

An apparatus used to measure high temperatures.

Q

(NO LISTINGS)

R

RNA (pronounce as separate letters) RNA

Ribo**n**ucleic **a**cid. A chemical that provides the structure for protein synthesis (building up).

RATE OF RETURN RATE OF RETURN

A value which indicates the return of funds received on the basis of the total equity capital used to finance physical facilities. Similar to the interest rate on savings accounts or loans.

REAGENT (re-A-gent) REAGENT

A pure chemical substance that is used to make new products or is used in chemical tests to measure, detect, or examine other substances.

RECEIVING WATER RECEIVING WATER

A stream, river, lake, ocean, or other surface or groundwaters into which treated or untreated wastewater is discharged.

REDUCING AGENT REDUCING AGENT

Any substance, such as base metal (iron) or the sulfide ion (S^{2-}), that will readily donate (give up) electrons. The opposite is an OXIDIZING AGENT.

REDUCTION (re-DUCK-shun) REDUCTION

Reduction is the addition of hydrogen, removal of oxygen, or the addition of electrons to an element or compound. Under anaerobic conditions (no dissolved oxygen present), sulfur compounds are reduced to odor-producing hydrogen sulfide (H_2S) and other compounds. The opposite of OXIDATION.

REFLUX REFLUX

Flow back. A sample is heated, evaporates, cools, condenses, and flows back to the flask.

REPRESENTATIVE SAMPLE REPRESENTATIVE SAMPLE

A sample portion of material or wastestream that is as nearly identical in content and consistency as possible to that in the larger body of material or wastestream being sampled.

RESIDUAL ANALYZER, CHLORINE RESIDUAL ANALYZER, CHLORINE

See CHLORINE RESIDUAL ANALYZER.

RESIDUAL CHLORINE RESIDUAL CHLORINE

The concentration of chlorine present in water after the chlorine demand has been satisfied. The concentration is expressed in terms of the total chlorine residual, which includes both the free and combined or chemically bound chlorine residuals.

RESPONSIBILITY RESPONSIBILITY

Answering to those above in the chain of command to explain how and why you have used your authority.

RIGHT-TO-KNOW LAWS RIGHT-TO-KNOW LAWS

Employee "Right-to-Know" legislation requires employers to inform employees (operators) of the possible health effects resulting from contact with hazardous substances. At locations where this legislation is in force, employers must provide employees with information regarding any hazardous substances which they might be exposed to under normal work conditions or reasonably foreseeable emergency conditions resulting from workplace conditions. OSHA's Hazard Communication Standard (HCS) (Title 29 CFR Part 1910.1200) is the federal regulation and state statutes are called Worker Right-to-Know Laws.

RIPRAP RIPRAP

Broken stones, boulders, or other materials placed compactly or irregularly on levees or dikes for the protection of earth surfaces against the erosive action of waves.

RISING SLUDGE RISING SLUDGE

Rising sludge occurs in the secondary clarifiers of activated sludge plants when the sludge settles to the bottom of the clarifier, is compacted, and then starts to rise to the surface, usually as a result of denitrification, or anaerobic biological activity that produces carbon dioxide and/or methane.

ROD (SEWER) ROD (SEWER)

A light metal rod, three to five feet long with a coupling at each end. Rods are joined and pushed into a sewer to dislodge obstructions.

ROTAMETER (RODE-uh-ME-ter) ROTAMETER

A device used to measure the flow rate of gases and liquids. The gas or liquid being measured flows vertically up a tapered, calibrated tube. Inside the tube is a small ball or bullet-shaped float (it may rotate) that rises or falls depending on the flow rate. The flow rate may be read on a scale behind or on the tube by looking at the middle of the ball or at the widest part or top of the float.

ROTATING BIOLOGICAL CONTACTOR (RBC) ROTATING BIOLOGICAL CONTACTOR (RBC)

A secondary biological treatment process for domestic and biodegradable industrial wastes. Biological contactors have a rotating "shaft" surrounded by plastic discs called the "media." The shaft and media are called the "drum." A biological slime grows on the media when conditions are suitable and the microorganisms that make up the slime (biomass) stabilize the waste products by using the organic material for growth and reproduction.

S

SDGS SYSTEM SDGS SYSTEM

See SMALL-DIAMETER GRAVITY SEWER (SDGS) SYSTEM.

STEF SYSTEM STEF SYSTEM

See SEPTIC TANK EFFLUENT FILTER (STEF) SYSTEM.

STEP SYSTEM STEP SYSTEM

See SEPTIC TANK EFFLUENT PUMP (STEP) SYSTEM.

SANITARY SEWER SANITARY SEWER

A pipe or conduit (sewer) intended to carry wastewater or waterborne wastes from homes, businesses, and industries to the treatment works. Storm water runoff or unpolluted water should be collected and transported in a separate system of pipes or conduits (storm sewers) to natural watercourses.

SAPROPHYTIC (SAP-row-FIT-ick) ORGANISMS SAPROPHYTIC ORGANISMS

Organisms living on dead or decaying organic matter. They help natural decomposition of the organic solids in wastewater.

SCUM SCUM

The material which rises to the surface of a liquid and forms a layer or film on the surface. A scum layer forms on the water surface of a septic tank.

SECCHI (SECK-key) DISC SECCHI DISC

A flat, white disc lowered into the water by a rope until it is just barely visible. At this point, the depth of the disc from the water surface is the recorded Secchi disc transparency.

SECONDARY CLARIFIER SECONDARY CLARIFIER

A wastewater treatment device which consists of a rectangular or circular tank that allows those substances not removed by previous treatment processes that settle or float to be separated from the wastewater being treated.

SECONDARY TREATMENT SECONDARY TREATMENT

A biological wastewater treatment process used to reduce the BOD of organic dissolved or suspended materials in the wastewater being treated. Usually the process follows primary treatment by sedimentation. The process commonly is followed by secondary clarifiers that allow the solids to settle out from the wastewater being treated.

SEDIMENTATION (SED-uh-men-TAY-shun) SEDIMENTATION

The process of settling and depositing of suspended matter carried by wastewater. Sedimentation usually occurs by gravity when the velocity of the wastewater is reduced below the point at which it can transport the suspended material.

SEIZING SEIZING

Seizing occurs when an engine overheats and a component expands to the point where the engine will not run. Also called freezing.

SEPTAGE (SEPT-age) SEPTAGE

The sludge produced in septic tanks.

SEPTIC (SEP-tick) SEPTIC

A condition produced by anaerobic bacteria. If severe, the sludge produces hydrogen sulfide, turns black, gives off foul odors, contains little or no dissolved oxygen, and the wastewater has a high oxygen demand.

SEPTIC TANK SEPTIC TANK

A system sometimes used where wastewater collection systems and treatment plants are not available. The system is a settling tank in which settled sludge and floatable scum are in intimate contact with the wastewater flowing through the tank and the organic solids are decomposed by anaerobic bacterial action. Used to treat wastewater and produce an effluent that flows into a subsurface leaching (filtering and disposal) system where additional treatment takes place. Also referred to as an INTERCEPTOR; however, the preferred term is *SEPTIC TANK*.

SEPTIC TANK EFFLUENT FILTER (STEF) SYSTEM SEPTIC TANK EFFLUENT FILTER (STEF) SYSTEM

A facility where effluent flows from a septic tank into a gravity flow collection system which flows to a gravity sewer, treatment plant or subsurface leaching system. The gravity flow pipeline is called an effluent drain.

SEPTIC TANK EFFLUENT PUMP (STEP) SYSTEM SEPTIC TANK EFFLUENT PUMP (STEP) SYSTEM

A facility where effluent is pumped from a septic tank into a pressurized collection system which may flow into a gravity sewer, treatment plant, or subsurface leaching system.

SEPTICITY (sep-TIS-it-tee) SEPTICITY

Septicity is the condition in which organic matter decomposes to form foul-smelling products associated with the absence of free oxygen. If severe, the wastewater produces hydrogen sulfide, turns black, gives off foul odors, contains little or no dissolved oxygen, and the wastewater has a high oxygen demand.

SEQUENCING BATCH REACTOR (SBR) SEQUENCING BATCH REACTOR (SBR)

A form of activated sludge treatment process.

SERIES OPERATION SERIES OPERATION

Wastewater being treated flows through one treatment unit and then flows through another similar treatment unit. Also see PARALLEL OPERATION.

SET POINT SET POINT

The position at which the control or controller is set. This is the same as the desired value of the process variable. For example, a thermostat is set to maintain a desired temperature.

SHOCK LOAD (TRICKLING FILTERS) SHOCK LOAD

The arrival at a plant of a waste which is toxic to organisms in sufficient quantity or strength to cause operating problems. Possible problems include odors and sloughing off of the growth or slime on the trickling filter media. Organic or hydraulic overloads also can cause a shock load.

SHOCK LOAD (ACTIVATED SLUDGE) SHOCK LOAD

The arrival at a plant of a waste which is toxic to organisms in sufficient quantity or strength to cause operating problems. Possible problems include odors and bulking sludge which will result in a high loss of solids from the secondary clarifiers into the plant effluent and a biological process upset that may require several days to a week to recover. Organic or hydraulic overloads also can cause a shock load.

SHORT-CIRCUITING SHORT-CIRCUITING

A condition that occurs in tanks or basins when some of the flowing water entering a tank or basin flows along a nearly direct pathway from the inlet to the outlet. This is usually undesirable since it may result in shorter contact, reaction, or settling times in comparison with the theoretical (calculated) or presumed detention times.

SHREDDING SHREDDING

Comminution. A mechanical treatment process which cuts large pieces of wastes into smaller pieces so they won't plug pipes or damage equipment. SHREDDING and COMMINUTION usually mean the same thing.

SIDESTREAM SIDESTREAM

Wastewater flows that develop from other storage or treatment facilities. This wastewater may or may not need additional treatment.

SLOUGHED SLOUGHED

Biological or biomass growths that break off from the rotating biological contactor (RBC) media. The "sloughed" biological growth is suspended in the RBC effluent and is later removed in the secondary clarifier as sludge.

SLUDGE (sluj) SLUDGE

The settleable solids separated from liquids during processing or the deposits of foreign materials on the bottoms of streams or other bodies of water.

SLUDGE AGE SLUDGE AGE

A measure of the length of time a particle of suspended solids has been retained in the activated sludge process.

$$\text{Sludge Age, days} = \frac{\text{Suspended Solids Under Aeration, lbs or kg}}{\text{Suspended Solids Added, lbs/day or kg/day}}$$

SLUDGE DENSITY INDEX (SDI) SLUDGE DENSITY INDEX (SDI)

This calculation is used in a way similar to the Sludge Volume Index (SVI) to indicate the settleability of a sludge in a secondary clarifier or effluent. The weight in grams of one milliliter of sludge after settling for 30 minutes. SDI = 100/SVI. Also see SLUDGE VOLUME INDEX (SVI).

SLUDGE VOLUME INDEX (SVI) SLUDGE VOLUME INDEX (SVI)

This is a calculation which indicates the tendency of activated sludge solids (aerated solids) to thicken or to become concentrated during the sedimentation/thickening process. SVI is calculated in the following manner: (1) allow a mixed liquor sample from the aeration basin to settle for 30 minutes; (2) determine the suspended solids concentration for a sample of the same mixed liquor; (3) calculate SVI by dividing the measured (or observed) wet volume (m*L*/*L*) of the settled sludge by the dry weight concentration of MLSS in grams/*L*.

$$\text{SVI, m}L\text{/gm} = \frac{\text{Settled Sludge Volume/Sample Volume, m}L/L}{\text{Suspended Solids Concentration, mg/}L} \times \frac{\text{1,000 mg}}{\text{gram}}$$

SLUGS SLUGS

Intermittent releases or discharges of wastewater.

SLURRY (SLUR-e) SLURRY

A thin, watery mud or any substance resembling it (such as a grit slurry or a lime slurry).

SMALL-DIAMETER GRAVITY SEWER (SDGS) SYSTEM SMALL-DIAMETER GRAVITY SEWER (SDGS) SYSTEM

A type of collection system in which a series of septic tanks discharge effluent by gravity, pump, or siphon to a small-diameter wastewater collection main. The wastewater flows by gravity to a lift station, a manhole in a conventional gravity collection system, or directly to a wastewater treatment plant.

SOLIDS CONCENTRATION SOLIDS CONCENTRATION

The solids in the aeration tank which carry microorganisms that feed on wastewater. Expressed as milligrams per liter of Mixed Liquor Volatile Suspended Solids (MLVSS, mg/*L*).

SOLUBLE BOD SOLUBLE BOD

Soluble BOD is the BOD of water that has been filtered in the standard suspended solids test. The soluble BOD is a measure of food for microorganisms that is dissolved in the water being treated.

SOLUTION SOLUTION

A liquid mixture of dissolved substances. In a solution it is impossible to see all the separate parts.

SPECIFIC GRAVITY SPECIFIC GRAVITY

(1) Weight of a particle, substance or chemical solution in relation to the weight of an equal volume of water. Water has a specific gravity of 1.000 at 4°C (39°F). Wastewater particles or substances usually have a specific gravity of 0.5 to 2.5.

(2) Weight of a particular gas in relation to the weight of an equal volume of air at the same temperature and pressure (air has a specific gravity of 1.0). Chlorine has a specific gravity of 2.5 as a gas.

SPLASH PAD SPLASH PAD

A structure made of concrete or other durable material to protect bare soil from erosion by splashing or falling water.

SPORE SPORE

The reproductive body of an organism which is capable of giving rise to a new organism either directly or indirectly. A viable (able to live and grow) body regarded as the resting stage of an organism. A spore is usually more resistant to disinfectants and heat than most organisms. Gangrene and tetanus bacteria are common spore-forming organisms.

STABILIZATION STABILIZATION

Processes that convert organic materials to a form that resists change. Organic material is stabilized by bacteria which convert the material to gases and other relatively inert substances. Stabilized organic material generally will not give off obnoxious odors.

STABILIZE STABILIZE

To convert to a form that resists change. Organic material is stabilized by bacteria which convert the material to gases and other relatively inert substances. Stabilized organic material generally will not give off obnoxious odors.

STABILIZED WASTE STABILIZED WASTE

A waste that has been treated or decomposed to the extent that, if discharged or released, its rate and state of decomposition would be such that the waste would not cause a nuisance or odors in the receiving water.

STAFF GAGE STAFF GAGE

A ruler or graduated scale used to measure the depth or elevation of water in a channel, tank, or stream.

STANDARD METHODS STANDARD METHODS

STANDARD METHODS FOR THE EXAMINATION OF WATER AND WASTEWATER, 20th Edition. A joint publication of the American Public Health Association (APHA), American Water Works Association (AWWA), and the Water Environment Federation (WEF) which outlines the accepted laboratory procedures used to analyze the impurities in water and wastewater. Available from Water Environment Federation, Publications Order Department, 601 Wythe Street, Alexandria, VA 22314-1994. Order No. S82010. Price to members, $155.00; nonmembers, $200.00; plus shipping and handling based on weight.

STANDARD SOLUTION STANDARD SOLUTION

A solution in which the exact concentration of a chemical or compound is known.

STANDARDIZE STANDARDIZE

To compare with a standard.

(1) In wet chemistry, to find out the exact strength of a solution by comparing it with a standard of known strength. This information is used to adjust the strength by adding more water or more of the substance dissolved.

(2) To set up an instrument or device to read a standard. This allows you to adjust the instrument so that it reads accurately, or enables you to apply a correction factor to the readings.

STATIC HEAD STATIC HEAD

When water is not moving, the vertical distance (in feet) from a specific point to the water surface is the static head. Also see DYNAMIC HEAD.

STATIC LIFT STATIC LIFT

Vertical distance water is lifted from upstream water surface up to downstream water surface (which is at a higher elevation) when no water is being pumped.

STATIC WATER HEAD STATIC WATER HEAD

Elevation or surface of water that is not being pumped.

STEP-FEED AERATION STEP-FEED AERATION

Step-feed aeration is a modification of the conventional activated sludge process. In step aeration, primary effluent enters the aeration tank at several points along the length of the tank, rather than all of the primary effluent entering at the beginning or head of the tank and flowing through the entire tank in a plug flow mode.

STERILIZATION (STARE-uh-luh-ZAY-shun) STERILIZATION

The removal or destruction of all microorganisms, including pathogenic and other bacteria, vegetative forms and spores. Compare with DISINFECTION.

STOP LOG STOP LOG

A log or board in an outlet box or device used to control the water level in ponds and also the flow from one pond to another pond or system.

SUBSURFACE LEACHING SYSTEM SUBSURFACE LEACHING SYSTEM

A method of treatment and disposal of septic tank effluent, sand filter effluent, or other treated wastewater. The effluent is applied to soil below the ground surface through open-jointed pipes or drains or through perforated pipes (holes in the pipes). The effluent is treated as it passes through porous soil or rock strata (layers). Newer subsurface leaching systems include chamber and gravelless systems, and also gravel trenches without pipe the full length of the trench.

SUCTION HEAD SUCTION HEAD

The *POSITIVE* pressure [in feet (meters) or pounds per square inch] on the suction side of a pump. The pressure can be measured from the centerline of the pump *UP TO* the elevation of the hydraulic grade line on the suction side of the pump.

SUCTION LIFT SUCTION LIFT

The *NEGATIVE* pressure [in feet (meters) or inches (centimeters) of mercury vacuum] on the suction side of the pump. The pressure can be measured from the centerline of the pump *DOWN TO* (lift) the elevation of the hydraulic grade line on the suction side of the pump.

SUMP SUMP

The term "sump" refers to a structure which connects an industrial discharger to a public sewer. The structure (sump) could be a sample box, a clarifier or an intercepting sewer.

SUPERNATANT (sue-per-NAY-tent) (ACTIVATED SLUDGE) SUPERNATANT

Liquid removed from settled sludge. Supernatant commonly refers to the liquid between the sludge on the bottom and the scum on the surface of a tank.

SUPERNATANT (sue-per-NAY-tent) (RBCs) SUPERNATANT

Liquid removed from settled sludge. Supernatant commonly refers to the liquid between the sludge on the bottom and the scum on the surface of an anaerobic digester or a septic tank (interceptor). This liquid from an anaerobic digester is usually returned to the influent wet well or to the primary clarifier. If the liquid (supernatant) is from a septic tank, the liquid is discharged by gravity or by a pump to a wastewater collection system.

SURCHARGE SURCHARGE

Sewers are surcharged when the supply of water to be carried is greater than the capacity of the pipes to carry the flow. The surface of the wastewater in manholes rises above the top of the sewer pipe, and the sewer is under pressure or a head, rather than at atmospheric pressure.

SURFACTANT (sir-FAC-tent) SURFACTANT

Abbreviation for surface-active agent. The active agent in detergents that possesses a high cleaning ability.

SUSPENDED SOLIDS SUSPENDED SOLIDS

(1) Solids that either float on the surface or are suspended in water, wastewater, or other liquids, and which are largely removable by laboratory filtering.

(2) The quantity of material removed from wastewater in a laboratory test, as prescribed in *STANDARD METHODS FOR THE EXAMINATION OF WATER AND WASTEWATER*, and referred to as Total Suspended Solids Dried at 103–105°C.

T

TOC (pronounce as separate letters) TOC

Total **O**rganic **C**arbon. TOC measures the amount of organic carbon in water.

TWA TWA

See TIME WEIGHTED AVERAGE (TWA).

TELEMETERING EQUIPMENT TELEMETERING EQUIPMENT

Equipment that translates physical measurements into electrical impulses that are transmitted to dials or recorders.

TERTIARY (TER-she-AIR-ee) TREATMENT TERTIARY TREATMENT

Any process of water renovation that upgrades treated wastewater to meet specific reuse requirements. May include general cleanup of water or removal of specific parts of wastes insufficiently removed by conventional treatment processes. Typical processes include chemical treatment and pressure filtration. Also called ADVANCED WASTE TREATMENT.

THIEF HOLE THIEF HOLE

A digester sampling well which allows sampling of the digester contents without venting digester gas.

THRESHOLD ODOR THRESHOLD ODOR

The minimum odor of a sample (gas or water) that can just be detected after successive odorless (gas or water) dilutions.

THRUST BLOCK THRUST BLOCK

A mass of concrete or similar material appropriately placed around a pipe to prevent movement when the pipe is carrying water. Usually placed at bends and valve structures.

TIME WEIGHTED AVERAGE (TWA) TIME WEIGHTED AVERAGE (TWA)

The average concentration of a pollutant (or sound) based on the times and levels of concentrations of the pollutant. The time weighted average is equal to the sum of the portion of each time period (as a decimal, such as 0.25 hour) multiplied by the pollutant concentration during the time period divided by the hours in the workday (usually 8 hours).

TITRATE (TIE-trate) TITRATE

To *TITRATE* a sample, a chemical solution of known strength is added drop by drop until a certain color change, precipitate, or pH change in the sample is observed (end point). Titration is the process of adding the chemical reagent in small increments (0.1 – 1.0 milliliter) until completion of the reaction, as signaled by the end point.

TOPOGRAPHY (toe-PAH-gruh-fee) TOPOGRAPHY

The arrangement of hills and valleys in a geographic area.

TOTAL CHLORINE TOTAL CHLORINE

The total concentration of chlorine in water, including the combined chlorine (such as inorganic and organic chloramines) and the free available chlorine.

TOTAL CHLORINE RESIDUAL TOTAL CHLORINE RESIDUAL

The total amount of chlorine residual (value for residual chlorine, including both free chlorine and chemically bound chlorine) present in a water sample after a given contact time.

TOTAL DYNAMIC HEAD (TDH) TOTAL DYNAMIC HEAD (TDH)

When a pump is lifting or pumping water, the vertical distance (in feet) from the elevation of the energy grade line on the suction side of the pump to the elevation of the energy grade line on the discharge side of the pump. The total dynamic head is the static head plus pipe friction losses. Static head is the vertical distance between water surfaces when the water is not moving.

TOTALIZER TOTALIZER

A device or meter that continuously measures and calculates (adds) a process rate variable in cumulative fashion; for example, total flows displayed in gallons, million gallons, cubic feet, or some other unit of volume measurement. Also called an INTEGRATOR.

TOXIC (TOX-ick) — TOXIC

A substance which is poisonous to a living organism.

TOXIC SUBSTANCES — TOXIC SUBSTANCES

Substances which are poisonous to living organisms and restrict their ability to treat wastes.

TOXICITY (tox-IS-it-tee) — TOXICITY

The relative degree of being poisonous or toxic. A condition which may exist in wastes and will inhibit or destroy the growth or function of certain organisms.

TRANSPIRATION (TRAN-spur-RAY-shun) — TRANSPIRATION

The process by which water vapor is released to the atmosphere by living plants. This process is similar to people sweating. Also see EVAPOTRANSPIRATION.

TRICKLING FILTER — TRICKLING FILTER

A treatment process in which the wastewater trickles over media that provide the opportunity for the formation of slimes or biomass which contain organisms that feed upon and remove wastes from the water being treated.

TURBIDITY (ter-BID-it-tee) — TURBIDITY

The cloudy appearance of water caused by the presence of suspended and colloidal matter. In the waterworks field, a turbidity measurement is used to indicate the clarity of water. Technically, turbidity is an optical property of the water based on the amount of light reflected by suspended particles. Turbidity cannot be directly equated to suspended solids because white particles reflect more light than dark-colored particles and many small particles will reflect more light than an equivalent large particle.

TURBIDITY METER — TURBIDITY METER

An instrument for measuring and comparing the turbidity of liquids by passing light through them and determining how much light is reflected by the particles in the liquid. The normal measuring range is 0 to 100 and is expressed as Nephelometric Turbidity Units (NTUs).

TURBIDITY UNITS (TU) — TURBIDITY UNITS (TU)

Turbidity units are a measure of the cloudiness of water. If measured by a nephelometric (deflected light) instrumental procedure, turbidity units are expressed in nephelometric turbidity units (NTU) or simply TU. Those turbidity units obtained by visual methods are expressed in Jackson Turbidity Units (JTU) which are a measure of the cloudiness of water; they are used to indicate the clarity of water. There is no real connection between NTUs and JTUs. The Jackson turbidimeter is a visual method and the nephelometer is an instrumental method based on deflected light.

U

UNIFORMITY COEFFICIENT (U.C.) — UNIFORMITY COEFFICIENT (U.C.)

The ratio of (1) the diameter of a grain (particle) of a size that is barely too large to pass through a sieve that allows 60 percent of the material (by weight) to pass through, to (2) the diameter of a grain (particle) of a size that is barely too large to pass through a sieve that allows 10 percent of the material (by weight) to pass through. The resulting ratio is a measure of the degree of uniformity in a granular material such as filter media.

$$\text{Uniformity Coefficient} = \frac{\text{Particle Diameter}_{60\%}}{\text{Particle Diameter}_{10\%}}$$

V

V-NOTCH WEIR — V-NOTCH WEIR

A triangular weir with a "V" notch calibrated in gallons per minute readings. The weir can be placed in a pipe or open channel. As the flow passes through the "V", the depth of water flowing over the weir can be measured and converted to a flow in gallons per minute.

VARIABLE COSTS — VARIABLE COSTS

Costs that a utility must cover or pay that are associated with the actual collection, treatment, and disposal of wastewater. The costs vary or fluctuate. Also see FIXED COSTS.

VAULT — VAULT

A small box-like structure that contains valves used to regulate flows.

VECTOR VECTOR

An insect or other organism capable of transmitting germs or other agents of disease.

VELOCITY HEAD VELOCITY HEAD

The energy in flowing water as determined by a vertical height (in feet or meters) equal to the square of the velocity of flowing water divided by twice the acceleration due to gravity ($V^2/2g$).

VOID VOID

A pore or open space in rock, soil or other granular material, not occupied by solid matter. The pore or open space may be occupied by air, water, or other gaseous or liquid material. Also called a void space or interstice.

VOLATILE (VOL-uh-tull) VOLATILE

(1) A volatile substance is one that is capable of being evaporated or changed to a vapor at relatively low temperatures. Volatile substances also can be partially removed from water or wastewater by the air stripping process.

(2) In terms of solids analysis, volatile refers to materials lost (including most organic matter) upon ignition in a muffle furnace for 60 minutes at 550°C. Natural volatile materials are chemical substances usually of animal or plant origin. Manufactured or synthetic volatile materials such as plastics, ether, acetone, and carbon tetrachloride are highly volatile and not of plant or animal origin. Also see NONVOLATILE MATTER.

VOLATILE ACIDS VOLATILE ACIDS

Fatty acids produced during digestion which are soluble in water and can be steam-distilled at atmospheric pressure. Also called organic acids. Volatile acids are commonly reported as equivalent to acetic acid.

VOLATILE LIQUIDS VOLATILE LIQUIDS

Liquids which easily vaporize or evaporate at room temperature.

VOLATILE MATTER VOLATILE MATTER

Matter in water, wastewater, or other liquids that is lost on ignition of the dry solids at 550°C.

VOLATILE SOLIDS VOLATILE SOLIDS

Those solids in water, wastewater, or other liquids that are lost on ignition of the dry solids at 550°C.

VOLUMETRIC VOLUMETRIC

A measurement based on the volume of some factor. Volumetric titration is a means of measuring unknown concentrations of water quality indicators in a sample *BY DETERMINING THE VOLUME* of titrant or liquid reagent needed to complete particular reactions.

VOLUTE (vol-LOOT) VOLUTE

The spiral-shaped casing which surrounds a pump, blower, or turbine impeller and collects the liquid or gas discharged by the impeller.

W

WASTEWATER WASTEWATER

A community's used water and water-carried solids that flow to a treatment plant. Storm water, surface water, and groundwater infiltration also may be included in the wastewater that enters a wastewater treatment plant. The term "sewage" usually refers to household wastes, but this word is being replaced by the term "wastewater."

WATER CYCLE WATER CYCLE

The process of evaporation of water into the air and its return to earth by precipitation (rain or snow). This process also includes transpiration from plants, groundwater movement, and runoff into rivers, streams and the ocean. Also called the HYDROLOGIC CYCLE.

WEIR (weer) WEIR

(1) A wall or plate placed in an open channel and used to measure the flow of water. The depth of the flow over the weir can be used to calculate the flow rate, or a chart or conversion table may be used to convert depth to flow.

(2) A wall or obstruction used to control flow (from settling tanks and clarifiers) to ensure a uniform flow rate and avoid short-circuiting.

WET CHEMISTRY WET CHEMISTRY

Laboratory procedures used to analyze a sample of water using liquid chemical solutions (wet) instead of, or in addition to, laboratory instruments.

WET WELL WET WELL

A compartment or tank in which wastewater is collected. The suction pipe of a pump may be connected to the wet well or a submersible pump may be located in the wet well.

X

(NO LISTINGS)

Y

(NO LISTINGS)

Z

ZOOGLEAL (ZOE-glee-al) MASS ZOOGLEAL MASS

Jelly-like masses of bacteria found in both the trickling filter and activated sludge processes. These masses may be formed for or function as the protection against predators and for storage of food supplies. Also see BIOMASS.

ZOOGLEAL (ZOE-glee-al) MAT ZOOGLEAL MAT

A complex population of organisms that form a "slime growth" on the sand filter media and break down the organic matter in wastewater. These slimes consist of living organisms feeding on the wastes in wastewater, dead organisms, silt, and other debris. On a properly loaded and operating sand filter these mats are so thin as to not be visible to the naked eye. "Slime growth" is a more common term.

SUBJECT INDEX

Q

R

S

T

NOTES

NOTES

NOTES

NOTES

NOTES